A Transition to Advanced Mathematics

A Transition to Advanced Mathematics

A Survey Course

William Johnston
Alex M. McAllister

OXFORD
UNIVERSITY PRESS
2009

OXFORD
UNIVERSITY PRESS

Oxford University Press, Inc., publishes works that further
Oxford University's objective of excellence
in research, scholarship, and education.

Oxford   New York
Auckland   Cape Town   Dar es Salaam   Hong Kong   Karachi
Kuala Lumpur   Madrid   Melbourne   Mexico City   Nairobi
New Delhi   Shanghai   Taipei   Toronto

With offices in
Argentina   Austria   Brazil   Chile   Czech Republic   France   Greece
Guatemala   Hungary   Italy   Japan   Poland   Portugal   Singapore
South Korea   Switzerland   Thailand   Turkey   Ukraine   Vietnam

Published by Oxford University Press, Inc.
198 Madison Avenue, New York, New York 10016
www.oup.com

Oxford is a registered trademark of Oxford University Press

Library of Congress Cataloging-in-Publication Data
Johnston, William, 1960–
A transition to advanced mathematics : a survey course / William Johnston,
Alex M. McAllister.
     p. cm.
Includes bibliographical references and index.
ISBN 978-0-19-531076-4
1. Mathematics—Textbooks. I. McAllister, Alex M. II. Title.
QA37.3.J65 2009
510—dc22   2009009644

9 8 7 6 5 4 3 2 1

Printed in the United States of America
on acid-free paper

For our teachers and our students

Preface

A Transition to Advanced Mathematics: A Survey Course promotes the goals of a "transition" course in mathematics, helping to lead students from courses in the calculus sequence to theoretical upper-level mathematics courses. The text simultaneously promotes the goals of a "survey" course, describing the intriguing questions and insights fundamental to many diverse areas of mathematics. Its only prerequisite is single variable calculus, and there are many chapters, such as chapters 1, 2, 3, and 6, that do not even require calculus. A hallmark of the book is its flexibility—an instructor may choose to use the text in a variety of ways. The standard adoption would be for a transition course, but this text could also be used in other settings.

A lack of diversity is perhaps the most noteworthy weakness in many institutions' current introductory mathematics curricula. A significant number of students (indeed, most people in the general population) have little understanding of the broad scope of mathematics. Since many promising students never even complete the calculus sequence, they drop out of mathematics before having had the opportunity to study some mathematical field they would have loved. Calculus doesn't stir everyone's imagination. Could a potential coding theory wizard have missed out on the fun of public key cryptography? Has a potential complex analyst who could have proven the Riemann hypothesis turned to another major? Has a potential logician who might have followed in the footsteps of Gödel decided math was purely computational? In addition, without a survey course, most mathematics majors do not possess an appreciation for the multifaceted aspects of the study of mathematics—at least not until after they have declared their major and taken a variety of upper-level courses. This situation in mathematics stands in marked contrast to virtually every other area of academics, where a survey course is among the regular course offerings. Surely our standard undergraduate course offerings can do better? But how can we succeed in showing our students the expansive vista of mathematics without a significant restructuring of the curriculum?

The answer can come in many forms, and this text can help. Combining the goals of a transition course with the desire to provide a survey of the subject, *A Transition to Advanced Mathematics: A Survey Course* teaches proof writing, reading, and understanding mathematics in the context of its many wonderful and interesting subfields. And so the text is written primarily for use in a one-semester transition course, enhancing that course by giving students a taste of the many areas of mathematics. Learning to read and write proofs is an important yet challenging process; by embedding it in the study of interesting and diverse mathematics, this text is designed to motivate and inspire students in their further studies.

Depending on how a department stresses theory at the junior-senior level, some instructors may also wish to use the text in a survey course at the upper level. Students who have seen only one or two advanced courses, such as differential equations or complex analysis, would benefit from this text's approach, as it promotes a training in the abstract nature of the subject. The text would be terrific at a large university that might offer many curricular tracks toward mathematical science majors. It also serves small colleges well, including those institutions whose resources limit the possibility of offering the full breadth of courses common in larger programs. The book could also be used as a training tool in independent studies, where a bright student could work through the sections by reading, answering questions, and working through selected exercises. Or it would make an inspirational gift for a young person who has expressed an interest in mathematics but is not yet a student in a four-year undergraduate program—anyone who loves mathematics and wants to know more about mathematical thinking would benefit from working through this text.

And so the main objective of the book is to bring about a deep change in the mathematical character of students—how they think and their fundamental perspectives on the world of mathematics. Instead of just calculating a derivative, we want students to enjoy the theory that Newton and Leibniz developed, especially as the theory leads to the techniques used in calculations. Instead of just knowing such facts as the first three primes are 2, 3, and 5, we want students to respond well to the variety of theoretical questions about primes, to formulate such questions on their own, and to be impressed by and to understand key elements of the mathematical theory of primes. In this way, we hope that working through the text will encourage students to become mathematicians in the fullest sense of the word.

How can we bring about this change in our students? We believe this text promotes three major mathematical traits in a meaningful, transformative way: to develop an ability to communicate with precise language, to use mathematically sound reasoning, and to ask probing questions about mathematics. These skills are the hallmarks of a good mathematician.

Mathematicians live in a unique world. Our language is the natural language of our culture (for most people in the United States and the United Kingdom this language is English), but a mathematician's use of this natural language is refined and specific. Through the common consensus of professional researchers and teachers, mathematical words and phrases are given precise, unambiguous interpretations, making it crucial for a mathematician to be able to work carefully with formal, rigorous definitions. With years of experience and practice, most mathematicians naturally express themselves in this formal language, but at the same time, this ability is an acquired skill that sometimes runs counter to the fluidity and adaptability of our natural language. With care and practice, students can develop the ability to write and speak well using the formal, explicit language of mathematics— its terminology and symbols, its expression of deductive and inductive reasoning, and its insistence on clarity and organizational neatness. *A Transition to Advanced Mathematics* offers engagement in the necessary experience to develop a mathematical voice.

Similarly, a mathematician's rational mode of thought is rooted in natural human reasoning, but it also differs from that of the mainstream, being uniquely refined and sophisticated. It searches for general truths that follow from deductive reasoning. The creation of new mathematics often follows from leaps of intuitive insight based on results gathered from examples. But examples are not enough. Centuries of experience and practice have led mathematicians to rely on logical deductive arguments as the litmus test for mathematical truth. These arguments are traditionally presented in formal mathematical proofs. The format of this text encourages students to develop the logical thought processes needed to reason through these proofs. The book introduces the fundamentals of mathematical thought by placing the study of logic (as a description of this formal deductive reasoning) up front. And it gives students practice in applying mathematical arguments and proofs in the context of the broad landscape of mathematical fields. A reviewer of this text recognized the strategy well and wrote, "The justification for axiomatic reasoning … is clearest when there are questions on the table that simply cannot be resolved in any way other than employing the logical precision of a mathematical argument." The text also encourages students to learn to write proofs not for the sake of writing proofs, but because they see the value of applying sound reasoning to intriguing mathematical questions. In short, the book invites students to enter the ongoing mathematical dialogue with mentors and colleagues.

Finally, mathematicians have an active curiosity and a constant desire to ask questions. Mathematicians perceive a world of ideas to be grappled with, research interests to be explored, and applications of theory to be determined. While much great mathematics is already known, students need to understand that there is so much more waiting to be discovered! Put succinctly, discovering patterns and forming conjectures are essential to the pursuit of mathematical truth. *A Transition to Advanced Mathematics* has many questions and exercises that promote the formulation of reasonable hypotheses; diverse examples throughout the text help students begin investigations of many different types of mathematical objects.

A Transition Course

A Transition to Advanced Mathematics nicely serves as a text for the "transition" course now so common in many institutions' undergraduate mathematics curriculum. Usually offered at the sophomore level, a transition course bridges the gap between computationally oriented lower-level courses and theoretically oriented upper-level courses.

Most mathematics students begin their college career in a calculus sequence that emphasizes computational problem-solving and applications of calculus methods. There are many good reasons for beginning the undergraduate curriculum with this sequence of courses. Students learn a lot of analysis and function theory by the end of their second year, which provides them with good depth in one area of mathematics and a great deal of experience in solving many problems at increasing levels of sophistication. In addition, calculus is the field of mathematics that is most useful as a prerequisite for the physical sciences, engineering, the social sciences, and business. Students majoring in these areas of study need to learn differential and integral calculus by the end of their first year of college, and mathematics teachers across the country do an excellent job of preparing these students for the rigors ahead.

On the other hand, as budding mathematicians our students should seek more than just knowing *what* mathematical truths hold; they should want to understand *why* mathematical truths hold. The good news is that computations and algorithms learned in lower-level courses often contain the kernel of the ideas behind the truth of certain mathematical statements. Thus, by working through calculations, students can develop an insightful intuition about many mathematical truths. The next step for a student to mature into a fully developed mathematician is to gain an ability to articulate precisely reasoned arguments that explain and justify the mathematical idea under scrutiny.

Unfortunately, as many in the mathematical community have recognized, a focus on the computational elements of calculus is not preparing students for this transition into theoretically oriented upper-level courses. Many students enter courses on abstract mathematics having minimal experience with either the deductive reasoning or the abstract thought processes that are characteristic of proofs. Furthermore, many have never been exposed to the experimentation and conjecture essential to the discovery and creation of mathematics. This text is designed to bridge the gap and improve the success of students in upper-level courses. By making mathematics enjoyable and manageable, and by serving the need to train students in mathematics well, this book is also intended to serve as the mathematical community's much sought after "pump" to bring more students into the mathematical fold. As they work through the text, students hopefully will recognize that they are learning the art of mathematics, and, like an apprentice artist, hopefully they will enjoy the resulting creations as they use their "mathematical palette."

In summary, as a text for a transition course, *A Transition to Advanced Mathematics* encourages students to:

- Develop careful reasoning skills as the student is transitioning from computationally oriented, algorithmic thinking to more sophisticated modes of reasoning;
- Learn to read mathematics, specifically definitions, examples, proofs, and counterexamples;
- Learn to write mathematics, primarily formal proofs, but also intuitive explanations and conjectures.

A Survey Course

More than just serving as a text for a transition course, *A Transition to Advanced Mathematics* is also designed to provide students with a broad survey of many fundamental areas of mathematics. Students completing a calculus sequence may not realize that mathematicians are a diverse lot with wide-ranging interests. Indeed, many different areas of mathematics suit individual skills and insights as well as personal interests and temperaments. With the calculus sequence serving as the primary point of entry to the mathematics major, many students are unaware of the marvelous variety inherent in mathematics.

A Transition to Advanced Mathematics responds in a positive way to the need to provide students with a broad survey of mathematical ideas and explorations, as it is intended to:

- Provide students with a broad and comprehensive introduction to mathematics, including both continuous and discrete mathematics;
- Introduce students to "upper-level" topics at an earlier stage in the mathematics major;
- Create greater continuity and flow in the mathematics major, introducing various topics, mathematical objects, and proof techniques multiple times at increasing levels of sophistication.

The text responds to the mathematical community's ambitious desire to show students a vast array of mathematical ideas. In its writing, we had to decide which topics to include and which to omit. Two questions guided the decision process: (1) What fundamental ideas should all mathematics majors know when they complete their undergraduate degree? (2) What ideas do mathematicians experience as intriguing, exciting, and central to mathematics? In some ways, these questions may be highly personal with subjective answers; very reasonable people may give very different and equally compelling answers. This text offers an answer in a way that we believe represents a thoughtful response of the full mathematical community. The answers have naturally been guided by our own experiences, but they have also been informed by discussions with many colleagues and friends, presentations and panels at national and regional mathematics meetings, published statements of professional mathematical societies, and our personal understanding of the consensus of the contemporary mathematical culture. Some people may wish that we had included additional areas, but we feel the text promotes mathematics in general and intends that students be able to make the jump into areas not discussed in the book (such as general and algebraic topology, differential geometry, non-Euclidean geometry, or relativity theory) by having discussed the mathematics presented.

The following general descriptions of the chapters, together with the detailed Table of Contents, present the balance we have struck between continuous and discrete mathematical topics in light of the "survey" aspect of the course. We hope that this panoramic view of the mathematics we know and love will intrigue, excite, and ultimately encourage students to take up a more thorough study of the topics in upper-level mathematics courses.

Suggestions for the Instructor

We wrote the text to give instructors options when using it in a one-semester course, although it is impossible to teach every topic from every section in such a short time. We intentionally provided plenty of material to allow for a follow-up study, such as an independent project for a student who might be excited about a mathematical problem described in the readings. In this way, the book offers flexibility within a mathematical curriculum; it will usually be used in a one-semester course, but some institutions may have short terms where a follow-up course would fit in well.

The one-semester offering is the standard fare, and the book is designed for this setting. An instructor can choose from the Contents in a variety of ways. Chapter 1 is required and needs to be discussed first, but then there are many options for the way this book can be used. Use will typically depend on the needs of the department and the curriculum, the interests of the instructor, the purpose of the course, and the backgrounds of the students. A reviewer for the text said it best, "The beauty of this type of text is that you can jump around, since … most of the chapters are self-contained." The flowchart, though it does not have to be followed, gives some guidance as to the rough logical dependence of the chapters.

We believe that the heart of the course is in chapters 1–4, and these four chapters could support a wonderful course in and of themselves. The first chapter is designed to teach students to think mathematically and to prove mathematical theorems in the context of mathematical logic. We chose to begin the book with symbolic logic because we have found that students' proof-writing skills improve tremendously when their approaches are grounded in proper logical thought. The study of logic is rightfully approached for its own sake as an interesting field of mathematics, and in this text it doubles as an important tool to develop theorem-proving skills.

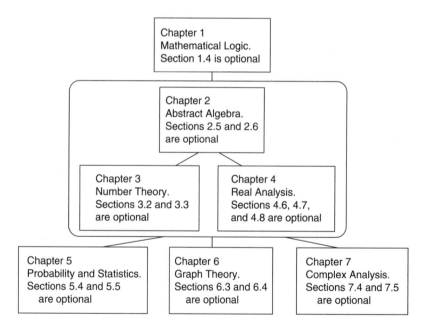

The last section of chapter 1 is the most important in the book, in the sense that it gathers the ideas from formal logic into a discussion of how to prove mathematical theorems. It sets the stage for proving mathematical results in all other chapters. Additional ideas introduced in chapter 1 include the sentential (or propositional) logic of connectives, truth tables, validity of arguments, Gödel's incompleteness theorems, and predicate logic. Nearly all of these topics are directly connected to learning about the fundamental proof techniques of mathematics, and the text intends for students to be motivated by seeing the value of symbolic logic throughout the study. An application section explores the design of computer circuits via sentential logic and Karnaugh maps. Not all of these topics need to be explored, and an instructor may choose to omit many of the sections. A streamlined approach to chapter 1, for example, could examine only sections 1.1, 1.2, 1.6, and 1.7. In the flowchart, we have listed section 1.4 as optional because we often choose to omit it, but a quick review of any chapter will indicate that an instructor may pick and choose from the many topics found within sections in a variety of ways.

Chapter 2 studies number systems as foundational to understanding mathematics. The chapter explores the integers and other basic number systems from the perspective of abstract algebraic properties and relations. These notions lead to important insights that are applied in later chapters, especially chapter 3. The fundamental ideas introduced in chapter 2 include a basic algebra of sets, Russell's paradox, the division algorithm, modular arithmetic, congruence of integers modulo n, equivalence relations, proofs of the uniqueness of mathematical objects, dihedral groups, and the basic notions of group theory. An application section explores a variety of check digit schemes.

Chapter 3 is meant to be a lot of fun. It expands on chapter 2's study of number systems from the perspective of examining abstract algebraic properties, including the exploration of solutions to polynomials. This theme is picked up on in many later chapters, especially in the study of polynomials as functions. An instructor can pick and choose from the many interesting, accessible, and historic topics from number theory, including ideas on the infinitude of primes, the prime number theorem, Goldbach's conjecture, the fundamental theorem of arithmetic, the Pythagorean theorem, solutions of basic Diophantine equations, fields, Fermat's last theorem (the proof is given for $n = 4$), the irrationality of the square root of two, the classical fundamental theorem of algebra, Abel's theorem, and the proof technique of mathematical induction. An application section explores public key encryption (via the RSA system) and Hamming codes, which require a short introduction to matrix multiplication.

The mathematics developed during the Age of Enlightenment sets the stage for the development of both calculus and the theory of transfinite numbers. Chapter 4 introduces the basic notions of real analysis that underlie calculus. An instructor can choose to cover all of the topics in any section or simply focus on the basic definitions provided. The ideas introduced in this chapter include Descartes' development of analytic geometry, the definition and properties of functions, the theory of inverse functions, the definition and basic properties of limits, derivatives, and Riemann integrals, the definitions of cardinality and countability, Cantor's diagonalization arguments to prove the countability of the rationals and the uncountability of the reals, and a brief introduction to L_2 spaces. An application section explores how differential equations can model physical processes such as the motion of a clock pendulum. The chapter assumes competency with topics found in a standard single-variable calculus course. As for any of the chapters from chapter 4 on, an instructor may choose to stop at any midway point through the list of sections. When we teach the one-semester course, we often decide to go on to chapter 5 after covering section 4.6.

Chapters 5–7 are offered as sweet desserts. There are two distinct approaches to these last three chapters of the text. When we teach the course, we like to choose at least two or three sections from each of these chapters in order to give the students a taste of the many different disciplines in mathematics. Our students value this exposure—they say it helps them choose which courses they might later select from the upper-level offerings. Alternatively, each of the chapters is a completely independent module and can be studied in greater depth or omitted. Chapter 5 explores the mathematics of likelihood and the long-term patterns in discrete events. The section on hypothesis testing provides a mathematical approach to inductive thinking, parallel to the way in which chapter 1 provides a mathematical approach to deductive reasoning. The fundamental ideas introduced in this chapter include basic combinatorics, Pascal's triangle, the binomial theorem, basic probability, hypothesis testing, and least squares regression. Many of the problems are computational, but the overriding framework of hypothesis testing and many of the abstract notions of probability theory are presented. This exposure is meant to assist greatly any student entering the corresponding upper level course.

Chapter 6 introduces the study of graphs by indicating how they model and solve real-world questions, beginning with the Königsberg bridge problem. In this way, the chapter describes the mathematics of adjacency and the abstract descriptions of networks of "connected" points (or objects). This chapter's fundamental ideas include the definition and basic properties of graphs, Eulerian and Hamiltonian circuits, trees and spanning trees, and weighted graphs. The chapter presents many algorithms for constructing shortest paths, spanning trees, Hamiltonian cycles, and minimum weight versions of these objects in a given graph.

Chapter 7 presents an introduction to the theory of complex-valued functions, teaching students about the basic algebra of complex numbers, single- and multivalued functions such as nth roots, exponential, trigonometric, and logarithmic functions and their graphical representation, analytic functions, partial differentiation and the Cauchy–Riemann equations, power series representations of analytic functions, harmonic functions, and the Laplacian. An application section explores the use of streamlines and equipotentials to understand and model fluid flow.

Key Elements of the Text

We hope *A Transition to Advanced Mathematics* will be recognized as a clear and cogent text in support of a transition course surveying mathematics. It is designed to serve ideally in collaboration with mathematics professors helping students to explore new mathematical vistas, to grow into the perspectives of the mathematician, and to successfully practice mathematics. The following elements of the text are intended to help facilitate this partnership between professor and text in the creation of a dynamic and interesting learning experience.

Embedded questions. In each section, after reading through the text and examples that illustrate and explain fundamental concepts, students are invited to create and display their personal understanding of the mathematical idea at hand by answering questions. Many of these queries are straightforward and useful in providing good introductory experiences with the new ideas at hand; as such, they can be assigned as homework in preparation for class or used during class in the spirit of active learning and engaged discussion. Some of them lead to a main idea of an upcoming proof. An example is question 3.1.9 in section 3.1, which asks students about computations of the form $p_1 \cdot p_2 \cdots p_n + 1$ where each p_k is prime—are integers of that form always prime? (It is still an open question whether there are infinitely many primes in this sequence.)

Reading questions. An effective pedagogical tool is to expect students to read the text before coming to class and to be able to answer a collection of basic questions. We always want our students to use a text more than as a reference for worked examples. Reading comprehension questions at the end of each section ask for definitions, examples, and the central ideas of the material, leading students to open the book and read.

There are many ways for an instructor to use the reading questions. We assign them before every class meeting and expect students to write their responses in complete English sentences. Our hope is that students both learn the value of reading the book and get practice in expressing mathematical concepts well. They also come better prepared for class. In this way, teachers can respond to students' questions and engage the mathematical ideas at a much deeper level during class, and the students develop the independent reading skills essential for more sophisticated mathematical studies.

Exercises. Every section is accompanied by 70 exercises that allow the professor considerable flexibility in assigning homework and that give the reader practice. As with any exercise set, the ultimate goal is to provide students needed practice to deepen their understanding of the corresponding mathematical concepts. Instructors can pick and chose from many different types of problems. The exercises are grouped according to topic; if the instructor has focused on just part of the section's material, it is easy to pick out corresponding problems to assign.

The end of each exercise set always contains a variety of more challenging exercises. These questions sometimes anticipate ideas in upcoming sections, require the study and use of a new definition or idea, or ask students to make conjectures based on some pattern arising from a collection of computations. Instructors could occasionally use them to motivate students to pursue a topic in more detail, or as a staging point for further investigations that might lead to a short paper or presentation.

An application section. Every chapter includes a section that explores an application of the theoretical ideas under study. All involve interesting "real-world" issues. Students are often surprised when theoretical notions find expression as a useful tool in life. The text intends to teach students, as they see a variety of applications, to view purely abstract, theoretical ideas as not antithetical to using mathematics to benefit society. The intent is for students to begin to perceive pure and applied mathematics as going hand in hand and strengthening one another in interplay: a search for applications often results in the development of new theoretical ideas, and theoretical mathematics often manifests itself as a critical underpinning of an applied tool.

None of the application sections are required for the text's other sections. When we teach the course, we sometimes treat a chapter's applied section in the same way as the others in the chapter, but at other times we might simply ask our students to read the section outside of class and submit the reading questions, or have them work in teams to answer some of the exercises. Depending upon the instructor's interests and the parameters of the course, any applied section may be skipped.

Embedded reflections on the history, culture, and philosophy of mathematics. Mathematics is a timeless study that has been gradually developed through the corporate efforts of diverse individuals and cultures. The historical origins of mathematical ideas and the accompanying cultural standards for definitions, examples, and proofs are worthwhile and interesting and contribute to a student's ability to understand and appreciate contemporary mathematics. Throughout the text, we tell stories about the struggles, the insights, and the people and events that helped shape mathematics. Our hope is for students to enjoy the drama, getting a sense of the eureka of mathematical breakthroughs and connecting proofs and mathematical statements (so often presented as devoid of human emotion), and relating to the human lives of the men and women who first presented them.

Acknowledgments

We thank our many friends and colleagues for supporting this work and our efforts. Through Faculty Development Committee grants, Centre College helped fund work during several summer months of writing. In 2001 we received a generous grant from the Associated Colleges of the South, funded by the Mellon Foundation, to assist with the incorporation of the technology-based portions of the text.

Our colleagues in the mathematics departments at Centre College and at Randolph–Macon College supported our efforts by implementing a transitions course in the major curriculum and by allowing us to use early drafts of the work in the courses' initial offerings. In short, our departments' responses were universally supportive and meant a great deal to us. We are also grateful to Stan Perrine at Charleston Southern University, Deirdre Smeltzer at Eastern Mennonite University, and John Thompson at University of Pittsburgh at Johnstown for using an initial version of this text in their courses, and for their insightful suggestions that improved the text.

Our students welcomed our requests for them to use early versions of the text, making suggestions, giving us positive feedback, and expressing their enthusiasm for both the project and the course. Josh Smith provided valuable assistance, giving feedback and working on solutions to exercises in chapters 1 and 2. Tyshaun Lang and Morgan Smith worked every exercise and question in a late version of the text, providing valuable feedback.

Our families have sustained us with love, patience, and advice. We thank them, especially Susan and Julie, for being supportive as we worked and typed.

Our work together has been greatly satisfying and enriching for both of us. A project of this magnitude would have been very difficult without a mutual respect and admiration between coauthors. Our friendship and professional esteem for each other's efforts helped make this venture enjoyable throughout, and we each publicly acknowledge our thanks for the other's contributions. We feel very blessed to have a collegial relationship with one another, and we hope and sense that this good fortune is reflected in our writing.

We especially thank Phyllis Cohen, Edward Sears, Michael Penn, and those at Oxford University Press for their advice, support, and coordination of the book's publication. The reviewers' comments were extremely helpful. Any errors in the text are the responsibility of the authors, but a host of people around us should share in whatever compliments the text receives. There are too many to name individually, but we sincerely thank them all very much.

Contents

A Transition to Advanced Mathematics

1 Mathematical Logic

The formal study of logic is ancient, going back to at least the fourth century B.C.E., when Aristotle and his Greek compatriots sought to identify those forms of human reasoning that are correct (or valid) and those that are not. Our motivation is similar. In this text, we explore diverse areas of mathematics, identify new mathematical objects, investigate the relationships among them, and develop algorithms to facilitate their study. In short, we pursue mathematical truth. But more than just the "what" of mathematical truth, we seek the "why" of mathematical truth. We develop an ability to understand and prove theoretical mathematical results, including those that derive the computational tools so useful in applied mathematics. Successful insight into this theory of mathematics is essentially dependent on the use of correct reasoning.

And so our study of mathematical logic has two goals. The first is the study of logic for its own sake, as a field of mathematics with interesting objects, algorithms, and insights. The second goal is the study of logic as a tool and a language for understanding legitimate forms of human reasoning; in this way, logic will facilitate our study of the theory of mathematics in many different settings.

In writings such as *Prior Analytics*, Aristotle developed the insight that human reasoning can itself be studied via reasoning: we can turn inward and examine how we think. In fact, Aristotle believed that logic should be studied before pursuing any other branch of knowledge. The next significant step forward in the study of logic did not occur until 2,200 years later in the heady aftermath of the Scientific Revolution. In the middle of the nineteenth century, the Irish mathematician George Boole introduced the notion of a formal language with an accompanying algebra of logic. Beginning with the seminal paper *An Investigation of the Laws of Thought, on Which Are Founded the Mathematical Theories of Logic and Probabilities*, Boole and his fellow mathematicians described how formal languages overcome the ambiguity of natural languages and provide a more precise analysis of both our natural languages and our reasoning processes.

Less than 40 years later, the Austrian mathematician Kurt Gödel's study of formal languages illuminated both the potential and the essential limitations of the human mind as it operates within a formal system of logic. Gödel's incompleteness theorems demonstrate that some true mathematical statements are not provable (that is, they can never be proven in a suitable formal system) and are among the most significant mathematical and philosophical insights of the twentieth century. Within another

40 years, the use of formal languages began playing a key role in the design of the computer chips that are so essential to our technologically based society.

In this chapter, we develop a formal language in the spirit of Aristotle and Boole known as "sentential" logic. We examine the interaction of sentential logic with our natural language and our intuitive notion of truth. We develop an algebra of sentential logic, explore the expressiveness of this language, consider an application to the design of computer chips, and study common rules of natural deductive reasoning that are valid. We also consider an extension of sentential logic known as "predicate" logic that incorporates a finer analysis of sentence structure. We end this chapter with a discussion of the fundamental proof techniques widely utilized by mathematicians. By developing some sophistication in our ability to work with these techniques, we assume the role of a theoretical mathematician as we apply formal reasoning to prove the truth of mathematical statements.

Why should we begin this book with a chapter on logic? Most of you have recently finished studying the intricacies of calculus, and (in high school) the ins and outs of geometry, trigonometry, and advanced algebra. Perhaps this chapter may strike you as the study of odd-looking symbols that seem to have little relevance to your previous mathematics courses. But mathematics is, after all, the study of "mathematical objects" such as numbers, which are only symbols—meaningless, except in their definitions and relationships. And yet these objects become powerful tools in making sense of our world. Proving statements about such objects is the primary concern of theoretical mathematicians and forms the basis for any rational, deductive investigation of mathematics. And so in this chapter we get down to basics: True, False, or Maybe. We present mathematical logic as an essential tool that you can use in your attempts to determine the truth of mathematical statements.

In the study of more advanced mathematical ideas, one can go off on tangents that either have no basis in sound logic and are irrelevant, or that lead to incorrect conclusions and are counterproductive. Mathematical logic can keep us on track, and this chapter is then essential as the basis for your continuing study of mathematics.

1.1   The Formal Language of Sentential Logic

The goal of Aristotle's logic was the analysis of arguments constructed as a combination of sentences in our natural language. There is a great deal of consistency across human cultures and languages in how we reason; there is little difference between representing the idea of argument with the word "logos" in Aristotle's natural language of ancient Greek and the word "argument" in our natural language of modern English. Rather, the way that we construct and reason through arguments shares much in common with the way Aristotle and others reasoned. This universality enables the success of sentential logic as a fundamental tool in the study of human reasoning.

In common usage, the word "argument" carries a host of connotations, including fights or emotional outbursts that may not involve any rational thought. In this study, we are interested in arguments in the precise logical sense of the word. For us, an *argument* is a list of sentences. The last sentence is the *conclusion* and the previous sentences include some *premises* or *assumptions* with some intermediate steps included for clarity. Ideally, the conclusion follows from the set of premises via some valid means of logical reasoning.

The first systematic study of arguments and deductive reasoning was undertaken by the eminent Greek philosopher Aristotle in the third century B.C.E. Aristotle was a philosopher who made important contributions to the development of all areas of knowledge. The son of a physician, he was orphaned at a young age and raised by his uncle. At 17 he became a student at Plato's Academy in Athens, and he soon joined the faculty of the Academy. Political unrest in Greece and Macedonia eventually forced him to leave Athens, although he eventually returned to found his own school, the Lyceum. Aristotle died at the age of 62—his legacy, in thinking of the way rational thought is structured and in the workings of the mind on philosophical issues, continues to have an important influence to this day.

In studying arguments, Aristotle focused his attention on a special type of argument known as a *syllogism*, which consists of two premises and a conclusion, and he developed the theory of syllogisms in his book *Prior Analytics*. The following two examples of syllogisms are based on the work of the scholastic logicians of the Middle Ages.

Every Greek is a person.   Every Greek is a person.
Every person is mortal.   Every Trojan is a person.
Thus, every Greek is mortal.   Thus, every Greek is a Trojan.

On the surface, these two arguments appear quite similar, but your intuition may identify an important difference between them. The first argument should seem "right" (in fact, it is a valid argument), while the second should seem "wrong" (in fact, it is invalid). We seek to understand and clarify this distinction between valid and invalid arguments and to develop various approaches for identifying the validity of arguments without having to rely exclusively on intuition.

Natural languages are often ambiguous and at times misleading. You may be able to think of moments in your life when something you said was misinterpreted by another person; storylines, from Shakespearean plays to soap operas, are often driven by misunderstandings among characters. In real life we rely on context, voice inflection, and further conversation to provide clarity, but these tools of extended communication are not available in formal, written mathematical settings. The inherent ambiguity of natural languages is one of the factors that prompted the creation of formal languages. As mathematicians, we need to eliminate ambiguities in the language we use, which we will do by controlling the structure allowed within the sentences we consider. The first structure we will discuss is the "connective." Without such control, ambiguity can play havoc. The following questions verify that English really is an ambiguous natural language.

Question 1.1.1   Consider the poorly written sentence "I am going to bike and run or swim."

(a) Identify two distinct interpretations of the given sentence by inserting a pair of parentheses in two different ways.

(b) State English sentences that express the two distinct interpretations from part (a) by inserting the pairs of words "either–or" and "both–and" into the given sentence.

■

In this book, the symbol "■" indicates the completion of the task at hand, including the end of either an example or a question, and the successful completion of a proof. As we can see from the results of question 1.1.1, the words "and" and "or" play a crucial role in determining our interpretation of a sentence. These words are *connectives* in the natural language of English and are essential to obtaining an unambiguous formal language. We typically think of connectives linking simple subject–verb–object sentences, such as "I am going to bike" or "Bailey hires Andy," to create more complex, compound sentences, such as "I am going to bike, and Bailey hires Andy."

Question 1.1.2   In each sentence, try to identify the connective linking the simple component sentences.

(a) Bailey hires Alex or Alex becomes a telemarketer.
(b) Alex does not become a telemarketer.
(c) Alex becomes a telemarketer if and only if Bailey hires Alex.
(d) Both Alex graduates from college and Bailey hires Alex.
(e) Bailey hires Alex if Alex graduates from college.
(f) Bailey hires Alex when Alex graduates from college.
(g) If Alex does not graduate from college, then Alex becomes a telemarketer.

■

Did you find all the connectives? Probably the most difficult to identify is the one-place connective "not" in sentences (b) and (g). The five most common connectives are "not," "and," "or," "if–then," and "if and only if." These connectives are expressed in many different ways in written and spoken English and you need to become familiar with the corresponding variants. For example, the "if and only if" connective is often expressed as "precisely when" or "exactly when" and the phrase "*A* if and only if *B*" succinctly expresses "both if *A*, then *B* and if *B*, then *A*." The formal language of sentential logic uses the following symbols, known as *logical connectives*, to represent the given English connectives and their variants.

English connectives	Logical connectives	Formal names
not	$\sim$	negation
and, both–and, but	$\wedge$	conjunction
or, either–or	$\vee$	disjunction
if–then, implies, if, when, only if	$\rightarrow$	implication
if and only if, precisely when	$\leftrightarrow$	biconditional

Most of the time we can directly substitute a logical connective's symbol for the identified English words; the most important exceptions are "if" and "when" (expressions of implication) as discussed in example 1.1.1 below. Plainly put, "*A* when *B*" is the same as "*B* implies *A*." Similarly, "*A* if *B*" means "*B* implies *A*." In the formal language, simple sentences are represented by upper case letters.

For example, in one context, we might define A to represent "Alex graduates from college," while in another, we might define P to represent "The number n is prime." We refer to a collection of definitions assigning sentence symbols to particular English statements as a *dictionary*. With these basics of our formal language in mind, we reconsider the sentences from question 1.1.2.

Example 1.1.1 We use the given dictionary to translate each English sentence into sentential logic.

A: Alex graduates from college.
B: Bailey hires Alex.
C: Alex becomes a telemarketer.

(a) Bailey hires Alex or Alex becomes a telemarketer. $B \vee C$
(b) Alex does not become a telemarketer. $\sim C$
(c) Alex becomes a telemarketer if and only if Bailey hires Alex. $C \leftrightarrow B$
(d) Both Alex graduates from college and Bailey hires Alex. $A \wedge B$
(e) Bailey hires Alex if Alex graduates from college. $A \rightarrow B$
(f) Bailey hires Alex when Alex graduates from college. $A \rightarrow B$
(g) If Alex does not graduate from college, then Alex becomes a telemarketer. $(\sim A) \rightarrow C$

■

For sentence (d), an attempted translation of $\wedge A \wedge B$ is incorrect since the logical connective $\wedge$ expresses the complete phrase "both–and" and is only written once as $A \wedge B$. Sentences (e) and (f) illustrate the correct translation of the implication expressed by "if" and "when"; for these sentences $B \rightarrow A$ would be an incorrect translation. We must include parentheses in sentence (g) to avoid the potential ambiguity that accompanies the use of multiple connectives. In this example, we must clarify the correct translation as $(\sim A) \rightarrow C$, rather than the incorrect translation of $\sim (A \rightarrow C)$, which actually expresses: "It's not the case that if Alex graduates from college, then Alex becomes a telemarketer." Sentences can also be translated in the other direction, from sentential logic into English.

Example 1.1.2 We use the dictionary from example 1.1.1 to translate each formal sentence into English.

- $A \vee (\sim B)$: Alex graduates from college or Bailey does not hire her.
- $A \wedge (\sim C)$: Alex graduates from college, but does not become a telemarketer.
- $B \leftrightarrow [A \vee (\sim C)]$: Bailey hires Alex precisely when either she graduates from college or she does not become a telemarketer.
- $A \wedge (B \wedge C)$: Alex graduates from college, and both Bailey hires her and she becomes a telemarketer.
- $(A \wedge B) \wedge C$: Both Alex graduates from college and Bailey hires her, and she becomes a telemarketer.
- $A \rightarrow (B \wedge C)$: If Alex graduates from college, then both Bailey hires her and she becomes a telemarketer.

■

In light of the variety of English connectives, there are many possible translations of these formal sentences. For example, we could translate the last sentence in example 1.1.2 as: "Alex is hired by Bailey and becomes a telemarketer when she graduates from college." As translators, we are free to give such alternate renditions, provided we carefully obtain an English sentence accurately expressing the precise meaning of the formal sentence.

Question 1.1.3   Use the given dictionary to translate each English sentence into sentential logic and each formal sentence into English. Some of these sentences are true and some are false. We'll grapple with those issues soon, but for the moment we focus on the process of translation.

P: The number n is prime.
Q: The number n is rational. (Q is for "quotient.")
S: The number n is the square root of an integer.
Z: The number n is an integer. (*Zahlen* is German for "count.")

(a) The number n is a prime integer.
(b) The number n is rational exactly when n is the square root of an integer.
(c) If the number n is the square root of an integer but n is also an integer, then n is prime.
(d) The number n is the square root of an integer and if n is an integer, then n is prime.
(e) The number n is rational when n is the square root of an integer.
(f) Either the number n is prime and n is an integer, or n is rational.
(g) The number n is prime and either n is an integer or n is rational.
(h) $P \lor Q$
(i) $Q \rightarrow [(\sim P) \land S]$
(j) $P \leftrightarrow (\sim Q)$
(k) $(\sim P) \lor [Z \land (\sim Q)]$

■

In the preceding examples and questions, dictionaries have been given to facilitate the process of translation. Eventually you will create your own dictionary when translating English sentences into the formal language of sentential logic. In such cases, you must first identify the connectives in the sentences you are analyzing and then represent the corresponding simple sentence components with appropriate sentence symbols. For example, in the sentence "Two is even and two is prime," we identify the connective "and" and represent "Two is even" with E and "Two is prime" with P to obtain the sentential logic rendition $E \land P$.

We now turn to a precise definition of *sentential logic*, also known as *propositional logic* or *statement logic*. This formal language has two components: an *alphabet* identifying the legal symbols that may be used, and *sentences* consisting of legal strings of symbols from the alphabet. Throughout this section we have used uppercase letters as sentence symbols. The formal language of sentential logic also uses lowercase letters to represent unspecified simple sentences; that is, lowercase letters p, q, r represent unspecified simple sentences in the same way that the variables x, y, z represent unspecified numbers in algebra.

Definition 1.1.1 *The formal* **alphabet** *of sentential logic consists of exactly the following symbols.*

sentence symbols:	$A, B, \ldots, Z$
sentence variables:	$a, b, \ldots, z$
logical connectives:	$\sim, \wedge, \vee, \rightarrow, \leftrightarrow$
grouping symbols:	(,), [,], {, }

In a formal sentence, we may use only these symbols; any other symbols are "illegal" and should not be used. A single exception allows the indexing of sentence symbols and sentence variables with subscripts if the situation warrants. For example, we work with sentence variables $p_1, \ldots, p_n$ when describing generic sentences.

Sentential logic also identifies a collection of "legal" sentences consisting of certain strings of symbols from the alphabet. In the following definition, the symbols $\mathbb{B}$ and $\mathbb{C}$ denote generic sentences and may be sentence symbols, sentence variables, or compound sentences.

Definition 1.1.2 *A* **sentence** *of sentential logic is a string of symbols from the alphabet of sentential logic that satisfies the following:*

(a) *A single sentence symbol or a single sentence variable is a sentence;*
(b) *If $\mathbb{B}$, $\mathbb{C}$ are sentences, then so are $(\sim\mathbb{B})$, $(\mathbb{B} \wedge \mathbb{C})$, $(\mathbb{B} \vee \mathbb{C})$, $(\mathbb{B} \rightarrow \mathbb{C})$, and $(\mathbb{B} \leftrightarrow \mathbb{C})$;*
(c) *Only strings of symbols obtained by finitely many applications of (a) and (b) are sentences.*

When building up formal sentences, we carefully include parentheses as grouping symbols at each step. However, for the sake of readability, we often abbreviate sentences by omitting the outermost pair of parentheses. We also utilize the other grouping symbols from the alphabet to facilitate clarity of expression; for example, we may write $[(\sim A) \rightarrow C]$ for $((\sim A) \rightarrow C)$.

Example 1.1.3 Every string of formal symbols introduced thus far in this section has been a sentence. In contrast, the following strings of symbols are not sentences, since they do not satisfy any of the forms given in definition 1.1.2.

Nonsentence	Reason
$A \sim$	$\sim$ must precede, not follow sentence symbols
$\sim \wedge A$	connectives cannot be adjacent
$\vee\, p \wedge q$	$\vee$ must be between sentences
I LOVE MATH	sentence symbols cannot be adjacent

■

In contrast to the nonsentence $\vee\, p \wedge q$ in example 1.1.3, the string $(\sim p) \wedge q$ is a sentence since $(\sim p)$ is a sentence of the form $(\sim\mathbb{A})$, and if we label $(\sim p)$ as $\mathbb{B}$ and q as $\mathbb{C}$, then $(\sim p) \wedge q$ is of the form $(\mathbb{B} \wedge \mathbb{C})$.

Question 1.1.4 Identify each string of symbols as a sentence or as a nonsentence. Give reasons justifying your answer.

(a) $A \wedge (p \vee A)$

(b) $A \wedge p \vee A$

(c) $A \sim \wedge B \rightarrow 6$

(d) $((\sim A) \vee B) \leftrightarrow (A \rightarrow B)$

(e) $(p \wedge q) \leftrightarrow (q \vee p)$

(f) $(p \,\&\, q) \leftrightarrow (q \vee p)$

∎

We end this section with a few thoughts about mathematical definitions. While mathematics is a language rich in expression, mathematics is also quite focused and precise in its use of words. In contrast to the adaptability and fluidity of word use in natural languages, mathematicians generally assign one meaning to each technical word in a given context via a formal definition. For example, in definition 1.1.2, we specified the meaning of the word "sentence" in the context of sentential logic. In mathematical conversation, the word "sentence" now identifies exactly the objects specified in the definition—no more and no less—and mathematicians restrict the use of the word "sentence" to precisely these objects.

While at times definitions may seem somewhat arbitrary, they are most often the result of months (if not years and centuries) of discussion and reflection by researchers and teachers of mathematics. The definitions we use in this text are consistent with the common consensus of the mathematical community and should be learned and used with care. Perhaps some day soon you will choose to join in the ongoing conversation about mathematical ideas and craft definitions that arise in research ventures.

1.1.1 Reading Questions for Section 1.1

1. State two goals in studying mathematical logic.
2. Define an argument. What two types of sentences appear in an argument?
3. Give an example of an argument and identify the premises and conclusion.
4. Give an example of a syllogism.
5. What motivates our interest in developing formal languages?
6. Specify a natural language sentence with two distinct interpretations.
7. State both English and formal versions of the five connectives.
8. Discuss the relationship between the sentences "if A, then B," "B if A," and "B when A."
9. Identify the two components of a formal language.
10. State the symbols in the alphabet of sentential logic.
11. Define a sentence in the context of sentential logic.
12. Give an example of a sentence and a nonsentence of sentential logic.

1.1.2 Exercises for Section 1.1

In exercises 1–11, use the given dictionary to translate each English sentence into sentential logic and each formal sentence into English.

C: Taylor is a college student.

L: Taylor is a natural leader.

M: Taylor is a math major.

Q: Taylor will be qualified for a high-paying job.

1. Taylor is a college student.
2. Taylor is not a math major.
3. If Taylor is a math major, then she will be qualified for a high-paying job.
4. Taylor is not in college, but she is a natural leader.
5. Since Taylor is not in college, she will not be qualified for a high-paying job.
6. $(L \wedge M) \rightarrow Q$
7. $(L \vee M) \rightarrow Q$
8. $\sim(C \rightarrow Q)$
9. $(\sim C) \rightarrow (\sim Q)$
10. $(\sim M) \rightarrow [(\sim L) \vee (\sim Q)]$
11. $C \leftrightarrow L$

In exercises 12–22, use the given dictionary to translate each English sentence into sentential logic and each formal sentence into English. Note that some of these assertions are mathematically true and some are false.

A: X is associative.

C: X is commutative.

F: X is a field.

G: X is a group.

12. If X is a group, then X is associative.
13. X is a group but is not commutative.
14. X is associative or commutative, but not both.
15. X is associative and commutative when X is a field.
16. X is a group does not imply that X is a field.
17. $F \vee G$
18. $C \wedge (\sim A)$
19. $\sim(A \rightarrow G)$
20. $(\sim C) \rightarrow (\sim F)$
21. $F \rightarrow [(C \wedge A) \wedge G]$
22. $(F \vee G) \rightarrow A$

In exercises 23–33, use the given dictionary to translate each English sentence into sentential logic and each formal sentence into English. Note that some of these assertions are mathematically true and some are false.

B: A sequence $\{a_n\}$ is bounded.

C: A sequence $\{a_n\}$ converges.

D: A sequence $\{a_n\}$ diverges.

M: A sequence $\{a_n\}$ is monotone.

23. A sequence $\{a_n\}$ converges or diverges.
24. A sequence $\{a_n\}$ diverges exactly when it does not converge.
25. If a sequence $\{a_n\}$ is bounded and monotone, then it does not diverge.
26. If a sequence $\{a_n\}$ is not bounded and not monotone, then it does not converge.
27. A sequence $\{a_n\}$ diverging does not imply it is unbounded.

28. $C \rightarrow B$
29. $(\sim D) \leftrightarrow C$
30. $\sim[(\sim M) \rightarrow D]$
31. $[(\sim B) \wedge M] \rightarrow D$
32. $\sim[(M \rightarrow B) \vee (B \rightarrow M)]$
33. $D \rightarrow [(\sim B) \vee (\sim M)]$

In exercises 34–53, translate each English sentence into sentential logic.

34. A if and only if B, but not C.
35. R if both P and Q.
36. Either U or T, otherwise Q.
37. Neither L nor R, but not Z.
38. D or both Q exactly when S and X.
39. A otherwise not B.
40. C or not D.
41. Neither E nor F, or G.
42. Either not H or both I and if J then K.
43. Y if and only if both Z and W implies X.

44. If H, then either J or both K and L.
45. Either if H, then J or both K and L.
46. If H, then either J or K, but not L.
47. If either H or J, then both K and L.
48. If H, then either both J and K or L.
49. p if both q and r.
50. p when q, or r.
51. If either p or q, then r exactly when s.
52. p or q, if and only if not r.
53. Neither p nor q, but not r.

In exercises 54–63, identify each string of symbols as a sentence or as a nonsentence. Recall that the outermost pair of parentheses may be dropped. Give reasons justifying your answer.

54. $\sim A \rightarrow B$
55. $(\sim A) \rightarrow B$
56. $\sim(A \rightarrow B)$
57. $\sim\sim A \rightarrow \sim\sim A$
58. $(A \leftrightarrow A) \vee [\sim(B \wedge C)]$

59. $A \leftrightarrow \sim[B \wedge (\sim C)]$
60. $A \rightarrow [p \wedge (\sim B)]$
61. $p \leftrightarrow P \wedge B$
62. *MATH IS AWESOME*
63. *logic is fun*

Exercises 64–66 explore the ambiguity of the English language with respect to connectives.

64. State a natural language sentence with exactly five distinct interpretations.
65. How many connectives are necessary to create a natural language sentence with exactly 14 distinct interpretations?
66. State a natural language sentence with exactly three distinct interpretations.

Exercises 67–70 outline a "proof by induction" that the number of left parentheses in any sentence is the same as the number of right parentheses. The technique of proof by induction will be studied in section 3.6. In this context, do not drop the outermost pair of parentheses from a formal sentence.

67. How many left parentheses and how many right parentheses appear in an individual sentence symbol (for example, A by itself) or in an individual sentence variable (for example, p by itself)?
68. Assume that $\mathbb{B}$ has m left parentheses and m right parentheses. How many left parentheses appear in $(\sim \mathbb{B})$? How many right parentheses appear in $(\sim \mathbb{B})$?

69. Assume that $\mathbb{B}$ has m left parentheses and m right parentheses and that $\mathbb{C}$ has n left parentheses and n right parentheses. How many left parentheses appear in $(\mathbb{B} \wedge \mathbb{C})$? How many right parentheses appear in $(\mathbb{B} \wedge \mathbb{C})$?

70. Following the model given in exercise 69, argue that $(\mathbb{B} \vee \mathbb{C})$, $(\mathbb{B} \rightarrow \mathbb{C})$, $(\mathbb{B} \leftrightarrow \mathbb{C})$ each have the same number of left and right parentheses. Conclude from exercises 67–70 that *any* sentence has the same number of left and right parentheses.

1.2   Truth and Sentential Logic

Mathematicians seek to discover and to understand mathematical truth. The five logical connectives of sentential logic play an important role in determining whether a mathematical statement is true or false. Specifically, the truth value of a compound sentence is determined by the interaction of the truth value of its component sentences and the logical connectives linking these components. In this section, we learn a *truth table algorithm* for computing all possible truth values of any sentence from sentential logic.

In mathematics we generally assume that every sentence has one of two truth values: *true* or *false*. As we discuss in later chapters, the reality of mathematics is far less clear; some sentences are true, some are false, some are neither, while some are unknown. Many questions can be considered in one of the various interesting and reasonable multi-valued logics. For example, philosophers and physicists have successfully utilized multi-valued logics with truth values "true," "false," and "unknown" to model and analyze diverse real-world questions. In this book, we keep our study immediately relevant to the most common needs in mathematics by assuming a two-valued logic with truth values "true" denoted by T, and "false" denoted by F. In a given setting, one of these two truth values is assigned to each sentence symbol $(A, B, \ldots, Z)$, while sentence variables $(a, b, \ldots, z)$ are free to assume either truth value. We use *truth tables* to determine the truth value of sentences built up from sentence symbols, sentence variables, and logical connectives.

We begin by stating the distinct truth table for each logical connective. In defining these basic truth tables, an intuitive understanding of connectives in our natural language drives the interpretation of connectives in the formal language of sentential logic, and so we appeal to our intuition in motivating our formal definitions.

First, consider negation, the "not" connective denoted by $\sim$. Negation switches truth values. For example, if "The number n is prime" is true, then "The number n is not prime" is false; that is, if P is true, then $\sim P$ is false. Similarly, if "The number n is prime" is false, then "The number n is not prime" is true; that is, if P is false, then $\sim P$ is true. We express this analysis both as a phrase to aid memorization and as a truth table.

	p	$\sim p$
$\sim$ swaps truth values	T	F
	F	T

This basic truth table uses the sentence variable p, since p (as a variable) is free to assume either truth value T or F, enabling a complete analysis of the negation connective. In addition, the truth table has only two rows, since p is the only sentence variable in the sentence $\sim p$.

With this definition in hand, we no longer need to rely on intuition when interpreting the negation connective in a sentence. Instead, the truth table for negation has mathematically formalized the interpretation of negation when computing the truth of sentences. We refer to this truth table when a negation appears in a sentence, an approach which is particularly helpful when working with elaborate compound sentences. By developing similar truth tables for the other logical connectives and capturing our natural intuitions about these connectives, we establish the complete tools for developing an algebra of truth for sentential logic.

Turning to the other connectives, consider conjunction, the "and" connective denoted $\wedge$. We interpret $p \wedge q$ as true exactly when both p and q are true. If p is false or if q is false or if both p and q are false, then $p \wedge q$ is false. As above, we gather this analysis (and the results of a similar analysis for the other connectives) into a collection of phrases and truth tables.

$\wedge$ is T if	p	q	$p \wedge q$		$\vee$ is F if	p	q	$p \vee q$
both T and	T	T	T		both F and	T	T	T
F otherwise	T	F	F		T otherwise	T	F	T
	F	T	F			F	T	T
	F	F	F			F	F	F

$\rightarrow$ is F if	p	q	$p \rightarrow q$		$\leftrightarrow$ is T if	p	q	$p \leftrightarrow q$
$T \rightarrow F$ and	T	T	T		the same and	T	T	T
T otherwise	T	F	F		F otherwise	T	F	F
	F	T	T			F	T	F
	F	F	T			F	F	T

Since each sentence in the above chart has two sentence variables, there are four rows in each truth table. In particular, each sentence variable can be either true or false, resulting in the four possible permutations of truth values: TT, TF, FT, FF. The left columns in each truth table list these four possibilities. We think of a truth table with permutations TT, TF, FT, FF (in this order) as the *standard truth table* for a sentence with two variables. You should mirror this pattern in your truth table computations to facilitate comparisons among sentences.

The truth tables for disjunction and implication warrant further comment. For the disjunction $p \vee q$, note that there are two standard yet very different usages of the word "or" in our natural language of English. For example, suppose you are eating at your favorite fast food restaurant and the server asks you two questions:

- Would you like french fries or onion rings with your value meal?
- Would you like cream or sugar with your coffee?

In response to the fries–rings question, you can ask for fries or for onion rings, but not both, and you would not be upset that you can only have one; we refer to this use of disjunction as an *exclusive-or*. In contrast, in response to the cream–sugar question, you can ask for cream or sugar or both, and opting for both is a common choice among coffee lovers; we refer to this use of disjunction as an *inclusive-or*. In everyday life, context and social norms typically clarify this potential ambiguity in the use of "or." However, for our formal language, we must avoid such ambiguity and choose just one of these two options as the standard for all disjunctions. Over time, mathematicians and philosophers have adopted the inclusive-or as the standard interpretation of "or," and so we define $p \vee q$ as true when p is true, when q is true, or when *both* p and q are true.

In standard mathematical practice, the implication $p \rightarrow q$ is the most important logical connective. Mathematics is essentially a science of implications in which we explicitly identify assumptions and establish the conditional truth of mathematical statements. The first two lines of the truth table for implication match most people's intuitions: "true implies true" is true and "true implies false" is false. But, why should "false implies true" or "false implies false" be defined as a true statement?

A couple of examples may clarify this choice. First, consider a common "bribe" offered by parents to their children: "If you behave in the store, then we will stop for ice cream." If the child does not behave in the store, the parents' statement would be considered true not only if they do not stop for ice cream (the "false implies false" case), but even if, in a moment of benevolent generosity, they do stop for ice cream (the "false implies true" case). In particular, the parent's statement is false only when the child behaves in the store, but they do not stop for ice cream (the "true implies false" case). Similar situations arise quite frequently in mathematics. For example, consider the assertion "If $n \geq 3$, then $n^2 \geq 4$." This statement is true even for $n = 1$, when $n \geq 3$ is false and $n^2 \geq 4$ is false (the "false implies false" case); similarly, it is true for $n = 2$, when $n \geq 3$ is false and $n^2 \geq 4$ is true (the "false implies true" case). In short, both "false implies true" and "false implies false" are considered true.

We now focus on the mechanics of using the five basic truth tables to compute the truth of compound sentences. This analysis is based on both the truth value of the component sentences and the logical connectives linking them.

Example 1.2.1 We compute the truth table for $(\sim p) \vee q$.

The two sentence variables p and q generate the $2 \times 2 = 2^2 = 4$ permutations of truth values TT, TF, FT, FF in the corresponding truth table. After listing these permutations, we begin with the innermost connective (the connective farthest inside the parentheses—in this case the negation $\sim$ on p) and work our way out through any other connectives (in this case, the disjunction $\vee$). We compute one row at a time, applying the corresponding basic truth tables to the particular truth values given in the appropriate columns of the truth table. For this sentence, the operation of the innermost connective (the negation of p with truth values in the first column) is given in the third column. The effect of the next connective (the disjunction of the third and second columns) follows in the final (fourth) column.

p	q	$\sim p$	$(\sim p) \vee q$
T	T	F	T
T	F	F	F
F	T	T	T
F	F	T	T

■

Example 1.2.2   We compute the truth table for $(\sim p) \wedge p$.

The one sentence variable p generates the two rows of the corresponding truth table. As in example 1.2.1, the innermost connective is $\sim$ and the outermost is $\wedge$. First, the operation of the innermost connective (the negation of p with truth values in the first column) is given in the second column. The effect of the next connective (the conjunction of the second and first columns) follows in the final (third) column.

p	$\sim p$	$(\sim p) \wedge p$
T	F	F
F	T	F

■

Example 1.2.3   We compute the truth table for $(p \wedge q) \rightarrow r$.

The three distinct statement variables p, q and r generate the $2 \times 2 \times 2 = 2^3 = 8$ permutations of truth values in the corresponding truth table. For this sentence, the innermost connective is $\wedge$ and the outermost is $\rightarrow$. The construction of the truth table proceeds as above, starting with the computation for the innermost connective (the conjunction of the first and second columns) in the fourth column and working outward to the next connective (the implication of the fourth and third columns) in the final (fifth) column.

p	q	r	$p \wedge q$	$(p \wedge q) \rightarrow r$
T	T	T	T	T
T	T	F	T	F
T	F	T	F	T
T	F	F	F	T
F	T	T	F	T
F	T	F	F	T
F	F	T	F	T
F	F	F	F	T

■

As can be seen from these three examples, the number of variables in a sentence determines the number of rows in the corresponding truth table. In fact, if a sentence has n variables, the truth table for the sentence has 2^n rows. The proof of this numerical relationship uses mathematical induction and is discussed in section 3.6.

Example 1.2.4   Another variation of the truth table question occurs in the context of sentences blending sentence symbols (which have a fixed, known truth value) with sentence variables (which are unspecified and may be either true or false). For example, if A has truth value T and B has truth value F, we compute the corresponding truth table for $(A \vee p) \rightarrow B$.

p	A	B	$A \vee p$	$(A \vee p) \rightarrow B$
T	T	F	T	F
F	T	F	T	F

■

Question 1.2.1   Compute the truth table for each formal sentence.

(a) $(\sim p) \vee p$ 　　　　　　　　　　　　　　　　　　　　(b) $(\sim p) \wedge (\sim q)$

■

Reflecting on the previous examples and questions, notice that some of the truth tables we have computed possess interesting and important features. In example 1.2.2, we found that the truth table for $(\sim p) \wedge p$ has all F's in the its final column. Similarly, in question 1.2.1, the truth table for $(\sim p) \vee p$ has all T's in its final column. These are special events for sentences and (as with many special events) such sentences are given distinctive names.

Definition 1.2.1   • *A **tautology** is a sentence that has truth value T for every assignment of truth values to its sentence variables.*
　　　　　　　• *A **contradiction** is a sentence that has truth value F for every assignment of truth values to its sentence variables.*
　　　　　　　• *A **contingency** is a sentence that has truth value T for at least one assignment of truth values to its sentence variables and truth value F for at least one assignment of truth values to its sentence variables.*

Example 1.2.5   From question 1.2.1, the truth table for $(\sim p) \vee p$ has all T's in its final column, and so $(\sim p) \vee p$ is a tautology. From example 1.2.2, the truth table for $(\sim p) \wedge p$ has all F's in the its final column, and so $(\sim p) \wedge p$ is a contradiction. From example 1.2.1, the truth table for $(\sim p) \vee q$ has both T's and F's in its final column, and so $(\sim p) \vee q$ is a contingency.

■

Question 1.2.2   Compute the truth table for each sentence and identify each as a tautology, a contradiction, or a contingency.

(a) $p \leftrightarrow (\sim p)$ 　　　　　　　　　　　　　(c) $p \leftrightarrow (p \vee q)$
(b) $p \leftrightarrow p$ 　　　　　　　　　　　　　　　(d) $p \leftrightarrow (p \wedge q)$

■

We finish this section by defining an important relationship between sentences based on their truth tables. When two sentences have identical final columns in their respective truth tables, we identify them as "the same" in the algebra of logic. This insight motivates the following definition.

Definition 1.2.2 *Sentences $\mathbb{B}$ and $\mathbb{C}$ are **logically equivalent** if the standard truth tables for $\mathbb{B}$ and $\mathbb{C}$ have the same final column. We write $\mathbb{B} \equiv \mathbb{C}$ to denote that $\mathbb{B}$ and $\mathbb{C}$ are logically equivalent.*

The use of the word "if" in mathematical definitions (as in the preceding definition of logical equivalence) is a common practice in mathematical discourse and is always interpreted to mean "if and only if." This broader interpretation of "if" is used only in the context of definitions, while for theorems, lemmas, and other mathematical statements, we adhere to the strict, formal interpretation of the if–then logical connective. Thus, when we are reading a mathematical *definition* and encounter the word "if," we read the definition as an "if and only if" statement asserting the exact meaning of the identified word, allowing us to move freely back and forth between the defined word and the definition.

For example, if two sentences are logically equivalent, then the two sentences have the same final column in their standard truth tables. In addition, if two sentences have the same final column in their standard truth tables, then the two sentences are logically equivalent. You will want to develop a facility in this process of transitioning back and forth between defined mathematical words and the corresponding formal definitions.

We develop a good understanding of logical equivalences by considering some pairs of sentences that are logically equivalent, and some that are not.

Example 1.2.6 We prove that $(p \to q) \equiv [(\sim p) \vee q]$.

The basic truth table for the implication $p \to q$ and the standard truth table for $(\sim p) \vee q$ given in example 1.2.1 have the same final columns, as demonstrated below.

p	q	$p \to q$
T	T	T
T	F	F
F	T	T
F	F	T

p	q	$\sim p$	$(\sim p) \vee q$
T	T	F	T
T	F	F	F
F	T	T	T
F	F	T	T

■

Example 1.2.7 We prove that both $[(\sim p) \vee p] \not\equiv [(\sim p) \vee q]$ and $[(\sim p) \vee p] \not\equiv (p \to q)$.

Using the result of example 1.2.6, neither $(p \to q)$ nor $[(\sim p) \vee q]$ is logically equivalent to a contradiction. A contradiction has truth value F in every row of the final column of its standard truth table, while both of these sentences have T in the first row (and also in the third and fourth rows) of their respective final columns. In example 1.2.2, we found that $(\sim p) \wedge p$ is a contradiction. Alternatively, observe that the first sentence in each pair has one sentence variable, while the second sentence has two sentence variables, and so they cannot be logically equivalent.

■

A particularly important pair of logical equivalences is referred to as De Morgan's laws in honor of the nineteenth century English mathematician Augustus De Morgan, who first identified the significance of these relations for mathematical logic, set theory, and general mathematical discourse. De Morgan was born in India while his father was serving as an officer in the military, and shortly after birth lost sight in his right eye. While a child, he showed no particular aptitude for academics or athletics, but in 1823 he entered Trinity College of Cambridge University. In 1827, while only 21 years old, De Morgan was appointed as the first professor of mathematics at the newly founded University College London. As a research mathematician, De Morgan is best known for his contribution to mathematical logic, mathematical induction, and the study of algebras. He was also a prolific writer and was a co-founder and the first president of the London Mathematical Society. De Morgan loved mathematical trivia, and noted that he was x years old in the year x^2 (he was 43 in 1849); people born in 1980 share this in common with De Morgan (they will be $x = 45$ in $x^2 = 45^2 = 2025$).

Question 1.2.3 **De Morgan's laws**   De Morgan's laws specify how negation distributes across conjunctions and disjunctions, changing the primary connective. Verify that the sentences in each of the following pairs are logically equivalent by computing the corresponding truth tables.

(a) $[\sim(p \wedge q)] \equiv [(\sim p) \vee (\sim q)]$ (b) $[\sim(p \vee q)] \equiv [(\sim p) \wedge (\sim q)]$ ■

1.2.1   Reading Questions for Section 1.2

1. State the two truth values of sentential logic. How are they represented?
2. Give an example of a setting in which a three-valued logic might prove useful.
3. State the basic truth tables for the five logical connectives $\sim$, $\wedge$, $\vee$, $\rightarrow$, and $\leftrightarrow$.
4. Define the standard truth table for a sentence with two variables.
5. What is the relationship between the number of variables in a sentence and the number of rows in the corresponding truth table?
6. Discuss the distinction between an inclusive-or and an exclusive-or.
7. Discuss the definition of the truth table for the implication $p \rightarrow q$.
8. Define and give examples of a tautology, a contradiction, and a contingency.
9. Give natural language examples of a tautology, a contradiction, and a contingency.
10. Define logically equivalent sentences.
11. Give an example of a pair of sentences that are logically equivalent and a pair that are not.
12. State De Morgan's laws in both sentential logic and English.

1.2.2   Exercises for Section 1.2

For exercises 1–20, compute the truth table for each sentence and identify each sentence as a tautology, a contradiction, or a contingency.

1. $p \leftrightarrow (\sim p)$
2. $p \wedge (p \rightarrow p)$
3. $\sim[(\sim p) \rightarrow p]$
4. $[p \rightarrow (\sim p)] \vee p$
5. $(\sim p) \rightarrow q$
6. $p \leftrightarrow (\sim q)$

7. $p \rightarrow (q \rightarrow p)$

8. $\sim[p \rightarrow (p \vee q)]$

9. $(p \leftrightarrow q) \leftrightarrow (\sim p)$

10. $(p \vee q) \vee (\sim p)$

11. $[(p \rightarrow q) \wedge (\sim q)] \rightarrow p$

12. $(p \vee q) \wedge [(\sim p) \wedge (\sim q)]$

13. $(p \vee r) \leftrightarrow \sim\{[(\sim p) \wedge (\sim r)]\}$

14. $[q \leftrightarrow r] \leftrightarrow [(\sim q) \wedge r]$

15. $(p \wedge q) \vee r$

16. $(p \wedge q) \rightarrow [(\sim q) \wedge r]$

17. $(p \leftrightarrow q) \leftrightarrow (\sim r)$

18. $(p \vee r) \rightarrow (q \wedge r)$

19. $\{p \rightarrow [\sim(q \wedge r)]\} \rightarrow (r \rightarrow p)$

20. $\{p \rightarrow [q \wedge (\sim r)]\} \rightarrow [(\sim q) \rightarrow (\sim p)]$

In exercises 21–42, determine if each pair of sentences is logically equivalent by computing the corresponding truth tables. Some pairs of sentences have names associated with them to facilitate their use later in the text.

21. Double negation: $\sim(\sim p)$; p

22. De Morgan's laws: $\sim(p \wedge q)$; $(\sim p) \vee (\sim q)$

23. De Morgan's laws: $\sim(p \vee q)$; $(\sim p) \wedge (\sim q)$

24. $p \wedge q$; p

25. $p \vee q$; p

26. Commutativity: $p \wedge q$; $q \wedge p$

27. Commutativity: $p \vee q$; $q \vee p$

28. Associativity: $(p \wedge q) \wedge r$; $p \wedge (q \wedge r)$

29. Associativity: $(p \vee q) \vee r$; $p \vee (q \vee r)$

30. $p \wedge (q \vee r)$; $(p \wedge q) \vee r$

31. Distributivity: $p \wedge (q \vee r)$; $(p \wedge q) \vee (p \wedge r)$

32. Distributivity: $p \vee (q \wedge r)$; $(p \vee q) \wedge (p \vee r)$

33. $p \vee (q \wedge r)$; $(p \vee q) \wedge r$

34. $(p \rightarrow q) \wedge p$; q

35. Contrapositive: $p \rightarrow q$; $(\sim q) \rightarrow (\sim p)$

36. Inverse: $p \rightarrow q$; $(\sim p) \rightarrow (\sim q)$

37. Converse: $p \rightarrow q$; $q \rightarrow p$

38. Implication expansion: $p \rightarrow q$; $(\sim p) \vee q$

39. $p \rightarrow q$; $\sim[p \wedge (\sim q)]$

40. $\sim(q \rightarrow p)$; $(\sim p) \rightarrow (\sim q)$

41. Biconditional expansion: $p \leftrightarrow q$; $(p \rightarrow q) \wedge (q \rightarrow p)$

42. $p \leftrightarrow q$; $(\sim p) \leftrightarrow (\sim q)$

In exercises 43–52, compute the truth table for each sentence under the assumption that sentence symbol A has truth value T and sentence symbol B has truth value F.

43. $A \rightarrow (\sim B)$

44. $(A \wedge B) \vee (\sim B)$

45. $A \rightarrow p$

46. $p \rightarrow B$

47. $p \rightarrow (A \vee B)$

48. $p \rightarrow (A \wedge B)$

49. $A \leftrightarrow [p \vee (\sim B)]$

50. $(B \wedge p) \rightarrow (\sim A)$

51. $[\sim(B \wedge q)] \rightarrow (A \leftrightarrow p)$

52. $(A \wedge p) \rightarrow (q \vee B)$

Exercises 53–55 show that logical equivalence is an "equivalence relation" (an important concept discussed in section 2.3) sharing three key properties in common with the standard equality relation $=$. Verify that $\equiv$ satisfies each property for formal sentences $\mathbb{B}$, $\mathbb{C}$, and $\mathbb{D}$ from sentential logic.

53. Prove $\mathbb{B} \equiv \mathbb{B}$.

54. Prove that if $\mathbb{B} \equiv \mathbb{C}$, then $\mathbb{C} \equiv \mathbb{B}$.

55. Prove that if $\mathbb{B} \equiv \mathbb{C}$ and $\mathbb{C} \equiv \mathbb{D}$, then $\mathbb{B} \equiv \mathbb{D}$.

In exercises 56–57, let $\mathbb{B}$ and $\mathbb{C}$ be formal sentences from sentential logic and use the definitions of tautology and logical equivalence to prove each statement.

56. If $\mathbb{B} \equiv \mathbb{C}$, then $\mathbb{B} \leftrightarrow \mathbb{C}$ is a tautology.
57. If $\mathbb{B} \leftrightarrow \mathbb{C}$ is a tautology, then $\mathbb{B} \equiv \mathbb{C}$.

Exercises 58–70 consider the truth functional rendition of the basic truth tables. The basic truth tables can be thought of as defining functions on truth values as illustrated in the following two examples.

$$f_\sim(T) = F \qquad f_\sim(F) = T$$
$$f_\wedge(T, T) = T \qquad f_\wedge(T, F) = F \qquad f_\wedge(F, T) = F \qquad f_\wedge(F, F) = F$$

In exercises 58–60, follow the model given for $f_\sim$ and $f_\wedge$ and define each truth function on the four distinct ordered pairs of Ts and Fs.

58. $f_\vee$
59. $f_\rightarrow$
60. $f_\leftrightarrow$

In exercises 61–66, use the examples and your answers from exercise 58–60, to compute the value of each composite function.

61. $f_\wedge(f_\sim(T), F)$
62. $f_\leftrightarrow(f_\sim(T), f_\wedge(T, T))$
63. $f_\rightarrow(f_\vee(T, F), f_\wedge(F, T))$

64. $f_\vee(f_\sim(T), f_\sim(F))$
65. $f_\sim(f_\leftrightarrow(T, F))$
66. $f_\sim(f_\rightarrow(f_\sim(T), F))$

In exercises 67–70, determine the function resulting from each composition or explain why the function is not defined.

67. $f_\sim \circ f_\wedge$
68. $f_\wedge \circ f_\sim$

69. $f_\vee \circ f_\sim$
70. $f_\sim \circ f_\vee$

1.3 An Algebra for Sentential Logic

In 1854 George Boole published his groundbreaking work *An Investigation of the Laws of Thought, on Which Are Founded the Mathematical Theories of Logic and Probabilities* [22]. In this book, Boole developed an algebra of logic for manipulating and simplifying formal sentences. Boole was born in Lincolnshire, England in 1815 and, due to financial constraints, was essentially a self-taught mathematician of extraordinary accomplishments. From the age of 16, Boole supported his parents and siblings by running a series of day and boarding schools. During this time he began studying and researching mathematics, eventually winning the Royal Society's Royal Medal in 1844 for a paper *On a general method of analysis* applying algebraic methods to solve differential equations. In 1849 Boole was appointed the first professor of mathematics at the newly founded Queen's College in Cork, Ireland. He taught in Cork for the rest of his life, earning a reputation as an outstanding teacher while remaining a prolific researcher. At the relatively young age of 49, Boole died of a fever after walking from his home to the College in a soaking rainstorm.

Boole's algebra enables an analysis of the reasoning processes fundamental to the pursuit of mathematical truth. Through the subsequent efforts of Augustus De Morgan, Gottlob Frege, Charles Pierce, and other logicians, Boole's initial work ultimately led to a variety of results, including Gödel's incompleteness theorems (which demonstrate that some true mathematical statements can never be proven in any formal system of logic). In this section, we describe an algebra of sentential logic in the spirit of Boole based on the notion of logical equivalence. We also develop an ability to manipulate the logical connectives appearing in sentences, transforming complex sentences into simpler sentences.

Working with this algebra of sentential logic, we also address a fundamental question about the "expressiveness" of our set of connectives. In the last section, our investigations focused on constructing truth tables to determine all possible truth values for a given formal sentence. In this section, we take up the question of turning this process around and ask, "Given a truth table, can we find a sentence satisfying the truth table?" We show the set of five basic logical connectives $\{\sim, \wedge, \vee, \rightarrow, \leftrightarrow\}$ is *adequate* in the sense of possessing enough expressive power to identify a sentence that satisfies any given truth table. We will see that this set of connectives is redundant in the sense that some proper subcollections are also adequate. Armed with the algebra of sentential logic, we identify new adequate sets of connectives by reducing them to known adequate sets of connectives. As we highlight in the exercises, we can even define new connectives "nand" and "nor" that, taken by themselves, form an adequate set of connectives! Besides being of academic interest, we will see in the next section that these ideas play an essential role in the design of computer circuits.

The algebra of sentential logic is based on the notion of logical equivalence and is really quite similar to the standard algebra of numbers and variables. For example, we can expand $(2x)^2 = 4x^2$ using either of the algebraic identities $(ab)^2 = a^2 b^2$ or $(a + b)^2 = a^2 + 2ab + b^2$. Similarly, in the setting of sentential logic, we utilize known logical equivalences to manipulate and simplify formal sentences. In this way, logical equivalence describes a relationship between formal sentences in sentential logic. Consider the following example.

Example 1.3.1   We simplify $[\sim(\sim p)] \vee p$.

$$
\begin{aligned}
\sim(\sim p) \vee p &\equiv p \vee p &&\text{since } [\sim(\sim p)] \equiv p \\
&\equiv p &&\text{since } p \vee p \equiv p
\end{aligned}
$$
∎

As can be surmised from example 1.3.1, we must know certain basic logical equivalences in order to perform such algebraic manipulations. In section 1.2, we began developing a familiarity with various logical equivalences. For ready reference, we gather together the most important and frequently used logical equivalences in the following table.

Formal name	Logical equivalence
Double negation	$\sim(\sim p) \equiv p$
De Morgan's laws	$\sim(p \wedge q) \equiv (\sim p) \vee (\sim q)$ $\sim(p \vee q) \equiv (\sim p) \wedge (\sim q)$

Formal name	Logical equivalence
Implication Expansion Contrapositive Biconditional Expansion	$p \rightarrow q) \equiv (\sim p) \vee q$ $(p \rightarrow q) \equiv (\sim q) \rightarrow (\sim p)$ $(p \leftrightarrow q) \equiv (p \rightarrow q) \wedge (q \rightarrow p)$
Commutativity	$(p \wedge q) \equiv (q \wedge p)$ $(p \vee q) \equiv (q \vee p)$
Associativity	$(p \wedge q) \wedge r \equiv p \wedge (q \wedge r)$ $(p \vee q) \vee r \equiv p \vee (q \vee r)$
Distributivity	$p \wedge (q \vee r) \equiv (p \wedge q) \vee (p \wedge r)$ $p \vee (q \wedge r) \equiv (p \vee q) \wedge (p \vee r)$
Tautology Contradiction	$p \vee (\sim p) \equiv T$ $p \wedge (\sim p) \equiv F$
Simplification	$p \wedge T \equiv p$ and $p \wedge F \equiv F$ $p \vee T \equiv T$ and $p \vee F \equiv p$

Many of the names assigned to these logical equivalences correspond to the names mathematicians have given to similar properties in other algebraic settings. We will want to become adept at referencing these properties and transitioning from one version of a logical equivalence to another.

Example 1.3.2   We prove that $[\sim (p \vee q)] \rightarrow (\sim q)$ is logically equivalent to the tautology T.

$$
\begin{array}{lll}
[\sim (p \vee q)] \rightarrow (\sim q) & \equiv & \{\sim [\sim (p \vee q)]\} \vee (\sim q) \quad & \text{Implication expansion} \\
& \equiv & (p \vee q) \vee (\sim q) & \text{Double negation} \\
& \equiv & p \vee (q \vee (\sim q)) & \text{Associativity} \\
& \equiv & p \vee T & \text{Tautology} \\
& \equiv & T & \text{Simplification}
\end{array}
$$

As with standard algebraic manipulations, there is often more than one path to an answer; the following is another approach to demonstrating this same logical equivalence.

$$
\begin{array}{lll}
[\sim (p \vee q)] \rightarrow (\sim q) & \equiv & [\sim (\sim q)] \rightarrow \{\sim [\sim (p \vee q)]\} \quad & \text{Contrapositive} \\
& \equiv & q \rightarrow (p \vee q) & \text{Double negation (twice)} \\
& \equiv & (\sim q) \vee (p \vee q) & \text{Implication expansion} \\
& \equiv & (\sim q) \vee (q \vee p) & \text{Commutativity} \\
& \equiv & [(\sim q) \vee q] \vee p & \text{Associativity} \\
& \equiv & T \vee p & \text{Tautology} \\
& \equiv & T & \text{Simplification}
\end{array}
$$

■

Example 1.3.3   We identify a sentence logically equivalent to $p \wedge q$ that uses only the logical connectives $\sim$ and $\vee$.

$$
\begin{array}{lll}
p \wedge q & \equiv & [\sim (\sim p)] \wedge [\sim (\sim q)] \quad & \text{Double negation (twice)} \\
& \equiv & \sim [(\sim p) \vee (\sim q)] & \text{De Morgan's laws}
\end{array}
$$

■

The two De Morgan's Laws express the relationship between $\wedge$ (conjunction) and $\vee$ (disjunction) using $\sim$ (negation). We make frequent use of this pair of logical equivalences in transitioning between conjunction and disjunction.

Question 1.3.1   Identify a formal sentence logically equivalent to each sentence that uses only the logical connectives $\sim$ and $\wedge$.

(a) $p \vee q$ Hint: Use double negation and De Morgan's laws.
(b) $p \rightarrow q$ Hint: Use implication expansion and De Morgan's laws.
(c) $p \leftrightarrow q$ Hint: Use biconditional expansion.

■

This algebra of logical equivalence enables us to examine the expressiveness of the connectives in the formal language of sentential logic. For example, since $(p \rightarrow q) \equiv [(\sim p) \vee q]$, can we drop the implication (the "if–then" connective denoted $\rightarrow$) from the set of connectives and make do with just using negation (the "not" connective denoted $\sim$) and disjunction (the "or" connective denoted $\vee$) whenever we need an implication? On the other hand, perhaps we would prefer to drop disjunction and express all disjunctions in terms of negations and implications? In the context of logical equivalence, the strongest rendition of this question of expressiveness is:

Can we find a sentence satisfying any given truth table?

In fact, we can produce such a sentence and, even better, we can accomplish this task for every given truth table using the same standard algorithm. First, we identify a collection of conjunctions (based on the truth values of the variable in each of the "true" rows of the given table) and then form the disjunction of these conjunctions to obtain the desired formal sentence. We illustrate this algorithm in the next two examples.

Example 1.3.4   We identify a formal sentence (using only the connectives $\sim$, $\wedge$, and $\vee$) satisfying the following truth table.

p	q	?
T	T	F
T	F	T
F	T	F
F	F	T

In the context of producing a formal sentence satisfying this truth table, only the "true" rows are important for implementing our algorithm. In particular, we use the two "true" rows of the truth table to identify conjunctions as follows.

p	q	?		
T	T	F		
T	F	T	$p \wedge (\sim q)$	since $p = T$ and $q = F$
F	T	F		
F	F	T	$(\sim p) \wedge (\sim q)$	since $p = F$ and $q = F$

In the work given next to the truth table, we are being a little "loose" in our use of the equality symbol. When we write $p = T$ and $q = F$ next to the second row, we are observing the particular assignment of truth values to the sentence variables in the second row of the truth table. Since $p = T$ in the second row, we take the positive instance p of the sentence variable p and, since $q = F$, we take the negative instance $(\sim q)$ of the sentence variable q to obtain the conjunction $p \wedge (\sim q)$. Similarly, in the fourth row, we have both $p = F$ and $q = F$ and, taking the negative instance $(\sim p)$ and $(\sim q)$ of each sentence variable, we obtain the conjunction $(\sim p) \wedge (\sim q)$. Finally, we take the disjunction of the sentences determined by the second and fourth rows to obtain the desired formal sentence.

$$? \quad \equiv \quad [p \wedge (\sim q)] \vee [(\sim p) \wedge (\sim q)]$$

A complete truth table computation verifies our solution.

p	q	$\sim p$	$\sim q$	$p \wedge (\sim q)$	$(\sim p) \wedge (\sim q)$	$[p \wedge (\sim q)] \vee [(\sim p) \wedge (\sim q)]$
T	T	F	F	F	F	F
T	F	F	T	T	F	T
F	T	T	F	F	F	F
F	F	T	T	F	T	T

We observe that each conjunction outputs exactly one T, while all the other rows are F; this T occurs in the row used to construct the conjunction. The final disjunction combines these various Ts into exactly the right rows needed to produce the given truth table. This method of focusing on the "true" rows and taking the disjunction of the resulting sentences works for every truth table.

Example 1.3.5   We identify a formal sentence (using only the connectives $\sim$, $\wedge$, and $\vee$) satisfying the following truth table.

p	q	r	$?$		
T	T	T	T	$p \wedge q \wedge r$	since $p = q = r = T$
T	T	F	F		
T	F	T	T	$p \wedge (\sim q) \wedge r$	since $p = r = T$ and $q = F$
T	F	F	F		
F	T	T	T	$(\sim p) \wedge q \wedge r$	since $p = F$ and $q = r = T$
F	T	F	F		
F	F	T	F		
F	F	F	F		

We are free to write $p \wedge q \wedge r$ without grouping symbols by the associativity of $\wedge$; recall that $p \wedge (q \wedge r) \equiv (p \wedge q) \wedge r$ from our table of logical equivalences. Taking the disjunction of the sentences determined by rows 1, 3, and 5, we obtain the desired formal sentence (as can be verified with a complete truth table computation).

$$? \quad \equiv \quad [p \wedge q \wedge r] \vee [p \wedge (\sim q) \wedge r] \vee [(\sim p) \wedge q \wedge r]$$

For the sake of completeness, we observe that a truth table without any true rows must have only false rows and is therefore a contradiction. If such a table has 2^n rows, the truth table is satisfied by the contradiction $[p_1 \wedge (\sim p_1)] \wedge p_2 \wedge \cdots \wedge p_n$.

Question 1.3.2 Find formal sentences (using only the connectives $\sim$, $\wedge$, and $\vee$) satisfying each truth table.

(a)

p	q	?
T	T	F
T	F	T
F	T	T
F	F	F

(b)

p	q	r	?
T	T	T	F
T	T	F	T
T	F	T	F
T	F	F	F
F	T	T	F
F	T	F	T
F	F	T	F
F	F	F	T

■

The algorithm illustrated in examples 1.3.4 and 1.3.5 enables us to find a formal sentence expressing *any* given truth table and *always* yields an "or" sentence (or disjunction) of several "and" sentences (or conjunctions). Since this algorithm requires only the logical connectives of negation, conjunction, and disjunction, it leads to the following definition and theorem.

Definition 1.3.1 *A set of connectives is* **adequate** *if every truth table is satisfied by a sentence using only the connectives in the set.*

Theorem 1.3.1 $\{\sim, \wedge, \vee\}$ *is an adequate set of connectives.*

Sketch of Proof Given a truth table, we identify the true rows. If there are no true rows (and so only 2^n false rows) the contradiction $[p_1 \wedge (\sim p_1)] \wedge p_2 \wedge \ldots \wedge p_n$ is the desired formal sentence. If there are true rows, we produce the corresponding conjunction for each true row, with sentence variable p conjoined if p has value T in the row and $(\sim p)$ conjoined if p has value F in the row. Recalling the discussion after example 1.3.4, we observe that the truth table for each conjunction is F on all rows of the corresponding truth table except for a single T in exactly the row used to construct the conjunction. Taking the disjunction of these various conjunctions combines all the various T's into exactly the right rows needed to produce the given truth table.

■

Observe that the output obtained by implementing the algorithm detailed in the preceding sketch of a proof is always an "or" sentence (a disjunction) of several "and" sentences (several conjunctions). We give a special name to sentences exhibiting this distinctive structure.

Definition 1.3.2 *A formal sentence is said to be in* **disjunctive normal form** *if the sentence is the disjunction of sentences consisting of conjunctions of sentence symbols, sentence variables, or their negations.*

Example 1.3.6 The following two sentences are in disjunctive normal form.

$$[p \wedge (\sim q)] \vee [(\sim p) \wedge (\sim q)] \qquad\qquad [p \wedge q \wedge r] \vee [p \wedge (\sim q) \wedge r]$$

■

A single sentence variable can be viewed as a "trivial" conjunction containing no $\wedge$'s. From this perspective, both p and q are trivial conjunctions, and so the sentence $p \vee q$ is in disjunctive normal form. Similarly, the conjunction $p \wedge q \wedge (\sim r)$ can be viewed as a trivial disjunction (containing no $\vee$'s), and so the sentence $p \wedge q \wedge (\sim r)$ is in disjunctive normal form.

Example 1.3.7   In contrast, the following two sentences are *not* in disjunctive normal form.

- $[p \vee (\sim q)] \wedge [(\sim p) \vee (\sim q)]$
  The conjunction $\wedge$ is the primary connective joining two disjunctions, failing to meet the requirements of disjunctive normal form. In the exercises, we consider such sentences which are said to be in *conjunctive normal form*.
- $p \rightarrow [p \vee (\sim q)]$
  Implication is not a negation, conjunction, or disjunction, which are the only logical connectives allowed for disjunctive normal form.

■

Since the proof of theorem 1.3.1 is the first in this text, we reflect briefly on the nature and role of theorems in mathematics. In the sense of using rational thought as a guide toward truth, theorems are the lifeblood of mathematics. A *theorem* is a declaration of mathematical truth that is supported by a *proof*, or a convincing mathematical argument. The truths of mathematics, as embodied in theorems, are of a distinctly different character than the truths of the other sciences or the truths of almost any other area of human endeavor. The theorems of mathematics have a universal character. When we have a proof that "*A* implies *B*," the truth of the theorem does not rely on a mechanical apparatus or a real world manifestation; instead, truth is understood more absolutely as a definite piece of knowledge. When we claim that every truth table is satisfied by a sentence using only the logical connectives $\sim$, $\wedge$, and $\vee$, we really mean *every* truth table. This claim is not only different from the declaration that "Everyone loves chocolate milk," but is also fundamentally different from scientific theories and hypotheses that are true based on the empirical data that is currently available.

A mathematical truth is only identified as a theorem once a thorough and convincing rational argument has been created justifying its truth. As scientists pursuing truth, mathematicians begin with a small collection of (hopefully self-evident) assumptions or properties known as *axioms*. Working from these axioms, further results are argued to be true using deductive reasoning; such an argument is referred to as a *proof* of the result. In practice, many different names are assigned to proven results, including theorem, lemma, corollary, and proposition. In addition, notice that we have identified the argument for theorem 1.3.1 as a "sketch" of a proof; further details must be provided for a complete proof of theorem 1.3.1. In a mathematics course, students study and learn the contemporary norms for mathematical proofs through example, practice, and feedback from the professor who is mentoring the learning experience.

We turn our attention back to the study of the adequacy of sets of connectives. The addition of further connectives to a known adequate set preserves adequacy. The definition of adequate does not require the use of every connective in the given set, and so every extension of an adequate set of connectives is adequate. Therefore, since $\{\sim, \wedge, \vee\}$ is adequate, each of the sets $\{\sim, \wedge, \vee, \rightarrow\}$, $\{\sim, \wedge, \vee, \leftrightarrow\}$, and $\{\sim, \wedge, \vee, \rightarrow, \leftrightarrow\}$ is also adequate.

Working in the other direction, we are led to ask if any smaller sets of connectives are adequate. For example, are any proper subsets of $\{\sim, \wedge, \vee\}$ adequate? The nonempty, proper subsets of $\{\sim, \wedge, \vee\}$ are $\{\sim, \wedge\}$, $\{\sim, \vee\}$, $\{\wedge, \vee\}$, $\{\sim\}$, $\{\wedge\}$, and $\{\vee\}$. The first two of these proper subsets of connectives are adequate, while the last four are not. A set of connectives is proven not adequate by finding a specific truth table that is not expressible by the given set of connectives. For example, $\{\sim\}$ is not adequate because no sentence using only negation satisfies the following truth table.

$$
\begin{array}{c|c}
p & ? \\
\hline
T & T \\
F & T \\
\end{array}
$$

A complete justification that $\{\sim\}$ is not adequate requires more work than simply stating this one observation; these further details are left for your later studies. Instead, we focus on the more positive goal of showing a given set of connectives is adequate.

The strategy employed to show the first two sets of connectives $\{\sim, \wedge\}$ and $\{\sim, \vee\}$ are adequate is common to many areas of mathematics. We reduce the mathematical object under study to another object that is already known to possess the desired property. In this setting, we reduce a given set of logical connectives to another set of connectives already known to be adequate, as modeled in the following example.

Example 1.3.8   We prove $\{\sim, \wedge\}$ is an adequate set of connectives.

From theorem 1.3.1, $\{\sim, \wedge, \vee\}$ is adequate. We show $\{\sim, \wedge\}$ is adequate by finding sentences logically equivalent to each of $\sim p$, $p \wedge q$, and $p \vee q$ using only the given connectives. Since $\sim p$ is logically equivalent to $\sim p$ and $p \wedge q$ is logically equivalent to $p \wedge q$, we just need a sentence logically equivalent to $p \vee q$ using only $\sim$ and $\wedge$. From De Morgan's laws, we know that $\sim (p \vee q) \equiv [(\sim p) \wedge (\sim q)]$. Negating both sides, we have $\sim [\sim (p \vee q)] \equiv \sim [(\sim p) \wedge (\sim q)]$. Thus, by double negation, $(p \vee q) \equiv \sim [(\sim p) \wedge (\sim q)]$ and we have the desired sentence logically equivalent to $p \vee q$ using only $\sim$ and $\wedge$. Thus, $\{\sim, \wedge\}$ is an adequate set of connectives. The following table summarizes this argument that $\{\sim, \wedge\}$ is adequate.

Given adequate $\{\sim, \wedge, \vee\}$		Proving adequate $\{\sim, \wedge\}$
$\sim p$	$\equiv$	$\sim p$
$p \wedge q$	$\equiv$	$p \wedge q$
$p \vee q$	$\equiv$	$\sim [(\sim p) \wedge (\sim q)]$

∎

Question 1.3.3 Prove $\{\sim, \vee\}$ is an adequate set of connectives. Emulate the template given in example 1.3.8, using double negation and the other half of De Morgan's laws to express $\sim p$, $p \wedge q$, and $p \vee q$ using only $\sim$ and $\vee$. ■

1.3.1 Reading Questions for Section 1.3

1. What relationship between formal sentences is the basis for an algebra of sentential logic?
2. State the logical equivalences double negation and De Morgan's laws.
3. State three additional logical equivalences between formal sentences.
4. What is an important question of expressiveness in the context of sentential logic?
5. Describe an algorithm for producing a formal sentence with a given truth table.
6. What is the default formal sentence for a truth table with only false rows?
7. What is the characteristic structure of a formal sentence in disjunctive normal form?
8. Discuss the nature of theorems and proofs in mathematics.
9. Define and give an example of an adequate set of connectives.
10. Describe a strategy for proving that a given set of logical connectives is adequate.
11. Give an example of an inadequate set of connectives.
12. How do we prove that a set of connectives is not adequate?

1.3.2 Exercises for Section 1.3

In exercises 1–6, identify a formal sentence logically equivalent to $(p \to q)$ that uses only the given connectives.

1. $\{\sim, \wedge, \vee, \to, \leftrightarrow\}$
2. $\{\sim, \wedge, \vee, \leftrightarrow\}$
3. $\{\sim, \wedge, \vee\}$
4. $\{\sim, \wedge\}$
5. $\{\sim, \vee\}$
6. $\{\sim, \to\}$

In exercises 7–12, identify a formal sentence logically equivalent to $(p \leftrightarrow q)$ that uses only the given connectives.

7. $\{\sim, \wedge, \vee, \to, \leftrightarrow\}$
8. $\{\sim, \wedge, \vee, \to\}$
9. $\{\sim, \wedge, \vee\}$
10. $\{\sim, \wedge\}$
11. $\{\sim, \vee\}$
12. $\{\sim, \to\}$

In exercises 13–24, identify a formal sentence logically equivalent to each sentence that uses only the connectives $\sim$ and $\wedge$.

13. $p \vee q$
14. $p \to q$
15. $p \leftrightarrow q$
16. $(\sim p) \to q$
17. $p \to (q \to p)$
18. $\sim((\sim p) \to p)$
19. $\sim(p \to (p \vee q))$
20. $(p \leftrightarrow q) \wedge (\sim p)$

21. $(p \wedge q) \leftrightarrow r$

22. $(p \vee q) \wedge r$

23. $(p \vee r) \rightarrow (q \wedge r)$

24. $(p \vee r) \leftrightarrow \{\sim[(\sim p) \wedge (\sim r)]\}$

In exercises 25–30, identify a formal sentence logically equivalent to each sentence that uses only the connectives $\sim$ and $\vee$.

25. $p \wedge q$

26. $p \rightarrow q$

27. $p \leftrightarrow q$

28. $(\sim p) \rightarrow q$

29. $p \rightarrow (q \rightarrow p)$

30. $\sim((\sim p) \rightarrow p)$

In exercises 31–42, identify a formal sentence in disjunctive normal form satisfying each truth table using the algorithm described in theorem 1.3.1.

31.

p	q	?
T	T	F
T	F	T
F	T	T
F	F	T

32.

p	q	?
T	T	T
T	F	F
F	T	T
F	F	F

33.

p	q	?
T	T	T
T	F	F
F	T	T
F	F	T

34.

p	q	?
T	T	T
T	F	T
F	T	F
F	F	T

35.

p	q	?
T	T	F
T	F	T
F	T	F
F	F	F

36.

p	q	?
T	T	T
T	F	T
F	T	T
F	F	T

37.

p	q	r	?
T	T	T	F
T	T	F	F
T	F	T	T
T	F	F	T
F	T	T	F
F	T	F	F
F	F	T	T
F	F	F	F

38.

p	q	r	?
T	T	T	F
T	T	F	T
T	F	T	T
T	F	F	T
F	T	T	F
F	T	F	F
F	F	T	T
F	F	F	T

39.

p	q	r	?
T	T	T	F
T	T	F	T
T	F	T	F
T	F	F	F
F	T	T	T
F	T	F	T
F	F	T	T
F	F	F	T

40.

p	q	r	?
T	T	T	T
T	T	F	T
T	F	T	F
T	F	F	F
F	T	T	T
F	T	F	F
F	F	T	F
F	F	F	F

41.

p	q	r	?
T	T	T	T
T	T	F	F
T	F	T	T
T	F	F	F
F	T	T	F
F	T	F	F
F	F	T	T
F	F	F	T

42.

p	q	r	?
T	T	T	F
T	T	F	F
T	F	T	F
T	F	F	F
F	T	T	T
F	T	F	T
F	F	T	F
F	F	F	T

In exercises 43–52, use the fact that $\{\sim, \vee, \wedge\}$ is adequate to prove each set of connectives is adequate.

43. $\{\sim, \vee\}$

44. $\{\sim, \wedge\}$

45. $\{\sim, \rightarrow\}$

46. $\{\sim, \vee, \rightarrow\}$

47. $\{\sim, \wedge, \rightarrow\}$

48. $\{\sim, \vee, \leftrightarrow\}$

49. $\{\sim, \wedge, \leftrightarrow\}$

50. $\{\sim, \wedge, \vee, \rightarrow\}$

51. $\{\sim, \wedge, \vee, \leftrightarrow\}$

52. $\{\sim, \wedge, \vee, \rightarrow, \leftrightarrow\}$

Exercises 53–56 consider some general properties of truth tables and adequate sets of connectives.

53. Using only the connectives in the set $\{\sim, \leftrightarrow\}$, there are four distinct types of four-row truth tables for two sentence variables. Identify these four truth tables. What does this tell us about the adequacy of $\{\sim, \leftrightarrow\}$?

54. Using truth tables, verify the logical equivalence $(p \vee q) \equiv [(p \leftrightarrow q) \rightarrow q]$.

55. Based on the logical equivalence in exercise 54, we need only show $(\sim p)$ is logically equivalent to a sentence using only $\rightarrow$ and $\leftrightarrow$ in order to prove that $\{\rightarrow, \leftrightarrow\}$ is adequate. However, there is no such sentence, and $\{\rightarrow, \leftrightarrow\}$ is not adequate. What row of the truth table for $(\sim p)$ is problematic and why?

56. Based on exercise 55, what common reason ensures that each of $\{\rightarrow, \wedge\}$, $\{\rightarrow, \vee\}$, and $\{\wedge, \vee\}$ is not an adequate set of connectives?

Exercises 57–58 consider logical connectives that are adequate by themselves. Two such connectives are defined by the following truth tables.

p	q	$p \mid q$
T	T	F
T	F	T
F	T	T
F	F	T

p	q	$p \downarrow q$
T	T	F
T	F	F
F	T	F
F	F	T

The connective $\mid$ is referred to as either the Scheffer stroke or the "nand" connective (since the truth table for $\mid$ is that of a negated and-sentence). Similarly, the connective $\downarrow$ is referred to as either the Pierce arrow or the "nor" connective (since the truth table for $\downarrow$ is that of a negated or-sentence).

In exercises 57–58, prove $\{\mid\}$ is adequate by using truth tables to verify each logical equivalence.

57. $\sim p \equiv p \mid p$

58. $p \wedge q \equiv (p \mid q) \mid (p \mid q)$

In exercises 59–60, prove $\{\downarrow\}$ is adequate by using truth tables to verify each logical equivalence.

59. $\sim p \equiv p \downarrow p$

60. $p \vee q \equiv (p \downarrow q) \downarrow (p \downarrow q)$

Exercises 61–70 consider implications. In technical discourse, the left side of an implication is referred to as the "antecedent" and the right side of an implication is referred to as the "consequent." In addition, the "contrapositive" of an implication $p \rightarrow q$ is the logically equivalent sentence $(\sim q) \rightarrow (\sim p)$.

In exercises 61–70, identify the antecedent, the consequent, and the contrapositive of each implication.

61. If p, then q.

62. If $\sim p$, then q.

63. If $p \vee q$, then $q \vee p$.

64. $q \wedge p$ when $p \wedge q$.

65. $(p \vee q) \equiv p$ when the variable $q = F$.

66. $(p \wedge q) \equiv p$ when the variable $q = T$.

67. If $n > 2$, then $n^2 > 4$.

68. If $n \leq 2$, then $n^2 \leq 4$.

69. $n^2 > 4$ when $n > 2$.

70. $n^2 \leq 4$ when $n \leq 2$.

1.4 Application: Designing Computer Circuits

In theorem 1.3.1 of section 1.3, we described an algorithm for constructing a formal sentence satisfying a given truth table. Recall that the resulting output is always an "or" sentence (a disjunction) of several "and" sentences (several conjunctions) and that we say such sentences are in *disjunctive normal form*.

In addition to the disjunctive normal form's connections with sentences satisfying given truth tables and with adequate sets of connectives, this form is also important for the design of computer circuits. The logical processes simulated by computers are described using a two-state system and correspond directly with sentential logic. The application of sentential logic to two-state systems actually predates computers and was first developed for the design of telephone systems. In the 1920s and 1930s, the first telephone networks were constructed using physical switches that were in either an "open" or a "closed" position. Similarly, the electric current in a computer circuit is either "on" or "off." These two-state systems can be modeled using sentential logic by identifying a correspondence with the truth values "true" and "false." The standard correspondence is given in the following table. The binary values 1 and 0 are traditionally used by computer scientists and computer engineers in the design of computer circuits and so these binary values are included in this table.

Truth value	Phone switch	Electric current	Binary value
true	closed	on	1
false	open	off	0

We study the design of basic computer circuits using our familiarity with sentential logic and with disjunctive normal form. Recall that $\{\sim, \wedge, \vee\}$ is an adequate set of connectives, and so every truth table is satisfied by a formal sentence using only these connectives. Therefore, when designing computer circuits, we utilize three basic circuits or *gates*, where these gates correspond to the connectives for negation, conjunction, and disjunction. We assign the values 1 to T and 0 to F and transform the connectives' truth tables into *input–output tables*. The following chart gives the input–output tables defining the three basic gates along with their standard circuit diagram symbols.

Gate	Input–output table		Diagram symbol
NOT-gate (for ~)	input	output	
	1	0	
	0	1	
AND-gate (for ∧)	input	output	
	1 1	1	
	1 0	0	
	0 1	0	
	0 0	0	
OR-gate (for ∨)	input	output	
	1 1	1	
	1 0	1	
	0 1	1	
	0 0	0	

Any adequate set of connectives can be used to determine a collection of basic gates since every truth table (and so every input–output table) is expressible by an adequate set of connectives. For example, {~, ∧} is an adequate set of connectives, and we could design computer circuits using just a NOT-gate and an AND-gate. However, as we have also seen (particularly in the exercises for section 1.3), using only two connectives can significantly increase the complexity of formal sentences. Therefore, we choose to utilize all three of the basic gates. Interestingly enough, the "nand" and "nor" connectives (introduced in the exercises for section 1.3) are utilized in actual practice. Both nand and nor are adequate by themselves and only require the physical manufacture of a single basic circuit rather than three basic circuits.

A computer circuit is presented as a diagram of wires and gates. The diagram begins on the left with several input wires (these correspond to sentence variables), and a single output wire terminates the circuit on the right of the diagram. Four fundamental rules are followed when creating these circuit diagrams.

- Any single wire can split and provide input wires to two or more gates.
- Input wires cannot combine.
- An output wire from one gate can serve as an input wire for another gate.
- An output wire from a gate cannot loop back to serve as an input wire for the same gate, neither directly nor after passing through any number of intermediate gates.

Before we dive into designing computer circuits, we first trace the computation of a given circuit on some sample inputs. In this section we are generous in using the equal sign to denote both a particular assignment of truth values to sentence variables and the correspondence between truth values and binary values. We also freely utilize our knowledge of truth tables from sentential logic.

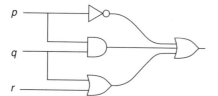

Figure 1.1 The computer circuit for example 1.4.1

Example 1.4.1   We trace some computations of the computer circuit given in figure 1.1.

We first trace the computation of this circuit on the inputs $p = 1$, $q = 0$, $r = 1$.

- From the top gate (the NOT-gate), we have $\sim p =\, \sim 1 =\, \sim T = F = 0$.
- From from the middle gate (the AND-gate), we have $p \wedge q = 1 \wedge 0 = T \wedge F = F = 0$.
- From the bottom gate (the OR-gate on the left), we have $q \vee r = 0 \vee 1 = F \vee T = T = 1$.

Taking the final disjunction (the OR-gate on the right), we combine the output values of the first three gates and obtain $0 \vee 0 \vee 1 = F \vee F \vee T = T = 1$. Therefore, the circuit computes 1 from the given inputs of $p = 1$, $q = 0$, and $r = 1$.

We now trace the computation of this circuit on the set of inputs $p = 0, q = 0$, $r = 1$.

- From the top gate, we have $\sim p =\, \sim 0 =\, \sim F = T = 1$.
- From from the middle gate, we have $p \wedge q = 0 \wedge 0 = F \wedge F = F = 0$.
- From the bottom gate, we have $q \vee r = 0 \vee 1 = F \vee T = T = 1$.

Taking the final disjunction, the circuit computes $1 \vee 0 \vee 1 = T \vee F \vee T = T = 1$.

From tracing these computations, we recognize that the top gate computes $(\sim p)$, the middle gate computes $(p \wedge q)$, and the bottom gate computes $(q \vee r)$. Taking the final disjunction, we see that the given circuit computes the formal sentence $(\sim p) \vee (p \wedge q) \vee (q \vee r)$. Based on this analysis, we can determine a complete input–output table for the given circuit by computing the truth table for this sentence and expressing the result in binary notation (using 1 for T and 0 for F).

■

In the circuit diagram given in figure 1.1 for example 1.4.1, we have three input wires entering the right OR-gate, rather than just two input wires as specified in the original definition of the OR-gate. We are free to adopt this shorthand notation for both the OR-gate and the AND-gate by the associativity of conjunction and disjunction provided by the logical equivalences $p \wedge (q \wedge r) \equiv (p \wedge q) \wedge r$ and $p \vee (q \vee r) \equiv (p \vee q) \vee r$, respectively. Similar logical equivalences hold for an arbitrary number of sentence variables in multiple disjunctions and conjunctions. Therefore, we can increase the number of input wires into a single OR-gate and a single AND-gate as needed in circuit diagrams.

Question 1.4.1 Trace the computation of the circuit given in figure 1.1 on each set of inputs.

    (a) $p = 1$, $q = 1$, $r = 1$ (b) $p = 0$, $q = 1$, $r = 0$

                                                                                  ■

We now study the design of a computer circuit with a given input–output table. In light of the natural correspondence between truth tables and input–output tables suggested by our work, we follow an approach suggested by section 1.3. Specifically, we produce a sentence in disjunctive normal form satisfying the given table, and then use this sentence to design the computer circuit using NOT-gates, AND-gates, and OR-gates.

Example 1.4.2 We use disjunctive normal form to design a computer circuit with the following input–output table.

p	q	output
1	1	1
1	0	0
0	1	1
0	0	1

    Identifying 1 with T and 0 with F, we implement the standard algorithm for disjunctive normal form to produce the sentence $[p \wedge q] \vee [(\sim p) \wedge q] \vee [(\sim p) \wedge (\sim q)]$. This sentence serves as a guide in designing the corresponding computer circuit. First, we compute each of the three conjunctions using NOT-gates and AND-gates. Then, we take the disjunction of the resulting outputs in a rightmost OR-gate (in this case, with three input wires). This process produces the desired circuit diagram given in figure 1.2.

                                                                                    ■

The computer circuit produced in example 1.4.2 is not the simplest possible circuit computing the given input–output table. You may be able to design a simpler circuit using fewer gates based on your familiarity with the algebra of sentential logic. After another example and question, we study "Karnaugh maps" as a means of reducing circuit complexity.

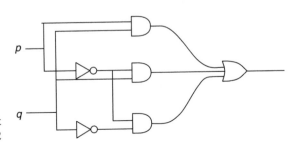

Figure 1.2 The computer circuit for example 1.4.2

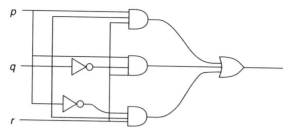

Figure 1.3 The computer circuit for example 1.4.3

Example 1.4.3 We use disjunctive normal form to design a computer circuit with the following input–output table.

p	q	r	output
1	1	1	1
1	1	0	0
1	0	1	1
1	0	0	0
0	1	1	1
0	1	0	0
0	0	1	0
0	0	0	0

The input–output table is satisfied by $[p \wedge q \wedge r] \vee [p \wedge (\sim q) \wedge r] \vee [(\sim p) \wedge q \wedge r]$ by our standard algorithm for disjunctive normal form. First, we compute each of the three conjunctions using NOT-gates and AND-gates, and then we take the disjunction of the resulting outputs in a rightmost OR-gate (again with three input wires). This process produces the desired circuit diagram given in figure 1.3.

■

Question 1.4.2 Use disjunctive normal form to design computer circuits with each input–output table.

(a)

p	q	output
1	1	0
1	0	1
0	1	1
0	0	0

(b)

p	q	r	output
1	1	1	0
1	1	0	1
1	0	1	0
1	0	0	0
0	1	1	0
0	1	0	1
0	0	1	0
0	0	0	1

■

Figure 1.4 The computer circuit for $(\sim p) \vee q$

Recall from example 1.4.2 that the formal sentence $[p \wedge q] \vee [(\sim p) \wedge q] \vee [(\sim p) \wedge (\sim q)]$ was used to design a circuit computing the following input–output table.

p	q	output
1	1	1
1	0	0
0	1	1
0	0	1

You may recognize the pattern of 1s in this input–output table from another setting: if we substitute T for 1 and F for 0, we obtain the basic truth table for the implication $p \rightarrow q$. Example 1.2.6 in section 1.2 proved implication expansion, the logical equivalence $(p \rightarrow q) \equiv [(\sim p) \vee q]$. Based on this logical equivalence, the circuit for $(\sim p) \vee q$ (figure 1.4) also computes the input–output table given in example 1.4.2.

This second circuit is much simpler than one produced in example 1.4.2 and would certainly be favored by engineers and manufacturers because of this relative simplicity. However, the design of this simpler circuit hinged on a bit of clever insight. In contrast, the algorithmic approach of disjunctive normal form guarantees a solution (that is, a circuit diagram) for every given input–output table. Fortunately, this example is not an isolated event and there exists an algorithm that enables the simplification of many formal sentences without requiring too much cleverness. The algorithm involves searching a *Karnaugh map* representation of a given input–output table for patterns of 1's. We illustrate this approach in the next two examples.

Example 1.4.4 We use a Karnaugh map to design a circuit with the input–output table from example 1.4.2.

p	q	output
1	1	1
1	0	0
0	1	1
0	0	1

For two sentence variables p and q, the corresponding 2×2 Karnaugh map is determined by reorganizing the input–output table into a two-row by two-column table. We label the rows and columns with the possible values of p and q and list the output value of 1 in each interior square for which the input–output table has an output value 1. For example, when $p = 1$ and $q = 1$, the output is 1, so the entry in the first-row, first-column interior square of the following

Karnaugh map is also 1. The following is the complete Karnaugh map for the input–output table.

		p	
		1	0
q	1	1	1
	0		1

We now inspect the Karnaugh map for adjacent pairs of 1's in either rows or columns (but not diagonals). In two-variable settings, we are interested in an adjacent pair of 1's since they can always be represented by a *single* sentence variable or its negation. In this example, we find two distinct adjacent pairs of 1's.

- The first row of the interior square is represented by q, since the variable q only takes on the value 1 and the variable p takes on both values 1 and 0. The corresponding formal sentence in disjunctive normal form is $(p \wedge q) \vee [(\sim p) \wedge q]$ and (using logical equivalences), we have

$$(p \wedge q) \vee [(\sim p) \wedge q] \equiv [p \vee (\sim p)] \wedge q \equiv T \wedge q \equiv q.$$

Sentence variables that take on both values 1 and 0 are referred to as *free variables* and always "factor out" of the corresponding formal sentence. These variables represent extraneous data and so they are designated as "free" variables that can take on any value and do not impact the outcome of the computation.

- The second column of the interior square is represented by $(\sim p)$, since the variable p only takes on value 0 and the variable q takes on both values 1 and 0. The corresponding formal sentence in disjunctive normal form and the resulting simplification are

$$[(\sim p) \wedge q] \vee [(\sim p) \wedge (\sim q)] \equiv (\sim p) \wedge [q \vee (\sim q)] \equiv (\sim p) \wedge T \equiv (\sim p).$$

Since every 1 in the Karnaugh map appears in at least one of these adjacent pairs, we move onto the final step. We take a disjunction $q \vee (\sim p) \equiv (\sim p) \vee q$ to obtain the final sentence satisfying the given input–output table. When working with a Karnaugh map, we always take this final disjunction of the component sentences determined by adjacent pairs of 1's. This process produces the desired circuit diagram given in figure 1.5.

■

Karnaugh maps were developed in the early 1950s by the telecommunications engineer Maurice Karnaugh while he was working at Bell Laboratories. In 1953, Karnaugh published his results on what have become known as Karnaugh maps in *The Map Method for Synthesis of Combinational Logic Circuits* [137] and these diagrams

Figure 1.5 The computer circuit for example 1.4.4

are a standard component of computer science and engineering curricula. We are using Karnaugh maps to simplify circuits and formal sentences and this process can be refined to obtain minimal circuits and sentences (minimal in terms of the number of connectives appearing in the final sentence). The success of Karnaugh maps hinges on humans' natural affinity for identifying certain patterns, and this approach works quite well for up to six variables. More sophisticated and subtle algorithms have been developed for simplifying sentences with more than six variables. Consider the following use of a Karnaugh map in the three-variable setting.

Example 1.4.5   We use a Karnaugh map to design a circuit with the input–output table from example 1.4.3.

p	q	r	output
1	1	1	1
1	1	0	0
1	0	1	1
1	0	0	0
0	1	1	1
0	1	0	0
0	0	1	0
0	0	0	0

In this three-variable setting, the two sentence variables p and q are grouped together and the original input–output table is reorganized into a 2×4 Karnaugh map. As in example 1.4.4, we list a 1 in the interior square corresponding to each output of 1 in the input–output table and so obtain the following complete Karnaugh map.

		| 11	10	00	01
r	1	1	1		1
	0				

pq

The column labeling of the 2×4 Karnaugh map is particularly important. We use what is known as **grayscale** labeling, in which exactly one bit changes from one column to the next; this labeling permits adjacent pairs of 1's and 2×2 squares of 1's to "wrap around" the ends of the map. We first inspect the Karnaugh map for 2×2 squares of 1's; such squares can be represented by a single variable (or its negation). We then look for adjacent pairs of 1's in rows and columns (but not diagonals); such pairs can be represented by a conjunction of just two variables (or their negations). In this example, we do not find any 2×2 squares of 1's, but we do find two distinct adjacent pairs of 1's in the first row.

- The first two columns of the first row are represented by $(p \wedge r)$ since p takes on 1, q takes on both 1 and 0, and r takes on 1. We eliminate the free variable q that takes on both 1 and 0 and, since $p = 1 = T$ and $r = 1 = T$, we have $(p \wedge r)$.
- The last and first column of the first row are represented by $(q \wedge r)$. This pair of 1's is adjacent because 2×2 squares and adjacent pairs of 1's can "wrap around" the ends of the map under the given grayscale labeling of columns.

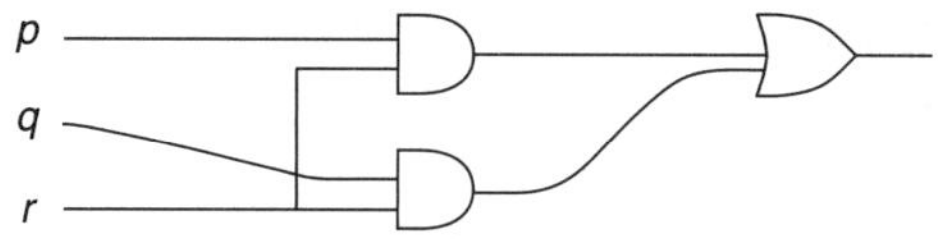

Figure 1.6 The computer circuit for example 1.4.5

For this adjacent pair, p takes on both 1 and 0, q takes on 1, and r takes on 1. We eliminate the free variable p and, since $q = 1 = T$ and $r = 1 = T$, we have $(q \wedge r)$.

Since every 1 in the Karnaugh map appears in at least one of these adjacent pairs, we take the disjunction $(p \wedge r) \vee (q \wedge r)$ to obtain the final sentence satisfying the given input–output table. This process produces the desired circuit diagram given in figure 1.6.

■

Question 1.4.3 Use a Karnaugh map to design a computer circuit with the following input–output table.

p	q	output
1	1	0
1	0	1
0	1	0
0	0	1

■

Thus far, we have used our knowledge of sentential logic to inform and guide our work with computer circuits. However, we now find ourselves in a position to reverse this relationship. Just as we have used Karnaugh maps to simplify computer circuit diagrams, we can also use Karnaugh maps to simplify formal sentences.

Example 1.4.6 We use a Karnaugh map to simplify the following formal sentence.

$$[p \wedge q \wedge (\sim r)] \vee [p \wedge (\sim q) \wedge r] \vee [(\sim p) \wedge (\sim q) \wedge r] \vee$$
$$[p \wedge (\sim q) \wedge (\sim r)] \vee [(\sim p) \wedge (\sim q) \wedge (\sim r)]$$

First, we construct the corresponding 2×4 Karnaugh map, using the standard grayscale labeling of columns as noted in example 1.4.5 and listing the value 1 in the interior square corresponding to each conjunction of the given formal sentence.

		pq 11	10	00	01
r	1		1	1	
	0	1	1	1	

We search the Karnaugh map for instances of 2×2 squares of 1's and adjacent pairs of 1's, which corporately include all 1's in the map. We find one square and one adjacent pair.

- The center 2×2 square of 1's is represented by $(\sim q)$. In particular, p takes on both 1 and 0, q takes on 0, and r takes on both 1 and 0. We eliminate the free variables p and r and, since $q = 0 = F$, we have $(\sim q)$.
- The adjacent pair of 1's determined by the first two columns of the second row is represented by $[p \wedge (\sim r)]$. The variable p takes on 1, q takes on both 1 and 0, and r takes on 0. We eliminate the free variable q and, since $p = 1 = T$ and $r = 0 = F$, we have $[p \wedge (\sim r)]$.

Taking the final disjunction, the given formal sentence is logically equivalent to the much simpler sentence $(\sim q) \vee [p \wedge (\sim r)]$.

■

Question 1.4.4   Use a Karnaugh map to simplify the following formal sentence.

$$[p \wedge q \wedge r] \vee [p \wedge q \wedge (\sim r)] \vee [(\sim p) \wedge q \wedge r] \vee$$

$$[(\sim p) \wedge (\sim q) \wedge (\sim r)] \vee [(\sim p) \wedge q \wedge (\sim r)]$$

In addition, sketch the corresponding circuit diagram. As you work with the Karnaugh map, keep in mind that squares and adjacent pairs can wrap around the ends of the map.

■

For completeness, we mention that isolated 1's in Karnaugh maps do not allow the elimination of any free variables. For example, consider a given input–output table with the following Karnaugh map.

		11	10	00	01
		pq			
r	1	1			
	0			1	

Since no 2×2 squares of 1's or adjacent pairs of 1's appear in this map, there are no free variables to eliminate. Therefore, the corresponding formal sentence is $[p \wedge q \wedge r] \vee [(\sim p) \wedge (\sim q) \wedge (\sim r)]$.

Finally, we should note that a Karnaugh map with all 1's corresponds to a tautology utilizing an appropriate number of sentence variables, while a Karnaugh map with no 1's corresponds to a contradiction utilizing an appropriate number of sentence variables.

1.4.1   Reading Questions for Section 1.4

1. Define disjunctive normal form. Give an example of a sentence that is in disjunctive normal form and a sentence that is not.

2. Define and sketch the symbol for the three basic gates used in designing computer circuits.
3. Discuss the relationship between input–output tables and truth tables. Give an example to facilitate your discussion.
4. State four fundamental rules for designing computer circuits.
5. Why are we free to use just three gates when designing computer circuits?
6. What is the first step in identifying a computer circuit for a given input–output table?
7. What role do Karnaugh maps play in designing computer circuits?
8. What role do Karnaugh maps play in simplifying formal sentences?
9. State the dimensions of a Karnaugh map representing an input–output table with two inputs. What are the dimensions for three inputs?
10. What configuration of 1's do we look for in a 2×2 Karnaugh map? in a 2×4 Karnaugh map?
11. How many variables are needed to represent an adjacent pair of 1's in a 2×2 Karnaugh map? in a 2×4 Karnaugh map?
12. How many variables are needed to represent a square of 1's in a 2×2 Karnaugh map? in a 2×4 Karnaugh map?

1.4.2   Exercises for Section 1.4

In exercises 1–4, trace the computation of the computer circuit given in figure 1.7 on each set of inputs.

1. $p = 1$, $q = 1$ 3. $p = 0$, $q = 1$
2. $p = 1$, $q = 0$ 4. $p = 0$, $q = 0$

In exercises 5–8, trace the computation of the computer circuit given in figure 1.8 on each set of inputs.

5. $p = 1$, $q = 1$ 7. $p = 0$, $q = 1$
6. $p = 1$, $q = 0$ 8. $p = 0$, $q = 0$

In exercises 9–14, trace the computation of the computer circuit given in figure 1.9 on each set of inputs.

9. $p = 1$, $q = 1$, $r = 1$ 12. $p = 0$, $q = 1$, $r = 1$
10. $p = 1$, $q = 1$, $r = 0$ 13. $p = 0$, $q = 0$, $r = 1$
11. $p = 1$, $q = 0$, $r = 1$ 14. $p = 0$, $q = 0$, $r = 0$

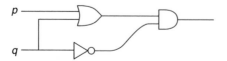

Figure 1.7 The computer circuit for exercises 1–4

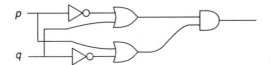

Figure 1.8 The computer circuit for exercises 5–8

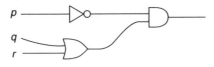

Figure 1.9 The computer circuit for exercises 9–14

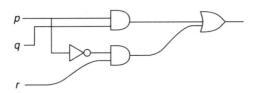

Figure 1.10 The computer circuit for exercises 15–20

In exercises 15–20, trace the computation of the computer circuit given in figure 1.10 on each set of inputs.

15. $p = 1$, $q = 1$, $r = 1$ 18. $p = 0$, $q = 1$, $r = 1$

16. $p = 1$, $q = 1$, $r = 0$ 19. $p = 0$, $q = 0$, $r = 1$

17. $p = 1$, $q = 0$, $r = 1$ 20. $p = 0$, $q = 0$, $r = 0$

In exercises 21–32, use disjunctive normal form to design a computer circuit with each input–output table.

21.

p	q	output
1	1	0
1	0	1
0	1	1
0	0	1

23.

p	q	output
1	1	0
1	0	1
0	1	1
0	0	0

25.

p	q	output
1	1	0
1	0	1
0	1	0
0	0	0

22.

p	q	output
1	1	1
1	0	0
0	1	1
0	0	0

24.

p	q	output
1	1	1
1	0	1
0	1	0
0	0	1

26.

p	q	output
1	1	1
1	0	1
0	1	1
0	0	1

27.

p	q	r	output
1	1	1	0
1	1	0	0
1	0	1	1
1	0	0	1
0	1	1	0
0	1	0	0
0	0	1	1
0	0	0	0

29.

p	q	r	output
1	1	1	0
1	1	0	1
1	0	1	0
1	0	0	0
0	1	1	1
0	1	0	1
0	0	1	1
0	0	0	1

31.

p	q	r	output
1	1	1	1
1	1	0	0
1	0	1	1
1	0	0	0
0	1	1	0
0	1	0	0
0	0	1	1
0	0	0	1

28.

p	q	r	output
1	1	1	0
1	1	0	0
1	0	1	1
1	0	0	1
0	1	1	0
0	1	0	0
0	0	1	1
0	0	0	1

30.

p	q	r	output
1	1	1	1
1	1	0	1
1	0	1	0
1	0	0	0
0	1	1	1
0	1	0	0
0	0	1	0
0	0	0	0

32.

p	q	r	output
1	1	1	0
1	1	0	0
1	0	1	0
1	0	0	0
0	1	1	1
0	1	0	1
0	0	1	0
0	0	0	1

In exercises 33–44, state a formal sentence satisfying the input–output table represented by each Karnaugh map.

33.

		p	
		1	0
q	1		1
	0	1	1

36.

		p	
		1	0
q	1		1
	0	1	

34.

		p	
		1	0
q	1	1	
	0	1	1

37.

		p	
		1	0
q	1	1	1
	0	1	1

35.

		p	
		1	0
q	1	1	1
	0		1

38.

		p	
		1	0
q	1	1	
	0		1

39.

		pq			
		11	10	00	01
r	1	1		1	1
	0	1			1

40.

		11	10	00	01
				pq	
r	1		1	1	
	0	1			1

41.

		11	10	00	01
				pq	
r	1	1		1	
	0	1		1	

42.

		11	10	00	01
				pq	
r	1	1	1	1	
	0	1	1	1	

43.

		11	10	00	01
				pq	
r	1	1	1	1	1
	0	1			1

44.

		11	10	00	01
				pq	
r	1	1	1		1
	0	1			1

In exercises 45–56, use a Karnaugh map to design a computer circuit with each input–output table.

45. The input–output table from exercise 21.

46. The input–output table from exercise 22.

47. The input–output table from exercise 23.

48. The input–output table from exercise 24.

49. The input–output table from exercise 25.

50. The input–output table from exercise 26.

51. The input–output table from exercise 27.

52. The input–output table from exercise 28.

53. The input–output table from exercise 29.

54. The input–output table from exercise 30.

55. The input–output table from exercise 31.

56. The input–output table from exercise 32.

In exercises 57–64, use a Karnaugh map to simplify each formal sentence.

57. $[p \wedge q] \vee [(\sim p) \wedge q]$

58. $[p \wedge q] \vee [p \wedge (\sim q)]$

59. $[p \wedge q] \vee [p \wedge (\sim q)] \vee [(\sim p) \wedge (\sim q)]$

60. $[p \wedge q] \vee [p \wedge (\sim q)] \vee [(\sim p) \wedge (\sim q)]$

61. $[p \wedge q \wedge r] \vee [p \wedge (\sim q) \wedge (\sim r)] \vee [p \wedge (\sim q) \wedge r]$
62. $[(\sim p) \wedge (\sim q) \wedge r] \vee [(\sim p) \wedge (\sim q) \wedge (\sim r)] \vee [(\sim p) \wedge q \wedge (\sim r)]$
63. $[(\sim p) \wedge (\sim q)] \vee [p \wedge q \wedge (\sim r)] \vee [p \wedge (\sim q) \wedge (\sim r)] \vee [(\sim p) \wedge q \wedge (\sim r)]$
64. $[(\sim p) \wedge (\sim q)] \vee [(\sim p) \wedge q] \vee [p \wedge (\sim q) \wedge (\sim r)]$

Exercises 65–70 consider an alternative to disjunctive normal form. A formal sentence is in conjunctive normal form when the sentence is a conjunction of sentences consisting of disjunctions of sentence symbols, sentence variables, or their negations. Given a truth table, a sentence in conjunctive normal form satisfying the truth table is obtained by stating a conjunction based on the truth values of the sentence variable in each "false" row of the given table and then forming the disjunction of all these conjunctions. Finally, we take the negation of the resulting sentence and apply De Morgan's laws and double negation to obtain the desired sentence in conjunctive normal form. Consider the following example.

p	q	?		
T	T	F	$p \wedge q$	since $p = T, q = T$
T	F	T		
F	T	F	$(\sim p) \wedge q$	since $p = F, q = T$
F	F	T		

Taking the negation of the resulting disjunction and applying De Morgan's laws and double negation, we have.

$$
\begin{aligned}
\sim [(p \wedge q) \vee ((\sim p) \wedge q)] &\equiv [\sim (p \wedge q)] \wedge \{\sim [(\sim p) \wedge q]\} \\
&\equiv [(\sim p) \vee (\sim q)] \wedge \{[\sim (\sim p)] \vee (\sim q)\} \\
&\equiv [(\sim p) \vee (\sim q)] \wedge [p \vee (\sim q)].
\end{aligned}
$$

Thus, $[(\sim p) \vee (\sim q)] \wedge [p \vee (\sim q)]$ is a formal sentence in conjunctive normal form satisfying the given truth table.

In exercises 65–70, use conjunctive normal form to design a circuit with each input–output table. For the last three exercises use the generalized De Morgan's law $\sim (p \wedge q \wedge r) \equiv (\sim p) \vee (\sim q) \vee (\sim r)$, which can be verified via a truth table computation.

65. The input–output table from exercise 21.
66. The input–output table from exercise 22.
67. The input–output table from exercise 23.
68. The input–output table from exercise 27.
69. The input–output table from exercise 28.
70. The input–output table from exercise 29.

1.5   Natural Deductive Reasoning

In this section we discuss natural language arguments and deductive reasoning—the very topics that motivated Aristotle's original study of logic. There are many different kinds of arguments, and some of the most effective arguments are those blending rational with irrational (yet persuasive) elements. We focus on the logical content of arguments, identifying arguments as valid or invalid based purely on their formal structure. In this context, an *argument* is a list of sentences. The last sentence is the *conclusion* and the previous sentences include some *premises* or *assumptions* with some intermediate steps often included for clarity. Ideally, the conclusion follows from the set of premises by some valid means of logical reasoning. The process of determining the validity of an argument is quite important, particularly when we must decide whether we believe an argument and when our choices carry profound consequences.

The first step in analyzing arguments is to translate natural language arguments into formal language arguments. This translation enables us to focus on the logical structure of a given argument and to determine its validity. Good reasoning uses a blend of various different argument forms or *rules of deduction*. We define "valid" rules of deduction and then learn some particular rules of deduction. The two most common templates for rules of deduction are

$$\mathbb{A} \quad \therefore \mathbb{B} \qquad \text{and} \qquad \mathbb{C}, \quad \mathbb{D} \quad \therefore \mathbb{E}.$$

The symbol " $\therefore$ " is commonly translated as "therefore" or "thus" and denotes the conclusion of the rule of deduction, and so the first template is read "$\mathbb{A}$, therefore $\mathbb{B}$." Similarly, the second template is read "$\mathbb{C}$ and $\mathbb{D}$, therefore $\mathbb{E}$." In such arguments, $\mathbb{A}$, $\mathbb{C}$, and $\mathbb{D}$ represent formal sentences that serve as premises or assumptions. Given these assumptions, we deduce the conclusions $\mathbb{B}$ and $\mathbb{E}$, respectively.

There are many different rules of deduction, some of which are valid, or correct, and some of which are invalid, or incorrect. The following two rules of deduction are valid.

$$\begin{aligned} \text{Double negation:} &\quad \sim(\sim p) \quad \therefore p \\ \text{Modus ponens:} &\quad p \rightarrow q, \ p \ \therefore q \end{aligned}$$

Double negation assumes the premise $\sim(\sim p)$ and deduces the conclusion p. Similarly, modus ponens assumes the two premises $p \rightarrow q$ and p, and deduces the conclusion q. The name "modus ponens" is Latin for "mode that affirms" and was given to this rule of deduction by the logicians of the Scholastic period during the Middle Ages, when the study of Aristotle's logic flourished in European monasteries.

Before engaging in a broad study of many different rules of deduction, we first define what it means for a rule of deduction to be *valid*, or correct. The fundamental guiding principle that motivates our formal definition asserts, *"an argument is incorrect if it can have true premises and a false conclusion."* We also work with the positive rendition of this guideline: *"an argument is correct if it can never have true premises and a false conclusion."*

With these principles in mind, we develop an algorithm identifying arguments that can *never* have both true premises and a false conclusion. Thinking in terms of the five original logical connectives and their basic truth tables, implication (the "if-then" connective denoted $\rightarrow$) provides the key tool. In particular, implication returns the value true, except in the case "true implies false." Therefore, we define an argument to be valid when a corresponding implication is a tautology. If an argument has multiple premises, we are only interested in the cases when all the premises are simultaneously true, and so we combine the argument's multiple premises using a conjunction. Consider the following definition.

Definition 1.5.1 *A rule of deduction of the form* $\mathbb{A}_1, \ldots, \mathbb{A}_n \therefore \mathbb{C}$ *is* **valid** *if* $(\mathbb{A}_1 \wedge \cdots \wedge \mathbb{A}_n) \to \mathbb{C}$ *is a tautology. Thus, a rule of deduction of the form* $\mathbb{A} \therefore \mathbb{B}$ *is* **valid** *if* $\mathbb{A} \to \mathbb{B}$ *is a tautology; similarly, a rule of deduction of the form* $\mathbb{C}, \mathbb{D} \therefore \mathbb{E}$ *is* **valid** *if* $(\mathbb{C} \wedge \mathbb{D}) \to \mathbb{E}$ *is a tautology.*

Example 1.5.1 We prove double negation: $\sim(\sim p) \therefore p$ is a valid rule of deduction.

Proof Using the definition of validity (that is, definition 1.5.1), we compute the truth table for the corresponding implication $\sim(\sim p) \to p$.

p	$(\sim p)$	$\sim(\sim p)$	$\sim(\sim p) \to p$
T	F	T	T
F	T	F	T

Since this implication is a tautology, the given argument can never have both a true premise and a false conclusion. Therefore, double negation is a valid rule of deduction.

∎

Double negation is a special case of a more general principle: if $\mathbb{B}$ and $\mathbb{C}$ are formal sentences and $\mathbb{B} \equiv \mathbb{C}$, then both $\mathbb{B} \therefore \mathbb{C}$ and $\mathbb{C} \therefore \mathbb{B}$ are valid rules of deduction. In particular, since $\mathbb{B}$ and $\mathbb{C}$ are logically equivalent, they have the same final column in their respective standard truth tables. Therefore, both $\mathbb{B} \to \mathbb{C}$ and $\mathbb{C} \to \mathbb{B}$ are tautologies, and the corresponding rules of deduction $\mathbb{B} \therefore \mathbb{C}$ and $\mathbb{C} \therefore \mathbb{B}$ are valid.

Example 1.5.2 We prove modus ponens: $(p \to q), p \therefore q$ is a valid rule of deduction.

Proof Using the definition of validity, we compute the truth table for the corresponding implication $[(p \to q) \wedge p] \to q$.

p	q	$p \to q$	$(p \to q) \wedge p$	$[(p \to q) \wedge p] \to q$
T	T	T	T	T
T	F	F	F	T
F	T	T	F	T
F	F	T	F	T

Since this implication is a tautology, the given argument can never have both a true premise and a false conclusion. Therefore, modus ponens is a valid rule of deduction.

∎

Question 1.5.1 Prove each rule of deduction is valid.

(a) Conjunctive simplification: $p \wedge q \therefore p$
(b) Modus tollens: $p \to q, \sim q \therefore (\sim p)$ (Latin for "mode that denies")
(c) Disjunctive syllogism: $p \vee q, \sim p \therefore q$

∎

Now that we have a good handle on proving that a rule of deduction is valid, we consider the dual goal of proving a rule of deduction is *not* valid. We show that the corresponding implication is *not* a tautology by demonstrating that one row of the implication's truth table is false (although certainly more than one row—and perhaps even every row—may be false).

Example 1.5.3   We prove the converse error: $p \to q, \ q \ \therefore p$ is *not* a valid rule of deduction.

Proof   Using the definition of validity, we compute the truth table for the corresponding implication $[(p \to q) \wedge q] \to p$.

p	q	$p \to q$	$(p \to q) \wedge q$	$[(p \to q) \wedge q] \to p$
T	T	T	T	T
T	F	F	F	T
F	T	T	T	F
F	F	T	F	T

The final column has truth value F in the third row when $p = F$ and $q = T$. Since the corresponding implication is not a tautology, the given argument can have true premises, but a false conclusion. Therefore, the converse error is not a valid rule of deduction.

■

Question 1.5.2   Prove each rule of deduction is not valid.

  (a)   Inverse error: $p \to q, \ (\sim p) \ \therefore \sim q$ (b) $p \vee q, \ p \ \therefore \sim q$

■

While this truth table algorithm provides a complete approach to determining the validity of arguments, the size of the resulting truth tables and the corresponding computational requirements make this approach unreasonable when the number of variables is large. Since the truth table of a sentence with n variables has 2^n rows, checking the validity of an argument involving multiple premises with many distinct sentence variables can quickly become impractical.

In addition, we are interested in Aristotle's goal of modeling natural deductive reasoning. Natural reasoning often proceeds in incremental steps using basic rules of deduction, including double negation, modus ponens, and conjunctive simplification. For example, we soon verify the validity of the rule of deduction: $(\sim A) \to B$, $(\sim C) \to (\sim A)$, $\sim B$, $C \to D \ \therefore D$. Rather than computing the corresponding 16-row truth table, we proceed from the premises to the conclusion in several steps, justifying each step with a known rule of deduction. The underlying idea is that an argument obtained by composing valid rules of deduction must also be valid, providing us a model of the step-by-step processes inherent in natural reasoning. The next several examples illustrate this approach to proving the validity of complex arguments.

Example 1.5.4   We prove $A \wedge B$, $A \rightarrow C$ $\therefore$ C is a valid argument by justifying each step in the given deduction with a known rule of deduction.

1.	$A \wedge B$	premise
2.	A	1—conjunctive simplification with $p = A$ and $q = B$
3.	$A \rightarrow C$	premise
4.	C	2,3—modus ponens with $p = A$ and $q = C$

In the justification for line 2, the number 1 identifies $A \wedge B$ from line 1 as the premise allowing the deduction of A. Similarly, in the justification for line 4, the numbers 2 and 3 identify the lines containing the premises for the particular implementation of modus ponens that yield the conclusion C given in line 4.

■

Example 1.5.5   We prove $A \rightarrow (\sim B)$, A, $C \rightarrow B$ $\therefore \sim C$ is a valid argument by justifying each step in the given deduction with a known rule of deduction.

1.	A	premise
2.	$A \rightarrow (\sim B)$	premise
3.	$\sim B$	1,2—modus ponens with $p = A$ and $q = \sim B$
4.	$C \rightarrow B$	premise
5.	$\sim C$	3,4—modus tollens with $p = C$ and $q = B$

■

Example 1.5.6   We prove $A \vee B$, $\sim A$, $B \rightarrow C$ $\therefore$ C is a valid argument by justifying each step in the given deduction with a known rule of deduction.

1.	$A \vee B$	premise
2.	$\sim A$	premise
3.	B	1,2—disjunctive syllogism with $p = A$ and $q = B$
4.	$B \rightarrow C$	premise
5.	C	3,4—modus ponens with $p = B$ and $q = C$

■

As seen in these examples, our study focuses on the formal rendition of the component sentences of an argument. When applying this approach to analyze natural language arguments, we would employ the skills and techniques from section 1.1 to create an appropriate dictionary and express the natural language sentences as formal sentences. While one would probably never engage in such a careful, explicit analysis, the habits and patterns of correct reasoning and a careful, implicit analysis of logical arguments are essential to the creation of proofs of mathematical truth.

As can be surmised from these examples, we must know some basic rules of deduction in order to justify the steps in a given deduction. For ready reference, the most frequently used rules of deduction are gathered together in the following chart.

Formal name	Rule of deduction
Modus ponens	$p \rightarrow q,\ p\ \therefore\ q$
Modus tollens	$p \rightarrow q,\ \sim q\ \therefore \sim p$
Double negation	$\sim(\sim p)\ \therefore\ p$ $p\ \therefore \sim(\sim p)$
Conjunctive simplification	$p \wedge q\ \therefore\ p$ $p \wedge q\ \therefore\ q$
Conjunctive addition	$p,\quad q\quad \therefore\ p \wedge q$ $p,\ q\ \therefore\ q \wedge p$
Disjunctive syllogism	$p \vee q,\ \sim p\ \therefore\ q$ $p \vee q,\ \sim q\ \therefore\ p$
Disjunctive addition	$p\ \therefore\ p \vee q$ $q\ \therefore\ p \vee q$
Hypothetical syllogism	$p \rightarrow q,\ q \rightarrow r\ \therefore\ p \rightarrow r$
Dilemma	$p \vee q,\ p \rightarrow r,\ q \rightarrow r\ \therefore\ r$
Contradiction	$(\sim p) \rightarrow [q \wedge (\sim q)],\quad \therefore\ p$
Logical equivalence	if $\mathbb{B} \equiv \mathbb{C}$, then $\mathbb{B}\ \therefore\ \mathbb{C}$
De Morgan's law	$\sim(p \wedge q)\ \therefore\ (\sim p) \vee (\sim q)$ $(\sim p) \vee (\sim q)\ \therefore \sim(p \wedge q)$ $\sim(p \vee q)\ \therefore\ (\sim p) \wedge (\sim q)$ $(\sim p) \wedge (\sim q)\ \therefore \sim(p \vee q)$

This collection of rules of deduction is commonly used in both mathematical and philosophical courses in logic and has been isolated over centuries of study and practice as essential guidelines for correct reasoning. We have already seen proofs of the validity of some of these rules in the preceding examples and questions. The proofs of the new rules of deduction appearing in the bottom half of the chart are given as exercises at the end of this section.

Question 1.5.3 Prove $(\sim A) \rightarrow B,\ (\sim C) \rightarrow (\sim A),\ (\sim B),\ C \rightarrow D\ \therefore\ D$ is a valid argument by justifying each step in the given deduction with a known rule of deduction from the above table, as in the previous examples.

1. $(\sim A) \rightarrow B$ 3. $\sim(\sim A)$

2. $\sim B$ 4. $(\sim C) \rightarrow (\sim A)$

5. $\sim(\sim C)$	7. $C \rightarrow D$
6. C	8. D

∎

Question 1.5.4 Prove $[(\sim A) \wedge B] \rightarrow C, \ \sim A, \ \sim(A \vee C) \ \therefore \sim B$ is a valid argument by justifying each step in the given deduction with a known rule of deduction.

1.	$[(\sim A) \wedge B] \rightarrow C$	5.	$\sim[(\sim A) \wedge B]$
2.	$\sim(A \vee C)$	6.	$[\sim(\sim A)] \vee (\sim B)$
3.	$(\sim A) \wedge (\sim C)$	7.	$A \vee (\sim B)$
4.	$\sim C$	8.	$\sim A$
		9.	$\sim B$

∎

We end this section with an alternative approach to proving a rule of deduction is invalid. A rule of deduction is valid if every entry is T in the final column of the truth table for the corresponding implication, while a rule of deduction is invalid if at least one entry is F in the final column of the truth table for the corresponding implication. The alternative approach is based on the observation that we do not need to produce a complete truth table in order to prove that a rule of deduction is invalid—the T rows of the truth table are irrelevant to demonstrating the invalidity of an argument. Instead, we just need to identify one assignment of truth values to the sentence variables (i.e., one row in the corresponding truth table) for which all the premises are true, but the conclusion is false. Consider the following example.

Example 1.5.7 We prove the converse error: $p \rightarrow q, \ q \ \therefore \ p$ is an invalid rule of deduction using the alternative approach discussed above.

Proof We assume the premises are T and the conclusion is F, resulting in the following collection of assignments of truth values to the sentences appearing in the argument.

$$
\begin{array}{ccc}
p \rightarrow q & q & p \\
T & T & F
\end{array}
$$

Working from these assumptions, we identify the possible assignments of truth values to the corresponding sentence variables as determined by the basic truth tables for the logical connectives.

$$
\begin{array}{lll}
(p \rightarrow q) = T & \Rightarrow & p = F \text{ or } q = T \\
q = T & \Rightarrow & q = T \\
p = F & \Rightarrow & p = F
\end{array}
$$

From the second and third conditions, we have $q = T$ and $p = F$. The first line is now satisfied, since we have both $p = F$ and $q = T$ (recall that mathematicians work with the inclusive-or). Therefore, $p = F$ and $q = T$ is an assignment of truth values to the sentence variables that is consistent with our assumptions and results in the premises all being true while the conclusion is false. We say that $p = F$

and $q = T$ serves both as a *witness* to the invalidity of the converse error and as a *counterexample* to the validity of the converse error.

■

As demonstrated in example 1.5.7, we first assume that the premises are T and the conclusion is F when proving the invalidity of a given argument. We then work with the basic truth tables to obtain a corresponding assignment of truth values to sentence variables that is consistent with all of these assumptions and serves as a witness to the invalidity of the given argument.

Example 1.5.8   We use the alternative approach to prove $(\sim p)$, $(p \vee q)$, $(r \to q)$ ∴ $(p \wedge r)$ is an invalid rule of deduction.

Proof   We assume the premises are T and the conclusion is F, resulting in the following collection of assignments of truth values to the sentences appearing in the argument.

$$
\begin{array}{cccc}
(\sim p) & p \vee q & r \to q & p \wedge r \\
T & T & T & F
\end{array}
$$

Working from these assumptions, we identify the possible assignments of truth values to the corresponding sentence variables as determined by the basic truth tables for the logical connectives.

$$
\begin{array}{lcl}
(\sim p) = T & \Rightarrow & p = F \\
(p \vee q) = T & \Rightarrow & p = T \ \text{ or } \ q = T \\
(r \to q) = T & \Rightarrow & r = F \ \text{ or } \ q = T \\
(p \wedge r) = F & \Rightarrow & p = F \ \text{ and } \ r = F
\end{array}
$$

From the first and last conditions, we have $p = F$ and $r = F$. Based on the second line, we must choose $q = T$, since we already have $p = F$. Under this assignment of truth values, the third line is now satisfied, since we have both $r = F$ and $q = T$ (again, we use the inclusive-or). Therefore, $p = F$, $q = T$, and $r = F$ is an assignment of truth values to sentence variables that results in the premises all being true while the conclusion false, witnessing the invalidity of the given argument.

■

Question 1.5.5   Use the alternative approach to prove the invalidity of each rule of deduction from question 1.5.2.

(a) Inverse error: $p \to q$, $(\sim p)$ ∴ $\sim q$ (b) $p \vee q$, p ∴ $\sim q$

■

1.5.1   Reading Questions for Section 1.5

1. State the definition of an argument for natural deductive reasoning.
2. State one of the general forms for a rule of deduction, and give an example.

3. What is the guiding principle for identifying an argument as valid?
4. How many rows are there in the truth table of a sentence with n distinct sentence variables?
5. Define and give an example of a valid rule of deduction.
6. Define and give an example of an invalid rule of deduction.
7. Describe two distinct approaches to proving that a rule of deduction is valid.
8. Discuss the positive and negative aspects of each approach to proving that a rule of deduction is valid.
9. State four basic rules of deduction.
10. Describe two distinct approaches to proving that a rule of deduction is invalid.
11. Discuss the positive and negative aspects of the two approaches to proving that a rule of deduction is invalid.
12. Discuss the relationship between a witness and a counterexample.

1.5.2   Exercises for Section 1.5

In exercises 1–18, use the definition of validity (that is, an appropriate truth table computation) to prove each rule of deduction is valid. In exercises 1–11, also state the name associated with the rule of deduction.

1. $p \rightarrow q, \quad \sim q \quad \therefore \sim p$

2. $p \wedge q \quad \therefore p$

3. $p \wedge q \quad \therefore q$

4. $p, \quad q \quad \therefore p \wedge q$

5. $p \vee q, \quad \sim p \quad \therefore q$

6. $p \vee q, \quad \sim q \quad \therefore p$

7. $p \quad \therefore p \vee q$

8. $q \quad \therefore p \vee q$

9. $p \rightarrow q, \quad q \rightarrow r \quad \therefore p \rightarrow r$

10. $p \vee q, \quad p \rightarrow r, \quad q \rightarrow r \therefore r$

11. $[p \wedge (\sim q)] \rightarrow [r \wedge (\sim r)]$
    $\therefore p \rightarrow q$

12. $p \rightarrow q, \quad p \rightarrow r$
    $\therefore p \rightarrow (q \wedge r)$

13. $p \leftrightarrow q, \quad p \quad \therefore q$

14. $p \leftrightarrow q, \quad q \quad \therefore p$

15. $p \leftrightarrow q, \quad \sim p \quad \therefore \sim q$

16. $p \leftrightarrow q, \quad \sim q \quad \therefore \sim p$

17. $p \leftrightarrow q, \quad p \leftrightarrow r$
    $\therefore (\sim q) \vee r$

18. $p \leftrightarrow q, \quad p \leftrightarrow r \quad \therefore q \leftrightarrow r$

In exercises 19–32, use the definition of validity (that is, an appropriate truth table computation) to prove each rule of deduction is *invalid*.

19. $p \rightarrow q, \quad \sim p \quad \therefore \sim q$

20. $p \rightarrow q, \quad \sim q \quad \therefore p$

21. $p \vee q, \quad p \quad \therefore \sim q$

22. $p \vee q, \quad p \quad \therefore q$

23. $p \vee q \quad \therefore p$

24. $p \wedge q \quad \therefore \sim r$

25. $p \wedge q \quad \therefore \sim q$

26. $(p \wedge q) \rightarrow r, \quad \sim q \quad \therefore r$

27. $(p \wedge q) \rightarrow r, \quad p \quad \therefore r$

28. $(p \wedge q) \rightarrow r, \quad r \quad \therefore p \wedge q$

29. $(p \vee q) \vee r, \quad \sim p \quad \therefore r$

30. $p \vee q, \quad p \leftrightarrow q \quad \therefore (\sim p) \wedge q$

31. $p \leftrightarrow q, \quad q \leftrightarrow r \quad \therefore (\sim p) \wedge r$

32. $p \leftrightarrow q, \quad \sim q \quad \therefore p$

In exercises 33–42, prove each argument is valid by justifying each step in the given deduction with a known rule of deduction. In addition, state the number of rows in the corresponding truth table proof of validity; do not compute these truth tables, just state the number of rows.

33. $\sim(\sim B), \quad B \rightarrow A \quad \therefore A$
    1. $\sim(\sim B)$
    2. B
    3. $B \rightarrow A$
    4. A

34. $A \wedge B, \quad B \rightarrow C \quad \therefore C$
    1. $A \wedge B$
    2. B
    3. $B \rightarrow C$
    4. C

35. $A \rightarrow B, \quad (\sim B) \wedge C \quad \therefore \sim A$
    1. $(\sim B) \wedge C$
    2. $\sim B$
    3. $A \rightarrow B$
    4. $\sim A$

36. $(A \vee B) \rightarrow C, \quad B \quad \therefore C$
    1. B
    2. $A \vee B$
    3. $(A \vee B) \rightarrow C$
    4. C

37. $A \rightarrow B, \quad B \rightarrow C, \quad A \vee D,$
    $\sim D \quad \therefore C$
    1. $A \rightarrow B$
    2. $B \rightarrow C$
    3. $A \rightarrow C$
    4. $A \vee D$
    5. $\sim D$
    6. A
    7. C

38. $(\sim A) \wedge (\sim B), \quad (\sim C) \rightarrow B,$
    $C \rightarrow D \quad \therefore D$
    1. $(\sim A) \wedge (\sim B)$
    2. $\sim B$
    3. $(\sim C) \rightarrow B$
    4. $\sim(\sim C)$
    5. C
    6. $C \rightarrow D$
    7. D

39. $A \rightarrow (B \wedge C), \quad A \wedge D,$
    $(\sim B) \vee E \quad \therefore E$
    1. $A \wedge D$
    2. A
    3. $A \rightarrow (B \wedge C)$
    4. $B \wedge C$
    5. B
    6. $\sim(\sim B)$
    7. $(\sim B) \vee E$
    8. E

40. $(A \wedge B) \rightarrow C, \quad B \vee D, \quad \sim D,$
    $A \wedge E \quad \therefore C$
    1. $A \wedge E$
    2. A
    3. $B \vee D$
    4. $\sim D$
    5. B
    6. $A \wedge B$
    7. $(A \wedge B) \rightarrow C$
    8. C

41. $[A \vee (\sim B)] \rightarrow C,$
     $(\sim B) \vee D, \sim C,$
     $E \rightarrow A, \quad \therefore \ D \wedge (\sim E)$
     1. $[A \vee (\sim B)] \rightarrow C$
     2. $\sim C$
     3. $\sim [A \vee (\sim B)]$
     4. $(\sim A) \wedge \sim (\sim B)$
     5. $\sim A$
     6. $E \rightarrow A$
     7. $\sim E$
     8. $\sim (\sim B)$
     9. $D \vee (\sim B)$
     10. D
     11. $D \wedge (\sim E)$

42. $(\sim A \vee B) \rightarrow C,$
     $D \vee (\sim B), \sim E, A \rightarrow E,$
     $[(\sim A) \wedge C] \rightarrow (\sim D) \therefore \sim B$
     1. $A \rightarrow E$
     2. $\sim E$
     3. $\sim A$
     4. $(\sim A) \vee B$
     5. $[(\sim A) \vee B] \rightarrow C$
     6. C
     7. $(\sim A) \wedge C$
     8. $[(\sim A) \wedge C] \rightarrow (\sim D)$
     9. $\sim D$
     10. $D \vee (\sim B)$
     11. $\sim B$

In exercises 43–56, use the alternative approach (illustrated in example 1.5.7) to prove each rule of deduction is invalid.

43. The invalid rule from exercise 19.
44. The invalid rule from exercise 20.
45. The invalid rule from exercise 21.
46. The invalid rule from exercise 22.
47. The invalid rule from exercise 23.
48. The invalid rule from exercise 24.
49. The invalid rule from exercise 25.
50. The invalid rule from exercise 26.
51. The invalid rule from exercise 27.
52. The invalid rule from exercise 28.
53. The invalid rule from exercise 29.
54. The invalid rule from exercise 30.
55. The invalid rule from exercise 31.
56. The invalid rule from exercise 32.

In exercises 57–64, classify each argument as an example of modus ponens, modus tollens, converse error, or inverse error.

57. If Socrates is human, Socrates is mortal.
     Socrates is human.
     Therefore, Socrates is mortal.

58. If Socrates is human, Socrates is mortal.
     Socrates is mortal.
     Therefore, Socrates is human.

59. If Socrates is human, Socrates is mortal.
     Socrates is not human.
     Therefore, Socrates is not mortal.

60. If Socrates is human, Socrates is mortal.
     Socrates is not mortal.
     Therefore, Socrates is not human.

61. If n is an even prime, then $n = 2$.
     $n \neq 2$.
     Therefore, n is not an even prime.

62. If n is an even prime, then $n = 2$.
    The number n is an even prime.
    Therefore, $n = 2$.

63. If n is an even prime, then $n = 2$.
    The number n is not an even prime.
    Therefore, $n \neq 2$.

64. If n is an even prime, then $n = 2$.
    $n = 2$.
    Therefore, n is an even prime.

In exercises 65–70, let $\mathbb{B}$ and $\mathbb{C}$ be formal sentences and use the definitions of tautology, logical equivalence, and valid argument to prove each claim.

65. If $\mathbb{B} \equiv \mathbb{C}$, then $\mathbb{B} \leftrightarrow \mathbb{C}$ is a tautology.
66. If $\mathbb{B} \leftrightarrow \mathbb{C}$ is a tautology, then $\mathbb{B} \equiv \mathbb{C}$.
67. If $\mathbb{B} \leftrightarrow \mathbb{C}$ is a tautology, then $\mathbb{B} \,\therefore\, \mathbb{C}$ is a valid argument.
68. $\mathbb{B} \,\therefore\, \mathbb{C}$ is a valid argument does not imply that $\mathbb{B} \leftrightarrow \mathbb{C}$ is a tautology.
    Hint: Give an example of $\mathbb{B}$ and $\mathbb{C}$ such that $\mathbb{B} \,\therefore\, \mathbb{C}$ is a valid argument, but $\mathbb{B} \leftrightarrow \mathbb{C}$ is not a tautology.
69. If $\mathbb{B} \equiv \mathbb{C}$, then $\mathbb{B} \,\therefore\, \mathbb{C}$ is a valid argument.
70. $\mathbb{B} \,\therefore\, \mathbb{C}$ is a valid argument does not imply that $\mathbb{B} \equiv \mathbb{C}$.
    Hint: Give an example of $\mathbb{B}$ and $\mathbb{C}$ such that $\mathbb{B} \,\therefore\, \mathbb{C}$ is a valid argument, but $\mathbb{B} \not\equiv \mathbb{C}$.

1.6   The Formal Language of Predicate Logic

Aristotle and Boole developed formal logic to facilitate the use of human reasoning to study itself. In pursuing this objective, we have defined sentential logic and studied the fundamental connectives of our natural language: not, and, or, if–then, and if and only if. As we have seen through our work with translations, truth, expressibility, computer circuits, and natural deductions, this endeavor has been quite successful. But there is still more to be done.

For example, we cannot yet analyze the validity of the following syllogism (which is another variation on Aristotle's original work):

Every Greek is mortal.
There exists a Greek.
Thus, there exists a mortal.

Intuitively, this argument appears valid, but it can not be verified using sentential logic. In particular, noun–verb–object sentences without connectives are the fundamental "units" or "building blocks" of sentential logic; this syllogism is formalized as p, $q \,\therefore\, r$, where, p is "Every Greek is mortal," q is "There exists a Greek," and r is "There exists a mortal." From the perspective of sentential logic, this argument is invalid, since

$(p \wedge q) \rightarrow r$ is not a tautology under the assignment of truth values $p = T$, $q = T$, and $r = F$.

This example illustrates a need to expand the expressive power of sentential logic to capture more sophisticated forms of valid reasoning. We work toward overcoming the limitations of sentential logic by delving more deeply into the sentence structure of our natural language. In the end, we make two significant extensions of sentential logic to define what is known as *predicate logic*.

The first extension is the addition of *predicates*, which express the verb–object portion of a sentence and identify a property of the subject. Some examples of predicate phrases include: "x is Greek," "x loves y," "x is even," "$x > y$," and the distinguished identity predicate "$x = y$." Adding predicates to the formal logic also leads us to consider the various *names* that can be substituted into predicates, including constants (such as "b" representing Bailey and "5" representing five), variables (such as "x" and "y"), and functions (such as addition and differentiation).

The second extension is the addition of *quantifiers*, which express the notions of "every" and "exists." In the above syllogism, "every" and "exists" are central to the argument about Greeks and mortals. In mathematics as a whole, quantifiers play an important role as we seek to understand and express general truths about mathematical objects.

This section begins with the study of predicates and then develops quantifiers. We start by examining a collection of sentences from the perspective of sentential logic.

Example 1.6.1   We translate each English sentence into sentential logic. In translating, we implicitly define a dictionary; for example, based on the first sentence, C represents "Bailey loves Chris".

- Bailey loves Chris, but not Morgan. $\qquad\qquad\qquad\qquad C \wedge (\sim M)$
- Bailey loves Chris only if Dakota loves Morgan. $\qquad\qquad C \rightarrow D$
- Chris loves neither Bailey nor Dakota. $\qquad\qquad\qquad \sim(B \vee A)$
- Five is either even or odd. $\qquad\qquad\qquad\qquad\qquad\qquad E \vee O$
- Five is even if and only if five squared is even. $\qquad\qquad E \leftrightarrow S$
- Since five is odd and two is even, five plus
  two is odd. $\qquad\qquad\qquad\qquad\qquad\qquad\qquad (O \wedge T) \rightarrow F$

∎

Throughout this section the examples and questions explore both natural language and mathematical translations. Both types of translations are of interest and reveal the power and the versatility of predicate logic in the study of human reasoning.

Question 1.6.1   Translate each English sentence into sentential logic. Notice how this level of translation has become much easier compared to our initial work in section 1.1.

(a) Chris loves Bailey or she loves Dakota.
(b) If Chris loves Dakota, then Dakota does not love Morgan.
(c) Bailey and Chris do not love each other.
(d) Two is both prime and even.

(e) Five is odd only if five plus five is even.

(f) Five is not even, but two is even.

■

Now that we have a good handle on the sentential logic analysis of these English sentences, we develop the perspective of predicate logic. We delve more deeply into the structure of sentences by means of predicates and names. For example, in the sentence "Bailey loves Chris, but not Morgan," the relevant predicate is "x loves y," and the relevant names are "Bailey," "Chris," and "Morgan." In the spirit of formal logic, we represent these English phrases and words with the symbols $L(x, y)$, b, c, and m. At this point, we make an important transition in our practice: we apply the word *predicate* exclusively to strings of symbols of the form $P(x_1, \ldots, x_n)$ that can be interpreted as English predicate phrases and the word *names* exclusively to constants, variables, and functions applied to names.

Definition 1.6.1   *A* **predicate** *is a string of symbols of the form* $P(x_1, \ldots, x_n)$ *where* $x_1, \ldots, x_n$ *are variables. A* **name** *is a string consisting of a single constant, a single variable, or a function applied to names. A predicate has a finite number of variables and is interpreted as true or false in a given context, when nonvariable* **names** *are substituted for the variables appearing in the predicate. We refer to a predicate with n distinct variables as an* **n-place predicate**.

Example 1.6.2   We give some examples of predicates, along with one of the many possible interpretations of each predicate.

1-place predicates	$G(x)$: x is Greek
	$P(x)$: x is prime
2-place predicates	$L(x, y)$: x loves y
	$G(x, y)$: x is greater than y
	$x = y$: x is equal to y
3-place predicates	$T(x, y, z)$: x thinks y is z
	$A(x, y, z)$: $x + y = z$

We also give some examples of nonvariable names, along with one of many possible interpretations of each name.

b : Bailey	c : Chris	$a(2, 2)$: two plus two
d : Dakota	2 : two	$a[a(2, 3), 5]$: two plus three, plus five

■

Question 1.6.2   Give an additional example of a 1-place predicate, a 2-place predicate, a name that is a constant, and a name that is a function applied to a constant.

■

When using predicates, we may *not* compose two or more predicates to assert that a single object possesses multiple properties. Instead, we must translate such sentences using a conjunction of the corresponding predicates. For example, working with the predicates given in example 1.6.2, the predicate logic translation of the (nonsensical) English sentence "x is a Greek prime" as $G[P(x)]$ is *incorrect*. Instead, we translate this sentence into predicate logic as $G(x) \wedge P(x)$, using the conjunction to express that x is both Greek and prime. With these ideas in hand, we reconsider the sentences previously translated into sentential logic.

Example 1.6.3   We use the given dictionary to translate each English sentence from example 1.6.1 into predicate logic.

b : Bailey	c : Chris	d : Dakota	m : Morgan
2 : two	5 : five	$a(x, y) = x + y$	$s(x) = x^2$
$L(x, y)$: x loves y	$E(x)$: x is even	$O(x)$: x is odd	$P(x)$: x is prime

- Bailey loves Chris, but not Morgan. $L(b, c) \wedge [\sim L(b, m)]$
- Bailey loves Chris only if Dakota loves Morgan. $L(b, c) \rightarrow L(d, m)$
- Chris loves neither Bailey nor Dakota. $\sim [L(c, b) \vee L(c, d)]$
- Five is either even or odd. $E(5) \vee O(5)$
- Five is even if and only if five squared is even. $E(5) \leftrightarrow E[s(5)]$
- Since five is odd and two is even, five plus two is odd.

$$[O(5) \wedge E(2)] \rightarrow O[a(5, 2)]$$

∎

We use traditional notation for familiar functions and predicates. From studying algebra and calculus, we recognize that functions are often expressed using the generic notation $a(x, y)$ or $s(x)$, as in example 1.6.3. However, some common functions are usually expressed differently. For example, we typically write $x + y$ rather than $+(x, y)$ and x^2 rather than $^2(x)$. Using this traditional notation, the last two sentences from example 1.6.3 can be translated into predicate logic as follows.

$E(5) \leftrightarrow E[s(5)]$	as	$E(5) \leftrightarrow E(5^2)$
$[O(5) \wedge E(2)] \rightarrow O[a(5, 2)]$	as	$[O(5) \wedge E(2)] \rightarrow O[5 + 2]$

This same practice is also followed when translating familiar mathematical predicates. For example, we usually write $x < y$, rather than $<(x, y)$. Thus, we generally prefer to use traditional mathematical notation for functions and predicates for the sake of readability. However, we do require the strict notation for functions and predicates when crafting proofs about sentences from predicate logic.

Question 1.6.3   Use the dictionary from example 1.6.3 to translate each English sentence from question 1.6.1 into predicate logic.

(a) Either Chris loves Bailey or she loves Dakota.
(b) If Dakota loves Chris, then Dakota does not love Morgan.
(c) Bailey and Chris do not love each other.
(d) Two is both prime and even.
(e) Five is odd only if five plus five is even.
(f) Five is not even, but two is even.

∎

The second significant extension of sentential logic is the introduction of *quantifiers*. The words "every," "all," "exists," and "some" make regular appearances in Aristotle's syllogisms, in our natural language, and in the common language of mathematics, and these words play a central role in much of human reasoning.

The formal language of predicate logic uses the following symbols to represent the given English quantifiers and their variants.

English quantifiers	Formal quantifiers	Formal names
for all, for every, for each	∀	universal
there exists, there is, for some	∃	existential

Just these two quantifiers (expressing the complementary notions of "every" and "exists") capture the full range of human expression and reasoning about quantity in our formal language. As illustrated in the following example, these quantifiers enable the translation of much more sophisticated English sentences and ideas into predicate logic.

Example 1.6.4   We use the given dictionary to translate each English sentence into predicate logic.

b : Bailey c : Chris d : Dakota m : Morgan

2 : two 5 : five $a(x, y) = x + y$ $s(x) = x^2$

$L(x, y)$: x loves y $E(x)$: x is even $O(x)$: x is odd $P(x)$: x is prime

- Someone loves Bailey. $\exists x L(x, b)$
- Everyone loves Chris and Morgan. $\forall x[L(x, c) \wedge L(x, m)]$
- Everyone who doesn't love Bailey, loves Chris. $\forall x\{[\sim L(x, b)] \rightarrow L(x, c)\}$
- There exists an even prime. $\exists x[E(x) \wedge P(x)]$
- If n is even, then n is not odd. $\forall x\{E(x) \rightarrow [\sim O(x)]\}$
- The square of a non-even integer is not even. $\forall x\{[\sim E(x)] \rightarrow [\sim E(x^2)]\}$

■

Even though the words "for all" (or their equivalent) did not explicitly appear in the last two sentences of example 1.6.4, these sentences are still translated as universal sentences (the formal quantifier ∀ is the universal quantifier expressing "for all"). Both of these sentences *implicitly* assert that all numbers satisfy the stated property. In our formal language, we must *explicitly* identify this implicit content of the sentence using the universal quantifier. Many mathematical statements make such implicit claims about all mathematical objects within some context. As we translate sentences into predicate logic, and later as we work on proving mathematical statements, we must be conscious of the frequent occurrence of such implicit universal assertions.

The implication is also crucial for the correct translation of the last two sentences of example 1.6.4. In particular, the last sentence is sometimes *mistakenly* translated as $\forall x\{[\sim E(x)] \wedge [\sim E(x^2)]\}$, which claims that "every number is not even and the square of every number is not even." This is not the sentence we have been asked to translate. Instead, we use an implication to specify a context for the claim made in this sentence. For ease of reference in technical discourse, the left side of an implication is referred to as the *antecedent*, and the right side of an implication is referred to as the *consequent*. In this last sentence of example 1.6.4, the antecedent ($\sim E(x)$ expressing "x is not even") frames the context, and the consequent ($\sim E(x^2)$ expressing "x^2 is not even") makes an assertion about the numbers in the specified context. This process of "setting the stage" is crucial in mathematics, since context determines the truth of mathematical statements.

Question 1.6.4   Using the dictionary from example 1.6.4, translate each English sentence into predicate logic.

(a) Bailey loves someone.
(b) Someone loves themselves and Dakota.
(c) Everyone who loves Chris also loves Morgan.
(d) Some primes are odd.
(e) If n is even, then n^2 is even.
(f) The square of every even integer is even.

∎

Predicates and quantifiers provide the tools needed to make a more careful analysis of the syllogism introduced at the beginning of this section.

Example 1.6.5   We state the sentential logic translation and the predicate logic translation of the syllogism given at the beginning of this section. For this translation, we use the dictionary $G(x)$: "x is Greek," and $M(x)$: "x is mortal."

The syllogism	Sentential logic	Predicate logic
Every Greek is mortal.	p	$\forall x\, [G(x) \rightarrow M(x)]$
There exists a Greek.	q	$\exists x\, G(x)$
Thus, there exists a mortal.	$\therefore\ r$	$\therefore\ \exists x\, M(x)$

A comparison of these translations clearly illustrates the finer analysis provided by predicate logic. The sentential logic is simply insufficient for analyzing valid arguments of this complexity, while the predicate logic enables us to verify the validity of this syllogism (see example 1.7.7 at the end of section 1.7).

∎

Thus far, we have only translated sentences that require the use of a single quantifier. Many interesting and important mathematical statements must be expressed with multiple quantifiers. Recall the following definition of the limit from calculus:

$$\lim_{x \to c} f(x) = L \qquad \text{iff} \qquad \text{for every } \varepsilon > 0, \text{ there exists } \delta > 0, \text{ such that}$$
$$0 < |x - c| < \delta \text{ implies } |f(x) - L| < \varepsilon$$

Considering this definition from the perspective of predicate logic, we see the definition begins with "every ε," followed by "there exists δ," and then an implicit "for all x." Therefore the formal translation of this definition begins $\forall \varepsilon\ \exists \delta\ \forall x$. In other words, we must work with multiple quantifiers to express precisely the definition of the limit of a function at a point. This alternation of quantifiers is why this notion is challenging for many calculus students.

Example 1.6.6   We use the given dictionary to translate each English sentence into predicate logic.

$$x + y \quad x^2 \quad L(x, y): x \text{ loves } y \quad E(x): x \text{ is even} \quad P(x): x \text{ is prime}$$

- Everyone loves someone. $\forall x \exists y L(x, y)$
- There is someone whom everyone loves. $\exists x \forall y L(y, x)$
- Everyone who loves someone does not love
  everyone. $\forall x [\exists y L(x, y) \rightarrow \sim \forall z L(x, z)]$
- If x and y are even, then the sum of x and
  y is even. $\forall x \forall y \{[E(x) \wedge E(y)] \rightarrow E(x + y)\}$
- The sum of two odds is even. $\forall x \forall y \{[O(x) \wedge O(y)] \rightarrow E(x + y)\}$
- There exist x, y, z such that $x^2 + y^2 = z^2$. $\exists x \exists y \exists z [x^2 + y^2 = z^2]$

■

Question 1.6.5   Use the dictionary from example 1.6.6 to translate each English sentence into predicate logic. Assume that $=$ is in the dictionary.

(a) Everyone is loved by someone.

(b) Someone loves everyone.

(c) Everyone who loves someone is loved by someone.

(d) The sum of two evens is not prime.

(e) There do not exist even x and odd y with an even sum.

(f) For some x, y, we have $(x + y)^2 = x^2 + y^2$.

■

As we have seen in calculus, the definition of limit is essential to calculus; the subtleties in the notion of a limit provide one example of the need for multiple quantifiers. Limits also arise in the context of infinite sequences of numbers. Recall that *sequences* are infinite lists of numbers; some basic examples include $1, 2, 3, \ldots, n, \ldots$ and $1, -1, 2, -2, 3, -3, \ldots$. In the following example, we consider the predicate logic rendition of various limit definitions associated with sequences.

Example 1.6.7   We use the given dictionary to translate each mathematical definition into predicate logic.

$$d(x, y) = |x - y| \quad \text{and} \quad x > y: x \text{ is greater than } y$$

The function $d(x, y)$ provides a measure of "distance" on the real line based on the absolute value function. The predicate $x > y$ is the standard "greater than" relation; we also use the "less than" $(y < x)$ version of this predicate for the sake of readability.

- The sequence $\{a_n\}$ *converges* to L if for every ε greater than 0, there exists an N such that for every n greater than N, a_n is within ε of L. Thus, $\lim_{n \to \infty} a_n = L$
  is translated as

$$\forall \varepsilon \, \exists N \, \forall n \, \{ \, [\, (\varepsilon > 0) \wedge (n > N) \,] \rightarrow [\, d(a_n, L) < \varepsilon \,] \, \}.$$

- The sequence $\{f(a_n)\}$ converges to $f(L)$ is translated as

$$\forall \varepsilon \; \exists N \; \forall n \; \{ \; [\; (\varepsilon > 0) \wedge (n > N) \;] \; \rightarrow \; [\; d[f(a_n), f(L)] < \varepsilon \;] \; \}.$$

- The sequence $\{a_n\}$ is *Cauchy* (named after a famous mathematician latter in the text) if for every ε greater than 0, there exists an N such that for every m, n greater than N, a_m and a_n are within ε of each other. Thus, we translate $\{a_n\}$ is Cauchy as

$$\forall \varepsilon \; \exists N \; \forall m \; \forall n \{ \; [\; (\varepsilon > 0) \wedge (m > N) \wedge (n > N) \;] \; \rightarrow \; [\; d(a_m, a_n) < \varepsilon \;] \; \}.$$

- We translate the sequence $\{f(a_n)\}$ is Cauchy as

$$\forall \varepsilon \; \exists N \; \forall m \; \forall n \{ \; [\; (\varepsilon > 0) \wedge (m > N) \wedge (n > N) \;] \; \rightarrow \; [\; d[f(a_m), f(a_n)] < \varepsilon \;] \}.$$

∎

We end this section with a precise definition of the formal language of predicate logic. This definition parallels our work with sentential logic, but has additional elements because of the introduction of names. The formal language of predicate logic has three components: an *alphabet* identifying the legal symbols that may be used; *names* consisting of strings of symbols from the alphabet that may be substituted into predicates; and *sentences* consisting of legal strings of symbols from the alphabet that make assertions. Consider the following definition of these three components.

Definition 1.6.2 *The formal* **alphabet** *of predicate logic consists of exactly the following symbols.*

constants:	$a, b, \ldots, o$
functions:	$f(x_1, \ldots, x_n)$ for all n
predicates:	$P(x_1, \ldots, x_n)$ for all n
identity predicate:	$=$
variables:	$p, q, \ldots, z$
logical connectives:	$\sim, \wedge, \vee, \rightarrow, \leftrightarrow$
quantifiers:	$\forall, \exists$
grouping symbols:	$(,), [,], \{, \}$

As with sentential logic, constants, functions, predicates, and variables may be indexed, and so we have infinitely many such symbols. For example, there exist infinitely many constants $a_1, a_2, a_3, \ldots$, infinitely many variables $x_1, x_2, x_3, \ldots$, infinitely many functions $f_1(x_1), f_2(x_2), f_3(x_3), \ldots$, and infinitely many predicates $P_1(x_1), P_2(x_2), P_3(x_3), \ldots$. In addition, the choice of constants from roughly the first half of the English alphabet and variables from roughly the second half of the English alphabet is not a strict distinction. We also use other symbols for functions and predicates. For example, g and h are commonly used for functions, and L, E, and O are used to identify predicates as in the various examples and questions of this section. Finally, the phrase "for all n" in the specification of functions and predicates indicates the availability of functions with any finite number of variables. Thus, we have functions $f(x_1), f(x_1, x_2), f(x_1, x_2, x_3), \ldots$, so on, and an *n-place function* is a function with n variables; similarly, we have predicates $P(x_1), P(x_1, x_2), P(x_1, x_2, x_3), \ldots$ and so on, and an *n-place predicate* is a predicate with n variables.

Definition 1.6.3 *A **name** of predicate logic is a string of symbols from the alphabet of predicate logic that satisfies the following:*

- *a single constant or a single variable is a name;*
- *an n-place function applied to n names is a name.*

In this section, we have used multiple names, including the constants b for Bailey and 2 for two, the variables x and y, and the functions $x + y$ and x^2. As suggested by the second condition of this definition, $x + 2$, $(x + y) + 2$, and $[(2 + y) + 2]^2$ are all names, since we are allowed to compose, or "layer," functions multiple times to produce new names. In contrast, $\sim(x + y)$, $x \wedge y$, $T(x, y)$, $2^2 < 5$, and $\forall x\, x^2$ are not names, since we may not use connectives, predicates, or quantifiers in stating names—such symbols make assertions rather than identify objects.

Definition 1.6.4 *A **sentence** of predicate logic is a string of symbols from the alphabet of predicate logic that satisfies the following:*

- *if $A_1, \ldots, A_n$ are names and $P(x_1, \ldots, x_n)$ is an n-place predicate, then $P(A_1, \ldots, A_n)$ is a sentence;*
- *if A_1 and A_2 are names, then $A_1 = A_2$ is a sentence;*
- *if x is a variable and $\mathbb{B}, \mathbb{C}$ are sentences, then so are $(\sim\mathbb{B})$, $(\mathbb{B} \wedge \mathbb{C})$, $(\mathbb{B} \vee \mathbb{C})$, $(\mathbb{B} \to \mathbb{C})$, $(\mathbb{B} \leftrightarrow \mathbb{C})$, $\forall x(\mathbb{B})$, $\exists x(\mathbb{B})$.*

For the sake of readability, we often abbreviate sentences by omitting the outermost pair of parentheses, and we utilize the other grouping symbols given in the alphabet. We also drop the numeric subscripts from variables; for example, we may write $\forall x \exists y L(x, y)$ for $\forall x_1 \exists x_2 L(x_1, x_2)$. We have examined and produced many different examples of sentences in this section, including $E(5) \leftrightarrow E(5^2)$ and $\forall x \exists y L(x, y)$. In contrast, the following chart provides strings of symbols that are not sentences.

Non-sentence	Reason
$(x + y) \wedge 5$	Connectives do not apply to names such as 5, only sentences
$5 = (2 < 5)$	A name and a sentence cannot be equal, only two names can be equal
$\exists \sim x[P(x) \wedge E(x)]$	The ordering of connectives and quantifiers is critical
$P(E(x))$	Predicates cannot be composed

1.6.1   Reading Questions for Section 1.6

1. What motivates our interest in predicate logic?
2. Define an n-place predicate and give examples for $n = 1, 2, 3$.
3. Define and give an example of each type of name.
4. Define an n-place function and give examples for $n = 1, 2, 3$.
5. Define and give examples of the two quantifiers.
6. Give an example of an implication and identify its antecedent and consequent.
7. Discuss the role of implication in translating mathematical statements into sentential logic.
8. Discuss the role of quantifiers in the definition of a limit.
9. Define and give an example of a sequence.
10. What are the three components of the formal language of predicate logic?
11. Give an example for each clause in the definition of a sentence of predicate logic.
12. Give an example of an expression using symbols from the alphabet of predicate logic that is not a sentence.

1.6.2   Exercises for Section 1.6

In exercises 1–16, use the given dictionary to translate each English sentence into predicate logic.

$$c : \text{Chris} \qquad p : \text{Pat} \qquad L(x, y) : x \text{ loves } y$$

1. Chris loves Pat or Pat loves Chris.
2. Chris and Pat do not love each other.
3. Chris and Pat love each other.
4. If Chris loves someone, then Chris loves Pat.
5. Pat loves both Chris and himself.
6. Pat loves someone.
7. Pat loves everyone, except Chris.
8. If anyone loves Pat, then Chris does.
9. If someone loves Chris, then Pat loves Chris.
10. Someone loves both Chris and Pat.
11. Everyone loves themselves.
12. Everyone loves someone who is loved by someone.
13. Chris doesn't love anyone who doesn't love someone.
14. No one loves Chris, but everyone loves Pat.
15. Since everyone loves themselves, everyone loves someone.
16. Not everyone who loves someone also loves themselves.

In exercises 17–32, use the given dictionary to translate each English sentence into predicate logic.

$$0, 2, 5 \qquad E(x) : x \text{ is even} \qquad x > y : x \text{ is greater than } y$$
$$x + y \qquad P(x) : x \text{ is prime}$$

17. Two is both prime and even.
18. Zero is even, but not prime.
19. Five is not even, but two is even.
20. Either zero or two is prime, but not both.
21. The sum of two and five is greater than two and greater than five.
22. The sum of two and five is prime, but not even.
23. If the sum of two and five is prime, then the sum is not even.
24. Some even numbers are not prime.
25. If a number is greater than zero, then the number is not zero.
26. There exists an even number and an odd number.
27. The sum of two even numbers is even.
28. The sum of two odd numbers is even.
29. The sum of three even numbers is not prime.
30. If n is a prime that is not even, then n is greater than two.
31. For every even number, there is a greater even number.
32. There exists an even prime.

In exercises 33–40, use the two given dictionaries to translate each English sentence into predicate logic in two different ways.

Dictionary A: 0 $E(x)$: x is even $x > y$: x is greater than y
Dictionary B: $Z(x)$: x is zero $E(x)$: x is even $x > y$: x is greater than y

33. Zero is even.
34. If n is zero, then n is even.
35. There exists a number greater than zero.
36. There exists an even number greater than zero.
37. Every number is greater than zero.
38. Every even number is greater than zero.
39. Some number is less than zero.
40. If a number is greater than zero, then it is not zero.

In exercises 41–50, use the given dictionary to translate each definition of a mathematical property into predicate logic.

$$x + y \qquad x < y : x \text{ is less than } y$$

41. Commutativity: The sum of x and y is the same as the sum of y and x.
42. Non-commutativity: For some x and y, the sum of x and y is different from the sum of y and x.
43. Identity: There exists an element e such that for all x, the sum of x and e is x and the sum of e and x is x.
44. Reflexivity: For every x, x is less than x.
45. Irreflexivity: For every x, x is not less than x.
46. Symmetry: If x is less than y, then y is less than x.
47. Asymmetry: If x is less than y, then y is not less than x.
48. Transitivity: If x is less than y and y is less than z, then x is less than z.

49. Density: If x is less than y, then there exists z such that z is between x and y.

50. Comparability: For every x and y, x is less than y or x is the same as y or y is less than x.

In exercises 51–56, use the given dictionary to translate each mathematical definition into predicate logic (as in example 1.6.7).

$0, c, \delta, \varepsilon, f(x)$	$d(x, y) = \|x - y\|$	$x > y :\ x$ is greater than y
L, N, M	$x = y$	$x < y :\ x$ is less than y

51. $\lim\limits_{x \to c} f(x) = L$ if for every $\varepsilon > 0$, there exists $\delta > 0$ such that for all x within δ of c, but not equal to c, then $f(x)$ is within ε of L.

52. $\lim\limits_{x \to c} f(x) \neq L$ if for some $\varepsilon > 0$, for every $\delta > 0$, there exists x within δ of c such that $x \neq c$ and $f(x)$ is more than ε from L.

53. $\lim\limits_{x \to c} f(x) = +\infty$ if for every $M > 0$, there exists $\delta > 0$ such that for all x within δ of c, but not equal to c, then $f(x)$ is greater than M.

54. $\lim\limits_{x \to c} f(x) = -\infty$ if for every $M < 0$, there exists $\delta > 0$ such that for all x within δ of c, but not equal to c, then $f(x)$ is less than M.

55. $\lim\limits_{x \to +\infty} f(x) = L$ if for every $\varepsilon > 0$, there exists $N > 0$ such that for all x greater than N, then $f(x)$ is within ε of L.

56. $\lim\limits_{x \to +\infty} f(x) = L$ if for every $\varepsilon > 0$, there exists $N < 0$ such that for all x less than N, then $f(x)$ is within ε of L.

Exercises 57–70 focus on the identity predicate. Recall that this predicate is denoted by "$x = y$" and is automatically included in every dictionary for predicate logic. While the identity predicate is just one among many 2-place predicates, some sentences make essential use of the distinguished identity predicate. For example, "Chris loves only Pat," "No one loves everyone else," and "There exist at least two even numbers," can only be expressed via the notion of identity. The words "only," "besides," "else," "at least," and "exactly" refer to relations among objects in terms of being the same or different than some other object and must be expressed using the identity.

In exercises 57–70, use the given dictionary to translate each English sentence into predicate logic.

$c :$ Chris	$p :$ Pat	$L(x, y) :\ x$ loves y	$x > y :\ x$ is greater than y
0	4	$E(x) :\ x$ is even	$P(x) :\ x$ is prime

57. Chris loves only Pat.
58. Only Chris loves Pat.
59. Everyone loves someone.
60. Everyone loves someone else.
61. No one loves everyone.
62. No one loves everyone else.
63. There is exactly one even prime.
64. Every other prime is greater than the even prime.
65. There are at least two numbers greater than zero.

66. There are at least two primes greater than zero.
67. There are at least three numbers greater than zero.
68. There are at least three primes greater than zero.
69. There are at least three numbers between zero and four.
70. There are exactly three numbers between zero and four.

1.7   Fundamentals of Mathematical Proofs

In this chapter we have studied logic partly for its own sake, but also to facilitate our pursuit of mathematical truth; we expect the study of human reasoning to enhance our intuition when we turn to the broad study of mathematics. In the last section, we translated well-known mathematical statements into predicate logic, including such assertions as "The sum of two even numbers is even" and "There exists an even prime." We now turn our attention to exploring when and why these statements are true. While there is great value in knowing what mathematical statements are true, there is even greater value in understanding why they are true. Among other things, knowing why a statement is true often allows us to understand a whole host of other mathematical statements. This general understanding then promotes our creative efforts in extending the body of known mathematical results. In this section we explore five fundamental approaches to proving mathematical truths: *direct proof*; *proof by contradiction*; *proof by contrapositive*; *proof by example*; and *proof by counterexample*. These are the most widely used tools of mathematical reasoning, and mastering these proof techniques will enable your increasingly more sophisticated forays into mathematical truth.

Before we dive into working with these proof techniques, we briefly discuss an overall perspective of mathematics as a science. Mathematics is a "deductive science of the conditional" in which the objects of study are ideas, and the fundamental tool of study is logical reasoning. Many would argue that mathematics extends beyond rational thought—that intuition and unconscious insight are key to mathematical creativity, and that the thirst to uncover new truths is the driving motivation behind mathematical endeavor. Furthermore, the many surprising and deep connections between mathematics and our physical world are of great significance and interest. However, creative intuitions, physical results, and specific examples do not ensure mathematical truth. While these are indeed essential elements supporting the continuing exploration and development of mathematics, logical reasoning alone is the final arbiter of mathematical truth, and we rely on logical argument in extrapolating mathematical truth from the evidence provided by specific examples.

Traditionally, mathematicians have looked to *Elements* [73] as a first and ideal model of mathematics as a deductive science. A geometry text written by the Greek mathematician Euclid in 300 B.C.E, copies of the *Elements* have played an important role in sparking periods of intense mathematical creativity—from the Arabic world in the ninth century to Italy in the sixteenth century and even to this day, as a continuing source of inspiration for mathematicians. *Elements* consists of a collection of logical arguments for geometric truths deduced from a few *axioms*, or assumptions. More than just being convincing, many of these proofs have a constructive flavor, and so hopefully enable a deep understanding of why the corresponding mathematical statements are true. *Elements* was the first book to develop the goals of proof in mathematics: to convince; to explain; to illuminate; and to inspire—in short, to make us wiser.

With these reflections in mind, we turn toward developing our skills in crafting proofs. The proof techniques of direct proof, proof by contradiction, and proof by contrapositive are approaches to arguing the truth of conditional statements of the form $(p \rightarrow q)$. Most mathematical statements have this form, with the antecedent p determining the context in which the property expressed by the consequent q is true. We begin with the *premises* or *assumptions* expressed by p, and we attempt to demonstrate the truth of the *conclusion* expressed by q with a logical argument. In this section, we consider an illuminating example of each proof technique and provide an opportunity to work with each in an accompanying question. You will want to reflect carefully on these examples as you start crafting your own proofs and as you begin growing into your own style of "doing mathematics."

As we develop the ability to work with these proof techniques, we consider some very basic mathematical notions: even, odd, rational, and irrational numbers. In subsequent chapters we study more sophisticated mathematical ideas and use these proof techniques to establish more elegant and subtler mathematical truths. Mastering these techniques in this simpler setting will enable you to grapple with more sophisticated notions. We formally define the ideas used in constructing the proofs in this section. In the definitions we use the set-theoretic notation "$\in$" to denote "is an element of" or "in the set." Set theory is studied more fully in section 2.1.

Definition 1.7.1 • *The **integers** are the numbers* $\ldots, -2, -1, 0, 1, 2, \ldots$; *the set of all integers is denoted by* $\mathbb{Z}$. *("Zahlen" is German for "count.")*
  • *An integer* $n \in \mathbb{Z}$ *is **even** if there exists an integer* $k \in \mathbb{Z}$ *such that* $n = 2k$.
  • *An integer* $n \in \mathbb{Z}$ *is **odd** if there exists an integer* $k \in \mathbb{Z}$ *such that* $n = 2k + 1$.
  • *The **reals** are the numbers on the continuum of the real line. They are directed distances from a designated point zero; the set of all real numbers is denoted by* $\mathbb{R}$.
  • *A real* $r \in \mathbb{R}$ *is **rational** if there exist integers* $p, q \in \mathbb{Z}$ *with* $q \neq 0$ *such that* $r = p/q$; *the set of all rational numbers is denoted by* $\mathbb{Q}$.
  • *A real* $r \in \mathbb{R}$ *is **irrational** if* r *is not rational.*

These number systems and adjectives for numbers may be familiar from previous mathematics courses; for example, $\sqrt{2}$ and π are examples of irrational numbers. Before moving on, you should think of a specific example of each type of number identified in definition 1.7.1. In this section we also use the fact that the sum, difference, and product of two integers yields an integer. In addition, we need two more theorems. We prove these claims about the integers and reals at appropriate points later in the text, but for now we just state and use these results.

Theorem 1.7.1 The parity property of the integers *Every integer is either even or odd.*

Theorem 1.7.2 **The zero product property of the reals** *The product of two nonzero real numbers is nonzero.*

With these definitions and theorems in hand, we focus on developing an ability in writing proofs. In a *direct proof* of a conditional statement ($p \rightarrow q$), we assume the premise p (or multiple premises) and work toward the conclusion q. Mathematicians often give direct proofs of mathematical statements that are not phrased as implications; in such settings it is often helpful to first phrase the statement as an implication. Typically, definitions and previously established results enable the transition from p to q as we argue the truth of the desired sentence. In the following examples, we present two versions of each proof. The first is an "expanded" proof in which the arguments are fully described to transparently indicate the thought processes essential to the proof. We then present a "succinct" proof in a more elegant style expressing the essential details of the argument. Each style of proof has its pros and cons. As you begin writing your own proofs, you should probably emulate the "expanded" proofs to help ensure that you don't miss any important details; eventually your proofs will evolve to mirror more closely the "succinct" style of proofs.

Example 1.7.1 We prove the sum of two even integers is even.

An Expanded Proof We first phrase this mathematical statement as an implication: if two integers are even, then the sum of these integers is even. Since the goal is to prove something about every pair of even integers, we identify two arbitrary even integers. Let m and n be even integers. The implication can now be phrased in terms of m and n as: if m and n are even, then $m + n$ is even. Thus, the goal is now to prove that $m + n$ is even.

By the definition of an even integer, we must show that $m + n = 2k$ for some integer $k \in \mathbb{Z}$. The only information available to help us achieve this goal is the fact that m and n are even, so we apply the definition of even to these two integers. Since m is even, there exists an integer $i \in \mathbb{Z}$ such that $m = 2i$. Similarly, since n is even, there exists an integer $j \in \mathbb{Z}$ such that $n = 2j$. Computing the sum and making the appropriate substitutions, we have

$$m + n = 2i + 2j = 2(i + j) = 2k, \quad \text{where } k = i + j.$$

Thus, by the definition of even integers, $m + n$ is even.

■

A More Succinct Proof Let m and n be even integers. We prove that $m + n$ is even. Since m and n are even, there exist integers i and j such that $m = 2i$ and $n = 2j$. Therefore, $m + n = 2i + 2j = 2(i + j)$, and so $m + n$ is even.

■

In both the expanded and the succinct versions of the proof, we explicitly state not only the premises we are using, but also the conclusion toward which we are working. As you develop your own proofs, you should follow this same practice of explicitly identifying both where you are starting from and where you are going.

Often, the bridge linking the premises to the conclusion becomes apparent from the corresponding definitions and other known results.

Question 1.7.1   Prove (directly) that the product of two even integers is even; that is, prove that if m and n are even integers, then $m \cdot n$ is an even integer.

■

We now consider establishing the truth of a conditional mathematical statement based on *proof by contradiction*. The approach taken in a proof by contradiction is justified by the logical equivalence $(p \to q) \equiv \{ [p \wedge (\sim q)] \to [r \wedge (\sim r)] \}$; that is, if we can prove that $p \wedge (\sim q)$ implies a contradiction, then the implication $p \to q$ must also be true. The set-up for a proof by contradiction is apparent from this logical equivalence: we assume both the premise and the negation of the conclusion. The next step can be less obvious—we work toward a contradiction of the form $r \wedge (\sim r)$. While r can be any mathematical statement, we often have the option of using either $r = p$ or $r = q$.

Example 1.7.2   Prove that the sum of a rational number and an irrational number is irrational.

An Expanded Proof   We are asked to prove something about a rational number and an irrational number, so we give ourselves an arbitrary number of each type. Let x be rational, so $x = p/q$ for some integers p and q with $q \neq 0$, and let y be irrational, so y is not equal to such a quotient. Phrasing the mathematical statement we are proving as an implication, we have: if x is rational and y is irrational, then $x + y$ is irrational.

We proceed by contradiction, assuming the premises and the negation of the conclusion and working toward a contradiction. So, in addition to our assumptions that x is rational and y is irrational, we also assume that $x + y$ is not irrational. Since $x + y$ is not irrational, we know that $x + y$ is rational and there exist integers r and s with $s \neq 0$ such that $x + y = r/s$. By substituting and algebraically manipulating the sum, we obtain the following.

$$x + y = \frac{r}{s} \quad \Rightarrow \quad \frac{p}{q} + y = \frac{r}{s} \quad \Rightarrow \quad y = \frac{r}{s} - \frac{p}{q} = \frac{rq - ps}{sq}$$

The product and difference of integers yield an integer and, by the zero product property, the product $sq \neq 0$ since $s \neq 0$ and $q \neq 0$. Therefore, y is a rational number. We have just shown that under our assumptions, the irrational number y must be rational, which is a contradiction. Therefore, the sum of a rational and an irrational is irrational.

■

A More Succinct Proof   We proceed by contradiction. Assume that $x = p/q$ is rational where p and q are integers with $q \neq 0$ and assume that y is irrational. In addition, we assume that $x + y$ is not irrational and work toward a contradiction. Since $x + y$ is not irrational, the sum is rational, and so $x + y = r/s$, where r and s are integers with $s \neq 0$. Substituting and algebraically manipulating the sum, we obtain the following.

$$x + y = \frac{r}{s} \quad \Rightarrow \quad \frac{p}{q} + y = \frac{r}{s} \quad \Rightarrow \quad y = \frac{r}{s} - \frac{p}{q} = \frac{rq - ps}{sq}$$

The product and difference of integers yields an integer and, by the zero product property, the product $sq \neq 0$ since $s \neq 0$ and $q \neq 0$. Thus, the irrational number y is rational, which is a contradiction. Therefore, the sum of a rational and an irrational is irrational.

■

Question 1.7.2 Prove (by contradiction) that the product of a nonzero rational number and an irrational number is irrational. Why do we need a *nonzero* rational number in the product?

■

The last proof technique we consider for establishing the truth of a conditional mathematical statement is *proof by contrapositive*. This third proof technique for implications is justified by the logical equivalence $(p \to q) \equiv [(\sim q) \to (\sim p)]$; that is, in a proof by contrapositive, we swap and negate the premises and conclusion, and then proceed to give a direct proof of the resulting implication. Note that a proof by contrapositive begins in the same fashion as a proof by contradiction—we assume the negation of a conclusion.

Example 1.7.3 We prove that for every integer $n \in \mathbb{Z}$, if n^2 is even, then n is even.

An Expanded Proof Taking the contrapositive of "if n^2 is even, then n is even" by swapping and negating the premise and conclusion, we obtain "if n is not even, then n^2 is not even." By the parity property of the integers, every integer is either even or odd, and so an integer that is not even must be odd. Therefore, the contrapositive is equivalent (by the parity property) to the implication "if n is odd, then n^2 is odd." We give a direct proof of this contrapositive.

We assume n is odd and prove that n^2 is odd. By the definition of an odd integer, we must show $n^2 = 2k + 1$ for some integer $k \in \mathbb{Z}$. The only information we have available to help us achieve this goal is the fact that n is odd, so we apply the definition of an odd integer. Since n is odd, there exists an integer $i \in \mathbb{Z}$ such that $n = 2i + 1$. Computing the square and algebraically manipulating the resulting sum, we have

$$n^2 = (2i + 1)^2 = 4i^2 + 4i + 1 = 2(2i^2 + 2i) + 1$$
$$= 2k + 1 \quad \text{where } k = 2i^2 + 2i.$$

Therefore, n^2 is odd by the definition of an odd integer. Since n is odd implies n^2 is odd, by contrapositive, we know that if n^2 is even, then n is even.

■

A More Succinct Proof We give a proof by contrapositive that n^2 is even implies n is even. Taking the contrapositive and applying the parity property of the integers, we prove that n is odd implies n^2 is odd. We assume n is an odd integer and prove that n^2 is odd. Since n is odd, there exist an integer $i \in \mathbb{Z}$ such that $n = 2i + 1$. Therefore, $n^2 = (2i + 1)^2 = 4i^2 + 4i + 1 = 2(2i^2 + 2i) + 1$ and n^2 is odd. Since n is odd implies n^2 is odd, by contrapositive, we know that if n^2 is even, then n is even.

■

At this point, you might ask the natural question: "When and how do you recognize that proof by contrapositive is appropriate?" We identify a few positive indicators for proof by contrapositive that you should watch for as you craft proofs of implications.

- A direct proof becomes complicated or subtle. In example 1.7.3, if we had tried to prove directly that n^2 is even implies n is even, we would have assumed $n^2 = 2i$ and then worked with the more complicated properties of factors and primes. While continuing in this vein may prove necessary, we should at least consider an alternate approach.
- The negation of the premises and the conclusion are easy to state, enabling us to readily give a proof of the contrapositive.
- The more complicated computational component of an implication is embedded in the theorem's premises rather than in the conclusion. In example 1.7.3, the squaring of n is in the premise. In general, we prefer that computations appear in the conclusion because they provide us something to work with as we craft a proof.

Question 1.7.3 Prove (by contrapositive) that for every integer $n \in \mathbb{Z}$, if n^2 is odd, then n is odd.

■

The proof of a biconditional mathematical statement of the form $(p \leftrightarrow q)$ requires us to consider two implications. Recall that the biconditional is expressed in English by such phrases as "if and only if," "exactly when," and "precisely when." In addition, mathematicians often use a standard abbreviation of "*iff*" for the biconditional phrase "if and only if" in mathematical exposition (and so we must be careful to watch for the second "f" when reading mathematical statements). The logical equivalence of biconditional expansion $(p \leftrightarrow q) \equiv [(p \rightarrow q) \wedge (q \rightarrow p)]$ provides the strategy for proving biconditionals: we prove a biconditional $(p \leftrightarrow q)$ by proving the two corresponding implications $(p \rightarrow q)$ and $(q \rightarrow p)$, each by a direct proof, a proof by contradiction, or a proof by contrapositive. We illustrate this strategy for proving a biconditional in the following example, which links together a couple of different pieces of our work in this section.

Example 1.7.4 We prove n is even iff n^2 is even.

Proof We prove the result by proving the two corresponding conditionals:

- If n is even, then n^2 is even.
- If n^2 is even, then n is even.

We have already proven some results that facilitate working with these conditional statements. The first conditional follows from question 1.7.1, which states that the product of two even integers is even. We assume n is even and so, by question 1.7.1, the product $n \cdot n = n^2$ is even. The second conditional was proven in example 1.7.3. Thus, n is even iff n^2 is even.

■

If we did not have the results of question 1.7.1 and example 1.7.3 in hand, then the proof of the biconditional in example 1.7.4 would follow the same outline given above, only we would need to work from scratch and fill in the details of the proof of the two corresponding conditionals. We also note that in mathematical discourse, we commonly say the first conditional is a *corollary* of question 1.7.1 because the corollary follows directly from the question. The word "corollary" is derived from the Latin word for "gift" and refers to such an immediate consequence of a known theorem.

Once we learn the definitions and basic properties of a few more mathematical objects, we are ready to prove a whole host of mathematical truths using these three proof techniques for implications. By continuing to work with these proof techniques, you will learn to recognize which approach is most useful in a given situation. When you are considering a new mathematical statement and are unsure how to proceed, start by trying to construct a direct proof. If for some reason you run into difficulty, don't despair—*every* mathematician has shared this experience. Often, important insights are gained at such mathematical roadblocks. However, when you encounter such difficulties, you should also be prepared to "bail out" and attempt a different approach. If a direct proof is not working, try a proof by contradiction or a proof by contrapositive. The ability to move fluidly between these proof techniques is one key to success in understanding and creating new mathematics.

We now consider proving existential statements and negated universal statements. For mathematical statements of these forms, producing a single example is sufficient to prove the statement. For existentials this should be clear; an existential statement claims that a certain object exists and so it suffices to produce at least one such object to show that the statement is true.

Example 1.7.5   We prove each of the following existential mathematical statements.

  • There exists an even integer.

Proof   The number $2 = 2 \cdot 1$ is an even integer. There are infinitely many different examples to prove this statement: $0, 2, 4, \ldots$ are all even.

■

  • There exists an even prime.

Proof   The number $2 = 2 \cdot 1$ is an even prime. This value is the only example that will prove this statement.

■

  • There exists a rational number.

Proof   As in the case of the even integers, there are infinitely many different rational numbers. Examples of rational numbers include: $0, \frac{1}{2}, \frac{3}{4}$ and 2 (since it can be written in the form $\frac{2}{1}$).

■

Question 1.7.4   Prove the following existential mathematical statements.

   (a)   There exists an odd prime.   (b)   There exists an irrational.
                                                                                ∎

   A similar approach works for proving negated universal statements; they are logically equivalent to existentials as witnessed by the following pair of logical equivalences.

$$\exists x(\mathbb{B}) \equiv \sim \forall x \sim (\mathbb{B}) \qquad\qquad \forall x(\mathbb{B}) \equiv \sim \exists x \sim (\mathbb{B})$$

   A precise rendition of $\sim \forall x \sim (\mathbb{B})$ would be written as $[\sim (\forall x[\sim (\mathbb{B})])]$; we omit all but the innermost parentheses for the sake of readability. We do not formally prove these logical equivalences, but reading the English renditions of them may provide some intuitive justification. For example, the left logical equivalence claims that "there exists x such that $\mathbb{B}$ holds" is equivalent to "it is not the case that for all x not $\mathbb{B}$ holds." Similarly, the right logical equivalence claims that "for all x $\mathbb{B}$ holds" is equivalent to "it is not the case that there exists x such that not $\mathbb{B}$ holds." Most often, negated universals arise in the context of *disproving* a universal mathematical statement. In such a setting, the object produced to disprove a universal sentence is referred to as a *counterexample*.

Example 1.7.6   We illustrate the equivalence of proving a negated universal statement and disproving a universal statement.

   • We prove that not every integer is even.

   Proof $3 = 2 \cdot 1 + 1$ is an integer that is not even.
                                                                                ∎

   • We disprove the claim that every integer is even.

   Proof $3 = 2 \cdot 1 + 1$ is an integer that is not even (and so, 3 is a counterexample).
                                                                                ∎

Question 1.7.5   Disprove the following universal mathematical statements.

   (a)   Every prime is even.   (b)   For all $n > 2$, $n^2 \geq 25$.
                                                                                ∎

Question 1.7.6   Prove the negated universal statement: Not every square root is rational.
                                                                                ∎

   We end this chapter by demonstrating how the mathematical proof techniques developed in this section, along with the formal language of predicate logic, are powerful enough to prove the validity of the syllogism given at the beginning of section 1.6. In this way, the mathematical ideas we have studied in this chapter fulfill Aristotle's fundamental desire: they provide a rational framework for determining the truth of arguments.

Example 1.7.7 We prove the validity of the following syllogism.

> Every Greek is mortal.
> There exists a Greek.
> Thus, there exists a mortal.

Proof We give a direct proof of this syllogism. We assume that "Every Greek is mortal" and "There exists a Greek," and we show that "There exists a mortal." We begin with the second assumption that "There exists a Greek," which we translated into predicate logic as $\exists x G(x)$. From this assumption, there must exist a Greek and we let g denote a Greek; working with the predicate notation, $G(g)$ is true. We now consider the first assumption that "Every Greek is mortal," which we translated into predicate logic as $\forall x[G(x) \rightarrow M(x)]$. This statement is assumed true for all objects (including g) and so we have $G(g) \rightarrow M(g)$; that is, if g is Greek, then g is mortal. We now have the truth of both $G(g)$ and $G(g) \rightarrow M(g)$. Applying modus ponens, we deduce $M(g)$, which asserts g is mortal. Since g is mortal, "There exists a mortal" is true. Thus, if the two assumptions are true, the conclusion must be true, and the given syllogism is a valid argument.

■

1.7.1 Reading Questions for Section 1.7

1. State the five proof techniques discussed in this section.
2. Define and give an example of an even integer and an odd integer.
3. State the parity property of the integers.
4. Define and give an example of a rational number and an irrational number.
5. State the zero product property of the reals.
6. Compute the truth table verifying that $(p \rightarrow q) \equiv \{[p \wedge (\sim q)] \rightarrow [r \wedge (\sim r)]\}$. What proof technique is justified by this logical equivalence?
7. Compute the truth table verifying that $(p \rightarrow q) \equiv [(\sim q) \rightarrow (\sim p)]$. What proof technique is justified by this logical equivalence?
8. Compute the truth table verifying that $(p \leftrightarrow q) \equiv [(p \rightarrow q) \wedge (q \rightarrow p)]$. What proof technique is justified by this logical equivalence?
9. What logical connective is abbreviated "iff" in mathematical exposition?
10. Why can we prove an existential sentence with an example?
11. Why can we not prove a universal sentence with an example?
12. What types of mathematical statements can be proven by counterexamples?

1.7.2 Exercises for Section 1.7

In exercises 1–18, give a direct proof of each mathematical statement.

1. The sum of an odd integer and an even integer is odd.
2. The difference of two even integers is even.
3. The product an odd integer and an even integer is even.
4. The cube of an even integer is even.

5. The sum of two odd integers is even.
6. The difference of two odd integers is even.
7. The product of two odd integers is odd.
8. The square of an odd integer is odd.
9. If n is odd, then $n^2 = 8i + 1$ for some integer i.
10. If the sum of two integers is even, their difference is even.
11. If the sum of two integers is odd, their difference is odd.
12. The sum of two rational numbers is rational.
13. The difference of two rational numbers is rational.
14. The product of two rational numbers is rational.
15. The quotient of two nonzero rational numbers is nonzero.
16. The square of a rational number is rational.
17. The double of a rational number is rational.
18. Every integer is rational.

In exercises 19–30, give a proof by contradiction of each mathematical statement.

19. The square of an even integer is even.
20. The square of an odd integer is odd.
21. The cube of an even integer is even.
22. The cube of an odd integer is odd.
23. If r is an irrational number, then $\sqrt{r}$ is irrational.
24. The double of a rational number is rational.
25. The square of a rational number is rational.
26. The product of a nonzero rational number and an irrational number is irrational.
27. There does not exist a greatest integer.
28. There does not exist a greatest even integer.
29. There does not exist a least positive rational number.
30. There does not exist a least positive real number.

In exercises 31–35, give a proof by contrapositive of each mathematical statement.

31. If n^3 is even, then n is even.
32. If n^3 is odd, then n is odd.
33. If mn is odd, then both m and n are odd.
34. If the unit digit of an integer is nonzero, then the integer is not a multiple of 10.
35. If r is an irrational number, then $\sqrt{r}$ is an irrational number.

In exercises 36–40, prove each biconditional mathematical statement.

36. n^3 is odd iff n is odd.
37. n^2 is odd iff n is odd.
38. n is even iff $n + 1$ is odd.
39. n is odd iff $n + 1$ is even.
40. n is even iff n can be written as the sum of two odds.

In exercises 41–48, prove each existential mathematical statement.

   41. There exists an odd integer.
   42. There exists an odd rational.
   43. There exists an even integer that can be written as a sum of two distinct primes.
   44. There exists an even integer that can be written as a sum of two primes in two different ways.
   45. There exists an irrational number.
   46. There exists a rational number.
   47. There exists a rational integer.
   48. There exists an even rational number.

In exercises 49–59, disprove each universal mathematical statement.

   49. Every prime is odd.
   50. For all $n > 4$, $n^2 \geq 36$.
   51. The ratio of the circumference of a circle to its radius is rational.
   52. The sum of two evens is odd.
   53. The sum of two odds is odd.
   54. The sum of an even and an odd is even.
   55. Every odd integer is irrational.
   56. Every even integer is irrational.
   57. The sum of two irrational numbers is irrational.
   58. The sum of a rational and irrational is rational.
   59. For every pair of reals r and s, if $r^2 = s^2$, then $r = s$.

In exercises 60–67, prove each negated universal mathematical statement.

   60. Not every square root of a positive integer is rational.
   61. Not every square root of a positive integer is irrational.
   62. Not every rational number is even.
   63. Not every rational number is odd.
   64. Not every integer is even.
   65. Not every integer is odd.
   66. Not every square root is greater than zero.
   67. Not every square is positive.

In exercises 68–70, identify the error in each of the following *incorrect* "proofs" that the sum of two even integers is even.

   68. False proof 1: Let x be even. Since $x + x = 2x$ is even, the sum of two evens is even.
   69. False proof 2: Let $x + y$ be even. Therefore the sum of two evens is even.
   70. False proof 3: If x is even and y is even, then $x + y$ is even. Therefore, $x + y$ is even and the sum of two evens is even.

Notes

The history of Western thought traces its roots to the Greeks in the third and fourth century B.C.E. The works of Socrates, Plato, and Aristotle are among the preeminent accomplishments of this time, and their thoughts and insights continue to influence Western thought and culture to this day. Cahill [35] provides a generally accessible description of the impact of ancient Greek ideas on contemporary Western mathematical culture (as passed down from Socrates to Plato to Aristotle and, finally, to us), and also describes the Greek influence on how we feel, how we rule, how we party, and how we see. Both Kline [142] and Jacobs [126] provide a similar, but more thorough, introductory survey of the historical progression of mathematical thought, including the contributions of the Greeks. Aristotle's analysis of natural reasoning in the *Prior Analytics* remains the foundation of the philosophical and mathematical approaches to logic; Smith [220] is a contemporary English translation of this treatise. This vein of analyzing natural language arguments continues in philosophy departments across the U.S.A. in logical reasoning courses; such courses are supported by texts by Browne and Keeley [32], Kelley [138], and McInerny [173].

The more formal approach of the sentential and predicate logic traces its roots to George Boole in his seminal treatise *An Investigation of the Laws of Thought, on Which are Founded the Mathematical Theories of Logic and Probabilities* (a reprint [22] is available from Dover Publications), and the German philosopher Gottlob Frege in his treatise *The Foundations of Arithmetic: A Logico-Mathematical Enquiry into the Concept of Number* (see the translation by Austin [91]). Davis [54] traces the development of rational thought from the perspective of the historical development of computing and contemporary understanding of algorithms, including a discussion of Boole's and Frege's work in this context. MacHale [161] has interwoven the story of Boole's personal and intellectual life in a biographical format.

The sentential and predicate logic are standard elements of formal logic courses offered by both philosophy and mathematics departments. The philosophical rendition of these courses focuses on working with translations and developing students' capability to create natural deductive arguments (rather than just identifying the steps in such arguments as we have done in this text). Copi and Cohen [46], Gustason and Ulrich [105], and Jacquette [127] are widely used texts that support such philosophy courses and enable students to develop their skills in this more computational approach to these logics. There is also a *Schaum's Outline* [181] and a more recent abridgment by McAllister [182] that explore this line of development. Another approach adopted in some philosophical contexts is a "tree method proof system" that captures the essence of the natural deductive system, while also enabling a relatively quick approach to the theoretical aspects of logic; Jeffrey [131] is a respected text supporting this approach to logic.

Mathematicians typically adopt a more abbreviated approach to the sentential and predicate logic in the interest of exploring the theoretical aspects of logic. Hamilton [109], Leary [152], and Mendelson [175] are all fine introductions to this mathematical approach to logic; Enderton [71] is the classic introduction to mathematical logic at the advanced undergraduate and beginning graduate level. There are also philosophical texts that adopt this approach, including Hunter [123]. Mathematicians and philosophers have collaborated for many years, and there is an important cross-fertilization of insights and ideas between these complementary approaches to logic.

Among the most significant intellectual accomplishments of the twentieth century are the incompleteness theorems of the accomplished Austrian–American mathematician Kurt Friedrich Gödel, which assert that some true mathematical statements are not provable (that is, they can never be proven) in sufficiently powerful formal systems. There are many enjoyable books describing various aspects of these results. Nagel and Newman [178] is the classic text in this area, providing a marvelous description of the intellectual background and the content of Gödel's *On Formally Undecidable Propositions of Principia Mathematica and Related Systems* (reprinted by Dover Publications [99]). Davis et al. [55] explores the context and content of Gödel's work, as well as a host of other interesting mathematical miscellany. More technical explorations of Gödel's incompleteness theorems can be found in Enderton [71], Jeffrey [131], and the graduate level text by Smullyan [223]. Crossley et al. [49] contains an abbreviated outline of the proof of these results and, more recently, Franzen [90] has written a book exploring the "uses and abuses" of Gödel's theorems. A number of more playful explorations of Gödel's insights have been written by Smullyan (see both [225] and [222]), using puzzles and riddles to illuminate various facets and applications of these theorems. Several of Gödel biographies have been written, including the recent definitive book by Dawson [52]. The Pulitzer Prize winning work of Hofstadter [118] also explores the interconnections between Gödel's results, music, art, and biology.

Since we have touched on one of the many connections between mathematics and computer science in the design of computer circuits, we also mention the foundational work of the English mathematician Alan Turing in the mid-twentieth century. As discussed by Davis [54], Turing isolated the notions at the heart of our understanding of "computability," developing abstract and practical tools for grappling with many questions of computer science. The Turing test remains the gold-standard in efforts to assess and describe artificial intelligence, and the Turing Award is the computer science equivalent of the Nobel Prize. Hodges [117] has written the definitive biography of the personal and intellectual life of Turing. Hamilton [109] and Cutland [50] both contain accessible descriptions of Turing's model for computing, which has become known as a Turing machine.

This chapter's application of logic to computer science traces its historical roots to telephone circuit design in the early 1930s and the continuation of this work at Bell Laboratories. Karnaugh was working as a telecommunications engineer in the 1950s when he published his results on what has become known as Karnaugh maps in *The Map Method for Synthesis of Combinational Logic Circuits* [137]. These diagrams have become a standard component of computer science and engineering curricula, and our presentation is based on the introductory courses in these disciplines. Further details and applications of these ideas can be found in Comer [44], Kerns and Irwin [139], and Mano [167].

Finally, the results presented in section 1.7 are widely known among mathematicians (and others!), and so they are contained in many different mathematical textbooks. These ideas are often studied in Discrete Mathematics courses, which are supported by such texts as those by Epp [72], Richmond and Richmond [193], and Scheinerman [209]. Alternatively, there are a growing number of "Foundations of Mathematics" textbooks that consider these notions, including those by Barnier and Feldman [10], D'Angelo and West [51], and Smith et al. [219]. Two fun books about mathematical ideas and proofs put in the form of stories, rhymes, and enjoyable explanations of problems are Fadiman [77] and [78].

2 Abstract Algebra

An important goal of mathematics is to understand the concept of quantity. This concept has an ancient anthropological origin; among the first mathematical steps made by humans was the ability to distinguish between one and many and, eventually, between one, two, three, and so on. In modern English "quantity" is synonymous with such words as amount, number, size, and magnitude. As we have seen in previous mathematics courses, the term "quantity" is also frequently used to refer to objects in mathematical expressions (for example, "substitute the quantity $x = 2$ into ..."). We seek a more precise mathematical understanding of this concept.

The development of number systems has a rich and interesting history. In Zaire, Africa, archeologists have discovered what appear to be number representations etched into fossilized bone fragments that have been carbon-dated to 20000 B.C.E. By 2000 B.C.E., the Babylonians, who lived in the area of modern Turkey, Iraq, and Iran, had developed symbols to represent quantities and to perform basic arithmetic operations. The Babylonians also began to analyze the general properties of numbers in connection with the study of astronomy. Around 300 B.C.E., the Hindu mathematicians in India developed our modern notation for numbers, which eventually reached the Middle East through the interaction of merchants along ancient trade routes. By 1000 C.E., Islamic settlers and traders had brought this Hindu–Arabic numeral system to southern Europe, along with the mathematical advances of the Babylonians and the ancient Greeks. These ideas found fertile ground in Italian academic circles and fostered a renaissance in mathematical interests and studies.

Abstract algebra grew out of an interest in polynomial equations. The Italian mathematicians of the 1500s studied not just individual numbers, but collections of numbers identified as solutions of polynomial equations. By the mid-1800s, European mathematicians had made significant progress in understanding both the power and the essential limitations of our ability to solve polynomials (as discussed in section 3.5). This sophisticated study used the abstract properties of numbers and formal number systems. In this way, abstract algebra was born. This area of study remains a lively theoretical field in which mathematicians continue to make significant progress and contributions. Abstract algebra is also widely applicable and is used in a variety of ways by mathematicians, physicists, chemists, computer scientists, mineralogists, artists, and many others.

We begin to study abstract algebra by developing an algebra of sets. Intuitively, a set is a collection of objects known as elements; set theory is the study of the properties, relations, and operations for sets. While set theory is not technically part of abstract algebra, sets are fundamental to all areas of mathematics and we need to establish a precise language for sets. We also explore operations on sets and relations between sets, developing an "algebra of sets" that strongly resembles aspects of the algebra of sentential logic. In addition, as we discussed in chapter 1, a fundamental goal in mathematics is crafting articulate, thorough, convincing, and insightful arguments for the truth of mathematical statements. We continue the development of theorem-proving and proof-writing skills in the context of basic set theory.

After exploring the algebra of sets, we study two number systems denoted $\mathbb{Z}_n$ and $U(n)$ that are closely related to the integers. Our approach is based on a widely used strategy of mathematicians: we work with specific examples and look for general patterns. This study leads to the definition of modified addition and multiplication operations on certain finite subsets of the integers. We isolate key axioms, or properties, that are satisfied by these and many other number systems and then examine number systems that share the "group" properties of the integers. Finally, we consider an application of this mathematics to check digit schemes, which have become increasingly important for the success of business and telecommunications in our technologically based society. Through the study of these topics, we engage in a thorough introduction to abstract algebra from the perspective of the mathematician—working with specific examples to identify key abstract properties common to diverse and interesting mathematical systems.

2.1 The Algebra of Sets

Intuitively, a *set* is a "collection" of objects known as "elements." But in the early 1900's, a radical transformation occurred in mathematicians' understanding of sets when the British philosopher Bertrand Russell identified a fundamental paradox inherent in this intuitive notion of a set (this paradox is discussed in exercises 66–70 at the end of this section). Consequently, in a formal set theory course, a set is defined as a mathematical object satisfying certain axioms. These axioms detail properties of sets and are used to develop an elegant and sophisticated theory of sets. This "axiomatic" approach to describing mathematical objects is relevant to the study of all areas of mathematics, and we begin exploring this approach later in this chapter. For now, we assume the existence of a suitable axiomatic framework for sets and focus on their basic relationships and operations. We first consider some examples.

Example 2.1.1   Each of the following collections of elements is a set.

- $V = \{\text{cat, dog, fish}\}$
- $W = \{1, 2\}$
- $X = \{1, 3, 5\}$
- $Y = \{n : n \text{ is an odd integer}\} = \{\ldots, -5, -3, -1, 1, 3, 5, \ldots\}$

■

In many settings, the upper case letters $A, B, \ldots, Z$ are used to name sets, and a pair of braces $\{, \}$ is used to specify the elements of a set. In example 2.1.1, V is a finite set of three English words identifying common household pets. Similarly, W is finite set consisting of the integers 1 and 2, and X is a finite set consisting of the integers 1, 3, and 5. We have written Y using the two most common notations for an infinite set. As finite beings, humans cannot physically list every element of an infinite set one at a time. Therefore, we often use the descriptive set notation $\{n : P(n)\}$, where $P(n)$ is a predicate stating a property that characterizes the elements in the set. Alternatively, enough elements are listed to define implicitly a pattern and ellipses "..." are used to denote the infinite, unbounded nature of the set. This second notation must be used carefully, since people vary considerably in their perception of patterns, while clarity and precision are needed in mathematical exposition.

Certain sets are of widespread interest to mathematicians. Most likely, they are already familiar from your previous mathematics courses. The following notation, using "barred" upper case letters, is used to denote these fundamental sets of numbers.

Definition 2.1.1   • $\emptyset$ *denotes the* **empty set** $\{\ \}$, *which does not contain any elements.*
   • $\mathbb{N}$ *denotes the set of* **natural numbers** $\{\ 1, 2, 3, \ldots\ \}$.
   • $\mathbb{Z}$ *denotes the set of* **integers** $\{\ \ldots, -3, -2, -1, 0, 1, 2, 3, \ldots\ \}$.
   • $\mathbb{Q}$ *denotes the set of* **rational numbers** $\{\ p/q\ :\ p, q \in \mathbb{Z}\ with\ q \neq 0\ \}$.
   • $\mathbb{R}$ *denotes the set of* **real numbers** *consisting of directed distances from a designated point zero on the continuum of the real line.*
   • $\mathbb{C}$ *denotes the set of* **complex numbers** $\{\ a + bi\ :\ a, b \in \mathbb{R}\ with\ i = \sqrt{-1}\ \}$.

In this definition, various names are used for the same collection of numbers. For example, the natural numbers are referred to by the mathematical symbol "$\mathbb{N}$," the English words "the natural numbers," and the set-theoretic notation "$\{1, 2, 3, \ldots\}$." Mathematicians move freely among these different ways of referring to the same number system as the situation warrants. In addition, the mathematical symbols for these sets are "decorated" with the superscripts "$*$," "$+$," and "$-$" to designate the corresponding subcollections of nonzero, positive, and negative numbers, respectively. For example, applying this symbolism to the integers $\mathbb{Z} = \{\ldots, -3, -2, -1, 0, 1, 2, 3, \ldots\}$, we have

$$
\begin{aligned}
\mathbb{Z}^* &= \{\ldots, -3, -2, -1, 1, 2, 3, \ldots\}, \\
\mathbb{Z}^+ &= \{1, 2, 3, \ldots\}, \\
\mathbb{Z}^- &= \{-1, -2, -3, \ldots\}.
\end{aligned}
$$

There is some discussion in the mathematics community concerning whether or not zero is a natural number. Many define the natural numbers in terms of the "counting" numbers $1, 2, 3, \ldots$ (as we have done here) and refer to the set $\{0, 1, 2, 3, \ldots\}$ as the set of *whole numbers*. On the other hand, many mathematicians think of zero as a "natural" number. For example, the axiomatic definition of the natural numbers introduced by the Italian mathematician Giuseppe Peano in the late 1800s includes zero. Throughout this book, we use definition 2.1.1 and refer to the natural numbers as the set $\mathbb{N} = \{\ 1, 2, 3, \ldots\ \}$.

Our study of sets focuses on relations and operations of sets. The most fundamental relation associated with sets is the "element of" relationship that indicates when an object is a member of a set.

Definition 2.1.2 *If a is an element of set A, then a ∈ A denotes "a is an element of A."*

Example 2.1.2 As in example 2.1.1, let $W = \{1, 2\}$ and recall that $\mathbb{Q}$ is the set of rationals.

- 1 is in W, and so $1 \in W$.
- 3 is not in W, and so $3 \notin W$.
- $\frac{1}{2}$ is rational, and so $\frac{1}{2} \in \mathbb{Q}$.
- $\sqrt{2}$ is not rational (as we prove in section 3.4), and so $\sqrt{2} \notin \mathbb{Q}$.

■

Question 2.1.1 Give an example of a finite set A with $2 \in A$ and an infinite set B with $2 \notin B$.

■

We now consider relationships between sets. We are particularly interested in describing when two sets are identical or equal. As it turns out, the identity relationship on sets is best articulated in terms of a more primitive "subset" relationship describing when all the elements of one set are contained in another set.

Definition 2.1.3 *Let A and B be sets.*

- *A is a **subset** of B if every element of A is an element of B. We write $A \subseteq B$ and show $A \subseteq B$ by proving that if $a \in A$, then $a \in B$.*
- *A is **equal** to B if A and B contain exactly the same elements. We write $A = B$ and show $A = B$ by proving both $A \subseteq B$ and $B \subseteq A$.*
- *A is a **proper subset** of B if A is a subset of B, but A is not equal to B. We write either $A \subset B$ or $A \subsetneq B$ and show $A \subset B$ by proving both $A \subseteq B$ and $B \nsubseteq A$.*

Formally, the notation and the associated proof strategy are not part of the definition of these set relations. However, these facts are fundamental to working with sets and you will want to become adept at transitioning freely among definition, notation, and proof strategy.

Example 2.1.3 As in example 2.1.1, let $W = \{1, 2\}$, $X = \{1, 3, 5\}$, and $Y = \{n : n \text{ is an odd integer}\}$. We first prove $X \subseteq Y$ and then prove $W \nsubseteq Y$.

Proof that $X \subseteq Y$ We prove $X \subseteq Y$ by showing that if $a \in X$, then $a \in Y$. Since $X = \{1, 3, 5\}$ is finite, we prove this implication by exhaustion; that is, we consider every element of X one at a time and verify that each is in Y. Since $1 = 2 \cdot 0 + 1$, $3 = 2 \cdot 1 + 1$, and $5 = 2 \cdot 2 + 1$, each element of X is odd; in particular, each element of X has been expressed as $2k + 1$ for some $k \in \mathbb{Z}$). Thus, if $a \in X$, then $a \in Y$, and so $X \subseteq Y$.

■

Proof that $W \nsubseteq Y$ We prove $W \nsubseteq Y$ by showing that $a \in W$ does not necessarily imply $a \in Y$. Recall that $(p \to q)$ is false precisely when $[p \wedge (\sim q)]$ is true; in this case, we need to identify a counterexample with $a \in W$ and $a \notin Y$. Consider $2 \in W$. Since $2 = 2 \cdot 1$ is even, we conclude $2 \notin Y$. Therefore, not every element of W is an element of Y.

■

Question 2.1.2   As in example 2.1.1, let $X = \{1, 3, 5\}$ and $Y = \{n : n$ is an odd integer $\}$. Prove that X is a proper subset of Y.

∎

Example 2.1.4   The fundamental sets of numbers from definition 2.1.1 are contained in one another according to the following proper subset relationships.

$$\varnothing \subset \mathbb{N} \subset \mathbb{Z} \subset \mathbb{Q} \subset \mathbb{R} \subset \mathbb{C}$$

∎

When working with relationships among sets, we must be careful to use the notation properly so as to express true mathematical statements. One common misuse of set-theoretic notation is illustrated by working with the set $W = \{1, 2\}$. While it is true that $1 \in W$ since 1 is in W, the assertion that $\{1\} \in W$ is *not* true. In particular, W contains only numbers, not sets, and so the set $\{1\}$ is not in W. In general, some sets do contain sets—W is just not one of these sets. Similarly, we observe that $\{1\} \subseteq W$ since $1 \in \{1, 2\} = W$, but $1 \subseteq W$ is *not* true; indeed, $1 \subseteq W$ is not a sensible mathematical statement since the notation $\subseteq$ is not defined between an element and a set, but only between sets.

Despite these distinctions, there is a strong connection between the "element of" relation $\in$ and the subset relation $\subseteq$, as you are asked to develop in the following question. In this way, we move beyond discussing relationships among specific sets of numbers to exploring more general, abstract properties that hold for every element and every set.

Question 2.1.3   Prove that $a \in A$ if and only if $\{a\} \subseteq A$.
Hint: Use definitions 2.1.2 and 2.1.3 to prove the two implications forming this "if-and-only-if" mathematical statement.

∎

We now turn our attention to six fundamental operations on sets. These set operations manipulate a single set or a pair of sets to produce a new set. When applying the first three of these operations, you will want to utilize the close correspondence between the set operations and the connectives of sentential logic.

Definition 2.1.4   *Let A and B be sets.*

- *A^C denotes the **complement** of A and consists of all elements not in A, but in some prespecified **universe** or **domain** of all possible elements including those in A; symbolically, we define $A^C = \{x : x \notin A\}$.*
- *$A \cap B$ denotes the **intersection** of A and B and consists of the elements in both A and B; symbolically, we define $A \cap B = \{x : x \in A$ and $x \in B\}$.*
- *$A \cup B$ denotes the **union** of A and B and consists of the elements in A or in B or in both A and B; symbolically, we define $A \cup B = \{x : x \in A$ or $x \in B\}$.*
- *$A \setminus B$ denotes the **set difference** of A and B and consists of the elements in A that are not in B; symbolically, we define $A \setminus B = \{x : x \in A$ and $x \notin B\}$. We often use the identity $A \setminus B = A \cap B^C$.*

- $A \times B$ denotes the **Cartesian product** *of A and B and consists of the set of all ordered pairs with first-coordinate in A and second-coordinate in B; symbolically, we define* $A \times B = \{(a, b) : a \in A \text{ and } b \in B\}$.
- $\mathbb{P}(A)$ *denotes the* **power set** *of A and consists of all subsets of A; symbolically, we define* $\mathbb{P}(A) = \{X : X \subseteq A\}$. *Notice that we always have* $\emptyset \in \mathbb{P}(A)$ *and* $A \in \mathbb{P}(A)$.

Example 2.1.5 As above, we let $W = \{1, 2\}$, $X = \{1, 3, 5\}$ and $Y = \{n : n \text{ is an odd integer }\}$. In addition, we assume that the set of integers $\mathbb{Z} = \{\ldots, -2, -1, 0, 1, 2, \ldots\}$ is the universe and we identify the elements of the following sets.

- $W^C = \{\ldots, -2, -1, 0, 3, 4, 5, \ldots\}$
- $Y^C = \{n : n \text{ is an even integer }\}$ by the parity property of the integers
- $W \cap X = \{1\}$, since 1 is the only element in both W and X
- $W \cup X = \{1, 2, 3, 5\}$, since union is defined using the inclusive-or
- $W \setminus X = \{2\}$
- $X \setminus W = \{3, 5\}$
- $\mathbb{Z}^* = \mathbb{Z} \setminus \{0\} = \{\ldots, -3, -2, -1, 1, 2, 3, \ldots\}$
- $W \times X = \{(1, 1), (1, 3), (1, 5), (2, 1), (2, 3), (2, 5)\}$
- $\mathbb{P}(W) = \{\, \emptyset, \{1\}, \{2\}, \{1, 2\} \,\}$

■

The last two sets given in example 2.1.5 contain mathematical objects other than numbers; the power set is also an example of a set containing other sets. As we continue exploring mathematics, we will study sets of functions, matrices, and other more sophisticated mathematical objects.

Question 2.1.4 Working with W, X, and Y from example 2.1.5, identify the elements in the sets X^C, $W \cap Y$, $W \cup Y$, $W \setminus Y$, $Y \setminus W$, $X \times W$, $W \times W$, $W \times Y$, and $\mathbb{P}(X)$. In addition, state six elements in $\mathbb{P}(Y)$; that is, state six subsets of Y.

■

The use of symbols to represent relationships and operations on mathematical objects is a standard feature of mathematics. Good choices in symbolism can facilitate mathematical understanding and insight, while poor choices can genuinely hinder the study and creation of mathematics. Historically, the symbols $\in$ for "element of," $\cap$ for "intersection," and $\cup$ for "union" were introduced in 1889 by the Italian mathematician Giuseppe Peano. His work in formalizing and axiomatizing set theory and the basic arithmetic of the natural numbers remains of central importance. The Cartesian product $\times$ is named in honor of the French mathematician and philosopher René Descartes, who first formulated "analytic geometry" (an important branch of mathematics discussed in section 4.1).

Although we have presented the Cartesian product $A \times B$ as an operation on pairs of sets, this product extends to any finite number of sets. Mathematicians work with ordered triples $A \times B \times C = \{(a, b, c) : a \in A, b \in B, \text{ and } c \in C\}$, ordered quadruples $A \times B \times C \times D = \{(a, b, c, d) : a \in A, b \in B, c \in C, \text{ and } d \in D\}$, and even ordered n-tuples $A_1 \times \cdots \times A_n = \{(a_1, \ldots, a_n) : a_i \in A_i \text{ for } 1 \leq i \leq n\}$. While the use of n-tuples may at first seem to be of purely academic interest, models for science and business with tens (and even hundreds and thousands) of independent variables have become more common as computers have extended our capacity to analyze increasingly sophisticated events.

Along with considering the action of set-theoretic operations on specific sets of numbers, we are also interested in exploring general, abstract properties that hold for all sets. In this way we develop an algebra of sets, comparing various sets to determine when one is a subset of another or when they are equal. In developing this algebra, we adopt the standard approach of confirming informal intuitions and educated guesses with thorough and convincing proofs.

Example 2.1.6   For sets A and B, we prove $A \cap B \subseteq A$.

Proof   We prove $A \cap B \subseteq A$ by showing that if $a \in A \cap B$, then $a \in A$. We give a direct proof of this implication; we assume that $a \in A \cap B$ and show that $a \in A$. Since $a \in A \cap B$, both $a \in A$ and $a \in B$ from the definition of intersection. We have thus quickly obtained the goal of showing $a \in A$.

■

In example 2.1.6 we used a direct proof to show that one set is a subset of another. This strategy is very important: we prove $X \subseteq Y$ by assuming $a \in X$ and showing $a \in Y$. In addition, the process of proving $a \in X$ implies $a \in Y$ usually involves "taking apart" the sets X and Y and characterizing their elements based on the appropriate set-theoretic definitions. Once X and Y have been expanded in this way, our insights into sentential logic should enable us to understand the relationship between the two sets and to craft a proof (or disproof) of the claim. We illustrate this approach by verifying another set-theoretic identity.

Example 2.1.7   For sets A and B, we prove $A \setminus B = A \cap B^C$.

Proof   In general, we prove two sets are equal by demonstrating that they are subsets of each other. In this case, we must show both $A \setminus B \subseteq A \cap B^C$ and $A \cap B^C \subseteq A \setminus B$.

$A \setminus B \subseteq A \cap B^C$: We assume $a \in A \setminus B$ and show $a \in A \cap B^C$. Since $a \in A \setminus B$, we know $a \in A$ and $a \notin B$. The key observation is that $a \notin B$ is equivalent to $a \in B^C$ from the definition of set complement. Since $a \in A$ and $a \notin B$, we have both $a \in A$ and $a \in B^C$. Therefore, by the definition of intersection, $a \in A \cap B^C$. Thus, we have $A \setminus B \subseteq A \cap B^C$, completing the first half of the proof.

$A \cap B^C \subseteq A \setminus B$: We assume $a \in A \cap B^C$ and show $a \in A \setminus B$. From the definition of intersection, we know $a \in A \cap B^C$ implies both $a \in A$ and $a \in B^C$. Therefore, both $a \in A$ and $a \notin B$ from the definition of complement. This is exactly the definition of set difference, and so $a \in A \setminus B$. Thus, $A \cap B^C \subseteq A \setminus B$, completing the second half of the proof.

The proof of these two subset relationships establishes the desired equality $A \setminus B = A \cap B^C$ for every set A and B.

■

Question 2.1.5   Prove that if A and B are sets with $A \subseteq B$, then $B^C \subseteq A^C$.

■

A whole host of set-theoretic identities can be established using the strategies illustrated in the preceding examples. As we have seen, the ideas and identities of sentential logic play a fundamental role in working with the set-theoretic operations. Recall that De Morgan's laws are among the most important identities from sentential logic; consider the following set-theoretic version of these identities.

Example 2.1.8 De Morgan's laws for sets   We prove one of De Morgan's laws for sets: If A and B are sets, then both $(A \cap B)^C = A^C \cup B^C$ and $(A \cup B)^C = A^C \cap B^C$.

Proof   We prove the identity $(A \cap B)^C = A^C \cup B^C$ by arguing that each set is a subset of the other based on the following biconditionals:

$$
\begin{array}{lll}
a \in (A \cap B)^C & \text{iff} \quad a \notin A \cap B & \text{Definition of complement} \\
& \text{iff} \quad a \text{ is not in both } A \text{ and } B & \text{Definition of intersection} \\
& \text{iff} \quad \text{either } a \notin A \text{ or } a \notin B & \text{Sentential De Morgan's laws} \\
& \text{iff} \quad \text{either } a \in A^C \text{ or } a \in B^C & \text{Definition of complement} \\
& \text{iff} \quad a \in A^C \cup B^C & \text{Definition of union}
\end{array}
$$

Working through these biconditionals from top to bottom, we have $a \in (A \cap B)^C$ implies $a \in A^C \cup B^C$, and so $(A \cap B)^C \subseteq A^C \cup B^C$. Similarly, working through these biconditionals from bottom to top, we have $a \in A^C \cup B^C$ implies $a \in (A \cap B)^C$, and so $A^C \cup B^C \subseteq (A \cap B)^C$. This proves one of De Morgan's laws for sets, $(A \cap B)^C = A^C \cup B^C$ for every set A and B.   ■

Question 2.1.6   Prove the other half of De Morgan's laws for sets; namely, prove that if A and B are sets, then $(A \cup B)^C = A^C \cap B^C$.   ■

We end this section by discussing proofs that certain set-theoretic relations and identities do *not* hold. From section 1.7, we know that (supposed) identities can be disproved by finding a counterexample, exhibiting specific sets for which the given equality does not hold. To facilitate the definition of sets A, B, C with the desired properties, we introduce a visual tool for describing sets and set operations known as a *Venn diagram*. In a Venn diagram, the universe is denoted with a rectangle, and sets are drawn inside this rectangle using circles or ellipses. When illustrating two or more sets in a Venn diagram, we draw overlapping circles to indicate the possibility that the sets may share some elements in common. The Venn diagrams for the first four set operations from definition 2.1.4 are given in figure 2.1.

Example 2.1.9   We disprove the *false* claim that if A, B, and C are sets, then $A \cap (B \cup C) = (A \cap B) \cup C$. This demonstrates that union and intersection operations are not associative when used together, and so we must be careful with the order of operation when "mixing" union and intersection.

The Venn diagrams given in figure 2.2 illustrate the sets we are considering in this example. We use three circles to denote the three distinct sets A, B, and C. In addition, the circles overlap in a general way so as to indicate all the various possibilities for sets sharing elements.

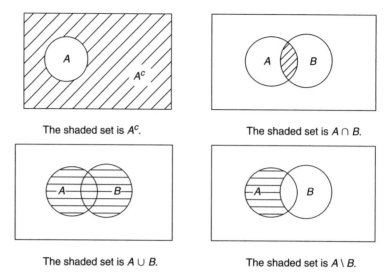

The shaded set is A^c.

The shaded set is $A \cap B$.

The shaded set is $A \cup B$.

The shaded set is $A \setminus B$.

Figure 2.1   Venn diagrams for basic set operations

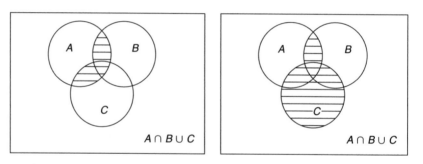

Figure 2.2   The Venn diagram for example 2.1.9 showing $A \cap (B \cup C) \neq (A \cap B) \cup C$

Examining the Venn diagrams, we see that if A, B, C are defined so that C contains an element that is in neither A nor B, the sets $A \cap (B \cup C)$ and $(A \cap B) \cup C$ will be different. Alternatively, we could define A, B, C so that $B \cap C$ contains an element that is not in A. Following the first approach, we choose to define the sets $A = \{1\}$, $B = \{1, 2\}$, and $C = \{1, 2, 3\}$ and verify the desired inequality with the following computations.

$$A \cap (B \cup C) \quad = \quad \{1\} \cap \{1, 2, 3\} = \{1\}$$
$$(A \cap B) \cup C \quad = \quad \{1\} \cup \{1, 2, 3\} = \{1, 2, 3\}$$

Therefore these three sets provide a counterexample demonstrating that sometimes $A \cap (B \cup C) \neq (A \cap B) \cup C$.

∎

In example 2.1.9, the choice of sets $A, B,$ and C is just one choice among many. We are certainly free to make other choices, and you might even think of constructing counterexamples as providing an opportunity to express your "mathematical personality."

Question 2.1.7   Guided by example 2.1.9, give another counterexample disproving the *false* claim that $A \cap (B \cup C) = (A \cap B) \cup C$ for all sets A, B, C.

          ■

We highlight one subtlety that arises in this setting. In example 2.1.9 and question 2.1.7, the counterexamples only disprove the general claim that $A \cap (B \cup C) = (A \cap B) \cup C$ for all sets A, B, C. However, these counterexamples do *not* prove that we have inequality for every choice of sets. In fact, there exist many different cases in which equality does hold. For example, both $A = \emptyset, B = \emptyset, C = \emptyset$ and $A = \{1, 2\}, B = \{1, 3\}, C = \{1\}$ produce the equality $A \cap (B \cup C) = (A \cap B) \cup C$, but only because we are working with these specific sets. We therefore cannot make any general claims about the equality of $A \cap (B \cup C)$ and $(A \cap B) \cup C$, but must consider each possible setting on a case-by-case basis. In short, if we want to prove that a set-theoretic identity does not always hold, then a counterexample accomplishes this goal; if we want to prove that a set-theoretic identity never holds, then we must provide a general proof and not just a specific (counter)example.

Question 2.1.8   Sketch the Venn diagram representing the following sets.

     (a) $(A \cup B) \cap C$ (b) $A^C \setminus B$

          ■

Question 2.1.9   Following the model given in example 2.1.9, disprove the *false* claim that the following identities hold for all sets A, B, C.

     (a) $(A \cup B) \cap C = A \cup (B \cap C)$ (b) $A^C \setminus B = (A \setminus B)^C$

          ■

2.1.1   Reading Questions for Section 2.1

1. What is the intuitive definition of a set?
2. What is the intuitive definition of an element?
3. Describe two approaches to identifying the elements of an infinite set.
4. Name six important sets and the symbolic notation for these sets.
5. Define and give an example of the "element of" relation $a \in A$.
6. Define and give an example of the set relations: $A \subseteq B$, $A = B$, and $A \subset B$.
7. If A and B are sets, what strategy do we use to prove that $A \subseteq B$?
8. If A and B are sets, what strategy do we use to prove that $A = B$?
9. Define and give an example of the set operations: A^C, $A \cap B$, $A \cup B$, $A \setminus B$, $A \times B$, and $\mathbb{P}(A)$.
10. Define and give an example of a generalized Cartesian product $A_1 \times A_2 \times \cdots \times A_n$.
11. State both the sentential logic and the set-theoretic versions of De Morgan's laws.
12. Discuss the use of a Venn diagram for representing sets.

2.1.2   Exercises for Section 2.1

In exercises 1–14, identify the elements in each set, assuming $A = \{w, x, y, z\}$ is the universe, $B = \{x, y\}$, $C = \{x, y, z\}$, and $D = \{x, z\}$.

1. B^C
2. C^C
3. $B \cap C$
4. $B \cap D$
5. $B \cup C$
6. $B \cup D$
7. $B \cap (C \cup D)$

8. $(B \cap C) \cup D$
9. $B \setminus D$
10. $D \setminus B$
11. $B \times C$
12. $B \times D$
13. $\mathbb{P}(B)$
14. $\mathbb{P}(C)$

In exercises 15–22, identify the elements in each set, assuming $A = (0, 2) = \{x : 0 < x \le 2\}$ and $B = [1, 3) = \{x : 1 \le x < 3\}$ are subsets of the real line $\mathbb{R}$.

15. A^C
16. B^C
17. $A \cap B$
18. $A \cup B$

19. $A \setminus B$
20. $B \setminus A$
21. $A^C \cap B^C$
22. $A^C \cup B^C$

In exercises 23–27, give an example proving each subset relationship is proper.

23. $\emptyset \subset \mathbb{N}$
24. $\mathbb{N} \subset \mathbb{Z}$
25. $\mathbb{Z} \subset \mathbb{Q}$

26. $\mathbb{Q} \subset \mathbb{R}$
27. $\mathbb{R} \subset \mathbb{C}$

In exercises 28–41, prove each set-theoretic identity for sets A, B, and C.

28. $\{2, 2, 2\} = \{2\}$
29. $\{1, 2\} = \{2, 1\}$
30. $\{1\} \in \mathbb{P}(\{1\})$
31. $A \subseteq A$ (and so $A \in \mathbb{P}(A)$)
32. $\emptyset \subseteq A$ (and so $\emptyset \in \mathbb{P}(A)$)
33. $A \setminus \emptyset = A$
34. $[A^C]^C = A$
35. $A \cap B \subseteq A$

36. $A \cap \emptyset = \emptyset$
37. $A \subseteq A \cup B$
38. If $A \subseteq B$ and $B \subseteq C$, then $A \subseteq C$.
39. If $A \subseteq B$ and $A \subseteq C$, then $A \subseteq B \cap C$.
40. $(A \cup B) \setminus C = (A \setminus C) \cup (B \setminus C)$
41. If $A \subseteq B$, then $\mathbb{P}(A) \subseteq \mathbb{P}(B)$.

In exercises 42–45, disprove each *false* set-theoretic identity.

42. $1 = \{1\}$
43. $1 \subseteq \{1\}$

44. $\{1\} \in \{1\}$
45. $\{1\} \subseteq \mathbb{P}(\{1\})$

For exercises 46–53, disprove the *false* claim that the following hold for all sets A, B, C by describing a counterexample.

46. If $A \not\subseteq B$ and $B \not\subseteq C$, then $A \not\subseteq C$.
47. If $A \subseteq B$, then $A^C \subseteq B^C$.
48. If $A^C = B^C$, then $A \cup B = \emptyset$.
49. If $A^C = B^C$, then $A \cap B = \emptyset$.

50. If $A \cup C = B \cup C$, then $A = B$.
51. If $A \cap C = B \cap C$, then $A = B$.
52. If $B = A \cup C$, then $A = B \setminus C$.
53. $(A \setminus B) \cup (B \setminus C) = A \setminus C$

Exercises 54–57 consider "disjoint" pairs of sets. We say that a pair of sets X and Y is disjoint when they have an empty intersection; that is, when $X \cap Y = \emptyset$.

In exercises 54–57, let $B = \{x, y\}$, $C = \{x, y, z\}$, $D = \{x, z\}$, $E = \{y\}$, and $F = \{w\}$ and identify the sets in this collection that are disjoint from the following sets.

54. B 56. D

55. C 57. E

Exercises 58–62 explore numeric properties of the power set operation.

58. State every element in $\mathbb{P}(\emptyset)$. How many elements are in $\mathbb{P}(\emptyset)$?
59. State every element in $\mathbb{P}(\{1\})$. How many elements are in $\mathbb{P}(\{1\})$?
60. State every element in $\mathbb{P}(\{1, 2\})$. How many elements are in $\mathbb{P}(\{1, 2\})$?
61. State every element in $\mathbb{P}(\{1, 2, 3\})$. How many elements are in $\mathbb{P}(\{1, 2, 3\})$?
62. Based on your answers to exercises 58–61, make a conjecture about how many elements are in $\mathbb{P}(\{1, 2, 3, 4\})$. Extend your conjecture to $\mathbb{P}(\{1, 2, \ldots, n\})$.

Exercises 63–65 consider how mathematicians have utilized set theory as a tool for defining the natural numbers. In particular, a correspondence between the nonnegative integers $\{0, 1, 2, 3, \ldots\}$ and certain sets is defined, beginning as follows.

$$0 = \emptyset$$
$$1 = \{0\} = \{\emptyset\}$$
$$2 = \{0, 1\} = \{\emptyset, \{\emptyset\}\}$$
$$3 = \{0, 1, 2\} = \{\emptyset, \{\emptyset\}, \{\emptyset, \{\emptyset\}\}\}$$

63. Using this model as a guide, state the set corresponding to the integer 4.
64. Using this model as a guide, state the set corresponding to the integer 5.
65. For each natural number from 0 to 5, how many elements are in the corresponding set? Based on this observation make a conjecture of how many elements are in the set for the natural number 50.

Exercises 66–67 consider the Barber paradox that was introduced by Bertrand Russell in an effort to illuminate Russell's paradox (discussed in the exercises 68–70). The Barber paradox is based on the following question.

If the barber shaves everyone who doesn't shave themselves and only those who don't shave themselves, who shaves the barber?

66. Assume the barber does not shave himself and find a contradiction.
67. Assume the barber shaves himself and find a contradiction.

Exercises 68–70 consider Russell's paradox. A set N is said to be normal if the set does *not* contain itself; symbolically, we write $N \notin N$. Examples of normal sets include the set of all even integers (which is itself not an even integer) and the set of all cows (which is itself not a cow). An example of a set that is not normal is the set of all thinkable things (which is itself thinkable).

68. Give two more examples of normal sets and an example of a set that is not normal.

69. Let N be the set of all normal sets. Assume N is a normal set and find a contradiction.

70. Let N be the set of all normal sets. Assume N is not a normal set and find a contradiction.

Bertrand Russell pointed out this paradox in our intuitive understanding of sets in a letter to Gottlob Frege in 1903. This paradox holds when a set is defined as "any collection" of objects and highlights the interesting observation that not every collection is a set.

2.2   The Division Algorithm and Modular Addition

Our study of abstract algebra begins with the system of whole numbers known more formally as the integers. Recall that $\mathbb{Z}$ denotes the set of *integers* $\{\ldots, -3, -2, -1, 0, 1, 2, 3, \ldots\}$. From previous mathematics courses, we are already familiar with several operations on the integers, including addition, subtraction, multiplication, division, and exponentiation. In this chapter, we "push the boundaries" on these operations by studying certain subsets of the integers along with a modified addition operation known as *modular addition*. We use the *division algorithm* to define this new addition operation.

The division algorithm is actually the name of a theorem, but the standard proof of this result describes the long division algorithm for integers. The ancient Greek mathematician Euclid included the division algorithm in Book VII of *Elements [73]*, a comprehensive survey of geometry and number theory. Traditionally, Euclid is believed to have taught and written at the Museum and Library of Alexandria in Egypt, but otherwise relatively little is known about him. And yet *Elements* is arguably the most important mathematics book ever written, appearing in more editions than any book other than the Christian Bible.

By the time *Elements* had appeared in 300 B.C.E., Greek mathematicians had recognized a duality in the fundamental nature of geometry. On the one hand, geometry is empirical, at least to the extent that it describes the physical space we inhabit. On the other hand, geometry is deductive because it uses axioms and reasoning to establish mathematically certain truths. Mathematicians and others continue to wonder at this duality. As Albert Einstein questioned, "How can it be that mathematics, being after all a product of human thought independent of experience, is so admirably adapted to the objects of reality?"

Mathematicians have a special affection for Euclid's book because *Elements* is the first known comprehensive exposition of mathematics to utilize the deductive, axiomatic method. In addition, a Latin translation of Euclid's *Elements* played a fundamental role in fostering the European mathematical renaissance of the sixteenth and seventeenth centuries. We now formally state the division algorithm.

Theorem 2.2.1 Division algorithm *If $m, n \in \mathbb{Z}$ and n is a positive integer, then there exist unique integers $q \in \mathbb{Z}$ and $r \in \{0, 1, \ldots, n-1\}$ such that $m = n \cdot q + r$. We refer to n as the **divisor**, q as the **quotient**, and r as the **remainder** when m is **divided** by n.*

The division algorithm makes two distinct claims about the quotient q and the remainder r. First of all, the division algorithm is an *existence* result guaranteeing that when we divide an integer m by a positive integer n, then we must obtain values q and r that are also integers. In addition, the division algorithm is a *uniqueness* result, ensuring that for each pair of integers m and n (with n positive) there is exactly one such quotient q and remainder r (when $r \in \{0, \ldots, n-1\}$). The uniqueness aspect of this theorem is in many ways just as significant as the existence, although perhaps highlighting uniqueness might seem a bit strange. Can you think of some setting in which a mathematical question does not have a unique answer? Perhaps thinking about the notion of "an" antiderivative in calculus is helpful. Many important results in mathematics make these dual claims of existence and uniqueness, and so the idea of uniqueness is something to watch for when studying mathematics.

Although the division algorithm guarantees the existence and uniqueness of an equation relating any two integers m and n with n positive, the theorem itself does not provide any information about how to actually *find* the equation. In most cases, the choice of integers q and r for the quotient and remainder are not immediately obvious, requiring a "behind the scenes" calculation using long division. The following example illustrates the intimate connection between the process of long division and the division algorithm.

Example 2.2.1 We use long division to specify the quotient q and the remainder r from the division algorithm when $m = 29$ is divided by $n = 12$. The grade school approach to long division produces the following result.

$$
\begin{array}{r}
2 \\
12 \overline{\big)\ 29} \\
-24 \\
\hline
5
\end{array}
$$

In the notation of the division algorithm, the quotient $q = 2$ and the remainder $r = 5$ when $m = 29$ is divided by $n = 12$. Alternatively, this result can be written in the division algorithm's $m = n \cdot q + r$ equation form as $29 = 12 \cdot 2 + 5$. ∎

Even though the division algorithm does not explicitly state an algorithm for finding $m = n \cdot q + r$, the value of this result lies in the guarantees of existence and uniqueness. When working with particular integers m and n with n positive, we are assured that a long division calculation will not be in vain. Also, in general settings, we can confidently work with the quotient q, the remainder r, and the equation $m = n \cdot q + r$ for any choice of integers m and n with n positive.

Example 2.2.2 Continuing to work with the divisor $n = 12$ from example 2.2.1, we use long division to verify that three more integers m have remainder $r = 5$ when m is divided by 12.

$$
\begin{array}{r}
3 \\
12 \overline{\big)\ 41} \\
-36 \\
\hline
5
\end{array}
\qquad
\begin{array}{r}
-2 \\
12 \overline{\big)\ -19} \\
-(-24) \\
\hline
5
\end{array}
\qquad
\begin{array}{r}
0 \\
12 \overline{\big)\ 5} \\
-0 \\
\hline
5
\end{array}
$$

Based on these computations, $m = 41$, $m = -19$, and $m = 5$ all have remainder $r = 5$ when divided by $n = 12$. As may be apparent, infinitely many integers m have remainder $r = 5$ under division by $n = 12$.

When creating this example, we chose $m = 41$, $m = -19$, and $m = 5$ by substituting different integer values q into the equation $m = q \cdot 12 + 5$. Distinct integers $q \in \mathbb{Z}$ produce distinct $m = q \cdot 12 + 5$ with remainder $r = 5$ under division by $n = 12$. Since there exist infinitely many integers $q \in \mathbb{Z}$, infinitely many integers m have remainder $r = 5$ under division by $n = 12$.

■

As we'll see throughout this chapter, the remainders identified by the division algorithm are key to many important mathematical insights and applications. One important property of these remainders is that they must be nonnegative—that is, r must be zero or positive. Thus, for a negative m, the nonnegative requirement for r also results in a negative quotient q. In example 2.2.2, we divided $m = -19$ by $n = 12$ and obtained the negative quotient $q = -2$ since the remainder r had to be nonnegative. Notice that the product of the quotient q and divisor n (in this case, $n \cdot q = (-2) \cdot 12 = -24$) had to be less than or equal to $m = -19$ so that adding the nonnegative reminder r produced the desired equality; and so, we must be careful (as always) when working with negative integers.

Question 2.2.1 Use long division to compute the quotient q and the remainder r when each m is divided by $n = 6$.

(a) $m = 39$ (c) $m = -9$

(b) $m = 195$ (d) $m = -603$

■

Question 2.2.2 State two positive integers and two negative integers $m \in \mathbb{Z}$ with each remainder r under division by $n = 6$.

(a) $r = 1$ (b) $r = 5$

■

Now that we have a good handle on the computations associated with the division algorithm, we outline a proof of this theorem. This argument discusses a constructive, algorithmic approach to identifying the quotient and the remainder that will hopefully provide some insight into why this theorem is true. As we begin, note that this is only a sketch of a proof rather than a complete proof with full details. A complete proof of the division algorithm is often discussed in abstract algebra and number theory courses and is left for your later studies.

A Sketch of a Proof of the division algorithm As highlighted above, the division algorithm makes two claims about the quotient and the remainder: an existence claim and a uniqueness claim. And so a proof of this theorem has two parts.

Existence Consider the case where m and n are both positive integers. The basic idea is to compare multiples of n with the integer m. First we check to see if $n \cdot 0 \le m < n \cdot 1$. If so, we are done, using $q = 0$ and $r = m$. If not, we check to see if $n \cdot 1 \le m < n \cdot 2$;

if so, we are done, using $q = 1$ and $r = m - n \cdot 1$. Continuing in this manner, we eventually find the desired quotient $q \in \mathbb{Z}$ such that $n \cdot q \le m < n \cdot (q + 1)$ with remainder $r = m - n \cdot q$. If m is negative, the idea is the same, only consider negative quotients $q \in \mathbb{Z}$. The following illustration may provide further insight into the process outlined here.

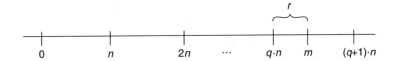

A complete, formal proof of the existence portion of the division algorithm typically uses an axiom known as the well-ordering principle of the integers to prove that there exist integers q and r satisfying $m = n \cdot q + r$. The well-ordering principle implies that every set of positive integers contains a least integer; the least integer in an appropriately defined set is used to obtain q and r.

Uniqueness The uniqueness portion of the division algorithm is proven by assuming that there exist two distinct quotient–remainder pairs for m and n and proving that the two quotients and the two remainders must actually be equal. Symbolically, we assume $m = n \cdot q_1 + r_1 = n \cdot q_2 + r_2$ where q_1, q_2 and r_1, r_2 satisfy the division algorithm conditions for the quotient and remainder, respectively. We then prove $q_1 = q_2$ and $r_1 = r_2$. This strategy is used in many different settings in proving uniqueness: we assume there exist two distinct mathematical objects with a given set of properties and show that they must actually be equal to each other.

■

We now use the division algorithm to identify certain subsets of the integers that are of particular interest to mathematicians—these sets allow for the definition of a modified addition operation. The division algorithm serves as a tool for partitioning the integers into a finite number of subsets, allowing us to "lump together" various integers into disjoint, or nonintersecting, sets of integers. The idea is to fix a specific positive integer n as a divisor and place all integers with the same remainder under division by n into the same subset of $\mathbb{Z}$.

For example, fix $n = 6$ as the divisor (as in question 2.2.1). Then, as you can verify, the integers $\ldots, -15, -9, -3, 3, 9, 15, \ldots$ each have remainder $r = 3$ under division by $n = 6$. Adopting the approach suggested above, we place all of these integers into the same subset $\{\ldots, -15, -9, -3, 3, 9, 15, \ldots\}$ of $\mathbb{Z}$. Similarly, the integers $\ldots, -13, -7, -1, 5, 11, 17, \ldots$ each have remainder $r = 5$ under division by $n = 6$, and we place all of these integers into the same subset $\{\ldots, -13, -7, -1, 5, 11, 17, \ldots\}$ of $\mathbb{Z}$. Continuing in this manner produces the following six disjoint subsets of integers.

$$
\begin{aligned}
\{\ldots, -18, -12, -6, \mathbf{0}, 6, 12, \ldots\} &= \{q \cdot 6 + 0 : q \in \mathbb{Z}\} \quad \text{for remainder} \quad r = 0 \\
\{\ldots, -17, -11, -5, \mathbf{1}, 7, 13, \ldots\} &= \{q \cdot 6 + 1 : q \in \mathbb{Z}\} \quad\quad\quad\quad\quad\quad r = 1
\end{aligned}
$$

$$
\begin{aligned}
\{\ldots, -16, -10, -4, \mathbf{2}, 8, 14, \ldots\} &= \{q \cdot 6 + 2 : q \in \mathbb{Z}\} \quad \text{for remainder} \quad r = 2 \\
\{\ldots, -15, -9, -3, \mathbf{3}, 9, 15, \ldots\} &= \{q \cdot 6 + 3 : q \in \mathbb{Z}\} \quad\quad\quad\quad\quad\quad r = 3 \\
\{\ldots, -14, -8, -2, \mathbf{4}, 10, 16, \ldots\} &= \{q \cdot 6 + 4 : q \in \mathbb{Z}\} \quad\quad\quad\quad\quad\quad r = 4 \\
\{\ldots, -13, -7, -1, \mathbf{5}, 11, 17, \ldots\} &= \{q \cdot 6 + 5 : q \in \mathbb{Z}\} \quad\quad\quad\quad\quad\quad r = 5
\end{aligned}
$$

In order to facilitate working with such sets of integers for different divisors n, we introduce a notation that conveniently relates integers sharing a common remainder. Specifically, we define $m \bmod n$ to be the remainder of m under division by n. For example, using this notation, we have $15 \bmod 6 = 3$ (since 3 is the remainder when 15 is divided by 6) and we have $17 \bmod 6 = 5$ (since 5 is the remainder when 17 is divided by 6).

Definition 2.2.1 *For integers m and n, we write $\boldsymbol{m} \textbf{ mod } \boldsymbol{n} = \boldsymbol{r}$ when $m = n \cdot q + r$ for integers q and r satisfying the division algorithm, and say "m mod n is r" or "m modulo n is r." For $a, b \in \mathbb{Z}$, a mod n = b mod n exactly when a and b have the same remainder r under division by n; in this case, we often write $\boldsymbol{a} \equiv \boldsymbol{b} \textbf{ mod } \boldsymbol{n}$ and say that "a is congruent to b mod n."*

Example 2.2.3 Continuing to work with the divisor $n = 6$, we illustrate the modular operation and the congruence relation by observing the following congruence relationships.

- $39 \bmod 6 = 3$, since $39 = 6 \cdot 6 + 3$
- $195 \bmod 6 = 3$, since $195 = 6 \cdot 32 + 3$
- $-9 \bmod 6 = 3$, since $-9 = 6 \cdot (-2) + 3$
- $-603 \bmod 6 = 3$, since $-603 = 6 \cdot (-101) + 3$

Furthermore, since all four of these integers have the same remainder of $r = 3$ under division by $n = 6$, the following congruence relations hold.

$$39 \equiv 195 \bmod 6, \quad 39 \equiv (-9) \bmod 6, \quad 39 \equiv (-603) \bmod 6, \quad \text{and so on} \ldots$$

∎

In light of the computations in example 2.2.3, the partition of the integers determined by remainders under division by $n = 6$ places all four of 39, 195, -9, and -603 into the same subset $\{q \cdot 6 + 3 : q \in \mathbb{Z}\}$ of $\mathbb{Z}$—the subset identified for remainder $r = 3$ above. The next question considers modular computations for various divisors $n \in \mathbb{N}$.

Question 2.2.3 Using the division algorithm, determine the value of each expression.

(a) $39 \bmod 3$, $195 \bmod 3$, $(-9) \bmod 3$, $(-603) \bmod 3$

(b) $39 \bmod 10$, $195 \bmod 10$, $(-9) \bmod 10$, $(-603) \bmod 10$

(c) $39 \bmod 2$, $195 \bmod 2$, $(2k + 1) \bmod 2$, $(2k) \bmod 2$ (where k is an arbitrary integer)

∎

The relation of congruence modulo n on the integers was first defined by Carl Friedrich Gauss in the late 1700s and played a significant role in Gauss's seminal treatise on number theory, *Disquisitiones Arithmeticae [97]*. Gauss was a German mathematician, physicist, astronomer, and surveyor who was born in 1777 in Brunswick and died in 1855 in Göttingen. Along with Archimedes and Newton, Gauss is widely regarded as one of the three most important mathematicians in recorded human history. Gauss developed a tremendous number of mathematical insights during his career and supervised the doctoral work of many active research mathematicians of the 1800s. He was particularly interested in number theory and in 1801 published *Disquisitiones Arithmeticae*, which included his work with congruence modulo n and modular arithmetic. These ideas have remained foundational in today's study of abstract algebra.

As described above, remainders play an important role in utilizing the division algorithm. Our study of modular arithmetic will benefit from determining all possible values of the remainder under division by a given positive integer n. For example, the possible remainders satisfying the division algorithm under division by $n = 2$ are the values $r = 0$ and $r = 1$.

Question 2.2.4   State a set consisting of all possible remainders that can result from the division algorithm under division by each n.

    (a) $n = 3$ (c) $n = 9$

    (b) $n = 6$ (d) an arbitrary $n \in \mathbb{N}$

                                                                                            ■

In light of the answers to question 2.2.4 and the abstract power and generality of the division algorithm, the the integers $0, 1, 2, \ldots, n - 1$ are the only possible values for the remainder r under division by a fixed positive integer $n \in \mathbb{N}$. For example, when dividing by $n = 3$, the only possible remainders are $0, 1,$ and 2 and, when dividing by $n = 4$, the only possible remainders are $0, 1, 2,$ and 3. To facilitate our work with the sets of integers sharing a common remainder under division by n, we identify each set of congruent integers with the corresponding remainder. For example, when $n = 3$, the integers are partitioned into the following three sets, which are then identified with their corresponding remainders.

$$\{\ldots, -6, -3, \mathbf{0}, 3, 6, \ldots\} \quad \text{with remainder } r = 0$$
$$\{\ldots, -5, -2, \mathbf{1}, 4, 7, \ldots\} \quad \text{with remainder } r = 1$$
$$\{\ldots, -4, -1, \mathbf{2}, 5, 8, \ldots\} \quad \text{with remainder } r = 2$$

The remainders serve as "representatives" for their corresponding sets and are the objects of study for much of this chapter. In the next section, we begin referring to these sets as *equivalence classes* of integers and develop the theory of equivalence classes and equivalence relations. For the moment, we define the set of these *remainders*.

Definition 2.2.2 $\mathbb{Z}_n$ *denotes the set of **integers mod n** consisting of the remainders* $\{0, 1, 2, \ldots, n - 1\}$ *under division by* $n \in \mathbb{N}$. *We refer to the set* $\mathbb{Z}_n$ *as "Z mod n."*

Question 2.2.5   State the elements in each of the sets $\mathbb{Z}_3$, $\mathbb{Z}_6$, and $\mathbb{Z}_{10}$.

                                                                                                           ■

These sets $\mathbb{Z}_n$ are important because they contain all the possible remainders under division by a fixed positive integer $n \in \mathbb{N}$. The remainders serve as representatives of the infinite sets of integers obtained by partitioning the integers (based on division by a fixed positive integer n) so that every integer appears in exactly one set. As Gauss and others since have found, this perspective has profound implications when studying the algebraic properties of integers.

The algebraic properties of number systems depend not just on the numbers, but also on the operations performed on them. This dependence leads us to consider possible operations on these sets of numbers. Since $\mathbb{Z}_n$ is a finite set of integers, we naturally think of applying the standard addition operation for integers (and other operations, including standard multiplication). This practice of implementing well-understood operations and techniques in new settings is common in mathematics. Unfortunately, the standard addition of integers suffers an important shortcoming in $\mathbb{Z}_n$, which is addressed by modifying the standard addition operation into a new modular addition operation. Before defining this operation, we explore the limitations of standard addition in the context of $\mathbb{Z}_6$.

Question 2.2.6   Consider the *closure* of the set $\mathbb{Z}_6$ under standard addition. Identify which of the following sums (computed using standard addition), are in $\mathbb{Z}_6 = \{0, 1, 2, 3, 4, 5\}$ and which are not in $\mathbb{Z}_6$.

(a) $0 + 3$ (c) $0 + 4$

(b) $3 + 3$ (d) $3 + 4$

∎

As question 2.2.6 indicates, when $a, b \in \mathbb{Z}_6$, sometimes $a + b \in \mathbb{Z}_6$ and sometimes $a + b \notin \mathbb{Z}_6$. This behavior is described by saying $\mathbb{Z}_6$ is *not closed under addition*. We typically want to work with sets and operations where the set is closed under the given operation. Fortunately, there is a modified "modular" addition operation $\oplus$ such that $\mathbb{Z}_6$ is closed under $\oplus$. The idea is to define $a \oplus b$ equal to the remainder r when $a + b$ (the standard sum of a and b) is divided by a fixed, given integer $n \in \mathbb{N}$. The following definition makes this notion precise.

Definition 2.2.3   *If $a, b \in \mathbb{Z}_n$ and $(a + b) \bmod n = r$, then $a \oplus b = r$. This operation is called* **addition mod n** *and we refer to $\mathbb{Z}_n$* **under addition mod n.**

The notation $\oplus$ for modular addition is traditionally used for all sets $\mathbb{Z}_n$. There are actually infinitely many different modular addition operations $\oplus_n$, one for each positive integer $n \in \mathbb{N}$ that can serve as a divisor. Mathematicians often rely on context to identify the particular operation in use, and the designation of a set $\mathbb{Z}_n$ automatically determines the value of the divisor n and the corresponding modular addition operation.

Example 2.2.4   We compute two sums in $\mathbb{Z}_6 = \{0, 1, 2, 3, 4, 5\}$ under addition mod 6.

- $1 \oplus 2 = 3$ since $(1 + 2) \bmod 6 = 3 \bmod 6 = 3$
- $3 \oplus 5 = 2$ since $(3 + 5) \bmod 6 = 8 \bmod 6 = 2$

∎

Question 2.2.7   Verify the following sums in $\mathbb{Z}_6 = \{0, 1, 2, 3, 4, 5\}$ under addition mod 6.

(a) $1 \oplus 4 = 5$ (c) $2 \oplus 4 = 0$

(b) $4 \oplus 5 = 3$ (d) $3 \oplus 4 = 1$

∎

The sums computed in example 2.2.4 and question 2.2.7 provide some *evidence* that $\mathbb{Z}_6$ is closed under addition mod 6; that is, $a \oplus b$ appears to be in $\mathbb{Z}_6$ for all $a, b \in \mathbb{Z}_6$. However, we have not yet proven this general fact—only that $a \oplus b$ is in $\mathbb{Z}_6$ for the specific pairs of numbers given in example 2.2.4 and question 2.2.7. In this simple setting, mathematicians often use exhaustion to provide a complete justification that $\mathbb{Z}_6$ is closed under addition mod 6, computing the mod 6 sum for every possible pair of integers in $\mathbb{Z}_6$. In the next section, we describe a "Cayley table" for $\mathbb{Z}_6$ that performs all of these computations. In section 2.4, we also study the abstract properties of integers under modular arithmetic; this approach shows that every set $\mathbb{Z}_n$ is closed under the corresponding modular addition operation.

For now, we explore two important algebraic properties of the set $\mathbb{Z}_n$ under addition mod n. As in the context of standard addition on the integers, the number zero plays a special role in $\mathbb{Z}_n$ under modular addition.

Question 2.2.8 Directly compute the value of $0 \oplus a$ and $a \oplus 0$ for every $a \in \mathbb{Z}_6$.

∎

The computations in question 2.2.8 demonstrate that the element $0 \in \mathbb{Z}_6$ "fixes" every element of $\mathbb{Z}_6$ and so preserves the "identity" of every element. Therefore, we refer to 0 as the *identity* element of $\mathbb{Z}_6$. As you might surmise, 0 plays a similar role in every set $\mathbb{Z}_n$ under addition mod n.

We also observe that some pairs of integers, when summed together using modular addition, produce the identity 0. This property is also found in the setting of the integers $\mathbb{Z}$ under standard addition. For example, $2 + (-2) = 0$ and $(-2) + 2 = 0$; in general, $a + (-a) = 0$ and $(-a) + a = 0$ for every integer $a \in \mathbb{Z}$. This process of identifying the *additive inverse* $(-a)$ of an integer $a \in \mathbb{Z}$ can also be carried out in the context of $\mathbb{Z}_n$. Whenever the sum of a and b produces the identity (in this case, when $a \oplus b = 0$), we say that a and b are *inverses*, or that b is the *additive inverse* of a.

Question 2.2.9 Answer the following questions about inverses in $\mathbb{Z}_6 = \{0, 1, 2, 3, 4, 5\}$ under addition mod 6.

(a) Prove that 2 and 4 are inverses under addition mod 6 by showing both $2 \oplus 4 = 0$ and $4 \oplus 2 = 0$.

(b) State the inverse of 3 under addition mod 6; that is, find $b \in \mathbb{Z}_6$ such that both $3 \oplus b = 0$ and $b \oplus 3 = 0$. The answer is an element of $\mathbb{Z}_6 = \{0, 1, 2, 3, 4, 5\}$.

(c) State the inverse of 5 under addition mod 6.

∎

Our discussion of closure, identity, and inverses in the context of $\mathbb{Z}_6$ is really just a preliminary investigation of the algebraic properties common to many different number systems. We continue to explore these properties in a variety of different number systems throughout this chapter. For example, in the next section we consider a modified version of another familiar operation: standard multiplication of integers. Perhaps you can already hazard a conjecture about how we will modify this operation.

For now, we end this section with a proof of one of the theorems that was introduced in section 1.7, using the division algorithm to prove the parity property of the integers.

Theorem 2.2.2 The parity property of the integers *Every integer is either even or odd.*

Proof   Assume $m \in \mathbb{Z}$ is an integer. We prove that m is either even or odd by showing that $m = 2k$ or $m = 2k + 1$ for some $k \in \mathbb{Z}$. Applying the division algorithm to m using the divisor $n = 2$ produces $m = 2q + r$ with $r \in \{0, 1\}$. There are two cases to consider. If $r = 0$, then $m = 2q + 0 = 2q$ and m is even. If $r = 1$, then $m = 2q + 1$ and m is odd. Since every integer falls into one of these two cases, every integer must be either even or odd. ∎

2.2.1   Reading Questions for Section 2.2

1. State the division algorithm.
2. Discuss the two distinct claims made by the division algorithm.
3. What quotient q and remainder r satisfy the division algorithm when $m = 7$ and $n = 3$?
4. Discuss why infinitely many integers m have remainder $r = 1$ under division by $n = 3$.
5. Define $a \equiv b \bmod n$ and give an interesting example for $n = 3$.
6. Define $\mathbb{Z}_n$ and give an example.
7. Why are we interested in the elements of $\mathbb{Z}_n$?
8. Define $a \oplus b = r$ and give an example.
9. When is a set G is closed under standard addition? Give an example of a set that is closed under addition and a set that is not.
10. How does closure motivate the definition of modular addition?
11. Define and give an example of an identity.
12. Define and give an example of an additive inverse.

2.2.2   Exercises for Section 2.2

In exercises 1–4, determine the quotient q and the remainder r from the division algorithm when each m is divided by $n = 7$.

1. $m = 39$
2. $m = 195$
3. $m = -9$
4. $m = -603$

In exercises 5–8, identify three integers $m \in \mathbb{Z}$ that produce each remainder r when m is divided by $n = 7$.

5. $r = 1$
6. $r = 2$
7. $r = 4$
8. $r = 5$

In exercises 9–12, determine the quotient q and the remainder r from the division algorithm when each m is divided by $n = 8$.

9. $m = 33$

10. $m = 198$

11. $m = -11$

12. $m = -612$

In exercises 13–16, identify three integers $m \in Z$ that produce each remainder r when m is divided by $n = 8$.

13. $r = 1$

14. $r = 4$

15. $r = 5$

16. $r = 7$

In exercises 17–22, determine the three smallest positive integers $m \in Z$ with remainder r under division by n.

17. $n = 3$ and $r = 1$

18. $n = 3$ and $r = 2$

19. $n = 7$ and $r = 3$

20. $n = 7$ and $r = 4$

21. $n = 10$ and $r = 3$

22. $n = 10$ and $r = 7$

In exercises 23–32, find the value of each expression.

23. $39 \bmod 7$, $195 \bmod 7$, $(-9) \bmod 7$, $(-603) \bmod 7$

24. $33 \bmod 8$, $198 \bmod 8$, $(-11) \bmod 8$, $(-608) \bmod 8$

25. $34 \bmod 9$, $199 \bmod 9$, $(-13) \bmod 9$, $(-606) \bmod 9$

26. $36 \bmod 10$, $197 \bmod 10$, $(-10) \bmod 10$, $(-605) \bmod 10$

27. $35 \bmod 11$, $196 \bmod 11$, $(-12) \bmod 11$, $(-607) \bmod 11$

28. $36 \bmod 2$, $197 \bmod 2$, and both $(2k + 1) \bmod 2$ and $(2k) \bmod 2$ for $k \in Z$

29. $0^2 \bmod 2$, $1^2 \bmod 2$

30. $0^3 \bmod 3$, $1^3 \bmod 3$, $2^3 \bmod 3$

31. $0^5 \bmod 5$, $1^5 \bmod 5$, $2^5 \bmod 5$, $3^5 \bmod 5$, $4^5 \bmod 5$.

32. $0^7 \bmod 7$, $1^7 \bmod 7$, $2^7 \bmod 7$, $3^7 \bmod 7$, $4^7 \bmod 7$, $5^7 \bmod 7$, $6^7 \bmod 7$.

33. *(Fermat's little theorem)* Based on the answers to exercises 29–32, formulate a conjecture about the value of $a^p \bmod p$ when $a \in Z_p$ and p is a prime integer.

In exercises 34–37, state the elements in each set.

34. $\mathbb{Z}_5$

35. $\mathbb{Z}_8$

36. $\mathbb{Z}_{11}$

37. $\mathbb{Z}_{15}$

In exercises 38–41, identify the infinite subset of the integers represented by each remainder r in the given set $\mathbb{Z}_n$.

38. $r = 2$ in $\mathbb{Z}_5$

39. $r = 4$ in $\mathbb{Z}_5$

40. $r = 2$ in $\mathbb{Z}_8$

41. $r = 4$ in $\mathbb{Z}_8$

In exercises 42–53, compute each modular sum in the given set $\mathbb{Z}_n$.

42. in $\mathbb{Z}_2$: $0 \oplus 1$ and $1 \oplus 1$

43. in $\mathbb{Z}_3$: $1 \oplus 2$ and $2 \oplus 2$

44. in $\mathbb{Z}_4$: $0 \oplus 3$ and $1 \oplus 3$

45. in $\mathbb{Z}_4$: $3 \oplus 2$ and $3 \oplus 3$

46. in $\mathbb{Z}_5$: $4 \oplus 3$ and $3 \oplus 2$

47. in $\mathbb{Z}_5$: $3 \oplus 4$ and $2 \oplus 2$

48. in $\mathbb{Z}_7$: $4 \oplus 5$ and $3 \oplus 2$

49. in $\mathbb{Z}_7$: $6 \oplus 5$ and $3 \oplus 6$

50. in $\mathbb{Z}_9$: $7 \oplus 5$ and $3 \oplus 6$ 52. in $\mathbb{Z}_{11}$: $4 \oplus 5$ and $7 \oplus 5$

51. in $\mathbb{Z}_9$: $2 \oplus 7$ and $8 \oplus 8$ 53. in $\mathbb{Z}_{11}$: $8 \oplus 9$ and $1 \oplus 10$

Exercises 54–55 consider identities and inverses in $\mathbb{Z}_4 = \{0, 1, 2, 3\}$.

54. Prove 0 is the additive identity of $\mathbb{Z}_4$ by directly computing $0 \oplus a = a$ and $a \oplus 0 = a$ for every $a \in \mathbb{Z}_4$.

55. The addition mod 4 inverse of 2 is 2 since $(2 + 2) \bmod 4 = 4 \bmod 4 = 0$. Identify the addition mod 4 inverse for each of the four elements in $\mathbb{Z}_4$.

In exercises 56–59, identify the inverse under addition mod n for every element of each set.

56. $\mathbb{Z}_5$ 58. $\mathbb{Z}_{11}$

57. $\mathbb{Z}_8$ 59. $\mathbb{Z}_{15}$

In exercises 60–64, prove each mathematical statement. For exercises 60 and 61, assume that $a, b, c, d \in \mathbb{Z}$ with $a \equiv b \bmod n$ and $c \equiv d \bmod n$.

60. $(a + c) \equiv (b + d) \bmod n$ 63. If a is odd, then $a^2 \equiv 1 \bmod 4$.

61. $(a - c) \equiv (b - d) \bmod n$ 64. If a is odd, then $a^2 \equiv 1 \bmod 8$.

62. If a is even, then $a^2 \equiv 0 \bmod 4$.

Exercises 65–68 consider the commutativity of set-theoretic operations. You may have noticed that identity computations require both $a \oplus 0 = a$ and $0 \oplus a = a$ for all $a \in \mathbb{Z}_n$. This requirement of commutativity (adding 0 on both the left and the right) may seem a bit mysterious. While many mathematical operations commute, some do not, as illustrated by considering the following (possible) identities.

In exercises 65–68, let A and B be sets. Either prove or disprove (with a counterexample) each set-theoretic identity.

65. $A \cap B = B \cap A$ 67. $A \setminus B = B \setminus A$

66. $A \cup B = B \cup A$ 68. $A \times B = B \times A$

Exercises 69 and 70 consider further algebraic properties of sets.

69. Prove that $\emptyset$ is the identity for union of sets; that is, prove that if A is a set, then both $A \cup \emptyset = A$ and $\emptyset \cup A = A$.

70. (*Unique empty set theorem*) Use exercise 69 to justify the two equalities in the following proof that the empty set is unique.

Proof   Assume that both $\emptyset_1$ and $\emptyset_2$ are empty sets. Then the following two equalities hold:

$$\emptyset_1 = \emptyset_1 \cup \emptyset_2 = \emptyset_2$$

∎

2.3   Modular Multiplication and Equivalence Relations

We continue our study of number systems by considering a "modular multiplication" operation on $\mathbb{Z}_n$. When working with products of elements of $\mathbb{Z}_n$, we must modify the standard multiplication operation for the same reason we modified the standard addition operation—the set $\mathbb{Z}_n$ is not closed under either standard addition or standard multiplication (see question 2.2.6). For example, $5, 6 \in \mathbb{Z}_7$, but the standard product $5 \cdot 6 = 30 \notin \mathbb{Z}_7$. As you might expect, the definition of multiplication mod n closely parallels the definition of addition mod n. The idea is to set $a \odot b$ equal to the remainder r when $a \cdot b$ (the standard product of a and b) is divided by a fixed, given integer $n \in \mathbb{N}$. The following definition makes this notion precise.

Definition 2.3.1 *If $a, b \in \mathbb{Z}_n$ and $(a \cdot b) \bmod n = r$, then $a \odot b = r$. This operation is called* **multiplication mod n**, *and we refer to $\mathbb{Z}_n$* **under multiplication mod n**.

As with modular addition, the notation $\odot$ for modular multiplication is traditionally used for all sets $\mathbb{Z}_n$ with context indicating the particular divisor n. The designation of a set $\mathbb{Z}_n$ automatically determines the value of the divisor n and the corresponding modular multiplication operation.

Example 2.3.1 We compute two products in $\mathbb{Z}_7 = \{0, 1, 2, 3, 4, 5, 6\}$ under multiplication mod 7.

- $2 \odot 3 = 6$ since $(2 \cdot 3) \bmod 7 = 6 \bmod 7 = 6$
- $2 \odot 5 = 3$ since $(2 \cdot 5) \bmod 7 = 10 \bmod 7 = 3$

■

Question 2.3.1 Verify the following products in $\mathbb{Z}_7 = \{0, 1, 2, 3, 4, 5, 6\}$ under multiplication mod 7.

(a) $1 \odot 4 = 4$

(b) $3 \odot 5 = 1$

(c) $4 \odot 5 = 6$

(d) $4 \odot 6 = 3$

■

In the previous section, we began investigating abstract algebraic properties of $\mathbb{Z}_n$ under modular addition. We identified three particularly important properties: closure, identity, and inverses under addition mod n. These three properties, together with associativity, describe the notion of a "group" and are fundamental to our work with number systems. We continue this study of abstract algebra in the context of $\mathbb{Z}_n$ under modular multiplication.

We first determine the *identity* for $\mathbb{Z}_n$ under multiplication mod n. Recall that an identity "fixes" every element of the number system under the given operation. For example, 0 was the identity element of $\mathbb{Z}_n$ under addition mod n since both $0 \oplus a = a$ and $a \oplus 0 = a$ for every $a \in \mathbb{Z}_n$. In the multiplicative setting, we are similarly interested in identifying an element $e \in \mathbb{Z}_n$ such that both $e \odot a = a$ and $a \odot e = a$ for every $a \in \mathbb{Z}_n$. Based on our experience with the integers under standard multiplication, we can readily identify the element that serves as this identity.

Question 2.3.2 Compute the value of $1 \odot a$ and $a \odot 1$ for every $a \in \mathbb{Z}_7$.

■

The computations in question 2.3.2 demonstrate that 1 "fixes" every element of $\mathbb{Z}_7$ under multiplication mod 7 and is therefore the identity of $\mathbb{Z}_7$ under multiplication mod 7. Similar computations verify that 1 is the *identity* for $\mathbb{Z}_n$ under multiplication mod n for any choice of $n \in \mathbb{N}$.

We now turn our attention to determining multiplication mod n *inverses* of the elements of $\mathbb{Z}_n$. For each element $a \in \mathbb{Z}_n$, the goal is to find $b \in \mathbb{Z}_n$ such that the modular product of a and b is equal to the identity under modular multiplication; symbolically, we need both $a \odot b = 1$ and $b \odot a = 1$. When this happens, we say that a and b are *inverses under multiplication mod n* or that b is the *multiplicative inverse* of a.

Example 2.3.2   We verify that 3 and 5 are multiplicative inverses in $\mathbb{Z}_7$.

As indicated in question 2.3.1 above, $3 \odot 5 = 15 \bmod 7 = 1$ in $\mathbb{Z}_7$. Similarly, $5 \odot 3 = 15 \bmod 7 = 1$. Since both $3 \odot 5$ and $5 \odot 3$ are equal to the identity of $\mathbb{Z}_7$, we know that 3 is the inverse of 5 under multiplication mod 7 and that 5 is the inverse of 3 under multiplication mod 7.

∎

Question 2.3.3   Determine the multiplicative inverse of each element from $\mathbb{Z}_7 = \{0, 1, 2, 3, 4, 5, 6\}$ under multiplication mod 7.

(a) $a = 1$ (c) $a = 3$
(b) $a = 2$ (d) $a = 6$

In addition, identify the unique element of $\mathbb{Z}_7$ that does not have a multiplicative inverse; this same element has "inverse issues" under standard multiplication of integers.

∎

As highlighted in question 2.3.3, not every element of $\mathbb{Z}_n$ has an inverse under multiplication mod n. For $\mathbb{Z}_7$, the unique element without a multiplicative inverse is 0. This situation is the best possible result we can hope for since 0 does not have a multiplicative inverse in any $\mathbb{Z}_n$. In general, most $\mathbb{Z}_n$'s contain several elements not having multiplicative inverses. However, having an inverse is a desirable property. Therefore, we modify the set under examination to guarantee every element has an inverse. In this way, we follow of the practice of mathematicians—making slight, incremental changes to known objects to enable a fruitful analysis.

In the context of studying abstract algebraic properties, we are particularly interested in the subset of $\mathbb{Z}_n$ that consists of exactly those elements of $\mathbb{Z}_n$ with multiplicative inverses; this set is denoted $U(n)$. We refer to $U(n)$ as the set of *units* of $\mathbb{Z}_\mathbf{n}$. In the next section, we prove that the set $U(n)$ of invertible elements under the operation of multiplication mod n satisfies the desirable algebraic properties that define a "group"; in fact, the sets $U(n)$ under multiplication mod n provide the standard examples of finite multiplicative groups.

As demonstrated in question 2.3.2, we can observe that the multiplicative identity is always its own inverse (since $1 \odot 1 = 1$), and so the element 1 is always in $U(n)$. But what about the nonidentity elements of $\mathbb{Z}_n$? One approach to determining which of these elements are also in $U(n)$ is to check for inverses one element at a time. Unfortunately, this exhaustive approach is too computationally intense for large $n \in \mathbb{N}$. Instead, there exists an algorithmic approach for identifying the elements of $U(n)$. The algorithm is based on determining the elements of $\mathbb{Z}_n$ that are relatively prime to n; those elements satisfy the axiom for inverses.

Working in this direction, we say that a positive integer $p \in \mathbb{N}$ is a *factor* of an integer $n \in \mathbb{Z}$ if p divides n evenly; using the notation of the division algorithm, p is a factor of n if $n = q \cdot p$ for some quotient $q \in \mathbb{Z}$. For example, 7 is a factor of 21 since $21 = 3 \cdot 7$. We say that $m \in \mathbb{Z}$ *shares a common factor* with $n \in \mathbb{Z}$ if there exists an integer $p \in \mathbb{Z}$ such that p is a factor of both m and n. For example, 36 shares a common factor with 21, since 3 is a factor of both. Finally, we say that $m \in \mathbb{Z}$ is *relatively prime* to $n \in \mathbb{Z}$ if m and n do not share a common factor $p > 1$. For example, 14 and 15 are relatively prime, since 2 and 7 (the factors of 14) are not factors of 15. These ideas enable the following characterization of the set of units.

Definition 2.3.2 *U(n) denotes the set of **units** of $\mathbb{Z}_n$ consisting of the nonzero elements of $\mathbb{Z}_n$ that are relatively prime to $n \in \mathbb{N}$; these elements share no common factors with n greater than 1. We typically work with $U(n)$ under the operation of multiplication mod n.*

Example 2.3.3 We identify the elements in $U(9)$. From definition 2.3.2, the set $U(9)$ is a subset of $\mathbb{Z}_9$ and membership in $U(9)$ is determined by identifying which nonzero elements of $\mathbb{Z}_9 = \{0, 1, 2, 3, 4, 5, 6, 7, 8\}$ share a common factor with 9 greater than 1. We first observe that the factors of 9 are 1, 3, and 9.

- For 0:   Only nonzero elements of $\mathbb{Z}_9$ can be elements of $U(9)$, and so $0 \notin U(9)$.
- For 1: The only factor of 1 is 1 itself, so 1 is relatively prime to 9 and $1 \in U(9)$.
- For 2: The factors of 2 are 1 and 2, so 2 is relatively prime to 9 and $2 \in U(9)$.
- For 3: The factors of 3 are 1 and 3, and 3 is a common factor of both 3 and 9 that is greater than 1. Therefore, 3 and 9 are not relatively prime, and we have $3 \notin U(9)$.

Continuing in this fashion, we find that $U(9) = \{1, 2, 4, 5, 7, 8\}$.

■

Question 2.3.4 Identify the elements in the sets $U(5) \subset \mathbb{Z}_5$ and $U(6) \subset \mathbb{Z}_6$.

■

As you have perhaps surmised in answering question 2.3.4, the elements of $U(n)$ are readily identified when the integer n is prime. Prime numbers are studied more fully in section 3.1, but for the moment, perhaps you can recall that an integer n is prime if the only factors of n are 1 and n. Therefore, when n is a prime integer, no positive integer between 1 and n shares a common factor with n. In this case, $U(n)$ is equal to the set of all nonzero elements of $\mathbb{Z}_n$; symbolically, $U(n) = \mathbb{Z}_n \setminus \{0\} = \{1, 2, \ldots, n - 1\}$ when n is prime.

We now examine the property of closure under modular multiplication. In the previous section, we asserted that $\mathbb{Z}_n$ is closed under addition mod n since for every pair $a, b \in \mathbb{Z}_n$, we have $a \oplus b \in \mathbb{Z}_n$. As we will see, $\mathbb{Z}_n$ and $U(n)$ are both closed under multiplication mod n. Since we are interested in knowing this mathematical truth for every integer n, we eventually consider a general proof of this result using abstract properties of integers. For the moment, we use the method of exhaustion for relatively small integers n. The direct exhaustion approach to verifying the closure of $U(n)$ and $\mathbb{Z}_n$ under modular multiplication (and $\mathbb{Z}_n$ under modular addition) requires the computation of all possible products of pairs of elements from these sets. Fortunately, a *Cayley table* provides a systematic approach to computing all of these various products.

Definition 2.3.3 *Let G be a finite set that is closed under an operation $\circ$; that is, for every $a, b \in G$, we have $a \circ b \in G$. A* **Cayley table for G under operation** $\circ$ *is an operation table that displays the result of applying the operation $\circ$ to each pair of elements $a, b \in G$, and so identifies the element $a \circ b \in G$ for every pair $a, b \in G$.*

These computational tables are named in honor of the talented and prolific English mathematician Arthur Cayley. In the 1800s, relatively few professorships were available at the handful of colleges and universities in England. Therefore, despite exhibiting tremendous mathematical talent during and immediately after his undergraduate work at Trinity College in Cambridge, Cayley worked as a lawyer for 14 years in order to make enough money to support his mathematical "hobby." During this time, Cayley shared strong friendships and deep mathematical conversations with a number of other well-regarded mathematicians (many of whom were also lawyers and actuaries). In just this relatively short time span, Cayley published approximately 250 research papers. Even today this publication record would be considered a highly prolific career, let alone just a few years worth of work while otherwise employed. In the 1850s, Cayley first extended the notion of a "group" from the setting of finite functions to a variety of other number systems. In the paper developing these ideas, Cayley gave the first computational tables of the type illustrated below for $\mathbb{Z}_3$ and $U(3)$; in this way they have come to be known as Cayley tables.

Example 2.3.4   We compute the Cayley table for $\mathbb{Z}_3 = \{0, 1, 2\}$ under addition mod 3.

The Cayley table is constructed so the interior table position determined by the row with element a and the column with element b contains the element $a \oplus b$, with a in the left position of the sum and b in the right position of the sum.

$\oplus$	0	1	2
0	$0 \oplus 0$	$0 \oplus 1$	$0 \oplus 2$
1	$1 \oplus 0$	$1 \oplus 1$	$1 \oplus 2$
2	$2 \oplus 0$	$2 \oplus 1$	$2 \oplus 2$

$\Rightarrow$

$\oplus$	0	1	2
0	0	1	2
1	1	2	0
2	2	0	1

Examining the Cayley table on the right, we can see that the set $\mathbb{Z}_3$ is closed under addition mod 3 since only elements of $\mathbb{Z}_3$ appear in the table. The algebraic properties of identity and inverses are readily observed. Since 0 fixes every element of $\mathbb{Z}_3$ under addition mod 3 (as witnessed in the first column and first row of the Cayley table), 0 is the additive identity . In addition, the additive inverse of a given element can be determined by searching for 0 in the appropriate row and column. For example, 1 and 2 are additive inverses since 0 appears both in the second row, third column and in the third row, second column.

∎

Example 2.3.5   We compute the Cayley table for $U(3) = \mathbb{Z}_3 \setminus \{0\} = \{1, 2\}$ under multiplication mod 3.

$\odot$	1	2
1	$1 \odot 1$	$1 \odot 2$
2	$2 \odot 1$	$2 \odot 2$

which is

$\odot$	1	2
1	1	2
2	2	1

As in example 2.3.4, $U(3)$ is closed under multiplication mod 3 since only elements of $U(3)$ appear in the Cayley table. In addition, 1 is the identity for multiplication mod 3 and both 1 and 2 are their own inverses under multiplication mod 3.

■

As discussed in these examples, the Cayley table for a set under an associated operation contains a great deal of algebraic information, which can play a key role in analyzing the set under the given operation. Significant patterns can appear; for example, in the Cayley tables of examples 2.3.4 and 2.3.5, notice that each element of the set appears *exactly once* in each row and each column. Such squares of n symbols in an array of size n with each symbol occurring exactly once in each row and in each column are known as *Latin squares*. The Cayley tables for $\mathbb{Z}_n$ under modular addition and $U(n)$ under modular multiplication are *always* Latin squares, and the Cayley table for any "group" is a Latin square; in practice, we often use this fact when computing the Cayley tables of such sets.

Question 2.3.5   Compute the Cayley table for each set under the given operation.

(a) $\mathbb{Z}_4$ under addition mod 4

(b) $\mathbb{Z}_6$ under addition mod 6

(c) $U(6)$ under multiplication mod 6

(d) $U(7)$ under multiplication mod 7

■

Another important conceptual key to developing a deep understanding of $\mathbb{Z}_n$ and $U(n)$ is the notion of an *equivalence relation*. As you may recall from the previous section, the elements of $\mathbb{Z}_n$ and $U(n)$ are not just numbers, but also sets (even though we work with them in much the same way that we work with numbers). Specifically, the elements of $\mathbb{Z}_n$ and $U(n)$ represent certain subsets of integers known as *equivalence classes*. In preparation for studying equivalence relations and equivalence classes, we revisit some consequences of the division algorithm.

Question 2.3.6   (a) Using the division algorithm, identify the remainder r when each of $m = 7$, $m = 4$, and $m = -5$ is divided by $n = 3$.

(b) Determine an infinite set consisting of every integer m with remainder $r = 1$ under division by $n = 3$.

(c) Similarly, describe the two infinite sets of integers with corresponding remainders $r = 0$ and $r = 2$ under division by $n = 3$.

■

As we have seen, the elements of $\mathbb{Z}_3$ are not just integers, but also represent sets of integers. In particular, an element $r \in \mathbb{Z}_3$ represents the set consisting of all integers with remainder r under division by $n = 3$; that is, an integer a is in the set represented by r if $a \bmod 3 = r$. The full correspondence is given by the following identification of sets and remainders.

Identify the set $\{\ldots, -6, -3, \mathbf{0}, 3, 6, \ldots\}$ with remainder $r = 0$.
Identify the set $\{\ldots, -5, -2, \mathbf{1}, 4, 7, \ldots\}$ with remainder $r = 1$.
Identify the set $\{\ldots, -4, -1, \mathbf{2}, 5, 8, \ldots\}$ with remainder $r = 2$.

More generally, this correspondence extends to any $n \in \mathbb{N}$ with each element $r \in \mathbb{Z}_n$ representing the corresponding infinite set of integers with remainder r under division by n; symbolically, $a \in \mathbb{Z}$ is in the set represented by $r \in \mathbb{Z}_n$ iff $a \bmod n = r$.

Gathering together the elements of these various subsets recognizes a relationship that exists among integers. In the particular case of $\mathbb{Z}_n$, the elements are related because they share a common remainder under division by n. Frequently in mathematics, we are interested in defining and working with relationships among numbers and other mathematical objects, and we identify (or "equate") objects based on these relationships. The following definition states the key properties of such relationships.

Definition 2.3.4   *Let S be a nonempty set. A **relation on S** is a set of ordered pairs (a, b) with $a, b \in S$. In this setting, we write the ordered pair (a, b) as $a \sim b$ and say "a is related to b." An **equivalence relation** $\sim$ on a set S is a relation satisfying the following three properties for all $a, b, c \in S$.*

- **Reflexivity:** *$a \sim a$; that is, every element is related to itself;*
- **Symmetry:** *$a \sim b$ implies $b \sim a$; that is, if a is related to b, then b is related to a;*
- **Transitivity:** *$a \sim b$ and $b \sim c$ imply $a \sim c$; that is, if a is related to b and b is related to c, then a is related to c.*

Recall that we have already used the symbol $\sim$ to denote the negation connective of sentential logic. The context in which $\sim$ is used will indicate the intended meaning of this symbol. In addition, mathematicians often use other, traditional symbols to denote equivalence relations. For example, we use "$\equiv$" to denoted the equivalence relation of logical equivalence in the context of sentential logic.

Even if the name "equivalence relation" is new, we are already familiar with several equivalence relations. Exercises 53–55 from section 1.2 demonstrate that logical equivalence is an equivalence relation on the sentences of sentential logic. More importantly, the standard identity or equality relation "$a = b$" is an equivalence relation on every set of mathematical objects. This should be clear from the definition of an equivalence relation and your mathematical experience with equality. In fact, the definition of equivalence relation is motivated in large part by the key properties of the standard equality relation. Another important example of an equivalence relation is provided by modular arithmetic on integers as detailed in the following example.

Example 2.3.6   Recall from definition 2.2.1 in section 2.2 that if $n \in \mathbb{Z}$ is a divisor and $a, b \in \mathbb{Z}$, then $a \equiv b \bmod n$ iff $a \bmod n = b \bmod n$, and we say that a **is congruent to** b mod n. We often use the fact that $a \bmod n = b \bmod n$ if and only if $a - b = n \cdot q$

(see exercise 41 at the end of this section). We prove that congruence modulo n is an equivalence relation on $\mathbb{Z}$ by verifying that each of the three properties from definition 2.3.4 hold for $a \equiv b \bmod n$ on $\mathbb{Z}$. In this example, we use the traditional mathematical notation $a \equiv b \bmod n$ to denote that a is related to b (rather than $a \sim b$).

- **Reflexivity:** We show $a \equiv a \bmod n$. Since $a \bmod n = r \in \{0, 1, \ldots, n-1\}$, we have $a \bmod n = a \bmod n$. Therefore, since standard equality is reflexive on the integers, $a \equiv a \bmod n$ and congruence mod n is reflexive.

- **Symmetry:** We assume that $a \equiv b \bmod n$ and show that $b \equiv a \bmod n$. Since $a \equiv b \bmod n$, we know $a \bmod n = b \bmod n$, where $a \bmod n = b \bmod n = r \in \{0, 1, 2, \ldots, n-1\}$ is an integer. Since standard equality is symmetric on the integers, $b \bmod n = a \bmod n$ and so $b \equiv a \bmod n$. Therefore, congruence modulo n is symmetric.

- **Transitivity:** We assume $a \equiv b \bmod n$ and $b \equiv c \bmod n$, and show $a \equiv c \bmod n$. Since $a \equiv b \bmod n$, we know $a \bmod n = b \bmod n$, where $a \bmod n = b \bmod n = r \in \{0, 1, 2, \ldots, n-1\}$ is an integer. Similarly, since $b \equiv c \bmod n$, we know $b \bmod n = c \bmod n$, where $b \bmod n = c \bmod n = s \in \{0, 1, 2, \ldots, n-1\}$ is an integer. Since standard equality is transitive on the integers, the following string of equalities yields $a \bmod n = c \bmod n$.

$$a \bmod n = r = b \bmod n = s = c \bmod n$$

Therefore, $a \equiv c \bmod n$ and congruence modulo n is transitive.

Since congruence modulo n is reflexive, symmetric, and transitive, congruence modulo n is an equivalence relation.

■

With this example in hand, we consider the identification of set elements induced by an equivalence relation. One good example is the "equality" of the fractions $1/2$, $5/10$, and $10/20$, which are (formally) distinct elements of $\mathbb{Q}$ and are only identified as the same by an appropriate equivalence relation on $\mathbb{Q}$ (see questions 2.3.8–2.3.10 below). We express this identification of elements in set-theoretic terms by gathering together all elements that are equivalent to one another into the same set.

Definition 2.3.5 *Let $\sim$ be an equivalence relation on a nonempty set S. For every $a \in S$, the **equivalence class of a** is the subset of S consisting of all $b \in S$ such that $a \sim b$. Symbolically, the equivalence class of a is often denoted by $[a] = \{b : a \sim b\}$. Every element of S appears in exactly one equivalence class.*

As discussed in the exercises at the end of this section, the equivalence classes partition the set S on which $\sim$ is defined. In particular, equivalence classes are disjoint; that is, they have an empty intersection. In addition, the union of all equivalence classes is the entire set S. We highlight these features of equivalence classes in the following example.

Example 2.3.7 The equivalence relation $a \equiv b \bmod n$ partitions the set of integers $\mathbb{Z}$ into equivalence classes that are represented by the elements of $\mathbb{Z}_n$. These equivalence classes are labeled according to the different possible remainders r that result from applying the division algorithm with divisor n. For example, the element 0 in $\mathbb{Z}_n$ is not just the number 0, but the set of all integers $a \in \mathbb{Z}$ such that $a \equiv 0 \bmod n = 0$. Integers with remainder 0 under division by n are all multiples of n, and so the equivalence class of 0 is $[0] = \{\ldots, -2n, -n, 0, n, 2n, \ldots\}$. ∎

Question 2.3.7   Working in $\mathbb{Z}_4$ under addition mod 4, identify the set of integers that form the corresponding equivalence classes for 0 and for 2. ∎

We end this section with an example of an equivalence relation on the set $\mathbb{Q}$ of rational numbers, consisting of quotients of integers with nonzero denominators. For example, $1/2, 3/1$ and $3/2$ are rational numbers, while $3/0$ and π are not. As mentioned above, we consider a relation on $\mathbb{Q}$ describing the identification of fractions as equal to one another. For example, the fraction $1/2$ is typically identified with the fractions $2/4, 3/6, 4/8$, and so on. Informally, we say that m/n is related to s/t when we can cross multiply and obtain equal integers. Formally, we define this relation $\sim$ on the rationals $\mathbb{Q}$ by

$$\frac{m}{n} \sim \frac{s}{t} \quad \text{iff} \quad m \cdot t = n \cdot s.$$

Example 2.3.8   We consider the relation $\sim$ on $\mathbb{Q}$ defined by $m/n \sim s/t$ iff $m \cdot t = n \cdot s$.

- $\dfrac{4}{5} \sim \dfrac{12}{15}$, since $4 \cdot 15 = 60$ and $5 \cdot 12 = 60$
- $\dfrac{1}{2} \nsim \dfrac{4}{5}$, since $1 \cdot 5 = 5$, $2 \cdot 4 = 8$, and $5 \neq 8$

∎

Question 2.3.8   Explain why the following statements are true or false for the relation $\sim$ on $\mathbb{Q}$ defined by $m/n \sim s/t$ iff $m \cdot t = n \cdot s$,

(a) $\dfrac{7}{3} \sim \dfrac{28}{12}$ (c) $\dfrac{3}{4} \sim \dfrac{6}{8}$

(b) $\dfrac{7}{3} \sim \dfrac{28}{3}$ (d) $\dfrac{5}{4} \sim -\dfrac{10}{8}$

∎

Now that we have some computational experience with this relation $\sim$ on $\mathbb{Q}$, we consider the proof that $\sim$ is an equivalence relation on $\mathbb{Q}$. A given relation is an equivalence relation if the relation is reflexive, symmetric, and transitive, and so this proof has three parts. Since this relation $\sim$ on $\mathbb{Q}$ is computationally defined using familiar arithmetic operations on the integers, we lead you through the proof in the three steps of the following question.

Question 2.3.9   Prove that the relation $\sim$ on the rationals $\mathbb{Q}$ defined by $m/n \sim s/t$ iff $m \cdot t = n \cdot s$ is an equivalence relation by showing $\sim$ satisfies the three properties of an equivalence relation. While arguing that these properties hold for $\sim$, assume standard equality on the integers is an equivalence relation (as we did in example 2.3.6).

(a) Reflexivity: Assume $m, n \in \mathbb{Z}$ with $n \neq 0$. Prove that $m/n \sim m/n$ by direct computation.

(b) Symmetry: Assume $m, n, s, t \in \mathbb{Z}$ with $n, t \neq 0$ and that $m/n \sim s/t$. Prove that $s/t \sim m/n$ by direct computation and the symmetry of equality.

(c) Transitivity: Assume $m, n, s, t, u, v \in \mathbb{Z}$ with $n, t, v \neq 0$ and that both $m/n \sim s/t$ and $s/t \sim u/v$. Prove that $\frac{m}{n} \sim \frac{u}{v}$.

■

Question 2.3.10 Continue the study of the relation $\sim$ on $\mathbb{Q}$ defined by $m/n \sim s/t$ iff $m \cdot t = n \cdot s$ by considering the equivalence classes of rational numbers. For example, the equivalence class of $\frac{1}{3} \in \mathbb{Q}$ is the set $\left[\frac{1}{3}\right] = \left\{ \frac{n}{3n} : n \in \mathbb{Z}^* \right\}$. Determine the equivalence classes of the following rational numbers.

(a) $\dfrac{2}{4}$

(b) $\dfrac{5}{3}$

■

2.3.1 Reading Questions for Section 2.3

1. When is a set G closed under an operation $\circ$? Give an example of a set that is closed under some operation and a set that is not.
2. Define $a \odot b$ and give an example.
3. Define multiplicative inverse mod n and give an example.
4. Define $U(n)$ and give an example.
5. Why are we interested in the elements of $U(n)$?
6. What is distinctive about $U(n)$ when n is prime?
7. Define the Cayley table for G under operation $\circ$ and give an example.
8. What is a relation on a set of numbers? What notation identifies a relation?
9. Give an example of a relation and a set that is not a relation.
10. State and define the three properties satisfied by an equivalence relation.
11. Give three examples of equivalence relations.
12. Define and give an example of an equivalence class.

2.3.2 Exercises for Section 2.3

In exercises 1–10, compute each modular product in the given set $\mathbb{Z}_n$.

1. in $\mathbb{Z}_2$: $0 \odot 1$ and $1 \odot 1$
2. in $\mathbb{Z}_3$: $1 \odot 2$ and $2 \odot 2$
3. in $\mathbb{Z}_5$: $4 \odot 3$ and $3 \odot 2$
4. in $\mathbb{Z}_5$: $1 \odot 4$ and $2 \odot 2$
5. in $\mathbb{Z}_7$: $4 \odot 5$ and $3 \odot 2$
6. in $\mathbb{Z}_7$: $6 \odot 5$ and $3 \odot 6$
7. in $\mathbb{Z}_9$: $7 \odot 5$ and $3 \odot 6$
8. in $\mathbb{Z}_9$: $2 \odot 7$ and $8 \odot 8$
9. in $\mathbb{Z}_{11}$: $4 \odot 5$ and $7 \odot 5$
10. in $\mathbb{Z}_{11}$: $8 \odot 9$ and $1 \odot 10$

In exercises 10–16, state the elements in each set.

11. $\mathbb{Z}_2$ under addition mod 2

12. $\mathbb{Z}_8$ under addition mod 8

13. $\mathbb{Z}_{11}$ under addition mod 11

14. $U(2)$ under multiplication mod 2

15. $U(8)$ under multiplication mod 8

16. $U(11)$ under multiplication mod 11

In exercises 17–22, compute the Cayley table for each set under the natural corresponding modular operation.

17. $\mathbb{Z}_2$

18. $\mathbb{Z}_5$

19. $\mathbb{Z}_8$

20. $U(2)$

21. $U(5)$

22. $U(8)$

In exercises 23–24, identify the multiplication mod n inverse for every element in the given set using the Cayley tables from exercises 21 and 22.

23. $U(5)$ 24. $U(8)$

Exercises 25–28 consider the "associativity" of modular multiplication. Working in $U(11)$ under multiplication mod 11, verify the property $(a \odot b) \odot c = a \odot (b \odot c)$ for each triple of elements by directly computing each pair of products.

25. $2, 3, 4 :\ (2 \odot 3) \odot 4 = 2 \odot (3 \odot 4)$

26. $5, 3, 6 :\ (5 \odot 3) \odot 6 = 5 \odot (3 \odot 6)$

27. $4, 8, 10 :\ (4 \odot 8) \odot 10 = 4 \odot (8 \odot 10)$

28. $7, 8, 9 :\ (7 \odot 8) \odot 9 = 7 \odot (8 \odot 9)$

Exercises 29–31 consider a case in which nonzero elements of $\mathbb{Z}_n$ do not have multiplicative inverses by studying $\mathbb{Z}_8 \setminus \{0\} = \{1, 2, 3, 4, 5, 6, 7\}$ under multiplication mod 8.

29. Compute the complete operation table for $\mathbb{Z}_8 \setminus \{0\}$ under multiplication mod 8; the result is not a Cayley table because this set is not closed under multiplication mod 8.

30. Give an example of $a, b \in \mathbb{Z}_8 \setminus \{0\}$ such that $a \odot b = c$ with $c \notin \mathbb{Z}_8 \setminus \{0\}$.

31. Identify the elements of $\mathbb{Z}_8 \setminus \{0\}$ with inverses under multiplication mod 8 and state the corresponding inverse. Do you note anything special about the elements with multiplicative inverses?

Exercises 32–36 explore a basic version of the Chinese remainder theorem, which relates standard and modular multiplication. The following computations use both standard multiplication (denoted $\cdot$) and multiplication mod 11 (denoted $\odot$).

33. Compute both $12 \odot 5 = (12 \cdot 5) \bmod 11$ and $(12 \bmod 11) \cdot (5 \bmod 11)$.

34. Compute both $101 \odot 48 = (101 \cdot 48) \bmod 11$ and $(101 \bmod 11) \cdot (48 \bmod 11)$.

35. In light of the answers to exercises 33 and 34, formulate a conjecture about the relationship between $a \odot b = (a \cdot b) \bmod 11$ and $(a \bmod 11) \cdot (b \bmod 11)$.

36. Compute both $14 \odot 10 = (14 \cdot 10) \bmod 11$ and $(14 \bmod 11) \cdot (10 \bmod 11)$. Was your conjecture correct?

37. Compute $[(14 \bmod 11) \cdot (10 \bmod 11)] \bmod 11$. If necessary, formulate a revised conjecture.

Exercises 37–38 consider Wilson's theorem, which describes factorials under congruence modulo n. Recall that $n!$ denotes the standard product of integers $1 \cdot 2 \cdot 3 \cdots n$.

37. Compute $(2!) \bmod 3$, $(4!) \bmod 5$, and $(6!) \bmod 7$. Based on these computations, formulate a conjecture about $(n!) \bmod (n+1)$.

38. Compute $(3!) \bmod 4$, $(5!) \bmod 6$, and $(10!) \bmod 11$. Was your conjecture correct? What is distinctive about the numbers for which your conjecture works? If necessary, formulate a revised conjecture.

In exercises 39–41, prove each mathematical claim about congruence modulo **n**, where $a,b,c,d \in \mathbb{Z}$ and $n \in \mathbb{N}$.

39. If $a \equiv b \bmod n$ and $c \equiv d \bmod n$, then $(a \times c) \equiv (b \times d) \bmod n$.

40. $a^2 \equiv b^2 \bmod n$ does not imply $a \equiv b \bmod n$. Hint: Give a counterexample.

41. $a \equiv b \bmod n$ if and only if $a - b$ is divisible by n.

In exercises 42–45, use the biconditional from exercise 41 to prove each mathematical claim about congruence modulo n, where $a,b,c,d \in \mathbb{Z}$ and $n \in \mathbb{N}$.

42. If $a \equiv b \bmod n$, then $(a+c) \equiv (b+c) \bmod n$. Hint: Use the biconditional from exercise 41.

43. If a is even, then $a^2 \equiv 0 \bmod 4$.

44. If a is odd, then $a^2 \equiv 1 \bmod 4$.

45. If a is odd, then $a^2 \equiv 1 \bmod 8$.

Exercises 46–48 consider definition 2.3.4 of an equivalence relation.

46. Express each property of an equivalence relation in predicate logic.

47. Express the negation of each property of an equivalence relation in both predicate logic and English.

48. Prove that if $\sim$ is an equivalence relation on a set S and $[a]$ denotes the equivalence class of a in S under $\sim$, then $a \sim b$ if and only if $[a] = [b]$.

In exercises 49–54, let S be the set of humans and determine whether or not each relation $\sim$ is an equivalence relation. If not, state the properties of an equivalence relation that fail.

49. Define $a \sim b$ iff a is the same age as b.

50. Define $a \sim b$ iff a loves b.

51. Define $a \sim b$ iff a is a full-brother of b.

52. Define $a \sim b$ iff a is a sibling of b.

53. Define $a \sim b$ iff a is a first cousin of b.

54. Define $a \sim b$ iff a is an ancestor of b.

In exercises 55–56, let $S = \{$ Alex, Andy, Bailey, Chris, Dakota, Morgan $\}$. List the pairs of elements in each equivalence relation and identify the corresponding distinct equivalence classes.

55. For $a, b \in S$, define $a \sim b$ iff a and b begin with the same letter.

56. For $a, b \in S$, define $a \sim b$ iff a and b contain the same number of letters.

In exercises 57–65, prove each relation is an equivalence relation and identify two distinct equivalence classes.

58. For $a, b \in \mathbb{R}$, define $a \sim b$ iff $a - b \in \mathbb{Z}$.
59. For $a, b \in \mathbb{Z} \setminus \{0\}$, define $a \sim b$ iff $ab \geq 0$.
60. For $a, b \in \mathbb{Z}$, define $a \sim b$ iff $a + b$ is even.
61. For $a, b \in \mathbb{Z}$, define $a \sim b$ iff $a - b$ is even.
62. For $(a, b), (x, y) \in \mathbb{R}^2$, define $(a, b) \sim (x, y)$ iff $a = x$.
63. For $(a, b), (x, y) \in \mathbb{R}^2$, define $(a, b) \sim (x, y)$ iff $a - x \in \mathbb{Z}$.
64. For differentiable functions f, g on $\mathbb{R}$, define $f \sim g$ iff $f' = g'$.
65. For points $(a, b), (x, y)$ on the Cartesian plane $\mathbb{R}^2$, define $(a, b) \sim (x, y)$ iff (a, b) and (x, y) are equidistant from the origin.
66. For lines J, K on the plane $\mathbb{R}^2$, define $J \sim K$ iff the slope of J is equal to the slope of K.

In exercises 66–70, each relation is *not* an equivalence relations. Determine the properties of an equivalence relation that hold and the properties that fail for each relation.

66. For $a, b \in \mathbb{Z}$, define $a \sim b$ iff $ab > 0$.
67. For $a, b \in \mathbb{Z}$, define $a \sim b$ iff $a > b$.
68. For $a, b \in \mathbb{Z}$, define $a \sim b$ iff $a \geq b$.
69. For $a, b \in \mathbb{Z}$, define $a \sim b$ iff a divides b.
70. For $(a, b), (x, y) \in \mathbb{R}^2$, define $(a, b) \sim (x, y)$ iff either $a = x$ or $b = y$.

2.4 An Introduction to Groups

We continue our exploration of abstract algebra by developing the mathematical notion of a group. In this chapter, we have focused on algebraic properties when studying numbers systems with their corresponding operations and equivalence relations. This focus has enabled us to identify and ferret out fundamental properties from well-understood settings and has given us the ability to look at these properties in similar, though new, settings. For example, equivalence relations highlight the key properties of the standard equality relation and help us recognize these properties in a whole host of quite diverse settings. In this section, we focus on four of the most important algebraic properties of the set of integers under standard addition and refer to every set–operation pair satisfying these four properties as a *group*. After carefully articulating the definition of a group, we study several different number systems satisfying the four group axioms, most of which will be familiar from your previous mathematics courses.

The notion of a group traces its origins to Évariste Galois, a French student who attempted to prove the nonexistence of general algorithms for solving polynomials of sufficient complexity. Galois considered functions mapping the solutions of a polynomial equation to other solutions of this same equation, and so the first groups studied by mathematicians were essentially functions on finite sets (since every polynomial equation over the reals has finitely many solutions). Galois' mathematical insights were so far ahead of his time that even some of the best mathematicians of that era failed to recognize and understand his ideas until years after his death. In one of the more dramatic tales in the history of mathematics, Galois' love life led him into a duel in 1832 and he died at the young age of 20. Realizing that he would probably not survive the duel, Galois wrote a long letter to his friend, Auguste Chevalier, in which he scribbled down his mathematical inspirations on solving polynomial equations, including his insights into groups. Fortunately for us, Chevalier preserved Galois' work and passed his manuscripts along to Joseph Liouville, who published them in 1846.

Over the next 50 years, mathematicians gradually recognized the power of Galois' ideas, and his initial work was developed and refined into a sophisticated and powerful mathematical theory of groups. The application of groups to questions about the solvability of polynomials examines sets of finite functions under composition, but the axiomatic group properties have proven essential to understanding the algebraic properties of many different sets and number systems. The English mathematician Arthur Cayley gave what might be considered the first general definition of a group in the 1850s. At this time, Cayley was studying matrix groups and the quaternions, a number system that extends the complex numbers. His work played a pivotal role in broadening mathematicians' understanding of abstract number systems and helped open doors to applications of group theory in many different areas of mathematics and the physical sciences. To this day, the study of groups (or group theory) remains an active and fertile area of research.

Group theory is also a widely applicable field of mathematics and has been used in a variety of essential ways in many different arts and sciences. For example, crystallographers have used group theory in the study of natural crystal structures and in their efforts to design synthetic crystals with certain desirable properties. Physicists have recognized that subatomic particles satisfy the properties of groups, and group theory allowed them to predict the existence of the "top quark" subatomic particle shortly before its discovery in superaccelerator experiments.

We soon define a group as a set under an operation satisfying four particular algebraic properties. From among the various familiar number systems, we choose to begin our study of groups with the important example of the integers $\mathbb{Z} = \{\ldots, -2, -1, 0, 1, 2, \ldots\}$ under standard addition. The following question recalls four algebraic properties already identified in the previous two sections. Studying these properties deepens our understanding of the integers and helps motivate the formal definition of a group.

Question 2.4.1   Consider the integers $\mathbb{Z}$ under the standard addition operation $+$.

    (a) If we add two integers from $\mathbb{Z}$, can we obtain a number that is not an integer in $\mathbb{Z}$?

    (b) Compute both $1 + (2 + 3)$ and $(1 + 2) + 3$. What is the relationship between these two sums? Does this relationship hold whenever we add three integers? Formulate a general statement.

    (c) In the context of the integers under standard addition, what is distinctive about zero?

    (d) What integer can we add to 3 and obtain 0? Given an arbitrary integer n, what integer can we add to n and obtain 0? Formulate a general statement.

These properties are familiar from our work with modular addition and modular multiplication. What name is given to each of these four algebraic properties?

∎

Motivated by the answers to question 2.4.1 and our work with other number systems, we formally define the notion of a group as follows.

Definition 2.4.1   *A* **binary operation** *on a set G is a function that maps each ordered pair in $G \times G$ to a unique element of a set containing G. A set G under binary operation $\circ$ is a* **group** *when the following four properties hold.*

(1) Closure: *For every $a, b \in G$, we have $a \circ b \in G$;*
(2) Associativity: *For every $a, b, c, \in G$, we have $a \circ (b \circ c) = (a \circ b) \circ c$;*
(3) Identity: *There exists an element $e \in G$ such that for every $a \in G$, we have both $e \circ a = a$ and $a \circ e = a$; we call e the* **identity** *for G under $\circ$ ("einheit" is German for "identity");*
(4) Inverses: *For every $a \in G$, there exists $b \in G$ such that both $a \circ b = e$ and $b \circ a = e$; we call b the* **inverse** *of a and we often write $b = a^{-1}$.*

Mathematicians refer to these four properties as the *axioms* of group theory or as the *group axioms*. As with many abstract concepts in mathematics, the definition of a group evolved gradually. Galois isolated this notion while studying functions on solution sets of polynomials in 1832. Cayley's contributions in the 1850s extended the notion of a group to the context of matrices and the quaternions. These insights eventually led Walter von Dyck to articulate definition 2.4.1 of a group in 1882. In parallel to this work, the eminent French mathematician Augustin-Louis Cauchy implicitly defined the notion of a group in the 1850s. His work appears to have influenced Heinrich Weber, who independently articulated definition 2.4.1. of a group in 1882. Even so it was not until the early 1900s that the abstract definition of a group gained widespread understanding and acceptance by the mathematics community.

These four axioms serve as fundamental assumptions when proving various theorems, or mathematical truths, in group theory. One of the primary characteristics of group theory is that the axioms are sufficiently weak so that many different number systems satisfy the four axioms, and yet they are sufficiently strong so as to enable the proof of many results and the development of a rich mathematical theory. We begin to describe this richness through examples that refine and bring into focus an understanding of the definition of a group. We have already pointed out one very important example of a group: the integers under standard addition.

Example 2.4.1   We observe that each of the four group axioms holds for the set of integers $\mathbb{Z}$ under the standard addition operation $+$.

1. Closure: For every $n, m \in \mathbb{Z}$, we have $n + m \in \mathbb{Z}$.
2. Associativity: For every $n, m, k \in \mathbb{Z}$, we have $(n + m) + k = n + (m + k)$.
3. Identity: The additive identity of the integers is 0, since for every $n \in \mathbb{Z}$, we have both $0 + n = n$ and $n + 0 = n$.
4. Inverses: For every $n \in \mathbb{Z}$, the additive inverse of n is $(-n) \in \mathbb{Z}$, since we have both $n + (-n) = 0$ and $(-n) + n = 0$.

A more complete, rigorous proof that the four group axioms are satisfied by the integers under standard addition uses a formal, axiomatic description of the integers. The details of such a proof are beyond the scope of this book and are left for your later studies.

■

In definition 2.4.1, notice that a group consists of both a set *and* an operation. For example, the set of integers, by itself, is not a group. However, the set of integers under standard addition is a group, as we observed in example 2.4.1. The particular operation associated with a given set is central to determining whether or not we have a group, since some sets can be a group under one operation but not under another. There are infinitely many different examples of groups. The next question identifies an infinite group with a multiplicative operation.

Question 2.4.2 Consider the set of nonzero rational numbers $\mathbb{Q}^* = \mathbb{Q} \setminus \{0\} = \{\frac{p}{q} : p, q \in \mathbb{Z}$ and $p, q \neq 0\}$. Prove $\mathbb{Q}^*$ is a group under the standard multiplication operation by considering each of the four group axioms as outlined below.

(a) Closure: Let $p/q, r/s \in \mathbb{Q}$ and prove that the product of p/q and r/s is rational. Assume the closure of the integers under multiplication and use the zero product property of the reals; in particular, assume the product of two nonzero reals is nonzero and recall that every integer is real.

(b) Associativity: Show that the three rationals $1/2$, $3/5$, and $8/7$ satisfy associativity under standard multiplication. Prove that any three rationals m/n, p/q, $r/s \in \mathbb{Q}$ satisfy associativity under standard multiplication, assuming the associativity of the integers under standard multiplication.

(c) Identity: State the identity of the nonzero rationals under standard multiplication.

(d) Inverses: Determine the inverse of the rational $3/2$ under multiplication. Identify the inverse of an arbitrary nonzero rational m/n.

(e) Zero is omitted from $\mathbb{Q}$ in this question because one of the four group axioms fails to hold if 0 is included. Which group axiom fails?

■

As we might hope and expect, many widely used number systems under their standard operations are groups. For some of these number systems, we need to exclude an element (as with 0 for the rational numbers under standard multiplication), but these special cases are often widely known or easily identified. The next example attempts a naive, straightforward approach to identifying a finite group. In its failure, we see how a given set and operation may not completely satisfy the group axioms.

Example 2.4.2 We prove the set $\{0, 1\}$ under the standard addition operation $+$ is *not* a group.

- Closure: The closure axiom fails for $\{0, 1\}$ under standard addition. The unique counterexample is provided by working with the element $1 \in \{0, 1\}$. In particular, $1 \in \{0, 1\}$, but $1 + 1 = 2 \notin \{0, 1\}$. Therefore, $\{0, 1\}$ is not closed under standard addition.

- Associativity: We observe that standard addition is associative on $\{0, 1\}$ since addition is associative on the set of all integers $\mathbb{Z}$, which includes the elements

0 and 1. Often we find that associativity is "inherited" from the integers or some other appropriate, ambient base set that is already known to satisfy associativity.

- Identity: We observe that 0 is the additive identity. In particular, we have $0 + 0 = 0$ and both $0 + 1 = 1$ and $1 + 0 = 1$.
- Inverses: The inverse axiom fails for $\{0, 1\}$ under standard addition. The element $0 \in \{0, 1\}$ has an inverse since $0 + 0 = 0$, and so 0 is its own inverse. However, the element $1 \in \{0, 1\}$ does not have an inverse since the only possible candidates for inverses in this set are 0 and 1, but $0 + 1 = 1 \neq 0$ and $1 + 1 = 2 \neq 0$. The inverse of 1 under standard addition is -1, but $-1 \notin \{0, 1\}$.

Since the group axioms of closure and inverses do not hold, the set $\{0, 1\}$ is not a group under standard addition.

∎

Only one group axiom needs to fail in order for a given set not to be a group under a given operation. In example 2.4.2, we considered all four axioms for the sake of developing a thorough understanding of the group axioms. As you study further examples of set–operation pairs that are not groups, feel free to identify any one group axiom that fails (unless directed otherwise).

In disproving the closure axiom, example 2.4.2 utilized the methods developed in section 1.7 for disproving universal statements. For example, a proof that the closure axiom is not satisfied involves identifying a counterexample in the given set; that is, specific elements $a, b \in G$ such that $a \circ b \notin G$. Similarly, a proof that the inverse axiom is not satisfied requires us to produce (at least) one concrete counterexample $a \in G$ such that for every $b \in G$, the inverse property $a \circ b = e = b \circ a$ does not hold for a and b. Many given sets under an operation are not groups because of the failure of closure and inverses. Associativity and identity fail less frequently; we consider a few such cases in the exercises at the end of this section.

Despite the failure of our first attempt to identify a finite group, some groups are finite. Modular arithmetic provides examples of finite groups: both $\mathbb{Z}_n$ under modular addition and $U(n)$ under modular multiplication are groups. For example, the closure axiom is satisfied in both $\mathbb{Z}_n$ and $U(n)$ under their respective modular operations, since we identify every modular sum and product with a remainder under division by n, and since these remainders are precisely the elements of $\mathbb{Z}_n$ and $U(n)$. The following question carefully considers the ideas behind the proof that $\mathbb{Z}_n$ under addition mod n satisfies the four group axioms. While this discussion does not constitute a complete, formal proof, your answers should include sufficient detail to convince you of the validity of the claim.

Question 2.4.3   Verify that $\mathbb{Z}_n$ under addition mod n is a group by considering the four group axioms as outlined below.

(a) Closure: For $a, b \in \mathbb{Z}_n$, we defined $a \oplus b = (a + b) \bmod n$; that is, we compute $a \oplus b$ by applying the division algorithm to $a + b$ to obtain $r \in \mathbb{Z}$ where $a + b = n \cdot q + r$ for some integer q. According to the division algorithm, what are the possible values of r? What are the possible values of $a \oplus b$? Is $a \oplus b$ always in $\mathbb{Z}_n$?

(b) Associativity: Associativity for $\mathbb{Z}_n$ under modular addition follows from repeated use of the division algorithm and the associativity of standard addition on the integers $\mathbb{Z}$. As an example (and just an example, not a proof), verify that the following equalities hold in $\mathbb{Z}_6$ under addition mod 6.

- $(4 \oplus 5) \oplus 3 = 4 \oplus (5 \oplus 3)$ $\bullet$ $(2 \oplus 5) \oplus 1 = 2 \oplus (5 \oplus 1)$

(c) Identity: What is the identity of $\mathbb{Z}_n$ under addition mod n? Justify your answer.

(d) Inverses: For $a \in \mathbb{Z}_n$, show that 0 is the inverse of $a = 0$, and otherwise $n - a$ is the inverse of a by arguing that both $(n - a) \oplus a = 0$ and $a \oplus (n - a) = 0$.

■

The answers to question 2.4.3 form the heart of a complete proof that $\mathbb{Z}_n$ under addition mod n is a group. We leave the remaining details for your later studies and summarize this important result in the following theorem.

Theorem 2.4.1 *The set $\mathbb{Z}_n$ under the operation of addition mod n is a group.*

We now turn our attention to the sets $U(n)$ under multiplication mod n. We provide a more complete proof that $U(n)$ is a group under multiplication mod n, particularly for the closure and inverse axioms. The proof that $U(n)$ under multiplication mod n satisfies these group axioms uses a pair of results from number theory (that is, the abstract study of the properties of integers and the solutions to polynomial equations). As we did with the parity property of the integers and the zero product property of the reals in section 1.7, we just state and use these results without proof for the moment.

Theorem 2.4.2 (a) *For $a, b \in \mathbb{Z}$, if p is a prime factor of $a \cdot b$, then either p is a factor of a or p is a factor of b.*

(b) *If $a, n \in \mathbb{Z}$ are relatively prime, then there exist $h, k \in \mathbb{Z}$ such that $a \cdot h + n \cdot k = 1$.*

We give some specific examples illustrating this pair of results. These examples do not constitute proofs but are intended to facilitate an intuitive understanding.

Example 2.4.3 • 2 is a prime factor of $6 \cdot 15$, and 2 is a factor of $6 = 2 \cdot 3$.
- 2 is a prime factor of $6 \cdot 20$, and 2 is a factor of both $6 = 2 \cdot 3$ and $20 = 2 \cdot 10$ (this example shows that the first part of theorem 2.4.2 uses the inclusive-or).
- 3 and 5 are relatively prime because they have no common factor greater than 1. Letting $h = 2$ and $k = -1$, we have $3 \cdot 2 + 5 \cdot (-1) = 1$.
- 4 and 9 are relatively prime. Letting $h = -2$ and $k = 1$, we have $4 \cdot (-2) + 9 \cdot (1) = 1$.

■

Question 2.4.4 Identify $h, k \in \mathbb{Z}$ satisfying $a \cdot h + n \cdot k = 1$ for the following pairs of relatively prime numbers.

(a) $a = 3$ and $n = 7$ (b) $a = 8$ and $n = 5$

■

With the results of theorem 2.4.2 in hand, we prove that $U(n)$ under multiplication mod n is a finite group for every integer $n \in \mathbb{Z}$.

Theorem 2.4.3   *The set $U(n)$ under multiplication mod n is a group.*

Proof   We verify that each of the four group axioms holds for $U(n)$ under multiplication mod n.

- Closure: We assume $a, b \in U(n)$ and show $a \odot b \in U(n)$. First, since $a \odot b = (a \cdot b) \bmod n$, we know that $a \odot b \in \mathbb{Z}_n$. Therefore we need only prove that $(a \cdot b) \bmod n$ is relatively prime to n in order to show $a \odot b \in U(n)$. We proceed by contradiction; we assume $a \odot b \notin U(n)$ and work toward a contradiction. Since $a \odot b \notin U(n)$, we know $a \odot b$ is not relatively prime to n and so shares a common factor with n greater than 1. Let $p \in \mathbb{Z}$ with $1 < p \le a \odot b$ denote such a common factor. In addition, assume that p is prime. By the division algorithm, there exist $q, r \in \mathbb{Z}$ such that $a \cdot b = n \cdot q + r$ where $r = a \odot b \in \mathbb{Z}_n$. Since p is a factor of both n and r, we know that p is a factor of $a \cdot b$ based on the equation $a \cdot b = n \cdot q + r$. At this point, we use theorem 2.4.2(a): if p is a prime factor of $a \cdot b$, then either p is a factor of a or p is a factor of b. However, if p is a factor of a, then a and n share a common factor greater than 1, contradicting our assumption that $a \in U(n)$. Similarly, if p is a factor of b, then b and n share a common factor greater than 1, contradicting our assumption that $b \in U(n)$. In either case, we have a contradiction. Therefore, $a \odot b$ must be relatively prime to n, and so $a \odot b \in U(n)$.
- Associativity: Associativity for $U(n)$ under multiplication mod n follows from repeated use of the division algorithm and the associativity of the integers $\mathbb{Z}$ under standard multiplication. Further details are left to the reader.
- Identity: The identity of $U(n)$ under multiplication mod n is 1 since for every $a \in U(n)$, we have $1 \odot a = (1 \cdot a) \bmod n = a \bmod n = a$ and, similarly, $a \odot 1 = a$.
- Inverses: The inverse axiom is proven using theorem 2.4.2(b): if two integers $a, n \in \mathbb{Z}$ are relatively prime, then there exist integers $h, k \in \mathbb{Z}$ such that $a \cdot h + n \cdot k = 1$. We let $a \in U(n)$ and identify the inverse of a, including the justification that the inverse is actually in $U(n)$. Since $a \in U(n)$, we know that a and n are relatively prime, and so there exist $h, k \in \mathbb{Z}$ such that $a \cdot h + n \cdot k = 1$. We now determine the mod n value of both sides of this equation. On the left side $(a \cdot h + n \cdot k) \bmod n = (a \cdot h) \bmod n$, since $n \cdot k \bmod n = 0$. On the right side, $1 \bmod n = 1$. Therefore, $(a \cdot h) \bmod n = 1$, and so $h \bmod n \in U(n)$ is the multiplicative inverse of $a \in U(n)$. We note that $h \bmod n \in U(n)$ since otherwise h and n share a common factor greater than 1; this common factor would then divide 1 since $a \cdot h + n \cdot k = 1$, which is a contradiction. Since $a \in U(n)$ was an arbitrary element of $U(n)$, every element of $U(n)$ has a multiplicative inverse, and the inverse axiom holds.

∎

We end this section with two theorems that make claims about all groups. As we have seen in this section (and as will continue to see in the next two sets of exercises), many diverse number systems satisfy the group axioms. In the proofs of the following theorems, we use only the group axioms as justifications in our proofs, and so these results hold for every number system that is a group. This general approach is the source of the far-reaching scope of group theory and is a powerful feature of abstract mathematics. In general, mathematicians seek to utilize the weakest possible assumptions to determine truths for broad classes of number systems; in this case, we use the four axioms of groups to prove two theorems that hold for all of the many different groups simultaneously. These two theorems are really just the tip of the proverbial group-theoretic iceberg. Whole courses in the upper-level undergraduate and graduate mathematics curriculum are devoted to continuing and extending the first steps taken here. If you enjoy these investigations, then you can look forward to more advanced work in group theory and other courses in abstract algebra.

Theorem 2.4.4 Unique inverses theorem for groups *If G is a group under operation $\circ$ and $a \in G$, then the $\circ$-inverse of a is unique.*

Proof Mathematicians typically prove the uniqueness of a mathematical object by assuming there exist two objects with the properties under consideration and then showing these two objects must actually be the same. With this strategy in mind, assume that G is a group under operation $\circ$, that $a \in G$, and that both $x, y \in G$ are $\circ$-inverses of a. We then use the axioms of group theory to prove that $x = y$. Letting e denote the $\circ$-identity of G, observe the following string of equalities.

$$
\begin{aligned}
x &= e \circ x & \text{via the identity axiom} \\
&= (y \circ a) \circ x & \text{via the inverse axiom for } y; \text{ in particular, } e = y \circ a \\
&= y \circ (a \circ x) & \text{via associativity} \\
&= y \circ e & \text{via the inverse axiom for } x; \text{ in particular, } e = a \circ x \\
&= y & \text{via the identity axiom}
\end{aligned}
$$

Since the two $\circ$-inverses $x, y \in G$ for a are equal, the $\circ$-inverse of a is unique.

■

In light of theorem 2.4.4, we are now free to follow our intuitive instinct of referring to "the" inverse of an element $a \in G$, rather than just "an" inverse of a group element. In addition, theorem 2.4.4 ensures that the notation a^{-1} for an inverse is unambiguous. Sometimes we also use the notation $(-a)$ for the inverse of a group element when the group operation is expressed using additive notation rather than multiplicative notation. Most students become comfortable with this diversity in notation as they spend more time studying groups. Now consider the second of the two promised theorems; this result extends a familiarity with cancellation in equations to the group-theoretic setting.

Theorem 2.4.5 Left cancellation theorem for groups *If G is a group under operation $\circ$ and $a, b, c \in G$, then $a \circ b = a \circ c$ implies $b = c$.*

Proof Assuming the hypotheses of this theorem, we give a direct proof that $b = c$. Since $a \in G$, and G is a group under operation $\circ$, the $\circ$-inverse of a exists and

$a^{-1} \in G$. Multiplying both sides of the given equation $a \circ b = a \circ c$ by a^{-1} produces

$$a^{-1} \circ (a \circ b) = a^{-1} \circ (a \circ c).$$

Applying the axioms of associativity, inverses, and identity for $\circ$ on elements of G (in this order) to first the left side and then the right side of this equality produces the following two strings of equalities.

$$a^{-1} \circ (a \circ b) = (a^{-1} \circ a) \circ b = e \circ b = b$$
$$a^{-1} \circ (a \circ c) = (a^{-1} \circ a) \circ c = e \circ c = c$$

Based on this collection of equalities, $b = a^{-1} \circ (a \circ b) = a^{-1} \circ (a \circ c) = c$. We also give the condensed version of these equalities, in which we integrate all of the preceding strings of equalities into one line:

$$b = e \circ b = (a^{-1} \circ a) \circ b = a^{-1} \circ (a \circ b) = a^{-1} \circ (a \circ c) = (a^{-1} \circ a) \circ c$$
$$= e \circ c = c.$$

Each equality in this string can be justified by the assumptions or by an axiom of group theory; you may find it interesting and helpful to justify each equality explicitly.

■

We finish this introduction to group theory with a comment on the choice of the name "left" cancellation theorem for groups. As stated, theorem 2.4.5 allows the cancellation of terms in an equality only when the same terms appear on the *left* side of the two expressions in the given equality. There is a similar, but distinct, "right" cancellation theorem for groups that allows the cancellation of the same terms appearing on the right side of two equal expressions. We actually need both a left and a right cancellation theorem because not every group satisfies the axiom of commutativity (that is, the assertion that $a \circ b = b \circ a$ for every pair of elements a and b in the group). Even though most number systems introduced in lower level mathematics courses satisfy commutativity, there are important groups that do not; we consider such groups in the exercises to follow. This fact might also lead us to wonder what other familiar properties might be satisfied in some number systems but not in others. From these subtle differences we can develop a keen and insightful understanding of diverse mathematical objects.

2.4.1   Reading Questions for Section 2.4

1. Define and give an example of a binary operation.
2. State the four axioms satisfied by every group.
3. Give an example of an infinite group.
4. Give an example of an infinite set with a binary operation that is not a group. Identify the group axioms that fail.
5. State theorem 2.4.1. How is this result helpful when studying groups?
6. Give two examples of finite groups.

7. Give an example of a finite set with a binary operation that is not a group. Identify the group axioms that fail.
8. State theorem 2.4.2 and give examples illustrating the two parts of this theorem.
9. How do we use the number-theoretic results of theorem 2.4.2 in our study of groups?
10. State theorem 2.4.3. How is this result helpful when studying groups?
11. State two general properties that hold for all groups.
12. Discuss the distinction between "an" inverse and "the" inverse of a group element.

2.4.2 Exercises for Section 2.4

Exercises 1–3 consider definition 2.4.1 of a group.

1. Express each property of a group in predicate logic.
2. State the negation of each property of a group in English.
3. Express the negation of each property of a group in predicate logic.

In exercises 4–15, each set is a group under the given binary operation. State both the identity element of the group and the inverse of an arbitrary element of the group.

4. The rational numbers $\mathbb{Q}$ under standard addition $+$. Recall that: $m/n + p/q = (mq + np)/(nq)$. Also, verify associativity by proving the following are equal:

$$\frac{mq + np}{nq} + \frac{r}{s} \quad \text{and} \quad \frac{m}{n} + \frac{ps + rq}{qs}$$

5. The nonzero rational numbers $\mathbb{Q}^* = \mathbb{Q} \setminus \{0\}$ under standard multiplication. Why omit 0?
6. The real numbers $\mathbb{R}$ under standard addition.
7. The nonzero real numbers $\mathbb{R}^* = \mathbb{R} \setminus \{0\}$ under standard multiplication. Why omit 0?
8. The complex numbers $\mathbb{C}$ under standard addition: $(a + b \cdot i) + (c + d \cdot i) = (a + c) + (b + d) \cdot i$.
9. The nonzero complex numbers $\mathbb{C} \setminus \{0\}$ under standard multiplication: $(a + b \cdot i) \cdot (c + d \cdot i) = (ac - bd) + (bc + ad) \cdot i$. Why omit 0?
10. The ordered pairs of integers $\mathbb{Z}^2 = \{(m, n) : m, n \in \mathbb{Z}\}$ under componentwise addition: $(m, n) + (j, k) = (m + j, n + k)$.
11. The ordered pairs of rational numbers $\mathbb{Q}^2 = \{(r, s) : r, s \in \mathbb{Q}\}$ under componentwise addition: $(r, s) + (p, q) = (r + p, s + q)$.
12. The ordered pairs of reals $\mathbb{R}^2 = \{(r, s) : r, s \in \mathbb{R}\}$ under componentwise addition: $(r, s) + (p, q) = (r + p, s + q)$.
13. The ordered pairs of nonzero rational numbers $[\mathbb{Q}^*]^2 = \{(r, s) : r, s \in \mathbb{Q}$ and $r, s \neq 0\}$ under componentwise multiplication: $(r, s) \cdot (p, q) = (r \cdot p, s \cdot q)$.
14. The ordered pairs of nonzero reals $[\mathbb{R}^*]^2 = \{(r, s) : r, s \in \mathbb{R}$ and $r, s \neq 0\}$ under componentwise multiplication: $(r, s) \cdot (p, q) = (r \cdot p, s \cdot q)$.

15. The ordered pairs of nonzero reals $[\mathbb{R}^*]^2 = \{(r, s) : r, s \in \mathbb{R} \text{ and } r, s \neq 0\}$ under the operation $(r, s) * (p, q) = (rp - sq, rq + sp)$. Verify that $(1, 0)$ is the identity element. For inverses, consider the similarities between this $*$-multiplication operation and multiplication of complex numbers.

In exercises 16–27, each set is *not* a group under the given operation. Identify the group axioms that do not hold and state corresponding counterexamples.

16. The natural numbers $\mathbb{N}$ under standard addition.
17. The integers $\mathbb{Z}$ under standard multiplication.
18. The rational numbers $\mathbb{Q}$ under standard multiplication.
19. The real numbers $\mathbb{R}$ under standard multiplication.
20. The set $\{0, 1, 2\}$ under standard addition.
21. The set $\{0, 1, 2, 3\}$ under standard addition.
22. The set $\{0, \ldots, n\}$ under standard addition for integers $n > 2$.
23. The set $\{-1, 0, 1\}$ under standard addition.
24. The set $\{-2, -1, 0, 1, 2\}$ under standard addition.
25. The set $\{-n, \ldots, -1, 0, 1, \ldots, n\}$ under standard addition.
26. The ordered pairs of integers $\mathbb{Z}^2 = \{(m, n) : m, n \in \mathbb{Z}\}$ under componentwise multiplication: $(m, n) \cdot (j, k) = (m \cdot j, n \cdot k)$.
27. The ordered pairs of reals $\mathbb{R}^2 = \{(r, s) : r, s \in \mathbb{R}\}$ under componentwise multiplication: $(r, s) \cdot (p, q) = (rp, sq)$.

In exercises 28–31, prove each mathematical statement.

28. The set natural numbers $\mathbb{N}$ is not closed under subtraction. Give a counterexample, identifying two natural numbers witnessing the nonclosure of $\mathbb{N}$ under subtraction.
29. Subtraction is not associative on the integers $\mathbb{Z}$. Give a counterexample, identifying three integers witnessing the nonassociativity of subtraction.
30. *(Unique identity theorem for groups)* If G is a group under operation $\circ$, then the $\circ$-identity in G is unique. Hint: By way of contradiction, suppose there are two identities e and f, and consider $e \circ f$.
31. *(Right cancellation theorem for groups)* If G is a group under operation $\circ$ and $a, b, c \in G$, then $a \circ b = c \circ b$ implies $a = c$. Hint: See theorem 2.4.5.

Exercises 32–45 study "zero divisors" in the context of modular groups. A nonzero element $a \in \mathbb{Z}_n$ is a zero divisor of $\mathbb{Z}_n$ if there exists a nonzero $b \in \mathbb{Z}_n$ such that $a \odot b = 0$.

32. Prove that $0 \in \mathbb{Z}_6$ under multiplication mod 6 is *not* a zero divisor.
33. Prove that $1 \in \mathbb{Z}_6$ under multiplication mod 6 is *not* a zero divisor by computing all products $1 \odot b$ for all nonzero $b \in \mathbb{Z}_6$.
34. Prove that $2 \in \mathbb{Z}_6$ under multiplication mod 6 is a zero divisor by identifying a nonzero $b \in \mathbb{Z}_6$ such that $2 \odot b = 0$.
35. Prove that $3 \in \mathbb{Z}_6$ under multiplication mod 6 is a zero divisor by identifying a nonzero $b \in \mathbb{Z}_6$ such that $3 \odot b = 0$.
36. Prove that $4 \in \mathbb{Z}_6$ under multiplication mod 6 is a zero divisor by identifying a nonzero $b \in \mathbb{Z}_6$ such that $4 \odot b = 0$.

37. Prove that $5 \in \mathbb{Z}_6$ under multiplication mod 6 is *not* a zero divisor by computing all products $5 \odot b$ for all nonzero $b \in \mathbb{Z}_6$.

38. Based on exercises 32–37, what property identifies the zero divisors of $\mathbb{Z}_6$? Conjecture a relationship between $\mathbb{Z}_6$, $U(6)$, and the set of zero divisors of $\mathbb{Z}_6$.

39. Identify the zero divisors of $\mathbb{Z}_8$ under multiplication mod 8.

40. Identify the zero divisors of $\mathbb{Z}_{12}$ under multiplication mod 12.

41. Identify an odd integer n such that $\mathbb{Z}_n$ under multiplication mod n contains at least one zero divisor. State the zero divisors in this $\mathbb{Z}_n$.

42. Prove that $\mathbb{Z}_3$ does not contain any zero divisors based on the Cayley table for $\mathbb{Z}_3$ under multiplication mod 3. What property of 3 results in $\mathbb{Z}_3$ not having any zero divisors?

43. Prove that $\mathbb{Z}_5$ does not contain any zero divisors based on the Cayley table for $\mathbb{Z}_5$ under multiplication mod 5. What property of 5 results in $\mathbb{Z}_5$ not having any zero divisors?

44. Prove that if $\mathbb{Z}_n$ contains zero divisors, then left cancellation does *not* hold for all elements of $\mathbb{Z}_n$ by identifying $a, b, c \in \mathbb{Z}_n$ such that $a \odot b = a \odot c$ but $b \neq c$.

45. Prove that if $\mathbb{Z}_n$ contains zero divisors, then right cancellation does *not* hold for all elements of $\mathbb{Z}_n$ by identifying $a, b, c \in \mathbb{Z}_n$ such that $a \odot b = c \odot b$ but $a \neq c$.

Exercises 46–54 consider "idempotent" elements in modular groups. An element $a \in \mathbb{Z}_n$ is an *idempotent* of $\mathbb{Z}_n$ if $a \cdot a = a$ (often written as $a^2 = a$).

46. Identify the idempotents of $\mathbb{Z}_3$ under multiplication mod 3.

47. Identify the idempotents of $\mathbb{Z}_5$ under multiplication mod 5.

48. Identify the idempotents of $\mathbb{Z}_7$ under multiplication mod 7.

49. Identify the idempotents of $\mathbb{Z}_6$ under multiplication mod 6.

50. Identify the idempotents of $\mathbb{Z}_{10}$ under multiplication mod 10.

51. Identify the idempotents of $\mathbb{Z}_{14}$ under multiplication mod 14.

52. Based on exercises 46–51, conjecture a relationship between the divisor n for $\mathbb{Z}_n$ and the number of idempotents of $\mathbb{Z}_n$. What if n is prime? What if $n = 2p$, where p is prime?

53. Prove that if $a, b \in \mathbb{Z}_n$ are idempotents, then $a \odot b$ is also an idempotent under multiplication mod n.

54. Prove that if $\mathbb{Z}_n$ has no zero divisors (as defined before exercise 32), then 0 and 1 are the only idempotents of $\mathbb{Z}_n$ under multiplication mod n.

Exercises 55–70 consider symmetric groups. Symmetric groups, also known as permutation groups, were the first groups studied by Galois and others in the mid-1800s. In fact, the symmetric groups can be considered the most fundamental finite groups since every finite group can be identified with a subset of a symmetric group that is itself a group. The symmetric group S_n consists of all one-to-one, onto functions on the integers $\{1, \ldots, n\}$ under the operation of composition. Such functions are studied in section 4.2, but for now we consider examples on the finite sets $\{1, \ldots n\}$. The elements of S_n are functions α represented in arrays of the form

$$\alpha = \begin{bmatrix} 1 & 2 & \cdots & n \\ \alpha(1) & \alpha(2) & \cdots & \alpha(n) \end{bmatrix}.$$

The top row of the array denotes the inputs of the function and the bottom row consists of the corresponding, distinct outputs, which are also elements of $\{1, \ldots n\}$. For example, the following are three elements of S_3 represented in array form.

$$\epsilon = \begin{bmatrix} 1 & 2 & 3 \\ 1 & 2 & 3 \end{bmatrix} \qquad \alpha = \begin{bmatrix} 1 & 2 & 3 \\ 2 & 3 & 1 \end{bmatrix} \qquad \beta = \begin{bmatrix} 1 & 2 & 3 \\ 1 & 3 & 2 \end{bmatrix}$$

For this $\alpha \in S_3$, we have $\alpha(1) = 2$, $\alpha(2) = 3$, and $\alpha(3) = 1$. Elements of S_n are combined under the operation of composition, denoted by $\circ$. For example, for this $\alpha, \beta \in S_3$, the composition $\alpha \circ \beta$ is

$$\begin{array}{ccccccc} \alpha \circ \beta(1) & = & \alpha[\beta(1)] & = & \alpha(1) & = & 2, \\ \alpha \circ \beta(2) & = & \alpha[\beta(2)] & = & \alpha(3) & = & 1, \\ \alpha \circ \beta(3) & = & \alpha[\beta(3)] & = & \alpha(2) & = & 3. \end{array}$$

Representing the composition $\alpha \circ \beta$ in array form, we have

$$\alpha \circ \beta = \begin{bmatrix} 1 & 2 & 3 \\ 2 & 1 & 3 \end{bmatrix}.$$

In exercises 54–60, express each function from S_4 in array form.

55. $\alpha(1) = 1$, $\alpha(2) = 3$, $\alpha(3) = 4$, $\alpha(4) = 2$
56. $\alpha(1) = 3$, $\alpha(2) = 1$, $\alpha(3) = 2$, $\alpha(4) = 4$
57. $\alpha(1) = 2$, $\alpha(2) = 3$, $\alpha(3) = 4$, $\alpha(4) = 1$
58. $\alpha(1) = 3$, $\alpha(2) = 2$, $\alpha(3) = 1$, $\alpha(4) = 4$
59. $\alpha(1) = 4$, $\alpha(2) = 1$, $\alpha(3) = 2$, $\alpha(4) = 3$
60. $\alpha(1) = 4$, $\alpha(2) = 2$, $\alpha(3) = 3$, $\alpha(4) = 1$

In exercises 61–66, identify the function from S_5 in array form resulting from each composition of the given functions.

$$\alpha = \begin{bmatrix} 1 & 2 & 3 & 4 & 5 \\ 2 & 1 & 4 & 3 & 5 \end{bmatrix} \quad \beta = \begin{bmatrix} 1 & 2 & 3 & 4 & 5 \\ 3 & 2 & 1 & 5 & 4 \end{bmatrix} \quad \gamma = \begin{bmatrix} 1 & 2 & 3 & 4 & 5 \\ 5 & 4 & 3 & 2 & 1 \end{bmatrix}$$

61. $\alpha \circ \beta$ 64. $\beta \circ \gamma$
62. $\alpha \circ \gamma$ 65. $\gamma \circ \alpha$
63. $\beta \circ \alpha$ 66. $\gamma \circ \beta$

Exercises 67–70 consider the symmetric group $S_2 = \{\epsilon, \alpha\}$ on $\{1, 2\}$ under composition. The two elements of S_2 are

$$\epsilon = \begin{bmatrix} 1 & 2 \\ 1 & 2 \end{bmatrix} \qquad \text{and} \qquad \alpha = \begin{bmatrix} 1 & 2 \\ 2 & 1 \end{bmatrix}.$$

67. Express the function $\epsilon \circ \epsilon \in S_2$ in array form.
68. Express the functions $\alpha \circ \epsilon \in S_2$ and $\epsilon \circ \alpha \in S_2$ in array form.
69. Express the function $\alpha \circ \alpha \in S_2$ in array form.
70. Based on exercises 67–69, state the Cayley table for S_2 under composition.

2.5 Dihedral Groups

We continue our study of number systems with a radically different type of group called the "dihedral group." Their description uses Euclidean geometry and is based on an understanding of motions in the two-dimensional space of the Euclidean plane. Over the centuries, such motions have inspired artisans when creating beautiful patterns for decorating pottery, clothing, and buildings. In the early middle ages, Islamic mathematicians and artists decorated mosques throughout southern Spain and northern Morocco with intricate designs based on these motions. In the early 1900s, the artist Maurits Cornelis Escher created many fantastical and diverse images using planar motions. Posters of his work often decorate the walls of mathematicians' classrooms and offices and even a few dormitory rooms. When mathematicians learned that Escher had identified a complete classification of certain groups of motions, he was invited to share his work with group theorists at a national mathematics conference!

We are interested in planar motions with certain distinctive features. There are many different "types" of these transformations of the plane: we can *translate* the plane in a uniform direction; *rotate* the plane about some fixed point; *reflect* the plane across some given line; or implement a *glide reflection* combining a translation and a reflection. These four types of motions are referred to as *Euclidean plane isometries* since they preserve the geometric property of distance (or length) between points in the plane. In contrast, *non-isometries* are transformations such as bending, stretching, twisting, folding, cutting, or otherwise distorting the plane so that the distance between some points is changed.

The dihedral groups are certain subcollections of Euclidean plane isometries—namely, those that preserve the orientation of a given regular polygon. These sets of isometries have many fundamental group-theoretic properties. We begin our study of the dihedral groups with the definition of a regular polygon and identify some examples and nonexamples of polygons.

Definition 2.5.1 *A **polygon** is a closed geometric figure in the plane with three or more (but only finitely many) straight sides. A polygon is said to be a **regular polygon** if every side of the polygon has the same length and every interior angle of the polygon has the same magnitude.*

We often encounter polygons in our everyday lives from the rectangular buildings we live in, to octagonal stop signs, to the hexagonal honeycombs of bees. Most students study polygons from the earliest days in school and learn the familiar, distinctive names of many, including triangles, squares, and pentagons. The definition of a regular polygon guarantees that there exists exactly one regular polygon with n sides for every positive integer $n \geq 3$ (once the length of the sides is determined).

Example 2.5.1 The sides and the interior angles of each of the polygons in figure 2.3 are equal, and so each polygon is regular.

■

As we have seen, a good understanding of a formal definition is developed by considering not only examples satisfying the definition, but also a collection of examples exploring the boundaries and limitations of the definition. The two

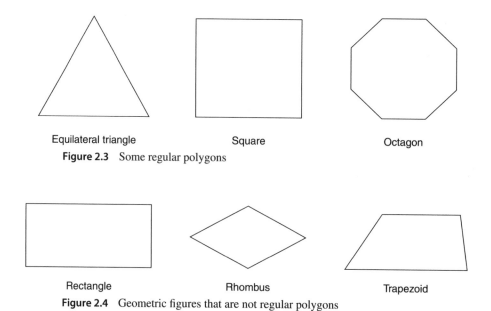

Equilateral triangle   Square   Octagon

Figure 2.3   Some regular polygons

Rectangle   Rhombus   Trapezoid

Figure 2.4   Geometric figures that are not regular polygons

requirements that a closed figure with straight sides must satisfy in order to be a regular polygon are equality of sides and equality of interior angles. From sentential logic, we know that these requirements are not satisfied if one or the other condition fails or when both fail simultaneously.

Example 2.5.2   We identify a figure in the plane that is not a polygon. We also describe the three different ways a four-sided polygon can fail to be regular as illustrated in figure 2.4.

- A circle is not a polygon because the circle does not have at least three straight sides.
- A nonsquare rectangle is not regular. Even though all the interior angles are equal, the sides have different lengths.
- A nonsquare rhombus (popularly known as a "diamond") is not regular. Even though the lengths of the sides are all equal, the interior angles are different from one another.
- A trapezoid whose sides and interior angles are different from each other is not regular. ∎

Dihedral groups do not include all motions of regular polygons, but just the orientation-preserving, distance-preserving motions of regular polygons. An isometry *preserves the orientation* of a polygon if the motion consists of "picking up" the polygon and moving it around so that when the motion is completed, every vertex is mapped to the initial location of one of the vertices of the polygon. For example, consider the two motions of an equilateral triangle illustrated in figure 2.5. The left diagram depicts a motion that is orientation-preserving, while the right diagram depicts a motion that is not orientation-preserving.

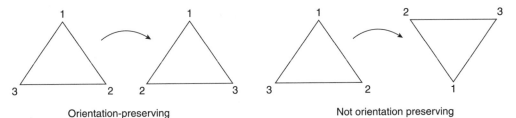

Orientation-preserving   Not orientation preserving

Figure 2.5 Two motions of an equilateral triangle

From these specific examples of motions, we can see that there exist infinitely many different isometries of an equilateral triangle that are not orientation-preserving. Recall that all isometries are translations, rotations, reflections, or glide reflections. The translations and glide reflections that involve shifting the figure to a different location in the plane are clearly not orientation preserving. Many of these motions that do not preserve orientations are interesting in their own right and have served as the basis for the work of the Islamic artists and of Escher, among others. However, some rotations and reflections are orientation-preserving. In fact, only a relative handful of the isometries of the plane are orientation-preserving for an equilateral triangle (the regular three-sided polygon) or for any regular n-sided polygon, but they are enough to yield an interesting set of mathematical objects. As it turns out, the set of orientation-preserving isometries of an equilateral triangle is a group under composition. The following example and questions explore the dihedral group consisting of these motions under composition.

Example 2.5.3 The *dihedral group* D_3 is the set of orientation-preserving isometries of an equilateral triangle under the operation of composition. In the notation D_3, the "D" stands for the "dihedral group" and the subscript "three" indicates that we are working with a regular three-sided polygon. To facilitate the discussion, we number the vertices of the equilateral triangle in its original position, where the top vertex is 1, the bottom right vertex is 2, and the bottom left is 3; this assignment of numbers to vertices is illustrated in the left triangle of figure 2.6.

The orientation-preserving isometries of an equilateral triangle may change the position of the vertices (with their corresponding numberings) but always return the triangle to the space in the plane that the triangle originally occupied. For example, when the triangle is rotated counterclockwise 120 degrees, the 1 vertex

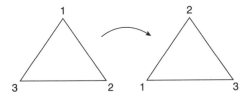

A counterclockwise, 120-degree rotation.   **Figure 2.6** The motion R_{120}

moves to the original position of the 3 vertex, the 3 vertex moves to 2's original position, and the 2 vertex moves to 1's original position. This motion is denoted by R_{120} and is illustrated in figure 2.6.

The dihedral group D_3 consists of *all* orientation-preserving isometries of an equilateral triangle. There are exactly six such motions, since vertex 1 can be placed at any of the three original vertex positions, which leaves two choices for placing the vertex 2 and just one choice for the vertex 3. Comparing the original labeling of vertices with the post-motion labeling of vertices enables us to identify three of these motions as *rotations* about the center of the equilateral triangle and the other three motions as *flips* (or reflections) across the three lines of symmetry of the equilateral triangle. The six possible motions are identified using the following suggestive labels.

- R_0 = rotate counterclockwise 0 degrees, moving $1 \rightarrow 1$, $2 \rightarrow 2$, $3 \rightarrow 3$;
- R_{120} = rotate counterclockwise 120 degrees, moving $1 \rightarrow 3$, $2 \rightarrow 1$, $3 \rightarrow 2$ (as illustrated in figure 2.6, vertex 1 has moved to the position originally occupied by vertex 3, etc.);
- R_{240} = rotate counterclockwise 240 degrees, moving $1 \rightarrow 2$, $2 \rightarrow 3$, $3 \rightarrow 1$;
- F_T = flip, or reflect, across the axis drawn from the top vertex to the center of the opposite side, moving $1 \rightarrow 1$, $2 \rightarrow 3$, $3 \rightarrow 2$;
- F_R = flip, or reflect, across the axis drawn from the right vertex to the center of the opposite side, moving $1 \rightarrow 3$, $2 \rightarrow 2$, $3 \rightarrow 1$;
- F_L = flip, or reflect, across the axis drawn from the left vertex to the center of the opposite side, moving $1 \rightarrow 2$, $2 \rightarrow 1$, $3 \rightarrow 3$.

Since these six motions are all the orientation-preserving, distance-preserving motions of an equilateral triangle, $D_3 = \{R_0, R_{120}, R_{240}, F_T, F_R, F_L\}$. The standard operation on these elements of D_3 is composition of motions, read from right to left. For example, the composition $R_{120} \circ R_{240}$ means that we first apply R_{240}, and then R_{120}; figure 2.7 illustrates this composition.

Intuitively, we recognize that $R_{120} \circ R_{240} = R_0$, since a counterclockwise rotation of 120 degrees followed by a counterclockwise rotation of 240 degrees returns every vertex to its original position; that is, the net effect of composing this pair of motions is a counterclockwise rotation of 0 degrees. This observation provides one piece of evidence that D_3 is closed under composition, as must be verified in a proof that D_3 is a group.

■

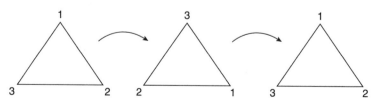

Figure 2.7   The motion of R_{240} followed by R_{120}

The following questions continue our study of D_3, particularly in light of the four group axioms of closure, associativity, identity, and inverses. We also pay close attention to the further algebraic property of commutativity. As a first step in this direction, you are invited to practice a bit more with the operation of composition on the elements of D_3.

Question 2.5.1 Compute each composition of elements from $D_3 = \{R_0, R_{120}, R_{240}, F_T, F_R, F_L\}$.

(a) $R_0 \circ R_{240}$ (d) $F_T \circ F_T$

(b) $R_{120} \circ R_{120}$ (e) $F_T \circ F_R$

(c) $R_{240} \circ R_{240}$ (f) $F_R \circ F_T$

                                                             ■

The last two computations in question 2.5.1 are of particular interest because they prove that composition on D_3 is not commutative. Recall that an operation $\circ$ is *commutative* on a set S if for every $a, b \in S$, we have $a \circ b = b \circ a$. The results of question 2.5.1 show that $F_T \circ F_R \neq F_R \circ F_T$, and so composition is not commutative on D_3. At the same time, we observe that *some* elements of D_3 do commute with each other; for example, $R_0 \circ F_T = F_T \circ R_0$ (in fact, R_0 commutes with every element of D_3). Thus, $F_T \circ F_R \neq F_R \circ F_T$ only proves that the universal property of *all* elements commuting with each other does not hold in D_3.

This behavior in D_3 is striking. Every number system we have studied thus far has been commutative, but D_3 behaves quite differently. As it turns out, many groups are noncommutative, and so we must be careful not to appeal to commutativity until we verify that the group in question actually satisfies this condition. As mentioned in the previous section, the issue of commutativity is why we considered both "left cancellation" and "right cancellation" separately in section 2.4. Motivated by the dihedral groups and other similar groups, mathematicians have come to recognize the importance of the algebraic property of commutativity and (as with most notions of importance) a special name is given to commutative groups.

Definition 2.5.2 *A group G under operation $\circ$ is* **Abelian** *if the operation $\circ$ is commutative on G; that is, G is Abelian if for every $a, b \in G$, we have $a \circ b = b \circ a$.*

Abelian groups are named in honor of the insightful Norwegian mathematician Niels Henrik Abel. In the late 1700s, Abel found quite limited opportunities to study mathematics in the schools of his native country. Despite (or because of) these limitations, Abel was a primarily self-educated mathematician who learned directly from the books and the research papers of the greatest mathematicians in history. In his early twenties, Abel solved one of the fundamental research questions of his time. He proved that there does not exist a general solution (in radicals) for polynomial equations of degree five or higher, in striking contrast to the existence of the quadratic equation for solving quadratic polynomials, and similar equations for solving cubics and quartics (such solutions of polynomial equations are discussed more fully in section 3.5). Abel's work preceded Galois' work on this same question, although Galois' approach was more general and is more widely known among contemporary mathematicians. Sadly, Abel died of tuberculosis at the age of 24, shortly before receiving word of an offer of a position at one of the leading research universities in Germany.

Every group we have studied thus far has been Abelian (except D_3), including $\mathbb{Z}$ under addition, $\mathbb{Q}^*$ under multiplication, $\mathbb{Z}_n$ under modular addition, and $U(n)$ under modular multiplication. Question 2.5.1 (together with the proof that D_3 is a group) show that D_3 is a nonAbelian group, as stated in theorem 2.5.1 below.

Working toward a proof that D_3 is a group, we first consider the closure of D_3 under composition—for every $a, b \in D_3$, we want to show that $a \circ b \in D_3$. Recall from section 2.3 that a Cayley table provides a thorough and systematic approach to computing the action of the binary operation on all possible pairings of elements. If only elements from the given set appear in the corresponding Cayley table, the set is closed under the given operation.

Question 2.5.2   Simplify the following Cayley table for D_3 under composition.

$\circ$	R_0	R_{120}	R_{240}	F_T	F_R	F_L
R_0	$R_0 \circ R_0$	$R_0 \circ R_{120}$	$R_0 \circ R_{240}$	$R_0 \circ F_T$	$R_0 \circ F_R$	$R_0 \circ F_L$
R_{120}	$R_{120} \circ R_0$	$R_{120} \circ R_{120}$	$R_{120} \circ R_{240}$	$R_{120} \circ F_T$	$R_{120} \circ F_R$	$R_{120} \circ F_L$
R_{240}	$R_{240} \circ R_0$	$R_{240} \circ R_{120}$	$R_{240} \circ R_{240}$	$R_{240} \circ F_T$	$R_{240} \circ F_R$	$R_{240} \circ F_L$
F_T	$F_T \circ R_0$	$F_T \circ R_{120}$	$F_T \circ R_{240}$	$F_T \circ F_T$	$F_T \circ F_R$	$F_T \circ F_L$
F_R	$F_R \circ R_0$	$F_R \circ R_{120}$	$F_R \circ R_{240}$	$F_R \circ F_T$	$F_R \circ F_R$	$F_R \circ F_L$
F_L	$F_L \circ R_0$	$F_L \circ R_{120}$	$F_L \circ R_{240}$	$F_L \circ F_T$	$F_L \circ F_R$	$F_L \circ F_L$

∎

The simplified Cayley table in question 2.5.2 contains only elements of D_3 and exhibits the distinctive Latin square feature common to every Cayley table of a finite group. Therefore, D_3 is closed under composition and satisfies the first of the four group axioms.

The verification that composition is an associative operation on D_3 requires significant work, since D_3 is not contained in any set whose associativity is already known. A proof by exhaustion that composition is associative on D_3 requires the computation of the two possible orderings of composition on the $6^3 = 216$ distinct triples of elements from D_3. Rather than offer this tedious proof, the next question highlights a couple of representative examples.

Question 2.5.3   Verify each equality in D_3 under composition.

(a) $R_{120} \circ (F_T \circ R_{240}) = (R_{120} \circ F_T) \circ R_{240}$

(b) $F_T \circ (R_{120} \circ F_R) = (F_T \circ R_{120}) \circ F_R$

∎

We now investigate the identity and inverse axioms for D_3 under composition. Question 2.5.2 asked for the simplified Cayley table for D_3. Since the following analysis makes essential use of this Cayley table, we provide the simplified version here; you may wish to use it to double-check your answer to question 2.5.2.

$\circ$	R_0	R_{120}	R_{240}	F_T	F_R	F_L
R_0	R_0	R_{120}	R_{240}	F_T	F_R	F_L
R_{120}	R_{120}	R_{240}	R_0	F_R	F_L	F_T
R_{240}	R_{240}	R_0	R_{120}	F_L	F_T	F_R
F_T	F_T	F_L	F_R	R_0	R_{240}	R_{120}
F_R	F_R	F_T	F_L	R_{120}	R_0	R_{240}
F_L	F_L	F_R	F_T	R_{240}	R_{120}	R_0

Recall from section 2.4 that the Cayley table for a group readily allows us to identify the identity element of the group and the inverse of each element of the group. The identity is determined by finding the group element that "preserves the identity" of every element of D_3 under composition. We can then determine inverses by locating the identity element in each row (or in each column) of the Cayley table.

Question 2.5.4   Working with the above Cayley table for D_3 under composition, state the identity of D_3 and the inverse of each element of D_3.

■

The answers to questions 2.5.1–2.5.4 above constitute a proof of the following result.

Theorem 2.5.1   *D_3 under composition is a nonAbelian group.*

The next step in studying dihedral groups is to consider an arbitrary regular n-sided polygon. In D_3, the orientation-preserving isometries of a regular three-sided polygon are three rotations (each through a multiple of $360/3 = 120$ degrees) and three flips (each over an axis through a vertex and the center of the opposite side). A similar pattern occurs in the more general setting for the set D_n of orientation-preserving isometries of a regular n-sided polygon. The set D_n contains n rotations, one for each a multiple of $360/n$ degrees or $2\pi/n$ radians. In addition, D_n contains n flips (or reflections) across the n axes of symmetry that bisect the regular n-sided polygon, where the description of the axes depends on whether the polygon has an even or an odd number of sides. If n is odd, then each flip is over an axis through a vertex and the center of the opposite side (as for the equilateral triangle). If n is even, then half the flips are over an axis through a pair of opposite vertices and the other half of the flips are over an axis through the centers of two opposite sides. To help clarify this general description in a more concrete setting, think about a square with its four axes of symmetry. The dihedral group for a square is studied in exercises 11–17 at the end of this section.

In general, the number of elements in a finite group is referred to as the *order* of the group. Since $D_3 = \{R_0, R_{120}, R_{240}, F_T, F_R, F_L\}$, we say that D_3 has order six. Similarly, the set D_n contains n rotations and n flips, and so D_n has order $2n$. This notion of order extends to other groups, including $\mathbb{Z}_n$ under addition mod n (which has order n) and $U(n)$ under multiplication mod n (which has order $n-1$ when n is prime).

As with D_3, the operation on D_n of composition from right to left provides a group structure and satisfies the four group axioms. As you might expect, the proof that D_n is a group closely mirrors the argument that D_3 is a group. We formally state the theorem for this general case and provide a sketch of a proof.

Theorem 2.5.2   *The set D_n of orientation-preserving isometries of a regular n-sided polygon is a group under the operation of composition. We refer to D_n as the **dihedral group of order** $2n$.*

Sketch of proof Closure is satisfied since the aggregate result of implementing one orientation-preserving isometry and then another is itself an orientation-preserving isometry. Associativity follows from an exhaustive consideration of all combinations of group elements using appropriate general descriptions of the orientation-preserving isometries of a regular n-sided polygon. The identity element of D_n under composition is R_0. Finally, every flip is its own inverse and the inverse of a rotation of $360/n$ degrees is a rotation of $360 - 360/n$ degrees (which is also an orientation-preserving isometry and so in D_n). Thus, D_n under composition is a group.

∎

For the remainder of this section, we examine patterns that arise when composing elements of D_n and take a more careful look at the issue of commutativity and noncommutativity in this group-theoretic setting. We focus our discussion around the study of the Cayley table for D_3 under composition computed in question 2.5.2 and given just before question 2.5.4. We first look for general patterns in computing compositions in D_3.

Question 2.5.5   Examining the Cayley table for D_3, we observe that the orientation-preserving isometries of the equilateral triangle are bunched together in four distinct blocks: two blocks of rotations in the upper left and lower right quadrants and two blocks of flips in the lower left and upper right quadrants. Based on the computations given in this table, identify the following general statements as true or false. For the false statements, give a counterexample.

(a) A rotation followed by a rotation is a rotation; that is, $R \circ R = R$.
(b) A rotation followed by a rotation is a flip; that is, $R \circ R = F$.
(c) A rotation followed by a flip is a rotation; that is, $F \circ R = R$.
(d) A rotation followed by a flip is a flip; that is, $F \circ R = F$.
(e) A flip followed by a rotation is a rotation; that is, $R \circ F = R$.
(f) A flip followed by a rotation is a flip; that is, $R \circ F = F$.
(g) A flip followed by a flip is a rotation; that is, $F \circ F = R$.
(h) A flip followed by a flip is a flip; that is, $F \circ F = F$.

∎

We now consider the issue of commutativity and noncommutativity. In addition to observing the noncommutative behavior of some elements of D_3, we also note that some elements of D_3 do commute with each other. Every group actually has some elements that commute with one another. For example, e is the identity of a group G under operation $\circ$ iff $e \circ a = a = a \circ e$ for every element $a \in G$. Therefore, by definition, the identity always commutes with every element of a group. In the specific setting of D_3, the identity element R_0 commutes with every element of D_3. Similarly, group elements always commute with their inverses since (by definition) $a \circ a^{-1} = e = a^{-1} \circ a$ for every $a \in G$. Furthermore, every element commutes with itself since $a \circ a = a \circ a$ by the reflexivity of standard equality. Based on these observations, we conclude that a great deal of commutativity still must exist among individual group elements, even when a group is nonAbelian.

In studying the role of commutativity in a particular group, mathematicians sometimes gather together the elements that commute with every other element of the group into a set called the *center* of the group. In addition to considering such global behavior, mathematicians also study more locally defined behavior. With regard to commutativity, we consider the set of group elements that commute with a single given group element $a \in G$. This set is called the *centralizer* of a. Consider the following definition.

Definition 2.5.3 *If a is an element of a group G under operation $\circ$, then the **centralizer** of a in G is the set of elements in G that commute with a. Symbolically, the centralizer of a in G is denoted by $C(a) = \{g : g \in G$ and $a \circ g = g \circ a\}$.*

Example 2.5.4 We identify the centralizers of R_0, R_{120}, and F_T in D_3 under composition.

Since the identity element R_0 commutes with every element of D_3, we immediately have $C(R_0) = D_3$. In contrast, the centralizer of every other element of D_3 is a proper subset of D_3, as can be verified by checking the (lack of) symmetry between the row and column of a given element in the Cayley table for D_3. For example, for the rotation R_{120}, we find $C(R_{120}) = \{R_0, R_{120}, R_{240}\}$ since R_{120} commutes with every rotation but with no flips. Similarly, for the flip F_T, we find $C(F_T) = \{R_0, F_T\}$.

■

The conclusion that $C(F_T) = \{R_0, F_T\}$ is a special case that warrants further discussion. For groups containing two or more elements, the least possible order of the centralizer of a group element is two, since every element commutes with at least the identity of the group and itself. As mentioned above, every group element also commutes with its inverse, and so a centralizer has order two exactly when a group element is its own inverse. Thus, $C(F_T) = \{R_0, F_T\}$ since F_T commutes with precisely the identity (R_0), with itself (F_T), and with its inverse (which is also F_T).

Question 2.5.6 Determine the centralizer of R_{240}, F_R, and F_L using the Cayley table for D_3 under composition.

■

Finally, we say a bit more about cancellation in the context of the dihedral groups. From theorem 2.4.5 and exercise 31 in section 2.4, the left cancellation and the right cancellation properties hold for any group, and so they hold for D_n. For example, working in D_3 we know that $F_T \circ R_{120} = F_T \circ (R_{240} \circ R_{240})$ implies $R_{120} = R_{240} \circ R_{240}$ by left cancellation and that $R_{120} \circ F_T = (R_{240} \circ R_{240}) \circ F_T$ implies $R_{120} = R_{240} \circ R_{240}$ via right cancellation.

In many other familiar number systems, we are perhaps used to using "cross" cancellation. For example, working with products of integers and rationals, we may say that $2 \cdot 1 = (3 \cdot 1/3) \cdot 2$ implies $1 = 3 \cdot 1/3$ or that $2x = (3 + x) \cdot 2$ implies $x = 3 + x$ by cancelling the 2 on both sides of the original equations. However, such cross-cancellation makes implicit use of the commutativity of elements in the underlying number systems for these equations. As we have seen, commutativity fails for several pairs of elements in D_3, and we will soon see that commutativity also fails to hold in a number of other important groups. Therefore, in general we cannot utilize cross-cancellation when manipulating equations unless we first verify and reference that we are working in an Abelian group. The following question explicitly highlights the failure of cross-cancellation in D_3.

Question 2.5.7   Verify that cross-cancellation is not valid in D_3 under composition by identifying $A \in D_3$ such that $F_T \circ A = R_{120} \circ F_T$, but $A \neq R_{120}$.

∎

2.5.1   Reading Questions for Section 2.5

1. What is a regular polygon? Give examples of polygons that are regular and irregular.
2. Give an example of a geometric figure that is not a polygon.
3. What is an isometry of the plane? Give an example of an isometry and a motion of the plane that is not an isometry.
4. What is an orientation-preserving isometry? Give an example of an orientation-preserving isometry of a square and an isometry of a square that is not orientation-preserving.
5. List the elements of D_3.
6. Define when an operation ∘ is commutative on set S.
7. Give an example of a commutative operation.
8. Give an example of an operation that is not commutative.
9. Define and give an example of an Abelian group.
10. Define the order of a group and give examples of groups of orders $2, 5$, and 6.
11. Define the center of a group G and give an example.
12. Define the centralizer of an element α in a group G and give an example.

2.5.2   Exercises for Section 2.5

In exercises 1–10, sketch a geometric figure satisfying each condition, or explain why no such figure exists.

1. a regular two-sided polygon
2. a regular four-sided polygon
3. a regular five-sided polygon
4. a regular eight-sided polygon
5. an irregular five-sided polygon with equal sides
6. an irregular five-sided polygon with equal interior angles
7. an irregular five-sided polygon with neither equal sides nor equal interior angles

8. an irregular six-sided polygon with equal sides
9. an irregular six-sided polygon with equal interior angles
10. an irregular six-sided polygon with neither equal sides nor equal interior angles

Exercises 11–17 consider the dihedral group D_4 under composition. The set D_4 consists of the eight orientation-preserving isometries of a square (a regular *four*-sided polygon). The set D_4 contains four rotations, denoted R_0, R_{90}, R_{180}, R_{270} with the subscript indicating the angle of counterclockwise rotation. In addition, D_4 contains four flips (denoted F_V, F_H, F_R, F_L), where F_V is the flip across the vertical axis through the center of the top and bottom sides, F_R is the flip across the axis from the lower right vertex to the upper left vertex, and F_H, F_L are the horizontal and lower left variations on these flips.

11. As modeled in example 2.5.3, give a written description of the eight isometries in D_4.
12. Compute the Cayley table for D_4 under composition.
13. State the identity of D_4 under composition.
14. State the inverse of each element of D_4 under composition.
15. Based on exercise 14, describe any patterns that exist for inverse pairs.
16. Prove that D_4 is not an Abelian group by identifying two elements of D_4 that do not commute under composition.
17. Determine the centralizer of each element in D_4.

Exercises 18–28 consider the dihedral group D_5 under composition. The set D_5 consists of the ten orientation-preserving isometries of a regular pentagon (a regular 5-sided polygon). In particular, D_5 contains five rotations (denoted R_0, R_{72}, R_{144}, R_{216}, R_{288}) and five flips (denoted F_1, F_2, F_3, F_4, F_5). Note that each flip is across an axis through a vertex and the opposite side's midpoint.

18. As modeled in example 2.5.3, give a written description of the ten isometries in D_5.
19. Compute the Cayley table for D_5 under composition.
20. State the identity of D_5 under composition.
21. State the inverse of each element of D_5 under composition.
22. Based on exercise 21, describe any patterns that exist for inverse pairs.
23. Prove that D_5 is not an Abelian group by identifying two elements of D_5 that do not commute under composition.
24. Determine the elements in the centralizer of an arbitrary nonidentity rotation in D_5.
25. Determine the elements in the centralizer of an arbitrary flip in D_5.
26. Describe the isometry that results from the composition of two rotations in D_5. Justify your answer.
27. State the numeric relationship that ensures two rotations in D_5 are inverses of each other.
28. Describe the isometry that results from the composition of two flips in D_5. Justify your answer.

Exercises 29–34 consider the general dihedral group D_n of order $2n$ consisting of the orientation-preserving isometries of a regular n-sided polygon under the operation of composition.

29. How many rotations are in D_n? Justify your answer.
30. As modeled in example 2.5.3, give a written description of the rotation elements of D_n.
31. If n is odd, how many flips are in D_n? Give a geometric justification of your answer.
32. As modeled in example 2.5.3, give a written description of the flip elements of D_n when n is odd.
33. If n is even, how many flips are in D_n? Give a geometric justification of your answer.
34. As modeled in example 2.5.3, give a written description of the two distinct types of flip elements of D_n when n is even.

In exercises 35–44, identify the order of each group by determining the number of elements in the group.

35. $\mathbb{Z}_5$ 40. D_8
36. $U(5)$ 41. $\mathbb{Z}_{11}$
37. D_5 42. $U(11)$
38. $\mathbb{Z}_8$ 43. $U(14)$
39. $U(8)$ 44. $\mathbb{Z}_{14}$

In exercises 45–54, identify the elements in the group of orientation-preserving isometries of each geometric figure.

45. a nonsquare rectangle   50. a valentine's heart (or a cardioid)
46. a nonrectangular parallelogram   51. an addition symbol $+$
47. a rhombus   52. a multiplication symbol $\times$
48. a circle   53. a division symbol $\div$
49. a noncircular ellipse   54. a subtraction symbol $-$

Exercises 55–65 consider the $M_2(\mathbb{Z})$ of 2×2 matrices on the set of integers. Matrix groups were studied by Cayley in the 1850s and are of interest in this section since they provide another important example of noncommutativity. A 2×2 matrix on the integers is an array of integers with two rows and two columns. The following are some elements of $M_2(\mathbb{Z})$.

$$A = \begin{bmatrix} 1 & 0 \\ 0 & 1 \end{bmatrix} \quad B = \begin{bmatrix} -1 & 0 \\ 1 & -1 \end{bmatrix} \quad C = \begin{bmatrix} 1 & 3 \\ 2 & 4 \end{bmatrix}$$

Addition and multiplication of matrices are defined by combining the entries in a given pair of matrices in an appropriate fashion based on the corresponding operations on the integers. Consider the following formal definitions of matrix addition and multiplication.

$$\begin{bmatrix} a & b \\ c & d \end{bmatrix} + \begin{bmatrix} e & f \\ g & h \end{bmatrix} = \begin{bmatrix} a+e & b+f \\ c+g & d+h \end{bmatrix}$$

$$\begin{bmatrix} a & b \\ c & d \end{bmatrix} \cdot \begin{bmatrix} e & f \\ g & h \end{bmatrix} = \begin{bmatrix} ae + bg & af + bh \\ ce + dg & cf + dh \end{bmatrix}$$

In exercises 55–62, compute each sum and product.

55. $A + B$

56. $B + A$

57. $A + C$

58. $C + A$

59. $A \cdot B$

60. $B \cdot A$

61. $B \cdot C$

62. $C \cdot B$

Exercises 63–65 consider the commutativity of matrix addition and matrix multiplication in light of the computations in exercises 55–62.

63. What do exercises 55–58 suggest about the commutativity of matrix addition on $M_2(\mathbb{Z})$?

64. Using the commutativity of integer addition, prove that addition of matrices from $M_2(\mathbb{Z})$ is commutative.

65. What do exercises 59–62 prove about the commutativity of matrix multiplication on $M_2(\mathbb{Z})$?

Exercises 66–70 continue a study of the symmetric groups begun in in exercises 55–70 of section 2.4. The symmetric group S_n consists of all one-to-one, onto functions on the integers $\{1, \dots, n\}$ under the operation of composition. Recall that the elements of S_n are functions α represented in arrays of the form

$$\alpha = \begin{bmatrix} 1 & 2 & \cdots & n \\ \alpha(1) & \alpha(2) & \cdots & \alpha(n) \end{bmatrix}.$$

The top row of the array denotes the inputs of the function and the bottom row consists of the corresponding distinct outputs which are also elements of $\{1, \dots n\}$. For example, the following are three elements of S_3 represented in array form.

$$\epsilon = \begin{bmatrix} 1 & 2 & 3 \\ 1 & 2 & 3 \end{bmatrix} \quad \alpha = \begin{bmatrix} 1 & 2 & 3 \\ 2 & 3 & 1 \end{bmatrix} \quad \beta = \begin{bmatrix} 1 & 2 & 3 \\ 1 & 3 & 2 \end{bmatrix}$$

In exercises 66–69, compute each composition.

66. $\beta \circ \alpha$

67. $\alpha^2 = \alpha \circ \alpha$

68. $\alpha^2 \circ \beta$

69. $\beta \circ \alpha^2$

70. What do exercises 68–69 prove about the commutativity of composition on S_3?

2.6 Application: Check Digit Schemes

We complete this study of abstract algebra with an application of modular arithmetic and dihedral groups. Contemporary American society assigns numbers to just about everything, using "identification numbers" to represent such diverse objects as books, food, financial transactions, and even people. These assignments of numbers provide benefits in record-keeping and tracking information, but they also produce some new, associated problems. In this section, we grapple with the problem of incorrect digits appearing in these numbers when they are stored or transmitted. We can readily imagine how easily an accountant might accidentally move a decimal place and record $10.00 instead of $1000, or a teacher might erroneously record a grade of 89 instead of 98. In addition to human error, the physical devices of communication systems sometimes break down, resulting in signals that degrade or become garbled due to background noise. Such disruptions can also introduce errors in identification numbers. In short, we need some algorithmic process for verifying the accuracy of stored and transmitted numbers.

Fortunately, mathematicians have developed a variety of approaches to the problem of verifying the accuracy of identification numbers. In this section, we study *check digits*—values that enable us to verify whether or not a given identification number (such as a credit card number or an inventory record) is correct. In this way, a check digit acts as a flag for incorrect numbers, identifying when an error has occurred in record keeping. The ideas of abstract algebra, particularly those of modular arithmetic and dihedral groups, are among the most important tools for specifying such check digit schemes. A variety of different schemes or algorithms have been devised for assigning a check digit to a given identification number.

Definition 2.6.1   *If $n \in \mathbb{N}$ is a positive integer that serves as an* **identification number** *for an object, then a* **check digit** *for n is an integer $c \in \{0, \ldots, 9\}$ appended to n to produce the corresponding* **record number** $n^\wedge c = 10 \cdot n + c$. *Thus,*

$$record\ number\ =\ identification\ number\ {}^\wedge check\ digit.$$

Example 2.6.1   An identification number 542 with check digit 7 has record number 5427.

■

Many different approaches have been devised to compute a check digit for a given identification number. Modular arithmetic is used in most check digit schemes as a device for "reducing" multidigit numbers to a single digit. A check digit permits the verification of the identification number up to some degree of accuracy depending on the type of scheme employed. In this section, we introduce four different check digit schemes and consider the relative strengths and weaknesses of each approach. This discussion will show that not all check digit schemes are created equal. We begin our study with perhaps the simplest scheme, based on congruence modulo ten.

2.6.1   The Mod 10 Check Digit Scheme

For a given identification number $n \in \mathbb{N}$, the mod 10 check digit scheme determines the check digit for n using the formula

$$\text{check digit for } n\ =\ n \bmod 10.$$

Therefore, under the mod 10 check digit scheme, an identification number $n \in \mathbb{N}$ has the record number $n^\wedge(n \bmod 10) = 10 \cdot n + n \bmod 10$.

Example 2.6.2 We use the mod 10 check digit scheme to compute the record number for identification number $n = 1165$. The corresponding check digit is $1165 \bmod 10 = 5$, which produces the record number $1165^{\wedge}5 = 11655$. Several more examples are presented in the following table.

Identification number n	Check digit $n \bmod 10$	Record number $n^{\wedge}(n \bmod 10)$
1165	5	11655
23876	6	238766
1234	4	12344
1235	5	12355
1284	4	12844
2134	4	21344

■

Question 2.6.1 Use the mod 10 check digit scheme to compute the record number for each identification number.

(a) 2345

(b) 4675

(c) 345

(d) 345654765

■

As discussed above, we are not just interested in specifying record numbers; we are also interested in using check digits to determine the accuracy of the record numbers, at least to the extent possible for the given check digit scheme.

Example 2.6.3 We determine the validity of a few record numbers under the mod 10 check digit scheme.

- 1222: Record number 1222 has identification number 122 and check digit 2. Since $122 \bmod 10 = 2$, the identification number 1222 is valid with respect to the mod 10 check digit scheme.
- 1236: Record number 1236 has identification number 123 and check digit 6. Since $123 \bmod 10 = 3 \neq 6$, the identification number 1236 is not valid.
- 3574: Since $357 \bmod 10 = 7 \neq 4$, 3574 is not a valid record number.

■

Question 2.6.2 Explain why each number is or is not a valid record number under the mod 10 check digit scheme.

(a) 155

(b) 12532

■

The following question explores the strengths and weaknesses of the mod 10 check digit scheme. We also introduce two of the most common errors that arise in identification numbers: the single-digit error and the transposition error.

Question 2.6.3   At Record Keeping International (RKI), record specialist Morgan Smith is entering record numbers obtained using the mod 10 check digit scheme in an inventory list.

(a) Morgan enters the number 18983 in the inventory list and the computer flashes the error message: Invalid Record Number! Identify the three possible errors that Morgan could have made in entering the number 18983 and that could be detected by the computer using the mod 10 check digit scheme.
Hint: One possible error is that Morgan could have typed the last digit wrong.

(b) Several record numbers later, Morgan is trying to enter record number 12344, but incorrectly types 12844 instead. Explain why the mod 10 check digit scheme does or does not detect this *single-digit error in the third position*.

(c) Suppose Morgan is trying to enter record number 12344, but incorrectly types 21344 instead. Explain why the mod 10 check digit scheme does or does not detect this *transposition error of the first two digits*.

(d) Should RKI terminate Morgan Smith's employment or should they explore the possibility of using another check digit scheme?

■

Statistical data indicate that single-digit errors and transposition errors are the two most common errors in storing and communicating numbers. Therefore, check digit schemes that detect these errors are particularly valuable. For the sake of clarity and precision, these two types of common errors are defined as follows.

Definition 2.6.2   *Let $a_1 \cdots a_n$ be a record number with $a_k \in \{0, \ldots, 9\}$ for every k with $1 \le k \le n$.*

- *A **single-digit error in the kth position** occurs if the record number stored is $a_1 \cdots a_{k-1} \, b_k \, a_{k+1} \cdots a_n$ where $b_k \ne a_k$ for some k with $1 \le k \le n$.*
- *A **transposition error** occurs if two adjacent digits are switched and the record number stored is $a_1 \cdots a_{k-1} \, a_{k+1} \, a_k \, a_{k+2} \cdots a_n$ for some k with $1 \le k \le n$.*

In summary, the mod 10 check-digit scheme detects single-digit errors in the last two positions of a record number. However, this scheme fails to detect any other errors and so has quite limited practical usefulness. Fortunately, mathematicians have developed other, more discerning check-digit schemes. We turn our attention to a second check-digit scheme, also based on modular arithmetic. As we will see, this "mod 9 check-digit scheme" does a better job of detecting single-digit errors than the mod 10 check-digit scheme.

2.6.2   The Mod 9 Check Digit Scheme

For a given identification number $n \in \mathbb{N}$, the mod 9 check-digit scheme determines the check digit for n using the formula

$$\text{check digit for } n \; = \; n \bmod 9.$$

Therefore, under the mod 9 check digit scheme, an identification number $n \in \mathbb{N}$ has the record number $n^{\wedge}n \bmod 9 = 10 \cdot n + n \bmod 9$.

Example 2.6.4 We use the mod 9 check digit scheme to compute the record number for identification number $n = 1165$. The corresponding check digit is $1165 \bmod 9 = 4$ (since $1165 = 9 \cdot 124 + 4$), producing record number $1165^{\wedge}4 = 11654$. Several more examples are presented in the following table.

Identification number n	Check digit $n \bmod 9$	Record number $n^{\wedge}(n \bmod 9)$
$1165 = 9 \cdot 129 + 4$	4	11654
$23876 = 9 \cdot 2652 + 8$	8	238768
$1234 = 9 \cdot 137 + 1$	1	12341
$1235 = 9 \cdot 137 + 2$	2	12352
$1284 = 9 \cdot 142 + 6$	6	12846
$2134 = 9 \cdot 237 + 1$	1	21341

■

Before considering some questions exploring the capabilities of the mod 9 check digit scheme, we point out an algorithm that facilitates mod 9 computations. One approach to computing $n \bmod 9$ is the standard long division technique of dividing n by 9 to determine the quotient q and then computing $n - 9 \cdot q = n \bmod 9$. Alternatively, if $n = a_1 \cdots a_k \in \mathbb{N}$ is a positive integer with decimal digits $a_1, \ldots, a_k$, then $n \bmod 9$ is equal to the mod 9 value of the digits' sum; that is, we can compute $n \bmod 9$ using the following formula:

$$n \bmod 9 = (a_1 \cdots a_k) \bmod 9 = (a_1 + \cdots + a_k) \bmod 9.$$

Reconsidering example 2.6.4 in which we determined the mod 9 check digit for identification number $n = 1165$, we can either determine the check digit $1165 \bmod 9 = 4$ directly (as in the example), or use this new formula to obtain

$$1165 \bmod 9 \;=\; (1 + 1 + 6 + 5) \bmod 9 \;=\; 13 \bmod 9 = 4.$$

This algorithm only works for mod 9 arithmetic and does not extend to division by other positive integers. This approach may be helpful in the following questions.

Question 2.6.4 Use the mod 9 check digit scheme to compute the record number for each identification number.

(a) 2345   (c) 345

(b) 4675   (d) 345654765

■

Example 2.6.5 We determine the validity of two record numbers under the mod 9 check-digit scheme.

- 1222: Record number 1222 has identification number 122 and check digit 2. Since $122 \bmod 9 = 5 \neq 2$, the identification number 1222 is not valid with respect to the mod 9 check digit scheme.

- 1236: For the record number 1236, we have identification number 123 and check digit 6. Since 123 mod 9 = 6, the identification number 1236 is valid with respect to the mod 9 check digit scheme.

■

Question 2.6.5   Explain why each number is or is not a valid record number under the mod 9 check digit scheme.

   (a) 155   (b) 12532

■

Question 2.6.6   After sufficient experience with the shortcomings of the mod 10 check-digit scheme, RKI has converted its inventory system to the mod 9 check-digit scheme. Record specialist Morgan Smith is once again entering record numbers in an inventory list for RKI. For each case, explain why the mod 9 check digit-scheme does or does not detect the single-digit error in the third position.

   (a) Morgan is trying to enter 12341, but incorrectly types 12841.
   (b) Morgan is trying to enter 18988, but incorrectly types 18888.
   (c) Morgan is trying to enter 18988, but incorrectly types 18088.
   (d) Morgan is trying to enter 12047, but incorrectly types 12947.

What is distinctive about cases (c) and (d) in which the mod 9 check digit scheme fails to detect the single-digit error?

■

The mod 9 check digit scheme detects all single-digit errors except for when 0 is substituted for 9 or 9 is substituted for 0. Since 9 mod 9 = 0 = 0 mod 9, mod 9 arithmetic is simply not capable of detecting such errors. Unfortunately, the mod 9 check digit scheme does not detect any transposition errors nor any other errors involving rearrangement of the digits.

Example 2.6.6   We verify that the mod 9 check-digit scheme does not detect transposition errors.
     Consider the identification numbers 1234 and 2134 on lines 3 and 6 of the table given in example 2.6.4. Both of these identification numbers have check digit 1, indicating that the mod 9 check-digit scheme does not detect the transposition of the first two digits.

■

Question 2.6.7   (a) Determine the check digit for three more rearrangements of the identification number 1234.
      (b) Using the algorithm explained before question 2.6.4, explain why the mod 9 check digit scheme does not detect transposition errors in any identification number.
      (c) Prove that the mod 9 check-digit scheme computes the same check digit when $a_1 \cdots a_n$ is rearranged as $a_{k_1} \cdots a_{k_n}$. This shows that that the mod 9 scheme cannot detect errors involving any type of rearrangement of digits in any given identification number.

■

In summary, the mod 9 check digit scheme detects single-digit errors in any position, except for single-digit errors in which the numbers 0 and 9 are substituted for each other. In addition to not detecting some single-digit errors, the mod 9 check digit scheme fails to detect any transposition errors, indicating the need for a more subtle, discerning, and ultimately more useful check digit scheme. In fact, of the four different check digit schemes studied in this section, only the last scheme (which is a dihedral check digit scheme) completely satisfies the goal of detecting all single-digit errors and all transposition errors.

2.6.3 The Codabar Check Digit Scheme

In preparation for working the dihedral check digit scheme, we first consider the Codabar check digit scheme. Codabar is a highly effective check digit scheme that detects all single-digit errors and 98% of other common errors (including, but not limited to transposition errors). In light of this high level of effectiveness, the Codabar check digit scheme is widely used in many diverse settings. Every major credit company uses Codabar, as well as many banks, libraries, universities, and a variety of other commercial enterprises. We are particularly interested in the Codabar check digit scheme because of the manner in which this algorithm blends standard arithmetic and modular arithmetic operations.

We present the Codabar check digit scheme in the context of determining the check digit for a credit card number. A credit card number is a 16-digit record number consisting of a 15-digit identification number assigned administratively by the issuing financial institution and a single check digit determined by the Codabar check digit scheme. In this context, the algorithm for the Codabar check digit scheme has the following four steps.

(a) Sum the digits in the odd positions, $1, 3, 5, \ldots, 15$, and double the sum.
(b) Determine how many digits in the odd positions *exceed* four (so, are either 5, 6, 7, 8, or 9) and add this number to the result of step (a).
(c) Add the digits in the even positions, $2, 4, 6, \ldots, 14$, to the result of step (b).
(d) Determine the check digit using the formula: [step (c) + check digit] mod $10 \equiv 0$.

The final credit card number consists of the original 15-digit identification number with the appended check digit.

Before working through some examples and questions, we note that credit card numbers are often presented in four blocks of four digits. Often spaces or hyphens are inserted between each block of digits; hyphens are used in this text. This presentation of credit card numbers facilitates the reading, recording, and recollection of these numbers by human beings, since most minds are not readily able to work with continuous blocks of sixteen digits.

Example 2.6.7 We determine the check digit and the complete credit card number for the identification number 8479-2675-3419-241.

Working from the left, we sum the digits in the odd positions and double the sum to obtain

$$2 \cdot (8 + 7 + 2 + 7 + 3 + 1 + 2 + 1) = 62.$$

For each 5, 6, 7, 8, or 9 in an odd position, we add one to the result of the first step.

$$\text{we have } 8, 7, 7 \quad \Rightarrow \quad 62 + 3 = 65$$

We add the remaining digits to the result of the second step to obtain

$$65 + 4 + 9 + 6 + 5 + 4 + 9 + 4 = 106.$$

We choose the check digit so that [step (c) + check digit] mod 10 $\equiv$ 0; in this case, we have

$$(106 + c) \bmod 10 \equiv 0 \quad \Rightarrow \quad c = 4.$$

Therefore, the Codabar check digit for the given identification number is 4, and the complete credit card number is 8479-2675-3419-2414.

∎

Perhaps the most interesting mathematical step in the algorithm for the Codabar check digit scheme is the final step in which we solve a modular equation for the unknown check digit c. The dihedral check digit scheme also requires us to solve a modular equation, as do many other sophisticated check digit schemes.

As with our other check digit schemes, we are interested in the Codabar check digit scheme as a tool for seeking errors in a given credit card number. As before, we compute the check digit for the given identification number and compare the resulting value with the given check digit. If the two digits are the same, the credit card number is accepted as valid (at least up to the error detection capabilities of the check digit scheme); if the two digits differ, then the credit card number is declared invalid.

Question 2.6.8   Use the Codabar check digit scheme to compute the credit card number with identification number 8479-2642-1937-847.

∎

Question 2.6.9   Explain why 9479-2675-3419-2414 is or is not a valid credit card number under the Codabar check digit scheme. Compare your result with example 2.6.7. These examples provide some evidence for what property of the Codabar check digit scheme?

∎

In summary, the Codabar check digit scheme combines standard integer arithmetic with modular arithmetic to obtain some pretty impressive error-detecting capabilities. As mentioned above, the Codabar check digit scheme detects all single-digit errors and 98% of other common errors. This check digit scheme has also provided a nice introduction to one of the essential elements of the fourth check digit scheme presented in this section.

2.6.4   The D_5 Check Digit Scheme

We finish this section on check digits with a "dihedral check digit scheme" that uses the dihedral group D_5 of order 10 in an essential way. The D_5 check digit scheme was developed by the Dutch mathematician Jacobus Verhoeff in 1969. Since the order of D_5 is 10, this group is a natural fit for the standard Hindu–Arabic numeral system with its 10 distinct digits. This D_5 check digit scheme is very effective in detecting the common errors in storing and communicating record numbers and is used by many banks and other financial institutions. This scheme detects all single-digit errors in any position and all transposition errors, and so is a significant improvement over the mod 9 and mod 10 check digit schemes presented above.

Recall that the dihedral group D_5 consists of all orientation-preserving isometries of a regular pentagon under the operation of composition. In order to facilitate our computations, the 10 digits of the Hindu–Arabic numeral system are associated with the 10 orientation-preserving isometries of the regular pentagon. Roughly speaking, the digits 0–4 with the five rotations in D_5 and the digits 5–9 with the five flips in D_5. While we do not give the exact mapping, the corresponding numeric rendition of the Cayley table for D_5 given below is used when carrying out computations in the D_5 check digit scheme.

∘	0	1	2	3	4	5	6	7	8	9
0	0	1	2	3	4	5	6	7	8	9
1	1	2	3	4	0	6	7	8	9	5
2	2	3	4	0	1	7	8	9	5	6
3	3	4	0	1	2	8	9	5	6	7
4	4	0	1	2	3	9	5	6	7	8
5	5	9	8	7	6	0	4	3	2	1
6	6	5	9	8	7	1	0	4	3	2
7	7	6	5	9	8	2	1	0	4	3
8	8	7	6	5	9	3	2	1	0	4
9	9	8	7	6	5	4	3	2	1	0

In this numeric rendition of the Cayley table for D_5, note the characteristic computational patterns for the dihedral group. In particular, the composition of two rotations (represented by 0–4) or two flips (represented by 5–9) produces a rotation (0–4), as witnessed in the upper left and lower right quadrants of this Cayley table. Similarly, the composition of a rotation (0–4) and a flip (5–9) produce a flip (5–9), as is apparent in the lower left and upper right quadrants of the table.

The D_5 check digit scheme also utilizes a collection of functions f_k defined on the elements of the dihedral group. In this text, we focus on applications of the D_5 check digit scheme to identification numbers with at most four digits, and so we give just the first four of these functions. Continuing to associate the digits 0–9 with the elements of D_5, these functions are presented in the following table; the inputs appear along the top row and the outputs of each function are listed beneath the corresponding inputs.

n	0	1	2	3	4	5	6	7	8	9
$f_1(n)$	1	5	7	6	2	8	3	0	9	4
$f_2(n)$	5	8	0	3	7	9	6	1	4	2
$f_3(n)$	8	9	1	6	0	4	3	5	2	7
$f_4(n)$	9	4	5	3	1	2	6	8	7	0

From the first row of this table, $f_1(0) = 1$, $f_1(1) = 5$, $f_1(2) = 7, \ldots, f_1(9) = 4$, and similarly for the other three functions. Given the function f_1, the other three functions are computed by repeated compositions. In particular, the function f_2 is equal to the composition $f_1 \circ f_1$, $f_3 = f_1 \circ f_1 \circ f_1$, and so on. Therefore, when a check digit is needed for a number with more than four digits, we can obtain the needed additional functions by taking an appropriate number of compositions of f_1 (see exercises 45–50 at the end of this section).

For a given identification number with digits $a_1 \cdots a_n$, the D_5 check digit is the digit c satisfying]

$$f_1(a_1) \circ f_2(a_2) \circ \cdots \circ f_n(a_n) \circ c = 0.$$

From an algorithmic perspective, the D_5 check digit scheme first applies the function f_k to the kth digit of the identification number, composes the resulting digits with the check digit c (as elements of D_5), sets the result equal to 0, and finally solves for the unknown check digit c in the resulting equation using the Cayley table for D_5 given above. The order of composition is critical here, since the dihedral group D_5 is a nonAbelian group and commutativity does not hold. For a four-digit identification number $a_1a_2a_3a_4$, the check digit c is chosen to satisfy

$$f_1(a_1) \circ f_2(a_2) \circ f_3(a_3) \circ f_4(a_4) \circ c = 0.$$

We illustrate the D_5 check digit scheme in the following example and then have you continue the study of the D_5 scheme in some questions.

Example 2.6.8   We use the D_5 check digit scheme to determine the record number for identification number 1165.

We proceed by substituting $a_1 = 1$, $a_2 = 1$, $a_3 = 6$, and $a_4 = 5$ into the D_5 check digit equation. The check digit c is determined by simplifying and solving this equation for c using the given table of functions and the Cayley table for D_5. In the following sequence of compositions, we use the fact that the composition operation $\circ$ is associative on D_5 and so we are free to work our way from left to right in simplifying each expression.

$$
\begin{aligned}
f_1(1) \circ f_2(1) \circ f_3(6) \circ f_4(5) \circ c &= 0 & \\
(5 \circ 8) \circ 3 \circ 2 \circ c &= 0 & \\
(2 \circ 3) \circ 2 \circ c &= 0 & \text{since } 5 \circ 8 = 2 \text{ in } D_5 \\
(0 \circ 2) \circ c &= 0 & \text{since } 2 \circ 3 = 0 \text{ in } D_5 \\
2 \circ c &= 0 & \text{since } 0 \circ 2 = 2 \text{ in } D_5
\end{aligned}
$$

Finally, we search for a 0 in the third row (the row for 2 in the D_5 Cayley table) and find that $2 \circ 3 = 0$, and so $c = 3$. Appending this check digit to the identification number, the desired record number is 11653.

■

**Question
2.6.10**   Use the D_5 check digit scheme to compute the record number for each identification number.

(a) 1234   (c) 1284

(b) 1235   (d) 2134

■

Notice that the D_5 check digits are different for each of the four identification numbers given in question 2.6.10. What evidence do these examples provide for the D_5 check digit scheme detecting single-digit errors and transposition errors? Knowing that the D_5 scheme succeeds in detecting these two types of errors, what recommendations do you have for Morgan Smith and the folks at RKI?

Question 2.6.11 Explain why each number is not a valid record number under the D_5 check digit scheme.

    (a) 45802   (c) 9873

    (b) 23943   (d) 123

For the third and fourth record numbers, we note that the D_5 check digit scheme does not require the use of four-digit identification numbers, and so we interpret the potential record number 9873 as consisting of an identification number 987 and a check digit 3.

■

2.6.5 Reading Questions for Section 2.6

   1. What motivates the development of check digit schemes?
   2. Define and give an example of a single-digit error.
   3. Define and give an example of a transposition error.
   4. Describe how the mod 10 check digit scheme determines a check digit.
   5. Discuss the relative strengths and weaknesses of the mod 10 check digit scheme.
   6. Describe how the mod 9 check digit scheme determines a check digit.
   7. Discuss the relative strengths and weaknesses of the mod 9 check digit scheme.
   8. Describe how the Codabar check digit scheme determines a check digit.
   9. Discuss the relative strengths and weaknesses of the Codabar check digit scheme.
  10. Describe how the D_5 check digit scheme determines a check digit.
  11. For the D_5 check digit scheme, state the definition of f_2 in terms of f_1.
  12. Discuss the relative strengths and weaknesses of the D_5 check digit scheme.

2.6.6 Exercises for Section 2.6

In exercises 1–4, use the mod 10 check digit scheme to compute the record number for each identification number.

   1. 1234   3. 1284

   2. 1235   4. 2134

In exercises 5–8, explain why each number is or is not a valid record number under the mod 10 check digit scheme.

   5. 45808   7. 9877

   6. 23944   8. 1236

In exercises 9–14, use the mod 9 check digit scheme to compute the record number for each identification number.

9. 1234

10. 1235

11. 1284

12. 2134

13. 2135

14. 2185

In exercises 15–20, explain why each number is or is not a valid record number under the mod 9 check digit scheme.

15. 45808

16. 23944

17. 9877

18. 1236

19. 345-936

20. 345-455

In exercises 21–26, use the Codabar digit scheme to compute the complete credit card number for each identification number.

21. 2181-2389-8824-398

22. 4566-3932-6858-147

23. 1234-7898-3243-311

24. 3577-1232-8098-294

25. 7678-1443-3425-768

26. 4556-7688-2345-355

In exercises 27–32, explain why each number is or is not a valid credit card number under the Codabar digit scheme.

27. 2345-4356-3112-7854

28. 9695-2859-8724-5659

29. 2456-5024-7695-4268

30. 0987-6568-3453-4452

31. 4586-7092-6795-4657

32. 0495-0256-6526-7096

In exercises 33–38, use the D_5 check digit scheme to compute the record number for each identification number.

33. 123

34. 132

35. 8345

36. 8435

37. 5345

38. 3545

In exercises 39–44, explain why each number is or is not a valid record number under the D_5 check digit scheme.

39. 2483

40. 8423

41. 45800

42. 54800

43. 45899

44. 55809

In exercises 45–46, we extend the D_5 check digit scheme to handle five- and six-digit identification numbers. Recall that f_1 is defined by the input-output table

n	0	1	2	3	4	5	6	7	8	9
$f_1(n)$	1	5	7	6	2	8	3	0	9	4

and that $f_2 = f_1 \circ f_1, f_3 = f_1 \circ f_1 \circ f_1 = f_1 \circ f_2$, and so on.

In exercises 45–46, follow this pattern to determine each functions.

45. f_5

46. f_6

In exercises 47–50, use the D_5 check digit scheme and the answers to exercises 45–46 to compute the the record number for each identification number.

47. 84765

48. 23987

49. 012345

50. 346589

Exercises 51–62 consider the check digits for ISBNs, or International Standard Book Numbers. Every published book is assigned a 10-digit ISBN, denoted by $a_1 - a_2a_3a_4a_5 - a_6a_7a_8a_9 - a_{10}$, where the hyphens are inserted for readability. For example, in the ISBN 0-7167-3817-1, we have $a_1 = 0$, $a_2 = 7$, $a_3 = 1$, ..., $a_{10} = 1$. The last digit a_{10} of the ISBN is a check digit and the ISBN check digit scheme determines the check digit using the following formula.

$$[10 \cdot a_1 + 9 \cdot a_2 + 8 \cdot a_3 + 7 \cdot a_4 + 6 \cdot a_5 + 5 \cdot a_6 + 4 \cdot a_7 + 3 \cdot a_8 + 2 \cdot a_9 + a_{10}] \bmod 11 = 0$$

We verify the check digit of the ISBN 0-7167-3817-1 by first computing:

$$10 \cdot 0 + 9 \cdot 7 + 8 \cdot 1 + 7 \cdot 6 + 6 \cdot 7 + 5 \cdot 3 + 4 \cdot 8 + 3 \cdot 1 + 2 \cdot 7 + a_{10} = 219 + a_{10}.$$

As with the Codabar check digit scheme and the D_5 check digit scheme, the ISBN check digit requires $[219 + a_{10}] \bmod 11 = 0$. Since $[219 + 1] \bmod 11 = 220 \bmod 11 = 0$, we have $a_{10} = 1$ and the given ISBN is correct. Since the ISBN check digit is determined using mod 11 arithmetic, the scheme sometimes needs an eleventh digit; the standard convention is to use X for the check digit when using $a_{10} = 10$. The ISBN check digit scheme detects all single-digit errors and all transposition errors of adjacent digits.

In exercises 51–56, use the ISBN check digit scheme to compute the ISBN for each identification number.

51. 2-3474-9129

52. 0-0823-7322

53. 0-7167-3818

54. 2-2343-6856

55. 3-3458-2134

56. 1-6987-5687

In exercises 57–62, explain why each number is or is not a valid ISBN under the ISBN check digit scheme.

57. 0-6181-2214-1

58. 0-5349-4422-3

59. 0-6181-4916-2

60. 0-9232-3140-4

61. 1-7365-4557-7

62. 2-8768-7698-5

Exercises 63–70 consider "isomorphic" finite groups. Intuitively, isomorphic groups are identical as groups (up to the particular choice of names used to identify group elements), and so share all group properties in common. Formally, two groups are isomorphic if there exists a one-to-one, onto function from one group to the other which preserves the group operation. The following exercises consider only finite groups and use Cayley tables to determine if two groups are isomorphic. Consider the following familiar Cayley tables for $\mathbb{Z}_2$, $U(3)$, and $\mathbb{Z}_3$ under the appropriate modular operations.

$\oplus$	0	1
0	0	1
1	1	0

$\odot$	1	2
1	1	2
2	2	1

$\oplus$	0	1	2
0	0	1	2
1	1	2	0
2	2	0	1

As we can see, the groups $\mathbb{Z}_2$ under addition mod 2 and $U(3)$ under multiplication mod 3 have identical Cayley tables provided we identify $0 \in \mathbb{Z}_2$ with $1 \in U(3)$ and $1 \in \mathbb{Z}_2$ with $2 \in U(3)$. We say that $\mathbb{Z}_2$ and $U(3)$ are isomorphic groups and write $\mathbb{Z}_2 \approx U(3)$; as mentioned above, we now know that $\mathbb{Z}_2$ and $U(3)$ share all group properties in common. On the other hand, the group $\mathbb{Z}_3$ under multiplication mod 3 is not isomorphic to either $\mathbb{Z}_2$ or $U(3)$; there are a different number of elements in $\mathbb{Z}_3$ than in either $\mathbb{Z}_2$ or $U(3)$, and so the Cayley tables are not identical. We can often prove that two groups are not isomorphic by comparing the number of elements in their respective sets, but sometimes we must carefully inspect the Cayley table to determine nonisomorphism.

In exercises 63–70, determine whether or not each pair of groups (under the appropriate operations) is isomorphic by inspecting Cayley tables. If the groups are isomorphic, state the identification of group elements witnessing the isomorphism.

63. $\mathbb{Z}_2$ and $U(4)$

64. $\mathbb{Z}_2$ and $U(5)$

65. D_3 and $U(7)$

66. $\mathbb{Z}_4$ and $U(8)$

67. $\mathbb{Z}_3$ and $U(8)$

68. $\mathbb{Z}_6$ and D_3

69. $\mathbb{Z}_6$ and S_3

70. $U(5)$ and $U(8)$

Notes

The ideas and results presented in section 2.1 are widely known among mathematicians (and others), and so they are contained in many different mathematical textbooks. Basic set theory is often studied in "Discrete Mathematics" courses, which are supported by such texts as those by Epp [72], Richmond and Richmond [193], and Scheinerman [209]. Alternatively, there are a growing number of "Foundations of Mathematics" textbooks that consider these notions, including those by Barnier and Feldman [10], D'Angelo and West [51], and Smith et al. [219].

A number of excellent books are devoted exclusively to the development and study of set theory as a rich mathematical field in its own right, including an undergraduate text by Halmos [108]; the standard graduate level texts exploring set theory are those written by Kunen [146] and Jech [130]. The rigorous, axiomatic study of set theory was initiated by a letter from Bertrand Russell to Gottlob Frege in which Russell outlined what has become known as Russell's paradox (see exercises 69–73 in section 2.1). Davis et al. [55] has a nice exploration of the context and content of this and other related paradoxes. Aside from his contributions to mathematics, Russell was a widely known and respected philosopher and wrote many different works exploring religion, happiness, and knowledge. Late in his life, Russell wrote his autobiography [204] reflecting on his intellectual and personal life in the context of the many events of the twentieth century.

The majority of this chapter was devoted to the development and study of abstract algebra. Most undergraduate mathematics majors take at least one course devoted exclusively to the study of abstract algebra. One of the most widely acclaimed undergraduate textbooks on abstract algebra is by Gallian [93]; two popular book are by Fraleigh [88] and Hillman et al. [116]; a standard graduate text in abstract algebra is by Hungerford [122].

In this chapter, we mentioned the work of several famous algebraists, including Abel, Galois, and Caley. The definitive biography of Niels Henrick Abel has been written by Stubhaug and Daly [236]. In addition, *Abel's Proof* by Pesic [187] provides an excellent exposition of the content and the historical results leading up to Abel's proof of the insolvability of the quintic, as does *The Equation That Couldn't Be Solved* by Livio [158]. In contrast, relatively little is known about Évariste Galois. Stewart [229], Livio [158], Bell [15], and Boyer and Merzbach [28] contain sketches about Galois' life. On the other hand, Galois' mathematical insights are well known, and there is a whole area of mathematics known as Galois theory. For undergraduates who have studied sufficient abstract algebra, Garling [96], Stewart [229] and Swallow [237] are excellent and accessible texts. Edwards [70] is a graduate text devoted exclusively to Galois theory, and Hungerford [122] is a standard graduate text in abstract algebra that also addresses Galois theory.

Arthur Caley had wide-ranging mathematical interests and was one of the most prolific mathematicians in known history; his collected works consist of over 2000 pages of published text. For those who are interested in learning more about Cayley, Crilly [48] has written a good and accessible biography. Emmy Noether is another mathematician who made important contributions to abstract algebra. Noether was a German mathematician who relocated to teach at Bryn Mawr College in Philadelphia shortly before the start of World War II. Noether was widely regarded as one of the most insightful algebraists of her time, and her work was praised by Einstein, Hilbert, and a host of other mathematicians. Van Der Waerden [244] has written an interesting book surveying the development of algebra from the contributions of the Islamic mathematicians in the Middle Ages to Noether's work in the twentieth century.

In the section on dihedral groups, we mentioned the connections between group theory and Maurits Cornelis Escher's work. In 1985, an International Congress was held exploring and discussing these interrelations; the proceedings of that conference can be found in [47]. In addition, those familiar with Escher's work will recognize that some of his pieces explore self-reference and self-perception, themes that are related to Gödel's mathematics and Bach's music, as explored in Hofstadter's Pulitzer Prize winning book *Gödel, Escher, Bach: An Eternal Golden Braid* (see [118]). Locher [159] is a good biographical account of Escher's life and work.

The application explored in this chapter concerned check digits for identification numbers. There are many different presentations of these notions. Perhaps the most accessible can be found in the textbooks written to support "Liberal Arts Mathematics" courses that focus on applications, including those texts by Burger and Starbird [34] and the Consortium for Mathematics and Its Applications [43]. In addition, Kirtland [141] has written a book exclusively devoted to developing a variety of check digit schemes.

The proliferation of electronic communication systems has not only generated a need for verification systems for transmitted information, but also an increasing need for secure communication systems. Hodges' biography of Turing [117] includes a description of the British effort to break the Enigma code utilized by the Germans during World War II. Such efforts continue to this day as federal agencies recruit mathematicians to contribute to more modern efforts to create and decipher code. This area of mathematical research is known as "cryptography" and is discussed further in section 3.2. Accessible introductions to coding theory include those by Bierbrauer [18], Hill [115], and Ling and Xing [156]. A young Irish mathematics student named Sarah Flannery has co-written an enjoyable autobiography [85], which includes a discussion of her development of a new coding scheme. Flannery was awarded both the 1998 Intel Fellows Achievement Award and the 1999 Ireland Young Scientist of the Year award for her work with this coding scheme.

3 Number Theory

In this chapter, we continue a study of numbers in all their grandness and diversity. As we have seen, there are many important and distinct number systems, including the natural numbers, the integers, the rationals, the reals, and the complex numbers. We consider each of these number systems from both a computational and a theoretical point of view, developing insights that provide a competent understanding of each one.

We begin with a special type of integer known as a prime number. Prime numbers can be thought of as the basic building blocks in the multiplicative structure of the integers. Many of the world's greatest mathematicians have devoted significant effort to exploring and understanding primes, including Euclid, Pierre de Fermat, Leonhard Euler, Carl Friedrich Gauss, Peter Lejeune Dirichlet, and Georg Bernhard Riemann. Through these collective efforts, mathematicians have developed a sound understanding of both the prime numbers and the integers. At the same time, many questions about these integers remain open (or unsolved); we introduce some of these intriguing questions that continue to inspire and challenge mathematicians.

Prime numbers have an important technological application in the fields of coding theory and cryptography. With the widespread use of electronic systems for sharing information, many people have become increasingly invested in secure and accurate means of communication. Cryptography is the field of mathematics devoted to the careful analysis and development of encryption algorithms that enable such communication. We first study RSA algorithms, which ingeniously ensure secure coding and decoding of messages using modular arithmetic and the group-theoretic properties of primes. We also study Hamming codes, a topic in coding theory (which is dedicated to detecting errors created during transmission). Hamming codes use modular matrix arithmetic to both detect and correct many such errors.

Mathematicians are also interested in certain relationships among ordered triples of integers. The Pythagorean theorem of geometry fame induces a relationship on triples of integers based on the possible side lengths of right triangles. Mathematicians have generalized this relation based on Diophantine equations, which are multivariable equations with integer solutions. Diophantus of Alexandria was a Greek mathematician from the third century C.E. who wrote the *Arithmetica*, a collection of 130 questions and solutions of linear and quadratic equations. Along with Euclid's *Elements*, Diophantus' *Arithmetica* and an important accompanying commentary by Hypatia were passed from the ancient Greeks to Islamic mathematicians, to Italian mathematicians, and eventually to the rest of western Europe.

The *Arithmetica* had a profound impact on European mathematicians in the sixteenth and seventeenth centuries, who extended Diophantus' work on specific linear and quadratic equations to a more general study of higher-order polynomial equations and multivariable equations with certain types of solutions. Aside from the Pythagorean theorem, perhaps the most famous Diophantine equations are those identified in Fermat's last theorem. This result claims that for every integer n greater than two, there are no integers a, b, c such that $a^n + b^n = c^n$. Fermat scribbled this claim in the margins of his personal copy of the *Arithmetica* around 1630, along with the tantalizing assertion that "I have discovered a truly remarkable proof which this margin is too small to contain." For more than three centuries, mathematicians sought to prove Fermat's last theorem. In 1995, the English mathematician Andrew Wiles from Princeton University gave the first complete proof of this result. We investigate the proof of one important case of Fermat's last theorem; the complete proof of the general theorem is quite advanced, using sophisticated mathematical ideas currently studied in graduate courses.

The chapter then takes up the study of the rational, real, and complex numbers. We develop definitions of these numbers in terms of the integers, and we prove the proper inclusions $\mathbb{Q} \subset \mathbb{R} \subset \mathbb{C}$. This discussion includes the classic proof that the square root of two is irrational, a startling insight in its time. Employing the abstract, algebraic approach of chapter 2, general properties are identified that hold in these number systems. This approach provides a coherent framework for studying these properties in the context of specific number systems. The primary algebraic object of interest is known as a "field," and extends the notion of a group to number systems with two operations. The rational, real, and complex numbers all have natural additive and multiplicative operations; the study of fields illuminates the interplay between these operations.

After developing a solid understanding of these number systems, the chapter then turns to polynomial equations. For centuries, mathematicians have studied polynomial equations and have developed polynomial models for physical and social behaviors. We consider the solvability and insolvability of polynomials over different number systems, seeing how the underlying numerical structure affects the set of solutions. We also take a closer look at the famous quadratic equation, its close cousins the cubic and the quartic equations, and the surprising result of the Norwegian mathematician Niels Abel, who proved that no such formula exists for polynomials of any degree greater than four.

The chapter ends with the study of "mathematical induction." Induction on the natural numbers (and other mathematical structures) is a useful proof technique of mathematics. We first use induction to prove general claims about all natural numbers. We also discuss the application of induction to verifying the truth of mathematical statements useful in other settings, including mathematical logic, abstract algebra, real analysis, and complex analysis.

3.1 Prime Numbers

Prime numbers serve as the basic building blocks in the multiplicative structure of the integers. As you may recall, an integer n greater than one is prime if its only positive integer multiplicative factors are 1 and n. Furthermore, every integer can be expressed as a product of primes, and this expression is unique up to the order of the primes in the product. This important insight into the multiplicative structure of the integers has become known as the *fundamental theorem of arithmetic*.

Beneath the simplicity of the prime numbers lies a sophisticated world of insights and results that has intrigued mathematicians for centuries. By the third century B.C.E., Greek mathematicians had defined prime numbers, as one might expect from their familiarity with the division algorithm. In Book IX of *Elements* [73], Euclid gives a proof of the infinitude of primes—one of the most elegant proofs in all of mathematics. Just as important as this understanding of prime numbers are the many unsolved questions about primes. For example, the Riemann hypothesis is one of the most famous open questions in all of mathematics. This claim provides an analytic formula for the number of primes less than or equal to any given natural number. A proof of the Riemann hypothesis also has financial rewards. The Clay Mathematics Institute has chosen six open questions (including the Riemann hypothesis)—a complete solution of any one would earn a \$1 million prize. Working toward defining a prime number, we recall an important theorem and definition from section 2.2.

Theorem 3.1.1 The division algorithm; theorem 2.2.1 in section 2.2 *If m, $n \in \mathbb{Z}$ and n is a positive integer, then there exist unique integers $q \in \mathbb{Z}$ and $r \in \{0, 1, \ldots, n-1\}$ such that $m = n \cdot q + r$. We refer to n as the* **divisor***, q as the* **quotient***, and r as the* **remainder** *when m is* **divided** *by n.*

Definition 3.1.1 *For m, $n \in \mathbb{Z}$, we say that n* **divides** *m when there exists $q \in \mathbb{Z}$ such that $m = n \cdot q$; that is, when the remainder r is 0 as the division algorithm is applied to m and n. In this context, n is called a* **divisor** *of m or a* **factor** *of m.*

The consideration of remainders from the division algorithm may bring to mind an important fact from the study of modular equivalence in chapter 2. Recall that n divides m exactly when $m \bmod n = 0$.

Example 3.1.1 We know that 3 divides 36 (or 3 is a factor of 36) because $36 = 3 \cdot 12$. On the other hand, 3 does not divide 37 (or 3 is not a factor of 37) because $37 = 3 \cdot 12 + 1$, and so the remainder from the division algorithm is 1 rather than 0.

■

Question 3.1.1 Determine if the following statements are true; if not, state the nonzero remainder.

(a) 5 divides 15   (c) 8 is a factor of 23

(b) 5 divides 24   (d) 8 is a factor of 32

■

In addition to determining if one particular integer divides another, mathematicians are often interested in identifying the complete set of *all* divisors of a given integer. For example, one important aspect of factoring polynomials is an ability to determine all possible divisors of the constant term. Similarly, divisors play a key role when determining the elements of a group $U(n)$.

Example 3.1.2   The positive integer divisors of 90 are 1, 2, 3, 5, 6, 9, 10, 15, 18, 30, 45, and 90.

∎

When compiling such lists, we typically consider only positive integer divisors (although certainly the negative of a divisor is also a divisor). It also helpful to notice that most divisors match up in pairs; for example, both 3 and 30 are divisors of $90 = 3 \cdot 30$.

Question 3.1.2   List every positive divisor of the integers 98 and 120.

∎

We now state the definition of a prime number.

Definition 3.1.2   *An integer $p \in \mathbb{Z}$ is* **prime** *when $p \geq 2$ and the only positive divisors of p are 1 and p itself. When an integer n is not prime, we say that n is* **nonprime***.*

Recall that nonprimes are also known as *composite* numbers. The following examples and questions highlight important examples and properties of these numbers.

Example 3.1.3   Using the definition of a prime, we observe that 3 is prime because its only divisors are 1 and 3. On the other hand, 4 is not prime because 2 divides 4, and so 1 and 4 are not the only positive divisors of 4.

∎

Question 3.1.3   Determine if each integer is prime; if not, state the positive integer divisors of the given number.

(a)   11

(b)   34

(c)   −3

(d)   1

(e)   83

(f)   6

∎

Question 3.1.4   How many prime numbers are even? Justify your answer.

∎

We often use prime numbers and their properties when working with integers. One approach to identifying if a given integer is prime is to try to factor it. A significant downside to this approach is its slowness—factoring arbitrary integers on the order of 100 digits can require up to 74 years of supercomputer time! As we discuss in the next section, this difficulty in quickly factoring large integers does have a positive side, enabling the security of certain encryption schemes. As an illustration of the relative slowness of this process, the next question asks you to distinguish among the first twenty primes and nonprimes.

Question 3.1.5   (a) List the first 10 prime numbers.

(b) List the first 10 positive integers that are nonprimes.

∎

The relationship between primes and nonprimes is expressed by the prime power factorization of integers as described in the fundamental theorem of arithmetic. A statement and proof of this result appear as Proposition 14 in Book IX of Euclid's *Elements [73]*. In this text, we state and use the fundamental theorem of arithmetic, leaving its proof for your later studies.

Theorem 3.1.2 Fundamental theorem of arithmetic *Every integer greater than 1 is either a prime or a product of primes; that is, every integer m can be written as*

$$m = p_1^{n_1} \cdot p_2^{n_2} \cdots p_k^{n_k}$$

where $p_1, p_2, \ldots, p_k \in \mathbb{Z}$ are prime numbers raised to positive integer powers $n_1, n_2, \ldots, n_k$. Furthermore, for a given integer, such a product of powers of primes is unique up to the order of the primes. We refer to such a product as the **prime power factorization** *of the integer m.*

Similar to the division algorithm, the fundamental theorem of arithmetic makes two distinct claims about every integer greater than one. First, the fundamental theorem of arithmetic is an *existence* result, guaranteeing that every integer greater than one can be expressed as a product of primes raised to powers. Second, the fundamental theorem of arithmetic is a *uniqueness* result, ensuring that every integer has exactly one such prime factorization up to order. As we will see, an integer's unique prime factors play a pivotal role in understanding and proving many insights into the properties of integers.

Example 3.1.4 We give the prime power factorizations of a few integers.

- $6 = 2 \cdot 3$
- $11 = 11$
- $1620 \;=\; 162 \cdot 10 \;=\; 2 \cdot 81 \cdot 10 \;=\; 2 \cdot 3^4 \cdot 2 \cdot 5 \;=\; 2^2 \cdot 3^4 \cdot 5$

■

Question 3.1.6 Find the prime power factorization of each integer.

(a) 30   (c) 12

(b) 5   (d) 27

■

Determining the prime power factorization of a relatively small integer is often straightforward. Number patterns help. For example, even numbers have a factor of two, and multiples of 10 have factors of two and five. Another simple pattern was introduced in section 2.6: if the sum of an integer's digits is divisible by nine, then the integer is divisible by nine (and so has a factor of 3^2). If no such pattern is apparent, then a factorization can be obtained by checking each integer up to $\sqrt{n}$ to find a divisor (if one exists); actually it is enough to check for divisibility by every prime number less than or equal to $\sqrt{n}$.

On the other hand, for sufficiently large integers, the prime power factorization can be extremely difficult to find. An exhaustive search for factors based on testing every integer (or prime) less than or equal to $\sqrt{n}$ can be extraordinarily time consuming and resource intensive. Computer scientists express this complexity by asserting that factoring integers requires nonpolynomial time computations; although Peter Shor has recently proven that a theoretical "quantum" computer is capable of polynomial time factorization of integers. The next example illustrates this complexity in producing prime power factorizations.

Example 3.1.5   We find the prime factorization of the integer 5,473,381,693.
Without sophisticated mathematical software, most people would have difficulty finding this integer's prime factorization. The square root of 5,473,381,693 is 73,983 (when rounded up), and so an exhaustive search may have to check for divisibility by every prime less than or equal to 73,983—which is a lot of dividing! Even the fastest of supercomputers can require a serious investment of time and space resources to factor this (and larger) integers. As it turns out, 5,473,381,693 can be factored by standard computer algebra systems as $13 \cdot 17^4 \cdot 71^2$. Interestingly enough, just a single change in the tens digit from 9 to 4 produces 5,473,381,643, which requires much more computer time to verify as prime.

■

When determining a prime power factorization by hand, it is often helpful to use intermediate steps and identify nonprime factors, which can in turn be factored. The following questions highlight the use of such intermediate steps.

Question 3.1.7   (a) Given that 6 divides 15,444,752,706, identify two primes that divide 15,444,752,706.
(b) Given that 6 divides an integer n, identify two primes that divide n.

■

Question 3.1.8   Determine the prime power factorization of 28,171,962,000 using direct computations. Hint: No prime greater than 13 divides this integer.

■

Computer algebra systems can be quite helpful when exploring divisibility and working with prime numbers. The following are useful commands from two widely used computer algebra systems (or CAS). The Maple function isprime(n) determines if n is a prime. The Mathematica function PrimepowerQ[n] determines if n is a power of a single prime.

CAS	Command	Example
Maple	[> m / n ;	[> 54 / 3
	[> ifactor(n) ;	[> ifactor(54) ;
	[> isprime(n) ;	[> isprime(54) ;
Mathematica	]: m / n	]: 54 / 3
	]: PrimeFactorList[n]	]: PrimeFactorList[54]
	]: PrimepowerQ[n]	]: PrimepowerQ[54]

We now shift attention from detailed computations with specific integers to more abstract, general questions about prime numbers. We know there exist infinitely many positive integers 1, 2, 3, …. But how many of these integers have the property of being prime? Could there be a "greatest" prime number, or are the primes unbounded in the set of integers (and so infinite)? As your intuition may suggest or as you've learned in other math courses, there are infinitely many distinct primes. In 300 B.C.E., Euclid gave an elegant proof of this result for Proposition 20 in Book IX of *Elements* [73]. To help motivate this proof, the following question considers integers obtained by adding one to a product of consecutive prime numbers.

Question 3.1.9 (a) Determine whether or not each integer is prime; if not, give a nontrivial divisor.

- $2 + 1$
- $2 \cdot 3 + 1$
- $2 \cdot 3 \cdot 5 + 1$
- $2 \cdot 3 \cdot 5 \cdot 7 + 1$

(b) Formulate a conjecture about the number $p_1 \cdot p_2 \cdots p_n + 1$ obtained by adding one to the product of the first n prime numbers.

(c) Find two primes greater than 50 that divide the number $2 \cdot 3 \cdot 5 \cdot 7 \cdot 11 \cdot 13 + 1 = 30,031$.

(d) If necessary, reformulate your conjecture from part (b). What can be said about the primes $p_1, \ldots, p_n$ (not) dividing $p_1 \cdot p_2 \cdots p_n + 1$?

■

Question 3.1.9 indicates that integers of the form $p_1 \cdot p_2 \cdots p_n + 1$ can be either prime or nonprime. This observation raises further questions. Do primes greater than 30,031 occur in the sequence of integers of this form? (Yes, they do—can you find one?) Is there something distinctive about the sixth prime 13 that leads to $p_1 \cdot p_2 \cdots p_6 + 1 = 30,031$ not being prime? How often do primes and nonprimes occur in this sequence? Mathematicians do not know if there exists an upper bound on the primes occurring in this sequence; in other words, it is an open question as to whether or not numbers of the form $p_1 \cdot p_2 \cdots p_n + 1$ are prime infinitely often.

Hopefully the insights gained from question 3.1.9 will help you understand and appreciate Euclid's proof. His argument is a classical proof by contradiction, assuming the negation of the desired result and working toward two mathematical statements that contradict each other.

Theorem 3.1.3 *There exist infinitely many prime numbers.*

Proof Assume that there are only finitely many prime numbers and that $p_1, \ldots, p_n$ is a complete list of these primes. This proof produces a "new" prime P that is not in the list, yielding a contradiction and leading to the conclusion that there are infinitely many primes.

Motivated by question 3.1.9, we define the desired integer as $P = p_1 \cdot p_2 \cdots p_n + 1$. Since P is greater than each of $p_1, \ldots, p_n$ in the complete list of primes, P is not prime. Therefore, by the fundamental theorem of arithmetic, P is a product of primes and so divisible by a prime. In particular, at least one of $p_1, \ldots, p_n$ must divide P, which in turn implies that one of $p_1, \ldots, p_n$ must divide

$$P \ - \ p_1 \cdot p_2 \cdots p_n \ = \ (p_1 \cdot p_2 \cdots p_n + 1) \ - \ p_1 \cdot p_2 \cdots p_n \ = \ 1.$$

However, the only positive divisor of 1 is 1, while every prime is greater than 1. Thus, none of $p_1, \ldots, p_n$ can divide 1. We have obtained the desired contradiction and conclude there exist infinitely many primes.

■

Since Euclid gave his proof of the infinitude of primes, many different and interesting proofs of this result have been given by various mathematicians, including one by the twentieth century Hungarian Paul Erdös. Most of Erdös's work was in discrete mathematics, particularly number theory and graph theory. In his lifetime, Erdös published more than 1,500 papers with at least 500 different coauthors. He liked to talk, in a jovial way, about "The Book" in which God had written a perfect proof for every mathematical theorem. *Proofs from THE BOOK* [3] is a recently published collection of theorems and proofs based on his suggestions and begins with six different proofs of the infinitude of primes, including Euclid's proof as well as proofs by Goldbach, Euler, and Erdös himself.

As you have perhaps surmised from our discussion in this section, there are many questions about prime numbers that remain open. We highlight three examples.

The Goldbach conjecture: In a 1742 letter to Leonhard Euler, the Russian mathematician Christian Goldbach conjectured that every even integer greater than two can be written as the sum of two primes. This claim is readily verified for small integers; for example, $4 = 2 + 2$, $6 = 3 + 3$, $8 = 3 + 5$, $10 = 5 + 5$, and so on. Despite the best efforts of many professional and amateur mathematicians, so far no proof of this conjecture has been pieced together. With the development of increasingly powerful and sophisticated supercomputers, the Goldbach conjecture has been verified for all even integers up to 12×10^{17} as of July 14, 2008. In addition, a number of "partial" Goldbach results have been proven, including independent proofs by Nikolai Chudakov, Theodor Estermann, and Johannes van der Corput in the 1930s that *"almost all" even numbers are the sum of two primes*, as well as Chen Jing-Run's proof in the 1960s that *every even number must be the sum of a prime and either a prime or a product of two primes* (such a product of two primes is known as a *semiprime*). While the Goldbach conjecture is widely believed to be true by mathematicians, evidence and intuition do not carry the same weight as a thorough, logical argument. And so mathematicians continue to seek a proof of the Goldbach conjecture in their quest for mathematical truth.

The twin primes conjecture: Pairs of prime numbers that differ by two are known as *twin primes*. For example, the first four pairs of twin primes are the primes 3 and 5, the primes 5 and 7, the primes 11 and 13, and the primes 17 and 19. The twin primes conjecture asserts that there are infinitely many pairs of twin primes. As with the Goldbach conjecture, the twin primes conjecture is generally believed to be true, but a complete proof continues to elude mathematicians.

The squares conjecture: The following examples help motivate this conjecture.

- $n = 1$: Consider $n^2 = 1$ and $(n + 1)^2 = 2^2 = 4$ and observe that 2 is a prime between 1 and 4.

- $n = 2$: Consider $n^2 = 4$ and $(n+1)^2 = 9$ and observe that 5 is a prime between 4 and 9.
- $n = 3$: Consider $n^2 = 9$ and $(n+1)^2 = 16$ and observe that 11 is a prime between 9 and 16.

The squares conjecture asserts that for every positive integer $n \in \mathbb{N}$, there exists a prime between n^2 and $(n+1)^2$. The squares conjecture is believed to be true, but mathematicians have not been able to prove the general result.

These three questions represent just a few of the many questions about primes that remain open. We hope your interests are piqued, and perhaps you will want to study such questions further. We end this section with a theorem stated in section 2.4. With the fundamental theorem of arithmetic in hand, we can now provide the proof.

Theorem 3.1.4 Theorem 2.4.2 in section 2.4 *For $a, b \in \mathbb{Z}$, if p is a prime factor of $a \cdot b$, then either p is a factor of a or p is a factor of b.*

Proof In proving a disjunction (an "or" statement), a standard strategy is to assume the hypothesis and the negation of one of the disjuncts, and then to argue for the truth of the other disjunct. The validity of this strategy is based on the logical equivalence of $p \rightarrow (q \vee r)$ and $(p \wedge \sim q) \rightarrow r$.

We assume $a, b \in \mathbb{Z}$, p is prime factor of $a \cdot b$, and p is not a factor of a. We show that p is a factor of b. Using the existence portion of the fundamental theorem of arithmetic, express $a \cdot b$, a, and b as the unique products of powers of primes (up to order)

$$a \cdot b = p_1^{n_1} \cdots p_i^{n_i}, \quad a = q_1^{m_1} \cdots q_j^{m_j}, \text{ and } \quad b = r_1^{l_1} \cdots r_k^{l_k},$$

where $p_1, \ldots, p_i, q_1, \ldots, q_j, r_1, \ldots, r_k$ are prime numbers and $n_1, \ldots, n_i$, $m_1, \ldots, m_j, l_1, \ldots, l_k$ are positive integers. Since p is a prime factor of $a \cdot b$, we know that p is one of $p_1, \ldots, p_i$ by the uniqueness of prime power factorizations. Without loss of generality, assume $p = p_1$, so that $a \cdot b = p^{n_1} \cdot p_2^{n_2} \cdots p_i^{n_i}$. Multiplying the prime power factorizations of a and b together, we also have $a \cdot b = q_1^{m_1} \cdots q_j^{m_j} \cdot r_1^{l_1} \cdots r_k^{l_k}$. Equating these two expressions for $a \cdot b$ produces

$$p^{n_1} \cdot p_2^{n_2} \cdots p_i^{n_i} = q_1^{m_1} \cdots q_j^{m_j} \cdot r_1^{l_1} \cdots r_k^{l_k}.$$

The uniqueness portion of the fundamental theorem of arithmetic implies that these primes and their powers are unique up to the order in which they appear. Since p is not a factor of a (by the assumption), p is not equal to any of $q_1, \ldots, q_j$. Therefore, p must be one of $r_1, \ldots, r_k$, and so p is a factor of $b = r_1^{l_1} \cdots r_k^{l_k}$. ∎

3.1.1 Reading Questions for Section 3.1

1. Define what is meant by the phrase "n divides m" and give an example.
2. What is the relationship between divides and modular equivalence?
3. Define and give an example of a prime number.
4. How many primes are there?
5. How many even primes are there?

6. State the fundamental theorem of arithmetic.
7. Discuss the nature of the two distinct claims made by the fundamental theorem of arithmetic.
8. What is the prime power factorization of an integer n? Determine the prime power factorization of $1,275$.
9. State and give an example for the Goldbach conjecture.
10. Define and give an example of a semiprime.
11. State and give an example for the twin primes conjecture.
12. State and give an example for the squares conjecture.

3.1.2   Exercises for Section 3.1

In exercises 1–4, verify each statement by finding the corresponding quotient q from the division algorithm.

1. 10 divides 30
2. 59 divides $7,729$

3. 34 is a factor of $2,414$
4. 23 is a factor of 161

In exercises 5–20, prove each mathematical statement for integers $m, n, k, p, a, b \in Z$.

5. The negative of a divisor of n is also a divisor of n.
6. The "divides" relation is reflexive; that is, m divides m.
7. The "divides" relation is transitive; that is, if m divides n and n divides k, then m divides k.
8. The "divides" relation is linear; that is, if m divides n and m divides k, then for every a, b, we have m divides $a \cdot n + b \cdot k$.
9. If m divides a and n divides b, then $m \cdot n$ divides $a \cdot b$.
10. If $m \cdot n$ divides k, then both m divides k and n divides k.
11. An integer m divides n if and only if $n \bmod m = 0$.
12. For every prime p and nonzero n, we have $n^2 \equiv n \bmod p$ if and only if either $n \equiv 1 \bmod p$ or $n \equiv 0 \bmod p$.
13. If 2 divides n, then 4 divides n^2.
14. If a prime p divides n, then p^2 divides n^2.
15. If a prime p divides n^2, then p divides n.   Note: This result is used in section 3.3.
16. If a prime p divides n^k, then p divides n.
17. If a prime p divides both m and n, then p^4 divides $m^4 - n^4$.
18. If no prime less than n divides n, then n is prime.
19. For any positive integer n, 3 divides $n^3 - n$.
20. The product of three consecutive integers is divisible by 6.

In exercises 21–25, disprove each *false* mathematical statement for $m, n, k, a, b \in Z$.

21. If m divides a and n divides b, then $m + n$ divides $a + b$.
22. If positive integers m and n both divide k, then $m \cdot n$ divides k. Hint: Consider $n = m^2$.

23. The "divides" relation is symmetric; that is, if m divides n, then n divides m.

24. The "divides" relation is asymmetric; that is, if m divides n, then n does not divide m.

25. The "divides" relation satisfies comparability; that is, for every m and n, m divides n, or m is equal to n, or n divides m.

In exercises 26–33, find the prime power factorization of each integer.

26. 1,045

27. 123

28. 61,600

29. 1,225

30. 2,103

31. 2,301

32. Every integer between 2 and 10 inclusive.

33. Every integer between 11 and 20 inclusive.

34. For a fixed, nonprime integer $n \in \mathbb{Z}$ with prime power factorization $n = p_1^{n_1} \cdot p_2^{n_2} \cdots p_k^{n_k}$, what is the largest possible integer that can appear in this factorization? Explain your answer.

Exercises 35–42 consider properties of *greatest common divisors*. A pair of integers has a greatest common divisor (gcd) (or factor) k when k is the greatest divisor of both. For example, 25 and 40 have a greatest common divisor 5 because 5 is a (common) divisor of both 25 and 40, and there is no common divisor of 25 and 40 greater than 5. In this case, we write $\gcd(25, 40) = 5$. In general, the primes shared by the prime power factorizations of two integers generate their greatest common divisor.

35. Find the prime power factorization of the integers 18 and 60.

36. Find all positive common divisors of 18 and 60. What is $\gcd(18, 60)$?

37. Determine $\gcd(12, 50)$.

38. Determine $\gcd(75, 100)$.

39. Determine $\gcd(31, 32)$.

40. Determine $\gcd(31, 62)$.

41. If p is prime and n is a positive integer, what are the two possible values of $\gcd(p, n)$?

42. Prove that if m and n are positive integers, then $[\gcd(m, n)]^2$ is a divisor of $m \cdot n$.

Exercises 43–46 consider greatest common divisors and linear combinations. The greatest common divisor $\gcd(m, n)$ is defined before Exercises 35–42. One consequence of the division algorithm is that for all positive integers m and n, there exist integers a, b such that $a \cdot m + b \cdot n = \gcd(m, n)$. We say that $\gcd(m, n)$ can be expressed as a *linear combination* of m and n.

In Exercises 43–46 find the greatest common divisor of each pair of integers and express this greatest common divisor as a linear combination of the two integers.

43. 3 and 8

44. 3 and 6

45. 12 and 16

46. 14 and 22

Exercises 47–52 consider relatively prime integers. Every pair of integers m and n have a common divisor of 1. When 1 is the greatest common divisor of m and n (that is, when $gcd(m, n) = 1$), we say that m and n are relatively prime.

In Exercises 47–52 determine if each pair of integers is relatively prime by finding their greatest common divisor.

47. 12 and 175

48. 31 and 67

49. 637 and 26,400

50. 164 and 25,83

51. 517 and 31,891

52. 517 and 51,183

In Exercise 53–58, prove each mathematical statement for $m, n, k, p \in \mathbb{Z}$. Relatively prime is defined before exercises 47–52.

53. A prime p is relatively prime to every integer $n < p$.

54. Positive integers n and n^2 are never relatively prime.

55. Positive integers n and $n + 1$ are always relatively prime.

56. If m is relatively prime to $n \cdot k$, then m is relatively prime to both n and k.

57. If m and n are relatively prime, then m^2 and n^2 are relatively prime.

58. Give a counterexample disproving the *false* assertion that "If m is relatively prime to $n \cdot p$, then $m \cdot n$ is relatively prime to p."

Exercises 59–63 consider a numerical approximation for the number of primes less than or equal to a given integer n, where $\pi(n)$ denotes this number. For example, $\pi(2) = 1$, $\pi(3) = 2$, $\pi(4) = 2$, and $\pi(5) = 3$. Mathematicians have long sought patterns and relations for primes—including the question of what percentage or ratio of integers are prime. The prime number theorem provides one answer, asserting the following limit.

$$\lim_{n \to \infty} \frac{\pi(n)}{\frac{n}{\ln(n)}} = 1$$

Exercises 59–63 consider numerical evidence supporting the *prime number theorem.*

59. Determine the value of $\pi(n)$ for every integer between 2 and 10 inclusive.

60. Determine the value of $\pi(n)$ for every integer between 11 and 20 inclusive.

61. Working with a table of primes (perhaps on the web), determine the value of $\pi(100)$ and $\pi(200)$.

62. Complete the following table. Does the resulting data support the assertion of the prime number theorem that

$$\lim_{n \to \infty} \frac{\pi(n)}{n/\ln(n)} = 1$$

n	$\pi(n)$	$\frac{n}{\ln(n)}$	$\frac{\pi(n)}{\frac{n}{\ln(n)}}$
10	4		
1,000	168		
100,000	9,592		
10,000,000	664,579		
1,000,000,000	50,847,534		

63. Complete the following table. Based on this data, what proportion of the integers less than or equal to n is also prime?

n	$\pi(n)$	$\frac{\pi(n)}{n}$	$\frac{1}{\ln(n)}$
10	4		
1,000	168		
100,000	9,592		
10,000,000	664,579		
1,000,000,000	50,847,534		

Exercises 64–70 consider questions related to Euclid's proof of the infinitude of the primes and the number-theoretic conjectures discussed at the end of this section.

64. What is the smallest composite positive integer of the form $p_1 \cdot p_2 \cdots p_n + 1$ with n greater than 1 and $p_1, p_2, \ldots, p_n$ distinct primes? Consider products of nonconsecutive primes such as $7 \cdot 13 + 1 = 92$.

65. Express every even number between 4 and 32 inclusive as a sum of two primes. For example, $12 = 5 + 7$. These sums verify the Goldbach conjecture up to 32.

66. Goldbach also made a conjecture about odd numbers and sums of primes: every odd positive integer greater than five is the sum of three primes. Verify this conjecture for every odd number between 7 and 31 inclusive.

67. State the first eight pairs of twin primes; this list begins with the pair $(3, 5)$.

68. Prove that an odd integer cannot be written as the sum of twin primes.

69. For every positive integer n between 2 and 20 inclusive, determine a prime between n^2 and $(n + 1)^2$. These primes verify the squares conjecture up to 20.

70. Mathematicians from Diophantus to Fermat thought that every positive integer can be expressed as the sum of four squares of integers. In the late 1700s, Lagrange gave the first rigorous proof of this result, based on work of Euler. Computationally verify this statement for every positive integer between 1 and 20 inclusive.

3.2 Application: Introduction to Coding Theory and Cryptography

This application of number theory involves prime numbers, modular arithmetic, and a bit of group theory in the context of sharing information. When two parties are communicating with one another, they (usually) seek an accurate exchange of information. In addition, the parties involved often have a strong interest in preserving the privacy of the shared information. Fortunately, mathematicians working in the fields of cryptography and coding theory have developed a variety of mathematical schemes that ensure both private and accurate communication.

These encryption schemes take a sensible string of characters and code them in some fashion so they appear to be garbled nonsense to everyone (except, hopefully, those who should be able to decode the message). Historically, nations and armed forces have invested significant resources in developing coding schemes, and (as you can readily imagine) such schemes have often provided entertaining subject matter for many a thrilling spy novel! The contemporary widespread use of the Internet for email, retail purchases, and other communication and financial transactions has also brought security concerns to the attention of people from various professions.

This section introduces two ingenious mathematical approaches to encoding and decoding messages. *RSA algorithms* are useful for preserving the privacy of transmitted information and are classic examples of a "public key" encryption scheme. Such schemes allow anyone to become a sender of a secure message, but permit only the publisher of the public key to decode the message. Public key encryption schemes are important; for example, they enable anyone to make purchases from on-line retailers' websites while preserving the privacy of the financial transaction. In general, public key encryption schemes rely on the relative ease of performing some mathematical operation coupled with the relative difficulty of undoing, or reversing, that operation. RSA codes rely on the ease of multiplying large prime numbers coupled with the difficulty of factoring large composite numbers. With the increasing sophistication of computers, RSA encryption has become a widely used algorithmic scheme for many aspects of electronic communication.

In contrast, *Hamming codes* address the issue of accurate communication. Hamming codes encrypt a message so the receiver can examine the transmitted result and determine if there was an error in the transmission. When an error occurs, Hamming codes enable the receiver to recover the original, correct message that had been intended for transmission. Coding schemes with the capability of detecting and fixing transmission errors are known as "error-correcting" codes. Hamming codes were the first coding schemes to incorporate error-detecting and error-correcting features and utilize a blend of matrix and modular arithmetic in the coding process.

3.2.1   RSA Cryptography

The algorithm embodied in RSA codes was developed by the cryptographers Ronald Rivest, Adi Shamir, and Leonard Adleman while working at the Massachusetts Institute of Technology in the late 1970s. The relatively recent declassification of Cold War era documents revealed that Clifford Cooks developed essentially the same scheme in 1973 at the British Government Communication Headquarters (a British intelligence agency). However, because Cooks' work was classified, the world at large first learned of this scheme from Rivest, Shamir, and Adleman in 1977. Their work with this encryption scheme was of such fundamental importance that they were jointly honored with the 2002 Turing Award by the Association for Computing Machinery (ACM). The Turing Award is given annually by the ACM for "contributions of lasting and major technical importance to computer science" and is the equivalent of the Nobel Prize for computer science.

The RSA coding scheme encodes characters of the alphabet by blending standard and modular arithmetic operations. We first identify each letter of the alphabet with a number; typically, we equate $A = 01, B = 02, \ldots, Z = 26$. The following chart may help your work with this correspondence.

A	B	C	D	E	F	G	H	I	J	K	L	M
01	02	03	04	05	06	07	08	09	10	11	12	13
N	O	P	Q	R	S	T	U	V	W	X	Y	Z
14	15	16	17	18	19	20	21	22	23	24	25	26

If we wish to encode more characters (perhaps the other symbols that appear on a standard keyboard), we appropriately extend this identification of symbols with positive integers. For the sake of human readability we use the vertical bar symbol (|) to separate codes for individual characters (for example when coding a word or a sentence).

Example 3.2.1   Using the identification $A = 01, B = 02, \ldots, Z = 26$ from above, we have

- "GO COLLEGE" identified with 07 | 15 | 03 | 15 | 12 | 12 | 05 | 07 | 05, and
- 16 | 08 | 15 | 14 | 05 | 08 | 15 | 13 | 05 identified with "PHONE HOME".

■

Question 3.2.1   State the alphabetic string and the numeric string identified with the following.

(a)   20 | 23 | 15 | 09 | 19 | 16 | 18 | 09 | 13 | 05   (b)   MATH IS FUN

■

Once the symbols in the alphabet have been correlated with numbers, select two prime numbers p and q. Almost any pair of primes works, and distinct choices of primes results in distinct RSA codes. The RSA code identifies the numbers $01, \ldots, 26$ with elements of $\mathbb{Z}_{p \cdot q}$, and so sufficiently large primes p and q are chosen to ensure $26 < p \cdot q$. In real-life applications, very large primes are chosen with the goal of producing a code that is difficult to break in any reasonable period of time.

Once the two primes p and q have been selected, one more choice is made: select a positive integer e less than $(p-1) \cdot (q-1)$ and relatively prime to $(p-1) \cdot (q-1)$. These two properties are exactly the defining features of the elements in $U[(p-1) \cdot (q-1)]$ (see section 2.4). The RSA code makes essential use of the multiplicative inverse of e in $U[(p-1) \cdot (q-1)]$; this choice of e ensures that e^{-1} exists.

How does an RSA code encode a letter? For $n = p \cdot q$, the "letter" L (which is really the corresponding number from $01, \ldots, 26$ identified with L) is encoded using the function

$$f(L) = L^e \bmod n.$$

Strings of letters are encoded one letter at a time.

How does an RSA code decode an encrypted letter? We first identify e^{-1} (the multiplicative inverse of e) in the group $U[(p-1)(q-1)]$, and then decode an encrypted letter $M = f(L)$ using the function

$$g(M) = M^{(e^{-1})} \bmod n.$$

Rivest, Shamir, and Adleman proved that $g(f(L)) = L$, and so g is the inverse of f under composition (also, $f(g(M)) = M$). In other words, this function g successfully decodes a letter encrypted by the function f given above.

Example 3.2.2 We use the RSA code with $p = 5$, $q = 7$, and $e = 5$ to encode and decode "GO COLLEGE".

For computational ease in this first example of an RSA code, we use relatively small primes p and q. For $p = 5$ and $q = 7$, we have $n = 35$ and $(p-1)(q-1) = 24$. There are many options for the choice of $e \in U(24)$. We chose the minimum nonidentity element $e = 5$ of $U(24)$ in creating this example, but any element of $U(24)$ will work when implementing this algorithm. Recall that the letters $A, \ldots, Z$ are identified with the numbers $01, \ldots, 26$; these numbers are less than $n = 35$ and are thought of as elements of $\mathbb{Z}_{35}$.

Encoding: We encode the message "GO COLLEGE" using the RSA code with $p = 5, q = 7$, $n = 35$, and $e = 5$. As in example 3.2.1, we identify this message with the list of numbers 07 | 15 | 03 | 15 | 12 | 12 | 05 | 07 | 05. For this RSA code, each two-digit number L in this list is encoded using the function: $f(L) = L^5 \bmod 35$. Using a calculator or computer, we obtain the following.

- $f(07) = 07^5 \bmod 35 = 16,807 \bmod 35 = (480 \cdot 35 + 07) \bmod 35 = 07$
- $f(15) = 15^5 \bmod 35 = 15$
- $f(03) = 03^5 \bmod 35 = 33$

Continuing in this fashion, the encoded message is

07 | 15 | 33 | 15 | 17 | 17 | 10 | 07 | 10.

Decoding: We now imagine this string of numbers 07 | 15 | 33 | 15 | 17 | 17 | 10 | 07 | 10 is transmitted and the receiver of the message is interested in decoding this string to obtain the original message. For the RSA code, each two-digit number in the message is decoded using the function: $g(M) = M^{(5^{-1})} \bmod 35$, where 5^{-1} is the multiplicative inverse of 5 in $U(24)$. Since $5 \cdot 5 = 25$ and $25 \bmod 24 = 1$, we have $5^{-1} = 5$. Using a calculator or computer algebra system, we obtain the following.

- $g(07) = 07^{(5^{-1})} \bmod 35 = 07^5 \bmod 35 = 07$
- $g(15) = 15^{(5^{-1})} \bmod 35 = 15^5 \bmod 35 = 15$
- $g(03) = 33^{(5^{-1})} \bmod 35 = 33^5 \bmod 35 = 03$

Continuing in this fashion, the decoded message is

07 | 15 | 03 | 15 | 12 | 12 | 05 | 07 | 05   or   "GO COLLEGE".

■

Question 3.2.2   Implement the RSA code with $p = 5$, $q = 7$, and $e = 5$. Recall that $e^{-1} = 5$.

    (a) Identify "PHONE HOME" with its corresponding list of two-digit numbers.

    (b) Encode the message "PHONE HOME".

    (c) Decode the message 06 | 10 | 23 | 13 | 01 | 20.

■

The algorithm for RSA cryptosystems is a classical example of a *public key encryption scheme*. When using a public key cryptography, the person interested in receiving a message sets up a *key pair* consisting of a *public key* and a *private key*. The public key is used for encoding messages and is announced widely, enabling anyone to code and transmit an encrypted message. The private key is used for decoding messages and is kept secret so that only the creator of the key can decode the transmitted message, even if the means of communication is vulnerable to eavesdropping.

When using RSA codes for public key cryptography, the person interested in receiving a message publishes the two integers $n = p \cdot q$ and e, and only these two integers. The numbers n and e serve as the public key, and anyone familiar with RSA codes can readily encode a message (using the function $f(L) = L^e \bmod n$) and transmit the result. The integers p, q and e^{-1} serve as the private key; only the publisher of the public key knows e^{-1} and can decode the encrypted message (using the function $g(M) = M^{(e^{-1})} \bmod n$).

If the primes p and q are kept secret and are sufficiently large, only the publisher of the key is able to decode the encrypted message. In order to determine e^{-1} as an element in $U[(p - 1)(q - 1)]$, the numbers $p - 1$ and $q - 1$ generally must be known; that is, the primes p and q must be known. Even though $n = p \cdot q$ is published publicly, the available algorithms for factoring large composite integers and the current state of computing technology render the determination of the primes p and q from n effectively impossible. In formal computer science terminology, integer factorizations require "nonpolynomial time" computations. In practical terms, if primes p and q are sufficiently large so that their product yields, say, a 100-digit number, then factoring p and q can require decades of time—which would provide no benefit to those seeking to eavesdrop. This difficulty in factoring integers makes RSA codes secure—even though the public keys for encoding strings are widely available; and so, if the corresponding private keys for decoding strings are kept secret, then the privacy of the transmitted message is ensured. In practical applications, composite numbers with more than 100 digits are routinely used for RSA codes.

As might be expected from the RSA codes' security features, federal agencies invested in national security are strongly interested in knowing large primes and in keeping these values secret in order to code and decode messages. As of November, 2008 there are 46 known Mersenne primes with the 46th equal to $2^{43,112,609} - 1$ which has 12,978,189 digits, identified through the collective efforts of hundreds of people participating in the Great Internet Mersenne Prime Search. This group continues in its efforts to identify large primes—perhaps you might be interested in joining in their ongoing work (see the Index of Online Resources).

Because of the scale of the numbers and computations involved, computers and computer algebra systems are essential to working with RSA codes in most practical settings. The following are some useful commands for two widely used computer algebra systems.

CAS	Command	Example
Maple	[> p * q ; [> L^e mod n ; [> e^(−1) mod [(p−1) * (q−1)] ;	[> 71 * 73 ; [> 18^11 mod 5183 ; [> 11^(−1) mod 5040 ;
Mathematica	]: p * q]: Mod[L^e , n]]: Mod[e^(−1) , (p−1) * (q−1)]	]: 71 * 73]: Mod[18^11 , 5183]]: Mod[11^(−1) , 5040]

Example 3.2.3 We use the RSA code with $p = 71$, $q = 73$, and $e = 11$ to encode and decode the message "RSA".

Encoding We first compute $n = p \cdot q = 71 \cdot 73 = 5183$ and identify the message "RSA" with the sequence of numbers 18 | 19 | 01. Each two-digit number L in this list is encoded using the function $f(L) = L^e \bmod n = L^{11} \bmod 5183$. Using a computer algebra system, each character is encoded as follows.

- $f(18) = 18^{11} \bmod 5183 = 4713$
- $f(19) = 19^{11} \bmod 5183 = 3685$
- $f(01) = 01^{11} \bmod 5183 = 1$

Since the computations use mod 5183 arithmetic, the resulting values appear as up to four-digit numbers. For the sake of uniformity, every encoded number is presented with the same number of digits; the encoded message is 4713 | 3685 | 0001.

Decoding We now imagine the string of numbers 4713 | 3685 | 0001 is transmitted, and the receiver decodes the message using the function $g(M) = g(M) = M^{(e^{-1})} \bmod n = M^{(11^{-1})} \bmod 5183$. Here 11^{-1} is the multiplicative inverse of 11 in

$$U[(p - 1) \cdot (q - 1)] = U(5040);$$

using a computer algebra system, this inverse is $11^{-1} = 2291$ because $(11 \cdot 2291) \bmod 5040 = 1$. Using a computer algebra system, each character is decoded as follows.

- $g(07) = 4713^{(11^{-1})} \bmod 5183 = 4713^{2291} \bmod 5183 = 18 = \text{R}$
- $g(15) = 3685^{(11^{-1})} \bmod 5183 = 3685^{2291} \bmod 5183 = 19 = \text{S}$
- $g(03) = 0001^{(11^{-1})} \bmod 5183 = 01 = \text{A}$

The decoded message is 18 | 19 | 01, or "RSA". ∎

Question 3.2.3 Implement the RSA code with $p = 31$, $q = 53$, and $e = 223$.

(a) Encode the message "I LOVE MATH".

(b) Decode the message 1188 | 0666 | 1391 | 0979 | 1502 | 0098 | 0850 | 0098 | 0586.

3.2.2   Hamming Codes

In 1950, the American mathematician Richard Hamming developed Hamming codes. For his work, Hamming received the 1968 Turing Award from the Association for Computing Machinery. Throughout his life, Hamming made many important contributions to coding theory, number theory, and numerical analysis; and so, in 1998, the Institute of Electrical and Electronics Engineers created an annual prize named the Hamming Medal for "exceptional contributions to information sciences, systems, and technology." Coding theory remains a rich and diverse area of mathematical research, and much of this work relies on a deep understanding of the notions of number theory and abstract algebra.

Hamming developed his error-correcting codes while working for Bell Laboratories. In the context of telecommunications, messages are transmitted in binary; that is, as strings of 0's and 1's. Such binary codes are the easiest to correct—just knowing the position of an error allows for its immediate correction by switching the digit from a 0 to 1, or vice-versa.

Hamming codes are not concerned with the security issues that were the primary focus of RSA encryption. Instead, these codes focus on the accuracy of transmitted information; once a message is received, how can we ensure that it was the message originally sent? This question may bring to mind the study of check digits in section 2.6, which addressed human error and breakdowns in physical communication devices introducing errors into messages. While check digits determined the accuracy of transmitted information, Hamming codes both determine if any single-digit error has occurred and the position of the error (and hence the correct binary digit for that position). Hamming codes can also detect double errors (occurring in two positions), but cannot automatically correct them.

Hamming codes append several check digits to the end of a message. These check digits are computed using matrix multiplication of the message (written as a row vector) by a *generating matrix*. Once the message has been transmitted, the receiver can then multiply by a second *parity check matrix*, where the result indicates if and where a single-digit error has occurred. Hamming codes are defined using matrix multiplication, and so our study begins with a description of matrices and matrix multiplication.

Example 3.2.4   An $m \times n$ *matrix* is an array of numbers with m rows and n columns. For the following, A is a 2×3 matrix, B is a 2×1 matrix (also called a *column vector*), and C is a 1×3 matrix (also called a *row vector*).

$$A = \begin{bmatrix} 1 & 2 & 3 \\ 4 & 5 & 6 \end{bmatrix} \qquad B = \begin{bmatrix} 1 \\ 4 \end{bmatrix} \qquad C = \begin{bmatrix} 2 & 3 & 4 \end{bmatrix}$$

■

The multiplication of 2×2 matrices was used in exercises 54–64 from section 2.5. For an arbitrary product, matrix multiplication is best studied after first learning how to multiply a row vector by a column vector of the same length.

Definition 3.2.1   *The product of a row vector $A = \begin{bmatrix} a_1 & a_2 & \cdots & a_n \end{bmatrix}$ with a column vector of equal length*

$$B = \begin{bmatrix} b_1 \\ b_2 \\ \vdots \\ b_n \end{bmatrix}$$

is the real number obtained by taking the sum of the products of the corresponding components: $a_1 \cdot b_1 + a_2 \cdot b_2 + \cdots + a_n \cdot b_n$. We write

$$\begin{bmatrix} a_1 & a_2 & \cdots & a_n \end{bmatrix} \cdot \begin{bmatrix} b_1 \\ b_2 \\ \vdots \\ b_n \end{bmatrix} = a_1 \cdot b_1 + a_2 \cdot b_2 + \cdots + a_n \cdot b_n = \sum_{i=1}^{n} a_i \cdot b_i.$$

Example 3.2.5   Using definition 3.2.1, we multiply

$$A = \begin{bmatrix} 1 & 5 & 2 \end{bmatrix}$$

and

$$B = \begin{bmatrix} 4 \\ 3 \\ 6 \end{bmatrix}.$$

These vectors have the same length $n = 3$ and can be multiplied together. First take the product of the corresponding components $1 \cdot 4 = 4$, $5 \cdot 3 = 15$, and $2 \cdot 6 = 12$ and then add these three products together to obtain $4 + 15 + 12 = 31$. Thus, we have

$$\begin{bmatrix} 1 & 5 & 2 \end{bmatrix} \cdot \begin{bmatrix} 4 \\ 3 \\ 6 \end{bmatrix} = 4 + 15 + 12 = 31.$$

       ■

Question 3.2.4   Using definition 3.2.1 of vector multiplication, answer the following.

(a) Compute each product of row and column vectors.

$$\bullet \ \begin{bmatrix} 2 & -10 \end{bmatrix} \cdot \begin{bmatrix} 13 \\ 1 \end{bmatrix} \qquad\qquad \bullet \ \begin{bmatrix} 2 & 8 & 10 \end{bmatrix} \cdot \begin{bmatrix} 3 \\ 5 \\ 1 \end{bmatrix}$$

(b) Explain why the product $\begin{bmatrix} 2 & 8 & 10 \end{bmatrix} \cdot \begin{bmatrix} 13 \\ 1 \end{bmatrix}$ is undefined.

       ■

     Vector multiplication is one step in the more general process of computing the product $A \cdot B$ of an $m \times n$ matrix A and an $n \times p$ matrix B. Such a product results in an $m \times p$ matrix, where the entry in the jth row and kth column is the vector product of the jth row of A with the kth column of B. The following definition describes this process.

Definition 3.2.2 *Let A be an m × n matrix and B be an n × p matrix with entries labeled as follows:*

$$A = \begin{bmatrix} a_{11} & a_{12} & \cdots & a_{1n} \\ a_{21} & a_{22} & \cdots & a_{2n} \\ \vdots & \vdots & & \vdots \\ a_{m1} & a_{m2} & \cdots & a_{mn} \end{bmatrix} \qquad B = \begin{bmatrix} b_{11} & b_{12} & \cdots & b_{1p} \\ b_{21} & b_{22} & \cdots & b_{2p} \\ \vdots & \vdots & & \vdots \\ b_{n1} & b_{n2} & \cdots & b_{np} \end{bmatrix}$$

The product $C = A \cdot B = AB$ is an m × p matrix where the entry c_{jk} in the jth row and kth column of C is the vector product of the jth row of A with with kth column of B; that is,

$$c_{jk} = \begin{bmatrix} a_{j1} & a_{j2} & \cdots & a_{jn} \end{bmatrix} \cdot \begin{bmatrix} b_{1k} \\ b_{2k} \\ \vdots \\ b_{nk} \end{bmatrix} = a_{j1} \cdot b_{1k} + a_{j2} \cdot b_{2k} + \cdots + a_{jn} \cdot b_{nk}.$$

Example 3.2.6 We use definition 3.2.2 to compute the product $C = A \cdot B$ of the 2 × 2 matrices

$$A = \begin{bmatrix} 1 & 2 \\ 3 & 4 \end{bmatrix}$$

and

$$B = \begin{bmatrix} 5 & 6 \\ 7 & 8 \end{bmatrix}.$$

- The entry c_{11} is the product of the first row of A with the first column of B.

$$c_{11} = \begin{bmatrix} 1 & 2 \end{bmatrix} \cdot \begin{bmatrix} 5 \\ 7 \end{bmatrix} = 1 \cdot 5 + 2 \cdot 7 = 19$$

- The entry c_{12} is the product of the first row of A with the second column of B.

$$c_{12} = \begin{bmatrix} 1 & 2 \end{bmatrix} \cdot \begin{bmatrix} 6 \\ 8 \end{bmatrix} = 1 \cdot 6 + 2 \cdot 8 = 22$$

- $c_{21} = \begin{bmatrix} 3 & 4 \end{bmatrix} \cdot \begin{bmatrix} 5 \\ 7 \end{bmatrix} = 3 \cdot 5 + 4 \cdot 7 = 43$

- $c_{22} = \begin{bmatrix} 3 & 4 \end{bmatrix} \cdot \begin{bmatrix} 6 \\ 8 \end{bmatrix} = 3 \cdot 6 + 4 \cdot 8 = 50$

Therefore, we have

$$C = A \cdot B = \begin{bmatrix} 1 & 2 \\ 3 & 4 \end{bmatrix} \begin{bmatrix} 5 & 6 \\ 7 & 8 \end{bmatrix} = \begin{bmatrix} 19 & 22 \\ 43 & 50 \end{bmatrix}.$$

■

Example 3.2.7 We compute the product of a 1 × 2 row vector and a 2 × 3 matrix:

$$\begin{bmatrix} 1 & 2 \end{bmatrix} \cdot \begin{bmatrix} 3 & 4 & 5 \\ 6 & 7 & 8 \end{bmatrix} = \begin{bmatrix} (1 \cdot 3 + 2 \cdot 6) & (1 \cdot 4 + 2 \cdot 7) & (1 \cdot 5 + 2 \cdot 8) \end{bmatrix}$$
$$= \begin{bmatrix} 15 & 18 & 21 \end{bmatrix}.$$

■

A matrix product $A \cdot B$ is only defined when the number of columns in the left matrix A is equal to the number of rows in the right matrix B. If these numbers differ, the corresponding vector products are undefined, and so the matrix product is undefined.

Example 3.2.8   The following product is undefined because the left matrix only has two columns, while the right matrix has three rows, and so we are unable to compute any of the corresponding vector products.

$$\begin{bmatrix} -1 & 12 \\ 7 & 3 \end{bmatrix} \cdot \begin{bmatrix} 15 & -6 & 8 \\ 1 & 1 & 2 \\ 0 & 1 & 0 \end{bmatrix} \text{ is undefined.}$$

∎

Question 3.2.5   Using definition 3.2.2 of vector multiplication, answer the following.

(a) Compute each product of matrices.

$$\bullet \begin{bmatrix} -1 & 10 & 3 \\ 0 & 4 & 1 \end{bmatrix} \cdot \begin{bmatrix} 3 & -2 \\ 12 & 5 \\ 2 & -2 \end{bmatrix}$$

$$\bullet \begin{bmatrix} -10 & 1 & 0 \end{bmatrix} \cdot \begin{bmatrix} 1 & 0 & 3 \\ -1 & 2 & 0 \\ 0 & 10 & -1 \end{bmatrix}$$

(b) Explain why the product $\begin{bmatrix} 0 & 1 \end{bmatrix} \cdot \begin{bmatrix} -1 & 10 \\ -12 & 1 \\ 0 & 0 \end{bmatrix}$ is undefined.

∎

As with RSA codes, there are many different Hamming codes. A Hamming code encodes "letters" that have been identified as row vectors of some uniform length r. Hamming codes are defined on vectors containing binary numbers, and so every entry in these row vectors is either 0 or 1. Row vectors with r entries may represent up to 2^r distinct "letters." For example, there are $2^4 = 16$ distinct row vectors available for coding letters when using vectors of length $r = 4$.

$$\begin{bmatrix} 0 & 0 & 0 & 0 \end{bmatrix} \quad \begin{bmatrix} 1 & 0 & 0 & 0 \end{bmatrix} \quad \begin{bmatrix} 0 & 1 & 0 & 0 \end{bmatrix} \quad \begin{bmatrix} 0 & 0 & 1 & 0 \end{bmatrix}$$

$$\begin{bmatrix} 0 & 0 & 0 & 1 \end{bmatrix} \quad \begin{bmatrix} 1 & 1 & 0 & 0 \end{bmatrix} \quad \begin{bmatrix} 1 & 0 & 1 & 0 \end{bmatrix} \quad \begin{bmatrix} 1 & 0 & 0 & 1 \end{bmatrix}$$

$$\begin{bmatrix} 0 & 1 & 1 & 0 \end{bmatrix} \quad \begin{bmatrix} 0 & 1 & 0 & 1 \end{bmatrix} \quad \begin{bmatrix} 0 & 0 & 1 & 1 \end{bmatrix} \quad \begin{bmatrix} 1 & 1 & 1 & 0 \end{bmatrix}$$

$$\begin{bmatrix} 1 & 1 & 0 & 1 \end{bmatrix} \quad \begin{bmatrix} 1 & 0 & 1 & 1 \end{bmatrix} \quad \begin{bmatrix} 0 & 1 & 1 & 1 \end{bmatrix} \quad \begin{bmatrix} 1 & 1 & 1 & 1 \end{bmatrix}$$

Hamming codes encrypt these "letters" by multiplying the corresponding row vectors by a *generating matrix* that is specific to each Hamming code. The entries of the generating matrix are 0's and 1's, and vector and matrix products are computed using modulo 2 arithmetic (since the code uses only binary vectors and matrices).

The generating matrix consists of two distinct components. The left component is the
$r \times r$ *identity matrix* with 1's on the main diagonal and 0's in all other entries. The
right component consists of additional *parity bit columns* that are cleverly selected
to produce the error-correction capability of this coding scheme. After multiplying
a given row vector by the generating matrix, the corresponding encoded row vector
consists of the original vector with multiple check digits appended. These encoded row
vectors are then transmitted.

The recipient can then check the received encoded row vectors for single-digit
errors in the message. The recipient multiplies the received vectors by a *parity
check matrix*, which is a variation on the generating matrix as defined below and
in the Exercises at the end of this section. When the product is the zero vector, the
Hamming code indicates that no single-digit errors have occurred. If the product is a
nonzero vector, the resulting vector identifies the location of any single-digit errors,
enabling their correction. With this general approach in mind, we study a specific
Hamming code.

3.2.3   The (7, 4) Hamming Code

The rest of this section details the well-known *(7,4) Hamming code*, which Hamming
first identified in 1950. In this setting, the row vectors representing "letters" all
have uniform length $r = 4$. The generating matrix for the $(7, 4)$ Hamming code
follows.

$$G = \begin{bmatrix} 1 & 0 & 0 & 0 & 1 & 1 & 1 \\ 0 & 1 & 0 & 0 & 1 & 1 & 0 \\ 0 & 0 & 1 & 0 & 1 & 0 & 1 \\ 0 & 0 & 0 & 1 & 0 & 1 & 1 \end{bmatrix}$$

Notice that the left four columns of this matrix form the 4×4 identity matrix (with
1's on the main diagonal and 0's elsewhere), and the right three columns consist of
the parity bits. Exercises 60–65 at the end of this section discuss the definition of an
arbitrary Hamming code's generating matrix.

Example 3.2.9   We use the generating matrix G for the $(7, 4)$ Hamming code to encode the
message "PI."

We first identify the message "PI" with the pair of integers $16 \mid 9$. We then
determine the binary representation of these integers; since $16 = 1 \cdot 2^3 + 1 \cdot
2^2 + 1 \cdot 2^1 + 1 \cdot 2^0$, and $9 = 1 \cdot 2^3 + 0 \cdot 2^2 + 0 \cdot 2^1 + 1 \cdot 2^0$, we identify
"PI" as $\begin{bmatrix} 1 & 1 & 1 & 1 \end{bmatrix} \begin{bmatrix} 1 & 0 & 0 & 1 \end{bmatrix}$. We now encode the message by
multiplying these row vectors by the generating matrix (using modulo 2 arithmetic)
as follows.

$$\begin{bmatrix} 1 & 1 & 1 & 1 \end{bmatrix} \cdot \begin{bmatrix} 1 & 0 & 0 & 0 & 1 & 1 & 1 \\ 0 & 1 & 0 & 0 & 1 & 1 & 0 \\ 0 & 0 & 1 & 0 & 1 & 0 & 1 \\ 0 & 0 & 0 & 1 & 0 & 1 & 1 \end{bmatrix} = \begin{bmatrix} 1 & 1 & 1 & 1 & 1 & 1 & 1 \end{bmatrix}$$

$$\begin{bmatrix} 1 & 0 & 0 & 1 \end{bmatrix} \cdot \begin{bmatrix} 1 & 0 & 0 & 0 & 1 & 1 & 1 \\ 0 & 1 & 0 & 0 & 1 & 1 & 0 \\ 0 & 0 & 1 & 0 & 1 & 0 & 1 \\ 0 & 0 & 0 & 1 & 0 & 1 & 1 \end{bmatrix} = \begin{bmatrix} 1 & 0 & 0 & 1 & 1 & 0 & 0 \end{bmatrix}$$

For example, the fifth entry in the first product is 1 because $(1 \cdot 1 + 1 \cdot 1 + 1 \cdot 1 + 1 \cdot 0) \bmod 2 = 1$. The code now transmits the following encoded version of the original message.

$$\begin{bmatrix} 1 & 1 & 1 & 1 & 1 & 1 & 1 \end{bmatrix} \begin{bmatrix} 1 & 0 & 0 & 1 & 1 & 0 & 0 \end{bmatrix}$$

■

Question 3.2.6   Using the $(7, 4)$ Hamming code, compute the encoded version of the following message.

$$\begin{bmatrix} 0 & 1 & 1 & 1 \end{bmatrix} \begin{bmatrix} 0 & 0 & 1 & 1 \end{bmatrix} \begin{bmatrix} 1 & 0 & 1 & 0 \end{bmatrix}$$

■

Comparing the original vectors with the corresponding encoded vectors in example 3.2.9 and question 3.2.6, we observe that the original vector appears as the first part of its encoded vector. Thus, the Hamming codes do not provide any measure of security against eavesdropping. Instead, Hamming codes are useful because they detect and correct any single-digit errors in the transmitted vectors. For example, suppose someone receives a message $\begin{bmatrix} 1 & 1 & 1 & 1 & 1 & 0 & 1 \end{bmatrix}$ encoded using the $(7, 4)$ Hamming coding scheme. This vector does not match any of the 16 possible correctly encoded vectors (as you can verify by multiplying each of the 16 possible binary vectors of length four by the generating matrix). Therefore, some error has occurred. If a transmission error occurred in at most one digit (a single-digit error), the Hamming code determines the correct, original message.

Single-digit errors are detected by multiplying the received row vector by the Hamming code's *parity check matrix*. The parity check matrix P is formed by attaching the $(r - 1) \times (r - 1)$ identity matrix (with all 1's on the main diagonal and 0's elsewhere) to the "bottom" of the parity bit columns from the generating matrix G. The $(7, 4)$ Hamming code has the following generating matrix G and parity check matrix P; for visual emphasis, the common parity bit portions of these matrices are printed in bold.

$$G = \begin{bmatrix} 1 & 0 & 0 & 0 & \mathbf{1} & \mathbf{1} & \mathbf{1} \\ 0 & 1 & 0 & 0 & \mathbf{1} & \mathbf{1} & \mathbf{0} \\ 0 & 0 & 1 & 0 & \mathbf{1} & \mathbf{0} & \mathbf{1} \\ 0 & 0 & 0 & 1 & \mathbf{0} & \mathbf{1} & \mathbf{1} \end{bmatrix} \qquad P = \begin{bmatrix} \mathbf{1} & \mathbf{1} & \mathbf{1} \\ \mathbf{1} & \mathbf{1} & \mathbf{0} \\ \mathbf{1} & \mathbf{0} & \mathbf{1} \\ \mathbf{0} & \mathbf{1} & \mathbf{1} \\ 1 & 0 & 0 \\ 0 & 1 & 0 \\ 0 & 0 & 1 \end{bmatrix}$$

If no single-digit error has occurred, then the product of the encoded vectors and the corresponding parity check matrix P results in the zero row vector (the row vector with 0 in every entry). On the other hand, if a single-digit error has occurred, then the product is one of the row vectors from P. The position of this row vector in P indicates the position of the single-digit error in the received encoded vector. The error is corrected by switching the parity of the entry (substituting 1 for 0 or 0 for 1).

Example 3.2.10 We check the following two encoded row vectors for single-digit errors using the parity check matrix P for the $(7, 4)$ Hamming code. Recall that all arithmetic is performed modulo 2.

- We check $\begin{bmatrix} 1 & 1 & 1 & 1 & 1 & 0 & 1 \end{bmatrix}$. Multiplying by P produces

$$\begin{bmatrix} 1 & 1 & 1 & 1 & 1 & 0 & 1 \end{bmatrix} \cdot \begin{bmatrix} 1 & 1 & 1 \\ 1 & 1 & 0 \\ 1 & 0 & 1 \\ 0 & 1 & 1 \\ 1 & 0 & 0 \\ 0 & 1 & 0 \\ 0 & 0 & 1 \end{bmatrix} = \begin{bmatrix} 0 & 1 & 0 \end{bmatrix}.$$

The result is not the zero vector, and so the Hamming code indicates that a single-digit error has occurred. Since the resulting vector $\begin{bmatrix} 0 & 1 & 0 \end{bmatrix}$ matches the sixth row of P, the error occurred in the sixth entry of the encoded vector. Reversing the parity of the sixth digit, the corrected version of the encoded vector is $\begin{bmatrix} 1 & 1 & 1 & 1 & 1 & 1 & 1 \end{bmatrix}$.

- We check $\begin{bmatrix} 0 & 1 & 0 & 1 & 1 & 0 & 1 \end{bmatrix}$. Multiplying by P produces

$$\begin{bmatrix} 0 & 1 & 0 & 1 & 1 & 0 & 1 \end{bmatrix} \cdot P = \begin{bmatrix} 0 & 0 & 0 \end{bmatrix}.$$

The resulting zero vector indicates that no single-digit error has occurred. ■

Question 3.2.7 Using the parity check matrix for the $(7, 4)$ Hamming coding scheme, identify the single-digit errors in the following received vectors and state the corrected version of the encoded vector.

(a) $\begin{bmatrix} 1 & 1 & 1 & 1 & 1 & 0 & 0 \end{bmatrix}$

(b) $\begin{bmatrix} 0 & 1 & 1 & 1 & 0 & 0 & 1 \end{bmatrix}$

■

3.2.4 Reading Questions for Section 3.2

1. Describe public key encryption.
2. How do we represent letters in the RSA encryption scheme?
3. How does an RSA code encode a letter?
4. How does an RSA code decode an encrypted letter?
5. What are the public and private keys for an RSA code?
6. What step in the RSA encryption process prevents a person who does not know the factorization of n into $n = p \cdot q$ from decoding an encrypted message?
7. Why are financial institutions and national security agencies interested in large primes?

8. Describe the process of vector multiplication and give an example.
9. Describe the process of matrix multiplication and give an example.
10. How do we represent letters when using a Hamming code?
11. How does a Hamming code encode a letter?
12. What kind of errors can a Hamming code both detect and correct?

3.2.5 Exercises for Section 3.2

In exercises 1–4, find the numeric string identified with each alphabetic string.

1. ALGEBRA
2. ANALYSIS

3. PEACE
4. TRUE LOVE

In exercises 5–11, find the alphabetic string identified with each numeric string.

5. 05 | 21 | 12 | 05 | 18
6. 07 | 01 | 21 | 19 | 19
7. 14 | 05 | 23 | 20 | 15 | 14
8. 12 | 05 | 09 | 02 | 14 | 09 | 26
9. 19 | 16 | 05 | 01 | 11 | 20 | 18 | 21 | 20 | 08
10. 01 | 18 | 03 | 08 | 09 | 13 | 05 | 04 | 05 | 19
11. 23 | 01 | 12 | 11 | 23 | 09 | 20 | 08 | 15 | 21 | 20 | 02 | 12 | 01 | 13 | 05

In exercises 12–15, identify how many digits the RSA code uses to represent encoded letters for the following values of $n = p \cdot q$.

12. $n = 33$
13. $n = 143$

14. $n = 1,919$
15. $n = 1,2533$

In exercises 16–19, encode each message using the RSA code with $p = 3$, $q = 11$, and $e = 7$.

16. ALGEBRA
17. ANALYSIS

18. PEACE
19. TRUE LOVE

In exercises 20–24, decode each encoded message using the RSA code with $p = 5$, $q = 13$, $e = 29$, and $e^{-1} = 5$.

20. 30 | 45 | 25
21. 08 | 45 | 61 | 05
22. 41 | 01 | 29 | 50 | 08

23. 48 | 08 | 01 | 18 | 29 | 50 | 25
24. 30 | 21 | 54 | 50 | 29 | 48 | 05

Exercises 25–28, encode each message using the RSA code with $p = 73$, $q = 103$, and $e = 2543$.

25. ALGEBRA
26. ANALYSIS

27. PEACE
28. TRUE LOVE

Exercises 29–32, decode each encoded message using the RSA code with $p = 73$, $q = 103$, and $e = 2543$.

29. 5758 | 2429 | 1318 | 4221
30. 1276 | 0001 | 4299 | 4208 | 5758

29. 3745 | 3949 | 4733 | 4208 | 4299 | 4502 | 4221
30. 4502 | 5758 | 0001 | 0036 | 4299 | 4208 | 4330

Exercises 33–37 consider the relative security of the RSA codes for various values of n.

33. Factor n into primes p and q when $n = 143$ and $(p - 1)(q - 1) = 120$.
34. Discuss the public encryption security of the RSA code that publishes $n = 143$.
35. Factor n into primes p and q if $n = 12,533$ and $(p - 1)(q - 1) = 12,300$.
36. Discuss the public encryption security of the RSA code that publishes $n = 12,533$.
37. Try to stump one of your classmates with an RSA encrypted message. Choose a large (at least four digits) integer value of n that factors into primes $n = p \cdot q$ and $e \in U[(p - 1)(q - 1)]$. Implement the RSA coding scheme with this p, q, and e to encode a three letter message (e.g., "RSA" or "ACE"). Publicly announce your values for n and e (but not p and q) and see how long it takes for your encrypted message to be deciphered. Are you satisfied with the security of your code?

Exercises 38–39 consider a type of prime number named for the French monk and mathematician Marin Mersenne. A *Mersenne prime* is a prime of the form $2^p - 1$, where p is prime. As of November 2008 there are 46 known Mersenne primes with the 46th equal to $2^{43,112,609} - 1$, which has 12,978,189 digits.

38. State the first three Mersenne primes.
39. Identify the first prime p such that $2^p - 1$ is not prime.

In exercises 40–47, compute each product, or explain why the product is undefined.

40. $\begin{bmatrix} -1 & 0 \end{bmatrix} \cdot \begin{bmatrix} -46 \\ 12 \end{bmatrix}$

41. $\begin{bmatrix} 3 & 1 \end{bmatrix} \cdot \begin{bmatrix} 15 \\ 1 \\ 27 \end{bmatrix}$

42. $\begin{bmatrix} -10 & 12 & 2 \end{bmatrix} \cdot \begin{bmatrix} -9 \\ 60 \\ 13 \end{bmatrix}$

43. $\begin{bmatrix} 1 & 5 & 2 & -3 & -2 \end{bmatrix} \cdot \begin{bmatrix} 4 \\ 1 \\ -2 \\ 0 \\ -1 \end{bmatrix}$

44. $\begin{bmatrix} -1 & 0 & 1 & 2 \end{bmatrix} \cdot \begin{bmatrix} 10 & 5 & 0 & 0 & 1 \\ 0 & 0 & -2 & 2 & 1 \\ 3 & -3 & 3 & -3 & 1 \end{bmatrix}$

45. $\begin{bmatrix} 6 & -10 & 1 \end{bmatrix} \cdot \begin{bmatrix} 0 & 17 \\ -8 & 8 \\ 30 & -30 \end{bmatrix}$

46. $\begin{bmatrix} a & b \\ c & d \end{bmatrix} \cdot \begin{bmatrix} e & f \\ g & h \end{bmatrix}$

47. $\begin{bmatrix} 1 & 0 & 0 \\ 0 & 1 & 0 \\ 0 & 0 & 1 \end{bmatrix} \cdot \begin{bmatrix} a & b & c \\ d & e & f \\ g & h & i \end{bmatrix}$

In exercises 48–51, encode each message using the $(7, 4)$ Hamming code.

48. $\begin{bmatrix} 1 & 1 & 0 & 0 \end{bmatrix}$

49. $\begin{bmatrix} 1 & 1 & 1 & 0 \end{bmatrix}$ $\begin{bmatrix} 1 & 1 & 0 & 1 \end{bmatrix}$

50. $\begin{bmatrix} 0 & 0 & 1 & 0 \end{bmatrix}$ $\begin{bmatrix} 0 & 1 & 1 & 0 \end{bmatrix}$

51. $\begin{bmatrix} 0 & 0 & 0 & 1 \end{bmatrix}$ $\begin{bmatrix} 1 & 0 & 0 & 1 \end{bmatrix}$ $\begin{bmatrix} 1 & 0 & 1 & 1 \end{bmatrix}$

In exercises 52–55, check each message encoded by the $(7, 4)$ Hamming code for single-digit errors. If multiplication by the parity check matrix indicates a single-digit error, state the corrected version of the vector.

52. $\begin{bmatrix} 0 & 1 & 1 & 1 & 1 & 0 & 0 \end{bmatrix}$

53. $\begin{bmatrix} 0 & 1 & 1 & 1 & 0 & 0 & 0 \end{bmatrix}$

54. $\begin{bmatrix} 1 & 0 & 1 & 0 & 1 & 1 & 0 \end{bmatrix}$ $\begin{bmatrix} 1 & 1 & 1 & 0 & 1 & 1 & 0 \end{bmatrix}$

55. $\begin{bmatrix} 1 & 0 & 1 & 0 & 0 & 1 & 0 \end{bmatrix}$ $\begin{bmatrix} 1 & 1 & 1 & 0 & 1 & 0 & 1 \end{bmatrix}$

Exercises 56–59 highlight the limitations of the $(7, 4)$ Hamming code for detecting multiple errors in transmitted messages and indicate the need for more sophisticated error-correcting schemes.

In exercises 56–59, verify that the $(7, 4)$ Hamming coding scheme does not detect the given errors. What correction does the $(7, 4)$ Hamming code recommend?

56. $\begin{bmatrix} 1 & 1 & 1 & 1 & 1 & 1 & 1 \end{bmatrix}$ received as $\begin{bmatrix} 1 & 1 & 1 & 1 & 1 & 0 & 0 \end{bmatrix}$

57. $\begin{bmatrix} 1 & 1 & 1 & 1 & 1 & 1 & 1 \end{bmatrix}$ received as $\begin{bmatrix} 1 & 0 & 1 & 1 & 1 & 0 & 1 \end{bmatrix}$

58. $\begin{bmatrix} 0 & 1 & 0 & 1 & 1 & 0 & 1 \end{bmatrix}$ received as $\begin{bmatrix} 0 & 1 & 0 & 1 & 1 & 1 & 0 \end{bmatrix}$

59. $\begin{bmatrix} 0 & 1 & 0 & 1 & 1 & 0 & 1 \end{bmatrix}$ received as $\begin{bmatrix} 1 & 1 & 0 & 1 & 1 & 1 & 1 \end{bmatrix}$

Exercises 60–65 introduce Hamming codes of higher dimension. For every pair of integers of the form $[(2^{r-1} - 1), r]$, we can determine a $[(2^{r-1} - 1), r]$ Hamming code that produces encrypted binary vectors of length $2^{r-1} - r$. Both the generating matrix G and a parity check matrix P for the $[(2^{r-1} - 1), r]$ Hamming code are determined by the parity bit columns. The parity bit columns consist of an array of all binary row vectors of length $r - 1$ with at least two 1's. In general, matrix G is obtained by attaching the $(2^{r-1} - r) \times (2^{r-1} - r)$ identity matrix to the front of this array and the parity check matrix P is obtained by attaching the $(r - 1) \times (r - 1)$ identity matrix to the bottom of this array. This procedure for the $(7, 4)$ Hamming coding scheme is illustrated below, with the parity bits in bold.

$$\begin{bmatrix} 1 & 1 & 1 \\ 1 & 1 & 0 \\ 1 & 0 & 1 \\ 0 & 1 & 1 \end{bmatrix} \Rightarrow G = \begin{bmatrix} 1 & 0 & 0 & 0 & 1 & 1 & 1 \\ 0 & 1 & 0 & 0 & 1 & 1 & 0 \\ 0 & 0 & 1 & 0 & 1 & 0 & 1 \\ 0 & 0 & 0 & 1 & 0 & 1 & 1 \end{bmatrix} \text{ and } P = \begin{bmatrix} 1 & 1 & 1 \\ 1 & 1 & 0 \\ 1 & 0 & 1 \\ 0 & 1 & 1 \\ 1 & 0 & 0 \\ 0 & 1 & 0 \\ 0 & 0 & 1 \end{bmatrix}$$

In exercises 60–65, consider Hamming codes of higher dimensions.

60. Determine a matrix consisting of the 11 binary column vectors of length $r - 1 = 5 - 1 = 4$ with at least two 1's.
61. What length vectors are encoded by the $(15, 5)$ Hamming coding scheme?
62. Find the generating matrix G for the $(15, 5)$ Hamming coding scheme.
63. Find the parity check matrix P for the $(15, 5)$ Hamming coding scheme.
64. How many vectors of length $r - 1 = 5$ have at least two 1's? What length vectors are encoded by the $(31, 6)$ Hamming code?
65. How many vectors of length $r - 1 = 6$ have at least two 1's? What length vectors are encoded by the $(63, 7)$ Hamming code?

In exercises 62–65, encode each message using the $(15, 5)$ Hamming code. The generating matrix G and parity check matrix P for the $(15, 5)$ Hamming code were identified in exercises 62 and 63 above.

66. $\begin{bmatrix} 0 & 0 & 1 & 1 & 0 & 1 & 0 & 1 & 1 & 0 & 0 \end{bmatrix}$
67. $\begin{bmatrix} 1 & 0 & 1 & 0 & 1 & 0 & 1 & 0 & 1 & 0 & 1 \end{bmatrix}$
68. $\begin{bmatrix} 1 & 1 & 1 & 0 & 1 & 1 & 0 & 1 & 1 & 0 & 1 \end{bmatrix}$

In exercises 69–70, check the following messages encoded by the $(15, 5)$ Hamming code for single-digit errors. If multiplication by the parity check matrix indicates a single-digit error, state the corrected version of the vector.

69. $\begin{bmatrix} 1 & 1 & 1 & 1 & 1 & 1 & 0 & 0 & 0 & 0 & 0 & 1 & 1 & 0 & 0 \end{bmatrix}$
70. $\begin{bmatrix} 1 & 1 & 1 & 1 & 0 & 0 & 0 & 0 & 0 & 0 & 0 & 1 & 0 & 0 \end{bmatrix}$

3.3   From the Pythagorean Theorem to Fermat's Last Theorem

Some relationships among numbers are based on equations containing two or more variables. This study of such relationships is focused around two of the most famous results in all of mathematics. The first is well-known to any student who has studied triangles—the Pythagorean theorem. For many centuries this result has been known to mathematicians around the world, from ancient Greece to ancient China, and has served as a staple of mathematical explorations. The second is an extension of the Pythagorean theorem, known as Fermat's last theorem. For more than three centuries,

Fermat's last theorem was one of the most significant open questions in mathematics, and mathematicians around the world rejoiced over its first complete proof in the 1990s. A study of these two results identifies some of the relations that do and do not exist among integers.

The Pythagorean theorem is named in honor of Pythagoras, a somewhat eccentric and yet brilliantly insightful Greek philosopher and mathematician. Pythagoras was born in 569 B.C.E. on the island of Samos, but later migrated to Croton on the south-eastern coast of Italy, where he established a semireligious, semiscientific society. Pythagoras and his followers immersed themselves in a study of numbers, becoming the first group in recorded history to work with numbers as abstract concepts, to identify the numeric relationships that exist among musical notes, and to explore the related geometric relations, including the Pythagorean theorem. At the same time, the Pythagoreans were well known for their mutual friendship, communal living, equal treatment of the sexes, and their intense secrecy. The society grew rapidly throughout Pythagoras' life but was violently suppressed shortly after his death. We now state the Pythagorean theorem and outline of a proof of this result.

Theorem 3.3.1 Pythagorean Theorem *For any right triangle, the square of the hypotenuse is equal to the sum of the squares of the other two sides, known as* **legs**. *If the hypotenuse has length c and the other two sides have lengths a and b, then we express this relation using the* **Pythagorean equation** $a^2 + b^2 = c^2$.

The Pythagorean theorem can also be expressed in terms of geometry, by attaching squares to a right triangle as indicated in figure 3.1. The Pythagorean theorem asserts that the area of the largest square is equal to the sum of the areas of the two smaller squares.

There are many different proofs of the Pythagorean theorem—some with a strong geometric flavor, and others that are more algebraic in nature. Many involve arranging right triangles in some clever fashion. The proof presented here considers an arrangement of four copies of the same generic right triangle in a square configuration, as illustrated in figure 3.2. Rather than simply stating a proof of this result, we

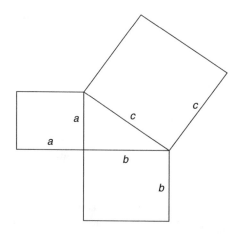

Figure 3.1 Geometric view of the Pythagorean theorem

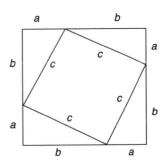

Figure 3.2 For Question 1's proof of the Pythagorean theorem

offer a series of questions that highlight the important characteristics of this figure, and invite you to piece together these relations and write a proof of this theorem yourself.

Question 3.3.1 Refer to figure 3.2, where the sides of the triangles are labeled with their corresponding lengths a, b, and c. Recall that the area of a square is the side length squared and the area of a triangle is one-half the product of the base and height.

    (a) What is the side length of the large, exterior square formed by the four triangles? Determine the area of this exterior square.

    (b) What is the side length of the small, interior square enclosed by the four triangles? Determine the area of this interior square.

    (c) Determine the area of the triangles.

    (d) Express the area of the exterior square as the sum of the area of the interior square and the area of the four triangles.

    (e) Set the expressions for the area of the exterior square from part (a) and part (d) equal to each other and algebraically simplify the result to obtain the Pythagorean equation $a^2 + b^2 = c^2$.

    (f) Based on your answers to these questions, write a proof of the Pythagorean theorem, using complete sentences and supportive algebraic computations at appropriate points in the argument.

■

While Pythagoras is recognized as the first mathematician to produce a general, abstract proof of this result, humanity's knowledge of numeric examples of the Pythagorean theorem predate his work by thousands of years. Megalithic monuments on the British Isles dating to 2500 B.C.E. are engraved with an example of integers satisfying the Pythagorean theorem. Numeric examples exist from ancient Egypt, Mesopotamia, India, and China (where the result is known as the Gougu theorem), including computational "proofs" with specific numbers that can be generalized. The oldest known written proof of the Pythagorean theorem is Proposition 47 in Book I of Euclid's *Elements* [73]. Since then more than 250 different proofs of the Pythagorean Theorem have been crafted, including an 1876 proof using trapezoids attributed to the twentieth President of the United States James Garfield while he was serving in the House of Representatives. The proof outlined in question 3.3.1

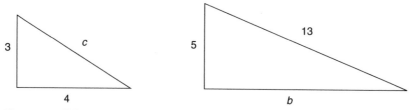

Figure 3.3   Triangles for example 3.3.1

can be traced back to the work of the Indian mathematician Bhaskara from the twelfth century C.E.

The Pythagorean theorem ensures that whenever the lengths of two sides of a right triangle are known, the length of the third side can be computed.

Example 3.3.1   We compute the length of the side identified with a variable for each right triangle.

- Applying the Pythagorean theorem to the triangle on the left in figure 3.3, $3^2 + 4^2 = c^2$, and so $25 = c^2$, which implies $c = 5$.
- For the triangle on the right in figure 3.3, $5^2 + b^2 = 13^2$, and so $b^2 = 169 - 25 = 144$, which implies $b = 12$.

∎

As mathematicians worked with the Pythagorean theorem, they recognized that right triangles with sides of integer length (such as those in example 3.3.1) are more the exception than the rule. When two sides of a triangle are integers, the third side is often not an integer. The following question provides some examples.

Question 3.3.2   Answer each question about a right triangle. Recall that the nonhypotenuse sides of a right triangle are called *legs*.

- (a) If the legs have lengths 3 and 4, what is the length of the hypotenuse?
- (b) If the legs have lengths 4 and 6.5, what is the length of the hypotenuse?
- (c) If the hypotenuse has length 17 and one leg has length 8, what is the length of the other leg?
- (d) If the hypotenuse has length 17 and one leg has length 11, what is the length of the other leg?

∎

Mathematicians are especially interested in integer solutions of the Pythagorean equation, and other similar multivariable equations known as *Diophantine equations*, as defined below. Motivated by this interest, mathematicians have defined certain distinguished types of triples of integers.

Definition 3.3.1   *A triple* (a, b, c) *of positive integers satisfying* $a^2 + b^2 = c^2$ *is called a* **Pythagorean triple**.

The next question considers various Pythagorean triples and explores the possibility that there exist infinitely many Pythagorean triples.

Question 3.3.3   The following questions show that some Pythagorean triples are multiples of
others.

   (a) Prove that $(5, 12, 13)$ is a Pythagorean triple.
   (b) Based on part (a), find a Pythagorean triple with $a = 10 = 5 \cdot 2$.
   (c) Find a Pythagorean triple with $a = 15 = 5 \cdot 3$.
   (d) If $n \in \mathbb{N}$ is an arbitrary positive integer, find a Pythagorean triple with $a = 5 \cdot n$.

   ▪

   As can be surmised from the answers to question 3.3.3, there exist infinitely many
Pythagorean triples; the following proof of this result is constructive.

Theorem 3.3.2   *There exist infinitely many Pythagorean triples; that is, there exist infinitely many
triples (a, b, c) of positive integers such that $a^2 + b^2 = c^2$.*

   Proof   The triple of positive integers $(3, 4, 5)$ is a Pythagorean triple, since $3^2 + 4^2 = 9 + 16 = 25 = 5^2$. Furthermore, for every positive integer $n \in \mathbb{N}$,

$$(3n)^2 + (4n)^2 \;=\; 3^2 n^2 + 4^2 n^2 \;=\; (3^2 + 4^2)n^2 \;=\; 5^2 n^2 \;=\; (5n)^2.$$

   Thus $(3n, 4n, 5n)$ is a Pythagorean triple for every positive integer n. Since there are
infinitely many positive integers, there exist infinitely many Pythagorean triples.

   ▪

   In addition to knowing that there exist infinitely many Pythagorean triples,
mathematicians have made another important step forward in describing Pythagorean
triples. The proof of theorem 3.3.2 simply manipulated the Pythagorean triple $(3, 4, 5)$
to obtain infinitely many others. But many different Pythagorean triples are not
multiples of $(3, 4, 5)$ (as we have seen in the preceding examples and questions). Faced
with this fact, mathematicians began seeking a pattern or formula that completely
classifies all Pythagorean triples—and they found one! Every Pythagorean triple is of
the form $(2mn, n^2 - m^2, n^2 + m^2)$, where $m, n \in \mathbb{N}$ are positive integers with $m < n$.
In addition to proving theorem 3.3.2, the formula shows that there exist an infinite
number of *primitive* Pythagorean triples; that is, Pythagorean triples whose values
(a, b, c) have no common divisor.
   The three integers $a = 3$, $b = 4$, and $c = 5$ are one solution of the Pythagorean
equation $a^2 + b^2 = c^2$. Many other multivariable equations have integer solutions;
such equations have come to be known as *Diophantine equations*. The ancient Greek
mathematician Diophantus, who lived in the third century, studied these types of
equations (especially linear ones), and wrote the famous text *Arithmetica* that was
the standard number theory reference (in fact the only thorough exposition on the
topic) as late as the seventeenth century! For the remainder of this section, we restrict
our attention to integer solutions of polynomial equations.

Definition 3.3.2   *A **Diophantine equation** is a polynomial equation in at least two variables with
integer solutions.*

   The Pythagorean equation $a^2 + b^2 = c^2$ is the best known example of a
Diophantine equation, but many other multivariable equations are Diophantine.

Example 3.3.2   We examine two equations—one is Diophantine with positive integer solutions, and one is not.

- $x - y = 0$ is a Diophantine equation with infinitely many solutions $x = y = 1$, $x = y = 2$, and so on; that is, every pair of integers with $x = y$ is a solution.
- $x + y = \pi$ is not a Diophantine equation because the sum of two integers cannot be an irrational number, and so this equation does not have integer solutions.

$\blacksquare$

Question 3.3.4   State an equation that is Diophantine and one that is not; give equations different from those in example 3.3.2

$\blacksquare$

Theorem 3.3.2 proves not only that the Pythagorean equation is Diophantine, but also that the Pythagorean equation is a Diophantine equation with infinitely many positive integer solutions. Not every Diophantine equation has this property. In fact, given any positive integer $n \in \mathbb{N}$, there exists a Diophantine equation with exactly n distinct positive integer solutions. The following example provides a linear Diophantine equation in two variables (so, both x and y are raised to the first power) with exactly two positive integer solutions.

Example 3.3.3   We identify the two pairs of positive integer solutions of $8x + 5y = 86$.
We can solve for either x or y in this equation; we solve for both obtaining

$$x = \frac{86 - 5y}{8} = 10 - \frac{5y - 6}{8} \quad \text{and} \quad y = \frac{86 - 8x}{5} = 17 - \frac{8x - 1}{5}.$$

Since x is a positive integer, the left equation implies $\frac{5y-6}{8}$ is a positive integer less than or equal to 9, and so y must be either 6 or 14. Substituting, we find the two pairs of solutions $(7, 6)$ and $(2, 14)$.

Alternatively, we can work with the right equation. Since y is a positive integer, the right equation implies $\frac{8x-1}{5}$ is a positive integer less than or equal to 16, and so x must be either 2 or 5. Substituting, we again find that $(7, 6)$ and $(2, 14)$ are the only two pairs of positive integer solutions of this Diophantine equation.

$\blacksquare$

Question 3.3.5   Identify the three pairs of positive integer solutions of $xy - 5x + 6y = 0$.
Hint: Solve for y and manipulate your solution to express y as the difference of an integer and a fraction in x as in example 3.3.3.

$\blacksquare$

The Pythagorean equation is just one example of a nonlinear Diophantine equation. Another famous example is the generalization of the Pythagorean equation to the form $a^3 + b^3 + c^3 = d^3$. Since $3^3 + 4^3 + 5^3 = 6^3$ and $1^3 + 6^3 + 8^3 = 9^3$, at least two quadruples of positive integers satisfy this equation. Following the same approach as the proof of the infinitude of Pythagorean triples for theorem 3.3.2, we can show that this Diophantine equation also has infinitely many positive integer solutions. Even though $a^3 + b^3 + c^3 = d^3$ has infinitely many solutions, mathematicians still do not have a general formula for generating every positive integer solution of this equation.

The infinite number of integer solutions to the Pythagorean equation $a^2 + b^2 = c^2$ and the above Diophantine equation $a^3 + b^3 + c^3 = d^3$ stands in marked contrast to what happens when we seek solutions to similar equations with greater integer exponents: $a^3 + b^3 = c^3$, $a^4 + b^4 = c^4$, $a^5 + b^5 = c^5$, and so on. The intensive study of this generalization of the Pythagorean theorem was initiated by the seventeenth century French mathematician Pierre de Fermat. Late in his life, Fermat claimed when the integer power n is greater than 2, there are no positive integer solutions to $a^n + b^n = c^n$. About this same time, Fermat proved this claim for $n = 4$ using a "method of infinite descent," and he announced this result to other mathematicians, inviting them to craft their own proofs.

Fermat's successes as a mathematician are even more impressive when we consider that he did not work full-time as a mathematician. A lawyer by day, Fermat had only his free time to study and develop mathematical insights. While in this sense only an amateur mathematician, Fermat came to be recognized as one of the greatest mathematical minds of his time. A short time after proving the $n = 4$ case, Fermat died without repeating his more general claim, nor providing any indication of how he was thinking of proving this result for all positive integers. Fortunately, mathematicians convinced Fermat's son to gather together and save Fermat's mathematical books and notes for later study. It became apparent that Fermat made many such claims with little or no proof, and mathematicians eagerly pursued complete proofs of these results. In the end, this one claim remained unproven, and so it came to be known as "Fermat's last theorem." For ease and clarity of reference, we formally state this result.

Theorem 3.3.3 Fermat's last theorem *For every integer n greater than* 2, *there are no positive integers a, b, c such that $a^n + b^n = c^n$.*

Many great mathematicians worked on proving Fermat's last theorem in its full generality. It turns out that the $n = 3$ case is much harder to prove than the $n = 4$ case. More than 100 years after Fermat's death, Leonhard Euler finally developed the mathematical insights enabling the complete proof that $a^3 + b^3 = c^3$ has no integer solutions. In subsequent years, Fermat's last theorem was proven for specific integer exponents one at a time, until in 1995 a complete proof was announced to the world. The English mathematician Andrew Wiles from Princeton University used sophisticated mathematical methods involving "elliptic curves" to prove the general version of Fermat's last theorem. For this work, Wiles was honored with numerous distinguished awards, including the presentation of a silver plate as a special tribute by the International Mathematical Union in 1998. The IMU is the organization responsible for awarding the Fields Medal, which is the equivalent of the Nobel Prize in mathematics. The Fields Medal is awarded every fourth year to at most four mathematicians "to recognize outstanding mathematical achievement for existing work and for the promise of future achievement." The Fields Medal is restricted to recipients who are at most 40 years of age; Wiles was 45 years old in 1998, or surely he would have been honored with the Fields Medal. While Wiles' proof is beyond the scope of this text, we can understand Fermat's proof for the $n = 4$ case.

3.3.1   Fermat's Last Theorem for $n = 4$

The rest of this section develops Fermat's proof that $a^4 + b^4 = c^4$ has no positive integer solutions. The first step is to modify the form of the equation. If a, b, c are positive integer solutions of $a^4 + b^4 = c^4$, then $(a^2)^2 + b^4 = c^4$. Thus, if $a^4 + b^4 = c^4$ has positive integer solutions, then so does $a^2 + b^4 = c^4$. Taking the contrapositive, if $a^2 + b^4 = c^4$ has no positive integer solutions, then neither does $a^4 + b^4 = c^4$. This proof therefore focuses on $a^2 + b^4 = c^4$, proving that it has no positive integer solutions.

Fermat's proof of this result uses the "method of infinite descent," which proceeds by contradiction. Assuming that there *does* exist a triple of positive integers (a, b, c) satisfying $a^2 + b^4 = c^4$, we prove the existence of another such solution (a^*, b^*, c^*) with the property that c^* is less than c. Applying this result to (a^*, b^*, c^*), there exists still another solution (a^{**}, b^{**}, c^{**}) with c^{**} less than c^*, and so on *ad infinitum*. This produces an infinite sequence of solutions with a corresponding infinite descending sequence of positive integers $c > c^* > c^{**} > c^{***} > \cdots$. But for any given positive integer c, there are only finitely many positive integers less than c, and we have the desired contradiction. Thus, $a^2 + b^4 = c^4$ has no positive integer solutions, and so $a^4 + b^4 = c^4$ has no positive integer solutions.

The main step in this argument is proving that if (a, b, c) is a triple of positive integers satisfying $a^2 + b^4 = c^4$, then there exists another such solution (a^*, b^*, c^*) with c^* less than c. There are many details to this proof, and we study some of them here; the exercises at the end of this section outline each part of the remaining portions of the proof.

The proof of Fermat's last theorem for $n = 4$ is quite long and involved, as are the proofs of many interesting and important mathematical results. Working through such proofs requires great care and determination, especially as they challenge a reader's mathematical understandings and intellectual abilities. These studies can result in frustration, but the rewards for perseverance can be tremendous. Furthermore, being able to follow sophisticated, involved proofs is an important step forward in learning to handle subtle mathematical ideas more easily and to craft such proofs yourself. And so, as we follow in the footsteps of Fermat through the rest of this section, be patient with yourself, be resolved to persevere through any tough spots—and enjoy a new understanding of these mathematical truths.

Since Fermat's proof proceeds by contradiction, the remainder of this section uses the assumption that (a, b, c) *is a triple of positive integers satisfying* $a^2 + b^4 = c^4$ *with c the least such positive integer.* To prove the existence of a solution (a^*, b^*, c^*) with c^* less than c, we consider three cases:

- b and c are both even;
- b and c have opposite parity; and
- b and c are both odd.

The case in which b and c are both even is taken care of quickly using the following theorem and question.

Theorem 3.3.4   *If (a, b, c) is a triple of positive integers satisfying $a^2 + b^4 = c^4$ with c the least such positive integer, then b and c have no common prime divisors.*

Proof   We proceed by contradiction, assuming that p is a prime divisor of both b and c (and working toward a contradiction of the "leastness" of c). Under this assumption, p^4 must divide $c^4 - b^4 = a^2$. From exercise 15 in section 3.1 or exercise 15 at the end of section 3.4, this fact implies p^2 divides a. Expressing these divisibility relations algebraically, we have $a = p^2 a^*$, $b = pb^*$, and $c = pc^*$. Substituting these expressions into the original equation and simplifying the result produces the following implications.

$$a^2 + b^4 = c^4 \quad \Rightarrow \quad (p^2 a^*)^2 + (pb^*)^4 = (pc^*)^4$$
$$\Rightarrow \quad p^4 (a^*)^2 + p^4 (b^*)^4 = p^4 (c^*)^4$$
$$\Rightarrow \quad (a^*)^2 + (b^*)^4 = (c^*)^4$$

Since $c = pc^*$ and p is prime (and so greater than 2), c^* must be less than c. But then, there exists a triple of positive integers (a^*, b^*, c^*) satisfying $(a^*)^2 + (b^*)^4 = (c^*)^4$ with c^* less than c. This contradicts the leastness of c, and so b and c must not share any common prime divisors.

■

Question 3.3.6   Use theorem 3.3.4 to prove that if (a, b, c) is a triple of positive integers satisfying $a^2 + b^4 = c^4$ with c the least such positive integer, then b and c cannot both be even.

■

We now turn our attention to the proof of Fermat's last theorem for $n = 4$ when b and c have opposite parity. The proof of this portion of the theorem relies on three lemmas, which we state and use without proof; exercises 62–69 at the end of this section suggestively outline the proofs of these Lemmas.

Lemma 3.3.1   *If (a, b, c) is a triple of positive integers satisfying $a^2 + b^4 = c^4$ with c the least such positive integer and with b and c of opposite parity, then there exist two odd positive integers s and t such that $\gcd(s, t) = 1$, $s^2 = c^2 + b^2$, and $t^2 = c^2 - b^2$.*

Lemma 3.3.2   *In the context of lemma 3.3.1, let $s + t = 2u$ and $s - t = 2v$. Then the following facts hold.*

- *The integer triple (u, v, c) is a Pythagorean triple.*
- *The integers u and v have opposite parity with $u = 2m^2$ and $v = (a^*)^2$ for positive integers m and a^*.*
- *$\gcd(u, v) = 1$.*

Lemma 3.3.3   *In the context of lemmas 3.3.1 and 3.3.2, there exist positive integers x and y such that the following hold:*

$$u = 2xy; \quad v = x^2 - y^2; \quad c = x^2 + y^2; \quad and \quad \gcd(x, y) = 1.$$

These lemmas lead to the proof of the following desired result.

Theorem 3.3.5   *In the context of lemmas 3.3.1, 3.3.2, and 3.3.3, there exist positive integers b^* and c^* with opposite parity and positive integer a^* such that $(a^*)^2 + (b^*)^4 = (c^*)^4$ with $0 < c^* < x < c$.*

Proof   Lemma 3.3.2 asserts that $u = 2m^2$, and lemma 3.3.3 asserts that $u = 2xy$; therefore, $m^2 = xy$. Replacing m with its prime power factorization $m = p_1^{n_1} \cdot p_2^{n_2} \cdots p_r^{n_r}$ produces $(p_1)^{2n_1} \cdot (p_2)^{2n_2} \cdots (p_r)^{2n_r} = xy$. Lemma 3.3.3 states that $\gcd(x, y) = 1$, and so each of the prime powers $(p_i)^{2n_i}$ appears intact within exactly one of the prime factorizations of either x or y. Thus, x and y have prime factorizations with even exponents on every prime, and so both x and y are perfect squares. Since the square roots of both x and y are integers, define $c^* = \sqrt{x}$ and $b^* = \sqrt{y}$. The rest of this proof verifies that a^* from lemma 3.3.2 with this b^* and c^* have the desired properties.

Lemma 3.3.2 states that $v = (a^*)^2$, and lemma 3.3.3 states that $v = x^2 - y^2$. Setting these equations equal to one another and substituting $c^* = \sqrt{x}$ and $b^* = \sqrt{y}$, produces the following equality

$$(a^*)^2 = v = x^2 - y^2 = ((c^*)^2)^2 - ((b^*)^2)^2 = (c^*)^4 - (b^*)^4$$

Simplifying, $(a^*)^2 + (b^*)^4 = (c^*)^4$. In addition, since $c^* = \sqrt{x}$, we have $0 < c^* < x$. Lemma 3.3.3 asserts that $c = x^2 + y^2$, which implies $x < c$. Thus, $0 < c^* < x < c$.

Finally, we argue that b^* and c^* have opposite parity. Lemma 3.3.2 asserts that u and v have opposite parity, and lemma 3.3.3 asserts that $u = 2xy$ is even, so $v = x^2 - y^2$ is odd. Therefore x and y have opposite parity, and so their respective square roots $c^* = \sqrt{x}$ and $b^* = \sqrt{y}$ must also have opposite parity.

We have arrived at the desired result: $(a^*)^2 + (b^*)^4 = (c^*)^4$ where b^* and c^* have opposite parity and $c^* < c$. This conclusion contradicts the original claim (that the triple of positive integers (a, b, c) with b and c of opposite parity is a solution of $a^2 + b^4 = c^4$ with the least possible value for c).   ∎

Lemmas 3.3.1, 3.3.2, and 3.3.3, and theorem 3.3.4 collectively prove that there does not exist a triple of positive integer solutions of $a^2 + b^4 = c^4$, which proves Fermat's last theorem for $n = 4$ when b and c have opposite parity. There is only one more case to consider: when b and c are both odd. The proof of this case is similar to the proof of theorem 3.3.4 and is outlined in exercise 70 at the end of this section.

3.3.2   Reading Questions for Section 3.3

1. State the Pythagorean theorem.
2. Give an example applying the Pythagorean theorem to a right triangle with one side of unknown length.
3. Sketch figure 3.2. How is this figure useful?
4. Define and give an example of a Pythagorean triple.
5. Give an example of a triple of integers that is not Pythagorean.
6. How many Pythagorean triples are there?
7. What is a primitive Pythagorean triple? Give an example.
8. Define and give an example of a Diophantine equation.

9. Explain why the Pythagorean equation is a Diophantine equation.

10. Give an example of a Diophantine equation with infinitely many solutions and a Diophantine equation with finitely many solutions.

11. State Fermat's last theorem.

12. Discuss the distinction between Fermat's and Euler's work on Fermat's last theorem and Wiles' work on Fermat's last theorem.

3.3.3   Exercises for Section 3.3

In exercises 1–6, determine if each triple is Pythagorean.

1. $(15, 36, 39)$ 4. $(48, 55, 73)$

2. $(10, 12, 22)$ 5. $(17, 144, 145)$

3. $(4, 3, 2)$ 6. $(18, 144, 146)$

In exercises 7–14, complete each Pythagorean triple by identifying the numeric value for the missing variable.

7. $(16, 63, c)$ 11. $(a, b, 5)$

8. $(48, 14, c)$ 12. $(a, b, 17)$

9. $(a, 40, 41)$ 13. $(a, 21, c)$

10. $(20, b, 52)$ 14. $(a, 45, c)$

Exercises 15–23 consider abstract questions and statements about the Pythagorean theorem and Pythagorean triples.

15. What is the least possible positive integer c for which there exist a and b such that (a, b, c) is a Pythagorean triple?

16. Prove that if n is a positive even integer, then $[n, (\frac{n}{2})^2 - 1, (\frac{n}{2})^2 + 1]$ is a Pythagorean triple.

17. Using the formula from exercise 16, list three distinct Pythagorean triples.

18. Prove that if n is a positive odd integer, then $[n, (n^2 - 1)/2, (n^2 + 1)/2]$ is a Pythagorean triple.

19. Using the formula from exercise 18, list three distinct Pythagorean triples.

20. Prove that if $m, n \in \mathbb{N}$, then $[2mn, n^2 - m^2, n^2 + m^2]$ is a Pythagorean triple.

21. Using the formula from Exercise 20, list three distinct Pythagorean triples.

22. Prove that if (a, b, c) satisfy the Pythagorean equation, then $(-a, -b, -c)$ also satisfies the Pythagorean equation. Conclude that a study of solutions of the Pythagorean equation can focus on positive solutions.

23. Suppose $(r/s, t/u, v/w)$ is a triple of rational numbers satisfying the Pythagorean equation. Determine a corresponding Pythagorean triple of integers expressed in terms of r, s, t, u, v, and w. Conclude that a study of rational solutions to the Pythagorean equation can focus on integer solutions.

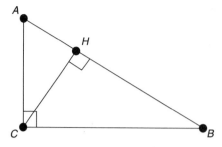

Figure 3.4   Figure for exercises 28–33

In exercises 24–27, consider the question of whether a Pythagorean triple can contain three prime numbers. Prove the following mathematical statements.

24. If (p, q, r) is a Pythagorean triple of primes, then one of p, q, r is even.

25. If p and q are prime, then $(p, q, 2)$ is not a Pythagorean triple.

26. If q and r are prime, then $(2, q, r)$ is not a Pythagorean triple.
    Hint: Manipulate the sum of squares in the Pythagorean equation to obtain a difference of squares and then factor.

27. A Pythagorean triple cannot contain three primes. Hint: Use exercises 24–26.

Exercises 28–33 develop Euclid's proof of the Pythagorean theorem given in Book I of *Elements*. Euclid added a line perpendicular to the hypotenuse (as shown in figure 3.4) and worked with similar triangles. Two triangles are said to be similar if their three interior angles are identical. Similar triangles are of interest in this setting because the side lengths of the corresponding sides share the same ratios. In exercises 28–33, let A, B, C, and H denote the four vertices determining the three triangles in figure 3.4, let pairs of adjacent letters (e.g., AB) denote the length of the side between the two vertices, and let triples of adjacent letters (e.g., ABC) denote the right triangle determined by the three vertices.

28. Prove that triangle ABC is similar to triangle ACH. Explain why this implies that $AC/AB = AH/AC$.

29. Algebraically manipulate the ratio from exercise 28 to find an expression for $(AC)^2$ in terms of AB and AH.

30. Prove that triangle ABC is similar to triangle BCH. Explain why this implies that $BC/AB = HB/BC$.

31. Algebraically manipulate the ratio from exercise 30 to find an expression for $(BC)^2$ in terms of AB and HB.

32. Working with the expressions from exercises 29 and 31, and with the equality $AH + HB = AB$, prove the Pythagorean identity $(AC)^2 + (BC)^2 = (AB)^2$.

33. Based on exercises 28–32, write a proof of the Pythagorean theorem, using complete sentences and supportive algebraic computations at appropriate points in the argument.

Exercises 34–40 develop an alternative proof for the existence of infinitely many Pythagorean triples.

34. Complete the following table of squares and differences of squares:

b	c	c^2	b^2	$c^2 - b^2$
0	1			
1	2			
2	3			
3	4			
4	5			
5	6			
6	7			
7	8			
12	13			

35. Describe the pattern in the right column of the table from exercise 34.
36. Prove that every odd number occurs in the right column of (the infinitely extended version of) the table from exercise 34 by considering the difference of squares of consecutive integers n and $n + 1$.
37. List two Pythagorean triples that occur in the table from exercise 34.
38. If the table from exercise 34 is extended to include rows with greater values of consecutive integers b and c, what is the next Pythagorean triple that appears in the table?
39. What integers have odd squares?
40. Based on exercises 34–39, write a proof that there exist infinitely many Pythagorean triples, using complete sentences and supportive algebraic computations at appropriate points in the argument.

In exercises 41–48, find all positive integer solutions x and y of the following Diophantine equations.

41. $3x + 5y = 12$

42. $6x + 15y = 30$

43. $8x + 5y = 1$

44. $20x + 25y = 125$

45. $10x + 11y = 320$

46. $xy - 7x + 6y = 0$

47. $xy + 3x - 2y = 0$

48. $xy - 3x + 4y = 0$

Exercises 49–52 consider the special category of Diophantine equations of the form $x^2 - Ny^2 = 1$, where N is a positive nonsquare integer. These equations are known as *Pell's equations* in honor of the seventeenth century English algebraist and number theorist John Pell.

49. Prove $x = 3$, $y = 2$ is a solution of Pell's equation $x^2 - 2y^2 = 1$.

50. Find a pair of positive integers (different from $x = 3$, $y = 2$) satisfying $x^2 - 2y^2 = 1$.
   Hint: There exists a solution with $x \leq 35$.

51. Prove that if a pair of positive integers (x, y) satisfies $x^2 - 8y^2 = 1$, then x is odd.

52. Prove that if a pair of positive integers (x, y) satisfies Pell's equation with N even, then x is odd.

Exercises 53–55 consider the special category of Diophantine equations of the form $x^n - Ny^n = 1$ where N and n are positive integers. These equations are known as *Thue's equations* in honor of the Norwegian mathematician Axel Thue. Thue made a number of important contributions to abstract algebra and number theory, and he identified conditions determining when these equations have a finite number of integer solutions.

53. Prove that $x = 2$, $y = 1$ is a solution of Thue's equation $x^3 - 7y^3 = 1$.

54. Find a pair of positive integers satisfying Thue's equation $x^4 - 5y^4 = 1$. Hint: There exists a solution with $x \leq 20$.

55. Prove that if a pair of positive integers (x, y) satisfies Thue's equation $x^3 - 7y^3 = 1$, then x and y must have opposite parity.

Exercises 56–61 consider the types of equations addressed by Fermat's last theorem.

56. Prove that $a = 2$, $b = \sqrt[3]{19}$, and $c = 3$ satisfy $a^3 + b^3 = c^3$. What does this tell us about $\sqrt[3]{19}$?

57. Find a triple of positive real numbers (a, b, c) satisfying $a^3 + b^3 = c^3$; identify an answer different from that given in exercise 56.

58. Find a pair of positive integers b and c (with $b \leq 10$) satisfying $1{,}701 + b^3 = c^3$.

59. Based on exercise 58, prove that $1{,}701$ cannot be expressed as the cube of a rational number; that is, prove $1{,}701 \neq (r/s)^3$ when r and s are positive integers

60. While visiting his friend and colleague Srinivasa Ramanujan in the hospital, Godfrey Harold Hardy remarked that he had arrived in taxi number $1{,}729$. Ramanujan immediately replied that "$1{,}729$ is an interesting number because it is the smallest integer that can be expressed as the sum of two cubes in two different ways." Find two distinct pairs of positive integers (a, b) (both with $a \leq 10$) satisfying $a^3 + b^3 = 1{,}729$.

61. Suppose $(r/s, t/u, v/w)$ is a triple of rational numbers satisfying $a^n + b^n = c^n$, where $n > 2$ is an integer. Find a corresponding triple of integers (x, y, z) expressed in terms of r, s, t, u, v, and w that would then satisfy $a^n + b^n = c^n$. Based on this insight and Fermat's last theorem, what do we know about the existence of triples of rational numbers satisfying $a^n + b^n = c^n$, where $n > 2$ is an integer?

Exercises 62–70 complete the proof of Fermat's last theorem for $n = 4$ by outlining the proofs of lemmas 3.3.1–3.3.3 and the proof of the setting where b and c are both odd. Specifically, you should assume in exercises 62–69 that b and c have opposite parity (as in lemmas 3.3.1–3.3.3). Then assume in exercise 70 that b and c are both odd (to prove the last case).

62. Working in the direction of proving lemma 3.3.1 in exercise 63, prove that $\gcd(c^2 + b^2, c^2 - b^2) = 1$. Develop a proof by contradiction, assuming there exists a prime divisor p of $c^2 + b^2$ and $c^2 - b^2$. Since $c^2 + b^2$ is odd, $p \neq 2$. Since p divides both $c^2 + b^2$ and $c^2 - b^2$, p divides their sum $2c^2$, and

so p divides c. Since p now divides both c^2 and $c^2 - b^2$, p also divides b. Explain why theorem 3.3.4 provides a contradiction and conclude that $\gcd(c^2 + b^2, c^2 - b^2) = 1$.

63. Assume the integer triple (a, b, c) solves $a^2 + b^4 = c^4$ with c the least such positive integer, and that b and c have opposite parity. Using the prime factorization of the integer a and the result from exercise 62, explain why the equation $(c^2 + b^2)(c^2 - b^2) = c^4 - b^4 = a^2$ implies that both of the factors must be squares whose greatest common divisor is 1; that is, there exist positive integers $s, t \in \mathbb{N}$ such that $\gcd(s, t) = 1$, $c^2 + b^2 = s^2$ and $c^2 - b^2 = t^2$. This completes the proof of lemma 3.3.1.

64. Working in the context of lemma 3.3.2, prove that the integer triple (u, v, c) is Pythagorean by substituting $u = (s+t)/2$ and $v = (s-t)/2$ into the expression $u^2 + v^2$ and expanding the resulting expression to obtain $(s^2 + t^2)/2 = c^2$.

65. Working in the context of lemma 3.3.2, prove that the integers u and v have opposite parity with $u = 2m^2$ and $v = (a^*)^2$ for positive integers m and a^* and that $\gcd(u, v) = 1$. Using the equation $2uv = (s^2 - t^2)/2 = b^2$ and the prime factorization of b, prove that either u or v has a factor of 2; without loss of generality, assume u is the term with this factor of 2. Then use the prime factorization of b and $2uv = b^2$ to prove the existence of the desired m and a^*. Finally, give a proof by contradiction that u and v have no common prime divisors: assume p divides both, prove that p would then divide both $2s = (s + t) + (s - t)$ and $2t = (s + t) - (s - t)$, and apply lemma 3.3.1 to conclude that $\gcd(u, v) = 1$.

66. Working in the context of lemma 3.3.3, use the prime factorization of v and the facts that $v^2 = (c - u)(c + u)$ and $\gcd(c - u, c + u) = 1$ to show that $c + u$ and $c - u$ are squares; in other words, prove that there exist positive integers e and f with $e^2 = c + u$ and $f^2 = c - u$.

67. Working in the context of lemma 3.3.3 and using the notation from exercise 66, let x and y be positive integers defined by $2x = e + f$ and $2y = e - f$. Using the fact that (u, v, c) is a Pythagorean triple and the results of exercise 66, prove that $v = x^2 - y^2$ (the second fact stated in lemma 3.3.3).

68. Working in the context of lemma 3.3.3 and exercises 66–67 and using the facts that $u = (e^2 - f^2)/2$ and $c = (e^2 + f^2)/2$, prove that $u = 2xy$ and $c = x^2 + y^2$ (the first and third facts stated in lemma 3.3.3).

69. Prove the last fact stated in lemma 3.3.3; that is, prove that $\gcd(x, y) = 1$. Develop a proof by contradiction, assuming p is a prime divisor of both x and y. Use the fact that $v = x^2 - y^2$ to conclude that p must divide v. But p also divides $e = x + y$ and $f = x - y$, and so p is a divisor of c, since $2c = (c + u) + (c - u) = e^2 + f^2$. Finally, p is a divisor of b, since $2b^2 = s^2 - t^2 = (s - t)(s + t) = 2u \cdot 2v$ and p divides v. This analysis shows p is a common divisor of c and b, which contradicts theorem 3.3.4.

70. Complete the proof of Fermat's last theorem for $n = 4$ by showing that if there exist positive integers a, b, c with b and c both odd such that $a^2 + b^4 = c^4$ (the reformulated version of $a^4 + b^4 = c^4$), then a contradiction results. Under the

assumption that such b and c are both odd (and so a is even) and $a^2 + b^4 = c^4$ with c the least such positive integer, prove each of the following statements.

(a) Theorem 3.3.4 asserts that b and c have no common prime divisors. Prove by contradiction that $c^2 + b^2$ and $c^2 - b^2$ have a unique common prime divisor of 2.

(b) Using the result from part (a), prove that $gcd(a, b, c) = 1$ by showing that a and b have no common prime divisors, and that a and c have no common prime divisors. Conclude that $gcd(c^2 + a, c^2 - a) = 1$.

(c) Algebraically manipulate $a^2 + b^4 = c^4$ to show that $b^4 = (c^2 + a)$ $(c^2 - a)$. Writing b in terms of its prime factorization and using $gcd(c^2 + a, c^2 - a) = 1$ from part (b), prove that there exist $s, t \in \mathbb{N}$ such that $c^2 + a = s^2$ and $c^2 - a = t^2$. Furthermore, note that $s > t$ and that both s and t are odd since c is odd and a is even.

(d) Working with s and t from part (c), define positive integers x and y such that $2x = s + t$ and $2y = s - t$ (note s and t are both odd); thus, $x > y$, $s = x + y$, and $t = x - y$. Using $gcd(c^2 + a, c^2 - a) = 1$ from part (b), prove that $gcd(x, y) = 1$.

(e) Prove that x and y from part (d) satisfy $a = 2xy$, $b^2 = x^2 - y^2$, and $c^2 = x^2 + y^2$.

(f) Using the identities from part (e), prove that $(bc)^2 + y^4 = x^4$ with $0 < x < c$, and explain why this is a contradiction of our assumption.

3.4 Irrational Numbers and Fields

This chapter's study of number systems continues by considering extensions of the integers $\mathbb{Z}$ to other well-known number systems, including the rationals $\mathbb{Q}$, the reals $\mathbb{R}$, and the complex numbers $\mathbb{C}$. Definition 2.1.1 in section 2.1 defined these sets of numbers, and example 2.1.4 in section 2.1 identified the proper subset relationships that exist among these sets of numbers: $\emptyset \subset \mathbb{N} \subset \mathbb{Z} \subset \mathbb{Q} \subset \mathbb{R} \subset \mathbb{C}$. We can think of these sets of numbers as arising from successive closure operations on each set to obtain the next in the sequence. This section examines such definitions of one set in terms of the next, and then extends the study of groups to the study of sets with two binary operations that form a "field."

We first consider the relationship between the integers $\mathbb{Z}$ and the rational numbers $\mathbb{Q}$. The extension of $\mathbb{Z}$ to $\mathbb{Q}$ follows from considering ratios of integers. The definition expresses the rational numbers as

$$ \mathbb{Q} \;=\; \left\{ \frac{m}{n} \;:\; m, n \in \mathbb{Z} \text{ with } n \neq 0 \right\}. $$

The next question asks you to verify that the set of $\mathbb{Z}$ is a proper subset of the set of $\mathbb{Q}$.

Question 3.4.1 (a) Express the integers 2 and -3 as a ratio of two integers m/n.

(b) Prove that $\mathbb{Z} \subseteq \mathbb{Q}$ by expressing an arbitrary $k \in \mathbb{Z}$ as a ratio m/n of two integers m and n with n nonzero.

(c) Prove that $\mathbb{Z} \subsetneq \mathbb{Q}$ by finding a ratio m/n of two integers that is not itself an integer.

(d) How many rational numbers are not integers?

■

We now consider the relationship between the rational numbers $\mathbb{Q}$ and the continuum of the real line $\mathbb{R}$. This study began with the Greeks, who were particularly enamored of the rationals and, at one time, believed that every number could be expressed as a ratio of two integers. The Pythagorean theorem asserts that for every right triangle, the square of the hypotenuse is equal to the sum of the squares of the other two sides. If we consider an isosceles right triangle with $a = b = 1$, then the length of the hypotenuse c is $c^2 = 1^2 + 1^2 = 1 + 1 = 2$, which implies $c = \sqrt{2}$. And so the Greeks thought of $\sqrt{2}$ in terms of geometry – as the distance determined by the hypotenuse of this simple triangle. In more recent centuries, as the sophistication of algebraic notation and manipulation has increased, mathematicians have also come to think of $\sqrt{2}$ algebraically—as the solution of the polynomial equation $x^2 - 2 = 0$; this perspective is developed in Section 3.5.

As the Pythagoreans investigated this number's properties, they recognized that $\sqrt{2}$ is not a rational number, but is instead *irrational*. In this setting, "irrational" simply means not rational; the other English usage of the word "irrational" is in reference to mental activities rather than numbers and is attributable to the common Greek linguistic root shared by "ratio" and "reason."

We consider a proof that $\sqrt{2}$ is irrational and explore some natural extensions of this result to prove that other real numbers are irrational. Over the centuries, many different proofs have been given for the irrationality of $\sqrt{2}$. In a famous book entitled *A Mathematician's Apology*, the English mathematician G. H. Hardy [112] praised the intrinsic beauty of mathematics and his love for mathematical results free of applications. Hardy highlighted the algebraic proof we study in this text that $\sqrt{2}$ is irrational as one of the most elegant proofs in mathematics. In preparation for presenting this proof, the following question considers common factors shared by numerators and denominators of rational numbers.

Question 3.4.2 For a rational number of the form $\frac{m}{n}$, the integer m is called the *numerator* and n is called the *denominator*. Sometimes the numerator and denominator are not relatively prime, but share a common factor; in this case the common factors can be cancelled and the rational number expressed in *lowest terms*. Express each rational number as a fraction in lowest terms.

(a) $\dfrac{2,965}{10,000}$

(c) 2.2965

(b) $\dfrac{10,505}{100}$

(d) 0.10505

■

With this terminology in hand, consider the following proof of the irrationality of $\sqrt{2}$.

Theorem 3.4.1   *The square root of two is irrational.*

Proof   This classical algebraic proof proceeds by contradiction, assuming that $\sqrt{2}$ is rational and working toward a contradiction.

From the assumption that $\sqrt{2}$ is rational, there exist integers m and n (with n nonzero) such that $\sqrt{2} = m/n$. Furthermore, we may assume that m and n share no common factors; any common divisor could be factored out and cancelled. A contradiction is obtained from these assumptions by showing that m and n must actually share a common factor of 2.

Since $\sqrt{2} = m/n$, we have $m = \sqrt{2} \cdot n$ and so $m^2 = 2n^2$. The fact that 2 divides m^2 implies 2 divides m (see exercise 15 in section 3.1 or exercise 15 at the end of this section). Therefore, $m = 2k$ for some $k \in \mathbb{Z}$. Substituting this term into the expression $m^2 = 2n^2$, we have $(2k)^2 = 2n^2$. Expanding gives us $4k^2 = 2n^2$, which implies that $2k^2 = n^2$. This same line of argument also proves that n is even. Since m and n are both even, they have a common divisor of 2, contradicting the assumption that m and n share no common factor.

■

Before considering extensions of this result, we pause to reflect on the significance of the proof that $\sqrt{2}$ is irrational and the integrity of the Pythagoreans in accepting the results of their intellectual explorations. For perhaps much of your mathematical life, you have been aware of the existence of irrational numbers such as π, e, and $\sqrt{2}$. In contrast, the Pythagoreans mistakenly first thought of *all* numbers as ratios of integers, and much of their pseudoreligious philosophy hinged on ratio relationships among integers. The realization that $\sqrt{2}$ is irrational must have initially upset their world view; dramatic, apocryphal stories describe the first Pythagorean to recognize this proof as being thrown overboard in the Mediterranean Sea to drown. And yet, the Pythagoreans recognized mathematical truth and proof as absolute, and the mathematicians of the sixth century B.C.E. accepted the veracity of this result and its consequences. In subsequent years, mathematicians eventually came to recognize just how fundamental the irrational numbers are to an understanding of the continuum of the real number line, as we will see in our continuing study of the reals.

The next question considers a natural extension of theorem 3.4.1 to the next prime number 3, and question 3.4 extends this result to any prime p. Both questions continue using the fact that if p is a prime number, n is an integer, and p divides n^2, then p divides n.

Question 3.4.3   The following questions develop a proof that $\sqrt{3}$ is irrational. The proof proceeds by contradiction, assuming $\sqrt{3}$ is rational and working toward the contradiction that the numerator and denominator of a rational expression for $\sqrt{3}$ in lowest terms must actually share a common factor of 3.

First assume that $\sqrt{3}$ is rational. By the definition, there exist integers m and $n \neq 0$ such that $\sqrt{3} = m/n$. Furthermore, assume m/n is in lowest terms, so that m and n do not share any common factors.

(a) Algebraically manipulate $\sqrt{3} = m/n$ and determine an expression for m^2 in terms of n.

(b) Working with the expression for m^2 from part (a), prove that m is a multiple of 3, and so $m = 3k$ for some integer $k \in \mathbb{Z}$.

(c) Substituting into the expression for m^2 from part (b), find an expression for n^2 in terms of k and prove that n is a multiple of 3.

(d) What common factor is share by m and n?

(e) State the resulting contradiction and conclusion.

(f) Based on your answers to these questions, write a proof that $\sqrt{3}$ is irrational, using complete sentences and supporting algebraic computations at appropriate points in the argument.

■

Question 3.4.4 Prove that if p is a prime number, then $\sqrt{p}$ is irrational.

■

Thus far, rational and irrational numbers have been defined in terms of the ability to represent them as fractions of integers. These numbers are also characterized by their decimal expansions. Consider the following theorem.

Theorem 3.4.2 *A real number $r \in \mathbb{R}$ is **rational** exactly when the decimal expansion of r terminates or repeats periodically, while a real number $r \in \mathbb{R}$ is **irrational** when the decimal expansion of r is nonterminating and nonrepeating.*

Theorem 3.4.2 may be familiar from your earlier studies in mathematics. The proof of this result is beyond the scope of this text and is left for your later studies. Even so, this description of rational and irrational numbers in terms of their decimal expansions is helpful in many settings.

Example 3.4.1 We consider some examples of decimal expansions of rational and irrational numbers.

Long division of integers produces the following decimal expansions.

$$\frac{1}{8} = 0.125 \qquad \frac{1}{9} = 0.111\ldots = 0.\overline{1}$$

In contrast, the following irrational numbers have infinite, nonrepeating decimal expansions, where

$$e = \lim_{n \to \infty} (1 + 1/n)^n$$

and π is defined as the circumference of a circle with diameter 1.

$$\sqrt{2} = 1.4142356\ldots \qquad e = 2.718281828\ldots \qquad \pi = 3.14159265\ldots$$

■

Although irrational numbers require infinite decimal expansions to be expressed exactly, humans and computers are only capable of manipulating finite decimal expansions. As such, identifying accurate finite approximations of irrational numbers can be an important goal in a study of the irrationals and has become a common topic in the sequence of calculus courses. On a less serious note, there even exist societies and contests devoted to memorizing initial parts of the decimal expansion of π, sometimes using mnemonics, such as "How I wish I could enumerate pi easily today" (the number of letters in each word is the first part of the expansion of π). How many digits of π do you know by heart?

The history of proving π and e irrational is interesting. While the ancient Greeks knew about the number π and "believed" it was irrational, it was not until 1768 that the first proof of the irrationality of π was given by the Johann Heinrich Lambert, a mathematician from the Alsace–Lorraine region on the Swiss–German border. In 1794, the French mathematician Adrien-Marie Legendre proved π^2 is irrational. The definition of e as the limit

$$e = \lim_{n \to \infty} (1 + 1/n)^n$$

was given in 1683 by the Swiss mathematician Jakob (Jacques) Bernoulli. Euler proved e is irrational in 1737, and a simple proof of this result is given with theorem 3.4.3 below. In addition to his work with π, Lambert also proved that e^n is irrational for any $n \in \mathbb{Z}^*$. In 1996, the Russian mathematician Yuri Nesterenko from Moscow State University proved that $\pi + e^\pi$ is irrational. At the same time, the (ir)rationality of many numbers is still an open question, including π^e, 2^e, and $\pi^{\sqrt{2}}$.

Theorem 3.4.3 *The real number e is irrational.*

Proof This result uses some results about power series from calculus. The strategy is to prove that the nonzero real number $1/e$ is not rational, and so e is not rational. Substituting $x = -1$ into the power series $e^x = \sum_{n=0}^{\infty} x^n/n!$, we obtain the following infinite series.

$$\frac{1}{e} = 1 - \frac{1}{1!} + \frac{1}{2!} - \frac{1}{3!} + \frac{1}{4!} - \frac{1}{5!} + \frac{1}{6!} - \cdots$$

Since this series is an alternating series, its sum $1/e$ is bounded by consecutive partial sums. Computing the third partial sum $S_3 = 1 - 1 + 1/2! = 1/2$ and the fourth partial sum $S_4 = 1 - 1/1! + 1/2! - 1/3! = 1/3$, we have the following inequalities (using a common denominator).

$$\frac{2}{6} < \frac{1}{e} < \frac{3}{6}$$

Considering further successive partial sums produces the following sequence of inequalities, which can be extended indefinitely.

$$\frac{2}{6} < \frac{1}{e} < \frac{3}{6}, \frac{8}{24} < \frac{1}{e} < \frac{9}{24}, \frac{44}{120} < \frac{1}{e} < \frac{45}{120}, \frac{264}{720} < \frac{1}{e} < \frac{265}{720}, \cdots$$

Notice that in each inequality, the numerators differ by 1 and the denominators are successive factorials $n!$ with $n = 2, 3, 4, 5, \ldots$.

With these observations in hand, we develop a proof by contradiction. Assume $1/e$ is rational, and so there exist integers $p, q \in \mathbb{Z}$ with $q \neq 0$ such that $1/e = p/q$. The first inequality $2/6 < 1/e < 3/6$ implies that the denominator q is not a divisor of 6. If q divides 6, then $1/e$ could be written as $m/6$ for some $m \in \mathbb{Z}$; but there does not exist such an integer m with $2 < m < 3$. Similarly, the second inequality $8/24 < 1/e < 9/24$ implies that the denominator q is not a divisor of $24 = 4!$, the third inequality implies that q is not a divisor of $120 = 5!$, and so on. In addition, q does not divide either $n = 1$ or $n = 2$, since $1/e$ is not equal to either 1 or 1/2.

Therefore, the denominator of $1/e$ is not a divisor of $n!$ for any $n \in \mathbb{N}$, and so $1/e$ is not rational. Since $1/e$ is not rational, its reciprocal e is not rational.

■

We now consider the relationship between the rational numbers $\mathbb{Q}$ and the continuum of the real number line $\mathbb{R}$. Every rational number appears in the reals (defined as signed distances); that is, we can identify every rational with the corresponding directed distance from the point zero on the real number line. Therefore, $\mathbb{Q} \subseteq \mathbb{R}$. Furthermore, this inclusion is proper; there exist irrational real numbers, include $\sqrt{2}$, $\sqrt{3}$, and e as shown above.

Just as the rationals can be defined in terms of the integers, we might seek to define the reals in terms of the rationals. Such a definition has been crafted, but it is quite subtle in nature and was developed only in the nineteenth century. The two approaches typically studied use either *Cauchy sequences* or *Dedekind cuts*, and both require the use of "infinite" objects to define irrational numbers. Cauchy sequences are named in honor of Augustin-Louis Cauchy, an insightful French mathematician who published an impressive 789 mathematical papers covering all areas of mathematics known at that time. When used to define irrational numbers in the reals, Cauchy sequences are infinite lists of rationals that converge to a given irrational number.

Dedekind cuts are named in honor of the German mathematician Richard Dedekind, who thought of this idea on November 24, 1858 and published its definition in *Stetigkeit und Irrationale Zahlen* in 1872. Dedekind was Gauss' last doctoral student and was a close friend and colleague of many of the leading mathematicians of the time, including Riemann, Dirichlet, and Cantor. A Dedekind cut defining a given irrational number consists of a pair of disjoint sets, one containing every rational number less than the given irrational and one containing every rational number greater. Both Cauchy sequences and Dedekind cuts provide a rigorous definition of the reals in terms of the rationals.

Finally, we consider the relationship between the real numbers $\mathbb{R}$ and the complex numbers $\mathbb{C} = \{a + bi : a, b \in \mathbb{R}\}$, where $i = \sqrt{-1}$. The history of mathematician's understanding and acceptance of the complex numbers is a bit checkered. The Babylonians and the ancient Greek mathematicians, including Diophantus in the third century C.E., labeled such equations as $x^2 + 1 = 0$ "meaningless." In 1572, the Italian mathematician Rafael Bombelli examined the equation $x^2 + 1 = 0$ and recognized the need to extend the reals to include "imaginary" numbers such as "$i = \sqrt{-1}$." This recognition was a significant step forward, although the mathematical community was not convinced of the legitimacy and value of studying complex numbers until the early 1800s. Section 7.1 provides more details about the history of complex numbers.

From the definition of $\mathbb{C}$, every real number $r \in \mathbb{R}$ is in the set of complex numbers since $r = r + 0i \in \mathbb{C}$. Furthermore, the reals are a proper subset of the complex numbers, since $i = \sqrt{-1}$ is not real. Among other things, i does not satisfy the property that $r^2 \geq 0$ for every $r \in \mathbb{R}$ (see exercises 67–70 at the end of this section). Thus, the real numbers are a proper subset of the complex numbers.

For the remainder of this section, we study these number systems from the perspective of their abstract algebraic properties, adopting the approach followed in chapter 2 when describing groups. In this setting, we consider sets of numbers with both additive and multiplicative operations and that satisfy the group properties for both operations. Such sets, together with their operations, are known as *fields*.

    The properties satisfied by a field are more extensive than the four group properties applied to both binary operations (as expressed in definition 2.4.1 of section 2.4). Perhaps this need for extra conditions seems natural since a field involves two operations on the base set, rather than the one operation of a group. These extra properties will be familiar, since they are satisfied by the standard addition and multiplication operations of the real numbers; the next question begins a study of the real numbers as a field.

Question 3.4.5   Recall that $\mathbb{R}$ under standard addition is an Abelian group.

    (a) Identify the additive identity of $\mathbb{R}$.
    (b) Identify the additive inverse of an arbitrary element $r \in \mathbb{R}$.

On the other hand, $\mathbb{R}$ under standard multiplication is not a group—although it comes very close. Only the inverse property is not satisfied, and this fails for just one real number.

    (c) What real number $r \in \mathbb{R}$ does not have a multiplicative inverse?

The nonzero reals $\mathbb{R}^* = \mathbb{R} \backslash \{0\}$ is a multiplicative group.

    (d) Identify the multiplicative identity of $\mathbb{R}^*$.
    (e) Identify the multiplicative inverse of an arbitrary element $r \in \mathbb{R}^*$.

                                                                 ■

    As question 3.4.5 indicates, standard addition and standard multiplication are both well-behaved operations on the real numbers, satisfying the properties of an Abelian group (except for the additive identity 0 under multiplication). The definition of a field reflects the properties of these two operations on the reals, and there exist many other important and familiar examples of fields. By articulating and working with these properties from a general, abstract perspective, mathematicians gain insight and prove results that can be applied to a host of other settings.

Definition 3.4.1   *A set F under two binary operations* $+, \times$ *is a* **field** *when the following three properties hold.*

    1. The set F is an Abelian group under the "addition" operation $+$*. The additive identity is called the* **zero** *and is denoted by* 0.
    2. The set $F^* = F \backslash \{0\}$ *is an Abelian group under the "multiplication" operation* $\times$*. The multiplicative identity is called the* **unity** *and is denoted by* 1.
    3. The multiplication operation $\times$ *distributes over the addition operation* $+$*; that is,* $a \times (b + c) = a \times b + a \times c$*, for every* $a, b, c \in F$.

    The references to addition $+$, multiplication $\times$, zero 0, and unity 1 in the above definition are notational. While some fields are composed of numbers with binary operations that correspond exactly with our intuitive understanding of these symbols, there also exist fields of vectors, functions, and other mathematical objects. In these contexts 0, 1, and the two binary operations are naturally different than the familiar operations that inspired their names. So why use this potentially ambiguous and confusing notation? In part, the answer lies in the context in which these notions were first isolated, the historical development of these ideas, and the ongoing tradition and culture of mathematicians. In addition, fields do behave like the real numbers under standard addition and multiplication, and so hopefully our intuitive understanding of the reals facilitates the study of other fields.

Given an abstract definition, mathematicians often immediately identify examples to sharpen their understanding. Working with concrete objects may throw further light on some mathematical behavior, perhaps leading to a modification of the definition. Furthermore, if a proposed definition is too strong and few objects satisfy the designated properties, then the scope of any results or insights is limited. On the other hand, if the proposed definition is too weak, then relatively few results may be provable. The definition of a field lies in the happy middle ground—strong enough to produce many interesting results and weak enough that many mathematical objects satisfy its conditions.

In addition to the real numbers, the set of rationals under standard addition and multiplication, the complex numbers under appropriately defined operations, and certain sets $\mathbb{Z}_n$ under modular addition and multiplication are all fields. For the rest of this section we consider the field properties of the complex numbers and the sets $\mathbb{Z}_n$.

An investigation of the field structure of the complex numbers requires the definition of addition and multiplication operations. If $a + bi$ and $c + di$ are complex numbers, then addition is defined componentwise as follows.

$$(a + bi) + (c + di) = (a + c) + (b + d)i$$

For example, $(1 + 2i) + (3 + 4i) = (1 + 3) + (2 + 4)i = 4 + 6i$. Multiplication of complex numbers uses the "F.O.I.L." method (multiplying First, Outer, Inner, and Last terms) and then applies the identity $i^2 = -1$ to simplify the resulting expression; the general algebraic formula follows.

$$(a + bi) \cdot (c + di) = ac + adi + bci + bdi^2 = (ac - bd) + (ad + bc)i$$

For example, $(1 + 2i) \cdot (3 + 4i) = 1 \cdot 3 + 1 \cdot 4i + 2 \cdot 3i + 2 \cdot 4i^2 = (1 \cdot 3 - 2 \cdot 4) + (1 \cdot 4 + 2 \cdot 3)i = -5 + 10i$.

Question 3.4.6 Add and multiply each pair of complex numbers.

(a) $1 + i$ and $3 + 5i$ (c) $2 - i$ and $-4 + 3i$

(b) 2 and i (d) i and $3 + 5i$

■

The following question outlines the proof that the set of complex numbers is a field under these addition and multiplication operations. In addition to the appropriate algebraic manipulations, the key insight is to recognize that complex numbers have two real components and to use the corresponding properties of the field of real numbers.

Question 3.4.7 This question considers the field properties of the set of complex numbers $\mathbb{C}$ under addition and multiplication.

The set of complex numbers is closed under both addition and multiplication and these two operations are both associative on the complex numbers. These properties hold because the real numbers are closed and associative under both addition and multiplication. The detailed computations supporting these claims are left for the reader. Verify that the other field properties hold for $(\mathbb{C}, +, \cdot)$ in response to the following questions.

(a) Prove that $0 = 0 + 0i$ is the additive identity of $\mathbb{C}$.

(b) Find the additive inverse of an arbitrary complex number $a + bi$.

(c) Prove that complex addition is commutative, using the fact that addition of real numbers is commutative.

(d) Prove that $1 = 1 + 0i$ is the multiplicative identity of $\mathbb{C}$.

(e) Find the multiplicative inverse of an arbitrary complex number $a + bi$ and express this inverse in the form $c + di$. Use the following identity in answering this question.

$$\frac{1}{a + bi} = \frac{a - bi}{(a + bi)(a - bi)}$$

(f) Prove that complex multiplication is commutative, using the fact that multiplication of real numbers is commutative.

(g) Prove that multiplication distributes across addition for arbitrary complex numbers $a + bi$, $c + di$, and $e + fi$.

∎

The fields studied thus far have all been infinite, but finite fields exist. Finite fields play a pivotal role in the abstract analysis of solutions of polynomials. As it turns out, many sets of the form $\mathbb{Z}_n$ are fields under modular addition and multiplication; the following theorem characterizes exactly which of these sets are fields.

Theorem 3.4.4 *The set $\mathbb{Z}_p$ under addition mod p and multiplication mod p is a field iff $p \in \mathbb{Z}$ is a prime number.*

Proof Assuming p is prime, we prove that $\mathbb{Z}_p$ is a field. Theorem 2.4.1 from section 2.4 asserts that $\mathbb{Z}_p$ under addition mod p is a group for every prime number p. Furthermore, because integer addition is commutative, so is modular addition.

$$a \oplus b = (a + b) \bmod p = (b + a) \bmod p = b \oplus a$$

Thus, $\mathbb{Z}_p$ under addition mod p is an Abelian group.

Similarly, theorem 2.4.3 from section 2.4 asserts that $U(p) = \{1, 2, \ldots, p - 1\} = \mathbb{Z}_p \backslash \{0\}$ under multiplication mod p is a group for every prime p. Furthermore, because integer multiplication is commutative, so is modular multiplication.

$$a \odot b = (a \cdot b) \bmod p = (b \cdot a) \bmod p = b \odot a$$

Thus, $\mathbb{Z}_p \backslash \{0\}$ under multiplication mod p is an Abelian group.

Finally, since integer multiplication distributes over integer addition, modular multiplication distributes over modular addition.

$$
\begin{aligned}
a \odot (b \oplus c) &= [\, a \cdot (b + c) \,] \bmod p \\
&= [\, a \cdot b + a \cdot c \,] \bmod p \\
&= (a \cdot b) \bmod p \oplus (a \cdot c) \bmod p \\
&= (a \odot b) \oplus (a \odot c)
\end{aligned}
$$

Thus $\mathbb{Z}_p$ under modular addition and multiplication is a field.

The other half of the biconditional is often proven using the contrapositive: if $n \in \mathbb{Z}$ is nonprime, then $\mathbb{Z}_n$ is not a field under modular addition and multiplication. Question 3.4.9 below considers a specific instance of this result, and exercise 64 at the end of the section asks for the general proof.

■

As with groups, Cayley tables play a role in the study and analysis of finite sets under binary operations. Recall that Cayley tables readily determine closure, identity, and inverses for finite sets, and so they can help prove or disprove that a given set with two binary operations is a field.

Question 3.4.8 Consider the field $\mathbb{Z}_5$ under addition mod 5 and multiplication mod 5.

(a) List the five elements of $\mathbb{Z}_5$.

(b) Compute the Cayley table for $\mathbb{Z}_5$ under addition mod 5.

(c) Using the Cayley table from part (b), argue that $\mathbb{Z}_5$ is closed under addition mod 5.

(d) Identify the additive identity of $\mathbb{Z}_5$.

(e) Determine the additive inverse of each element from $\mathbb{Z}_5$.

(f) Compute the Cayley table for $\mathbb{Z}_5 \setminus \{0\} = U(5)$ under multiplication mod 5.

(g) Using the Cayley table from part (f), argue that $\mathbb{Z}_5 \setminus \{0\}$ is closed under multiplication mod 5.

(h) Identify the multiplicative identity of $\mathbb{Z}_5$.

(i) Determine the multiplicative inverse of each element from $\mathbb{Z}_5 \setminus \{0\}$.

(j) Multiplication mod 5 distributes across addition mod 5. Demonstrate that this general property $a \odot (b \oplus c) = a \odot b \oplus a \odot c$ holds for the particular triple of elements $a = 2, b = 3, c = 4$ from $\mathbb{Z}_5$.

■

A Cayley table is also helpful for showing that a set with binary operations is not a field. The next question considers a number system for which the addition operation satisfies the field properties, but the multiplication operation does not.

Question 3.4.9 Consider the set $\mathbb{Z}_4$ under addition mod 4 and multiplication mod 4.

(a) List the four elements of $\mathbb{Z}_4$.

(b) Compute the Cayley table for $\mathbb{Z}_4$ under addition mod 4.

(c) Compute the Cayley table for $\mathbb{Z}_4 \setminus \{0\}$ under multiplication mod 4.

(d) Based on one of the Cayley tables from parts (b) and (c), argue that $\mathbb{Z}_4$ is not a field because the inverse property fails for one of these binary operations.

■

Finally, the next example identifies an infinite set with binary operations that is not a field. Again, the addition operation satisfies the field properties, but the multiplication operation does not.

Example 3.4.2 The set of integers under standard addition is an Abelian group. Furthermore, multiplication distributes over addition, as noted in the proof of theorem 3.4.4 above.

However, the nonzero integers are not an Abelian group under multiplication; in particular, only 1 and -1 have multiplicative inverses. Therefore, the set of integers under standard addition and multiplication is not a field.

∎

3.4.1 Reading Questions for Section 3.4

1. Define the rational numbers $\mathbb{Q}$ in terms of the integers $\mathbb{Z}$.
2. Give an example verifying that $\mathbb{Z}$ is a proper subset of $\mathbb{Q}$.
3. State theorem 3.4.1 and the generalization of this result to an arbitrary prime number.
4. If p is a prime number, then $\sqrt{p}$ is the solution of what algebraic equation?
5. Discuss the distinction between decimal representations of rational and irrational numbers.
6. Name two approaches to defining the real numbers $\mathbb{R}$ in terms of the rational numbers $\mathbb{Q}$.
7. Give an example verifying that $\mathbb{Q}$ is a proper subset of $\mathbb{R}$.
8. Define the complex numbers $\mathbb{C}$ in terms of the real numbers $\mathbb{R}$.
9. Give an example verifying that $\mathbb{R}$ is a proper subset of $\mathbb{C}$.
10. State the three properties satisfied by a field F under two binary operations.
11. Give four examples of a field—two infinite and two finite.
12. Provide two examples of a set under two binary operations that is not a field—one infinite and one finite.

3.4.2 Exercises for Section 3.4

In exercises 1–4, express each rational number as a fraction in lowest terms.

1. $\dfrac{346}{1,000}$

2. $\dfrac{783,552}{10,000}$

3. 0.953

4. 44.56423

Exercises 5–8 consider how to manipulate a repeating decimal expression to obtain an equivalent fraction. For example, $0.\overline{9} = 1$ by the following sequence of algebraic manipulations.

$$x = 0.\overline{9} \quad \Rightarrow \quad 10x = 9.\overline{9} \quad \Rightarrow \quad \begin{array}{rcl} 10x &=& 9.\overline{9} \\ - \quad x &=& 0.\overline{9} \\ \hline 9x &=& 9 \end{array} \quad \Rightarrow \quad x = 1$$

In exercises 5–8, use the technique demonstrated above to express the following rational numbers as a fraction in lowest terms.

5. $0.\overline{7}$

6. $0.\overline{79}$

7. $0.2\overline{5}$

8. $0.\overline{545}$

Exercises 9–12 consider decimal expansions and sequences of rational and irrational numbers.

9. As noted in example 3.4.1, the decimal expansion of e begins 2.718281828. Explain why the four digits 1828 cannot repeat indefinitely.

10. Decimal expressions that are nonterminating and nonrepeating represent irrational numbers. For example, Lindemann studied the irrational number $0.101001000100001\ldots$ on his way to proving that π is irrational. Using Lindemann's idea, find four distinct irrational numbers. Have some fun and think up some wild and wacky irrational numbers!

11. Prove that there exist infinitely many rational numbers by listing an infinite sequence of distinct rational numbers.

12. Prove that there exist infinitely many irrational numbers by listing an infinite sequence of distinct irrational numbers.

Exercises 13–17 consider divisibility properties of integers. Recall that for $m, n \in \mathbb{Z}$, we say that n divides m when there exists $q \in \mathbb{Z}$ such that $m = n \cdot q$.
In exercises 13–17, prove each mathematical statement.

13. If 2 divides an integer n, then 4 divides n^2.

14. If a prime p divides an integer n, then p^2 divides n^2.

15. If a prime p divides n^2, then p divides n.

16. If a prime p divides n^3, then p divides n.

17. If a prime p divides n^k for $k \in \mathbb{N}$, then p divides n.

Exercises 18–27 consider extensions and variations on the Pythagoreans' proof that the square root of two is irrational.

18. Prove that $\sqrt{5}$ is irrational.

19. Prove that $\sqrt{7}$ is irrational.

20. Prove that if p is a prime number, then $\sqrt{p}$ is irrational.

21. Prove that $\sqrt{6}$ is irrational.

22. Prove that if p is a prime number greater than 2, then $\sqrt{2p}$ is irrational.

23. Prove that if p and q are distinct prime numbers, then $\sqrt{pq}$ is irrational.

24. Prove that $\sqrt[3]{2}$ is irrational. Hint: Use exercise 16.

25. Prove that $\sqrt[k]{2}$ is irrational for every $k \geq 2$. Hint: Use exercise 17.

26. We know that $\sqrt{4} = 2$ is rational. In the Pythagorean proof that $\sqrt{2}$ is irrational identify the first place where the proof fails to hold for $\sqrt{4}$.

27. Give two examples demonstrating that if p and q are distinct prime numbers, then $\sqrt{p+q}$ may or may not be irrational.

In exercises 28–33, give an example of each type of number, or explain why such a number does not exist.

28. A rational number that is not an integer.

29. A real number that is not rational.

30. A rational number that is not real.

31. An irrational number that is not complex.

32. A complex number that is not real.

33. A number that is not complex.

In exercises 34–39, add and multiply each pair of complex numbers.

34. $3 + 2i$ and $4 + 5i$

35. $2 + i$ and $5 - 3i$

36. $6 - 2i$ and $4 + 2i$

37. 12 and $45i$

38. $1 + 5i$ and $1 - 5i$

39. $a + bi$ and $a - bi$, where $a, b \in \mathbb{R}$

In exercises 40–45, find the additive and multiplicative inverse of each complex number.

40. $8 + 3i$

41. $7 - 2i$

42. $-3 + 5i$

43. i

44. $-2i$

45. -2

Exercises 46–47 consider further computations with complex numbers.

46. For $u, v \in \mathbb{C}$, the fraction u/v can be thought of as $u \cdot v^{-1}$; in other words, as u times the multiplicative inverse of v. Using this approach, find the value of $(1 + i)/(2 - 3i)$.

47. Prove that $w = (1/\sqrt{2}) + (1/\sqrt{2})i$ is $\sqrt{i}$ by computing $w^2 = w \cdot w$.

In exercises 48–51, each set defined below is a field under the given pair of binary operations. Determine both the additive and multiplicative inverse of an arbitrary element of the field. (Hint: Cayley tables are helpful for answering these questions.)

48. $\{0, 2, 4, 6, 8\}$ under addition and multiplication mod 10.

49. $\mathbb{Z}_5\left[\sqrt{3}\right] = \{a + b\sqrt{3} : a, b \in \mathbb{Z}_5\}$ under addition and multiplication mod 5.

50. $\mathbb{Q}\left[\sqrt{2}\right] = \{a + b\sqrt{2} : a, b \in \mathbb{Q}\}$ under standard addition and multiplication
    Hint: Rationalize the denominator of $1/(a + b\sqrt{2})$ to find the multiplicative inverse.

51. $\mathbb{Q}\left[\sqrt{3}\right] = \{a + b\sqrt{3} : a, b \in \mathbb{Q}\}$ under standard addition and multiplication.

In exercises 52–56, compute the additive and multiplicative Cayley tables for each field.

52. $\mathbb{Z}_3$

53. $\mathbb{Z}_5$

54. $\mathbb{Z}_7$

55. $\{0, 2, 4, 6, 8\}$ under mod 10 operations

56. $\mathbb{Z}_3[i] = \{a + bi : a, b \in \mathbb{Z}_3\}$

In exercises 57–66, each set is *not* a field under the given pair of binary operations. Identify the field axioms that fail to hold and give a counterexample.

57. The natural numbers $\mathbb{N}$ under standard addition and multiplication.

58. The integers $\mathbb{Z}$ under standard addition and multiplication.

59. The set $n\mathbb{Z} = \{k \cdot n : k \in \mathbb{Z}\}$ (that is, the set of all multiples of an integer n) under standard addition and multiplication.

60. The set $\{0, 1\}$ under standard addition and multiplication.

61. The set $\{-1, 0, 1\}$ under standard addition and multiplication.

62. The set $\mathbb{Z}_4$ under addition mod 4 and multiplication mod 4.

63. The set $\mathbb{Z}_6$ under addition mod 6 and multiplication mod 6.

64. The set $\mathbb{Z}_n$ under addition mod n and multiplication mod n, when n is not a prime number.

65. The set of ordered pairs of integers $\mathbb{Z}^2 = \{(m, n) : m, n \in \mathbb{Z}\}$ under componentwise addition and multiplication.

66. The set of ordered pairs of real numbers $\mathbb{R}^2 = \{(r, s) : r, s \in \mathbb{R}\}$ under componentwise addition and multiplication.

Exercises 67–70 consider *ordered fields* and some of their basic properties. An ordered field $(F, +, \cdot)$ is a field with a binary relation $<$ satisfying the following properties.

- **comparability:** For all $x, y \in F$, exactly one of $x < y$, $x = y$, $y < x$ is true.
- **transitivity:** For all $x, y, z \in F$, if $x < y$ and $y < z$, then $x < z$.
- **addition preserves order:** For all $x, y, z \in F$, if $x < y$, then $x + z < y + z$.
- **multiplication preserves positives:** For all $x, y \in F$, if $x > 0$ and $y > 0$, then $x \cdot y > 0$.

In exercises 67–70, consider the ordered field of real numbers and prove the following mathematical statements.

67. If $r, s \in \mathbb{R}$, then $(-r) \cdot s = -(r \cdot s)$ and $r \cdot (-s) = -(r \cdot s)$, the unique additive inverse of $r \cdot s$.

68. If $r, s \in \mathbb{R}$, then $(-r) \cdot (-s) = r \cdot s$.

69. If $r \in \mathbb{R}$, then $r < 0$ implies $-r > 0$. Similarly, $r > 0$ implies $-r < 0$

70. If $r \in \mathbb{R}^* = \mathbb{R} \setminus \{0\}$, then $r^2 > 0$.

3.5　Polynomials and Transcendental Numbers

This section continues to study the basic number systems: the natural numbers; integers; rationals; reals; and complex numbers. The previous section verified the proper subset relationships among these sets of numbers, which clarified the numbers these sets do and do not share in common. This section revisits each number system, considering relationships among numbers that arise from variable expressions. Section 3.3 took the first steps in this direction with the study of the Pythagorean theorem and Fermat's last theorem.

The primary focus is polynomial equations. Polynomials are the simplest and most basic of variable expressions, and mathematicians have devoted many centuries of effort to the study of their solutions. Context plays a key role in this investigation—we must take care to identify the number system in which solutions are sought. The first step is the formal definition of a polynomial.

Definition 3.5.1　*A **polynomial over a set F** is an expression of the form $a_n x^n + \cdots + a_1 x + a_0$ with nonnegative integer powers and **coefficients** $a_n, \ldots, a_0 \in F$. The symbol x is a **variable** that can take on any value in a given variable domain. The value n is* *the greatest power of x with a nonzero coefficient and is called the* **degree** *of the polynomial. A nonzero constant polynomial is said to have degree* 0.

Polynomials are widely studied in mathematics courses. Most often, both the coefficient set F and the domain of the variable x are a field; when they are left unspecified, both of these sets are assumed to be the field $\mathbb{C}$.

Example 3.5.1   The expression $5x^3 + 6x + 1$ is a polynomial over $\mathbb{C}$ of degree 3. The coefficients of this polynomial are $a_3 = 5$, $a_2 = 0$, $a_1 = 6$, and $a_0 = 1$. Notice that this same expression could be interpreted as a polynomial over $\mathbb{Z}_7$; in this case, $F = \mathbb{Z}_7$ must be explicitly identified. Many familiar variable expressions are not polynomials, including $x^2 + \sqrt{x}$, $\sin(x)$, e^x, and $x + y$ (though $x + y$ is called a polynomial of two variables).

■

Question 3.5.1   Determine the degree and the coefficients of each polynomial over $\mathbb{C}$. For the first two polynomials, also identify a finite field that could serve as F.

(a) $3x^4 + 2x^3 - 7x^2 + 5x - 1$ (c) $(1 + i)x^3 + 2ix - 4$

(b) $2x^5 + 4x^2 + x$ (d) $2ix^7 + (1 + i)$

■

Question 3.5.2   Identify polynomials with degree 17, 2, 1, and 0.

■

Question 3.5.3   State three variable expressions that are not polynomials; give examples different from those in example 3.5.1.

■

Work with polynomials often focuses around the study of *zeros* of polynomial equations; that is, elements of the variable domain for which the polynomial's value equals 0. Zeros are often referred to as the *solutions* of the equation obtained by setting the polynomial equal to 0; zeros are also said to *satisfy* the corresponding polynomial equation. Solving polynomial equations is a common exercise in high school and early undergraduate mathematics courses, where zeros are shown to be significant in many settings. For example, zeros provide information in calculus about maximum and minimum polynomial function values. In practical applications, polynomials' zeros often identify the key features of mathematical models, providing insights into the physical and social world.

Definition 3.5.2   *A* **zero** *of the polynomial* $a_n x^n + \cdots + a_1 x + a_0$ *over a set F is an element w of the variable domain for which* $a_n w^n + \cdots + a_1 w + a_0 = 0$.

Example 3.5.2   The integer $x = 2$ is a zero of the polynomial $3x^2 - 12$ because $3 \cdot 2^2 - 12 = 12 - 12 = 0$. Similarly, $x = -2$ is also a zero of this same polynomial. On the other hand, $x = 1$ is not a zero of $3x^2 - 12$ because $3 \cdot 1^2 - 12 = 3 - 12 = -9 \neq 0$.

■

Question 3.5.4   (a) Verify that $x = 3$ is a zero of $2x^3 - 5x^2 - 9x + 18$ over $\mathbb{C}$.

(b) Verify that $x = 3$ is a zero of $2x^3 - 5x^2 - 9x + 18 = 2x^3 + x^2 + 3x$ over $\mathbb{Z}_6$.

(c) Verify that $x = 2$ is *not* a zero $2x^3 - 5x^2 - 9x + 18$ over $\mathbb{C}$.

(d) Verify that $x = i$ is a zero of $2x^3 + 3x^2 + 2x + 3$ over $\mathbb{C}$.

■

Example 3.5.3   The polynomial $2x^2 - 10x + 12$ has two zeros ($x = 2$ and $x = 3$), while the polynomial $x^2 - 2x + 1$ has only one zero ($x = 1$). If c is a zero of a polynomial, then $x - c$ is a factor of the polynomial. Thus, $x - 2$ and $x - 3$ are both factors of $2x^2 - 10x + 12$, and this polynomial can be written as $2x^2 - 10x + 12 = 2(x - 2)(x - 3)$. Similarly, $x^2 - 2x + 1 = (x - 1)^2$.

   The power of a linear term $x - c$ is referred to as the *multiplicity* of the zero c. Thus, $2x^2 - 10x + 12$ has two zeros each with multiplicity 1, and $x^2 - 2x + 1$ has a single zero with multiplicity 2.

■

   In this section, zeros of polynomials are considered as an alternative means of defining the basic number systems. This study leads to an important classification of real numbers. Recall that real numbers are either rational or irrational based on their expressibility as fractions of integers. Real numbers can also be viewed from the perspective of polynomial equations, with a classification of reals as either *algebraic* (expressible as zeros of polynomials) or *transcendental* (not expressible as zeros of polynomials).

   Working toward this classification of the reals, consider the natural numbers $\mathbb{N}$ as a given number system and generate the integers $\mathbb{Z}$ using polynomial equations over $\mathbb{N}$. In particular, elements of $\mathbb{Z}$ are either elements of $\mathbb{N}$ or are zeros of the polynomials over $\mathbb{N}$ of the form x or $x + n$, where $n \in \mathbb{N}$.

   In a similar way, $\mathbb{Q}$ can be obtained from $\mathbb{Z}$. Any rational number x is of the form m/n, where $m, n \in \mathbb{Z}$ and $n \neq 0$. For m/n, we have $nx = m$, which implies that $nx - m = 0$. In short, we obtain $\mathbb{Q}$ by starting with $\mathbb{Z}$ and identifying all zeros of the polynomials $n \cdot x - m$ over $\mathbb{Z}$.

   As suggested in the last section, the real numbers $\mathbb{R}$ are not so readily obtained from $\mathbb{Q}$. The definition of $\mathbb{R}$ in terms of $\mathbb{Q}$ is typically given using either Cauchy sequences or Dedekind cuts. Considering the possibility of defining $\mathbb{R}$ from $\mathbb{Q}$ using polynomials leads to some very interesting insights and results. Since the given base set is the rationals, the goal is to identify polynomials over the rationals whose zeros are irrational numbers. There are many such polynomials; for example, $\sqrt{2}$ is a zero of $x^2 - 2$, $\sqrt[3]{2}$ is a zero of $x^3 - 2$, $\sqrt[6]{2}$ is a zero of $x^6 - 2$, and so on. This handful of examples demonstrates that polynomials of every degree are needed just to define the nth roots of two, for every $n \in \mathbb{N}$. Similarly, polynomials of every degree are needed to obtain the nth roots of other integers, including prime numbers. Thus there does not exist a simple formulaic description for a subset of polynomials over $\mathbb{Q}$ with bounded, finite degree whose zeros generate the reals $\mathbb{R}$.

   Faced with this obstruction to the program of defining the reals as zeros of polynomials over the rationals, we can either give up or reframe the approach so as to accomplish as much as possible in terms of the original goal. When grappling with open questions, mathematicians often need to add limiting assumptions or rearticulate goals to proceed to a productive analysis. In this situation, we have been seeking a set of polynomials with bounded, finite degree over $\mathbb{Q}$ that generate $\mathbb{R}$. A natural redirection is to consider the zeros of *every* polynomial over $\mathbb{Q}$. Some polynomials over $\mathbb{Q}$ do not have real zeros; for example, $x^2 + 1$ has only complex zeros i and $-i$. But if we gather together every *real* zero of every polynomial over $\mathbb{Q}$, do we obtain the set of all real numbers? Surprisingly, the answer is no! In this way, there are fundamental limits to the expressiveness of polynomial equations. In light of this discussion, consider the following definition.

Definition 3.5.3   *Let $F \subseteq \mathbb{R}$. An element $a \in \mathbb{R}$ is called **algebraic over** F when a is the zero of a nonzero polynomial over F. An element $a \in \mathbb{R}$ is called **transcendental over** F when a is not algebraic over F. When $F = \mathbb{Q}$, the phrase "over F" is omitted and a is said to be either algebraic or transcendental.*

The eminent Swiss mathematician Leonhard Euler first stated this distinction among real numbers in 1744, giving the name "transcendental" to nonalgebraic reals because "they transcend the power of algebraic methods." Many familiar real numbers are algebraic over $\mathbb{Q}$.

Example 3.5.4   The real number 1 is algebraic because 1 is a zero of the polynomial $x - 1$ over $\mathbb{Q}$. The real number $\sqrt{2}$ is algebraic because $\sqrt{2}$ is a zero of the polynomial $x^2 - 2$ over $\mathbb{Q}$.

                                                                    ■

Extending the approach in the first half of example 3.5.4, every rational number is algebraic. Exercise 23 at the end of this section asserts the more general claim that every element of a field F is algebraic over F. Furthermore, example 3.5.4 also shows that some irrational numbers such as $\sqrt{2}$ are algebraic. However, not every real is algebraic, and some of the most famous of irrationals are transcendental. Several examples are presented in the following theorem, along with the mathematician who first proved the result and the year of its proof. Notice that these proofs followed over a century after Euler first isolated the distinction between algebraic and transcendental numbers in 1744.

Theorem 3.5.1   • *(Liouville, 1844) There exist (real) transcendental numbers .*
                 • *(Hermite, 1873) The real number e is transcendental.*
                 • *(Lindemann, 1882) The real number π is transcendental.*

Thus, some of the most significant numbers in mathematics are not only irrational, but also transcendental. The difficulty of these proofs is reflected by their relatively recent date of publication. Furthermore, there are many simply described real numbers whose classification is still undetermined; for example, mathematicians still do not know if $e + \pi$ is transcendental!

We pause to describe briefly the three mathematicians associated with theorem 3.5.1. Joseph Liouville was an accomplished French mathematician who worked in Paris for much of the nineteenth century. Liouville sought to prove that e is transcendental, and (while he did not succeed) he made important progress in this direction by proving the existence of (infinitely many) real transcendental numbers. In 1851, his work on this same question of proving e transcendental led to a constructive proof that the

"Liouvillian" number 0.11000100... (with one appearing in each $n!$ decimal place and zero appearing elsewhere) is transcendental.

Charles Hermite was another nineteenth century French mathematician. Hermite overcame many obstacles (including a physical disability and a somewhat limited and mediocre education) to become one of the great research mathematicians of his time. Hermite's proof that e is transcendental was followed closely by the German mathematician Ferdinand Lindemann's result a few years later. In essence, Lindemann coupled Hermite's proof with the fact that $e^{i\pi} = -1$ to prove that π is transcendental. This result also provided a negative answer to one of the classical open problems of ancient Greek mathematics: can we *square the circle* by constructing a square with the same area as a given circle using only a ruler and compass? Lindemann's proof that π is transcendental implies that the circle cannot be squared.

Despite the impossibility of defining $\mathbb{R}$ from $\mathbb{Q}$ using polynomials, the complex numbers $\mathbb{C}$ are defined from $\mathbb{R}$ readily enough. In general, an arbitrary element $a + bi \in \mathbb{C}$ is a zero of the quadratic polynomial $x^2 - 2ax + (a^2 + b^2)$ over $\mathbb{R}$, which can be verified by direct substitution as follows.

$$(a + bi)^2 - 2a(a + bi) + a^2 + b^2 = a^2 - b^2 + 2abi - 2a^2 - 2abi + a^2 + b^2 = 0.$$

Question 3.5.5   State polynomials over $\mathbb{R}$ with the following zeros.

(a) $2 + i$ (c) $3 - 4i$

(b) i^2 (d) i^4

■

The complex numbers possess another important property with respect to polynomials. Recall that the polynomial $x^2 + 1$ does not have a zero in any of the basic number systems except for $\mathbb{C}$. While $i \in \mathbb{C}$ provides a zero for $x^2 + 1$ (and many other polynomials), a natural question is whether or not there exists a polynomial over $\mathbb{C}$ that does not have a zero in $\mathbb{C}$. In fact, no such polynomial exists; that is, $\mathbb{C}$ is *algebraically closed* in the sense that every zero of every polynomial over $\mathbb{C}$ is an element of $\mathbb{C}$. This result implies that every polynomial over $\mathbb{C}$ can be expressed as a product of linear polynomials with complex coefficients. This key insight into the complex numbers has become known as the fundamental theorem of algebra.

Theorem 3.5.2 Fundamental theorem of algebra *A polynomial over $\mathbb{C}$ of degree $n > 0$ has n zeros in $\mathbb{C}$. These zeros need not be distinct but are counted according to multiplicity.*

Sometimes the fundamental theorem of algebra is stated as the assertion that every polynomial over $\mathbb{C}$ of degree $n > 0$ has a zero; theorem 3.5.2 follows from this claim by repeated applications of factoring and long division. A proof of the fundamental theorem of algebra is beyond the scope of this text and is left for your later studies.

Versions of the fundamental theorem of algebra were stated (but not proven) by the French mathematicians Albert Girard in 1629 and René Descartes in 1637 (in his seminal treatise *La Geometrie*). By the mid-1700s Jean Le Rond d'Alembert and Leonhard Euler had attempted proofs of this theorem, but their arguments were incomplete and so unsuccessful. The first known complete proof was given by

Carl Friedrich Gauss in 1796. Gauss published his first proof of the fundamental theorem of algebra at the age of 19 as part of his doctoral dissertation at the University of Helmstedt. Since then, more than 100 different proofs have been given for this result, and these proofs can be studied in such courses as abstract algebra, number theory, and complex analysis. During his lifetime, Gauss published four different proofs of the fundamental theorem of algebra, the last at the age of 70.

Question 3.5.6   Identify every zero and its multiplicity for each polynomial over $\mathbb{C}$.

(a) $x - 4$ (c) $x^3 - 1$
(b) $x^2 - 2x + 1$ (d) $x^4 - 1$

■

Knowing the existence of n complex zeros for a polynomial over $\mathbb{C}$ of degree n does not necessarily provide much insight into actually finding these zeros. And so it seems natural to ask, "How can one find the zeros of an arbitrary polynomial over $\mathbb{C}$?" This question has occupied the attention of mathematicians for centuries; the most significant breakthrough in this area occurred in the nineteenth century when Paolo Ruffini, Niels Henrik Abel, and Évariste Galois independently recognized the *impossibility* of finding a general formula providing the solutions of polynomials with sufficient complexity. Working in the direction of understanding these results, we consider the process of finding zeros of polynomials over $\mathbb{C}$ of increasing degree one case at a time.

3.5.1   Linear Polynomials

For a polynomial of degree $n = 1$, calculating the single zero is relatively straightforward. The unique zero of $a \cdot x + b$ is given by the following.

$$x = -\frac{b}{a}$$

When a and b are real numbers, then the meaning of $-b/a$ is apparent. But, what if a and b are complex? What complex number is associated with the fraction $-b/a$? For $a = c + di$ and $b = e + fi$ with $c, d, e, f, \in \mathbb{R}$, the fraction $x = -b/a$ is expressed as a complex number in standard form via the following computation.

$$x = -\frac{b}{a} = -\frac{e + fi}{c + di} = -\frac{e + fi}{c + di} \cdot \frac{c - di}{c - di} = -\frac{ce + df}{c^2 + d^2} + \frac{de - cf}{c^2 + d^2} \cdot i$$

In other words, $-b/a$ is computed by multiplying both the numerator and the denominator by the *complex conjugate* $c - di$ of the denominator $c + di$.

Example 3.5.5   We express the quotient $(2 + i)/(3 - 5i)$ as a complex number of the form $a + bi$, where $a, b \in \mathbb{R}$.

Based on the computation modeled above, we obtain the following.

$$\frac{2 + i}{3 - 5i} = \frac{2 + i}{3 - 5i} \cdot \frac{3 + 5i}{3 + 5i} = \frac{6 + 10i + 3i - 5}{9 + 25} = \frac{1 + 13i}{34} = \frac{1}{34} + \frac{13}{34}i$$

■

Question 3.5.7   Identify the zeros of the following polynomials over $\mathbb{C}$; express complex numbers that occur as solutions in the standard form $a + bi$, where $a, b \in \mathbb{R}$.

(a) $2x - 4$ (c) $e \cdot x + \pi + 1$

(b) $\sqrt{2} \cdot x - 8$ (d) $(2 + 3i)x - 12$

■

Before considering quadratic polynomials, we mention a bit of the interesting history of negative numbers. Many early mathematicians did not accept the idea of negative numbers, including Euclid and Diophantus. The ancient Greeks thought of numbers as physical lengths or magnitudes, and a negative length made little sense to them. By the seventh and ninth centuries, Indian and Islamic mathematicians were working with negative numbers. The first known example of a negative number being written in an equation is due to the French mathematician Nicolas Chuquet in 1484. But even as late as 1637, Descartes still referred to negative solutions of equations as "false roots." Ultimately, the practical use of negative numbers in commercial record-keeping brought negative numbers into widespread acceptance and use.

3.5.2   Quadratic Polynomials

For ease of calculation, the rest of this section examines only polynomials over $\mathbb{R}$, although all of the algorithms we introduce extend to the complex numbers. When the degree of a polynomial is $n = 2$, the familiar *quadratic formula* determines the two zeros of the polynomial. As early as the seventeenth century B.C.E., Babylonians understood how to apply the quadratic formula. A specific known example of their work includes the determination of the two roots of $x^2 - x - 87 = 0$ as $\frac{1}{2} \pm \sqrt{349}/2$. The following familiar description of the solutions of $ax^2 + bx + c = 0$ was not expressed until the late sixteenth century, by the French lawyer and mathematician Francois Viète.

$$x = \frac{-b \pm \sqrt{b^2 - 4ac}}{2a}$$

The term $b^2 - 4ac$ appearing inside the square root is called the *discriminant*. The next example illustrates the use of the quadratic formula.

Example 3.5.6   We identify the zeros of each quadratic.

- The zeros of $x^2 - 1$ are the two real values

$$x = \frac{0 \pm \sqrt{0 - 4 \cdot 1 \cdot (-1)}}{2 \cdot 1} = \pm 1$$

- The zeros of $x^2 - 2x + 1$ both equal

$$x = \frac{2 \pm \sqrt{4 - 4 \cdot 1 \cdot 1}}{2 \cdot 1} = 1,$$

which has multiplicity two.

- The zeros of $x^2 + x + 1$ are the complex values

$$x = \frac{-1 \pm \sqrt{1 - 4 \cdot 1 \cdot 1}}{2 \cdot 1} = -\frac{1}{2} \pm \frac{\sqrt{3}}{2} i.$$

The corresponding factorizations of these quadratics are therefore

$$x^2 - 1 = (x - 1)(x + 1), x^2 - 2x + 1 = (x - 1)^2,$$

and

$$x^2 + x + 1 = [x + (1 - \sqrt{3}i)/2][x + (1 + \sqrt{3}i)/2].$$

∎

Example 3.5.6 illustrates the only possible combinations of zeros for a quadratic polynomial with real coefficients: a pair of distinct real roots, a real root of multiplicity two, or a pair of complex conjugates. The form of the quadratic formula forces every polynomial with real coefficients to have roots that are either real or that occur as complex conjugate pairs.

Using the quadratic formula to identify the zeros of a quadratic polynomial with real coefficients is familiar and relatively straightforward, as illustrated above. However, for quadratics with complex coefficients (which result in a discriminant that is complex) the mathematics is more subtle. Mathematicians have developed a well-defined extension of the square root function on real numbers to a square root function on complex numbers that enables us to express such roots in the standard form $a + bi$, where $a, b \in \mathbb{R}$. This more advanced discussion of complex numbers is saved for chapter 7.

Question 3.5.8   Identify the two zeros of the following quadratic polynomials by either directly factoring the polynomials or using the quadratic formula.

(a) $x^2 + 4x + 4$ (c) $6x^2 - 3x + 7$
(b) $3x^2 - 5x - 2$ (d) $8x^2 - 8x + 1$

∎

3.5.3   Cubic Polynomials

Polynomials over the complex numbers with degree $n = 3$ also have a formula that generates its zeros—one that is more subtle and complicated. The development of the complete solution for cubic polynomials is due to several Italian mathematicians during the Renaissance. By 1515, Sciopione del Ferro had identified an approach to determining the zeros of every cubic polynomial of the form $x^3 + mx + n$; such polynomials are known as *depressed cubics*. Despite his position as the Chair of Arithmetic and Geometry at the University of Bologna, del Ferro did not publish or announce this breakthrough. However, others eventually recognized that such techniques must exist—del Ferro and his student Antonio Fior were winning multiple algebra contests, which were all the rage at that time. This recognition inspired Niccoló Fontana (better known as Tartaglia, or the "stammerer") in the 1530s to determine independently del Ferro's solution. This *del Ferro–Tartaglia solution* of the depressed cubic $x^3 + mx + n$ is the zero x_1 produced by the following formula.

$$x_1 = \sqrt[3]{\frac{-n}{2} + \sqrt{\frac{n^2}{4} + \frac{m^3}{27}}} - \sqrt[3]{\frac{n}{2} + \sqrt{\frac{n^2}{4} + \frac{m^3}{27}}}$$

Once x_1 is found, polynomial long division is used to express the cubic as

$$x^3 + mx + n = (x - x_1) \cdot (ax^2 + bx + c), \text{ where } a, b, c \in \mathbb{C}.$$

Finally, applying the quadratic formula to $ax^2 + bx + c$ produces the other two zeros.

Example 3.5.7   We identify the three zeros of the cubic polynomial $2x^3 - 6x + 4$.

The polynomial is not a depressed cubic, but this is easily remedied by dividing the original polynomial by two. The cubic $x^3 - 3x + 2$ has the same zeros as the original. The del Ferro–Tartaglia solution of this depressed cubic is computed as follows.

$$x_1 = \sqrt[3]{\frac{-2}{2} + \sqrt{\frac{2^2}{4} + \frac{(-3)^3}{27}}} - \sqrt[3]{\frac{2}{2} + \sqrt{\frac{2^2}{4} + \frac{(-3)^3}{27}}}$$

$$= \sqrt[3]{-1 + \sqrt{1-1}} - \sqrt[3]{1 + \sqrt{1-1}} = -1 - 1 = -2$$

Using long division, we factor $x^3 - 3x + 2$ into the linear term $(x + 2)$ and a quadratic term.

$$
\begin{array}{r}
x^2 - 2x + 1 \\
x + 2 \overline{\smash{)}\ x^3 - 3x + 2} \\
\underline{-(x^3 + 2x^2)} \\
-2x^2 - 3x + 2 \\
\underline{-(-2x^2 - 4x)} \\
x + 2 \\
\underline{-(x + 2)} \\
0
\end{array}
$$

Therefore, $x^3 - 3x + 2 = (x + 2)(x^2 - 2x + 1)$. Finally, applying the quadratic formula with $a = 1$, $b = -2$, and $c = 1$, we have $x_2 = x_3 = 1$. Thus, the three zeros of the original polynomial $2x^3 - 6x + 4$ are $x_1 = -2$, $x_2 = 1$, and $x_3 = 1$.

■

Question 3.5.9   Find the del Ferro solution to the cubic polynomial $3x^3 - 36x + 48$.

■

In the late 1530s, Tartaglia not only recovered the del Ferro solution for depressed cubics, he also obtained solutions for many types of other cubic polynomials (but not all). Like del Ferro, Tartaglia also did not publish or announce his results, although he did share them privately with Girolama Cardano, a well-established physician and mathematician from Milan. Just a few years later, Cardano extended Tartaglia's work to find a complete solution of general cubic polynomials, which he published in an important 1545 book entitled the *Ars Magna* (that is, *The Great Art, or the Rules of Algebra*).

Cardano's work identified a *transformation* that reduces a general cubic polynomial to a corresponding depressed cubic. For a cubic polynomial of the form $ax^3 + bx^2 + cx + d$, the following transformation produces a corresponding depressed cubic $y^3 + my + n$.

$$y = x + \frac{b}{3a} \qquad m = \frac{c}{a} - \frac{b^2}{3a^2} \qquad n = \frac{2b^3}{27a^3} - \frac{bc}{3a^2} + \frac{d}{a}$$

This mapping of variables and coefficients is known as *Cardano transformation*. The zeros of the general cubic are found by applying the Cardano transformation and then finding the del Ferro–Tartaglia solution y_1 that is a zero of $y^3 + my + n$. Computing the inverse transformation, $x_1 = y_1 - (b/3a)$ is a zero of the original cubic $ax^3 + bx^2 + cx + d$. The other two zeros are determined as before: use long division to factor the original polynomial into a linear term and a quadratic term, and finally apply the quadratic formula to produce the other two zeros. The next question carefully leads you through the process of using the Cardano transformation to find the zeros of a general cubic polynomial.

Question 3.5.10   Find the zeros of the cubic polynomial $x^3 - 6x^2 + 11x - 6$.

(a) The leading coefficient of this polynomial is $a = 1$. Identify the other three coefficients b, c, and d when viewing this polynomial from the perspective of the general cubic polynomial $ax^3 + bx^2 + cx + d$.

(b) Applying the Cardano transformation, find the coefficients m and n of the corresponding depressed cubic $y^3 + my + n$.

(c) Compute the del Ferro–Tartaglia solution y_1 that is a zero of $y^3 + my + n$.

(d) Implement the inverse transformation $x_1 = y_1 - (b/3a)$ to obtain a zero x_1 for $x^3 - 6x^2 + 11x - 6$.

(e) Find the other two zeros x_2 and x_3 for $x^3 - 6x^2 + 11x - 6$ using long division and the quadratic formula.

■

3.5.4 Quartic Polynomials

With this success in identifying zeros of a general cubic, mathematicians naturally turned their attention to polynomials of degree $n = 4$, known as quartics. In 1545, the Italian mathematician Lodovico Ferrari (a student of Cardano) developed an algorithm for finding the zeros of the general quartic polynomial $ax^4 + bx^3 + cx^2 + dx + e$. Ferrari's success in solving the general quartic polynomial led to a professorship in Bologna and was one of the factors that prompted Cardano's publication of the *Ars Magna*.

As you might expect, the Ferrari quartic algorithm is much more complicated than the del Ferro–Tartaglia solution; this text does not provide an example of a detailed Ferrari calculation. In general, the algorithm "completes the square" of the quartic and then finds zeros in a nesting of quartic, cubic, and square roots—similar in flavor to the nesting of roots used to find the zeros of cubics. For your interest, Ferrari's method is summarized here. The zeros of $ax^4 + bx^3 + cx^2 + dx + e$ are found by first calculating:

$$A = -\frac{3b^2}{8a^2} + \frac{c}{a}, \qquad B = \frac{b^3}{8a^3} - \frac{bc}{2a^2} + \frac{d}{a}, \qquad \text{and} \qquad C = -\frac{3b^4}{256a^4} + \frac{cb^2}{16a^3} - \frac{bd}{4a^2} + \frac{e}{a}.$$

If $B = 0$, the four roots are given by

$$x = -\frac{b}{4a} \pm \sqrt{\frac{-A \pm \sqrt{A^2 - 4C}}{2}}.$$

Otherwise, compute

$$D = -\frac{A^2}{12} - C, \qquad E = -\frac{A^3}{108} + \frac{AC}{3}, \qquad \text{and} \qquad F = \frac{E}{2} \pm \sqrt{\frac{E^2}{4} + \frac{D^3}{27}},$$

where plus or minus is chosen in the formula for F so that $F \neq 0$. Now set $G = \sqrt[3]{F}$ (in case F is complex, choose any one of the three cube roots) and define

$$H = \begin{cases} 0 & \text{if } G = 0 \\ \dfrac{D}{3G} & \text{if } G \neq 0. \end{cases}$$

Letting $y = -5A/6 - G + H$, the four roots are given by

$$x = -\frac{b}{4a} + \frac{\pm_1 \sqrt{A + 2y} \pm_2 \sqrt{-(3A + 2y \pm_1 2B/\sqrt{A + 2y})}}{2},$$

where the two symbols $\pm_1$ must have the same sign, and the other symbol $\pm_2$ acts independently.

The most important fact to understand here is that such an algorithm exists. Furthermore, the solution requires only certain types of algebraic operations applied to the coefficients of the given polynomial: addition and subtraction; multiplication and division; and nth roots. Mathematicians refer to such solutions as a *solutions by radicals*.

3.5.5 Quintic and Higher-Degree Polynomials

With so much success by the middle of the sixteenth century, mathematicians optimistically turned their attention to higher-degree polynomials. The next goal was to determine a solution by radicals for the zeros of a general quintic polynomial equation $ax^5 + bx^4 + cx^3 + dx^2 + ex + f$ of degree $n = 5$. Mathematicians in the sixteenth, seventeenth, and eighteenth centuries believed that all polynomials had solutions by radicals and expected it would be just a matter of time until a general algorithm would be identified. In 1771, the Italian–French mathematician Joseph-Louis Lagrange began studying "permutations," or mappings, of zeros of polynomials in his treatise *Reflections on the Algebraic Theory of Equations*. While he did not obtain a solution, Lagrange's work provided insight that helped the next generation of mathematicians obtain a complete analysis.

As it turns out, the general quintic polynomial is not solvable by radicals. The Italian mathematician Paolo Ruffini published his first attempt to prove the insolvability of the quintic in a 1799 book entitled *General Theory of Equations in which it is Shown that the Algebraic Solutions of the General Equation of Degree Greater than Four is Impossible*. Ruffini built on Lagrange's work and developed many new, important theorems (that are now interpreted as group-theoretic results). With the exception of one gap in his argument, Ruffini proved the insolvability of the quintic, but his work was not fully understood or accepted by the leading mathematicians of his time.

Independent of Ruffini's work, the Norwegian mathematician Niels Henrik Abel proved the insolvability of the quintic in 1824. His proof was similar in character to Ruffini's, but without the important gap that Ruffini had overlooked. Évariste Galois, who was unaware of the work of Ruffini and Abel, drafted a proof of the insolvability of the quintic by developing an abstract approach to studying solutions of polynomials. Galois' results give a condition that determines when a fifth degree polynomial can be solved by radicals and when one cannot. Though Galois died young, his insights were eventually developed into a subfield of abstract algebra that has become known as Galois theory, which remains an active area of research and study. Contemporary mathematicians typically prove the insolvability of the quintic using the powerful, general results of Galois theory. The next theorem formally states this insolvability of the quintic in what has become known as Abel's theorem.

Theorem 3.5.3 Abel's Theorem *If an integer $n \geq 5$, then there does not exist a solution by radicals that identifies every zero of an arbitrary polynomial over $\mathbb{C}$ of degree n; that is, there is no general formula using only the algebraic operations of addition, subtraction, multiplication, division, and integer roots of polynomial coefficients that provides every zero of an arbitrary polynomial over $\mathbb{C}$ of degree n.*

Although the proof of Abel's theorem is difficult, this result provides a complete understanding of a constructive approach to factoring an arbitrary polynomial into linear terms: such a uniform algorithm involving basic algebraic operations exists only for polynomials of degree 4 or less. In order to factor a polynomial of degree greater than 4, enough zeros must be identified by other means to reduce the polynomial's factors to degree less than or equal to four.

Abel's theorem does not imply the impossibility of finding zeros for all polynomials of degree greater than four (for example, we can readily solve $x^5 = 0$), only that there is no procedure using the basic algebraic operations that simultaneously solves every such polynomial of a given degree. In addition, this result ensures only that polynomials of degree greater than four are not uniformly solvable by radicals; other solutions are possible using other operations. Many positive results have been proven. In the 1850s, Charles Hermite, Leopold Kronecker, and Francesco Briosch independently proved that quintic polynomials are solvable using "elliptic modular" functions. In the following decades, solutions were found for polynomials of degree greater than four using such tools as modular functions, theta functions, and Mellin integrals. Furthermore, in the 1990s, mathematicians successfully found a solution by radicals for those quintics that are solvable by radicals.

3.5.6 Reading Questions for Section 3.5

1. What is a polynomial? Give an example of a polynomial and a variable expression that is not.
2. What is the degree of a polynomial? Give an example.
3. Define and give an example of a zero of a polynomial.
4. Define and give an example of an algebraic number over $\mathbb{Q}$.

5. Define and give an example of a transcendental number over $\mathbb{Q}$.

6. State the fundamental theorem of algebra.

7. What is the complex conjugate of a complex number? How do we use the complex conjugate to simplify fractions involving complex numbers?

8. State the formula that provides every zero of any given quadratic polynomial.

9. Define and give an example of a depressed cubic.

10. State the formula that provides the del Ferro–Tartaglia solution of a depressed cubic.

11. Define the Cardano transformation that maps a general cubic polynomial to a depressed cubic polynomial.

12. State Abel's theorem. Why is it interesting?

3.5.7   Exercises for Section 3.5

In exercises 1–6, state the degree of each polynomial or explain why the given variable expression is not a polynomial.

1. $2x^2 + 3x + 5$

2. $ix^5 + ex^3 + \pi$

3. $x + x^{-1}$

4. $x^{\frac{5}{2}} + x^{\frac{3}{2}} + x^{\frac{1}{2}}$

5. $x^{247} + \tan(x^2)$

6. $e^{x^2} + e^x + e$

In exercises 7–12, determine if the given number is a zero of the specified polynomial.

7. $x = 4$ for $x - 4$ over $\mathbb{C}$

8. $x = 4$ for $x - 4$ over $\mathbb{Z}_3$

9. $x = 3 + i$ for $x^2 + x - 3i$ over $\mathbb{C}$

10. $x = i$ for $x^4 - 1$ over $\mathbb{C}$

11. $x = 2$ for $x^2 + 3x + 2$ over $\mathbb{Z}_6$

12. $x = 3$ for $x^2 + 3x + 2$ over $\mathbb{Z}_6$

In exercises 13–16, prove each mathematical statement.

13. $x^2 + 3x + 2$ has four zeros over $\mathbb{Z}_6$

14. $x^2 + 2x + 2$ has no zeros over $\mathbb{Z}_4$

15. $x + 3$ has no zeros over $\mathbb{N}$

16. $2x + 1$ has no zeros over $\mathbb{Z}_8$

In exercises 17–30, prove each statement about algebraic and transcendental numbers.

17. $4/5$ is algebraic over $\mathbb{Q}$

18. $4/5$ is algebraic over $\mathbb{Z}$

19. $\sqrt[3]{5}$ is algebraic over $\mathbb{Z}$

20. $\sqrt[3]{5}$ is algebraic over $\mathbb{Q}$

21. $\sqrt[4]{14/5}$ is algebraic over $\mathbb{Z}$

22. $\sqrt[5]{3/4}$ is algebraic over $\mathbb{Q}$

23. If $a \in F$ a field, then a is algebraic over F.

24. $5e$ is transcendental over $\mathbb{Q}$

25. $\frac{\pi}{2}$ is transcendental over $\mathbb{Q}$

26. 4 is transcendental over $\mathbb{N}$

27. $\pi^2 \neq a\pi + b$, where $a, b \in \mathbb{Q}$

28. $\pi^2 = a\pi + b$, for some $a, b \in \mathbb{R}$

29. $e^3 = ae^2 + be + c$, for some $a, b, c \in \mathbb{R}$

30. $e^3 \neq ae^2 + be + c$, where $a, b, c \in \mathbb{Q}$

In exercises 31–54, identify every zero of each polynomial. Unless otherwise stated, assume all polynomials are over $\mathbb{C}$ and express your solutions in standard form.

31. $x - 2$

32. $x - 2$ over $\mathbb{Z}_5$

33. $x - 2$ over $\mathbb{Z}_7$

34. $4x + 5$

35. $(1 - 3i)x - (3 + 5i)$

36. $(-2 + 4i)x - (7 + 3i)$

37. $12x + (5 + 9i)$

38. $(2 + 10i)x - 6i$

39. $x^2 + 2x - 15$

40. $x^2 + 2x + 2$

41. $3x^2 + 2x + 2$

42. $x^2 + 10x - 39$

43. $x^3 - 8$

44. $8x^3 - 27$

45. $x^3 - 9x - 28$

46. $5x^3 - 45x - 140$

47. $x^3 - 12x + 16$

48. $x^3 - 12x - 16$

49. $x^3 - 3x^2 + 3x - 1$

50. $x^3 - 6x^2 + 11x - 6$

51. $x^3 - 7x^2 + x - 7$

52. $x^3 - 6x^2 + 14x - 15$

53. $x^3 - 15x^2 + 81x - 175$

54. $2x^3 - 6x^2 - 8x + 24$

In exercises 55–58, find the del Ferro–Tartaglia solution of each cubic.

55. $x^3 + 15x + 7$

56. $x^3 - 8x - 9$

57. $3x^3 + 27x - 6$

58. $2x^3 - 14x + 16$

In exercises 59–62, find the corresponding depressed cubic for each cubic under the Cardano transformation.

59. $3x^3 + 7x^2 - 6x - 4$

60. $17x^3 + 6x^2 - 4x + 11$

61. $3x^3 - 2x^2 + 19x - 8$

62. $4x^3 - 9x^2 + 16x - 1$

Exercises 63–70 introduce a number system extending the complex numbers known as the quaternions. This number system was first defined by the Irish mathematician Sir William Rowan Hamilton in 1843, and so $\mathbb{H}$ denotes the set of quaternions $\{a + bi + cj + dk : a, b, c, d \in \mathbb{R}\}$, where the three distinct quantities i, j, k satisfy the following relationships.

$$i^2 = j^2 = k^2 = ijk = -1, \qquad ij = -ji = k, \qquad jk = -kj = i, \qquad ki = -ik = j$$

The quaternions are a field under componentwise addition and a multiplication operation, where products of i, j, k simplified using the above identities. The following examples illustrate these operations.

$$
\begin{aligned}
(4 + 5i + 2j) + (2 + j + 4k) &= (4 + 2) + (5 + 0)i + (2 + 1)j + (0 + 4)k \\
&= 6 + 5i + 3j + 4k \\
(i + j) \cdot (2i + 3k) &= i \cdot 2i + i \cdot 3k + j \cdot 2i + j \cdot 3k \\
&= -2 + 3(-j) + 2(-k) + 3i = -2 + 3i - 3j - 2k
\end{aligned}
$$

In exercises 63–70, compute the following sums, differences, and products of quaternions, expressing the answers in the form $a + bi + cj + dk$ where $a, b, c, d \in \mathbb{R}$.

63. $(1 + 5i + 10k) + (2 - 3i + 3j - 2k)$

64. $(2 - 3i + 2j - 6k) + (3 - 3i - 2j + 5k)$

65. $(3 + 5i - 3k) - (4 + 3i + 7j + 3k)$

66. $(4 - 3i + 7j + k) - (1 + 3i - 2j - 4k)$

67. $(3 + 10k) \cdot (1 + 4i)$

68. $(4 + 2j) \cdot (5 - i)$

69. $(1 + 5i - 7j + 10k) \cdot (2 - 3i + 2j - 6k)$

70. $(3 + 5i + 2j - 3k) \cdot (1 + 4i)$

3.6 Mathematical Induction

This section studies a proof technique commonly used to demonstrate that mathematical statements are true for all elements of certain types of sets. The movement from considering particular examples and counterexamples to making claims about all mathematical objects in a given setting is fundamental to mathematics. As we have seen, intuition is key to developing new insights, but verifying the truth of a conjecture often requires considerably more effort. For certain infinite sets, we need a new proof technique known as *mathematical induction* to prove claims about all objects in a given set.

Before considering the infinite setting, recall that truths about finite sets can be established using *proof by exhaustion.* For a finite set (that is not too large), a mathematical statement can be proven true for every element in the set by verifying the statement holds for each element of the set one at a time. In this way, the set of possible counterexamples is "exhausted." For example, every element of the set $\{0, 2, 4, 6\}$ is even because $0 = 2 \cdot 0$, $2 = 2 \cdot 1$, $4 = 2 \cdot 2$, and $6 = 2 \cdot 3$; that is, each element of the given set is directly demonstrated to be even. As we can readily imagine, larger and larger finite sets can make proofs by exhaustion anything from tedious to impossible during a human's lifetime, although computers are of significant help when studying large finite sets that are readily amenable to algorithmic description. However, proof by exhaustion is not applicable when studying infinite sets. As finite beings, we are simply incapable of individually verifying that every element of an infinite set satisfies a given mathematical claim. Instead, in certain infinite settings, the proof technique known as mathematical induction is used to prove mathematical statements.

We begin the study of mathematical induction by considering infinite "end segments" of the integers $\mathbb{Z} = \{\ldots, -3, -2, -1, 0, 1, 2, 3, \ldots\}$. An *end segment* of $\mathbb{Z}$ is a set consisting of every element of $\mathbb{Z}$ that is greater than some designated integer $n \in \mathbb{Z}$. Notice that every end segment of $\mathbb{Z}$ is infinite. Induction is particularly suited to proofs about end segments of $\mathbb{Z}$, and most mathematicians identify induction with such sets of numbers.

Example 3.6.1 Each set given below is an end segment of the integers.

- $\mathbb{N} = \{1, 2, 3, \ldots\}$
- $\{n : n \geq 5\} = \{5, 6, 7, 8, \ldots\}$
- $\{n : n \geq -2\} = \{-2, -1, 0, 1, \ldots\}$

■

In previous mathematics courses, you may have studied mathematical claims about end segments of the integers, such as the following.

For every $n \in \mathbb{N}$, $\displaystyle\sum_{i=1}^{n} 2 = 2 + \cdots + 2 = 2n$.

For every $n \geq 5$, $n^2 < 2^n$.

If $a_1 = 1$, $a_2 = 3$, $a_k = a_{k-2} + 2a_{k-1}$, then for every $n \geq 1$, a_n is odd.

The truth of such statements is established using induction. There are two different renditions of induction: mathematical induction (also called "weak" induction or simply "induction"); and strong mathematical induction. The first two claims given above are proven via induction, while the third requires strong induction. The distinction between these two types of induction lies in how much information about a "base case" the proof must address, as well as how many elements we use in an "inductive" step of the proof. This distinction is discussed more carefully at an appropriate point as we consider various examples. The principles of mathematical induction are stated in the following theorem. Recall from section 1.6 that a predicate $P(k)$ is a string of symbols that can be interpreted as a predicate phrase about variables (such as k and n) or about specific numbers (such as a below).

Theorem 3.6.1 • **Principle of weak induction** *Let $P(k)$ be a predicate defined on integers $k \in \mathbb{Z}$ and let $a \in \mathbb{Z}$. If $P(a)$ is true and, for all $n \geq a$, we have $P(n)$ implies $P(n+1)$, then $P(n)$ is true for all integers $n \geq a$.*

• **Principle of strong induction** *Let $P(k)$ be a predicate defined on integers $k \in \mathbb{Z}$ and let $a, b \in \mathbb{Z}$ with $a \leq b$. If $P(a), P(a+1), \ldots, P(b)$ are true and, for all $n \geq a$, the conjunction $[P(a) \wedge P(a+1) \wedge \ldots \wedge P(n)]$ implies $P(n+1)$, then $P(n)$ is true for all integers $n \geq a$.*

These principles of induction are often described using the image of a line of dominoes falling down. For weak induction, we prove that $P(a)$ is true and that $P(n)$ implies $P(n+1)$ for all $n \geq a$. In the domino setting, this corresponds to knowing both that the first domino has fallen down and that when any domino falls down, the next domino in line must also fall down. These two observations about a line of dominoes leads to the conclusion that every domino in the line has fallen down. The principle of strong induction is similar except that we need to know that the first few dominos have fallen down, and that *all* the dominoes have fallen down up to some point in the line implies the next domino in line must also fall down. In some mathematical settings, the claim that a statement is true for a particular integer depends on knowing the statement is true for every preceding integer (rather than just the immediate predecessor).

While such intuitive descriptions of induction are helpful, a mathematician is also interested in a rational proof that induction is true. The choice of the word "Principle" in naming mathematical induction highlights the fact that these statements are typically not understood as theorems. Instead, induction is taken as an axiom (a fundamental belief) that helps define the natural numbers and end segments of the integers. In 1838, the English mathematician Augustus De Morgan introduced the term "mathematical induction" in *Induction Mathematics*, which provided the first clear statement of this fundamental proof technique. In 1887, the contemporary statement of induction was given by the German mathematician Richard Dedekind in *Was sind und Was sollen die Zahlen?* Furthermore, by 1889, the Italian mathematician Giuseppe Peano identified five axioms precisely defining the natural numbers $\mathbb{N}$ in *Arithmetices principia, nova methodo exposita*. The principle of weak induction is the fifth of Peano's axioms for arithmetic. We therefore do not *prove* that mathematical induction is valid, but instead choose to *accept* these principles because they are rooted in our intuitive understanding of the integers.

As a point of clarification, in some settings the principle of mathematical induction is proven using the well-ordering principle of the integers, which asserts that every subset of integers that is bounded below has a least element. However, one can also prove the well-ordering principle using the principle of mathematical induction, and so these two principles are actually equivalent. These proofs are often studied in courses in set theory and abstract algebra and are left for your later studies. Instead, this section focuses on developing a facility with using induction to prove the validity of mathematical statements. The first example proves a finite series (or finite summation) formula commonly used for directly computing Riemann sums when beginning a study of the integral in calculus.

Example 3.6.2   We use induction to prove that for every integer $n \geq 1$, we have

$$\sum_{i=1}^{n} i = \frac{n(n+1)}{2}.$$

Recall that the notation $\sum_{i=1}^{n}$ denotes a finite sum with an unspecified integer upper bound. For example, $\sum_{i=1}^{3} i = 1 + 2 + 3 = 6$ and $\sum_{i=1}^{n} i = 1 + 2 + 3 + \cdots + n$. Therefore, the above formula claims that an arbitrary sum of successive integers $\sum_{i=1}^{n} i$ is equal to $n(n+1)/2$.

Proof   This claim about every integer greater than or equal to 1 is proven using the principle of weak induction on the predicate $P(k)$ asserting "$\sum_{i=1}^{k} i = k(k+1)/2$." Referring to theorem 3.6.1, the principle of induction instructs us to begin with the *base case* $a = 1$; that is, we must prove $P(1)$ is true. After settling the base case, theorem 3.6.1 instructs us to prove the *inductive step* in which we assume $P(n)$ is true for some arbitrary, fixed integer $n \geq 1$ and demonstrate (under this assumption) that $P(n+1)$ must also be true. For this example, the assumption is that $\sum_{i=1}^{n} i = n(n+1)/2$, and this assumption is used to prove that

$$\sum_{i=1}^{n+1} i = \frac{(n+1)[(n+1)+1]}{2}$$

is true. In this context, $P(n)$ asserting that $\sum_{i=1}^{n} i = n(n+1)/2$ is known as the *induction hypothesis*.

■

Base case $a = 1$.   By direct computation, $\sum_{i=1}^{1} i = 1 = 1(1+1)/2$.

Inductive step   We assume $P(n)$, asserting that $\sum_{i=1}^{n} i = n(n+1)/2$. Under this assumption, we prove $P(n+1)$ asserting that

$$\sum_{i=1}^{n+1} i = \frac{(n+1)[(n+1)+1]}{2}.$$

The following string of equalities provides the desired conclusion.

$$\sum_{i=1}^{n+1} i = 1 + \cdots + n + (n+1) \qquad \text{Definition of series}$$

$$= [1 + \cdots + n] + (n+1) \qquad \text{Associativity of integer addition}$$

$$= \sum_{i=1}^{n} i + (n+1) \qquad \text{Definition of series}$$

$$= \frac{n(n+1)}{2} + (n+1) \qquad \text{Induction hypothesis}$$

$$= \frac{n(n+1)}{2} + \frac{2(n+1)}{2} \qquad \text{Basic algebra}$$

$$= \frac{n(n+1) + 2(n+1)}{2} \qquad \text{Common denominator}$$

$$= \frac{(n+1)(n+2)}{2} \qquad \text{Factor } (n+1)$$

$$= \frac{(n+1)[(n+1)+1]}{2} \qquad n+2 = (n+1)+1$$

Therefore, by induction, $\sum_{i=1}^{n} i = n(n+1)/2$ for every $n \geq 1$. ∎

In the inductive step of example 3.6.2 above, the use of the induction hypothesis is crucial to the success of the proof. As we work through several examples, observe how the inductive hypothesis is incorporated into the proof; understanding this element of these arguments will help you develop your own induction proofs. In addition, you'll want to observe the style and presentation of these inductive proofs and emulate these models when crafting your own proofs.

Question 3.6.1   Using induction, prove that $\sum_{i=1}^{n} 2 = 2n$ for every $n \geq 1$. ∎

The first known proof by induction was given by the Italian mathematician Francesco Maurolico in 1575. Maurolico proved that the sum of the first n odd integers is n^2 in *Arithmeticorum libri fuo*. However, the principle of mathematical induction is useful for proving a whole host of different results about infinite end segments of the integers, not just facts about series. In the next example and question, induction is used to prove the validity of certain inequalities among integers.

Example 3.6.3   We use induction to prove $2n + 1 < 2^n$ for every $n \geq 3$.

Proof   This inequality is proven using the principle of weak induction on the predicate $P(k)$ asserting that "$2k + 1 < 2^k$." The base case is $n = 3$, so we prove $2 \cdot 3 + 1 < 2^3$, and the inductive step assumes $2n + 1 < 2^n$ and proves $2(n+1) + 1 < 2^{n+1}$. ∎

Base case $n = 3$.   Direct computations produce the following equalities.

$$\begin{aligned} 2n + 1 &= 2 \cdot 3 + 1 &= 7 \\ 2^n &= 2^3 &= 8 \end{aligned}$$

Since $7 < 8$, we have $2n + 1 < 2^n$ when $n = 3$.

Inductive step   We assume that $2n + 1 < 2^n$ and prove that $2(n+1) + 1 < 2^{n+1}$. Some preparatory computations with the inequality help us identify the best use of the inductive hypothesis in this part of the proof. Direct computations produce the following equalities.

$$2(n+1)+1 \quad = \quad 2n+2+1 \quad = \quad (2n+1) \quad + \quad 2$$
$$2^{n+1} \qquad\qquad = \quad 2 \cdot 2^n \qquad = \quad 2^n \qquad\quad + \quad 2^n$$

The rightmost expressions indicate how we can use both the inductive hypothesis $2n+1 < 2^n$ and the fact that $2 < 2^n$ for $n \geq 1$ to prove that $2(n+1)+1 < 2^{n+1}$. We are now ready to piece these observations together into a fluid, articulate proof; that is, these "scratchwork" computations do not *prove* the inductive step, but only point us in the right direction. The proof of the inductive step follows from the following string of equalities and inequalities.

$$
\begin{aligned}
2(n+1)+1 \quad &= \quad 2n+2+1 & &\text{Distribution of } \times \text{ over } + \\
&= \quad (2n+1)+2 & &\text{Commutativity and associativity} \\
&< \quad 2^n + 2 & &\text{Induction hypothesis} \\
&< \quad 2^n + 2^n & &\text{Since } 2 < 2^n \text{ for } n \geq 1 \\
&= \quad 2^{n+1} & &\text{Exponentiation properties}
\end{aligned}
$$

Therefore, by induction, $2n+1 < 2^n$ for every $n \geq 3$. ∎

Question 3.6.2 Using induction, prove that $n^2 < 2^n$ for every $n \geq 5$. Hint: The base case is $n = 5$ since the claim asserts that the inequality holds for every $n \geq 5$. Also, the result of example 3.6.3 is useful in this proof's inductive step. ∎

The principle of induction is often applied to sequences of numbers defined by "recursion." Informally, a *sequence* is a list of numbers; for example, $2, 4, 6, 8, \ldots$. Such lists play an important role in the study and application of results in real analysis, complex analysis, topology, and computer science. Sequences arise quite naturally in both abstract mathematics and the real-world, and they exhibit a striking array of interesting and distinct behaviors. The following definition may be familiar from your previous studies.

Definition 3.6.1 *A sequence is a function defined on all integer inputs greater than or equal to some $n \in \mathbb{Z}$. A sequence is typically written using the subscript notation $a_1, a_2, a_3, \ldots$ where a_n denotes the **nth** term of the sequence and $\{a_n\}$ denotes the entire sequence.*

Example 3.6.4 Some examples of sequences include the following.

$$
\begin{aligned}
1, 1, 1, 1, \ldots \quad &\text{where} \quad a_n = 1 \\
1, -2, 4, -8, \ldots \quad &\text{where} \quad a_n = (-2)^n \\
1, 2, 6, 24, \ldots \quad &\text{where} \quad a_n = n!
\end{aligned}
$$

∎

In the context of studying induction, we are primarily interested in sequences that are obtained via recursion. A sequence is *defined by recursion* if the first few terms of the sequence are explicitly stated and if the value of later terms depends on the preceding terms in the sequence. Perhaps the most widely known sequence defined by recursion is the *Fibonacci sequence*: $1, 1, 2, 3, 5, 8, 13, \ldots$.

The first recorded definition of this sequence was given by Leonardo of Pisa, who is better known as Fibonacci or "son of Bonaccio." Fibonacci was an Italian merchant who wrote the classical mathematical work *Liber abaci* (or *Book of the Abacus*) in 1202. In this manuscript, Fibonacci developed a variety of algebraic methods, primarily in the context of commercial transactions. Most importantly for mathematics, this text played a key role in the dissemination and use of the Hindu–Arabic numeral system in Europe.

In the *Liber abaci*, Fibonacci asked: "How many pairs of rabbits will be produced in a year, beginning with a single pair, if in every month each pair bears a new pair which becomes productive from the second month on?" This question defines the Fibonacci sequence. More than just an amusing description of rabbit population growth, this sequence has been used to model a number of real-world processes, including the development of the arrangement of sunflower seeds and pineapple rinds, the distinctive spiral of nautilus shells, and the family trees of certain species. The following example studies the Fibonacci sequence using its contemporary formulation as a recursive sequence.

Example 3.6.5   The *Fibonacci sequence:* $1, 1, 2, 3, 5, 8, 13, \ldots$ is defined recursively as follows.

$$f_1 = 1, \qquad f_2 = 1, \qquad f_{n+2} = f_n + f_{n+1}, \text{ for } n \geq 1$$

The formula defining the general term f_{n+2} is referred to as a *recurrence relation* and indicates the dependence of the value f_{n+2} on the values of the previous two terms f_n and f_{n+1}. We illustrate the use of this recurrence relation by explicitly computing the first six terms of the Fibonacci sequence.

$$
\begin{aligned}
f_1 &= 1 \\
f_2 &= 1 \\
f_3 &= f_1 + f_2 = 1 + 1 = 2 \\
f_4 &= f_2 + f_3 = 1 + 2 = 3 \\
f_5 &= f_3 + f_4 = 2 + 3 = 5 \\
f_6 &= f_4 + f_5 = 3 + 5 = 8
\end{aligned}
$$

You can now see why the Fibonacci sequence begins $1, 1, 2, 3, 5, 8, \ldots$. Continuing to apply the recurrence relation in this way, we can find any desired number of terms of the Fibonacci sequence.

■

Question 3.6.3   State the first six terms of the sequence $\{a_n\}$ defined recursively as follows.

$$a_1 = 1, \qquad a_2 = 3, \qquad a_{n+2} = a_n + 2a_{n+1}, \text{ for } n \geq 1$$

■

Mathematical claims about sequences are usually proven using the principle of *strong* induction. The following example provides a first application of this proof technique.

Example 3.6.6 We prove that every term of the sequence $\{s_n\}$ defined recursively as follows is odd.

$$s_1 = 1, \qquad s_2 = 3, \qquad s_{n+1} = s_{n-1} + 2s_n, \text{ for } n \geq 2$$

Proof We use strong induction to prove the result. In this case, the predicate statement $P(k)$ asserts that "the term s_k is odd, for any $k \geq 1$." ■

Base case $n = 1$ and $n = 2$. Since the recursion relation for s_{n+1} is defined using the previous two terms, the base case must examine both $P(1)$ and $P(2)$. From the definition of the sequence, we immediately observe that $s_1 = 1 = 2 \cdot 0 + 1$ and $s_2 = 3 = 2 \cdot 1 + 1$ are both odd. Therefore, $P(1)$ and $P(2)$ are both true.

Inductive step The inductive hypothesis is the other key difference between strong induction and weak induction. In strong induction, $P(k)$ is assumed true for every integer from $k = 1$ to $k = n$, and $P(n + 1)$ is proved true using this collection of assumptions. In this proof, we assume that every term of the sequence $s_1, s_2, \ldots, s_n$ is odd, and prove that s_{n+1} is odd. Applying the inductive hypothesis to the terms s_{n-1} and s_n, there exist integers $i, j \in \mathbb{Z}$ such that $s_{n-1} = 2i + 1$ and $s_n = 2j + 1$. Substituting these values into the recurrence relation and simplifying, produces the following algebraic manipulations,

$$
\begin{aligned}
s_{n+1} &= s_{n-1} + 2s_n \\
&= 2i + 1 + 2(2j + 1) \\
&= 2(i + 2j + 1) + 1
\end{aligned}
$$

Therefore, s_{n+1} is an odd number and, by the principle of strong induction, every term of the sequence $\{s_n\}$ is odd. ■

Question 3.6.4 Prove that $b_n \leq 1$ for every term of the sequence $\{b_n\}$ defined recursively as follows.

$$b_1 = \frac{9}{10}, \qquad b_2 = \frac{10}{11}, \qquad b_{n+2} = b_n \cdot b_{n+1}, \text{ for } n \geq 1$$

Hint: The base case is $n = 1$ and $n = 2$ since the recursion relation for b_{n+2} is defined using the previous two terms. ■

While many proofs by induction are for end segments of the integers, this process can be extended to prove universal statements about other inductively defined mathematical structures. In addition to the integers, the definitions of some other mathematical objects are given using an inductive structure. Both Richard Dedekind and the Norwegian mathematician Thoralf Skolem made important contributions to understanding the role of induction in these more general settings. For example, the study of mathematical logic in chapter 1 involved several inductive definitions, including the following definition of a sentence of sentential logic.

Definition 3.6.2 (Definition 1.1.2 in Section 1.1) *A* **sentence** *of sentential logic is a string of symbols from the alphabet of sentential logic that satisfies the following:*

1. *a single sentence symbol or a single sentence variable is a sentence;*
2. *if* $\mathbb{B}$, $\mathbb{C}$ *are sentences, then so are* $(\sim\mathbb{B})$, $(\mathbb{B} \wedge \mathbb{C})$, $(\mathbb{B} \vee \mathbb{C})$, $(\mathbb{B} \rightarrow \mathbb{C})$, *and* $(\mathbb{B} \leftrightarrow \mathbb{C})$.

This inductive definition of sentences, first identifies a "base case" of sentences (the sentence symbols and variables), and then the other sentences are built up through repeated application of a restricted collection of operations on the base case sentences. This inductive definition of sentences directly parallels the inductive definition of the natural numbers, in which $n = 1$ is the base case and the other natural numbers are built up by repeatedly adding one. This inductive structure in the definition of a sentence enables us to prove mathematical claims about all sentences of sentential logic by using induction as illustrated in the following example.

Example 3.6.7   We prove that the number of left parentheses in any sentence of sentential logic is the same as the number of right parentheses.

Proof   We prove this claim by induction on definition 3.6.2 for a sentence of sentential logic.

Base case   We first verify the claim is true for sentence symbols and for sentence variables. A sentence symbol (for example, A) does not have any parentheses, and so the number of left parentheses is 0, as is the number of right parentheses. Similarly, a sentence variable does not have any parentheses, so again the number of left and right parentheses is 0.

Inductive step   The sentences considered in this inductive step, must include all parentheses (including the outermost pair) rather than omitting parentheses as has been our custom for the sake of readability. For the inductive hypothesis, assume that $\mathbb{B}$ and $\mathbb{C}$ are sentences, where $\mathbb{B}$ has m left and right parentheses and $\mathbb{C}$ has n left and right parentheses. We verify the claim is true for sentences built up from $\mathbb{B}$ and $\mathbb{C}$ by the operations identified in definition 3.6.2. The proof is therefore organized by examining these operations one at a time.

- $(\sim\mathbb{B})$: This sentence begins with a left parenthesis, which together with the m left parentheses from $\mathbb{B}$, yields a total of $1 + m$ left parentheses. Similarly, there are m right parentheses from $\mathbb{B}$, plus one additional right parenthesis at the end of the sentence, to yield a total of $m + 1$ right parentheses. From the commutativity of integer addition, $(\sim\mathbb{B})$ has the same number of left and right parentheses.

- $(\mathbb{B} \wedge \mathbb{C})$: This sentence begins with a left parenthesis, which together with the m left parentheses from $\mathbb{B}$ and n right parentheses from $\mathbb{C}$, yields a total of $1 + m + n = m + n + 1$ left parentheses. Similarly, there are m right parentheses from $\mathbb{B}$, n right parentheses from $\mathbb{C}$, plus one additional right parenthesis at the end of the sentence, to yield a total of $m + n + 1$

right parentheses. Therefore, $(\mathbb{B} \wedge \mathbb{C})$ has the same number of left and right parentheses.

- A similar argument shows that $(\mathbb{B} \vee \mathbb{C})$, $(\mathbb{B} \rightarrow \mathbb{C})$, and $(\mathbb{B} \leftrightarrow \mathbb{C})$ each have the same number of left and right parentheses. Further details are left for the next question.

Therefore, accepting these last details, this proof by induction shows that every sentence has the same number of left and right parentheses.

■

Question 3.6.5 Prove that if $\mathbb{B}$ has m left and right parentheses and $\mathbb{C}$ has n left and right parentheses, then $(\mathbb{B} \vee \mathbb{C})$, $(\mathbb{B} \rightarrow \mathbb{C})$, $(\mathbb{B} \leftrightarrow \mathbb{C})$ each have the same number of left and right parentheses.

■

Recall from section 1.3 that a set of connectives is "adequate" if every truth table is satisfied by a sentence using only the connectives in the set. The notion of an adequate set of connectives was introduced in a discussion of the expressibility of sentential logic. Another important reason for our interest in adequate sets of connectives is the simplification of inductive proofs. A complete proof of the inductive step in example 3.6.7 involved the consideration of five distinct cases, one for each of the connectives $\sim$, $\wedge$, $\vee$, $\rightarrow$, and $\leftrightarrow$.

However, if we (re)define sentences using a smaller adequate set of connectives, then a complete proof of the inductive step may involve just one or two distinct cases. For example, suppose we define sentences using only the connectives $\{\sim, \wedge\}$ (one of the sets of adequate connectives discussed in section 1.3). In this setting, the two cases considered in example 3.6.7 provides a complete proof of the claim that the number of left parentheses in any sentence of sentential logic is the same as the number of right parentheses.

3.6.1 Reading Questions for Section 3.6

1. Describe the process of proof by exhaustion.
2. Give an example of a proof by exhaustion.
3. Define and give an example of an end segment of the integers.
4. State the two principles of mathematical induction.
5. What is the distinction between weak and strong mathematical induction?
6. Give an example of mathematical claim that can be proven using weak mathematical induction.
7. Give an example of mathematical claim that can be proven using strong mathematical induction. Why is strong induction necessary?
8. Explain the role of the base case in a proof by induction.
9. Explain the role of the induction hypothesis in a proof by induction.
10. Describe recurrence relations and give an example.
11. State the first eight terms of the Fibonacci sequence.
12. What do we mean by induction on mathematical structures besides the integers?

3.6.2 Exercises for Section 3.6

In exercises 1–10, give a proof by exhaustion of each mathematical statement.

1. Every even integer between 4 and 20 inclusive can be written as the sum of two primes.
2. Every odd integer between 7 and 23 inclusive can be written as the sum of three primes.
3. Every even integer between 4 and 20 inclusive is composite.
4. Every odd integer strictly between 1 and 9 is prime.
5. There exist only two odd composite integers between 2 and 20.
6. There exists a prime between n^2 and $(n+1)^2$ for every integer between 2 and 10 inclusive.
7. Every integer between 2 and 4 inclusive satisfies the equation $2^n \leq n^2$.
8. Every positive integer n less than 7 satisfies the equation $n! < 3^n$.
9. Every element of $\mathbb{Z}_7$ has an inverse under addition mod 7.
10. Every nonzero element of $\mathbb{Z}_7$ has an inverse under multiplication mod 7.

In exercises 11–25, use induction to prove that each statement about finite series holds for every $n \in \mathbb{N}$.

11. $\displaystyle\sum_{i=1}^{n} 1 = n$

12. $\displaystyle\sum_{i=1}^{n} r = n \cdot r$, where $r \in \mathbb{R}$

13. $\displaystyle\sum_{i=1}^{n} 2^i = 2^{n+1} - 2$

14. $\displaystyle\sum_{i=1}^{n} r^i = \frac{r^{n+1} - r}{r - 1}$, where $r \in \mathbb{R}$

15. $\displaystyle\sum_{i=1}^{n} i \cdot (i!) = (n+1)! - 1$

16. $\displaystyle\sum_{i=1}^{n} i^2 = \frac{n(n+1)(2n+1)}{6}$

17. $\displaystyle\sum_{i=1}^{n} i^3 = \frac{n^2(n+1)^2}{4}$

18. $\displaystyle\sum_{i=1}^{n} 2i = n^2 + n$

19. $\displaystyle\sum_{i=1}^{n} (2i - 1) = n^2$

20. $\displaystyle\sum_{i=1}^{n} (2i - 1)^2 = \frac{n(2n-1)(2n+1)}{3}$

21. $\displaystyle\sum_{i=1}^{n} 4i - 3 = n(2n - 1)$

22. $\displaystyle\sum_{i=1}^{n} 3i - 2 = \frac{n(3n-1)}{2}$

23. $\displaystyle\sum_{i=1}^{n} i(i + 1) = \frac{n(n+1)(n+2)}{3}$

24. $\displaystyle\sum_{i=1}^{n} \frac{1}{i(i+1)} = \frac{n}{n+1}$

25. $\displaystyle\sum_{i=1}^{n} \frac{1}{(2i-1) \cdot (2i+1)} = \frac{n}{2n+1}$

In exercises 26–31, use induction to prove each statement about inequalities.

26. If $n > 2$, then $4 < n^2$.
27. If $n \geq 5$, then $n^2 < 2^n$.
28. If $n \geq 1$, then $n < 2^n$.

29. If $n \geq 2$, then $3n + 1 < 3^n$.
30. If $n \geq 0$, then $2^n < (n+2)!$.
31. If $n \geq 7$, then $3^n < n!$.

In exercises 32–39, use induction to prove each statement about divisibility relations. Recall that m is divisible by n iff there exists $q \in \mathbb{Z}$ such that $m = n \cdot q$.

32. If $n \geq 0$, then $2^{3n} - 1$ is divisible by 7.
33. If $n \geq 0$, then $3^{2n} - 1$ is divisible by 8.
34. If $n \geq 0$, then $4^n - 1$ is divisible by 3.
35. If $n \geq 0$, then $7^n - 2^n$ is divisible by 5.
36. If $n \geq 0$, then $n^2 - n$ is divisible by 2.
37. If $n \geq 1$, then $n^3 - n$ is divisible by 3.
38. If $n \geq 1$, then $4^n + 6n - 1$ is divisible by 9.
39. If $n \geq 1$, then $x^{2n} - y^{2n}$ is divisible by $x + y$.

In exercises 40–41, let $\{a_n\}$ be the sequence recursively defined by $a_1 = 1$, $a_2 = 2$, and $a_{n+2} = 2a_n + a_{n+1}$, for $n \geq 1$. Use induction to prove each mathematical statement.

40. For every $n \in \mathbb{N}$, $a_n \leq 2^n$.   41. For every $n \geq 2$, a_n is even.

In exercises 42–44, let $\{b_n\}$ be the sequence recursively defined by $b_1 = 4$, $b_2 = 8$, and $b_{n+2} = b_n + b_{n+1}$, for $n \geq 1$. Use induction to prove each mathematical statement.

42. For every $n \geq 5$, $b_n \leq 2^n$.   44. For every $n \in \mathbb{N}$, b_n is divisible
43. For every $n \in \mathbb{N}$, b_n is even.   by 4.

In exercises 45–46, let $\{c_n\}$ be the sequence recursively defined by $c_1 = 1$, $c_2 = 1$, $c_3 = 3$, and $c_{n+3} = c_n + c_{n+1} + c_{n+2}$, for $n \geq 1$. Use induction to prove each mathematical statement.

45. For every $n \in \mathbb{N}$, $c_n \leq 3^n$.   46. For every $n \in \mathbb{N}$, c_n is odd.

In exercises 47–49, let $\{d_n\}$ be the sequence recursively defined by $d_1 = 2$ and $d_{n+1} = 3 \cdot d_n$, for $n \geq 1$. Use induction to prove each mathematical statement.

47. For every $n \in \mathbb{N}$, $d_n \leq 3^n$.   49. For every $n \in \mathbb{N}$, d_n is even.
48. For every $n \in \mathbb{N}$, $d_n = 2 \cdot 3^{n-1}$.

In exercises 50–52, let $\{e_n\}$ be the sequence recursively defined by $e_1 = 3$ and $e_{n+1} = 2 + e_n$, for $n \geq 1$. Use induction to prove each mathematical statement.

50. For every $n \in \mathbb{N}$, $e_n \leq 2(n + 1)$.
51. For every $n \in \mathbb{N}$, $e_n = 3 + 2(n - 1)$.
52. For every $n \in \mathbb{N}$, e_n is odd.

In exercises 53–56, let $\{f_n\}$ be the Fibonacci sequence recursively defined by $f_1 = 1$, $f_2 = 1$ and $f_{n+2} = f_n + f_{n+1}$, for $n \geq 1$. Use induction to prove that each mathematical statement holds for every $n \in \mathbb{N}$.

53. $f_n \leq 2^n$
54. $f_1 + f_2 + \cdots + f_{n+1} = f_{n+3} - 1$
55. $f_1 + f_3 + \cdots + f_{2n+1} = f_{2n+2}$
56. there exist $a, b \in \mathbb{Z}$ so $a \cdot f_n + b \cdot f_{n+1} = 1$

In exercises 57–59, let $\{L_n\}$ be the *Lucas sequence* recursively defined by $L_1 = 2$, $L_2 = 1$ and $L_{n+2} = L_n + L_{n+1}$, for $n \geq 1$. This sequence was defined by Edouard Lucas as a generalization of the Fibonacci sequence. Use induction to prove that each mathematical statement holds for every $n \in \mathbb{N}$.

57. $L_n \leq 2^n$ 59. $L_{n+2} = 2 \cdot f_n + f_{n+1}$

58. $L_{n+2} = f_n + f_{n+2}$

In exercises 60–67, use induction to prove each mathematical statement.

60. For every $n \geq 4$, n is a linear combination of 2 and 3.

61. For every $n \geq 6$, n is a linear combination of 2 and 5.

62. For every $n \geq 8$, n is a linear combination of 2 and 7.

63. For every $n \in \mathbb{N}$, the product $\displaystyle\prod_{i=2}^{n} \left(1 - \frac{1}{i^2}\right) = \frac{n+1}{2n}$.

64. For every $n \in \mathbb{N}$, the product $\displaystyle\prod_{i=1}^{n} \frac{1}{(2i+1) \cdot (2i+2)} = \frac{2}{(2n+2)!}$.

65. For every $n \in \mathbb{N}$, $\sqrt{n} \leq \displaystyle\sum_{i=1}^{n} \frac{1}{\sqrt{i}}$.

66. For every $n \in \mathbb{N}$, if a set A contains n elements, then the power set $\mathbb{P}(A)$ of A contains 2^n elements.

67. *De Moivre's theorem* For every $n \in \mathbb{N}$ and $\theta \in \mathbb{R}$, $(\cos\theta + i\sin\theta)^n = \cos(n\theta) + i\sin(n\theta)$. Hint: In this setting $i = \sqrt{-1}$; also, consider the trigonometric identities $\cos(u \pm v) = \cos u \cos v \mp \sin u \sin v$ and $\sin(u \pm v) = \sin u \cos v \pm \cos u \sin v$.

68. *The power rule for differentiation* For every $n \in \mathbb{N}$, $(d/dx)(x^n) = n \cdot x^{n-1}$. Hint: Use the product rule for differentiation; see theorem 4.4.1 in section 4.4.

In exercises 69–71, give a complete proof by induction on the definition of sentence of sentential logic for each mathematical statement.

69. The number of left parentheses in any sentence is equal to the number of right parentheses.

70. The number of connectives in any sentence is equal to the number of right parentheses.

71. The number of connectives in any sentence is equal to the number of left parentheses.

Notes

Number theory is a lively area of ongoing study with many intriguing open questions. Andrews [6], Jones and Jones [134], Rosen [197], and LeVeque [155] introduce the contemporary study of number theory; a standard graduate text in number theory is Ireland and Rosen [124]. In addition, many discrete mathematics and abstract algebra texts present various aspects of the ideas studied in this chapter. Some standard textbooks used in discrete mathematics courses include Epp [72], Richmond and Richmond [193], and Scheinerman [209]. Fraleigh [88], Gallian [93], and Hungerford [122] provide excellent introductions to abstract algebra at the advanced undergraduate and graduate level.

As we have seen, ancient Greek mathematicians made many important contributions to the early development of number theory. Introductory surveys of the impact of Greek ideas on contemporary culture and mathematics are presented in Cahill [35], Jacobs [126], and Kline [142]. A more focused study of Pythagoras and the Pythagoreans can be found both in Kahn [135] and in Riedwig and Rendall [194]; Archimedes' work is still published and available in Archimedes [8]. Similarly, Bashmakova [11] and Heath [113] study the contributions of Diophantus to the development of algebra and discuss recent research into the solution of Diophantine equations and Fermat's last theorem.

As we have seen, Euclid's *Elements* presents many important number-theoretic results and has played an important role in the dissemination and development in the intervening centuries; Heath [73] is a fine contemporary translation of this ancient work. Similarly, Carl Friedrich Gauss' *Disquisitiones Arithmeticae* revolutionized the study of number theory; Clarke [97] is an available translation of this work. In addition, Dunnington [65] and Tent [238] are insightful biographies of this "Prince of Mathematicians."

The study of prime numbers remains an active and intriguing area of mathematical research. The number theory texts mentioned above can provide an excellent introduction to the study of these integers. In addition, Wells [253] provides an interesting survey of many "types" of primes and the sometimes startling relationships among primes. One of the most important open questions in mathematics is the "Riemann hypothesis," which conjectures that a certain analytic formula gives the number of primes less than or equal to a predesignated natural number. Du Sautoy [60] and Derbyshire [57] detail the historical development of mathematicians' efforts to resolve the Riemann hypothesis, and Rockmore [196] provides a more technical look at the recent work of contemporary mathematicians.

Prime numbers play an important role in the implementation of coding schemes to preserve the privacy and the accuracy of communication systems, particularly in the last few decades since the announcement of RSA codes in *A Method for Obtaining Digital Signatures and Public-Key Cryptosystems* [195]. Almost all number theory texts (and many discrete mathematics and abstract algebra texts) discuss RSA codes in some fashion. Hamming codes were first presented by Richard Hamming in *Error Detecting and Error Correcting Codes* [110]; Hamming codes are also discussed in both Gallian [93] and Grimaldi [103], as well as any text on error-correcting codes. For an introduction to the mathematical field of coding theory see Bierbrauer [18], Hill [115], and Ling and Xing [156]. More focused discussions of error-correcting codes are given by Huffman and Pless [120] and by MacWillians and Sloane [162]. Finally, Singh [215] provides an interesting historical survey of coding schemes from ancient Egypt to contemporary applications.

Fermat's last theorem is one of the most celebrated and famous of results in all of abstract mathematics. Andrew Wiles proof, with important contributions by Richard Taylor, appeared in "Modular elliptic curves and Fermat's Last Theorem" [257] in 1995. Naturally, the books on Fermat's last theorem written before and after this proof differ markedly. Ribenboim [192] is one interesting text that straddles this time period; most of the book discusses the proofs of specific cases of the theorem (including the $n = 4$ case studied in section 3.3), and an epilogue outlines Wiles' general approach to the complete proof. Both Hellegouarch [114] and Stewart and Tall [233] also provide interesting introductions to this area of study; and Singh and Lynch [216] detail the historical developments that led up to the complete solution of Fermat's last theorem by Wiles.

Throughout this chapter we have studied several very significant numbers in mathematics, most notably π, e, and i. Various books have been written about the mathematical history of these numbers and Mazur [172] has described ways of envisioning them. Beckmann [12] details the history of π; more recent books on this topic have been written by Eymard et al. [76] and by Posamentier and Lehmann [188]. Blatner's *The Joy of Pi* [20] is a playful collection of many intriguing facts and insights into this number. Maor [168] has written a history of e; Nahin [179] has written a history of i; and both Seife [212] and Kaplan and Kaplan [136] have traced the history of 0. Perhaps the story of 1 will be written before too much longer to complete the story of each constant appearing in the famous mathematical equation $e^{i\pi} + 1 = 0$!

We identified various resources for learning more about Abel, Galois, and the insolvability of the general quintic in the notes for chapter 2. Mathematicians currently prove Abel's theorem using Galois theory. For undergraduates who have studied sufficient abstract algebra, Garling [96], Stewart [229], and Swallow [237] are excellent and accessible texts introducing Galois theory; both Edwards [70] and Hungerford [122] are standard graduate texts.

Induction is used in significant ways in many areas of theoretical mathematics, and so is often discussed in standard undergraduate courses, including discrete mathematics, abstract algebra, and real analysis; the textbooks used in these courses (as identified above and in the notes for chapters 2 and 4) are good references for how induction impacts on these fields. A number of books have been written discussing the Fibonacci sequence and its generalizations. Both Garland [95] and Wahl [247] are directed toward a general audience; Benjamin and Quinn [16] detail the results of numerous undergraduate research projects exploring the Fibonacci sequence. Recently, Sigler has translated Fibonacci's *Liber abaci* [82].

Many of the theorems studied in this chapter appear in anthologies of the "fundamental" theorems of mathematics that detail both their proofs and some of the historical context of these results. Dunham's *Journey Through Genius* [64] and Davis et al.'s *The Mathematical Experience* [55] are two books in this genre. Aigner and Ziegler's *Proofs from THE BOOK* [3], inspired by their conversations and correspondence with Erdös, is a similarly intriguing collection of theorems and proofs. Finally, we mention G. H. Hardy's classic book *A Mathematician's Apology* [112], which reflects on the culture and nature of mathematics, and has been enjoyed by generations of mathematicians; you might also be interested in Stewart's recent *Letters to a Young Mathematician* [230], which was inspired by Hardy's book.

4 Real Analysis

The Renaissance and Baroque periods were times of profound change in the way western Europeans chose to explore and understand the universe. The humanistic intellectual and social movement in these eras was both encouraged by and contributed to insightful shifts in developing mathematics and applying it to the physical world. The "reawakening" in classical art and literature that defined the Renaissance (originating in Italy and spreading throughout Europe during the fourteenth, fifteenth, and sixteenth centuries) was accompanied by a movement to return to rational scientific investigation. Mathematics formed the core of this movement. And just as art, literature, and music flourished into the elaborate expressions of the Baroque period of the seventeenth century, so too mathematics began to flourish. New foundational ideas in mathematics and the mathematical community's increasing commitment to rigor during the seventeenth century paved the way for an "Age of Enlightenment" to follow—a time when the prevailing European culture strongly valued the rationalism that permeates Western culture to this day.

During the time between the ancient Greeks and the European Renaissance, the development of mathematics proceeded relatively slowly. A handful of Indian and Islamic mathematicians worked to preserve and extend the work of the ancient Greeks, and the dissemination of the Hindu–Arabic numeral system contributed to the mathematical achievements of the Renaissance and the Enlightenment. But the Age of Enlightenment in seventeenth and eighteenth century Europe was truly a unique time when a handful of exceptional individuals—mathematical geniuses, really—profoundly changed our approach to scientific investigation. Instead of being satisfied with calculations focused on practical problem solving, these mathematicians began developing a broad, methodological approach to mathematical and scientific thought. Many of the greatest mathematical minds in history worked at this time, including Galileo, Descartes, Fermat, Leibniz, and Newton. During this period, great advances occurred in the study of real-valued functions.

From your earlier courses in mathematics, you know that calculus characterizes properties of functions using limits, derivatives, and integrals. These mathematical tools help describe the way functions change as their independent, real-valued variables change. In this chapter we explore several insights into the theory of functions, describing these ideas in a mathematically rigorous fashion. We also investigate some related mathematical developments that occurred well past the dawning of the Age of Enlightenment, setting the stage for modern mathematical investigations into the theory of functions during the twentieth and twenty-first centuries.

4.1   Analytic Geometry

The modern understanding of the relationship between algebra and geometry can be traced back to the French philosopher René Descartes; in 1637 he published the treatise *Discours de la méthode pour bien conduire sa raison et chercher la vérité dans les sciences* (that is, *A Discourse on the method of rightly conducting the reason and seeking truth in the sciences*). In *La géometrie* (the third appendix of this work), Descartes explained a natural identification between algebraic equations and geometric curves in the real plane. Since you were taught such an identification years ago, this notion may seem relatively straightforward and obvious. But Descartes' insight was one of the key advances of the seventeenth century, initiating a new way of thinking about algebra and geometry, and ultimately contributing to the development of calculus.

We know this story well from previous mathematics courses. Descartes' correspondence between an algebraic equation and a geometric curve is obtained (in a modern way) by associating a unique mathematical label to each point on the plane. This unique name is known as an *ordered pair*. The values of an ordered pair are determined by identifying two perpendicular lines on the plane called *axes*; these two axes intersect in a single point called the *origin*. By convention, we visualize an *x-axis* with a horizontal orientation and a *y-axis* with a vertical orientation. The real plane with these axes is called the *coordinate plane* or sometimes the *Cartesian plane* in honor of Descartes. Scaling these axes, each point on the plane is now uniquely identified by an ordered pair of real numbers $(x, y) \in \mathbb{R}^2$. We determine the first coordinate $x \in \mathbb{R}$ by drawing a vertical line through the point and setting x equal to the directed distance along the x-axis from the origin to this vertical line. Similarly, $y \in \mathbb{R}$ is the directed distance along the y-axis from the origin to the horizontal line through the given point.

We can now describe Descartes' correspondence between an algebraic equation and a geometric curve: an equation in the variables x and y is identified with the curve whose points (x, y) satisfy the equation. Mathematicians gradually expanded on Descartes' identification of geometric points with algebraic ordered pairs into the mathematical field known as "analytic geometry."

Example 4.1.1   The equation $y = x^2$ is a parabola with vertex $(0, 0)$. The ordered pairs (x, y) satisfying this equation include $(-2, 4)$, $(-1, 1)$, $(0, 0)$, $(1, 1)$, $(2, 4)$, and $(3, 9)$, among others. The set of all ordered pairs (x, y) satisfying $y = x^2$ produces the parabola illustrated in figure 4.1.

■

Question 4.1.1   Identify three ordered pairs satisfying the equation $y = 2x + 5$ and sketch the curve identified by this equation. What do we call such a curve?

■

Analytic geometry can be developed for dimensions greater than two. Most often, we consider three-dimensional space $\mathbb{R}^3$. In this case, we identify three mutually perpendicular lines arranged as shown in figure 4.2; these lines are called x, y, and z *axes*. Scaling these axes as for $\mathbb{R}^2$, points in space are identified by *ordered triples* of

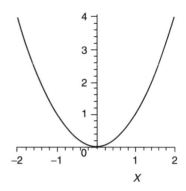

Figure 4.1　The coordinate plane with $y = x^2$

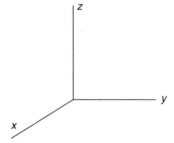

Figure 4.2　Nonnegative axes for a 3-dimensional coordinate system

the form (x, y, z), where each element of the ordered triple is the directed distance from the point to one of three coordinate planes. For example, the value a in the ordered triple (a, b, c) is the distance from the point to the yz-plane formed by the y-axis and the z-axis.

For much of this section, we determine the algebraic equations corresponding to many well-known curves in the plane. For most of these curves, the *distance formula* plays a key role in determining the equation. The distance formula follows from the Pythagorean theorem; further details of a proof are left to the reader. Also, the two-dimensional distance formula can be extended to obtain the formula for the distance between points in three-dimensional space.

Theorem 4.1.1 The Distance Formula *The distance D between any two points* (x, y) *and* (a, b) *in* $\mathbb{R}^2$ *is given by the formula*

$$D = \sqrt{(x - a)^2 + (y - b)^2}.$$

Similarly, the distance D between any two points (x, y, z) *and* (a, b, c) *in* $\mathbb{R}^3$ *is given by the formula*

$$D = \sqrt{(x - a)^2 + (y - b)^2 + (z - c)^2}.$$

The distance formula is readily applied to find the distance between points.

Question 4.1.2 Determine the distance between the following points.

(a) $(1, 2)$ and $(3, 4)$ (b) $(5, 6, 7)$ and $(10, 9, 8)$

■

An understanding of the relationship between geometric and algebraic structures provides a mathematical approach that enables us to prove many important results. For example, we can use the algebraic interpretation of a curve to prove rigorously that $y = x^2$ is a parabola in the classical sense of the ancient Greeks, who defined a parabola in terms of a fixed point called the *focus* and a fixed line called the *directrix*; a *parabola* consists of the set of all points that are equidistant from the focus and the directrix. The *vertex* of a parabola is the point on the curve closest to the focus (and the directrix). A parabola is one example of a "conic section." The following example identifies the standard algebraic equation for certain parabolas.

Example 4.1.2 We prove that the parabola with a focus on the positive y-axis at the point $(0, a)$ and with a directrix that is the horizontal line $y = -a$ has an algebraic equation of the form $y = cx^2$.

Proof Applying the distance formula, the distance from an arbitrary point (x, y) on this parabola to the focus $(0, a)$ is $\sqrt{(x - 0)^2 + (y - a)^2} = \sqrt{x^2 + (y - a)^2}$. Similarly, the distance from (x, y) to the directrix $y = -a$ is $\sqrt{(x - x)^2 + [y - (-a)]^2} = |y + a|$. Setting these distances equal to each other and squaring both sides, we obtain $x^2 + y^2 - 2ay + a^2 = y^2 + 2ay + a^2$. Algebraically manipulating this equation, $x^2 = 4ay$. Thus, $y = [1/(4a)]\, x^2 = cx^2$ when $c = 1/(4a)$.

■

Comparing this result with the equation $y = x^2$ from example 4.1.1, the curve corresponding to $y = x^2$ satisfies the classical definition of a parabola with focus at $(0, 1/4)$ and directrix $y = -1/4$. Note that for any given parabola, the axes of the coordinate plane can be positioned so that the x-axis is parallel to the directrix. If a parabola has vertex (h, k), focus $(h, k + a)$, and directrix $y = k - a$, then the corresponding equation is $(x - h)^2 = 4a(y - k)$. We obtain this equation using the distance formula and following the approach of example 4.1.2, which essentially amounts to replacing x with $x - h$ and y with $y - k$. If $c = 1/(4a)$, the equation can be algebraically manipulated to produce the following standard form for the equation of a parabola with vertex (h, k).

$$y = c(x - h)^2 + k$$

Descartes' identification of algebraic equations with curves on the plane ushered in a whole new era in the study of geometry. A geometric curve could now be studied in terms of the properties determined by its corresponding algebraic equation. In this way, the identification of algebraic equations expressing well-known curves became important.

Another example of a curve studied by the ancient Greeks is the circle. In the classical definition, a *circle* is the set of points equidistant from a fixed point called the *center*; the fixed distance from points on the circle to the center is called the *radius*.

For this discussion, let (h, k) denote the center and r denote the radius of a circle. Applying the distance formula to the center (h, k) and an arbitrary point on the circle, every point on the circle satisfies the equation $r = \sqrt{(x - h)^2 + (y - k)^2}$. Squaring both sides, the standard form for the equation of a circle with center (h, k) and radius r is

$$r^2 = (x - h)^2 + (y - k)^2.$$

Question 4.1.3   Graph each circle with center (h, k) and radius r, and state the corresponding algebraic equation.

(a) $(h, k) = (1, 1)$ and $r = 1$ (c) $(h, k) = (2, -4)$ and $r = 1$
(b) $(h, k) = (-2, 5)$ and $r = 8$ (d) $(h, k) = (-3, -4)$ and $r = 5$

▪

The *unit circle* has center $(h, k) = (0, 0)$ at the origin and radius $r = 1$. It is studied in many mathematics courses and is closely connected to trigonometric functions. The following question considers the set of points that make up the unit circle.

Question 4.1.4   (a) Graph the unit circle and state the corresponding algebraic equation.
(b) Using your equation, find the two points on the unit circle with $x = \sqrt{3}/2$ and label these points on your graph.
(c) Using your equation, find four other points on the unit circle and label these points on your graph.

▪

The unit circle helps define the trigonometric functions that are important in modeling many physical and social phenomena. The French mathematician Jean Baptiste Joseph Fourier proved early in the 1800s that many important functions can be expressed as a (possibly infinite) sum of sine and cosine functions; from this perspective, the trigonometric functions are the basic building blocks of a large class of functions. The first known study of trigonometric functions was undertaken by the Greek mathematician Hipparchus in the second century B.C.E., who applied mathematics to astronomy, which required the computation of chord lengths of certain circles. *Trigonometric* functions are defined on the unit circle by letting θ, measured in *radians*, be the angle formed between a ray emanating from the origin and the positive x-axis; a positive angle is measured counterclockwise up from the axis. The ray emanating from the origin intersects the unit circle at a point (x, y); the value $\cos \theta$ is the x-coordinate of this point, and $\sin \theta$ is the y-coordinate, as illustrated in figure 4.3. The remaining trigonometric functions are defined using the following ratios.

$$\tan \theta = \frac{\sin \theta}{\cos \theta} \qquad \sec \theta = \frac{1}{\cos \theta} \qquad \csc \theta = \frac{1}{\sin \theta} \qquad \cot \theta = \frac{\cos \theta}{\sin \theta}$$

Question 4.1.5   Complete the following table so that each ordered pair (x, y) is a point on the unit circle with both x and y nonnegative; the Pythagorean theorem may prove helpful. Use this table to answer the following questions.

$x = \cos \theta$	1		$\sqrt{2}/2$		0
$y = \sin \theta$		1/2		$\sqrt{3}/2$	

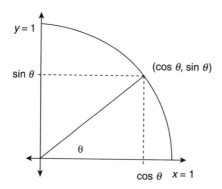

Figure 4.3 Defining $\cos\theta$ and $\sin\theta$ for a given angle θ

(a) Label these five points on a graph of the unit circle. For each point, draw a ray from the origin through the point and identify the angle θ between the ray and the x-axis in both degrees and radians.

(b) What are the values of $\cos(45°)$, $\sin(45°)$, $\cos(\pi/3)$, and $\sin(\pi/3)$?

(c) Calculate the value $\tan\theta$ for each column in the chart. What is the value of $\tan(\pi/6)$ and $\tan(\pi/2)$?

$\blacksquare$

As mentioned above, a parabola is one example of a conic section, as is a circle. Conic sections have been studied for at least the past 2300 years. In the third century B.C.E., the "Great Geometer" Appolonius of Perga wrote a comprehensive anthology *Conics* that greatly influenced the development of mathematics in subsequent centuries. Appolonius introduced the terms "parabola," "ellipse," and "hyperbola," and studied their properties based on the classical definitions discussed in this section. The fourth century mathematician Hypatia of Alexandria (the first woman known to contribute substantially to the development of mathematics) wrote important commentaries on Appolonius' *Conics* as well as Diophantus' *Arithmetica* and Ptolemy's *Almagest*.

The Greeks defined *conic sections* (or just *conics* for short) as those curves resulting from a plane intersecting a "double-napped" cone—one consisting of two cones joined at a common vertex and having axes (the "edges" of the cones) colinear. Parabolas, circles, ellipses, hyperbolas, a pair of intersecting lines, single lines, and points are the seven distinct types of conics. Lines and points are called *degenerate conic sections* because they are obtained when a plane intersects the cone's vertex. Each type of conic is identified with a standard equation; we are already familiar with the following correspondences between curves and equations.

Conic section	Standard algebraic equation
Parabola	$y = a(x-h)^2 + k$
Circle	$r^2 = (x-h)^2 + (y-k)^2$
Line	$y = mx + b$

In general, every linear equation corresponds to a line, and every quadratic equation corresponds to one of the nondegenerate conic sections. From the derivation of the equation for a parabola, we see that any polynomial in x and y with one variable of degree one and the other variable of degree two is a parabola. As we show below, polynomials in x and y with both variables of degree two are either circles, ellipses, or hyperbolas. For circles and ellipses, the coefficients of x^2 and y^2 have the same sign; for hyperbolas, the coefficients of x^2 and y^2 have opposite signs.

An *ellipse* is classically defined as the set of points (x, y) such that the sum of the distances from (x, y) to two fixed points called *foci* is equal to some fixed constant. Alternatively, an ellipse results from intersecting a "tilted" plane with a double-napped cone (although it cannot be tilted too much or it would generate either a parabola or a hyperbola). In the following example, we identify the standard algebraic equation for an ellipse.

Example 4.1.3   We prove that an ellipse with both foci on the x-axis equidistant from the origin has an algebraic equation of the form

$$\frac{x^2}{a^2} + \frac{y^2}{b^2} = 1.$$

Proof   Let $(-c, 0)$ and $(c, 0)$ denote the two foci of an ellipse with the sum of the distances equal to $2K$; note that in this setting $K > c$. For any point (x, y) on the ellipse, the sum of the distances from the point to the foci is $2K$; expressing this sum using the distance formula, we have

$$\sqrt{(x + c)^2 + (y - 0)^2} + \sqrt{(x - c)^2 + (y - 0)^2} = 2K.$$

Bringing the second square root term to the right side of the equation, squaring both sides, and simplifying yields the following equalities.

$$(x + c)^2 + y^2 \;=\; 4K^2 + -4K\sqrt{(x - c)^2 + y^2} + (x - c)^2 + y^2$$

$$x^2 + 2xc + c^2 + y^2 \;=\; 4K^2 + -4K\sqrt{(x - c)^2 + y^2} + x^2 - 2xc + c^2 + y^2$$

$$4xc \;=\; 4K^2 + -4K\sqrt{(x - c)^2 + y^2}$$

$$xc - K^2 \;=\; -K\sqrt{(x - c)^2 + y^2}$$

Now square both sides and simplify the result.

$$[xc - K^2]^2 \;=\; K^2[(x - c)^2 + y^2]$$

$$x^2c^2 - 2K^2xc + K^4 \;=\; K^2x^2 - 2xcK^2 + K^2c^2 + K^2y^2$$

$$(K^2 - c^2)x^2 + K^2y^2 \;=\; K^4 - K^2c^2 \;=\; K^2(K^2 - c^2)$$

$$\frac{x^2}{K^2} + \frac{y^2}{K^2 - c^2} \;=\; 1$$

Notice that since $K > c$, $K^2 - c^2$ is positive. The desired standard equation is obtained by defining $a = K$ and $b = \sqrt{K^2 - c^2}$.   ∎

The *hyperbola* is the last conic we consider in this section. Classically, a hyperbola is defined as the set of points (x, y) such that the difference of the distances from (x, y)

to two fixed points (called *foci*) is equal to a given fixed constant. Alternatively, a hyperbola results from the intersection of a double-napped cone with a plane "tilted" past the diagonal determined by the sides of the cone. The next question derives the standard form of a hyperbola's algebraic equation.

Question 4.1.6 Let $(-c, 0)$ and $(c, 0)$ denote the two foci on the x-axis equidistant from the origin, and consider the hyperbola consisting of points (x, y) such that the difference of the distances from (x, y) to the foci is $2K$, where $K < c$.

(a) Following a procedure similar to that of example 4.1.3, prove that the algebraic equation for such a hyperbola is of the following form.

$$\frac{x^2}{a^2} - \frac{y^2}{b^2} = 1$$

(b) Using the result from part (a), find the equation of the hyperbola with foci $(-4, 0)$ and $(4, 0)$ with a difference of distances equal to $2K = 6$.   ∎

While every curve on the xy-plane can be identified with a set of ordered pairs, not every curve in the plane is a conic section. The next few questions consider some patterns that can arise from studying finite sets of ordered pairs. This process is inductive since many different curves can satisfy a given set of points. Such curve-fitting is a delicate process, and mathematicians have developed sophisticated algorithms for finding the "best" fitting curves of a given type (for example, of polynomial type with least degree). The simplest possible curve that fits a given set of ordered pairs is typically chosen. The following questions present relatively simple patterns.

Question 4.1.7 Consider the ordered pairs presented in following table.

x	−3	−1	0	1	2	5	6	8
y	−6	−2	0	2	4	10	12	16

(a) Graph the ordered pairs identified in this table.
(b) Examining this graph, what geometric pattern exists among these points?
(c) Geometric patterns are often expressible as an algebraic equation. What equation describes the relationship between these x and y values?   ∎

Question 4.1.8 Consider the graph of points on the plane given in figure 4.4.

(a) State the ordered pairs that identify each point in figure 4.4.
(b) What two lines can be combined so that every point lies on their graph? Restrict the domains of the two lines to the negative and nonnegative real numbers, respectively.
(c) What single function expresses this pattern of points?   ∎

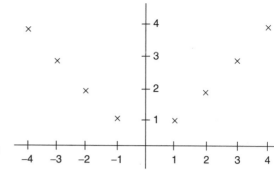

Figure 4.4 Points forming two
linear patterns

Question 4.1.9   Graph the ordered pairs in the set $\{(1, 2), (-5, 3), (0, 2), (4, -7), (-\sqrt{2}, -2),$
$(-3, 0)\}$ on a coordinate plane. These points may seem unrelated, but can you
draw a continuous curve with only two relative extrema (say, at $x = -\sqrt{2}$ and
$x = 1/2$) that passes through all six points? Continuity is defined in section 4.3,
but for now work with the intuitive notion of not lifting your pencil or pen as you
trace the curve.
                                                                                            ■

Basic geometric facts are often expressed in terms of a curve's ordered pairs (x, y).
For example, the perpendicular distance d from an arbitrary point $(x_0, y_0) \in \mathbb{R}^2$ in the
plane to a line determined by $y = mx + b$ is given by

$$d = \frac{|mx_0 + b - y_0|}{\sqrt{m^2 + 1}}.$$

The proof of this result is outlined in the exercises at the end of this section. The next
question uses this formula to study a triangle.

Question   Consider a triangle with vertices at $(1, 2)$, $(4, 7)$, and $(6, 3)$.
4.1.10
(a) Determine the equation of a line through $(4, 7)$ and $(6, 3)$.
(b) Use the formula presented above to find the perpendicular distance from the
     point $(1, 2)$ to the opposite side of the triangle; this distance is the triangle's
     height.
(c) Compute the area of the triangle using the standard formula

$$\text{Area} = \frac{1}{2} \cdot \text{base} \cdot \text{height}.$$

                                                                                            ■

We end this section by considering three-dimensional surfaces in $\mathbb{R}^3$. Every
point in three-dimensional space is identified with an ordered triple (x, y, z); algebraic
equations relating these variables define surfaces. A linear equation $ax + by + cz = d$
(where a, b, c and d are real constants) is the equation of a plane, which is classically
defined as the set of all points (x, y, z) that are equidistant from two distinct points
(e, f, g) and (p, q, r). We verify this algebraic equation using the distance formula

(for three dimensions), setting the corresponding distances equal, and performing some algebraic manipulations

$$\sqrt{(x-e)^2+(y-f)^2+(z-g)^2}=\sqrt{(x-p)^2+(y-q)^2+(z-r)^2}$$

$$x^2-2ex+e^2+y^2-2fy+f^2+z^2-2gz+g^2=x^2-2p+p^2+y^2-2qy+q^2+z^2-2rz+r^2$$

$$2(p-e)x+2(q-f)y+2(r-g)z=p^2+q^2+r^2-e^2-f^2-g^2$$

Letting $a = 2(p-e)$, $b = 2(q-f)$, $c = 2(r-g)$, and $d = p^2+q^2+r^2-e^2-f^2-g^2$, the standard form for the equation of a plane is $ax+by+cz = d$. The following question applies this result.

Question 4.1.11

(a) What is the equation of the plane whose points (x, y, z) are equidistant from the origin and from $(2, 2, 2)$?

(b) Find the real numbers $r, s, t \in \mathbb{R}$ such that the points $(r, 0, 0)$, $(0, s, 0)$, and $(0, 0, t)$ lie on the plane from part (a). These points are known as the x, y, and z *intercepts* of the plane, respectively.

(c) Produce a graph of the plane by plotting the three points from part (b) on the axes of a coordinate system in three-space and connecting these points with line segments.

∎

Just as we are interested in the three-dimensional analog of a line, we are similarly interested in the three-dimensional analog of a circle known as a sphere.

Question 4.1.12 Classically, a *sphere* is defined to be the set of points (x, y, z) that are a fixed *radius* r from a fixed *center* (h, j, k). Using the distance formula, show that the form for the algebraic equation of a sphere is $(x - h)^2 + (y - j)^2 + (z - k)^2 = r^2$.

∎

Geometric figures in two and three dimensions can be studied in terms of corresponding algebraic equations. Sometimes these equations may be viewed as functional expressions. For example, writing $y = f(x)$, the standard equation of a parabola may be written as $f(x) = c(x - h)^2 + k$. Similarly, the equation of a plane in three-space may be written as $f(x, y) = ax + by + c$, where $z = f(x, y)$ is a function of x and y. On the other hand, some algebraic equations cannot be viewed as functional expressions; for example, the hyperbola $x^2 - y^2 = 1$ does not have a single y-value corresponding to every x-value. The next section begins a detailed study of the theory of functions, since functions are special curves that lend themselves to mathematical operations such as composition, differentiation, and integration. A rigorous understanding of functional properties enables an exploration of the mathematical theory that underlies calculus. Ultimately this understanding allows us to study spaces of functions, in much the same way that we have studied spaces of points in this section.

4.1.1   Reading Questions for Section 4.1

1. Define axes and origin in the context of the two-space $\mathbb{R}^2$.
2. What point does the ordered pair $(1, 4)$ identify on the plane? Sketch a graph to facilitate your description.
3. What mathematical objects did Descartes identify with curves?
4. State the distance formula and give an example.
5. What is a conic section?
6. Name the seven distinct types of conics identified by the ancient Greeks.
7. What is the classical definition of a parabola, a circle, an ellipse, and a hyperbola?
8. State the standard algebraic equation of each conic.
9. Define $\cos\theta$ and $\sin\theta$ in terms of the unit circle.
10. Define tangent, cotangent, secant, and cosecant in terms of sine and cosine.
11. What is the classical definition of a plane and a sphere?
12. State the standard algebraic equation of a plane and a sphere.

4.1.2   Exercises for Section 4.1

In exercises 1–4, graph the points identified by each set of ordered pairs and tables.

1. $\{\,(2, 2),\ (\pi, e),\ (-1, 4),\ (-2, -2)\,\}$

2. $\{\,(5, 1),\ (-2, \sqrt{3}),\ (4, 0),\ (-5, -4),\ (1, -\pi)\,\}$

3.

x	-6	-4	-2	0	2	4	6
y	-6	-4	-1	0	2	4	6

4.

x	-3	-2	-1	0	1	2	3	4
y	6	4	1	0	2	4	6	8

In exercises 5–18, graph each curve on the coordinate plane and find an equation in x and y that corresponds to the curve.

5. The horizontal line two units above the x-axis.
6. The vertical line three units to the left of the y-axis.
7. The line passing through the points $(-2, 3)$ and $(5, -4)$.
8. The line passing through the points $(4, 7)$ and $(-1, -2)$.
9. The circle with center $(1, 2)$ and radius 6.
10. The circle with center $(3, -5)$ and radius 2.
11. The circle with center $(0, 2)$ and radius 2.
12. The circle with center $(2, 0)$ and radius 2.
13. The parabola with focus $(0, 1)$ and directrix $y = -1$.
14. The parabola with focus $(0, 3)$ and directrix $y = -1$.
15. The ellipse with foci $(-5, 0)$ and $(5, 0)$, and with sum of distances $2K = 20$.
16. The ellipse with foci $(-3, 0)$ and $(3, 0)$, and with sum of distances $2K = 20$.
17. The hyperbola with foci $(-5, 0)$ and $(5, 0)$, and with difference of distances $2K = 4$.

18. The hyperbola with foci $(-3, 0)$ and $(3, 0)$, and with difference of distances $2K = 2$.

In exercises 19–25, sketch the intersection of a double-napped cone (consisting of two cones joined at a common vertex with colinear axes) with a plane that yields each conic section.

19. A parabola.

20. A circle.

21. A ellipse.

22. A hyperbola.

23. A point.

24. A line.

25. Two intersecting lines.

Exercises 26–32 consider the general equations of conic sections.

26. Specify the focus and directrix of the parabola $y = 4x^2$.

27. Specify the focus and directrix of the parabola $y = x^2 + 1$.

28. What point on $y = 4x^2$ is closest to the focus of this parabola.

29. What point on $y = x^2 + 1$ is closest to the focus of this parabola.

30. The main cable hanging from the tops of the suspension towers of the Golden Gate Bridge is shaped as a parabola; the vertex of this parabola is the lowest point of the cable, lying six feet above the level of the road. The towers are approximately 520 feet above the level of the road and stand 4200 feet apart. Use these facts to find an equation representing the shape of the cable, placing the cable's vertex at the origin of your coordinate plane.

31. The *eccentricity* of an ellipse with foci at $(-c, 0)$ and $(c, 0)$ and with the sum of the distances equal to $2K$ is defined to be $e = c/K$. From the assumption that $K > c$, we have $0 < e < 1$. How do changes in e alter the shape of an ellipse? Contrast e close to 0 with e close to 1.

32. Show that the classical definition of a line as the set of points (x, y) equidistant from two fixed points (c, d) and (s, t) corresponds to an equation of the form $y = mx + b$.

In exercises 33–38, find a linear equation that corresponds to each set of points.

33. The line with slope 5 and y-intercept $b = -4$.

34. The line passing through points $(0, 0)$ and $(5, 3)$.

35. The line passing through points $(-2, 3)$ and $(6, -7)$.

36. The set of points (x, y) equidistant from the two points $(0, 0)$ and $(6, 4)$.

37. The set of points (x, y) equidistant from $(0, 0)$ and $(-2, 8)$.

38. The set of points (x, y) equidistant from $(-2, 3)$ and $(6, -7)$.

In exercises 39–44, simultaneously solve the equations of the given curves to determine all points of intersection of the curves.

39. The parabola $y = x^2 + 1$ and the line $y = x + 1$.

40. The parabola $y = x^2 + 1$ and the parabola $y = 1 - x^2$.

41. The parabola $y = x^2 + 1$ and the hyperbola $x^2 - 2y^2 = 1$.

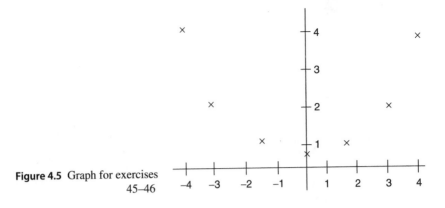

Figure 4.5 Graph for exercises 45–46

42. The ellipse $3x^2 + y^2 = 1$ and the line $y = x + 1$.
43. The ellipse $3x^2 + y^2 = 3$ and the parabola $2y = 3x^2$.
44. The ellipse $3x^2 + 2y^2 = 1$ and the hyperbola $5x^2 - 2y^2 = 1$.

In exercises 45–46, work with the following graph of points in the plane given in figure 4.5.

45. State the ordered pairs that correspond with each of the points in figure 4.5. Give your answer both as a set and as a table.
46. Sketch a quadratic curve that approximately fits the points given in figure 4.5 and state an algebraic equation defining this parabola.

In exercises 47–49, consider the set of points identified in the following table.

x	-3	-2	-1	0	1	2	3	4
y	-54	-16	-2	0	2	16	54	128

47. Graph the ordered pairs given in the above table on the coordinate plane.
48. Describe any geometric patterns you observe among the points graphed in exercise 47.
49. Based on exercises 47 and 48, state an algebraic equation expressing the relationship between the numbers in the x-row and the y-row in the above table.

In exercises 50–52, consider the set of points identified in the following table.

x	-153	-153	-100	-100	-50	-50	0	0	50	50	100	100	148
y	3	-3	115	-115	143	-143	150	-150	141	-141	110	-110	0

50. Graph the ordered pairs given in the above table on the coordinate plane.
51. Describe any geometric patterns you observe among the points graphed in exercise 50.

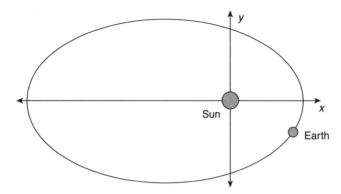

Figure 4.6 Earth's orbit for exercises 50–52

52. The collection of points in the above table corresponds with data on the Earth's orbit collected by astronomers before 1650. One of the major mathematical breakthroughs of the seventeenth century was Newton's determination that the Earth's orbit can be described by the equation: $x^2 + y^2 = (150.4716 - 0.0167x)^2$. Newton's study of planetary motion in terms of elliptical orbits was his first application of calculus, and it motivated his development of the theory describing how the position of objects change over time. The above equation corresponds to an ellipse, and Newton framed his equation so that the Sun (at one of the foci) is located at the origin as illustrated in figure 4.6. Based on this information, which half of the x-axis contains the Earth's perihelion (that is, its closest approach to the Sun)?

In exercises 53–58, find a linear equation that corresponds to each set of points in $\mathbb{R}^3$.

53. The set of points (x, y, z) equidistant from $(1, 2, 0)$ and the origin.
54. The set of points (x, y, z) equidistant from $(2, 0, 4)$ and the origin.
55. The set of points (x, y, z) equidistant from $(1, 1, 1)$ and $(2, 2, 2)$.
56. The set of points (x, y, z) equidistant from $(-2, -3, 4)$ and $(1, 3, -8)$.
57. The plane passing through $(2, 0, 0)$, $(0, 3, 0)$, and $(0, 0, 1)$.
58. The plane passing through $(4, 0, 0)$, $(0, 1, 0)$, and $(0, 0, 5)$.

Exercises 59–61 consider the intersection of the planes $ax + by + cz = 1$ with the unit sphere $x^2 + y^2 + z^2 = 1$. Answer each question by simultaneously solving the corresponding equations.

59. Prove that the nonempty intersection points of a plane with the unit sphere is always an ellipse, a circle, or a single point.
60. If $c = 1$, how are a and b related when the intersection of the plane and the unit sphere is a circle?
61. If $c = 1$, what values of a and b result in the intersection of the plane and unit sphere being a single point?

Exercises 62–64 develop a proof that the perpendicular distance d from a point (x_0, y_0) to a line $y = mx + b$ is

$$d = \frac{|mx_0 + b - y_0|}{\sqrt{m^2 + 1}}.$$

62. Sketch the line segment from (x_0, y_0) to $y = mx + b$ that is perpendicular to the line. Let $(\hat{x}, \hat{y})$ identify the point of intersection of this line segment with $y = mx + b$. Why is $-(1/m)(\hat{x} - x_0) + y_0 = m\hat{x} + b$? Solve this equation for $\hat{x}$ so that $m^2 + 1$ appears in the denominator.

63. As in exercise 62, explain why $(1/m)(\hat{y} - b) = -m(\hat{y} - y_0) + x_0$ is valid and solve this equation for $\hat{y}$ so that $m^2 + 1$ appears in the denominator.

64. Using the algebraic expressions for $\hat{x}$ and $\hat{y}$ from exercises 62 and 63, along with the distance formula $d = \sqrt{(\hat{x} - x_0)^2 + (\hat{y} - y_0)^2}$, prove that the perpendicular distance from the point (x_0, y_0) to the line $y = mx + b$ (that is, to the point $(\hat{x}, \hat{y})$) is given by the formula stated above.

Exercises 65–70 use similar triangles to prove the equivalence of the unit circle and right triangle definitions of the trigonometric functions. Recall from exercises 28–33 in section 3.3 that two triangles are similar if their three interior angles are identical; the side lengths of corresponding similar triangles share the same ratios.

In exercises 65–70, assume that ABC is an arbitrary right triangle with side lengths denoted by opp, adj, and hyp (greater than one). Use the similar triangles ABC and ADE in figure 4.7 and the ratio property of similar triangles to prove the equivalence of the definitions of the trigonometric functions. Note that edge AE is the radius of the unit circle and has length one.

65. The right triangle definition of cosine asserts that $\cos\theta = $ adj/hyp. Prove this quantity is equal to the unit circle definition of $\cos\theta$, which in this setting is $|AD|$.

66. The right triangle definition of sine asserts that $\sin\theta = $ opp/hyp. Prove this quantity is equal to the unit circle definition of $\sin\theta$, which in this setting is $|DE|$.

67. The right triangle definition of tangent asserts that $\tan\theta = $ opp/adj. Prove this quantity is equal to the unit circle definition of $\tan\theta$, which in this setting is $|DE|/|AD|$.

68. Prove the right triangle definition of secant equals the unit circle definition.

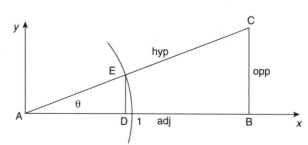

Figure 4.7 Similar right triangles for exercises 65–70

69. Prove the right triangle definition of cosecant equals the unit circle definition.
70. Prove the right triangle definition of cotangent is equal to the unit circle definition.

4.2 Functions and Inverse Functions

Functions are essential to the study of mathematics. As one might expect, humanity's understanding of functions developed gradually over centuries of reflection and dialogue. While the Babylonian, Greek, Indian, and Islamic mathematicians all worked with what we now understand as functions, the fourteenth century schools of natural philosophy at Oxford and Paris were the first to consider the more general notion of a function expressing dependence relations among quantities. The German philosopher and mathematician Gottfried Wilhelm von Leibniz is credited with introducing the word "function" into mathematical dialogue in the 1670s, and important refinements to the understanding of functions were made by Johann Bernoulli (or Jean Bernoulli), Euler, Cauchy, Fourier, and Dirichlet in the eighteenth and nineteenth centuries.

Functions enable mathematicians to think beyond computations with specific values to generalizations of algebraic rules, making it possible to apply a rule in one fell swoop to a whole set of values. The standard notation for a function is highly advantageous: abstractly writing $y = f(x)$ allows notational manipulations that are convenient, are easily understood, and prompt new insights into functions and operations on functions. One such operation is "composition" in which first one function and then another is applied to an input; this operation is defined more carefully in this section, but for the moment we observe the notational ease of expressing composition as $y = g(f(x))$. Even more, this notation allows us to name important functions and to describe easily properties of functions (for example, we can say "f is continuous" or "g is bounded").

This exploration of functions has two primary goals: to understand the abstract underpinnings of calculus; and to extend our investigations to spaces of functions. Along the way, set theory will play an important role as we characterize functions using domains and ranges (that is, using sets of inputs and sets of outputs).

This study begins with the rigorous, formal definition of a function that is commonly used by the mathematical community. Most often this definition (though its form is a bit modernized) is attributed to the work of the German mathematician Peter Lejeune Dirichlet in the 1830s. Dirichlet was trying to understand infinite sums of trigonometric functions known as Fourier series, and he needed to describe the notion of a "function" as something different from a "formula." The following definition rigorously expresses the intuitive notion that a function identifies every input with a unique output.

Definition 4.2.1 *Let D and Y be sets. A **function** from D to Y is a set of ordered pairs (x, y), where $x \in D$, $y \in Y$, and every $x \in D$ appears in exactly one ordered pair.*

We write $f : D \to Y$ to identify a function f from D to Y, and we write $f(x) = y$ to indicate that the ordered pair (x, y) appears in the function. The set D of x-values appearing in an ordered pair is called the **domain** *of f, and the set R of y-values appearing in an ordered pair is called the* **range** *of f. Finally, we say that f maps D to the* **target space** *Y and that f* **maps** *a to b whenever $f(a) = b$.*

In this definition of a function, notice that the range R is a subset of the target space Y. In some contexts the set Y is identified as the range. The next example considers a few simple finite functions on the integers.

Example 4.2.1   We first consider the function f defined as $\{(1, 2),\ (3, 2),\ (7, 3),\ (8, 12)\}$. This set is a function because every x-coordinate (that is, $1, 3, 7, 8$) appears in exactly one ordered pair, and so every x is mapped to exactly one y. There is no difficulty with 2 appearing as the y-coordinate for both 1 and 3; the definition prohibits only repeated x-coordinates. The domain of f is $D = \{1, 3, 7, 8\}$, and the range is $R = \{2, 3, 12\}$.

In contrast, consider the set of ordered pairs $\{(1, 2),\ (3, 2),\ (7, 3),\ (7, 12)\}$. This set is not a function because 7 appears as the x-coordinate in two distinct ordered pairs $(7, 3)$ and $(7, 12)$; hence every input does not have a unique output. We could delete one or both of these two ordered pairs to obtain a function; in a similar way, mathematicians often restrict the domain of an algebraic expression or geometric curve in order to obtain a function.

Finally, sets such as $\{1, 2\}$ and $\{4, (1, 2)\}$ are not functions since they are not sets of ordered pairs.

■

In an intuitive sense, a function is a correspondence (or a relationship) between two variables, where each possible value of the independent variable (the function's input) produces a single unique value of the dependent variable (the function's output). The sets that correspond to the possible input and output values are the domain and range sets, respectively. Recall from your previous math courses that often the domain D and the range R are not mentioned explicitly when defining or referring to a function. In these cases, D is understood to be the largest possible set on which the rule defining the function is defined. For example, without a domain or range specification, the real function $f(x) = 1/x$ is understood to have domain D consisting of all nonzero real numbers, since the equation is defined except when $x = 0$. The resulting range R is the set of nonzero reals, since these are the possible output values that could result.

As we saw in example 4.2.1, functions need not be defined in terms of an algebraic expression. Rather, any set of ordered pairs is a function (even infinite sets) if every x-value is identified with a unique y-value, even if an algebraic expression is not provided

Question 4.2.1   Explain why each set of ordered pairs is a function or not. For each function, identify the domain D and range R.

(a) $\{(0, 0),\ (1, 2),\ (2, 4),\ (3, 6),\ (3, 25)\}$
(b) $\{(0, 0),\ (1, 1),\ (2, 4),\ (3, 9),\ (4, 16)\}$
(c) $\{(x, x) : x \in \mathbb{N}\} \ = \ \{(1, 1),\ (2, 2),\ (3, 3), \ldots\}$

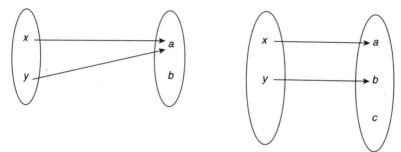

Figure 4.8   The left map is not one-to-one. The right map is not onto

(d) $\{(x, 2) : x \in \mathbb{N}\} \; = \; \{(1, 2), \; (2, 2), \; (3, 2), \ldots\}$
(e) $\{(2, y) : y \in \mathbb{N}\} \; = \; \{(2, 1), \; (2, 2), \; (2, 3), \ldots\}$

■

We verify that an algebraic equation relating variables x and y defines a function $y = f(x)$ in the following way: given any two distinct y-values y_1 and y_2 in the range R, prove the corresponding x values x_1 and x_2 are also distinct. Recall the geometric rendition of this property known as the "vertical line test": any vertical line can intersect the curve (corresponding to the algebraic equation) at most once.

Three important adjectives for functions are: one-to-one, onto, and one-to-one correspondence. As you may recall, one-to-one functions have "inverses." Intuitively, an inverse function "undoes" the work of a given function mapping outputs back to inputs. Many commonly used functions have inverses; for example, $y = x - 5$ is the inverse of $y = x + 5$, the natural logarithm function $\ln(x)$ is the inverse of $y = e^x$, $y = \arcsin(x)$ is the inverse of $y = \sin(x)$ with restricted domain $-\pi/2 < x < \pi/2$, and $y = ax + b$ is the inverse of $y = (x - b)/a$. One-to-one correspondences play a key role when studying the relative sizes of sets, and they are used to determine the equivalence of two algebraic structures such as groups.

An intuitive understanding of one-to-one and onto functions may be gained from simple illustrations. For one-to-one functions, every output comes from a unique input; the function illustrated on the left in figure 4.8 is not one-to-one because the element a is an output of two distinct inputs x and y. A function is onto a target set Y if every element of Y is an element of the range R; that is, if every element of Y is an output. The function on the right in figure 4.8 is not onto the illustrated set because the element c is not an output for any input. A one-to-one correspondence is both one-to-one and onto. The next definition precisely expresses these intuitive descriptions.

Definition 4.2.2 • *A function* $f : D \rightarrow Y$ *is* **one-to-one** *if for all* $x, y \in D, f(x) = f(y)$ *implies that* $x = y$; *equivalently, if for all* $x, y \in D, x \neq y$ *implies* $f(x) \neq f(y)$.
• *A function* $f : D \rightarrow Y$ *is* **onto** Y *if for every* $y \in Y$, *there exists an* $x \in D$ *such that* $f(x) = y$; *in other words, if the range of* f *is the target set.*
• *A function* $f : D \rightarrow Y$ *is a* **one-to-one correspondence** *if* f *is both one-to-one and onto. When this happens, we say that the sets* D *and* Y *are in one-to-one correspondence and we write* $|D| \; = \; |Y|$.

The next two examples consider functions in light of this definition, one that is a one-to-one correspondence and one that is neither one-to-one nor onto.

Example 4.2.2 We prove that the function f mapping the nonzero reals to the nonzero reals defined by $f(x) = 1/x$ is a one-to-one correspondence.

Proof We first show f is one-to-one, assuming that $a, b \in \mathbb{R} \setminus \{0\}$ with $f(a) = f(b)$, and proving that $a = b$. Since $f(a) = f(b)$, $1/a = 1/b$ from the definition of the function. Multiplying both sides of this equation by ab gives $a = b$. The function is therefore one-to-one.

We now show $f(x)$ is onto, assuming $b \in \mathbb{R} \setminus \{0\}$ (the target space) and finding a nonzero real value a such that $f(a) = b$. From the function's definition, this condition holds exactly when $1/a = b$, which identifies the corresponding domain value as $a = 1/b$. In conclusion, we have found a value a in the function's domain that is mapped to the given value b in the target set, and so f is onto.

Because f is both one-to-one and onto, it is a one-to-one correspondence from the set of nonzero reals to itself.

■

Example 4.2.3 We prove that the function $f : \mathbb{R} \to \mathbb{R}$ defined by $f(x) = x^2$ is neither one-to-one nor onto the reals.

Proof We identify counterexamples to each property. To disprove f is one-to-one, consider $a = 2$ and $b = -2$. For these values, $f(a) = f(2) = 4 = f(-2) = f(b)$, but $a = 2 \neq -2 = b$. Hence f is not one-to-one. To disprove f is onto the reals, consider $b = -1 \in \mathbb{R}$ (the target set). Since the square of every real number is positive, there does not exist $a \in \mathbb{R}$ such that $f(a) = a^2 = -1$. Thus f is not onto.

■

Restricting the domain of the function $f(x) = x^2$ in example 4.2.3 to the nonnegative reals would result in a one-to-one function. Similarly restricting the target space to the nonnegative reals would result in an onto function. Hence the function g from the nonnegative reals to the nonnegative reals defined by $g(x) = x^2$ is a one-to-one correspondence.

Question 4.2.2 (a) Prove that $f : \mathbb{R} \to \mathbb{R}$ defined by $f(x) = 12x - 10$ is a one-to-one correspondence.

(b) Prove that $f : \mathbb{R} \to \mathbb{R}$ defined by $f(x) = \sin x$ is neither one-to-one nor onto. Identify restrictions of the domain and range of $\sin x$ that yield a one-to-one correspondence.

■

Recall the importance of the composition operation on functions from your previous studies—most commonly used functional expressions are built up from compositions of more basic functions. For example, the function $y = \sqrt{4(x - 2)^2 + 1}$ is the composition of $y = x - 2$; $y = 4x^2 + 1$; and $y = \sqrt{x}$. Many properties of the

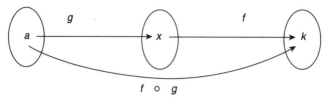

Figure 4.9   Composition of functions

composite function $(f \circ g)(x) = f(g(x))$ result from the corresponding properties of the component functions $f(x)$ and $g(x)$. Figure 4.9 graphically illustrates composition, where $f \circ g$ "combines" the process of first applying the function g and then the function f into a single functional operation; in this case, $f(g(a)) = f(x) = k$ (where $g(a) = x$) becomes $(f \circ g)(a) = k$.

The following definition of composition expresses this intuitive description.

Definition 4.2.3   *Let A, B, C be sets, and both $f : B \to C$ and $g : A \to B$ be functions. The* **composition** $f \circ g : A \to C$ *is defined by* $(f \circ g)(a) = f(g(a))$ *for every $a \in A$ such that $g(a) \in B$.*

In many different mathematical settings, a composite function $f \circ g$ inherits properties shared by its component functions f and g. Many of the functional properties we have studied in this section have this feature.

Theorem 4.2.1   *The composition of two one-to-one functions is one-to-one.*

     Proof   Suppose $f : B \to C$ and $g : A \to B$ are one-to-one, and assume $f(g(a)) = f(g(b))$. Since f is one-to-one, we conclude that $g(a) = g(b)$. Similarly, since g is one-to-one, we conclude that $a = b$. Hence the composition $f \circ g$ is one-to-one. ∎

Theorem 4.2.2   *The composition of two onto functions is onto.*

     Proof   The proof is left for exercise 56 at the end of the section. ∎

For the rest of this section we study inverse functions. Intuitively speaking, a function $f(x)$ maps each element of its domain to a unique element of its range. The inverse function (written $f^{-1}(x)$) reverses this process so that each output value in the range of f is mapped back to the corresponding input from which it came. In short, f and f^{-1} swap input and output values, as the following example illustrates.

Example 4.2.4   We consider the inverse of $f : \mathbb{Z} \to R$, where $R = \{2, 3, 4, \ldots\}$ defined by $f(x) = x + 2$. The following table represents this function on selected inputs.

f-input	0	1	2	3	4	5	...
f-output	2	3	4	5	6	7	...

The action of f^{-1} (which maps each f-output back to its corresponding f-input) reverses the rows of the table as follows.

f^{-1}-input $= f$-output	2	3	4	5	6	7	...
f^{-1}-output $= f$-input	0	1	2	3	4	5	...

Notice that the inverse function's output values are two less than the corresponding input; that is, $f^{-1}(x) = x - 2$.

■

The next definition expresses this intuitive description of the inverse of a function in terms of composition. As in the study of groups and fields in chapters 2 and 3, inverses should be "two-sided," which leads to both clauses in the definition.

Definition 4.2.4 *If $f : D \rightarrow Y$ is a function with range R, then $g : R \rightarrow D$ is the **inverse function** of f when both $(g \circ f)(x) = x$ for all $x \in D$ and $(f \circ g)(x) = x$ for all $x \in R$. We write $g = f^{-1}$ to identify the (unique) inverse of f.*

Example 4.2.5 As discussed in example 4.2.4, the inverse of $f(x) = x + 2$ is $f^{-1}(x) = x - 2$. Similarly, the inverse of $g(x) = 2x$ (which maps $\mathbb{R}$ to $\mathbb{R}$) is the function $g^{-1}(x) = x/2$; that is, we "undo" the work of multiplying by 2 by dividing by 2.

■

For all but a handful of functions, the inverse f^{-1} of a function $f(x)$ is *not* the same as its multiplicative inverse (or reciprocal) $[f(x)]^{-1} = 1/f(x)$. In the last example, we observed that $f(x) = x + 2$ has inverse function $f^{-1}(x) = x - 2$, while the multiplicative inverse of $f(x)$ is $[f(x)]^{-1} = 1/(x + 2)$. These functions are clearly not the same—they even have different domains! Therefore, we use the term "inverse function" exclusively to identify the function f^{-1} that results from inverting the operation of composition, rather than inverting the operation of multiplication.

As may be apparent, not all functions have inverses. Consider the familiar squaring function $f : \mathbb{R} \rightarrow \mathbb{R}$ defined by $f(x) = x^2$. Many pairs of numbers map to the same output; for example, both $f(2) = 2^2 = 4$ and $f(-2) = (-2)^2 = 4$. An inverse function for $f(x) = x^2$ would need to assign a unique value to $f^{-1}(4)$. But this task is impossible, since there were two values (2 and -2) that f maps to 4. Thus $f(x) = x^2$ does not have an inverse. Notice that $f(x) = x^2$ is not one-to-one, which is the reason it does not have an inverse. The following theorem extends this observation to all functions.

Theorem 4.2.3 *A function has an inverse iff the function is one-to-one.*

Comments on proof The proof of this theorem is a direct application of definitions, and is left for exercises 63–64 at the end of this section.

■

When a function is one-to-one, how do we find its inverse? A function f written in terms of basic algebraic operations can often be manipulated using the following two-step process to obtain an algebraic expression for its inverse f^{-1}.

- Switch the roles of x and y in the expression $y = f(x)$ to obtain $x = f(y)$.
- Solve the resulting new equation for y to obtain $y = f^{-1}(x)$.

The next example illustrates the process.

Example 4.2.6   We apply the two-step process to find the inverse f^{-1} of the one-to-one linear function $f(x) = 5x + 2$.

First switch the roles of x and y, writing $x = 5y + 2$. Then algebraically solve for y to obtain the inverse: $f^{-1}(x) = y = (x - 2)/5$. This expression is easily verified to be the inverse by directly computing $(f^{-1} \circ f)(x)$ and $(f \circ f^{-1})(x)$—they both should equal x. For example, $(f^{-1} \circ f)(x) = f^{-1}(f(x)) = f^{-1}(5x + 2) = [(5x + 2) - 2]/5 = 5x/5 = x$. Similarly, $(f \circ f^{-1})(x) = x$.

∎

Sometimes mathematicians modify the domain of a function that is not one-to-one (and so not invertible). For example, $f(x) = x^2$ is not invertible, but its domain can be restricted to $D = \{x \in \mathbb{R} : x \geq 0\}$ in order to generate an inverse. The function $h(x) = x^2$, where $x \in D$, has inverse function $h^{-1}(x) = \sqrt{x}$. This strategy is commonly employed by mathematicians—restrict a function's domain to make it one-to-one, and so invertible.

Question 4.2.3   Consider the one-to-one function $f : \{x \in \mathbb{R} : x \geq 0\} \to \mathbb{R}$ defined by $f(x) = 3x^2 - 5$.

(a) Find an algebraic expression for $f^{-1}(x)$.
(b) Verify that $(f^{-1} \circ f)(x) = x$.
(c) Verify that $(f \circ f^{-1})(x) = x$.

∎

Equations involving logarithms and exponentials are often solved by using inverse functions. The function $y = \log_a(x)$ is defined as the inverse of $f(x) = a^x$ (the natural logarithm function has base e, so $\ln(x) = \log_e(x)$). To solve for a variable x that is "trapped" in the power of an exponential function, we apply the logarithm to both sides, using the operation of composition.

Example 4.2.7   We solve the equation $y - 5 = e^{4x-8}$ for x. Applying the natural logarithm to both sides of this equation and simplifying (using the fact that $f^{-1}(f(x)) = x$) yields the following.

$$\ln(y - 5) = \ln(e^{4x-8})$$
$$\ln(y - 5) = 4x - 8$$
$$x = (\ln(y - 5) + 8)/4$$

∎

Question 4.2.4   Solve for x in the equation

$$y = \frac{\ln[(x - 3)^4]}{12}.$$

∎

We end this section with a discussion of the strong relationship between the graph of a function f and the graph of its inverse function f^{-1}. Understanding that an inverse function "swaps" the roles played by the x- and y-coordinates leads to the insight that the graphs of the two functions are related by switching the roles of the x- and y-axis. In other words, the graphs of a function and its inverse are reflections (or mirror images) of one another across the line $y = x$.

　　This observation is particularly useful when $f(x)$ is easy to graph, and when the graph of $f^{-1}(x)$ is much more difficult to identify. To find the graph of f^{-1}, simply reflect the graph of f across the line $y = x$. Mathematicians often use this graphical property to define new functions. Any one-to-one function f has an inverse whose graph can be obtained from the graph of f using this reflection property. For example, $\ln x$ is sometimes defined in terms of e^x in this fashion.

　　At the same time, an inverse function f^{-1} can be important in its own right, independent of any reference to the function f. For example, $y = \ln(x)$ has many independent uses; the fact that $\ln(x)$ is the area under the graph of $y = 1/t$ from $t = 1$ to $t = x$ is one such application. The last question of this section considers the graphical relationship between functions f and f^{-1} in the context of exponential and logarithmic functions having base two.

Question 4.2.5　Consider the exponential function $f : \mathbb{R} \to \mathbb{R}$ defined by $f(x) = 2^x$.

　　(a) Graph the six points of $f(x)$ identified by $x = -2, -1, 0, 1, 3, 5$.
　　(b) Extend the graph of points from part (a) to a complete graph of $f(x) = 2^x$.
　　(c) Graph the inverse $f^{-1}(x) = \log_2(x)$ of $f(x) = 2^x$.

　　　　　　　　　　　　　　　　　　　　　　　　　　　　　　　　　■

4.2.1　Reading Questions for Section 4.2

　　1. State both an intuitive description and the definition of a function.
　　2. Give an example of a set of ordered pairs that is a function and a set that is not.
　　3. State both an intuitive description and the definition of a one-to-one function.
　　4. Give an example of a function that is one-to-one and function that is not.
　　5. State both an intuitive description and the definition of an onto function.
　　6. State both an intuitive description and the definition of a one-to-one correspondence.
　　7. Define and give an example of a composition of functions.
　　8. State both an intuitive description and the definition of an inverse function.
　　9. Give an example of a function and its inverse.
　　10. What condition must a function satisfy to have an inverse?
　　11. Describe a process for algebraically identifying the inverse of an invertible function.
　　12. What is the relationship between the graphs of a function and its inverse?

4.2.2　Exercises for Section 4.2

In exercises 1–8, explain why each set of ordered pairs is a function or not. For each function identify the domain and range.

　　1. $\{(0, 5),\ (1, 2),\ (2, 4),\ (2, 6)\}$　　　3. $\{(0, 0),\ (1, 2),\ (2, 4),\ (3, 6)\}$
　　2. $\{(0, 2),\ (1, -1),\ (-1, 4),\ (-3, 9)\}$　　4. $\{(0, 0),\ (1, 1),\ (1, 4),\ (3, 9)\}$

5. $\{(x, 2) : x \in \mathbb{R}\}$ (or $y = 2$)

6. $\{(2, y) : y \in \mathbb{R}\}$ (or $x = 2$)

7. $\{(y^2, y) : y \in \mathbb{R}\}$ (or $x = y^2$)

8. $\{(y^2, y) : y \in \mathbb{N}\}$ (or $x = y^2$)

In exercises 9–12, explain why each set of ordered pairs from the unit circle is a function or not. For each function identify the domain and range.

9. The "top" half of the unit circle: $\{(x, y) : x^2 + y^2 = 1 \text{ and } y \geq 0\}$

10. The "bottom" half of the unit circle: $\{(x, y) : x^2 + y^2 = 1 \text{ and } y \leq 0\}$

11. The "right" half of the unit circle: $\{(x, y) : x^2 + y^2 = 1 \text{ and } x \geq 0\}$

12. The "left" half of the unit circle: $\{(x, y) : x^2 + y^2 = 1 \text{ and } x \leq 0\}$

In exercises 13–16, define functions on the finite sets $A = \{1, 2, 3\}$, $B = \{4, 5, 6\}$, and $C = \{7, 8\}$ with the following properties.

13. One-to-one and onto

14. One-to-one, but not onto

15. Not one-to one, but onto

16. Neither one-to-one nor onto

In exercises 17–22, prove each function is onto, or identify an element of the target that is not in the range.

17. $f : \mathbb{R} \to \mathbb{R}$ defined by $f(x) = 2x + 7$

18. $g : \mathbb{R} \to \mathbb{R}$ defined by $g(x) = x^2 - 5$

19. $h : \mathbb{R} \to \mathbb{R}$ defined by $h(x) = x^3 - 1$

20. $q : \mathbb{R} \to \mathbb{R}$ defined by $q(x) = 1/x$

21. $r : \mathbb{R} \to \mathbb{R}$ defined by $r(x) = \sin(x)$

22. $s : \mathbb{R} \to \mathbb{R}$ defined by $s(x) = \tan(x)$

In exercises 23–30, determine if each function has an inverse by proving or disproving the function is one-to-one. If so, specify the inverse function, including an explicit identification of its domain and range.

23. $f(x) = 5x - 2$, where $x \in \mathbb{R}$

24. $g(x) = x^2 + 2x + 3$, where $x \in \mathbb{R}$

25. $h(x) = (x + 3)^2 + 12$, where $x > 0$

26. $j(x) = \ln(x - 1)$, where $x > 1$

27. $k(x) = e^{x^2}$, where $x > 0$

28. $p(x) = e^{x^2}$, where $x \in \mathbb{R}$

29. $q(x) = 1/(2x + 3)$, where $x < -3/2$

30. $r(x) = 1/(x^2 + x)$, where $x > 1$

In exercises 31–34, state a restriction on the domain of each function to obtain a one-to-one function.

31. $f(x) = \sin(x)$

32. $g(x) = \cos(x)$

33. $r(x) = \tan(x)$

34. $s(x) = (x + 2)^4$

In exercises 35–40, graph each function and its inverse on the same axes. Assume appropriate domains so that all functions are defined and one-to-one, and so invertible.

35. $f(x) = x + 5$ and $f^{-1}(x) = x - 5$

36. $g(x) = 3x + 2$ and $g^{-1}(x) = 1/3x - 2/3$

37. $h(x) = x^2$ and $h^{-1}(x) = \sqrt{x}$

38. $q(x) = 1/x$ and $q^{-1}(x) = 1/x$

39. $r(x) = 5^x$ and $r^{-1}(x) = \log_5(x)$

40. $s(x) = (1/3)^x$ and $s^{-1}(x) = \log_{1/3}(x)$

In exercises 41–44, state the algebraic equation and identify the domain of each composition $f \circ g(x)$ and $g \circ f(x)$.

41. $f(x) = \sqrt{5x - 1}$ and
    $g(x) = 1/(2x + 1)$

42. $f(x) = x^3 - 1$ and $g(x) = \sqrt{x + 1}$

43. $f(x) = \ln(5x - 1)$ and
    $g(x) = 1/(x + 2)$

44. $f(x) = e^{1/x}$ and $g(x) = \sqrt{2x - 1}$

In exercises 45–55, identify each statement as *true* or *false*. For those statements that are false, provide an explanatory reason or a counterexample.

45. If $f(x) = x^3 + x$, then f is one-to-one, $f^{-1}(-2) = -1$, and $f^{-1}(2) = 1$.

46. If $f(x) = 2x^3 + 5x$, then f is one-to-one, $f^{-1}(7) = 1$, and $f^{-1}(14) = 2$.

47. The sum of two onto functions is onto.

48. The sum of two one-to-one functions is one-to-one.

49. The product of two onto functions is onto.

50. The product of two one-to-one functions is one-to-one.

51. If $f(x)$ is a function and $c \in \mathbb{R}$, then $f(c \cdot x) = c \cdot f(x)$.

52. Every polynomial function has an inverse.

53. The inverse of a polynomial is never a polynomial.

54. If $f(x)$ is invertible, the product $f(x) \cdot f^{-1}(x) = 1$ for every x in the domain of f.

55. If $f(x)$ is invertible, the composition $(f \circ f^{-1})(x) = 1$ for every x in the domain of f.

In exercises 56–70, prove each mathematical statement, assuming that $f : A \to B$ and $g : B \to C$ in your proofs.

56. If f and g are onto functions, then $g \circ f$ is an onto function.

57. If $g \circ f$ is onto, then g is onto.

58. If $g \circ f$ is onto, then f may or may not be onto.

59. If $g \circ f$ is one-to-one, then f is one-to-one.

60. If $g \circ f$ is one-to-one, then g may or may not be one-to-one.

61. If f and g are one-to-one correspondences, then $g \circ f$ is a one-to-one correspondence.

62. If $f \circ f$ is a one-to-one correspondence, then f is a one-to-one correspondence.

63. If f is one-to-one, then f has an inverse.

64. If f has an inverse, then f is one-to-one.

65. If f is invertible, then the inverse function of f is unique.

66. If f is invertible, then the inverse of f^{-1} is $(f^{-1})^{-1} = f$.

67. If f and g are invertible, then $(g \circ f)^{-1} = f^{-1} \circ g^{-1}$.

68. Any nonconstant linear function $f(x) = mx + b$ (with $m \neq 0$) is a one-to-one correspondence from $\mathbb{R}$ to $\mathbb{R}$.

69. The inverse of a nonconstant linear function $f(x) = mx + b$ exists and is a linear function.

70. The inverse of $f(x) = x^{2k+1}$ exists for any $k \in \mathbb{N} \cup \{0\}$.

4.3   Limits and Continuity

This section begins a study of the mathematical theory underlying calculus by exploring the concept of a limit. Hopefully your previous studies of calculus have provided a sense of how important limits are to the subject. For example, the property of continuity is defined in terms of limits, where the limits of continuous functions are evaluated by direct substitution. Similarly, the derivative is defined as a limit of difference quotients, and the Riemann integral is a limit of a sum. Hence the study of limits sets the stage for a study of the theory behind calculus.

In light of the dependence on limits of the definitions of derivative and integral, we might expect an orderly development of these mathematical ideas, with a rigorous understanding of limits historically preceding the notions of the derivative and the integral. But the actual historical development was not nearly so neat. The French mathematician Augustin-Louis Cauchy did not formulate the modern definition of a limit until the early 1800s—nearly 150 years after Sir Isaac Newton and Gottfried Wilhelm von Leibniz independently articulated the fundamental theorem of calculus linking the derivative and the integral.

How did Newton and Leibniz think about derivatives and integrals without a well-defined concept of a limit? Newton thought of the derivative in terms of small changes he called "moments of a fluent," and he called motion and the change in continuous variables over time "fluxions." He thought of the integral in terms of antiderivatives. In contrast, Leibniz thought of the derivative in terms of differences between successive terms in sequences with infinitely close values, and of the integral as a sum of infinitely many lines. These less-refined concepts based on infinitesimal quantities were adequate to handle many of the calculations that arose in the practical questions of optics, celestial mechanics, and astronomy that motivated their work. Sadly, the independent work of Newton in the 1660s and Leibniz in the 1670s resulted in a rather bitter argument over who should receive credit for calculus.

As the next generation of mathematicians continued to develop calculus, they recognized the importance of developing a rigorous, logical basis for the theory. A famous critique was given by the Irish philosopher George Berkeley in his 1734 tract *The analyst: or a discourse addressed to an infidel mathematician*; Berkeley asked piercing questions about the legitimacy of Newton's "fluxions." The Scottish mathematician Colin Maclaurin in his 1742 *Treatise of fluxions* and the French mathematician Jean Le Rond d'Alembert in his 1754 *Différential* both gave important responses to Berkeley's critique. However it was not until 1821 that Cauchy provided the first rigorous development of calculus (including the contemporary definition of the limit) in the text *Cours d'Analyse*.

Cauchy's definition of the limit provides an important foundation for studying many topics in the theory of functions. As you may have experienced, this definition can be difficult to understand because of its unfamiliar Greek letters used as symbols and the abstractness of its formulation. We discuss the definition of the limit and its interpretation in some detail. If you struggle with its logic, be patient, remembering that you are in good company—mathematics students have been struggling to master these notions for more than two centuries. In fact, Cauchy's own students *rioted* and marched to university officials in complaint when he first taught these ideas! And rest assured that mastering these ideas will provide deeper insights into the theoretical structure and behavior of functions.

We begin by discussing an intuitive description of a limit. In this discussion, all numbers and functions are interpreted in the context of the reals. Recall from calculus that the phrase "the *limit* of a function $f(x)$ as the variable x approaches a is equal to L," written as $\lim_{x \to a} f(x) = L$ means the following.

If x is close to a (but not equal to a), then $f(x)$ is close to L.

The following graphical example and question explore some of the subtleties in this intuitive understanding of limit.

Example 4.3.1 We identify the following limits based on the graph of the function $f(x)$ given in figure 4.10.

- $\lim_{x \to -4} f(x) = 2$.
- $\lim_{x \to -3} f(x) = 1$. The limit exists even though $f(-3)$ is undefined because we do not consider the value of the function *at $x = a = -3$* in determining the limit.
- $\lim_{x \to -2} f(x) = 0$. The limit is 0 even though $f(-2) = 1$ because we do not consider the value of the function *at $x = a = -2$* in determining the limit.
- $\lim_{x \to -1} f(x) = $ DNE (or Does Not Exist). The fact that x must be close to $a = -1$ from *either side* is important here. If $x < -1$ is close to $a = -1$, we have $f(x)$ close to 1, while if $x > -1$ is close to $a = -1$, we have $f(x)$ close to 0. Because these values differ the limit is undefined.

■

Question 4.3.1 State the value of each limit based on the graph of $f(x)$ given in figure 4.10.

(a) $\lim_{x \to 0} f(x)$ (d) $\lim_{x \to 2.5} f(x)$

(b) $\lim_{x \to 1} f(x)$ (e) $\lim_{x \to 3} f(x)$

(c) $\lim_{x \to 2} f(x)$ (f) $\lim_{x \to 4} f(x)$

■

The key to defining the limit $\lim_{x \to a} f(x) = L$ lies in articulating the mathematical meaning of the phrases "x is close (but not equal) to a" and "$f(x)$ is close to L." The word "close" is understood as a reference to distance. In the context of the real numbers, the distance between two numbers is measured using the absolute value metric; that is,

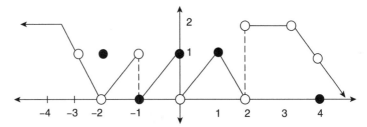

Figure 4.10 Graph for example 4.3.1 and question 4.3.1

the distance between two real numbers r, s is $|r - s|$. Therefore the phrase "x is close (but not equal) to a" is interpreted to mean that $|x - a|$ is a small, positive, nonzero real number; traditionally, this small number is denoted by "δ" (the lower case Greek letter "delta"), and we write $0 < |x - a| < \delta$. In a similar fashion, "$f(x)$ is close to L" is interpreted as meaning that $|f(x) - L|$ is a small, positive real number; traditionally, this small number is denoted by "ε" (the lower case Greek letter "epsilon"), and we write $|f(x) - L| < \varepsilon$. The next definition expresses the intuitive description of a limit.

Definition 4.3.1 Cauchy's definition of the limit *Let $f : D \to Y$ be a function and let $a \in \mathbb{R}$ such that an open interval containing a is a subset of D. Then, $\lim_{x \to a} f(x) = L$ means: for every $\varepsilon > 0$, there exists $\delta > 0$ such that $0 < |x - a| < \delta$ implies $|f(x) - L| < \varepsilon$. In this case, we say that the limit of f as x approaches a exists and is equal to L.*

We note that usually δ depends on ε, and can often be expressed as a function of ε. For example, the function $f(x) = x^2$ is graphed in figure 4.11, and the specific values for ε and δ are illustrated for the limit $\lim_{x \to 3} x^2 = 9$. In this case, the ε-interval around $L = 0$ determines a δ-interval around $a = 3$. In this case, it is possible to express $\delta = -3 + \sqrt{9 + \varepsilon}$. Several examples and questions will show how to find an appropriate value for δ.

Cauchy's definition of the limit phrases questions about the existence and value of a limit in terms of algebraic equations. Most students agree that it takes a little time to get used to working with this definition, and so we consider several examples. Again, be patient; the more you work with this definition, the more comfortable and intuitive it will become, and the more appreciative you will be of its sophisticated handling of a subtle concept.

Example 4.3.2 We use Cauchy's definition of the limit to prove $\lim_{x \to 2} 3x + 5 = 11$.

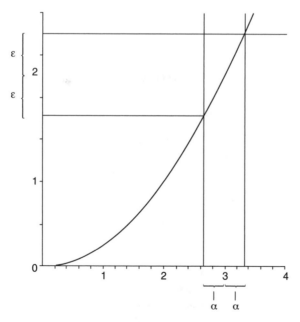

Figure 4.11 The delta-epsilon perspective on limits

Proof   Let $\varepsilon > 0$ be a real number. The proof requires us to identify a corresponding δ; we choose $\delta = \varepsilon/3$ for reasons soon discussed. Assuming that $0 < |x - a| < \delta$, we prove that $|f(x) - L| < \varepsilon$. Therefore, assuming $0 < |x - 2| < \delta$,

$$|f(x) - L| = |(3x + 5) - 11| = |3x - 6| = 3 \cdot |x - 2| < 3 \cdot \delta = 3 \cdot \frac{\varepsilon}{3} = \varepsilon.$$

By the formal of the limit, $\lim_{x \to 2} 3x + 5 = 11$.

■

Question 4.3.2   Prove each limit using the definition and the given δ.

(a) Using $\delta = \dfrac{\varepsilon}{4}$, prove $\lim_{x \to 3} 4x - 10 = 2$.

(b) Using $\delta = \dfrac{\varepsilon}{2}$, prove $\lim_{x \to 1} -2x + 5 = 3$.

(c) In light of parts (a) and (b), observe that $\delta = \dfrac{\varepsilon}{|m|}$ is a good candidate for proving that $\lim_{x \to a} mx + b = ma + b$. Using this hypothesis, prove that $\lim_{x \to 5} 4x + 15 = 35$.

■

Thus far we have only considered limits of linear functions. As you might expect, using the definition to verify limits of nonlinear functions can be subtler. The next two examples consider such functions.

Example 4.3.3   We use the definition of the limit to prove $\lim_{x \to 2} x^2 = 4$.

Proof   Assume $\varepsilon > 0$ is given; we must choose an appropriate value for δ. In this and other nonlinear settings, we begin by restricting the x-values to be considered. Our intuitive understanding of limits indicates the x-values must be "close to $a = 2$," and so we may reasonably assume the "restriction" that x is in an interval centered at $a = 2$, say $1 < x < 3$. Because δ is a small number that measures how close x is to $a = 2$ and because $1 < x < 3$ restricts the x-values to a distance at most one unit from $a = 2$, we are (at this point) assuming $\delta \leq 1$.

    To finish determining an appropriate choice for δ, we now examine $|f(x) - L|$, which equals

$$|f(x) - L| = |x^2 - 4| = |x + 2| \cdot |x - 2|.$$

The second term in this product (which is of the form $|x - a| = |x - 2|$) will be bounded above by δ under the hypothesis of the limit definition. But how do we specify an upper bound on the first term $|x + 2|$? Here is where the above restriction comes in; the inequality $1 < x < 3$ implies that $1 + 2 < x + 2 < 3 + 2$, and so we have $|x + 2| < 5$. Using this inequality, $|f(x) - L| = |x + 2| \cdot |x - 2| < 5 \cdot \delta$. Therefore, choosing δ so that $5 \cdot \delta \leq \varepsilon$, the above string of inequalities implies that $|f(x) - L| < \varepsilon$ whenever $0 < |x - a| < \delta$, provided both $\delta \leq 1$ (from the restriction) and $\delta \leq \varepsilon/5$ (from the choice $5 \cdot \delta \leq \varepsilon$). We ensure both of these conditions by defining δ as the minimum of 1 and $\varepsilon/5$; this minimum is denoted by $\delta = \min\{1, \varepsilon/5\}$. Informed by this analysis, we can now articulate the complete, formal proof.

Let $\varepsilon > 0$ and define $\delta = \min\{1, \varepsilon/5\}$. Assuming that $0 < |x - 2| < \delta$, we also know that $|x + 2| < 5$. Therefore,

$$|f(x) - L| = |x + 2| \cdot |x - 2| < 5 \cdot \delta < 5 \cdot \frac{\varepsilon}{5} = \varepsilon.$$

By the definition of the limit, $\lim_{x \to 2} x^2 = 4$. ∎

Example 4.3.4   We use the definition of the limit to prove

$$\lim_{x \to 4} \frac{1}{x - 2} = \frac{1}{2}.$$

Proof   As in example 4.3.3, first restrict the x-values under consideration to within one unit of $a = 4$, so that $3 < x < 5$ and (at this point) $\delta \leq 1$. We calculate

$$|f(x) - L| = \left| \frac{1}{x - 2} - \frac{1}{2} \right| = \left| \frac{2 - (x - 2)}{2(x - 2)} \right| = \left| \frac{4 - x}{2(x - 2)} \right| = |x - 4| \cdot \left| \frac{1}{2(x - 2)} \right|.$$

The first term of the rightmost expression is of the form $|x - a| = |x - 4|$ and is bounded by δ. A bound for the second term is based on the restriction that $3 < x < 5$; and so $1 < x - 2 < 3$, which implies $2 < 2(x - 2) < 6$, and so

$$\frac{1}{2} > \frac{1}{2(x - 2)} > \frac{1}{6}.$$

Substituting this upper bound for $1/2$ into the expression for $|f(x) - L|$, we have

$$|x - 4| \cdot \left| \frac{1}{2(x - 2)} \right| < \delta \cdot \frac{1}{2}.$$

Thus, choosing δ so that $\delta \cdot \frac{1}{2} \leq \varepsilon$ results in $|f(x) - L| < \varepsilon$ whenever $0 < |x - a| < \delta$, so long as $\delta \leq 1$ (from the restriction) and $\delta \leq 2\varepsilon$ (from the choice $\delta/2 \leq \varepsilon$). We therefore choose $\delta = \min\{1, 2\varepsilon\}$. The complete formal proof follows below.

Let $\varepsilon > 0$ and define $\delta = \min\{1, 2\varepsilon\}$. Assuming $0 < |x - 4| < \delta$, we have

$$\left| \frac{1}{2(x - 2)} \right| < \frac{1}{2}.$$

Thus,

$$|f(x) - L| = |x - 4| \cdot \left| \frac{1}{2(x - 2)} \right| < \delta \cdot \frac{1}{2} < 2\varepsilon \cdot \frac{1}{2} \leq \varepsilon.$$

By the definition of the limit,

$$\lim_{x \to 4} \frac{1}{x - 2} = \frac{1}{2}.$$ ∎

Question 4.3.3   Prove each limit using the definition of the limit and the restriction technique illustrated in examples 4.3.3 and 4.3.4.

(a) $\lim_{x \to 5} x^2 = 25$

(b) $\lim_{x \to 3} (x - 2)^2 = 1$

(c) $\lim_{x \to 3} \frac{1}{x - 1} = \frac{1}{2}$

(d) $\lim_{x \to 2} \frac{1}{x + 3} = \frac{1}{5}$ ∎

The ability to use Cauchy's definition of the limit is key to proving that a given limit exists and equals a specific value. More importantly, we use the formal definition to prove general theorems that hold for all limits. Among other things, these results help us to evaluate limits analytically, in the manner you are accustomed to from your calculus courses. We first consider a uniqueness result for limits (enabling us to refer to *the* limit of $f(x)$ as x approaches a), and then identify some familiar analytic results.

Theorem 4.3.1 *If $\lim\limits_{x \to a} f(x) = L$ exists, then L is unique.*

The next series of questions lead you through a proof of this result. Recall that a standard approach to proving the uniqueness of a mathematical object is to assume that two such objects exist and then either prove they are equal or obtain some other contradiction.

Question 4.3.4 The following steps develop a proof by contradiction for theorem 4.3.1.

(a) Assume $\lim\limits_{x \to a} f(x) = L$ is not unique. Formulate this assumption in terms of a mathematical statement about two limiting values, say L and M.

(b) Apply the definition of the limit to L and M for $\varepsilon = |L - M|/2$, writing out the resulting statement for each of these limiting values.

(c) The two statements from part (b) should involve two values for δ, which we refer to as δ_L and δ_M. Define $\delta = \min\{\delta_L, \delta_M\}$ and reexpress these two statements in terms of this δ.

(d) Explain why this last pair of statements yields a contradiction. Why can't $f(x)$ be "close" to both L and M as asserted in these statements?

■

The next theorem states the analytic properties of limits; using these results is a common exercise in calculus courses.

Theorem 4.3.2 *Let $a, c, L, M \in \mathbb{R}$, and let both f and g be functions on the reals with $\lim\limits_{x \to a} f = L$ and $\lim\limits_{x \to a} g = M$. Then the following equalities hold.*

- **Limit of a constant:** $\lim\limits_{x \to a} c = c$
- **Limit of a scalar multiple:** $\lim\limits_{x \to a} c \cdot f(x) = c \cdot L$
- **Limit of a sum:** $\lim\limits_{x \to a} f + g = L + M$
- **Limit of a difference:** $\lim\limits_{x \to a} f - g = L - M$
- **Limit of a square:** $\lim\limits_{x \to a} [f(x)]^2 = L^2$
- **Limit of a product:** $\lim\limits_{x \to a} f \cdot g = L \cdot M$
- **Limit of a quotient:** $\lim\limits_{x \to a} \dfrac{f}{g} = \dfrac{L}{M}$, *provided that $M \neq 0$*

You should develop a strong familiarity with both the linguistic and symbolic renditions of these results. For example, the third claim in theorem 4.3.2 is not only thought of as $\lim\limits_{x \to a} f + g = L + M$, but also as "the limit of a sum is the sum of the limits." A proof of each a statement in theorem 4.3.2 uses the definition of the limit. We present the proofs of several of these analytic properties here and leave some for the exercises at the end of this section. We highlight some important properties of the absolute value function $|x|$ that will be needed in these proofs.

Theorem 4.3.3   *Define the absolute value function on the reals by*

$$|x| = \begin{cases} x, & x \geq 0 \\ -x, & x < 0. \end{cases}$$

Then the following relationships hold for $x, y \in \mathbb{R}$.

- $|x+y| \leq |x|+|y|$ *(triangle inequality)*   • $|x| - |y| \leq |x - y|$
- $|x \cdot y| = |x| \cdot |y|$ • $\left| \dfrac{x}{y} \right| = \dfrac{|x|}{|y|}$

Example 4.3.5   We prove the limit of a scalar multiple rule (from theorem 4.3.2): If $a, c, L \in \mathbb{R}$ and f is a function on the reals with $\lim_{x \to a} f = L$, then $\lim_{x \to a} c \cdot f(x) = c \cdot L$.

Proof   If $c = 0$, then $c \cdot f(x) = 0$ is a constant function. Applying the limit of a constant rule in theorem 4.3.2; we have $\lim_{x \to a} 0 \cdot f(x) = \lim_{x \to a} 0 = 0 = 0 \cdot L$.

Assume $c \neq 0$. For any given $\varepsilon > 0$, consider $\varepsilon_1 = \varepsilon/|c|$. By Cauchy's definition of limit applied to the given value $\varepsilon_1 > 0$, there exists a value $\delta > 0$ such that $0 < |x - a| < \delta$ implies $|f(x) - L| < \varepsilon_1$. It follows that $0 < |x - a| < \delta$ implies

$$|c \cdot f(x) - c \cdot L| = |c| \cdot |f(x) - L| < |c| \cdot \varepsilon_1 = \varepsilon.$$

∎

Question 4.3.5   Prove the limit of a sum rule: If $a, L, M \in \mathbb{R}$ and both f and g are functions on the reals with $\lim_{x \to a} f = L$ and $\lim_{x \to a} g = M$, then $\lim_{x \to a} f + g = L + M$.
Hint: Let $\varepsilon > 0$. Since $\lim_{x \to a} f = L$, there exists $\delta_L > 0$ such that $0 < |x - a| < \delta_L$ implies $|f(x) - L| < \varepsilon/2$. The limit $\lim_{x \to a} g = M$ implies a similar condition for a $\delta_M > 0$. Choose $\delta = \min\{\delta_L, \delta_M\}$ so that both inequalities involving $\varepsilon/2$ are true when $0 < |x - a| < \delta$, and apply the triangle inequality from theorem 4.3.3 to complete the proof.

∎

The proofs of the product properties for limits identified in theorem 4.3.3 are a bit more complicated because of the algebra involved. Instead of jumping in to prove the general result, we first prove the simpler result for squares.

Example 4.3.6   We prove the limit of a square rule: If $a, L \in \mathbb{R}$ and f is a function on the reals with $\lim_{x \to a} f = L$, then $\lim_{x \to a} [f(x)]^2 = L^2$.

Proof   Without loss of generality, assume $L \geq 0$; a similar proof works if $L < 0$. For $\varepsilon > 0$, we identify $\delta > 0$ such that $0 < |x - a| < \delta$ implies $|[f(x)]^2 - L^2| < \varepsilon$. Working with $\varepsilon_1 = \sqrt{L^2 + \varepsilon} - L > 0$ (for reasons that become apparent in the following calculations) and with $\lim_{x \to a} f = L$, there exists

$\delta > 0$ such that $0 < |x - a| < \delta$ implies $|f(x) - L| < \varepsilon_1$. This δ is the needed value. If $0 < |x - a| < \delta$, the following relations hold.

$$
\begin{aligned}
|f(x)^2 - L^2| &= |f(x) - L| \cdot |f(x) + L| = |f(x) - L| \cdot |f(x) - L + 2L| \\
&\leq |f(x) - L| \cdot (|f(x) - L| + |2L|) \quad \text{by the triangle inequality} \\
&< \varepsilon_1 \cdot (\varepsilon_1 + 2L) = (\sqrt{L^2 + \varepsilon} - L) \cdot (\sqrt{L^2 + \varepsilon} + L) \\
&= L^2 + \varepsilon - L^2 = \varepsilon
\end{aligned}
$$

By the definition of the limit, $\lim_{x \to a} [f(x)]^2 = L^2$. ■

The proof of the limit of a product rule (which asserts that the limit of a product is the product of the limits) uses the following *polarization identity*:

$$
x \cdot y = \frac{1}{4}[(x + y)^2 - (x - y)^2].
$$

This algebraic identity simplifies a question about a product into a question about squares; the validity of the polarization identity is verified by simplifying the algebraic expression on the right-hand side. You can see how it is applied in the next example.

Example 4.3.7   We prove the limit of a product rule: If $a, L, M \in \mathbb{R}$ and both f and g be functions on the reals with $\lim_{x \to a} f = L$ and $\lim_{x \to a} g = M$, then $\lim_{x \to a} f \cdot g = L \cdot M$.

Proof   Applying the polarization identity and the limit rules from theorem 4.3.2 (that we have already verified) for scalar multiples, differences, and squares, we obtain the following equalities.

$$
\begin{aligned}
\lim_{x \to a} f(x) \cdot g(x) &= \lim_{x \to a} \frac{1}{4}\{ [f(x) + g(x)]^2 - [f(x) - g(x)]^2 \} \\
&= \frac{1}{4}\left\{ \lim_{x \to a}[f(x) + g(x)]^2 - \lim_{x \to a}[f(x) - g(x)]^2 \right\} \\
&= \frac{1}{4}\left\{ \left[\lim_{x \to a}f(x) + g(x)\right]^2 - \left[\lim_{x \to a}f(x) - g(x)\right]^2 \right\} \\
&= \frac{1}{4}\{ [L + M]^2 - [L - M]^2 \} = L \cdot M
\end{aligned}
$$
■

The final limit rule stated here is useful in a study and development of the integral. This famous "squeeze theorem" is studied in calculus; we state this result and leave the proof for exercise 56 at the end of this section.

Theorem 4.3.4 The squeeze theorem *If $a, L \in \mathbb{R}$ and f, g, h are functions on the reals such that $f(x) \leq g(x) \leq h(x)$ for every x (except possibly $x = a$) and both $\lim_{x \to a} f = L$ and $\lim_{x \to a} h = L$, then $\lim_{x \to a} g = L$.*

Now that we have investigated limits both from an intuitive perspective and in light of a mathematically rigorous definition, we are ready to study the second main topic of this section: continuity. A continuous function is sometimes informally described as a curve without any holes or gaps—one that can be drawn "without lifting the pencil." Mathematicians express continuity more carefully in terms of limits. A function $f(x)$ is continuous at a real value a if the limit of $f(x)$ as x approaches a is equal to $f(a)$; that is, if $\lim_{x \to a} f(x) = f(a)$. Not every function is continuous at every point; the value of the limit of f at a and the value of $f(a)$ are not guaranteed to be the same. In this sense, continuous functions are special.

We now define the continuity of a function $f(x)$ at $x = a$ to express the equality $\lim_{x \to a} f(x) = f(a)$. We state the definition of continuity in terms of δs and ε's; in this way, we can determine if a function is continuous in rigorous manner and prove mathematical truths about continuous functions. We first identify continuity as a pointwise property and then extend the definition from a point to sets of points, and ultimately to the entire real number line.

Definition 4.3.2 *A function $f : D \to Y$ is **continuous** at $a \in D$ if for every $\varepsilon > 0$, there exists $\delta > 0$ such that $|x - a| < \delta$ and $x \in D$ implies $|f(x) - f(a)| < \varepsilon$. We say that f is **discontinuous** at $x = a$ when f is not continuous at a. If f is continuous at every a in a set A, then we say f is **continuous on the set A**. If f is continuous on its domain, then we say f is a **continuous function**.*

When proving that a function is continuous at a point, we find that δ often depends on both ε and a. Furthermore, the definition no longer needs to insist that $0 < |x - a| < \delta$, but only that $|x - a| < \delta$, because the choice of $x = a$ automatically satisfies the limit definition: $|f(x) - f(a)| = |f(a) - f(a)| = 0 < \varepsilon$.

Example 4.3.8 We prove that $f(x) = x^3$ is continuous at $a = 2$.

Proof We begin by observing that $a = 2$ is in the domain of f and that $f(2) = 8$. Let $\varepsilon > 0$ and, (since the function is nonlinear) restrict the x-values to within one unit of 2, so that $1 < x < 3$. We are therefore (at this point) assuming that $\delta \leq 1$. Now consider $|x^3 - 8| = |x - 2| \cdot |x^2 + 2x + 4|$. Since δ bounds $|x - 2|$, we just need a bound on the second term $|x^2 + 2x + 4|$. Using the restriction $1 < x < 3$ and the fact that $x^2 + 2x + 4$ is increasing on $1 < x < 3$, we need only consider the endpoints; by direct substitution, we find $7 < x^2 + 2x + 4 < 19$. Then whenever $|x - 2| < \delta$, we have $|f(x) - f(a)| = |x - 2| \cdot |x^2 + 2x + 4| < \delta \cdot 19$. Choosing δ so that $19 \cdot \delta \leq \varepsilon$, we obtain $|f(x) - f(a)| < \varepsilon$. Thus for a given $\varepsilon > 0$, we choose $\delta = \min\{1, \varepsilon/19\}$. The complete formal proof follows below.

Let $\varepsilon > 0$ and define $\delta = \min\{1, \varepsilon/19\}$. Assuming $|x - 2| < \delta$, we know that $|x^2 + 2x + 4| < 19$. We now have

$$|f(x) - f(a)| = |x - 2| \cdot |x^2 + 2x + 4| < \delta \cdot 19 \leq \varepsilon.$$

By the definition of continuity, $f(x) = x^3$ is continuous at $a = 2$.

∎

Example 4.3.9 We prove that $f(x) = \sqrt{x}$ is continuous.

Proof By definition, $f(x) = \sqrt{x}$ is continuous if $\sqrt{x}$ is continuous at every element of its domain $D = \{x : x \geq 0\}$. First consider the domain value $a = 0$. Given any $\varepsilon > 0$, choose $\delta = \varepsilon^2$. Then whenever $|x - 0| < \delta$ and $x \in D$, we have $0 \le x < \varepsilon^2$, and so $|f(x) - f(a)| = |\sqrt{x} - 0| = \sqrt{x} < \varepsilon$. The function is therefore continuous at $a = 0$. Now let $a > 0$ be any other arbitrary element of this domain, and let $\varepsilon > 0$. We must identify an appropriate $\delta > 0$ so that $|x - a| < \delta$ implies $|f(x) - f(a)| < \varepsilon$. Since a may be between 0 and 1, we restrict the x-values to within $a/2$ units of a; this restriction gives us $a/2 < x < 3a/2$, and so we assume $\delta \le a/2$. Then, if $|x - a| < \delta$, we have

$$|f(x) - f(a)| = |\sqrt{x} - \sqrt{a}| \cdot \frac{|\sqrt{x} + \sqrt{a}|}{|\sqrt{x} + \sqrt{a}|} = \frac{|x - a|}{|\sqrt{x} + \sqrt{a}|} < \frac{\delta}{|\sqrt{\frac{a}{2}} + \sqrt{a}|}.$$

The restriction that $a/2 < x$ provides the right inequality in this last string of relations. Choosing δ so that $\delta \cdot (\sqrt{a/2} + \sqrt{a})^{-1} \le \varepsilon$; we obtain $|f(x) - f(a)| < \varepsilon$. Thus for a given $\varepsilon > 0$, we choose $\delta = \min\{a/2, (\sqrt{a/2} + \sqrt{a}) \cdot \varepsilon\}$. The complete formal proof follows below.

Let $\varepsilon > 0$ and define $\delta = \min\{\frac{a}{2}, (\sqrt{a/2} + \sqrt{a}) \cdot \varepsilon\}$. Assuming $|x - a| < \delta$, we have $a/2 < x$. Therefore,

$$|f(x) - f(a)| = \frac{|x - a|}{|\sqrt{x} + \sqrt{a}|} < \frac{\delta}{|\sqrt{\frac{a}{2}} + \sqrt{a}|} \le \varepsilon.$$

By the definition of continuity, $f(x) = \sqrt{x}$ is continuous.

■

Question 4.3.6   (a) Using the definition, prove that $f(x) = 2x - 3$ is continuous at $a = 5$.
(b) Using the definition, prove that $f(x) = 2x - 3$ is continuous.

■

In light of the analytic rules for computing limits presented in theorem 4.3.2, we might wonder if the basic algebraic operations "preserve" continuity; that is, if we combine two continuous functions using such operations, is the resulting function also continuous? For example, if we add two continuous functions, is the resulting sum continuous? The affirmative answer we expect for addition and the other arithmetic operations is given by the following theorem.

Theorem 4.3.5   *If $a, c \in \mathbb{R}$, $n \in \mathbb{N}$, and both f and g are continuous functions at $x = a$, then the following functions are also continuous at $x = a$: c, $c \cdot f$, $f + g$, $f - g$, $f \cdot g$, f/g (provided $g(x) \ne 0$ for all x in an open interval containing a), and $\sqrt[n]{f}$ (provided $f(x) \ge 0$ for all x in an open interval containing a). We say that these operations* **preserve** *continuity.*

The proof of theorem 4.3.5 closely resembles the proof of theorem 4.3.2 detailing the analytic rules for limit computations. The next example provides the proof that scalar multiplication preserves continuity; proofs of the other statements are left for the exercises at the end of this section.

Example 4.3.10   We prove that scalar multiplication preserves continuity (from theorem 4.3.5): If $a, c \in \mathbb{R}$ and f is a continuous function at $x = a$, then $c \cdot f$ is continuous at $x = a$.

Proof This proof follows the one given in example 4.3.5. If $c = 0$, then $c \cdot f(x) = 0$ is a constant function, and by the limit of a constant rule, the following equalities hold.

$$\lim_{x \to a} c \cdot f(x) = \lim_{x \to a} 0 = 0 = 0 \cdot f(a) = c \cdot f(a)$$

Therefore, $c \cdot f$ is a continuous function at $x = a$.

Now assume $c \neq 0$ and let $\varepsilon > 0$. We find $\delta > 0$ such that $|x - a| < \delta$ implies $|c \cdot f(x) - c \cdot f(a)| < \varepsilon$. From the definition of continuity applied to f for $\varepsilon_1 = \varepsilon / |c|$ (identified as in example 4.3.5), there exists $\delta > 0$ such that $|x - a| < \delta$ implies $|f(x) - f(a)| < \varepsilon_1$. This δ is the desired value: if $|x - a| < \delta$, then

$$|c \cdot f(x) - c \cdot f(a)| = |c| \cdot |f(x) - f(a)| < |c| \cdot \varepsilon_1 = |c| \cdot \frac{\varepsilon}{|c|} = \varepsilon.$$

By the definition of continuity, $c \cdot f$ is continuous at $x = a$. ∎

The limits of continuous functions are easily determined; we may simply use direct substitution to evaluate a limit at any point where a given function is continuous. In symbolic terms, if f is continuous at $x = a$, then $\lim_{x \to a} f(x) = f(a)$. In light of theorem 4.3.5, a whole host of (familiar) functions are easily identified as continuous, and so have limits that are readily evaluated.

Example 4.3.11 For each limit, the function is continuous on its domain, and we use direct substitution to evaluate the limit.

- $\lim_{x \to 7} 4\sqrt{3x^2 - 2x + 5} = 4\sqrt{3(7)^2 - 2(7) + 5} = 4\sqrt{138}$

- $\lim_{x \to -2} \dfrac{21x^5 - 4x^3 + 6}{\sqrt{11 - 3x} + 4} = \dfrac{21(-2)^5 - 4(-2)^3 + 6}{\sqrt{11 - 3(-2)} + 4} = \dfrac{-634}{\sqrt{17} + 4}$

- $\lim_{x \to 5} [3x^6 - 70x^4]\sqrt{2x^4 - 3x^2} = [3(5)^6 - 70(5)^4]\sqrt{2(5)^4 - 3(5)^2} = 15{,}625\sqrt{47}$ ∎

Question 4.3.7 Use the fact that each function is continuous and use substitution to evaluate each limit. Compare your answers with the results of question 4.3.2 and examples 4.3.3 and 4.3.4.

(a) $\lim_{x \to 3} 4x - 10$

(b) $\lim_{x \to 2} x^2$

(c) $\lim_{x \to 4} \dfrac{1}{x - 2}$

(d) $\lim_{x \to 0} \dfrac{1}{(x - 3)^2}$ ∎

The discussion thus far has focused on continuous functions from a positive perspective; the next example considers functions that are discontinuous.

Example 4.3.12 We prove that following function $f : \mathbb{R} \to \{0, 1\}$ is not continuous at any point in its domain. Dirichlet first highlighted the relevance of this function to the discussion of continuity. Mathematicians now refer to $f(x)$ as the *characteristic function* of the rationals $\mathbb{Q}$.

$$f(x) = \begin{cases} 1 & \text{if } x \in \mathbb{Q} \\ 0 & \text{if } x \notin \mathbb{Q} \end{cases}$$

Let a be an arbitrary element of the reals. Notice that any open interval containing a also contains both rational and irrational numbers, and so $f(x)$ takes on the values 0 and 1 in any interval containing a. Let $\varepsilon \in \mathbb{R}$ with $0 < \varepsilon < 1$. We prove that no $\delta > 0$ satisfies the requirements given in the definition of continuity. Let $\delta > 0$ be any positive real number. There are two cases to consider:

- If a is rational, then $f(a) = 1$. But there exists an irrational $x \in (a - \delta, a + \delta)$. For this value x, $|x - a| < \delta$, but also $|f(x) - f(a)| = |0 - 1| > \varepsilon$.
- If a is irrational, then $f(a) = 0$. But there exists a rational $x \in (a - \delta, a + \delta)$. For this value x, $|x - a| < \delta$, but also $|f(x) - f(a)| = |0 - 1| > \varepsilon$.

Therefore f is not continuous at any $a \in \mathbb{R}$.

■

One distinction between the definitions of limit and continuity is that the value $x = a$ need not be in the domain of a function f for the limit $\lim_{x \to a} f(x)$ to exist, but a must be in the domain of f for the function to be continuous at a. The next example and question highlight this distinction.

Example 4.3.13 We discuss limits and continuity for $f(x) = \dfrac{x - 2}{x^2 - x - 2}$.

First, note that f is not continuous at $a = 2$ because 2 is not in the domain of f. However, the limit of f as x approaches 2 does exist and can be computed as follows.

$$\lim_{x \to 2} \frac{x - 2}{x^2 - x - 2} = \lim_{x \to 2} \frac{x - 2}{(x - 2)(x + 1)} = \lim_{x \to 2} \frac{1}{x + 1} = \frac{1}{2 + 1} = \frac{1}{3}$$

In light of this computation, the function f can be redefined at $a = 2$ to produce a function that is identical to f (except at $x = 2$) and that is continuous at $a = 2$:

$$g(x) = \begin{cases} \dfrac{x - 2}{x^2 - x - 2} & \text{if } x \neq 2 \\ \dfrac{1}{3} & \text{if } x = 2. \end{cases}$$

When $f(a)$ may be redefined so that the function becomes continuous at a (as in this example), then a is called a *removable discontinuity* of the original function.

■

Question 4.3.8 The following functions $f(x)$ are not continuous at $a = 3$ because 3 is not in their domains. Prove that the limit of f as x approaches 3 exists and give a piecewise redefinition of f to obtain a function that is continuous at $a = 3$.

(a) $f(x) = \dfrac{x - 3}{x^2 + x - 12}$

(b) $f(x) = \dfrac{5x - 2}{5x^2 - 32x + 12}$

(c) $f(x) = \dfrac{x - 3}{x^2 - 9}$

(d) $f(x) = \dfrac{x - 3}{x^3 - 27}$

■

We finish this section with a brief discussion of infinite limits. As you may recall from your study of calculus, infinite limits describe important features of functions and graphs. For example, one of the distinguishing features of the function $f(x) = 1/x^2$ is that $\lim_{x \to 0} 1/x^2 = \infty$. Our interest here is not so much in the evaluation of such limits, but instead in developing its rigorous definition. The intuitive idea of an infinite limit $\lim_{x \to a} f(x) = \infty$ is that if x is close to a (but not equal to a), then $f(x)$ is greater than any prespecified real number (which is denoted by M). The following formal definition closely resembles Cauchy's definition of the limit.

Definition 4.3.3   *Let $f : D \to Y$ be a function whose domain D contains all points of an open interval around $a \in \mathbb{R}$, except for a itself. Then the expression $\lim_{x \to a} f(x) = \infty$ means: for every real $M > 0$, there exists $\delta > 0$ such that $0 < |x - a| < \delta$ implies $f(x) > M$. In this case, we say that $f(x)$ approaches infinity as x approaches a.*

Example 4.3.14   We use the definition of an infinite limit to prove that $\lim_{x \to 0} 1/x^2 = \infty$.

Proof   Let $M > 0$ be a given real number. We need to identify a corresponding δ so that $0 < |x - a| < \delta$ implies $f(x) > M$. We choose $\delta = 1/\sqrt{M}$ and assume $0 < |x - a| < \delta$, which implies $0 < |x - 0| = |x| < \delta = 1/\sqrt{M}$. The inequalities $0 < |x| < 1/\sqrt{M}$ imply $0 < x^2 < 1/m$, and so $1/x^2 > M$. Therefore, by the definition of an infinite limit, $\lim_{x \to 0} 1/x^2 = \infty$. ∎

■

Question 4.3.9   Prove each limit using the definition.

(a) $\lim_{x \to 3} \dfrac{1}{(x - 3)^4} = \infty$
(b) $\lim_{x \to 1} \dfrac{1}{x(x - 1)^2} = \infty$

■

4.3.1   Reading Questions for Section 4.3

1. State both an intuitive description and the definition of $\lim_{x \to a} f(x) = L$.
2. When using the definition to verify a limit, is δ always less than ε ? Consider question 4.3.2 when explaining your answer.
3. State theorem 4.3.1. How is this result helpful when studying limits?
4. Give an example of each analytic rule for computing limits identified in theorem 4.3.2.
5. State the triangle inequality. Give an example producing equality and an example producing a strict inequality.
6. Give an example for each property of the absolute value function identified in theorem 4.3.3.
7. State the polarization identity. How is this identity helpful when studying limits?
8. State the squeeze theorem and give an example of an application of this result.
9. State both an intuitive description and the definition of $f(x)$ is continuous at $x = a$.

10. Define and give an example of a continuous function.
11. Give an example of a function that is discontinuous at every element of its domain.
12. State both an intuitive description and the definition of $\lim\limits_{x \to a} f(x) = \infty$.

4.3.2 Exercises for Section 4.3

In exercises 1–6, use the graph of the function $f(x)$ given in figure 4.12 to identify the value of each limit, or to explain why the limit does not exist.

1. $\lim\limits_{x \to 0} f(x)$
2. $\lim\limits_{x \to 10} f(x)$
3. $\lim\limits_{x \to 20} f(x)$
4. $\lim\limits_{x \to 30} f(x)$
5. $\lim\limits_{x \to 40} f(x)$
6. $\lim\limits_{x \to 50} f(x)$

In exercises 7–12, prove each limit using the definition of a limit and the given δ.

7. $\lim\limits_{x \to 2} 7x - 8 = 6$ with $\delta = \dfrac{\varepsilon}{7}$

8. $\lim\limits_{x \to 4} -2x + 15 = 7$ with $\delta = \dfrac{\varepsilon}{2}$

9. $\lim\limits_{x \to a} mx + b = ma + b$ with $\delta = \dfrac{\varepsilon}{|m|}$

10. $\lim\limits_{x \to -2} x^2 - 2 = 2$ with $\delta = \min\{1, \varepsilon/5\}$

11. $\lim\limits_{x \to 0} ax^2 + bx + c = c$, where $a, b > 0$ with $\delta = \min\left\{\dfrac{b}{a}, \dfrac{\varepsilon}{2b}\right\}$

12. $\lim\limits_{x \to 5} \dfrac{3}{2x + 1} = \dfrac{3}{11}$ with $\delta = \min\left\{1, \dfrac{6\varepsilon}{121}\right\}$

In exercises 13–18, prove each limit using the definition.

13. $\lim\limits_{x \to 5} 2x + 3 = 13$
14. $\lim\limits_{x \to 3} -4x + 3 = -9$
15. $\lim\limits_{x \to 2} x^2 + 1 = 5$
16. $\lim\limits_{x \to 0} x^2 + 6x + 5 = 5$
17. $\lim\limits_{x \to 9} 3\sqrt{x} - 1 = 8$
18. $\lim\limits_{x \to 8} \dfrac{1}{x - 6} = \dfrac{1}{2}$

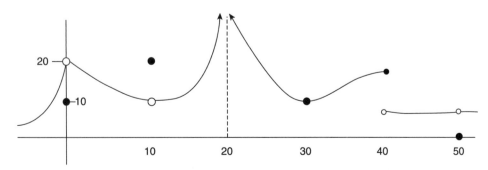

Figure 4.12 Graph of $f(x)$ for exercises 1–6

In exercises 19–26, evaluate the following limits using the analytic rules identified in theorem 4.3.2, or using other techniques that you learned in your calculus courses.

19. $\lim\limits_{x \to 2} 2x + 5$

20. $\lim\limits_{x \to 1} (8x^2 + 3) \cdot (3x + 1)$

21. $\lim\limits_{x \to 30} \sqrt[3]{x - 3}$

22. $\lim\limits_{x \to -3} \dfrac{9}{x + 1}$

23. $\lim\limits_{x \to 0} \dfrac{4x + 1}{5x + 2}$

24. $\lim\limits_{x \to 0} \dfrac{e^x}{x^2}$

25. $\lim\limits_{x \to 0} \dfrac{x}{\tan(x)}$

26. $\lim\limits_{x \to 0} \dfrac{\cos(x)}{4x^3}$

In exercises 27–32, determine the value of each limit under the following assumptions.

$$\lim\limits_{x \to 2} f(x) = 3 \qquad \lim\limits_{x \to 2} g(x) = 5 \qquad \lim\limits_{x \to 2} h(x) = 8$$

27. $\lim\limits_{x \to 2} f + g$

28. $\lim\limits_{x \to 2} f \cdot h$

29. $\lim\limits_{x \to 2} f - h$

30. $\lim\limits_{x \to 2} 3[f]^2 + 1$

31. $\lim\limits_{x \to 2} \sqrt{h + 6}$

32. $\lim\limits_{x \to 2} \dfrac{2f + 3h}{4g}$

In exercises 33–35, prove each function is continuous at $x = 0$ using the definition.

33. $f(x) = x$

34. $f(x) = |x|$

35. $f(x) = x^2$

In exercises 36–38, prove each function is continuous using the definition.

36. $f(x) = x$

37. $f(x) = |x|$

38. $f(x) = x^2$

In exercises 39–42, discuss the continuity of each function.

39. $f(x) = \dfrac{x^2 - 1}{x + 1}$

40. $f(x) = \dfrac{2x^2 - 4x}{x^3 - 4x}$

41. $f(x) = \begin{cases} x^2 & \text{if } x < 2 \\ \dfrac{(x+2)(x-2)}{x-2} & \text{if } x > 2 \end{cases}$

42. $f(x) = \begin{cases} 3x + 2 & \text{if } x < 0 \\ x^2 + 2 & \text{if } x > 0 \end{cases}$

In exercise 43–44, prove each limit using the definition of an infinite limit.

43. $\lim\limits_{x \to 3} \dfrac{2x}{(x - 3)^2} = \infty$

44. $\lim\limits_{x \to 2} \dfrac{3}{x(x - 2)^2} = \infty$

In exercises 45–56, prove each mathematical statement about limits.

45. The limit of a constant rule from theorem 4.3.2.

46. The limit of a sum rule from theorem 4.3.2.

47. The limit of a difference rule from theorem 4.3.2.

48. The limit of a quotient rule from theorem 4.3.2.

49. If $p(x)$ is a polynomial, then $\lim\limits_{x \to a} p(x) = p(a)$.

50. If $\lim\limits_{x \to a} f(x) = 0$, then $\lim\limits_{x \to a} |f(x)| = 0$.

51. Disprove the claim: If $\lim\limits_{x \to a} |f(x)| = L$, then either $\lim\limits_{x \to a} f(x) = L$ or $\lim\limits_{x \to a} f(x) = -L$.

52. If $\lim\limits_{x \to a} f(x) = \infty$ and $\lim\limits_{x \to a} g(x) = \infty$, then $\lim\limits_{x \to a} f + g = \infty$.

53. If $\lim\limits_{x \to a} f(x) = \infty$ and $\lim\limits_{x \to a} g(x) = L \in \mathbb{R}^*$, then $\lim\limits_{x \to a} \dfrac{f(x)}{g(x)} = \infty$.

54. If $\lim\limits_{x \to a} f(x) = L$, then $\lim\limits_{x \to a} (f(x) - L) = 0$.

55. Disprove the two claims:

    (a) If $\lim\limits_{x \to a} f(x) = L$, then $f(a) = L$.
    (b) If $f(a) = L$, then $\lim\limits_{x \to a} f(x) = L$.

56. The squeeze theorem (theorem 4.3.4).

In exercises 57–68, prove each mathematical statement about continuity.

57. The constant function is continuous (from theorem 4.3.5).
58. The sum of two continuous functions is continuous (from theorem 4.3.5).
59. The difference of two continuous functions is continuous (from theorem 4.3.5).
60. The product of two continuous functions is continuous (from theorem 4.3.5).
61. If f is continuous at $x = a$, then $-f$ is continuous at $x = a$.
62. If f is continuous at $x = a$, then $|f|$ is continuous at $x = a$.
63. Prove that $f(x) = x^n$ is continuous for all $n \in \mathbb{N}$ via induction (and theorem 4.3.5).
64. If $\lim\limits_{h \to 0} f(a + h) = f(a)$, then f is continuous at $x = a$.
65. Disprove the claim: If f and g are not continuous at $x = a$, then $f + g$ is not continuous at $x = a$.
66. Disprove the claim: If $|f|$ is continuous at $x = a$, then f is continuous at $x = a$.
67. Disprove the claim: If the composite function $f(g(x))$ is continuous, then $f(x)$ and $g(x)$ are both continuous.
68. The following variation on the characteristic function of $\mathbb{Q}$ is continuous at $x = 0$:

$$f(x) = \begin{cases} x & \text{if } x \in \mathbb{Q} \\ 0 & \text{if } x \notin \mathbb{Q} \end{cases}$$

In exercises 69–70, state both an intuitive description and a definition of each limit.

69. $\lim\limits_{x \to a} f(x) = -\infty$

70. $\lim\limits_{x \to \infty} f(x) = L$

4.4   The Derivative

Calculus is the study of change. While Sir Isaac Newton and Gottfried Leibniz are both credited for independently developing calculus in the late 1600s, mathematicians had already been working with derivatives for nearly a half century. The study of change as expressed by the derivative was motivated by a sixteenth and seventeenth century European reflection on and ultimate rejection of ancient Greek astronomy and physics. The European astronomers Nicolaus Copernicus, Tycho Brahe, and Johannes Kepler each had insights that challenged the theories of the ancient Greeks, setting the stage for the ground-breaking work of the Italian scientist Galileo Galilei in the early 1600s. Many of the questions about a moving object (that is, an object *changing* position and velocity) that these scientists were studying are readily answered by considering lines tangent to curves.

A number of mathematicians from many different countries made important contributions to the question of finding the equation of a tangent line. Pierre de Fermat studied maxima and minima of curves via tangent lines, essentially using the approach studied in contemporary calculus courses. This work prompted fellow French mathematician Joseph-Louis Lagrange to assert that Fermat should be credited with the development of calculus! The English mathematician Isaac Barrow, who was Newton's teacher and mentor, corresponded regularly with Leibniz on these mathematical ideas.

As we have mentioned, neither Newton nor Leibniz thought of the derivative as a measure of change in terms of our contemporary definition involving limits. Our study of the derivative follows more closely the work of Fermat and Barrow from the early 1600s, in which we think of a tangent line as a limit of secant lines. Naturally, the contemporary presentation is informed by an understanding of Cauchy's notion of the limit from the early 1800s.

The derivative enables the determination of the equation of a line tangent to a given curve at a given point. Given a function $y = f(x)$, the slope of a secant line joining two points $(c, f(c))$ and $(c + h, f(c + h))$ is

$$m = \frac{\text{rise}}{\text{run}} = \frac{\Delta y}{\Delta x} = \frac{f(c + h) - f(c)}{(c + h) - c} = \frac{f(c + h) - f(c)}{h}.$$

In this context, the symbol "m" is the first letter in the French word *montrer* which translates as "to climb."

To find the slope of the line tangent to a function f at the point $(c, f(c))$, we take a limit of the slopes of secant lines, letting h approach 0. Figure 4.13 illustrates why this limit process makes intuitive sense; you can see that the slopes of the secant lines get closer and closer to the slope of the tangent line as the point $(c + h, f(c + h))$ gets closer and closer to $(c, f(c))$. The definition of the derivative reflects these ideas.

The following definition expresses the real number c as a variable quantity x to identify a general formula for the derivative, enabling us to determine the slope of the line tangent to $f(x)$ whenever this slope is defined.

Definition 4.4.1   *Let $f(x)$ be a function with domain D. Then the* **derivative** *of $f(x)$ is*

$$f'(x) = \lim_{h \to 0} \frac{f(x + h) - f(x)}{h},$$

whenever this limit exists. We say that $f(x)$ is **differentiable at $x = c$** *when $f'(c)$ exists for $c \in D$, and that $f(x)$ is* **differentiable** *when $f'(x)$ exists for all $x \in D$. The ratio*

$$\frac{f(x + h) - f(x)}{h}$$

is called the **difference quotient** *of the derivative.*

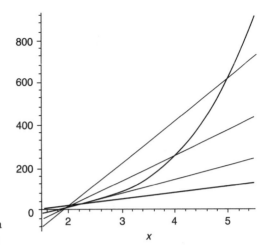

Figure 4.13 A tangent line at $x = 2$ as a
limit of secant lines

Recall from calculus that many different notations are used for the derivative of a
function $y = f(x)$, including

$$f'(x) \; = \; \frac{df}{dx} \; = \; \frac{d}{dx}(f) \; = \; y' \; = \; \frac{dy}{dx} \; = \; D_x(y) \; = \; \dot{y}.$$

Various phrases also refer to the derivative, including "f (or y) prime," "the derivative
of f (or y) with respect to x," the letters "d f d x" spoken individually, and "d y d x"
spoken individually. Most of this notation for the derivative is attributable to Leibniz,
who gave considerable thought to carefully identifying a useful symbolism and is
recognized as a genius in developing notation to make subtle concepts understandable.
The *alternate definition of the derivative* is sometimes helpful; if the limit exists, then

$$f'(x) = \lim_{t \to x} \frac{f(t) - f(x)}{t - x}.$$

The proof of the equivalence of this alternate definition and the one given in definition
4.4.1 is left for exercise 50 at the end of this section.

Example 4.4.1   We use the two definitions of the derivative to determine the equation of a line
tangent to $f(x) = x^2$ at $(2, 4)$.
    Applying the definition,

$$f'(x) = \lim_{h \to 0} \frac{(x+h)^2 - x^2}{h} = \lim_{h \to 0} \frac{x^2 + 2xh + h^2 - x^2}{h} = \lim_{h \to 0} \frac{2xh + h^2}{h} = \lim_{h \to 0} 2x + h = 2x.$$

Hence the slope of the tangent line at $(2, 4)$ is $f'(2) = 2 \cdot 2 = 4$, and the equation
of the line tangent to $f(x) = x^2$ at $(2, 4)$ is given by $y - 4 = 4(x - 2)$.
    Applying the alternate definition produces the same result:

$$f'(x) = \lim_{t \to x} \frac{t^2 - x^2}{t - x} = \lim_{t \to x} \frac{(t + x)(t - x)}{t - x} = \lim_{t \to x} t + x = 2x.$$

■

When using the definition (as in example 4.4.1), we often algebraically manipulate
the difference quotient so that h appears as a factor in the numerator. This factor

then cancels the denominator, simplifying the difference quotient so the limit can be evaluated. If the original function $f(x)$ is a rational function, then finding a common denominator will simplify the difference quotient in this way. If $f(x)$ contains a square root, multiplying by the *conjugate square root function* will simplify the difference quotient.

Example 4.4.2   We use the definition of the derivative to find the derivative of $f(x) = 5\sqrt{x+1}$.
Multiplying both the numerator and the denominator of the difference quotient by the conjugate square root function and then simplifying yields the following calculation.

$$
\begin{aligned}
f'(x) &= \lim_{h \to 0} \frac{5\sqrt{x+h+1} - 5\sqrt{x+1}}{h} \\[2mm]
&= \lim_{h \to 0} \frac{5\sqrt{x+h+1} - 5\sqrt{x+1}}{h} \cdot \frac{5\sqrt{x+h+1} + 5\sqrt{x+1}}{5\sqrt{x+h+1} + 5\sqrt{x+1}} \\[2mm]
&= \lim_{h \to 0} \frac{25[(x+h+1) - (x+1)]}{5h(\sqrt{x+h+1} + \sqrt{x+1})} = \lim_{h \to 0} \frac{5h}{h(\sqrt{x+h+1} + \sqrt{x+1})} \\[2mm]
&= \frac{5}{2\sqrt{x+1}}
\end{aligned}
$$

■

Question 4.4.1   Using the definition of the derivative, differentiate each function.

(a) $f(x) = 2x + 1$

(b) $g(x) = 7x^3$

(c) $s(x) = \dfrac{1}{x+5}$

(d) $t(x) = \dfrac{1}{3\sqrt{x}}$

■

While we can use the formal definition of the derivative to compute derivatives of a given function, theoretical applications of the definition are more important. Using the definition, we can prove general theorems that hold for all derivatives, making it easy to differentiate many familiar functions without explicitly applying the definition one function at at time. Many functions are so complicated in structure that directly using the difference quotient becomes unwieldy or impossible. The next theorem states analytic properties of derivatives to facilitate such computations. Using these results is a common exercise in calculus courses, but you may not have considered the underlying proofs that justify them. These proofs are the focus of the remainder of this section.

Theorem 4.4.1   *If $c \in \mathbb{R}$ and both f and g are differentiable functions, then the following hold.*

- *The constant rule:* $\dfrac{d}{dx}[c] = 0$

- *The scalar multiple rule:* $\dfrac{d}{dx}[c \cdot f(x)] = c \cdot f'(x)$

- *The sum rule:* $\dfrac{d}{dx}[f + g] = f' + g'$

- *The difference rule:* $\dfrac{d}{dx}[f - g] = f' - g'$

- *The power rule:* $\dfrac{d}{dx}[x^n] = n \cdot x^{n-1}$, *for* $n \in \mathbb{R}$

- *The product rule:* $\dfrac{d}{dx}[f \cdot g] = g \cdot f' + f \cdot g'$

- *The quotient rule:* $\dfrac{d}{dx}\left[\dfrac{f}{g}\right] = \dfrac{g \cdot f' - f \cdot g'}{g^2}$, *provided that* $g(x) \neq 0$

- *The chain rule:* $\dfrac{d}{dx}[f(g(x))] = f'(g(x)) \cdot g'(x)$

A standard goal of a calculus course is to develop a mastery in using these differentiation rules. Before diving into the proofs of various parts of this theorem, the next example provides the opportunity to revisit the skills you learned in calculus.

Question 4.4.2   Using theorem 4.4.1, differentiate each function.

(a) $f(x) = 10x^3 - 7x^2 + 5$ (d) $p(x) = (x^5 + x)\tan(2x)$

(b) $g(x) = \sqrt{5x + 2}$ (e) $q(x) = \ln(4x^2 + 1) \cdot \sin^2(5x + 3)$

(c) $h(x) = \dfrac{\cos(3x + 1)}{5x^2 + 2}$ (f) $r(x) = (2x^5 + 3)\sqrt[3]{4e^x + 6x}$

■

The next three examples give the proofs of some of these differentiation rules. As in the study of limits and continuity, we first consider the scalar multiple and sum rules, and then discuss a couple of different approaches to proving the power rule.

Example 4.4.3   We prove the scalar multiple rule from theorem 4.4.1: For any constant $c \in \mathbb{R}$ and differentiable function f,

$$\frac{d}{dx}[c \cdot f(x)] = c \cdot f'(x).$$

Proof   Apply the definition of the derivative and the limit of a scalar multiple rule.

$$\frac{d}{dx}[c \cdot f(x)] = \lim_{h \to 0} \frac{c \cdot f(x+h) - c \cdot f(x)}{h} = \lim_{h \to 0} \frac{c \cdot [f(x+h) - f(x)]}{h}$$

$$= c \cdot \lim_{h \to 0} \frac{f(x+h) - f(x)}{h} = c \cdot f'(x)$$

■

Example 4.4.4   We prove the sum rule: If f and g are differentiable functions, then $\dfrac{d}{dx}[f + g] = f'(x) + g'(x)$.

Proof   Apply the definition of the derivative and the limit of a sum rule.

$$\frac{d}{dx}[f + g] = \lim_{h \to 0} \frac{[f(x+h) + g(x+h)] - [f(x) + g(x)]}{h}$$

$$= \lim_{h \to 0} \frac{[f(x+h) - f(x)] + [g(x+h) - g(x)]}{h}$$

$$= \lim_{h \to 0} \frac{f(x+h) - f(x)}{h} + \lim_{h \to 0} \frac{g(x+h) - g(x)}{h} = f'(x) + g'(x)$$

∎

Example 4.4.5   We prove the power rule: If $n \in \mathbb{R}$, then

$$\frac{d}{dx}[x^n] = n \cdot x^{n-1}.$$

Proof   We prove the power rule in the case of the positive integers $n \in \mathbb{N}$ by using the binomial theorem to expand the term $f(x+h) = (x+h)^n$ in the difference quotient as follows:

$$(x+h)^n = x^n + n \cdot x^{n-1} \cdot h + \frac{n(n-1)}{2} \cdot x^{n-2} \cdot h^2 + \cdots + n \cdot x \cdot h^{n-1} + h^n.$$

Applying the definition of the derivative,

$$\frac{d}{dx}[x^n] = \lim_{h \to 0} \frac{(x+h)^n - x^n}{h}$$

$$= \lim_{h \to 0} \frac{[x^n + n \cdot x^{n-1} \cdot h + \frac{n(n-1)}{2} \cdot x^{n-2} \cdot h^2 + \cdots + n \cdot x \cdot h^{n-1} + h^n] - x^n}{h}$$

$$= \lim_{h \to 0} \frac{n \cdot x^{n-1} \cdot h + \frac{n(n-1)}{2} \cdot x^{n-2} \cdot h^2 + \cdots + n \cdot x \cdot h^{n-1} + h^n}{h}$$

$$= \lim_{h \to 0} \frac{h \cdot [n \cdot x^{n-1} + \frac{n(n-1)}{2} \cdot x^{n-2} \cdot h + \cdots + n \cdot x \cdot h^{n-2} + h^{n-1}]}{h}$$

$$= \lim_{h \to 0} n \cdot x^{n-1} + \frac{n(n-1)}{2} \cdot x^{n-2} \cdot h + \cdots + n \cdot x \cdot h^{n-2} + h^{n-1}$$

$$= n \cdot x^{n-1}.$$

Alternatively, the power rule for $n \in \mathbb{N}$ follows by induction (see exercise 67 in section 3.6). The definition of the derivative proves the base case $\frac{d}{dx}[x] = 1 \cdot x^0 = 1$, and the product rule applies in the inductive step (for $x^{n+1} = x^n \cdot x$).

A complete proof of the power rule must consider arbitrary real numbers $n \in \mathbb{R}$, not just positive integers $n \in \mathbb{N}$. The power rule extends to the negative integers via the quotient rule, to rational powers via implicit differentiation, and to all real numbers via logarithmic differentiation. The details of such a complete proof are left for your later studies.

∎

Question 4.4.3 The following steps outline a proof of the quotient rule: If $f(x)$, $g(x)$ are differentiable functions with $g(x) \neq 0$, then

$$\frac{d}{dx}\left[\frac{f(x)}{g(x)}\right] = \frac{g(x) \cdot f'(x) - f(x) \cdot g'(x)}{g(x)^2}.$$

(a) What is the difference quotient for the function $\dfrac{f(x)}{g(x)}$?

(b) Using the common denominator $g(x) \cdot g(x + h) \cdot h$, simplify the difference quotient from part (a).

(c) In the numerator from part (b), subtract and add the term $g(x) \cdot f(x)$. Now split the fraction into a difference of two differences, gathering together the two terms with $g(x)$ as a common factor and the two terms with $f(x)$ as a common factor.

(d) What is the limit of the difference of difference quotients from part (c) as h approaches 0?

(e) Based on parts (a)–(d), craft a complete proof of the quotient rule as modeled in examples 4.4.3, 4.4.4, and 4.4.5. ■

Question 4.4.4 The following steps outline a proof of the chain rule: If $f(x)$, $g(x)$ are differentiable functions, then

$$\frac{d}{dx}[f[g(x)]] = f'[g(x)] \cdot g'(x).$$

(a) What is the difference quotient

$$\frac{h(t) - h(x)}{t - x}$$

(from the alternate definition of the derivative) for the function $h(x) = f[g(x)]$?

(b) Assuming there are no values x for which $g(x) = g(t)$, multiply both the numerator and the denominator of the difference quotient from part (a) by $g(t) - g(x)$. Factor out the resulting difference quotient for $g(x)$.

(c) Take the limit of the product of difference quotients from part (b) as t approaches x to obtain the chain rule formula.

(d) Based on parts (a)–(d), craft a proof of the chain rule under the assumption that $g(x) \neq g(t)$ as modeled in examples 4.4.3, 4.4.4, and 4.4.5.

The assumption that there are no values for which $g(x)$ equals $g(t)$ may be unreasonable; a complete proof of the chain rule that does not use this assumption is outlined in exercises 67–70 at the end of this section. ■

We end this section by considering the relationship between two of the most significant properties of functions studied in this chapter: continuity and differentiability. Some properties of functions are completely independent of one another, as we saw in our discussion of one-to-one and onto functions; some functions are both, some are neither, while still others have just one of these properties. This observation leads us to ask if continuity and differentiability are independent of one another, or is there a connection between these two properties? As you may recall, every differentiable function is continuous, but not every continuous function is differentiable. We consider the theorem and its proof, along with a counterexample that together justify these assertions.

Theorem 4.4.2 *If a function f with domain D is differentiable at a $\in (b, c) \subseteq D$, then f is continuous at a.*

Proof   By the alternate definition of the derivative, given any $\varepsilon > 0$, there exists a value $\delta > 0$ so that

$$\left| \frac{f(x) - f(a)}{x - a} - f'(a) \right| < \varepsilon$$

whenever $0 < |x - a| < \delta$. Multiplying both sides by $|x - a|$, we see that

$$|f(x) - f(a) - f'(a)(x - a)| < \varepsilon |x - a|.$$

Applying the second inequality ($|x| - |y| \le |x - y|$) from theorem 4.3.3 in section 4.3, we have

$$|f(x) - f(a)| - |f'(a)(x - a)| \le |f(x) - f(a) - f'(a)(x - a)|.$$

This fact implies $|f(x) - f(a)| < |f'(a)(x - a)| + \varepsilon |x - a|$, and so

$$|f(x) - f(a)| < (|f'(a)| + \varepsilon) \cdot |x - a|.$$

The term on the right can be made arbitrarily small: we restrict values of x in that term so that $|x - a|$ is smaller than both δ (so that the first inequality holds) and $\varepsilon/(|f'(a)| + \varepsilon)$. Then $|f(x) - f(a)| < \varepsilon$, which proves the result. ∎

Theorem 4.4.2 asserts that every differentiable function is continuous. Are there continuous functions that are not differentiable? Perhaps you can recall from calculus examples of continuous functions that are not differentiable. The next example provides one such counterexample.

Example 4.4.6   We discuss the continuity and differentiability of $f(x) = |x|$ at $x = 0$.

We can show that $y = |x|$ is continuous at $x = 0$, using the definition. Let $\varepsilon > 0$ and choose $\delta = \varepsilon$. For any x such that $|x - 0| < \delta$, the following string of relations holds:

$$|f(x) - f(0)| = |\,|x| - |0|\,| = |\,|x|\,| = |x| < \varepsilon.$$

By the definition of continuity, $|x|$ is continuous at $x = 0$.

On the other hand, we can show that $|x|$ is not differentiable at $x = 0$, using the alternate definition of the derivative. The difference quotient for $f(x)$ at $x = 0$ is

$$\frac{f(x) - f(0)}{x - 0} = \frac{|x| - |0|}{x} = \frac{|x|}{x}.$$

Taking the limit of this difference quotient as x approaches 0,

$$\lim_{x \to 0^-} \frac{|x|}{x} = -1 \quad \text{and} \quad \lim_{x \to 0^+} \frac{|x|}{x} = 1.$$

Therefore the limit

$$\lim_{x \to 0} \frac{f(x) - f(0)}{x - 0}$$

does not exist, and $f(x) = |x|$ is not differentiable at $x = 0$.

■

Question 4.4.5 Give an example of a continuous function that is not differentiable at the following points:

(a) $x = 1$ (c) $x = n \cdot \pi$, for every $n \in \mathbb{Z}$

(b) both $x = 1$ and $x = -1$ (d) $x = 2n$, for every $n \in \mathbb{Z}$

■

These results show that (intuitively speaking) it is "more difficult" for a function to be differentiable than continuous. From an informal, graphical perspective, this fact is quite natural; at a point of discontinuity for a graph, we cannot draw a unique tangent line.

The results also provide another reason for the importance of studying continuity: the functions that are the most "well behaved" from the perspective of differential calculus are continuous. Section 4.6 will identify an important connection between continuity and Riemann integrability.

The derivative has transformed the way mathematicians think about functions. Many questions about mathematical objects and our real-world can be phrased in terms of the derivative's measure of change. In this way, the development of the derivative set the stage for much of the last three centuries of investigations into function theory. From your calculus courses, you know that these investigations include finding maxima and minima, and determining increasing and decreasing sections of curves, concavity, and points of inflection, as well as the construction of power series. In summary, the derivative flows through function theory in a useful and meaningful way.

4.4.1 Reading Questions for Section 4.4

1. Define and give an example of the slope of a line.
2. Describe an intuitive motivation for the definition of the derivative in terms of secant lines and tangent lines to a curve.
3. State the definition of the derivative $f'(x)$.
4. State the alternative definition of the derivative $f'(x)$.
5. Give an example of a differentiable function.
6. What is the distinction between a function being differentiable at a point $x = c$ and a function being differentiable?
7. State theorem 4.4.1. How is this result helpful when studying derivatives?
8. Give an example of each differentiation rule stated in theorem 4.4.1.
9. Define and give an example of a conjugate square root function.
10. State the binomial theorem. How is this result helpful when studying derivatives?
11. Discuss the relationship between continuity and differentiability.
12. Give two examples of functions that are continuous, but not differentiable.

4.4.2   Exercises for Section 4.4

In exercises 1–6, express the slope of a secant line to each function for the designated x-coordinates as a difference quotient, and sketch the corresponding graph.

1. $f(x) = x^2 + 2$ at $x = 3$ and $x = 4$
2. $f(x) = x^2 + 2$ at $x = 3$ and $x = 3.01$
3. $f(x) = x^2 + 2$ at $x = 3$ and $x = 3.0001$
4. $f(x) = x^3$ at $x = 0$ and $x = 1$
5. $f(x) = x^3$ at $x = 0$ and $x = 0.01$
6. $f(x) = x^3$ at $x = 0$ and $x = 0.0001$

In exercises 7–18, use the definition of the derivative to compute the derivative (if it exists) of each function.

7. $f(x) = 2x + 3$
8. $g(x) = 3x - 5$
9. $h(x) = x^2 + 1$
10. $j(x) = x^2 + x$
11. $p(x) = 1/x$
12. $q(x) = \dfrac{1}{x+1}$
13. $r(x) = \dfrac{1}{x-3}$

14. $s(x) = \sqrt{x}$
15. $t(x) = \sqrt{2x + 2}$
16. $u(x) = \dfrac{1}{\sqrt{x+7}}$
17. $v(x) = \begin{cases} 4x & \text{if } x \le 2 \\ 2x^2 & \text{if } x > 2 \end{cases}$
18. $w(x) = \begin{cases} 4x + 3 & \text{if } x \le 2 \\ 2x^2 & \text{if } x > 2 \end{cases}$

In exercises 19–28, compute the derivative of each function using the analytic differentiation rules from theorem 4.4.1, along with your recollection of the derivatives of functions from calculus.

19. $f(x) = (x^9 + x^6)^{37}$
20. $f(x) = (x + x^{-1})^4$
21. $f(x) = (3x^2 + \sqrt{6x+5} - 4) \cdot (2x + 1/x)$
22. $f(x) = (x^2 + 1)^3 \cdot \sqrt{(5x^3 + 2x)^2 + 1}$
23. $f(x) = \sin^5(x^3 + 2x)$
24. $f(x) = \ln(x) \cdot \cos(2x + 7)$

25. $f(x) = \log_3(\cot(2x))$
26. $f(x) = \ln(x^2 + 2) \cdot \log_5(\csc(x) + 2x)$
27. $f(x) = (k \cdot x^5 + 2x)\sqrt[3]{x}$, where $k \in \mathbb{R}$
28. $f(x) = \left(\sqrt{3x^2 + 2x} + \dfrac{1}{kx}\right)^{3n}$, where $k, n \in \mathbb{R}$

In exercises 29–34, determine the exact value of $h'(3\pi/4)$ and state the equation of the line tangent to $h(x)$ at $x = 3\pi/4$ using the information in the following table.

	$f(x)$	$f'(x)$	$g(x)$	$g'(x)$
$x = 3\pi/4$	4	2	5	3

29. $h(x) = 7 \cdot f(x) - \sec(x) + \pi^3$
30. $h(x) = g(x) \cdot \cos(x)$
31. $h(x) = \dfrac{g(x) + x}{f(x) + 2}$

32. $h(x) = \tan(x) + \pi \cdot \cot^2(g(x))$
33. $h(x) = \sin[\pi \cdot f(x)] + \cos[\pi \cdot g(x)]$
34. $h(x) = \dfrac{f(x)}{x} - \dfrac{x^2}{g(x)}$

In exercises 35–38, answer each question about $f(x) = \sqrt{x}$.

35. Using the definition of the derivative, find $f'(x)$.
36. Using the power rule, find $f'(x)$.
37. Determine the equation of the tangent line to $f(x) = \sqrt{x}$ at $(9, 3)$.
38. Determine the equation of the tangent line to $f(x) = \sqrt{x}$ that is perpendicular to the line determined by $2y + 8x = 16$.

Exercises 39–43 develop a proof that the derivative of $\sin \theta$ is $\cos \theta$.

39. Prove that $\sin \theta \cdot \cos \theta < \theta < \tan \theta$.

Hint: Compare the areas of the three nested regions in figure 4.14 and use the fact that a pie-shaped sector of the unit circle with central angle θ (in radians) has an area of $\theta/2$.

40. Identify upper and lower bounds on $\sin \theta/\theta$ using the inequalities from exercise 39.

Hint: Divide by $\sin \theta$ and take reciprocals.

41. Prove that $\lim\limits_{\theta \to 0} \sin \theta/\theta = 1$.

Hint: Apply the squeeze theorem (see theorem 4.3.4 from section 4.3) to the inequalities from exercise 40.

42. Prove that $\lim\limits_{\theta \to 0} (1 - \cos \theta)/\theta = 0$.

Hint: Multiply both the numerator and the denominator by $1 + \cos \theta$ and then use both the Pythagorean identity $\sin^2 \theta + \cos^2 \theta = 1$ and the limit from exercises 41.

43. Prove that the derivative of $\sin \theta$ is $\cos \theta$.

Hint: Working with the definition of the derivative, simplify the resulting difference quotient using the limits from exercises 41 and 42 along with the trigonometric identity $\sin(u + v) = \sin u \cos v + \sin v \cos u$.

In exercises 44–48, derive the formulas for the derivative of the other trigonometric functions; all but exercise 44 use the quotient rule.

44. Prove that the derivative of $\cos \theta$ is $-\sin \theta$.

Hint: Use the cofunction identity $\cos x = \sin(\pi/2 - x)$ and the derivative from exercises 43.

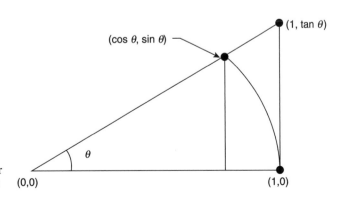

Figure 4.14 Figure for exercise 39

45. Prove that the derivative of $\tan\theta$ is $\sec^2\theta$.
46. Prove that the derivative of $\cot\theta$ is $-\csc^2\theta$.
47. Prove that the derivative of $\sec\theta$ is $\sec\theta\tan\theta$.
48. Prove that the derivative of $\csc\theta$ is $-\csc\theta\cot\theta$.

In exercises 51–66, prove each mathematical statement about derivatives.

49. The derivative of a differentiable function is unique. Hint: See the unique limit theorem (theorem 4.3.1 from section 4.3).
50. The two definitions of the derivative are equivalent. Hint: Let $h = a - x$.
51. The constant rule from theorem 4.4.1.
52. The difference rule from theorem 4.4.1.
53. The product rule from theorem 4.4.1. Hint: Add and subtract $f(x + h) \cdot g(x)$ in the numerator of the difference quotient for $f(x) \cdot g(x)$.
54. The quotient rule from theorem 4.4.1. Hint: See question 4.4.3.
55. The chain rule from theorem 4.4.1. Hint: See question 4.4.4.
56. Every polynomial is differentiable.
57. The derivative of a polynomial of degree n is a polynomial of degree $n - 1$.
58. The derivative of an even function is odd; that is, if $f(x) = f(-x)$, then $f'(x) = -f'(-x)$.
59. If f is a differentiable function on an interval $(x - h, x + h)$ for some $h \in \mathbb{R}$, then the derivative $f'(x)$ equals $\lim_{h\to 0} f'(x + h)$. Hint: Apply L'Hôpital's rule from calculus to the limit of the difference quotient.
60. Appling the alternative definition of the derivative at $x = 0$, the following function is not differentiable at $x = 0$.

$$f(x) = \begin{cases} x \cdot \sin\left[\dfrac{1}{x}\right] & \text{if } x \neq 0 \\ 0 & \text{if } x = 0 \end{cases}$$

61. The function $f(x)$ defined as follows has derivative $f'(0) = 0$.

$$f(x) = \begin{cases} x^2 \cdot \sin\left[\dfrac{1}{x}\right] & \text{if } x \neq 0 \\ 0 & \text{if } x = 0 \end{cases}$$

62. For every $k \in \mathbb{R}$, the function $f(x)$ defined as follows has derivative $f'(0) = 0$.

$$f(x) = \begin{cases} k \cdot x^2 & \text{if } x \in \mathbb{Q} \\ 0 & \text{if } x \notin \mathbb{Q} \end{cases}$$

63. If a function f is differentiable on (b, c) and $f'(a) = 0$ for $a \in (b, c)$, then it is not necessarily true that either a relative maximum or relative minimum for f occurs at $x = a$.
64. If f and g are differentiable functions on (a, b) with the same derivative, then $f(x) - g(x)$ is a constant for any $x \in (a, b)$.

65. If f and g are differentiable functions on (a, b) with $f - g$ a constant, then f and g have the same derivative at any $x \in (a, b)$.

66. Define a function f that is nowhere differentiable, while f^2 is everywhere differentiable. Hint: Consider a variation on the characteristic function of $\mathbb{Q}$.

Exercises 67–70 develop a proof of the chain rule in a fuller generality than was discussed in question 4.4.4. Throughout these exercises assume that $g(x)$ is differentiable at a point $x = a$ and that $f(x)$ is differentiable at $g(a)$.

67. Prove that the following function F is continuous at $h = 0$; intuitively, we think of F as the derivative of f with respect to $t = g(a)$.

$$f(h) = \begin{cases} \dfrac{f[g(a) + h] - f[g(a)]}{h} & \text{if } h \neq 0 \\ f'[g(a)] & \text{if } h = 0 \end{cases}$$

68. Prove that $f[g(a) + h] = f[g(a)] + h \cdot F(h)$ for sufficiently small values of h by taking the limit of these two expressions as h approaches 0.

69. In a parallel way, we can define a function G so that $G(0) = g'(a)$ and $g(a + k) = g(a) + k \cdot G(k)$ for sufficiently small values of k. Use this fact, the result from exercise 68, and the choice of $h = g(a + k) - g(a) = k \cdot G(k)$ to prove that:

$$f[g(a) + h] = f[g(a + k)] \quad \text{and} \quad h \cdot F(h) = k \cdot G(k) \cdot F(k \cdot G(k)).$$

70. Using the two equations obtained in exercise 69, substitute the first equation into the second to prove that

$$f[g(a + k)] = f[g(a)] + k \cdot G(k) \cdot F(k \cdot G(k)).$$

The last term on the right is continuous at 0 based on the definitions of F and G. Subtract $f[g(a)]$ on both sides of this equation, divide both sides by k and take the limit as k approaches 0 to obtain the chain rule.

4.5   Understanding Infinity

The notion of infinity has been an important element in many cultures' attempts to understand life: people refer to eternal time; an eternal spiritual afterlife; a boundless universe; an all-powerful deity. Mathematics has a unique and important perspective on infinity; the insights arising from mathematics' rigorous, logical approach to infinity have had an important influence on Western society's view of the world. But many advanced mathematical results on infinity (especially those that grew out of Georg Cantor's work in the late 1860s) are not widely known. In this section, we explore a mathematical understanding of the infinite.

We have already taken the first steps in this direction in our study of limits. One major breakthrough in the development of calculus is the harnessing of infinity in the very specific and powerful way expressed by the notion of limit to obtain the derivative (and the integral as discussed in section 4.6). As mathematicians developed and refined their understanding of limits, derivatives, and integrals in the eighteenth and nineteenth centuries, they began to recognize a need for a better understanding of the continuum of the real line. The German mathematician Georg Cantor made the most significant progress in this direction, ultimately developing a set-theoretic perspective of the reals that enabled him to study the basic properties and arithmetic of actual infinite numbers. While this study of infinity has a more discrete, set-theoretic feel, we immediately apply the results to the study of real analysis.

During the late 1600s, a European openness to infinite processes was a key element in the development of calculus. Based on the ancient Greek mathematician Archimedes' efforts to calculate the areas of closed plane figures via approximating polygons, it appears that he essentially understood the ideas of limits and integrals. Archimedes apparently held back because of the need to work with infinity. We can only wonder about the advancement of mathematical thought had Archimedes, Euclid, or Pappus of Alexandria successfully pursued these ideas further.

In our study of limits, we used the symbol "∞" to denote infinity. This notation was introduced in 1655 by the English mathematician John Wallis in his paper *De sectionibus conicus*. However, the symbol ∞ does not denote an actual infinite number, but indicates "unboundedness" or numbers growing without bound. For example, we know $\lim_{x \to 0} 1/x^2 = \infty$. This notation does *not* mean that we substitute 0 for x and the result is a number ∞; rather, this expression means that the values of the function $1/x^2$ grow larger and larger as we substitute numbers closer and closer to 0 for x. This notion of "growing large without bound" or of processes that continue indefinitely (such as counting positive integers) is referred to as *potential infinity*. The ancient Greeks were mostly accepting of the notion of potential infinity; in the third century B.C.E., Archimedes wrote about extremely large numbers growing without bound in his essay *The Sand Reckoner*. While such a perspective on infinity is useful (as evident in calculus), we are interested in more: we seek to define actual "infinite numbers."

Humans have been grappling with the notion of infinity for thousands of years. This struggle has led some of humanity's greatest thinkers to describe various "paradoxes of infinity," two of which are described in this section. Such paradoxes highlight the delicate issues at the heart of the notion of infinity–those that intuitively seem contradictory and without possibility of resolution, and that ultimately require new insights to unravel. The first was described during the fourth century B.C.E. by the Greek philosopher Zeno of Elea. Zeno proposed four paradoxes of infinity: the *Dichotomy*, *Achilles*, the *Arrow*, and the *Stadium*. Each paradox expresses concerns with the notion of a realized, *completed infinity*, as opposed to the familiar potential infinity from our studies of calculus.

Example 4.5.1 Zeno's paradox of Achilles and the tortoise   Achilles and the tortoise are racing, and Achilles has kindly given the tortoise a head start. After making some slow but steady progress, the tortoise decides to take a break and sit without moving on the race course. Achilles starts the race and begins covering the distance between himself and the stationary tortoise. First, Achilles covers half the distance between the starting line and the tortoise. He then covers half the remaining distance to the tortoise; at this point Achilles only needs to cover one-fourth of the distance between the starting line and the tortoise. Again, Achilles makes progress and covers half the remaining distance; now he has only one-eighth of the remaining distance left. He continues in this fashion: with 1/16th of the distance left to cover, then 1/32nd left, then 1/64th left, and so on. In this way, Achilles gets closer and closer to the tortoise. But he never catches the tortoise, because he has infinitely many of these half-distance intervals to pass through! And here is the paradox, as our "real-life" experience tells us that Achilles not only catches the tortoise (who remains at a standstill), but also passes him and wins the race (so long as Achilles continues to move toward the finish line).

A resolution of Zeno's Achilles–tortoise paradox. One resolution is often referred to as "irrelevant parametrization," which can be thought of as "keeping your eye on the wrong part of the problem." In this setting, the element of time is never considered in the statement of the paradox: How long does it take Achilles to pass through these increasingly small distance intervals?

A second resolution of this paradox follows from an understanding of infinite series (or sums) developed by European mathematicians in the eighteenth century. Perhaps you recall studying geometric series with $r = 1/2$, for which

$$\frac{1}{2} + \frac{1}{4} + \frac{1}{8} + \frac{1}{16} + \cdots = \frac{\frac{1}{2}}{1 - \frac{1}{2}} = 1.$$

In short, an infinite sum of positive terms can have finite value. In reference to Zeno's Achilles–tortoise paradox, if Achilles travels at a constant rate, then the series sums the times it takes him to travel each of the half-distances, and so the total distance between Achilles and the tortoise is covered in a finite amount of time.

■

While we may now understand how to respond to Zeno's paradoxes of infinity, these resolutions were not apparent for some 2,000 years. During this time, the notion of potential infinity continued to be accepted by philosophers and mathematicians, while the notion of an actual, complete infinity was regarded with skepticism.

A second paradox of infinity was stated by Galileo in the early 1600s. Contemporary mathematicians phrase Galileo's paradox in terms of one-to-one correspondences, which play a key role in the study of infinity. We defined a one-to-one correspondence (in definition 4.2.2 of section 4.2) as a function $f : X \to Y$ that is both one-to-one and onto. A one-to-one function satisfies the condition that every output comes from a unique input, and an onto function satisfies the condition that every element of the target set Y is in the range. Consider the following examples.

Example 4.5.2 We study functions on the sets $A = \{x, y, z\}$, $B = \{4, 5, 6\}$, and $C = \{a, b\}$.

- The function $f : C \to A$ defined by $f = \{(a, x), (b, z)\}$ is one-to-one; there are five other such one-to-one maps, but no maps from C onto A.
- The function $g : A \to C$ defined by $g = \{(x, a), (y, b), (z, b)\}$ is onto; there are five other such onto maps, but no one-to-one maps from A to C.
- The function $h : A \to B$ defined by $g = \{(x, 4), (y, 5), (z, 6)\}$ is both one-to-one and onto; there are five other such one-to-one correspondences between

A and *B*. We can think of these maps as expressing that *A* and *B* are the same size, while *C* has a different size.

■

Question 4.5.1   Prove each map on infinite sets has the specified property.

(a) The function $f : \mathbb{Z} \to \mathbb{Z}$ defined by $f(x) = x + 1$ is a one-to-one correspondence.
(b) The function $g : \mathbb{Z} \to \mathbb{Z}$ defined by $g(x) = 2x$ is one-to-one, but not onto.
(c) The function $h : \mathbb{R} \to \mathbb{R}$ defined by $h(x) = x^2$ is neither one-to-one nor onto.

■

One intriguing insight into one-to-one correspondences is that such maps between finite sets have different properties than such maps between infinite sets. For example, a finite set can never be placed in one-to-one correspondence with a proper subset of itself. A proper subset contains fewer elements than the original set, and so any map between the original set and a proper subset is either not one-to-one or not onto. As an intuitive illustration, consider the impossibility of matching four left shoes with five right shoes to form pairs of shoes without at least one right shoe left over. We can develop a careful proof that there does not exist a one-to-one correspondence between any finite set and a proper subset. This observation lies at the heart of Galileo's "paradox of infinity," as stated below.

Example 4.5.3 Galileo's paradox of squares   In the early 1600s, the Italian physicist and astronomer Galileo Galilei was reflecting on infinity and considering maps on infinite sets. He observed that the natural numbers $\mathbb{N} = \{1, 2, 3, ...\}$ can be placed in one-to-one correspondence with proper subsets of itself; for example, with the set of squares of natural numbers $S = \{1, 4, 9, 16, 25, ...\}$. Galileo identified the one-to-one correspondence $f : \mathbb{N} \to S$ defined by $f(n) = n^2$ (that is, by $f(1) = 1^2 = 1$, $f(2) = 2^2 = 4$, and so on). He identified this example as a paradox, (mistakenly) believing that infinite sets could not be placed in one-to-one correspondence with a proper subset.

A Resolution of Galileo's paradox of squares   Mathematicians now understand that finite and infinite sets have different properties with respect to one-to-one correspondences; in fact, a set is infinite exactly when the set can be placed in one-to-one correspondence with a proper subset. Rather then identifying a paradox, Galileo actually happened upon a characterization of infinity. We too must keep in mind that not every property of finite numbers extends to infinite numbers.

■

The definition of "number" grew out of the work of nineteenth century function theorists. In the early 1800s, Joseph Fourier modeled heat using trigonometric series, but it was unclear how widely this model could be applied. Peter Dirichlet and other mathematicians resolved this question by studying the points of discontinuity in domain sets, including infinite sets of points of discontinuity. Cantor's research efforts in this area led him to ask questions about the continuum of the real number line. Among other things, Cantor needed to know the relative size of the set of rational numbers vis-a-vis the set of real numbers. He originally conjectured that there was only one size of infinity; this intuition matched that of his colleagues and perhaps matches your own. But Cantor needed more than intuition—he needed proof. And his work led to some startling results.

Georg Cantor was a mathematician of Danish and Russian descent who spent most of his life working at the University of Halle in Germany. While his doctoral work was in number theory, the bulk of his early research at Halle was in real analysis. His study of Fourier series ultimately led him to explore perhaps one of the most fundamental questions of mathematics: What is a "number"? From the early 1870s to the late 1890s, Cantor published multiple papers that in aggregate develop set theory, precisely define one-to-one correspondences, explore finite and infinite numbers via such mappings, study an arithmetic of infinite numbers, and ask important questions that continue to guide mathematical research. This insightful work was greeted with mixed reactions, with some mathematicians expressing high praise and others strong disdain for Cantor's research program. The last couple of decades of Cantor's life were difficult; along with professional conflicts, Cantor struggled with the deaths of close family members, crippling mental illness, and the consequences to civilian life arising from the onset of World War I.

We now consider the question that Cantor asked: *What do we mean by the word "number"?* An intuitive description of "number" typically includes such phrases as "size" or "how many" (as in "How many elements are in a set?"). Cantor's insight was to think of "number" not as a property that is held in isolation, but as a relation. In this way, we can use sets and functions to develop a rigorous mathematical definition of "number." Instead of the words "number" or "size," mathematicians use the term "cardinality" to refer to the number of objects in a set. The following definition precisely expresses this concept and is at the heart of Cantor's theory of infinity.

Definition 4.5.1 *Sets X and Y have the same **cardinality** when there exists a one-to-one correspondence from X to Y, and the cardinality of a set X is denoted by $|X|$.*

- *$|X| = |Y|$ means that X and Y have the same cardinality.*
- *$|X| \leq |Y|$ means that there exists a one-to-one function from X to Y, which is not necessarily onto.*
- *$|X| < |Y|$ means that $|X| \leq |Y|$ but $|X| \neq |Y|$; that is, there exists a one-to-one map from X to Y, but no one-to-one correspondence.*

As you might expect, we denote the cardinality of finite sets using the nonnegative integers $0, 1, 2, 3, \ldots$. The "smallest" infinite cardinality is that of the natural numbers $\mathbb{N}$.

Example 4.5.4 We use the notation $|X|$ to express the cardinality of sets.

Considering the sets we studied in example 4.5.2, both $|\{x, y, z\}| = 3$ and $|\{4, 5, 6\}| = 3$; also $|\{a, b\}| = 2$. Since the empty set $\emptyset$ contains no elements, $|\emptyset| = 0$. In addition, we can consider the cardinalities of infinite sets. As observed in Galileo's paradox of squares, there exists a one-to-one correspondence between the set of natural numbers and the set of squares of natural numbers, and so $|\mathbb{N}| = |\{1, 4, 9, 16, 25, \ldots\}|$.

■

Question 4.5.2   Prove that $\mathbb{N}$ and the set E of even positive integers have the same cardinality by showing that the function $f : \mathbb{N} \rightarrow E$ defined by $f(n) = 2n$ is a one-to-one correspondence.   ∎

Cantor realized the importance of introducing symbols for infinite cardinalities and denoted the cardinality of the natural numbers $\mathbb{N}$ by $\aleph_0$, where $\aleph$ (called "aleph") is the first letter of the Hebrew alphabet. Since the symbol $\aleph_0$ is hard to write, we utilize the more familiar ω (the Greek letter "omega") in this text. In a more sophisticated study of infinite sets, ω denotes the set of natural numbers under the standard $\leq$ ordering of natural numbers; in such settings, we must use $\aleph_0$ when referring exclusively to the cardinality $|\mathbb{N}|$. However, in this text, we write $|\mathbb{N}| = \omega$, and we say that the set of natural numbers is *countably infinite* (or sometimes *denumerably infinite*). These conventions are formalized in the following definition.

Definition 4.5.2   *A set X is said to be* **countable** *when either X is finite or $|X| = |\mathbb{N}| = \omega$; that is, when X is finite or when there exists a one-to-one correspondence between X and $\mathbb{N}$. A set X is said to be* **uncountable** *when X is not countable.*

Observe that the elements of countable sets can be arranged in a sequence $a_1, a_2, a_3, \ldots$, where elements are allowed to be repeated in such a list. Writing the elements of a set X as a sequence implicitly identifies a one-to-one correspondence f between $\mathbb{N}$ and X. We use this approach to prove that many familiar sets are countable.

Example 4.5.5   We verify the following sets are countable; we consider examples of uncountable sets soon.

- The set of odd positive integers is countable as demonstrated by the sequence:

$$1, \ 3, \ 5, \ 7, \ 9, \ 11, \ 13, \ 15, \ 17, \ldots.$$

- The set of integers $\mathbb{Z}$ is countable as demonstrated by the sequence:

$$0, \ 1, \ -1, \ 2, \ -2, \ 3, \ -3, \ 4, \ -4, \ldots.$$

- The set Q_0 of rational numbers in the (real) interval $[0, 1)$ is countable. First, consider the sequence:

$$0, \ \frac{1}{2}, \ \frac{1}{3}, \ \frac{2}{3}, \ \frac{1}{4}, \ \frac{2}{4}, \ \frac{3}{4}, \ \frac{1}{5}, \ \frac{2}{5}, \ \frac{3}{5}, \ \frac{4}{5}, \ \frac{1}{6}, \ \frac{2}{6}, \ \frac{3}{6}, \ldots.$$

Do you see the pattern? Based on this sequence, we can define an onto map from $\mathbb{N}$ to Q_0. By omitting the repetitions in the sequence (for example, $1/2 = 2/4 = 3/6$, and so on) we obtain a one-to-one map, as suggested by the following sequence:

$$0, \ \frac{1}{2}, \ \frac{1}{3}, \ \frac{2}{3}, \ \frac{1}{4}, \ \frac{3}{4}, \ \frac{1}{5}, \ \frac{2}{5}, \ \frac{3}{5}, \ \frac{4}{5}, \ \frac{1}{6}, \ \frac{5}{6}, \ \frac{1}{7}, \ \frac{2}{7}, \ldots.$$

∎

Question 4.5.3   Using the sequence approach demonstrated in example 4.5.5, prove each set is countable; be careful to include every element of the given set in the sequence.

(a) the set E of even positive integers
(b) the set of even integers

(c) the set P of prime numbers

(d) the set of rational numbers in the (real) interval $[1, 2)$

■

Cantor also proved the following result about unions of countable sets; this result is useful in many settings, for example, when studying the theory of integration.

Theorem 4.5.1   *The union of a countable sequence of countable sets is countable.*

Proof   Let S_1, S_2, S_3, ... be a countable sequence of countable sets with $S_j = \{s_{j1}, s_{j2}, s_{j3}, \ldots\}$. Then $\bigcup_{j \in \mathbb{N}} S_j$ is a countable set as demonstrated by the sequence:

$$s_{11}, \ s_{12}, \ s_{21}, \ s_{13}, \ s_{22}, \ s_{31}, \ s_{14}, \ s_{23}, \ s_{32}, \ s_{41}, \ s_{15}, \ s_{24}, \ldots.$$

This sequence is constructed by following along the "diagonals" identified in the array of sequence elements given in figure 4.15; consequently, the technique used to define this sequence is called *Cantor's first diagonalization method*.

■

Theorem 4.5.1 can be used to prove the countability of many sets, as illustrated in the following example and question.

Example 4.5.6   We prove that the set $\mathbb{Q}$ of rational numbers is countable.

Proof   In light of theorem 4.5.1, we express the rationals as a union of a countable sequence of countable sets. Define Q_j to be the set of rational numbers in the interval $[j, j + 1)$. By appropriately extending the argument for Q_0 given in example 4.5.5, we can prove that each set Q_j is countable. We now observe that:

$$\mathbb{Q} = \bigcup_{j \in \mathbb{Z}} Q_j \, .$$

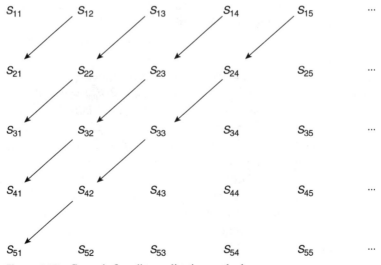

Figure 4.15   Cantor's first diagonalization method

Therefore, $\mathbb{Q}$ is a countable union of countable sets, and so $\mathbb{Q}$ is countable by theorem 4.5.1.

∎

Question 4.5.4 Using theorem 4.5.1, prove each set is countable.

(a) $\mathbb{N} \times \mathbb{N} = \{(m, n) : m, n \in \mathbb{N}\}$

(b) $\mathbb{Q}^n = \{(q_1, q_2, \ldots, q_n) : q_1, \ldots, q_n \in \mathbb{Q}\}$

(c) $\{(n_1, n_2, n_3, \ldots) : n_i \in \mathbb{Z} \text{ for all } i \in \mathbb{N}\}$

∎

Thus far, these results all seem to confirm a natural intuition that the countable cardinality ω of the set of natural numbers is the unique size of infinity. In addition, the rationals are *dense* in the set of reals in the sense that there exists a rational number strictly between any two real numbers. The countability of the rationals and the density of the rationals in the reals provides further evidence for the intuition that the real numbers are also countable. For many years, Cantor tried to prove that the set of real numbers is countable, but he was unsuccessful—for good reason. By December 1873, Cantor proved the astonishing, ground-breaking result that the set $\mathbb{R}$ of real numbers is uncountable! Even with the 1874 publication of his proof of this result, mathematicians continued to be surprised that there exists more than one infinite cardinality. The following elegant proof of the uncountability of the reals was published by Cantor in 1891.

Theorem 4.5.2 *The set $\mathbb{R}$ of real numbers is uncountable.*

Proof   We prove $[0, 1)$ is uncountable; because $[0, 1) \subset \mathbb{R}$, the uncountability of $[0, 1)$ implies the uncountability of $\mathbb{R}$. The proof proceeds by contradiction. First, assume that $[0, 1)$ is countable, and so its elements can be written as a sequence $\{a_1, a_2, a_3, a_4, \ldots\}$. Then produce a real number $r \in [0, 1)$ that is not one of the a_m's, contradicting the fact that the (supposed) one-to-one correspondence from $\mathbb{N}$ to $[0, 1)$ defined by the sequence is onto.

The real number r is defined using the decimal expansion of each $a_m \in [0, 1)$. First express each $a_m \in [0, 1)$ in an unambiguous way as a nonterminating decimal. Let b_{mn} represent the nth digit in the decimal expansion of a_m, so that we have: $a_1 = 0.b_{11}b_{12}b_{13}b_{14}\ldots$; $a_2 = 0.b_{21}b_{22}b_{23}b_{24}\ldots$; $a_3 = 0.b_{31}b_{32}b_{33}b_{34}\ldots$; and so on. In this setting, think of real numbers with terminating decimal expansions as having 0s appended to the end of their expansions; for example, think of $a_m = 1/2 = 0.5000\ldots$ as $b_{m1} = 5$, $b_{m2} = 0$, $b_{m3} = 0$, $b_{m4} = 0$, and so on.

We now apply what has become known as *Cantor's second diagonalization method* to obtain a real number $r \notin \{a_m : m \in \mathbb{N}\}$. Define $r = 0.b_1b_2b_3b_4\ldots$, where the nth digit b_n in the decimal expansion of r is determined by b_{nn}, the nth digit in the decimal expansion of a_n. If b_{nn} equals 0, define $b_n = 1$; if b_{nn} is not equal to 0, define $b_n = 0$. Based on this definition, observe that the nth digit of r differs from the nth digit of a_n. Thus, r is not an element of the sequence $\{a_m : m \in \mathbb{N}\}$, contradicting the fact that the (supposed) one-to-one correspondence from $\mathbb{N}$ to $[0, 1)$ defined by $\{a_m : m \in \mathbb{N}\}$ is onto. Thus $[0, 1)$ is uncountable, implying that $\mathbb{R}$ is uncountable.

∎

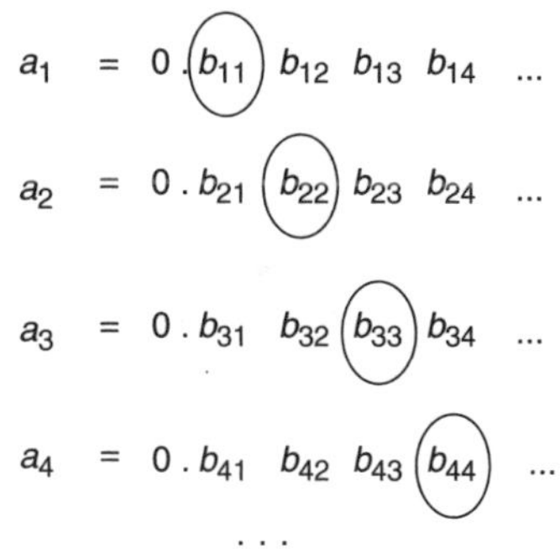

$$a_1 = 0.\boxed{b_{11}}\ b_{12}\ b_{13}\ b_{14}\ \ldots$$

$$a_2 = 0.b_{21}\ \boxed{b_{22}}\ b_{23}\ b_{24}\ \ldots$$

$$a_3 = 0.b_{31}\ b_{32}\ \boxed{b_{33}}\ b_{34}\ \ldots$$

$$a_4 = 0.b_{41}\ b_{42}\ b_{43}\ \boxed{b_{44}}\ \ldots$$

Figure 4.16 Cantor's second diagonalization method

This approach to proving the uncountability of the reals is called *Cantor's second diagonalization method* (or more simply a *diagonal argument*) because it obtains r by changing the digits in the decimal expansions of sequence elements along a diagonal. If we present these decimal expansions $a_m = 0.b_{m1}b_{m2}b_{m3}b_{m4}\ldots$ in an array and circle the digits that are modified (b_{nn} to obtain b_n) the "diagonal" of Cantor's argument is readily apparent, as illustrated in figure 4.16.

Theorem 4.5.2 leads to the inescapable conclusion that there are *different* sizes of infinity! This result surprised many mathematicians of the late nineteenth century, and it took years for Cantor's ideas to gain widespread acceptance. Cantor's methods and results provoked strong (positive and negative) reactions among mathematicians, from Leopold Kronecker's assertion that "God made the integers, all else is the work of man," to David Hilbert's claim that "No one will expel us from the paradise that Cantor has created." While this controversy continued into the early twentieth century, Cantor's approach has been invaluable to mathematicians' efforts to explore mathematical truth. As Cantor reflected on his proof that $|\mathbb{R}| \neq \omega$ and continued to develop his ideas, he was eventually able to identify a clever one-to-one correspondence that proved the precise relationship:

$$|\mathbb{R}| = 2^{\omega}.$$

This result led him to the following intriguing question:

Does there exist an infinite number between $\omega = |\mathbb{N}|$ and $2^{\omega} = |\mathbb{R}|$?

Despite literally decades of effort, Cantor was never able to answer this question, but he did conjecture what has become known as the *continuum hypothesis*: there is no infinite number between ω and 2^{ω} (just as there is no natural number between 1 and 2). The continuum hypothesis received a lot of attention during the twentieth century. At the 1900 International Congress of Mathematicians in Paris, the German mathematician David Hilbert gave a now famous address *The problems of mathematics* in which he presented a list of 23 questions that he considered to be the most significant of the time. These questions came to be called *Hilbert's problems*, and they profoundly influenced mathematical research throughout the twentieth century. Hilbert's first problem called for the resolution of the continuum hypothesis.

Surprisingly, there is no "right" answer to the continuum hypothesis. In 1938, the Austrian mathematician Kurt Friedrich Gödel showed that the continuum hypothesis cannot be proven false. In addition, Gödel conjectured that the continuum hypothesis also cannot be proven true; in 1963, the American mathematician Paul Cohen confirmed Gödel's conjecture, showing that continuum hypothesis is not provable. Mathematical statements that are neither provable nor disprovable are called *undecidable*. From the work of Gödel and Cohen, we know that the continuum hypothesis is undecidable (based on the standard axioms of set theory). Their results mean that we are "free" to assume the existence of an infinite number between ω and 2^ω, or the nonexistence of such an infinite number, provided we explicitly identify the use of such an assumption. These explorations of the infinite also tell us something about ourselves, as they highlight an essential limitation of our mathematical reasoning processes.

4.5.1 Cantor's Theorem

We now consider cardinality from an explicit set-theoretical perspective, including an analog of Cantor's proof of the uncountability of the reals based on the power set operation. We also study an "addition" operation and a "countable multiplication" operation for infinite numbers, along with what features of the finite versions of these operations carry over to the context of infinite numbers. Recall from definition 2.1.4 in section 2.1 that if A is a set, then $\mathbb{P}(A)$ denotes the *power set* of A and is the set of all subsets of A; symbolically, we define $\mathbb{P}(A) = \{X : X \subseteq A\}$. Notice that both $\emptyset \in \mathbb{P}(A)$ and $A \in \mathbb{P}(A)$.

Example 4.5.7 If $A = \{a, b\}$, then the 2-element subset of A is $\{a, b\}$, the 1-element subsets of A are $\{a\}$ and $\{b\}$, and the 0-element subset of A is $\emptyset = \{\}$. Therefore, $\mathbb{P}(A) = \{\ \{a, b\}, \{a\}, \{b\}, \emptyset\ \}$.

If $A = \emptyset$, then the only subset of $\emptyset$ is $\emptyset$, and so $\mathbb{P}(A) = \mathbb{P}(\emptyset) = \{\emptyset\}$. ∎

Question 4.5.5 Determine the power set of each set.

(a) $A = \{0\}$ (b) $A = \{0, 1\}$ ∎

In example 4.5.7 and question 4.5.5, if we compare the cardinality of a set A with the cardinality of its power set $\mathbb{P}(A)$, a distinctive pattern becomes apparent.

$$A \text{ has } 0 \text{ elements} \quad \Rightarrow \quad \mathbb{P}(A) \text{ has } 1 = 2^0 \text{ element}$$
$$A \text{ has } 1 \text{ element} \quad \Rightarrow \quad \mathbb{P}(A) \text{ has } 2 = 2^1 \text{ elements}$$
$$A \text{ has } 2 \text{ elements} \quad \Rightarrow \quad \mathbb{P}(A) \text{ has } 4 = 2^2 \text{ elements}$$

As it turns out, this numeric relation holds for every set A, including infinite sets: the cardinality of the power set of A is equal to two raised to the cardinality of the set A.

Theorem 4.5.3 *If A is a set, then $|\mathbb{P}(A)| = 2^{|A|}$.*

Comments on proof For finite sets, we can use induction on the nonnegative integers to prove theorem 4.5.3 (see exercise 65 in section 3.6). For infinite sets, we need to state a precise definition of 2 raised to an infinite power; this part of the proof is left for your later studies.

In light of theorem 4.5.3, Cantor wondered about the relationship between the cardinality of the set of natural numbers $|\mathbb{N}| = \omega$ and the cardinality of the power set of natural numbers $|\mathbb{P}(\mathbb{N})| = 2^\omega$; are ω and 2^ω equal, or not? As we have discussed already, not every property of finite numbers carries over to infinite numbers and so, even though $n < 2^n$ for every $n \in \mathbb{N}$ (see exercise 28 in section 3.6), Cantor insightfully recognized that perhaps ω could be equal to 2^ω. As you may recognize, this query is really another variation on the question of whether there are different sizes of infinity. Furthermore, from the uncountability of the reals and the fact that $|\mathbb{R}| = 2^\omega$, we know that $\omega \neq 2^\omega$. However, more than just working in this specific setting, Cantor proved a more general theorem with profound implications for a study of infinite cardinalities.

Theorem 4.5.4 Cantor's theorem *If A is a set, then the cardinality of A is strictly less than the cardinality of the power set of A; symbolically, we have $|A| < |\mathbb{P}(A)|$.*

Proof From definition 4.5.1, show that $|A| \leq |\mathbb{P}(A)|$ by defining a one-to-one function from A into $\mathbb{P}(A)$; we then need to prove that there is no one-to-one correspondence from A to $\mathbb{P}(A)$ to obtain the strict inequality $|A| < |\mathbb{P}(A)|$.

Define $f : A \to \mathbb{P}(A)$ by $f(x) = \{x\}$ for every $a \in A$. This function f is one-to-one; if $x, y \in A$ and $f(x) = f(y)$, then $\{x\} = \{y\}$ and so $x = y$ by the definition of set equality. Thus, $|A| \leq |\mathbb{P}(A)|$.

A diagonal argument (using Cantor's second diagonalization method) proves that there does not exist a one-to-one correspondence from A to $\mathbb{P}(A)$. As in the proof of the uncountability of the reals, this proof proceeds by contradiction. Assume $g : A \to \mathbb{P}(A)$ is a one-to-one correspondence and then define a "diagonal" set $D \in \mathbb{P}(A)$ such that D is not in the range of g (contradicting the assumption that g is onto). Given the (supposed) one-to-one correspondence g, define:

$$D = \{\, x \in A : x \notin g(x) \,\}.$$

By definition, $D \subseteq A$ and so $D \in \mathbb{P}(A)$. Therefore, since g is (supposedly) onto, there exists an element $d \in A$ satisfying $g(d) = D$. Either $d \in D$ or $d \notin D$, but both of these options lead us to a contradiction. If $d \in D$, then $d \notin g(d)$ by the definition of D; and so $d \notin g(d) = D$, giving a contradiction. On the other hand, if $d \notin D$, then $d \in g(d)$ by the definition of D; and so $d \in g(d) = D$, giving a contradiction. Thus g is not onto and there does not exist a one-to-one correspondence from A to $\mathbb{P}(A)$.

■

The proof of Cantor's theorem is relatively abstract, and so we consider an example in the context of the set of natural numbers to illustrate the diagonal set D.

Example 4.5.8 Suppose $g : \mathbb{N} \to \mathbb{P}(\mathbb{N})$ is a one-to-one correspondence with the first few output sets of g as follows.

$$g(1) = \{\, 1,\ 2,\ 3,\ 4, \ldots \}$$

$$g(2) = \{\, 1,\ 3,\ 5,\ 7, \ldots \}$$

$$g(3) = \{\, 2,\ 4,\ 6,\ 8, \ldots \}$$

$$g(4) = \{\,4,\ 5,\ 6,\ 7,\ 8,\ldots\,\}$$

$$g(5) = \{\,5\,\}$$

$$g(6) = \{\,5,\ 7\,\}$$

Working with the definition of $D = \{x \in \mathbb{N} : x \notin g(x)\}$, we have $D = \{2,\ 3,\ 6,\ldots\}$ because of the following relationships.

$$1 \in g(1),\quad 2 \notin g(2),\quad 3 \notin g(3),\quad 4 \in g(4),\quad 5 \in g(5),\quad 6 \notin g(6),\ \ldots$$

∎

Cantor's theorem provides another confirmation that there exist different infinite cardinalities. Since $|\mathbb{N}| = \omega$ and $|\mathbb{P}(\mathbb{N})| = 2^{\omega}$, we know that $\omega \neq 2^{\omega}$. Furthermore, we can apply the power set operation multiple times to obtain infinitely many distinct infinite numbers! The following sequence illustrates this process.

$$|\mathbb{N}| \ <\ |\mathbb{P}(\mathbb{N})| \ <\ |\mathbb{P}(\mathbb{P}(\mathbb{N}))| \ <\ |\mathbb{P}^3(\mathbb{N})| \ <\ \cdots$$
$$\omega \ <\ 2^{\omega} \ <\ 2^{2^{\omega}} \ <\ 2^{2^{2^{\omega}}} \ <\ \cdots$$

Recall the assertion of the (unprovable) continuum hypothesis that there does not exist an infinite number between ω and 2^{ω}. Gödel's and Cohen's results proving the undecidability of the continuum hypothesis extend to this broader context. The *generalized continuum hypothesis* states that for any infinite cardinal κ (the Greek letter "kappa"), there does not exist an infinite cardinal between κ and 2^{κ}, and is also undecidable.

4.5.2   Cardinal Arithmetic

We finish this section by studying cardinal arithmetic. Mathematicians have developed well-defined extensions of the operations of addition, multiplication, and exponentiation on the (finite) natural numbers to infinite cardinals, enabling us to associate meaning with such expressions as $1 + \omega$, $\omega + \omega$, and $3 \cdot \omega$. A first step in the direction of understanding cardinal arithmetic is the definition of addition and countable multiplication of cardinals; the complete definition of cardinal multiplication and cardinal exponentiation are left for your later studies. In this context, mathematicians commonly use κ and μ (the Greek letter "mu") to denote arbitrary cardinals (particularly infinite cardinals) in much the same way that m and n are used to denote arbitrary integers.

Definition 4.5.3   *If κ and μ are cardinals with $\kappa = |A|$ and $\mu = |B|$ for disjoint sets A and B, then*

- $\kappa + \mu = |A \cup B|$, *and*
- *if κ is finite or ω, then $\kappa \cdot \mu = |A \times B|$.*

As suggested by this definition, we perform cardinal operations by designating representative sets and defining one-to-one correspondences between these sets. In our discussion, we also informally illustrate these computations with imaginary

Hilbert Hotels, which have infinitely many hotel rooms. These illustrations can help us visualize solutions to some fairly complicated computations.

Example 4.5.9 We prove that $1 + \omega = \omega$.

An informal proof Study the identity $1 + \omega = \omega$ using an infinite Hilbert Hotel with ω many rooms. Suppose the traveler a arrives at a Hilbert Hotel and finds that every room is occupied. Despite having no vacancy, the staff can find a room for a. Doubling up is not allowed (as with any upscale hotel). But what if every occupant is willing to switch rooms? Can you see how to open up a room for traveler a? Figure 4.17 illustrates a process in which each occupant n moves one room up, and traveler a moves into room number 1, giving us $1 + \omega = \omega$. We now consider a formal proof of this result.

Proof The preceding informal discussion suggests how to prove that $1 + \omega = \omega$. We first choose representative sets for these cardinalities and then define a one-to-one correspondence between these sets. For this proof, let $A = \{a\} \cup \mathbb{N}$ represent $1 + \omega$ and let $\mathbb{N}$ represent ω; notice that these sets have the desired cardinalities. We define a one-to-one correspondence $f : A \to \mathbb{N}$ by

$$f(x) = \begin{cases} 1 & \text{if } x = a \\ x + 1 & \text{if } x \in \mathbb{N}. \end{cases}$$

A detailed proof that f is one-to-one and onto is left to the reader; given this one-to-one correspondence, we have $1 + \omega = \omega$.

■

Question 4.5.6 Prove each equality by giving both an informal Hilbert Hotel argument and a formal proof for a one-to-one correspondence between representative sets.

(a) $2 + \omega = \omega$

(b) $3 + \omega = \omega$

(c) $7 + \omega = \omega$

(d) $n + \omega = \omega$ for $n \in \mathbb{N}$

■

Example 4.5.10 We informally prove that $\omega + \omega = \omega$; that is, $2 \cdot \omega = \omega$.

An informal proof We consider a scenario in which the Hilbert Hotel chain is doing a bit of downsizing: suppose that one of two full Hilbert Hotels needs to close. To avoid a public relations nightmare, the Hilbert Hotel chain guarantees

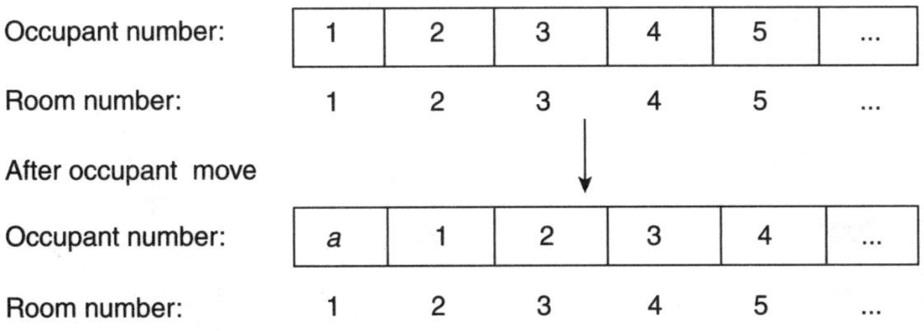

Figure 4.17 A vacancy is possible at a full Hilbert Hotel

every occupant of the closing hotel a room in the hotel remaining open. Now a full Hilbert Hotel needs to make room for countably infinite additional people. Can you see how the hotel might accomplish this feat? One strategy is to move everyone in the open hotel to even-numbered rooms and to move everyone in the closing hotel to the now vacant odd-numbered rooms of the open hotel; this approach is illustrated in figure 4.18. Thus, $\omega + \omega = \omega$, and so $2\omega = \omega$. The formal proof is left for the next question.

■

Question 4.5.7   Prove each equality by identifying a one-to-one correspondence between appropriate representative sets.

(a) $\omega + \omega = \omega$. Hint: Let $A = \{a_n : n \in \mathbb{N}\}$ and $B = \{b_n : n \in \mathbb{N}\}$. Define a one-to-one correspondence $f : A \cup B \to \mathbb{N}$ that gives a formula for the map depicted in example 4.5.10.

(b) $3\omega = \omega + \omega + \omega$ is equal to ω. Also, give an illustration of the corresponding informal Hilbert Hotel proof.

(c) $n \cdot \omega = \omega$ for $n \in \mathbb{N}$.

■

The ideas we have explored in this section are just the first steps in the mathematical study of infinite numbers—there is much more to learn about this topic. For example, besides the power set operation, there exist many other set-theoretic operations that yield larger and larger infinite numbers. These further explorations require a more formal and complete study of set theory, and are left for your later studies in mathematics. For now, we have identified many interesting results that play a role in the study of integral calculus in the next section.

4.5.3   Reading Questions for Section 4.5

1. Discuss the distinction between potential infinity and actual infinity.
2. Describe Zeno's paradox of Achilles and the tortoise. How do contemporary mathematicians resolve this paradox?
3. Define and give an example of a one-to-one correspondence.
4. Describe Galileo's paradox of squares. How do contemporary mathematicians resolve this paradox?
5. Define cardinality and give an example.

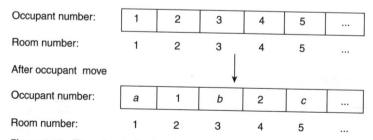

Figure 4.18   From the closing hotel, $a,b,c, \ldots$ move in among $1,2,3, \ldots$ from the open hotel

6. Discuss the distinction between $|X| = |Y|$ and $|X| < |Y|$.
7. Define and give an example of a countable set and an uncountable set.
8. State theorem 4.5.1. How is this result helpful when studying cardinality?
9. State theorem 4.5.2 and informally describe a diagonal argument.
10. State the continuum hypothesis. Explain what is meant by the assertion that the continuum hypothesis is undecidable.
11. Give an example for theorem 4.5.3.
12. State Cantor's theorem. How is this result helpful when studying cardinality?

4.5.4   Exercises for Section 4.5

In exercises 1–13, define a function with the indicated properties, or explain why such a function does not exist. The domains and targets of these functions are the following sets.

$$A = \{a,\ b\} \qquad B = \{1,\ 2,\ 3\} \qquad C = \{x\} \qquad D = \{y,\ z\}$$

1. A one-to-one function $f : A \to B$.
2. A one-to-one function $g : B \to A$.
3. An onto function $h : A \to B$.
4. An onto function $j : B \to A$.
5. A   one-to-one   correspondence $k : A \to B$.
6. A one-to-one function $f : A \to C$.
7. A one-to-one function $g : C \to A$.
8. An onto function $h : A \to C$.
9. An onto function $j : C \to A$.
10. A   one-to-one   correspondence $k : A \to C$.
11. A   one-to-one   correspondence $f : A \to D$.
12. A   one-to-one   correspondence $g : B \to D$.
13. A   one-to-one   correspondence $h : C \to D$.

In exercises 14–22, prove the following functions are one-to-one and onto, or identify counterexamples showing that one (or both) of these properties does not hold.

14. $f : \mathbb{N} \to \mathbb{N}$ defined by $f(x) = 4$.
15. $f : \mathbb{N} \to \mathbb{N}$ defined by $f(x) = 2x + 1$.
16. $f : \mathbb{R} \to \mathbb{R}$ defined by $f(x) = x^2 + 1$.
17. $f : \mathbb{R} \to \mathbb{R}$ defined by $f(x) = x^3 + 1$.
18. $f : \mathbb{R} \to \mathbb{R}$ defined by $f(x) = x^n + 1$, where $n \in \mathbb{N}$ is even.
19. $f : \mathbb{R} \to \mathbb{R}$ defined by $f(x) = x^n + 1$, where $n \in \mathbb{N}$ is odd.
20. $f : \mathbb{R} \to \mathbb{R}$ defined by $f(x) = |x|$.
21. $f : \mathbb{R}^+ \to \mathbb{R}$ defined by $f(x) = \sqrt{x}$.
22. $f : \mathbb{Z} \to \mathbb{R}$ defined by $f(x) = \begin{cases} \sqrt{x} & x \geq 0 \\ \sqrt{-x} & x < 0. \end{cases}$

In exercises 23–26, state all functions from the set A to the set B. Identify whether or not each function is one-to-one or onto.

23. $A = \{a\}$ and $B = \{1,\ 2\}$.
24. $A = \{a,\ b\}$ and $B = \{u,\ v\}$.
25. $A = \{1,\ 2,\ 3\}$ and $B = \{u,\ v\}$.
26. $A = \{a,\ b\}$ and $B = \{2,\ 4,\ 6\}$.

In exercises 27–34, define a one-to-one correspondence between the given pairs of sets, or explain why such a function does not exist. Hint: A one-to-one correspondence exists for the sets in 28, 29, 32, and 33.

27. $\{a, b, c, d\}$ and $\{1, 2\}$

28. $\{4, 5\}$ and $\{a, b\}$

29. $\mathbb{N}$ and $\mathbb{N} \cup \{a, b\}$

30. $\mathbb{Z}$ and $\mathbb{R}$

31. $\mathbb{Q}$ and $\mathbb{P}(\mathbb{N})$

32. $\mathbb{Z}$ and $\mathbb{N}$

33. real intervals $(0, 4)$ and $(0, 2)$

34. $\emptyset$ and $\{\emptyset\}$

In exercises 35–38, each sequence of real numbers might result from the (incorrect) assumption that $[0, 1)$ is countable (as in the proof of the uncountability of the reals for theorem 4.5.2). Identify the first few decimal digits of the real number r that would result from applying Cantor's second diagonalization method to each sequence.

35. $a_1 = 0.02345678 \ldots$
$a_2 = 0.22222222 \ldots$
$a_3 = 0.09009000 \ldots$
$a_4 = 0.14159261 \ldots$
$a_5 = 0.23580347 \ldots$
$a_6 = 0.77775555 \ldots$
$\vdots$

36. $a_1 = 0.65412389 \ldots$
$a_2 = 0.00111100 \ldots$
$a_3 = 0.98976649 \ldots$
$a_4 = 0.14159261 \ldots$
$a_5 = 0.90990999 \ldots$
$a_6 = 0.11149131 \ldots$
$\vdots$

37. $a_1 = 0.89114123 \ldots$
$a_2 = 0.00992211 \ldots$
$a_3 = 0.49149766 \ldots$
$a_4 = 0.61901592 \ldots$
$a_5 = 0.00339999 \ldots$
$a_6 = 0.31651491 \ldots$
$a_7 = 0.99887766 \ldots$
$a_8 = 0.10310310 \ldots$
$\vdots$

38. $a_1 = 0.23891114 \ldots$
$a_2 = 0.10109999 \ldots$
$a_3 = 0.66055504 \ldots$
$a_4 = 0.92619897 \ldots$
$a_5 = 0.25003322 \ldots$
$a_6 = 0.91316541 \ldots$
$a_7 = 0.11111111 \ldots$
$a_8 = 0.10000000 \ldots$
$\vdots$

In exercises 39–42, each sequence of sets might result from the (incorrect) assumption that $\mathbb{P}(\mathbb{N})$ is countable via a one-to-one correspondence $g : \mathbb{N} \to \mathbb{P}(\mathbb{N})$ (as in the proof of Cantor's theorem and example 4.5.8). Identify the first few elements of the set D that would result from applying Cantor's second diagonalization method to each sequence.

39. $g(1) = \{1, 2, 3, 4, \ldots\}$
$g(2) = \{1, 3, 5, 7, \ldots\}$
$g(3) = \{2, 4\}$
$g(4) = \{2, 3, 5, 7, \ldots\}$
$\vdots$

40. $g(1) = \{2, 4, 6, 8, \ldots\}$
$g(2) = \{5, 10, 15, 20, \ldots\}$
$g(3) = \{3\}$
$g(4) = \{4, 8, 12, 16, \ldots\}$
$\vdots$

41. $g(1) = \{1\}$
     $g(2) = \{3\}$
     $g(3) = \{5\}$
     $g(4) = \{2, 4, 6, 8, \ldots\}$
     $g(5) = \{1, 4, 7, 10, \ldots\}$
     $g(6) = \{4\}$
     $\vdots$

42. $g(1) = \{2, 3, 5, 7, \ldots\}$
     $g(2) = \{1\}$
     $g(3) = \{8, 16, 24, 32, \ldots\}$
     $g(4) = \{2\}$
     $g(5) = \{1, 2, 3, 4, 5, \ldots\}$
     $g(6) = \{3\}$
     $\vdots$

In exercises 43–52, prove each mathematical statement.

43. For sets A, B define $A \sim B$ when $|A| = |B|$. Prove that $\sim$ is an equivalence relation on sets.

44. If A and B are countable sets, then $A \times B = \{(a, b) : a \in A, b \in B\}$ is countable.
    Hint: Write $A \times B$ as a countable union, where the indexing is on the elements of A for sets of the form $\{(a, b) : b \in B\}$.

45. If $A \cup B$ is uncountable, then either A or B must be uncountable.
    Hint: Consider theorem 4.5.1.

46. The set of all infinite sequences of 0s and 1s is uncountable.
    Hint: Consider the proof of theorem 4.5.2 giving the uncountability of the reals.

47. Given a set A, the cardinality $|A| \geq 2$ iff there exists a one-to-one correspondence $f : A \rightarrow A$ that is not the identity function $f(x) = x$.

48. $|\mathbb{N} \times \mathbb{N}| = \omega$.
    Hint: Use Cantor's first diagonalization method as in the proof of theorem 4.5.1.

49. The function $f : \mathbb{N} \times \mathbb{N} \rightarrow \mathbb{N}$ defined by $f(m, n) = 2^{m-1} \cdot (2n - 1)$ is a one-to-one correspondence.
    Hint: When proving f onto, consider the prime power factorization of an arbitrary element from the range.

50. The function $f : \mathbb{N} \rightarrow \mathbb{Z}$ defined below is a one-to-one correspondence.

$$f(n) = \begin{cases} \dfrac{n}{2} & \text{if } n \text{ is even} \\ -\dfrac{n-1}{2} & \text{if } n \text{ is odd} \end{cases}$$

51. The function $f : (0, 1) \rightarrow \mathbb{R}$ defined below is a one-to-one correspondence.

$$f(x) = \begin{cases} \dfrac{x - \dfrac{1}{2}}{x} & \text{if } x \leq \dfrac{1}{2} \\ \dfrac{x - \dfrac{1}{2}}{1 - x} & \text{if } x > \dfrac{1}{2} \end{cases}$$

52. Every linear polynomial over $\mathbb{R}$ defines a one-to-one correspondence from $\mathbb{R}$ to $\mathbb{R}$.

In exercises 53–56, determine the power set $\mathbb{P}(A)$ of each set A, and state the cardinality of both A and $\mathbb{P}(A)$.

53. $A = \{\emptyset\}$.

54. $A = \{3, 5, 9\}$.

55. $A = \{w, x, y, z\}$.

56. $A = \{\text{car, bicycle, truck, bus}\}$.

In exercises 57–66, determine the cardinality of each set and its power set. For exercises 62 and 66, consider theorem 4.5.1 or exercise 45.

57. A set containing 6 elements.
58. $\mathbb{Z}$, the set of the integers.
59. $\mathbb{Q}$, the set of the rationals.
60. $\mathbb{R}$, the set of the reals.
61. $\mathbb{P}(\mathbb{R})$, the power set of the reals.
62. $\mathbb{R} \setminus \mathbb{Q}$, the set of the irrationals.

63. The set of linear polynomials over $\mathbb{Q}$.
64. The set of all polynomials over $\mathbb{Q}$.
65. The set of algebraic numbers over $\mathbb{Q}$. (Use 64.)
66. The set of transcendental numbers over $\mathbb{Q}$.

In exercises 67–70, prove each equality by giving an illustration of the corresponding informal Hilbert Hotel proof and by defining a one-to-one correspondence between appropriate representative sets.

67. $7 + \omega = \omega$.
68. $6 + \omega = \omega$.

69. $5\omega = \omega$.
70. $\omega + \omega = \omega$.

4.6   The Riemann Integral

Since ancient times, mathematicians, scientists, and engineers have struggled with the problem of computing the area enclosed by a curve on the plane. Many important practical questions are essentially area problems; the integral has played a vital role in answering these questions. Many aspects of our modern world are modeled and studied via integration, including an understanding of probability and statistics, the economic forecasts, building and monument design, manufacturing processes, space exploration, and transportation systems. In this section, we explore the theoretical ideas behind the familiar Riemann integral that is studied in introductory calculus courses. These courses focus practically on the computation of a given integral using the fundamental theorem of calculus and different integration techniques (integration by substitution, integration by parts, and so on). We assume a familiarity with such techniques and concentrate on developing a solid theoretical understanding of the definite integral.

This section's approach is based on the work of the German mathematician Bernhard Riemann in the early 1850s. Riemann was a doctoral student of Gauss at the University of Göttingen and made important contributions to many areas of mathematics, including real analysis (the foundational work on the integral that we study here), number theory (the Riemann hypothesis of chapter 3), geometry (Riemann's non-Euclidean geometry is a key element of Einstein's theory of relativity), and complex analysis (we study the Cauchy-Riemann equations in chapter 7). Sadly, Riemann died at the relatively young age of 39 as a result of complications arising from tuberculosis.

From your previous mathematics courses (including calculus), you know how to solve many area problems. Often such questions are phrased in terms of either polygons or functions, as in the following question.

Question 4.6.1 Determine a formula for the area enclosed by each region in the plane.

(a) A circle with diameter d.

(b) A rectangle with sides a and b.

(c) A trapezoid with parallel sides a and b and with perpendicular distance h between these sides.

(d) A regular pentagon with sides of length a.

(e) The triangular region bounded by $f(x) = x$, the x-axis, and the interval $[0, 3]$.

(f) The area on the plane bounded by $f(x) = 100 - 6x^2$, the x-axis, and the interval $[0, 4]$. Hint: This one is usually answered using calculus!

■

The development of the integral as a tool for computing area traces its roots back to the ancient Greeks. In the fourth century B.C.E., Eudoxus extended the work of his predecessors to articulate a precise "method of exhaustion" that measured areas by gradually expanding known areas to fill a given region. In the third century B.C.E., Archimedes made insightful use of the method of exhaustion. He computed the area of a regular polygon by "chopping" the figure into a collection of triangles and computing the area of the enclosed triangles; see figure 4.19 for an illustrative example. Archimedes extended this work to measure areas of parabolas (actually computing the sum of an infinite series!), to approximate π based on the area of a circle, and to measure areas, volumes, and surface areas of different geometric figures.

These computations become increasingly difficult when considering progressively more sophisticated geometric figures, and little progress was made on questions involving complicated figures for some 2,000 years. In the early 1600s, the Italian mathematician Bonaventura Cavalieri and the French mathematicians Gilles Persone de Roberval and Pierre de Fermat independently studied the idea of measuring area as sums of infinitely many, infinitely thin lines or rectangles. In the late 1600s, Leibniz also thought of areas as sums (introducing the notation "$\int$", which looks like the letter "S" in the word "Sum"), while Newton thought of areas in terms of antidifferentiation. Ultimately the work of Riemann in the 1850s provided a solid theoretical approach to understanding the computation of areas via definite integrals. Among Riemann's key insights was his choice to use Cauchy's notion of the limit (from the 1820s) in formulating a definition of the definite integral. When a region is bounded by a continuous function, computing a "Riemann sum" yields a close approximation of the actual area. Taking an appropriate limit yields the exact value of the bounded area.

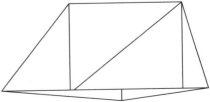

Figure 4.19 Archimedes' method for computing the area of a regular polygon

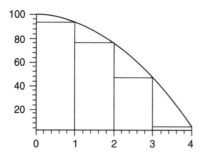

Figure 4.20 A four right-rectangle approximation of the bounded area

The Riemann sum uses rectangles to obtain the approximation, similar to the process illustrated in the next question.

Question 4.6.2   Approximate the area under the curve $f(x) = 100 - 6x^2$ between $a = 0$ and $b = 4$ using the four "right-rectangles" identified in figure 4.20.

(a) Compute the area of the four rectangles identified in figure 4.20.
Hint: We partition the interval $[0, 4]$ on the x-axis into subintervals of equal width $(b - a)/n$, where a and b are the endpoints of the interval and n is the number of rectangles (or subintervals). Figure 4.20 illustrates four subintervals of width $1 = (4 - 0)/4$. Each subinterval serves as the base for a rectangle whose height is determined by the subinterval's right endpoint x_i on the x-axis. In this case, the height of each rectangle is determined by the function $f(x) = 100 - 6x^2$; for example, the height of the leftmost rectangle in figure 4.20 is $f(1) = 100 - 6 \cdot 1^2 = 94$ and so the area of this rectangle is $1 \cdot 94 = 94$.

(b) What is the total area enclosed by the four rectangles?

(c) Similarly, compute the eight right-rectangle approximation of the bounded area.
∎

As you may recall from your study of calculus, using more and more rectangles leads to better and better approximations of the exact bounded area. One approach to defining the integral is to let the number of subintervals (or, equivalently, the number of rectangles) go to infinity. For any continuous function $f(x)$ on $[a, b]$, we could let n be the number of right-rectangles and find the area under f (as defined by the definite integral) to be the limit as n goes to infinity of the area enclosed by n right-rectangles. This approach works well when f is a continuous function, but fails in other more general settings. A more easily generalized approach involves taking sums that involve certain "upper bounds" and "lower bounds" for the function f on each subinterval. This technique follows from ideas Riemann developed in the 1850s and the French mathematician Jean Gaston Darboux refined in the 1870s.

Definition 4.6.1   *A real number M is an* **upper bound** *for a set $S \subseteq \mathbb{R}$ when $s \leq M$ for every $s \in S$; in this case, we say that S is* **bounded above***. If a real set S has an upper bound,* *then the* **supremum of S**, *denoted by* **sup S**, *is the "least upper bound" satisfying the following two properties:*

- sup *S is an upper bound for S,*
- *if M is any other upper bound for S, then* sup *S* < *M.*

Some sets have an upper bound and some do not. Also, for sets that are bounded above, the supremum may or may not be an element of the set. Consider the following example.

Example 4.6.1   The set $S = [0, \infty) = \{x : x \geq 0\}$ is not bounded above; for every positive real number M, we observe that $M + 1 \in S$ and $M + 1 \not\leq M$ (so M cannot be an upper bound). On the other hand, the sets $T = [0, 1]$ and $U = [0, 1)$ are both bounded above by any real number $M \geq 1$. You can see that the least of these upper bounds, which is the supremum of both these sets, is $\sup(T) = \sup(U) = 1$. The set T contains this supremum while U does not. Finally, the countable set of fractions $V = \{\frac{1}{2}, \frac{2}{3}, \frac{3}{4}, \frac{4}{5}, \dots\}$ is bounded above with supremum $\sup(V) = 1$. ■

Question 4.6.3   Give examples of sets with the following features.

(a) A set that is not bounded above.
(b) An uncountable set that is bounded above and contains its supremum of 4.
(c) An uncountable set that is bounded above, but does not contain its supremum of 5.
(d) A countable set that is bounded above and contains its supremum of 6.
(e) A countable set that is bounded above, but does not contain its supremum of 7. ■

As suggested in the discussion above, every nonempty set of real numbers that is bounded above has a supremum in the real numbers. Rather than proving this property of the reals, mathematicians have come to understand the existence of such least upper bounds as a defining, axiomatic property of the real numbers, similar in spirit to the principle of induction or the well-ordering principle discussed in chapter 3. This existence property is called the *axiom of completeness* and asserts that every nonempty set of real numbers that is bounded above has a supremum in the real numbers. Once the existence of a mathematical object is known, the question of uniqueness springs quite naturally to mind. In this case, the supremum of a given set bounded above is unique; the proof is left for exercise 55 at the end of this section. There are similar considerations for the existence and uniqueness of a set's lower bound and a greatest lower bound, which is known as the infimum. The next definition makes these terms precise.

Definition 4.6.2   *A real number m is a* **lower bound** *for a real set S when $s \geq m$ for every $s \in S$; in this case, we say that S is* **bounded below**. *If a real set S has a lower bound, then the* **infimum of S**, *denoted by* **inf S**, *is the "greatest lower bound" satisfying the following two properties:*

- inf *S is a lower bound for S,*
- *if m is any other lower bound for S, then* inf *S* > *m.*

Some sets of real numbers are bounded below while others are not; and sets that are bounded below may or may not contain their infimum. The existence of an infimum for a set bounded below follows from the axiom of completeness; the infimum of a set that is bounded below is also unique. For example, the set $S = \{1, \frac{1}{2}, \frac{1}{3}, \frac{1}{4}, \ldots\}$ is bounded below and has infimum inf $S = 0$, which is not an element of S. Similarly, the half-open interval $T = (0, 1]$ does not contain its infimum of 0, while the closed interval $U = [0, 1]$ contains its infimum of 0.

Question 4.6.4 Give examples of sets with the following features.

    (a) A set that is not bounded below.

    (b) An uncountable set that is bounded below and contains its infimum of 5.

    (c) An uncountable set that is bounded below, but does not contain its infimum of 4.

    (d) A countable set that is bounded below and contains its infimum of 3.

    (e) A countable set that is bounded below, but does not contain its infimum of 2.

                                                                  ■

Sets that are both bounded above and bounded below are said simply to be *bounded*. A rigorous study of integrals requires the characterization of sup S and inf S in terms of their relative distance from the elements of the given set S. The next lemma makes this notion precise for suprema and is useful for proving several important results later in this section. The proof follows directly from the definition and is left for exercise 57 at the end of this section.

Lemma 4.6.1 *If $S \subseteq \mathbb{R}$ is bounded above, then $M = \sup S$ iff both*

    • *M is an upper bound of S, and*

    • *for every $\varepsilon > 0$, there exists $s \in S$ such that $s > M - \varepsilon$.*

Question 4.6.5 Following the model given in lemma 4.6.1, state the corresponding characterization of inf S for a given real set S.

                                                                  ■

We now turn our attention to developing Riemann's definition of the definite integral. Intuitively, the area of a given region on the plane is found by summing the areas of a set of rectangles that approximate the given region. Taking an appropriate limit corresponds to filling up the given region with more and more rectangles; the limiting process gives the exact area. Riemann's great insight was to state these informal ideas in terms of the rigorous definitions discussed below.

Riemann's definition finds the area (which is denoted as the integral $\int_a^b f(x)\, dx$) of the region bounded by the graph of a function $f(x)$, the x-axis, and the vertical lines $x = a$ and $x = b$. The first step is to partition (or divide) the interval $[a, b]$ into n subintervals. In simple computations, these subintervals are often taken to have uniform length, but this condition is not strictly necessary. The subintervals form the bases of approximating rectangles, and so the base lengths needed for the corresponding area computations (area = base · height) are naturally the lengths of these subintervals. We state the definition of a partition and a refinement (needed for increasing the number of rectangles) and then we consider how to determine the rectangles' heights in this setting.

Definition 4.6.3 *A **partition** of a closed and bounded interval $[a, b] \subseteq \mathbb{R}$ is a finite set of reals $P = \{x_0, x_1, \ldots, x_n\}$ with $a = x_0 < x_1 < x_2 < \ldots < x_n = b$. If P and Q are two partitions of $[a, b]$, then Q is a **refinement** of P when $P \subset Q$.*

We intuitively think of a refinement as adding numbers to a partition, and so each original subinterval is either preserved or broken up into a finite number of smaller subintervals.

Example 4.6.2 We examine the interval $[a, b] = [1, 7]$. The set $P = \{1, 3, 5, 6, 7\}$ is a partition of $[1, 7]$. The set $Q = \{1, 2, 3, 4, 5, 6, 7\}$ is a refinement of P because Q is also a partition of $[1, 7]$ and $P \subset Q$. On the other hand $R = \{1, 2, 5, 6, 7\}$ is a partition of $[1, 7]$ that is not a refinement of P because $P \not\subset R$ (notice that $3 \in P$, but $3 \notin R$). Finally, $S = \{2, 5, 6, 7\}$ is not a partition of $[1, 7]$ because $1 \notin S$.

■

Question 4.6.6 Identify three distinct partitions P, Q, and R of $[0, 5]$ with the properties that Q is a refinement of P that contains at least one irrational number and R is not a refinement of P.

■

The Riemann definition of the integral considers both an upper and a lower approximation to the exact area beneath the curve. The two approximations use different rectangular heights (one upper and one lower) for each subinterval. A natural choice for these heights is the supremum and the infimum of the given function f on each subinterval. We consider only functions bounded on $[a, b]$; that is, functions for which there exists a real value M such that $|f(x)| \leq M$ for all $x \in [a, b]$.

Definition 4.6.4 *Let f be a bounded function defined on the interval $[a, b]$ and $P = \{x_0, x_1, \ldots, x_n\}$ be a partition of $[a, b]$. For each subinterval of $[a, b]$ of the form $[x_{i-1}, x_i]$, where $i = 1, 2, \ldots, n$, we define the following terms.*

- **The supremum of f on the ith subinterval** *is* $M_i(f) = \sup\{f(x) : x \in [x_{i-1}, x_i]\}$.
- **The infimum of f on the ith subinterval** *is* $m_i(f) = \inf\{f(x) : x \in [x_{i-1}, x_i]\}$.

Example 4.6.3 We identify the suprema $M_i(f)$ and the infima $m_i(f)$ for the function $f(x) = 3x^2 - 2x$ on the interval $[1, 7]$ under the partition $P = \{1, 3, 5, 6, 7\}$.

The function $f(x)$ is increasing on the interval $[1, 7]$. Therefore, each supremum $M_i(f)$ equals the value of the function at the right endpoint of the subinterval, and each infimum $m_i(f)$ equals the value at the left endpoint. For example, the supremum of f on $[x_0, x_1] = [1, 3]$ is $M_1(f) = f(3) = 3 \cdot 3^2 - 2 \cdot 3 = 21$, and the infimum of f on this subinterval is $m_1(f) = f(1) = 3 \cdot 1^2 - 2 \cdot 1 = 1$. Similar computations produce the following.

- For $[x_0, x_1] = [1, 3]$, $M_1(f) = 21$ and $m_1(f) = 1$.
- For $[x_1, x_2] = [3, 5]$, $M_2(f) = 65$ and $m_2(f) = 21$.
- For $[x_2, x_3] = [5, 6]$, $M_3(f) = 96$ and $m_3(f) = 65$.
- For $[x_3, x_4] = [6, 7]$, $M_4(f) = 133$ and $m_4(f) = 96$.

■

In more general settings, the suprema and infima may be the function's value at any point of the subinterval, or (for noncontinuous functions, they may not be realized as a function value). The examples in this section consider functions that lend themselves to relatively simple calculations of $m_i(f)$ and $M_i(f)$, and focus on developing a good theoretical understanding of these ideas.

Question 4.6.7   Identify the suprema $M_i(f)$ and the infima $m_i(f)$ for the function $f(x) = 3x^2 - 2x$ on the interval $[1, 7]$ under the partition $Q = \{\, 1,\ 2,\ 3,\ 4,\ 5,\ 6,\ 7 \,\}$. ∎

With this understanding of how to determine "bases" and "heights," we are now ready to define the upper and lower sums that enable us to approximate (and ultimately compute) the definite integral $\int_a^b f(x)dx$. Figure 4.21 provides a visual depiction of the use of upper sums and lower sums to approximate a given area.

Definition 4.6.5   *If f is a bounded function defined on the interval $[a, b]$ and $P = \{\, x_0,\ x_1, \ldots, x_n \,\}$ is a partition of $[a, b]$, then the **upper Riemann sum** (also called the **upper sum**) of f on $[a, b]$ with respect to P is*

$$U(f, P) \;=\; \sum_{i=1}^{n} M_i(f) \cdot (x_i - x_{i-1}).$$

*The **lower Riemann sum** (also called the **lower sum**) of f on $[a, b]$ with respect to P is*

$$L(f, P) \;=\; \sum_{i=1}^{n} m_i(f) \cdot (x_i - x_{i-1}).$$

Example 4.6.4   We compute the upper Riemann sum and the lower Riemann sum of $f(x) = 3x^2 - 2x$ on $[1, 7]$ with respect to partition $P = \{\, 1,\ 3,\ 5,\ 6,\ 7 \,\}$.

Example 4.6.3 computed the suprema $M_i(f)$ and the infima $m_i(f)$ for this function and partition. Based on these previous computations, the upper Riemann sum is

$$U(f,P) = \sum_{i=1}^{n} M_i(f) \cdot (x_i - x_{i-1})$$

$$= M_1(f) \cdot (x_1 - x_0) + M_2(f) \cdot (x_2 - x_1) + M_3(f) \cdot (x_3 - x_2) + M_4(f) \cdot (x_4 - x_3)$$

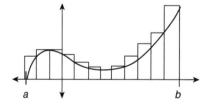
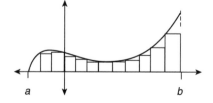

Upper sum geometry   Lower sum geometry

Figure 4.21   Computing upper sums and lower sums of a bounded area

$$= 21 \cdot (3-1) + 65 \cdot (5-3) + 96 \cdot (6-5) + 133 \cdot (7-6)$$

$$= 401.$$

Similarly, the lower Riemann sum is

$$
\begin{aligned}
L(f,P) &= \sum_{i=1}^{n} m_i(f) \cdot (x_i - x_{i-1}) \\
&= m_1(f) \cdot (x_1 - x_0) + m_2(f) \cdot (x_2 - x_1) + m_3(f) \cdot (x_3 - x_2) + m_4(f) \cdot (x_4 - x_3) \\
&= 1 \cdot (3-1) + 21 \cdot (5-3) + 65 \cdot (6-5) + 96 \cdot (7-6) \\
&= 205.
\end{aligned}
$$

■

Example 4.6.5 We compute the upper sum and the lower sum of $f(x) = 3x^2 - 2x$ on $[1,7]$ with respect to partition $Q = \{ 1, 2, 3, 4, 5, 6, 7 \}$, the refinement of P from question 4.6.7.

The calculations are similar to those in example 4.6.4, with the addition of two more subintervals contributing two additional terms to the sums (along with a corresponding modification of the subintervals that were split to create these new subintervals). The upper sum and lower sum in this setting are as follows.

$$
\begin{aligned}
U(f,Q) &= \sum_{i=1}^{n} M_i(f) \cdot (x_i - x_{i-1}) \\
&= 8 \cdot (2-1) + 21 \cdot (3-2) + 40 \cdot (4-3) + 65 \cdot (5-4) + 96 \cdot (6-5) + 133 \cdot (7-6) \\
&= 363
\end{aligned}
$$

$$
\begin{aligned}
L(f,Q) &= \sum_{i=1}^{n} m_i(f) \cdot (x_i - x_{i-1}) \\
&= 1 \cdot (2-1) + 8 \cdot (3-2) + 21 \cdot (4-3) + 40 \cdot (5-4) + 65 \cdot (6-5) + 96 \cdot (7-6) \\
&= 231
\end{aligned}
$$

■

Question 4.6.8 Find the upper sum and the lower sum for each function with respect to the partition $P = \{ 0, 1, 2, 3, 5 \}$ of $[0, 5]$.

(a) $f(x) = 4x^2 - 6$ (b) $f(x) = |x + 2|$

■

As suggested by their names, upper Riemann sums are upper bounds for the exact value of the corresponding integral, and lower Riemann sums are lower bounds. We obtain the exact value of the integral by determining the infimum of these upper bounds and the supremum of these lower bounds. When these infimum and supremum values are equal, Riemann defined this number as the value of the integral.

In this setting, the set of upper Riemann sums over which we take the infimum is the set of upper sums $U(f, P)$ over all possible partitions P of the given interval $[a, b]$. Similarly, the set of lower Riemann sums over which we take the supremum is the set of lower sums $L(f, P)$ over all possible partitions P of $[a, b]$. In many cases, we can think of the evaluation of the infimum and supremum in terms of a limit process, allowing us to consider taking more and more rectangles with smaller and smaller bases. The precise mathematical formulation of this intuitive description follows.

Definition 4.6.6   *If f is a bounded function defined on the interval $[a, b]$, then the **infimum of the upper Riemann sums** is*

$$U(f) = \inf\{\, U(f, P) : P \text{ is a partition of } [a, b] \,\}.$$

*Similarly the **supremum of the lower Riemann sums** is*

$$L(f) = \sup\{\, L(f, P) : P \text{ is a partition of } [a, b] \,\}.$$

The infimum of upper Riemann sums and the supremum of lower Riemann sums lead to a rigorous definition of integrability and the Riemann integral as follows.

Definition 4.6.7   *If f is a bounded function defined on the interval $[a, b]$, then f is **Riemann integrable on** $[a, b]$ exactly when $L(f) = U(f)$. When f is Riemann integrable **the Riemann integral of f on** $[a, b]$ is denoted by $\int_a^b f\, dx$ and is equal to $L(f) = U(f)$.*

We present a detailed computation of a Riemann integral using this definition. But as you might expect, computing $U(f)$ as the infimum over the upper sums with respect to *all* partitions of $[a, b]$ and $L(f)$ as the supremum over the lower sums with respect to *all* partitions of $[a, b]$ can be computationally difficult–even for many simple functions. This difficulty is one reason for the celebration of Newton's and Leibniz's genius in recognizing the fundamental theorem of calculus as providing a computationally straightforward approach to answering such questions. But a definitional approach can manage to compute Riemann integrals in light of the following result, which puts the computation in terms of partitions with certain features.

Theorem 4.6.1 Darboux's theorem   *Let f be a bounded function that is Riemann integrable on $[a, b]$ and $\{P_n : n \in \mathbb{N}\}$ be a sequence of partitions of $[a, b]$ such that the width of every subinterval of P_n is less than or equal to $1/n$. Then $U(f)$ and $L(f)$ can be expressed in terms of the sequence $\{P_n\}$, and both of the following equalities hold.*

$$U(f) = \inf\{\, U(f, P_n) : n \in \mathbb{N}\} = \lim_{n \to \infty} U(f, P_n)$$

$$L(f) = \sup\{L(f, P_n) : n \in \mathbb{N}\} = \lim_{n \to \infty} L(f, P_n)$$

We will soon see that any continuous function on an interval $[a, b]$ is integrable on $[a, b]$. Darboux's theorem therefore applies to any such continuous function f. The next example uses this fact.

Example 4.6.6   We evaluate $\displaystyle\int_0^1 x\, dx$ using the definition of the Riemann integral and Darboux's theorem.

Calculate $U(f) = \inf\{U(f, P) : P \text{ is a partition of } [a, b] \}$ using Darboux's theorem, which allows the choice of any sequence of partitions $\{P_n\}$ of $[0, 1]$ with the property that the width of each subinterval of P_n is less than or equal to $1/n$. For each $n \in \mathbb{N}$, we choose to define partitions that have subintervals of equal length $1/n$:

$$P_n = \left\{ 0, \frac{1}{n}, \frac{2}{n}, \dots, \frac{n-1}{n}, 1 \right\} = \left\{ \frac{i}{n} : 0 \le i \le n \right\}.$$

Now determine $U(f, P_n)$. Observe that $f(x) = x$ is increasing on $[0, 1]$, and so $M_i(f) = f(x_i) = f(i/n) = i/n$, where x_i is the right endpoint of the subinterval

$$[x_{i-1}, x_i] = \left[\frac{i-1}{n}, \frac{i}{n} \right].$$

Therefore, for $n \in \mathbb{N}$,

$$\begin{aligned}
U(f, P_n) &= \sum_{i=1}^{n} M_i(f) \cdot (x_i - x_{i-1}) = \sum_{i=1}^{n} x_i \cdot (x_i - x_{i-1}) = \sum_{i=1}^{n} \frac{i}{n} \cdot \frac{1}{n} \\
&= \frac{1}{n^2} \cdot \sum_{i=1}^{n} i = \frac{1}{n^2} \cdot \frac{n(n+1)}{2} \qquad \text{(see example 3.6 from section 3.6)} \\
&= \frac{n+1}{2n}.
\end{aligned}$$

Taking the limit as n goes to infinity (by Darboux's theorem),

$$U(f) = \lim_{n \to \infty} U(f, P_n) = \lim_{n \to \infty} \frac{n+1}{2n} = \frac{1}{2}.$$

By the continuity of f, this calculation shows that $\displaystyle\int_0^1 x \, dx = U(f) = \frac{1}{2}.$ ■

As we can see from example 4.6.6, computing Riemann integral using the definition is a complicated process, even with the help of Darboux's theorem. In many such cases, we are free to use the "right" sums when applying Darboux's theorem, as we did in the preceding example when the right sum happened to match the upper sum. When the Riemann integral does exist, Darboux's theorem allows us to use the following equation for direct computations:

$$\int_a^b f(x)dx = \lim_{n \to \infty} \sum_{i=1}^{(b-a)n} f\left(a + \frac{i}{n} \right) \cdot \frac{1}{n}$$

Also, the following summation formulas that are often useful for these computations:

$$\sum_{i=1}^{n} c = c \cdot n \qquad \sum_{i=1}^{n} i = \frac{n(n+1)}{2} \qquad \sum_{i=1}^{n} i^2 = \frac{n(n+1)(2n+1)}{6} \qquad \sum_{i=1}^{n} i^3 = \frac{n^2(n+1)^2}{4}$$

Question 4.6.9 Directly compute $L(f)$ for $f(x) = x$ on $[0, 1]$ using Darboux's theorem and the sequence of partitions $\{P_n\}$ of the form $P_n = \{i/n : 0 \le i \le n\}$ identified in example 4.6.6; you should find $L(f, P_n) = (n-1)/2n$.

■

Question   The function $f(x) = x^2 + 1$ is Riemann integrable on $[0, 2]$. Using the definition
4.6.10   of the Riemann integral and Darboux's theorem, evaluate $\int_0^2 x^2 + 1 \, dx$.   ∎

We have asserted that continuous functions are Riemann integrable on bounded
intervals $[a, b]$. This fact follows from the Riemann condition for integrability
and the Riemann–Lebesgue theorem, which we study in the rest of this section.
The next lemma works toward the Riemann condition for integrability, providing
information about the relationship between upper and lower Riemann sums with
respect to a partition and its refinements. While thinking about this result, you may
find it helpful to visualize rectangular approximations of a bounded finite area as in
figure 4.22.

Lemma   *If f is a bounded function on an interval $[a, b]$ and both P and Q are partitions of*
4.6.2   *$[a, b]$ with Q a refinement of P, then we have*

   (a) $U(f, Q) \geq L(f, Q)$,
   (b) $U(f, P) \geq U(f, Q)$,
   (c) $L(f, Q) \geq L(f, P)$.

Proof   We prove $U(f, Q) \geq L(f, Q)$ based on the definitions of upper and lower Riemann
sums. Since $M_i(f) \geq m_i(f)$, we have

$$U(f, Q) = \sum_{i=1}^{n} M_i(f) \cdot (x_i - x_{i-1}) \geq \sum_{i=1}^{n} m_i(f) \cdot (x_i - x_{i-1}) = L(f, Q).$$

The proof that $U(f, P) \geq U(f, Q)$ is by induction on the number of
points added to P to obtain the refinement Q. For the base case, assume
$P = \{x_0, \ x_1, \ x_2, \ldots, \ x_n\}$ and that the refinement Q is obtained by adding
one additional point x^* to P, so that $x_{k-1} < x^* < x_k$ for some $1 \leq k \leq n$ and
$Q = \{x_0, \ldots, \ x_{k-1}, \ x^*, \ x_k, \ldots, \ x_n\}$. Now focus on the intervals $[x_{k-1}, x_k]$,
$[x_{k-1}, x^*]$, and $[x^*, x_k]$, along with the corresponding suprema of f on these
intervals. In particular, define

$$M_1^*(f) \ = \ \sup\{f(x) : x \in [x_{k-1}, x^*]\} \text{ and } M_2^*(f) \ = \ \sup\{f(x) : x \in [x^*, x_k]\}.$$

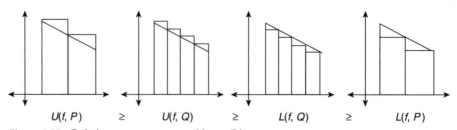

$U(f, P) \qquad \geq \qquad U(f, Q) \qquad \geq \qquad L(f, Q) \qquad \geq \qquad L(f, P)$

Figure 4.22   Relations among upper and lower Riemann sums

Because $M_k(f) = \sup\{f(x) : x \in [x_{k-1}, x_k]\}$, both $M_k(f) \geq M_1^*(f)$ and $M_k(f) \geq M_2^*(f)$. The definitions of $U(f, P)$ and $U(f, Q)$ imply

$$
\begin{aligned}
U(f,P) &= \sum_{i=1}^{n} M_i(f) \cdot (x_i - x_{i-1}) \\[2mm]
&= M_k(f) \cdot (x_k - x_{k-1}) + \sum_{i=1, i \neq k}^{n} M_i(f) \cdot (x_i - x_{i-1}) \\[2mm]
&= M_k(f) \cdot (x^* - x_{k-1}) + M_k(f) \cdot (x_k - x^*) + \sum_{i=1, i \neq k}^{n} M_i(f) \cdot (x_i - x_{i-1}) \\[2mm]
&\geq M_1^*(f) \cdot (x^* - x_{k-1}) + M_2^*(f) \cdot (x_k - x^*) + \sum_{i=1, i \neq k}^{n} M_i(f) \cdot (x_i - x_{i-1}) \\[2mm]
&= U(f,Q).
\end{aligned}
$$

Therefore if Q is a refinement of P with one additional point, then $U(f, Q) \leq U(f, P)$. If Q is an arbitrary refinement of P, then we consider a sequence of partitions beginning with P and adding one additional point at a time until we obtain Q, and the corresponding application from the base case to obtain:

$$
P \quad \subset \quad P_1 \quad \subset \quad \cdots \quad \subset \quad P_n \quad \subset \quad Q, \text{ and so}
$$

$$
U(f, P) \quad \geq \quad U(f, P_1) \quad \geq \quad \cdots \quad \geq \quad U(f, P_n) \quad \geq \quad U(f, Q).
$$

The proof $L(f, Q) \geq L(f, P)$ is similar to the proof just given for $U(f, P) \geq U(f, Q)$ and is left for exercise 59 at the end of this section. ∎

Lemma 4.6.2 enables the proof of a characterization of Riemann integrability.

Theorem 4.6.2 The Riemann condition for integrability *A function f is Riemann integrable on $[a, b]$ iff for every $\varepsilon > 0$, there exists a partition P of $[a, b]$ such that*

$$
U(f, P) - L(f, P) < \varepsilon.
$$

Proof Assume f is Riemann integrable on $[a, b]$ and prove that for every $\varepsilon > 0$, there exists a partition P of $[a, b]$ such that $U(f, P) - L(f, P) < \varepsilon$. Working in this direction, let $\varepsilon > 0$. Since f is Riemann integrable, $L(f) = U(f)$. Now apply the properties of suprema and infima identified in lemma 4.6.1 and question 4.6.5. By definition, $L(f) = \sup\{L(f, P) : P \text{ is a partition of } [a, b]\}$, and so there exists a partition S of $[a, b]$ such that $L(f, S) > L(f) - \varepsilon/2$ (by lemma 4.6.1). Similarly, $U(f) = \sup\{U(f, P) : P \text{ is a partition of } [a, b]\}$, and so there exists a partition T of $[a, b]$ such that $U(f, T) < U(f) + \varepsilon/2$ (by question 4.6.5). We combine these two partitions of $[a, b]$ to obtain the desired partition $P = S \cup T$. Applying these inequalities and the results on partitions produces the following.

$$U(f,P) - L(f,P) \le U(f,T) - L(f,P) \qquad \text{P is a refinement of T and lemma 4.6.2}$$

$$\le U(f,T) - L(f,S) \qquad \text{P is a refinement of S and lemma 4.6.2}$$

$$< [U(f) + \frac{\varepsilon}{2}] - [L(f) - \frac{\varepsilon}{2}] \quad U(f,T) < U(f) + \frac{\varepsilon}{2}, L(f,S) > L(f) - \frac{\varepsilon}{2}$$

$$= \varepsilon \qquad \text{$L(f) = U(f)$ since f is integrable}$$

To prove the converse, assume there exists a partition P of $[a, b]$ such that $U(f, P) - L(f, P) < \varepsilon$ and prove that f is Riemann integrable (that is, $L(f) = U(f)$). First note that $U(f, P) \ge L(f, P)$ for every partition P as observed in lemma 4.6.2; therefore, we always have $U(f) \ge L(f)$. The inequality $U(f) \le L(f)$ results from the following.

$$
\begin{aligned}
U(f) &\le U(f,P) & \text{Definition of infimum} \\
&< L(f,P) + \varepsilon & U(f,P) - L(f,P) < \varepsilon \\
&\le L(f) + \varepsilon & \text{Definition of supremum}
\end{aligned}
$$

Since ε is arbitrarily small, we have $U(f) \le L(f)$, which together with the previous inequality shows $U(f) = L(f)$. Hence f is Riemann integrable.

∎

We now consider a second condition for Riemann integrability, known as the *Riemann–Lebesgue theorem*, which is one of the most famous theorems of analysis. This result is named in honor of Bernhard Riemann and of Henri Lebesgue, who determined its straightforward characterization of Riemann integrable functions. The modern advanced theory of the integral took an important step forward in the early 1900s when the French mathematician Henri Lebesgue successfully analyzed infinite series constructed from sine and cosine functions. Such "Fourier series" represent bounded functions f using an infinite series whose terms are obtained by individually integrating sine and cosine expressions in f. In his initial work, Fourier assumed that the integrals of these expressions exist for every bounded function, but Lebesgue's more careful analysis revealed that only bounded Riemann integrable functions have this property. As part of this study, Lebsegue proved the Riemann–Lebsegue theorem characterizing Riemann integrable functions, resolving important questions about the existence of Riemann integrals. His work also highlighted some limitations of the Riemann integral in terms of Fourier series.

These limitations motivated Lebesgue to consider alternative formulations of the definite integral. In 1901, he developed an insightful "theory of measure" in an immediately famous paper *Sur une généralisation de l'intégrale définie*. Lebesgue offered a new definition of the definite integral that naturally generalizes the Riemann integral; this "Lebesgue integral" agrees with the Riemann integral whenever the Riemann integral exists, but many highly discontinuous functions have a well-defined and meaningful Lebesgue integral (even when their Riemann integral is undefined). The impact of his work was immediate. Among other things, Lebesgue helped determine when a Fourier series correctly represents its corresponding function. In this way, Lebesgue's accomplishments not only provided a whole new theoretical perspective on the concept of integration, but also approached the integral in a fashion that perfectly matched the practical requirements of many applications.

A first step toward appreciating Lebesgue's work is learning the Riemann–Lebesgue theorem, which characterizes Riemann integrable functions in terms of the function's continuity. While the proof of this result is beyond the scope of this text and is left for later studies, we seek to understand the statement and use of this theorem. We begin with two basic definitions from measure theory.

Definition 4.6.8   *An **interval open cover** for a set $S \subseteq \mathbb{R}$ is a countable collection of open intervals $\{I_n = (a_n, b_n) : n \in \mathbb{N}\}$ such that $S \subseteq \bigcup_{n=1}^{\infty} I_n$.*

The same set may have many different interval open covers as illustrated in the following example and question.

Example 4.6.7   We consider the set of all real numbers in the interval $(0, 1)$. Three distinct interval open covers of $(0, 1)$ include

$$\{(0, 1)\}, \quad \left\{ \left(-1, \frac{1}{2}\right), \left(\frac{1}{3}, \frac{2}{3}\right), \left(\frac{1}{2}, 1\right) \right\}, \quad \text{and}$$

$$\left\{ \left(\frac{1}{3}, \frac{2}{3}\right), \left(\frac{1}{4}, \frac{3}{4}\right), \left(\frac{1}{5}, \frac{4}{5}\right), \ldots, \left(\frac{1}{n}, \frac{n-1}{n}\right), \ldots \right\}.$$

Two distinct interval open covers of the irrationals $\mathbb{R} \setminus \mathbb{Q}$ are given by

$$\{(n, n+1) : n \in \mathbb{Z}\} \quad \text{and} \quad \left\{ \left(\frac{n}{2}, \frac{n+1}{2}\right) : n \in \mathbb{Z} \right\}.$$

■

Question 4.6.11   If possible, find both a finite and a countably infinite interval open cover of each set.

(a) $\{2, 4\}$ (c) $\mathbb{N}$

(b) $(1, 2) \cup [17, 19]$ (d) the set of transcendental real numbers

■

The general theory of Lebesgue measure is usually presented in graduate courses. However, it is easy to calculate the Lebesgue measure of any interval, since it is simply that interval's length. Some sets have Lebesgue measure zero and play an important role in the Riemann–Lebesgue theorem. The next definition explains when a set S has Lebesgue measure equal to 0.

Definition 4.6.9   *The **Lebesgue measure** of an interval $I = (a, b)$ is denoted by $m(I)$ or $m(a, b)$ and is defined as $m(I) = m(a, b) = b - a$. A set $S \subseteq \mathbb{R}$ has **measure zero** if for every $\varepsilon > 0$, there exists an interval open cover $\{I_n : n \in \mathbb{N}\}$ for S such that both*

$$S \subset \bigcup_{n=1}^{\infty} I_n$$

and

$$\sum_{n=1}^{\infty} m(I_n) \leq \varepsilon.$$

Example 4.6.8   The measure of an interval is easily computed by subtracting the endpoints; for example, the measure of the interval $[0, 1]$ is $m[0, 1] = 1 - 0 = 1$, and the measure of the half-open interval $[2, 46)$ is $m[2, 46) = 46 - 2 = 44$. Many infinite sets have nonzero Lebesgue measure, including the reals, the irrationals, and the transcendental numbers.

Any finite set has measure zero; for example, we prove that $S = \{1, 2\}$ has measure zero. For a given value $\varepsilon > 0$, define an interval of width $\varepsilon/2$ around each point in S. In particular,

$$\{ (1 - \frac{\varepsilon}{4}, 1 + \frac{\varepsilon}{4}), \ (2 - \frac{\varepsilon}{4}, 2 + \frac{\varepsilon}{4}) \}$$

is an open cover of S, since

$$S = \{1, 2\} \subseteq \left(1 - \frac{\varepsilon}{4}, 1 + \frac{\varepsilon}{4}\right) \cup \left(2 - \frac{\varepsilon}{4}, 2 + \frac{\varepsilon}{4}\right).$$

This cover shows S has measure zero, since

$$m\left(1 - \frac{\varepsilon}{4}, 1 + \frac{\varepsilon}{4}\right) + m\left(2 - \frac{\varepsilon}{4}, 2 + \frac{\varepsilon}{4}\right) = \frac{\varepsilon}{2} + \frac{\varepsilon}{2} = \varepsilon.$$

∎

Question 4.6.12   Prove that any countably infinite set of points $\{x_n : n \in \mathbb{N}\}$ has measure zero. Hint: Let $\varepsilon > 0$ and consider the interval open cover consisting of

$$I_n = \left(x_n - \frac{\varepsilon}{2 \cdot 2^n}, x_n + \frac{\varepsilon}{2 \cdot 2^n}\right).$$

Prove that $\sum_{n=1}^{\infty} m(I_n) = \varepsilon$ using the geometric series formula with $r = 1/2$ (see exercise 14 in section 3.6).

∎

The Riemann–Lebesgue theorem links the integrability of a function f with the measure of its set of discontinuities.

Theorem 4.6.3   The Riemann–Lebesgue theorem *If a function f is defined and bounded on an interval $[a, b]$, then f is Riemann integrable on $[a, b]$ iff the set of points in $[a, b]$ where f is discontinuous has measure zero.*

Example 4.6.9   An immediate consequence of the Riemann–Lebesgue theorem is that the Riemann integral always exists when $f(x)$ is continuous and bounded (in this case, the set of discontinuities is the empty set, which has measure zero). For example, $f(x) = x$ and $f(x) = x^2 + 1$ are continuous on $\mathbb{R}$ and thus Riemann integrable over any finite interval of $\mathbb{R}$.

In contrast, the characteristic function of the rationals (defined below) is discontinuous at every point in $\mathbb{R}$. Since every interval $[a, b] \subseteq \mathbb{R}$ has nonzero measure $b - a$, this function is not Riemann integrable on any interval of $\mathbb{R}$.

$$f(x) = \begin{cases} 1 & \text{if } x \in \mathbb{Q} \\ 0 & \text{if } x \notin \mathbb{Q} \end{cases}$$

■

Question   Prove that the following functions are Riemann integrable on $[0, 1]$ based on a
4.6.13   discussion of the measure of the set of discontinuities of f.

(a) $f(x) = \sqrt{5 \sin^2 x + 2}$

(b) $f(x) = \ln \sqrt{x + 1}$

(c) $f(x) = \begin{cases} \sin(1/x) & \text{if } x \neq 0 \\ 0 & \text{if } x = 0 \end{cases}$

(d) $f(x) = \begin{cases} 1 & \text{if } x = 1/2^k \text{ for } k \in \mathbb{N} \\ 0 & \text{otherwise} \end{cases}$

■

4.6.1 Reading Questions for Section 4.6

1. State an intuitive description of the idea motivating the definition of the Riemann integral.
2. Define and give an example of an upper bound and a supremum.
3. Define and give an example of a lower bound and an infimum.
4. Define and give an example of a partition and a refinement of $[0, 5]$.
5. Define and give an example of the supremum $M_i(f)$ of f on the subinterval $[x_i, x_{i+1}]$. For your example, use a two subinterval partition of the interval.
6. Define the upper Riemann sum $U(f, P)$ and the lower Riemann sum $L(f, P)$ for a bounded function f with respect to a partition P of $[a, b]$.
7. Define the infimum of the upper Riemann sum $U(f)$ and the supremum of the lower Riemann sum $L(f)$ for a bounded function f.
8. State Darboux's theorem. How is this result helpful when studying integration?
9. State the Riemann condition for integrability.
10. Give an example of an interval open cover.
11. Define and give an example of a set of measure zero.
12. State the Riemann–Lebesgue theorem. How is this result helpful when studying integration?

4.6.2 Exercises for Section 4.6

In exercises 1-4, consider the area on the plane bounded by the given function $f(x)$ and the x-axis on the interval $[2, 5]$. Hand-plot and compute the 4 right-rectangle approximation of this bounded area.

1. $f(x) = x - 2$
2. $f(x) = 2x + 4$
3. $f(x) = 2(x - 3)^2 + 1$
4. $f(x) = x^2 + x$

In exercises 5–12, find the supremum and infimum for the following bounded sets S. In addition, let $\varepsilon = 0.001$ and find a value $s \in S$ as guaranteed by lemma 4.6.1 such that $s > \sup S - \varepsilon$, as well as a value $t \in S$ such that $t < \inf S + \varepsilon$.

5. $S = \{x : 1 < x \le 3\}$
6. $S = \{x : -1 \le x < 3\}$
7. $S = \{x : 0 < x \le 5\} \cup \{x : 10 < x < 15\}$
8. $S = \{x : 0 < x \le 5\} \cup \{-1, -\frac{1}{2}, -\frac{1}{3}\}$
9. $S = \{\dfrac{1}{2^n} : n \in \mathbb{N}\}$
10. $S = \{1 - \dfrac{1}{n} : n \in \mathbb{N}\}$
11. $S = \{1 + \dfrac{1}{n} : n \in \mathbb{N}\}$
12. $S = \{\dfrac{2n-1}{2n+1} : n \in \mathbb{N}\}$

In exercises 13–20, give an example of sets with the following features, or explain why such a set does not exist.

13. A bounded set with infimum -1 and supremum 4.
14. A bounded set with infimum 4 and supremum -1.
15. A bounded set without an infimum.
16. A set that is bounded above without an infimum.
17. An uncountable set that is bounded below and contains its infimum of 0.
18. An uncountable set that is bounded above, but does not contain its supremum of 4.
19. A countable set that is bounded above and contains its supremum of 3.
20. A countable set that is bounded below, but does not contain its infimum of 8.

In exercises 21–24, give an example of a partition P of the following interval containing 4 points and a refinement Q of P containing 6 points.

21. $[3, 7]$
22. $[-1, 11]$
23. $[0, 8]$
24. $[1, 2]$

In exercises 25–30, find the upper sum and the lower sum for each function with respect to the partition $P = \{0, 2, 3, 5\}$ of $[0, 5]$.

25. $f(x) = 4x + 1$
26. $f(x) = 2x + 4$
27. $f(x) = -3x^2 + 4$
28. $f(x) = (x - 1)^2$
29. $f(x) = \sqrt{3x + 2}$
30. $f(x) = \ln(x + 1)$

In exercises 31–42, find the value of $M_i(f, P)$ and $m_i(f, P)$ for partition $P = \{1, 3, 4, 6\}$ of $[1, 6]$. In addition, use the definition of the definite integral and Darboux's theorem to determine the exact area bounded by $f(x)$ and the x-axis on the interval $[1, 6]$.

31. $f(x) = 4$
32. $f(x) = 2$
33. $f(x) = -x + 6$
34. $f(x) = x + 2$
35. $f(x) = -3x - 4$
36. $f(x) = 3x + 5$
37. $f(x) = x^2$
38. $f(x) = x^2 + 2$

39. $f(x) = x^2 + 3x - 2$

40. $f(x) = 2(x - 3)^2 + 1$

41. $f(x) = x^3$

42. $f(x) = x^3 + 1$

In exercises 43–50, use the definition of the definite integral and Darboux's theorem to evaluate each integral.

43. $\displaystyle\int_{2}^{5} 4\,dx$

44. $\displaystyle\int_{2}^{2} dx$

45. $\displaystyle\int_{0}^{3} x\,dx$

46. $\displaystyle\int_{2}^{5} x - 1\,dx$

47. $\displaystyle\int_{0}^{2} -x^2\,dx$

48. $\displaystyle\int_{0}^{3} 2x^2 + 4\,dx$

49. $\displaystyle\int_{0}^{2} x^3\,dx$

50. $\displaystyle\int_{0}^{2} x^3 - 4\,dx$

In exercises 51–54, prove each function is or is not Riemann integrable on the given interval using the Riemann–Lebesgue theorem.

51. For $[a, b] = [1, 4]$, the function $f(x) = \begin{cases} \dfrac{x - 2}{x^2 - 7x + 10} & \text{if } x \neq 2 \\ 32 & \text{if } x = 2 \end{cases}$.

52. For $[a, b] = [2, 6]$, the function $f(x) = \begin{cases} 1 & \text{if } x \in [2, 3] \\ x^2 & \text{if } x \in (3, 4] \\ x - 3 & \text{if } x \in (4, 5] \\ \sqrt{\ln(x + 10)} & \text{if } x \in (5, 6], \end{cases}$.

53. For $[a, b] = [0, 1]$, the function $f(x) = \begin{cases} x - 1 & \text{if } x = 1/2^n \text{ for } n \in \mathbb{N} \\ x + 1, & \text{otherwise} \end{cases}$.

54. For $[a, b] = [0, 1]$, the function $f(x) = \begin{cases} 1 & \text{if } x = m/2n \text{ for } m, n \in \mathbb{Z} \\ 0 & \text{otherwise} \end{cases}$.

    Hint: Prove that f is discontinuous at every $x \in [0, 1]$, as in example 12 from section 4.3.

In exercises 55–59, prove each mathematical statement.

55. If a set $S \subseteq \mathbb{R}$ is bounded above, then $\sup S$ is unique.

56. If a set $S \subseteq \mathbb{R}$ is bounded below, then $\inf S$ is unique.

57. Prove lemma 4.6.1 using the definition of supremum and upper bounds.

58. Prove the lower bound-infimum version of lemma 4.6.1: If $S \subseteq \mathbb{R}$ is bounded below, then $m = \inf S$ iff both

    - m is a lower bound of S, and
    - for every $\varepsilon > 0$, there exists $t \in S$ such that $t < M + \varepsilon$.

59. Prove lemma 4.6.2(c): If f is a bounded function on an interval $[a, b]$ and both P and Q are partitions of $[a, b]$ with Q a refinement of P, then $L(f, Q) \geq L(f, P)$.

Exercises 60–66 together prove that a bounded function f that is Riemann integrable on $[a, b]$ has a set of discontinuities $S = \{x \in [a, b] : f \text{ is discontinuous at } x\}$ of measure zero. The statements in many of these exercises depend on the following definition.

If $f : D \to Y$ and $x \in D$, then the *oscillation of f at x* is

$$\operatorname{osc}(f, x) = \lim_{h \to 0} \sup\{|f(x^*) - f(x^{**})| : x^*, x^{**} \in (x - h, x + h) \cap D\}.$$

A function f is continuous at a domain point $x \in D$ iff the oscillation $\operatorname{osc}(f, x)$ of f at x is zero.

In exercises 60–66, prove each mathematical statement.

60. If f is a bounded function that is Riemann integrable on $[a, b]$, why is the set of discontinuities S equal to $\{ x \in [a, b] : \operatorname{osc}(f, x) > 0 \}$?

61. Prove that $S = \bigcup_{i=1}^{\infty} S_n$, where $S = \{ x \in [a, b] : \operatorname{osc}(f, x) \geq \frac{1}{n} \}$.

62. We prove that $m(S_n) < \varepsilon$ for any given $\varepsilon > 0$, and so (since ε can be arbitrarily small), we have $m\{ S_n \} = 0$. To begin this proof, suppose $n \in \mathbb{N}$ is fixed; how do we know we can find a partition $P = \{x_0, x_1, ...x_n\}$ such that $U(f, P) - L(f, P) < \varepsilon/2n$?

63. Examine the interval (x_{i-1}, x_i) for the partition P and assume (x_{i-1}, x_i) contains a point $a \in S_n$. Using the fact that

$$\limsup_{h \downarrow 0}\{|f(x) - f(y)| : x, y \in (a - h, a + h)\} \leq \sup\{|f(x) - f(y)| : x, y \in [x_i, x_{i-1}]\}$$

$$\leq \sup\{f(x) : x \in [x_i, x_{i-1}]\} - \inf\{f(x) : x \in [x_i, x_{i-1}]\},$$

prove that $1/n \leq M_i(f) - m_i(f)$.

64. Prove that $\varepsilon/2n > \sum_{i=1}^{n}[M_i(f) - m_i(f)] \cdot (x_i - x_{i-1}) \geq 1/n \sum (x_i - x_{i-1})$, where the sums are taken only over those subintervals with $(x_{i-1}, x_i) \cap S_n \neq \emptyset$. Conclude that $\varepsilon/2 > \sum (x_i - x_{i-1})$, where the sum is taken over these same subintervals.

65. Prove that all of the points in S_n (except for possibly the values $x_0, x_1, ..., x_n$) are contained in a selection of the partition's subintervals whose total measure adds up to less than $\varepsilon/2$. Say why these subintervals, taken along with the set of open intervals $(x_i - \varepsilon/4n, x_i + \varepsilon/4n)$, for $i = 0, 1, 2, ..., n - 1$ must form an open cover for the set S_n whose total measure is less than ε.

66. Conclude from exercise 65 that $m(S_n) = 0$ for all n. Use the fact that "any countable union of sets of measure zero is also a set of measure zero" to explain why $m(S) = 0$.

Exercises 67–69 consider the famous *Cantor set*. The Cantor set is the subset of the interval $[0, 1]$ formed by repeatedly removing open subintervals. First, remove the interval $(\frac{1}{3}, \frac{2}{3})$, which is the middle third of the interval $[0, 1]$. Then two intervals $([0, \frac{1}{3}]$ and $[\frac{2}{3}, 1)$ remain. Remove the open interval forming the middle third of both of these intervals, so that four intervals $([0, \frac{1}{9}], [\frac{2}{9}, \frac{1}{3}], [\frac{2}{3}, \frac{7}{9}],$ and $[\frac{8}{9}, 1])$ remain. Continue this process indefinitely, removing the open interval forming the middle third of each remaining interval. The resulting set is called the Cantor set.

In exercises 67–69, prove each mathematical statement to show that the Cantor set is an uncountable set of measure zero.

67. Each step removes the open interval forming the middle third of each remaining interval. After two steps, the measure of the sets removed is the

sum of the interval lengths: $\frac{1}{3} + 2 \cdot \frac{1}{9} = \frac{5}{9}$. Calculate the measure of the sets removed after three steps, and then show that the total measure of the sets removed in the infinite number of steps it takes to create the Cantor set equals

$$\sum_{n=1}^{\infty} \frac{1}{3} \left(\frac{2}{3} \right)^{n-1}.$$

68. The measure of the Cantor set is the measure of the interval [0, 1] minus the total measure calculated in exercise 67. Use this fact and the geometric series formula (see exercise 14 in section 3.6) to find the measure of the Cantor set.

69. Each element a in the Cantor set turns out to have a *ternary expansion*

$$a = \frac{b_1}{3} + \frac{b_1}{3^2} + \frac{b_1}{3^3} + ...,$$

where every b_n equals either 0 or 2 (a finite expansion that ends in a string of zeros can be used). We write such an expansion as $a = 0.\, b_1\, b_2\, b_3\, ...\, [3]$, where the bracketed "3" indicates the expansion is ternary instead of base 10. Use this fact to show that the Cantor set is uncountable, employing a proof by contradiction that is reminiscent of the proof of theorem 4.5.3 in section 4.5.

4.7　The Fundamental Theorem of Calculus

This section studies the interrelationship between differential and integral calculus as expressed by Newton's and Leibniz's brilliantly insightful fundamental theorem of calculus. The Scottish mathematician James Gregory, who was in regular correspondence with Newton, published the first statement and proof of the fundamental theorem. As discussed in section 4.4, an understanding of the derivative had essentially been developed by the mid-1600s through the work of Pierre de Fermat and Isaac Barrow. At the time of this development, Bonaventura Cavalieri, Gilles de Roberval, and Fermat studied the integral as a measure of bounded area, and Evangalista Torricelli and Barrow studied the integral in the context of objects moving with variable speed. By 1660 the mathematical world was ready for the contributions of Newton, Leibniz, and Gregory.

The fundamental theorem of calculus asserts that differentiation (finding the slope of a tangent line) and integration (finding the area under a curve) are inverse operations. After working through examples computing the complex algebra of upper and lower sums connected to Riemann integrals, we can readily appreciate the advantage of using antidifferentiation to evaluate them. The fundamental theorem of calculus was first articulated by Newton in 1666 as part of his study of moving objects. His work reached full fruition in 1687 with the publication of *Philosophiae naturalis principia mathematica*. The *Principia* is recognized as one of the greatest scientific books ever written; it used calculus to develop and articulate Newton's fresh approach to physics and astronomy. Working independently of Newton, Leibniz developed a complete version of calculus by 1675 from an analytic perspective; he published his results on differential calculus in 1684 in *Nova methodus pro maximus et minimus itemque*

Tangentibus and on integral calculus in 1686. Sadly, delays in publishing results and in delivery of correspondence, personal misunderstandings, and nationalistic pride all played a role in a bitter priority dispute over who should receive credit for the development of calculus. Contemporary mathematicians credit both Newton and Leibniz for their brilliant insight; they independently described how the seemingly disparate operations of differentiation and integration are inverses. At the same time, both Newton and Leibniz freely acknowledged their appreciation for the contributions predecessors and mentors made to their own work. As Newton wrote in a letter to Robert Hooke in 1676, "If I have seen further it is by standing on the shoulders of giants."

This section's goal is to state and prove the fundamental theorem of calculus. Naturally enough, the proof relies on specific properties of the derivative and the integral. And so we begin by isolating a few key results about differentiation, antidifferentiation, and integration. These results lead to the fundamental theorem, which will provide a deeper understanding of continuous functions and of the differential and integral operations, as well as a way to to evaluate definite integrals.

Using Riemann's definition of the integral as a computational tool has serious limitations. In addition to the complexity of the computations (as illustrated in section 4.6), the calculations are impossible without additional closed-form expressions for series formulas working with broad classes of functions. The following question highlights the difficulty.

Question 4.7.1   (a) Using partitions P_n with subintervals of uniform width $1/n$, state (but do not evaluate) a limit of upper sums $U(f, P_n)$ whose value is the area bounded by $f(x) = 1/(x + 1)$ and the x-axis on the interval $[0, 1]$.

(b) Using partitions P_n with subintervals of uniform width $1/n$, state (but do not evaluate) a limit of upper sums $U(f, P_n)$ whose value is the area bounded by $f(x) = \sin(x)$ and the x-axis on the interval $[0, 1]$. ∎

If you think carefully about the two limits identified in answer to question 4.7.1 (with an eye toward actually computing these limits to find the values of the integrals), you can readily see the apparent shortcoming in using the Riemann definition to evaluate a definite integral. In order to evaluate these limits, we would need a closed-form expression for the sums:

$$\sum_{i=1}^{n} \frac{1}{i+1} \qquad \text{and} \qquad \sum_{i=1}^{n} \sin\left(\frac{i}{n}\right).$$

A need for more and more series formulas snowballs as we consider other functions. For example, integrating fourth, fifth, or sixth degree polynomials would require series formulas for $\sum_{i=1}^{n} i^4$, $\sum_{i=1}^{n} i^5$, and $\sum_{i=1}^{n} i^6$. Hence mathematicians quickly realized the importance of developing a straightforward approach to computing definite integrals—the approach ultimately provided by the fundamental theorem of calculus.

Before considering the inverse relationship between integration and differentiation expressed by the fundamental theorem, we lay important groundwork by considering two differentiation topics. The first is the mean value theorem, which provides a theoretical foundation used in later proofs; the second is antidifferentiation, or the process of running differentiation backwards. These notions will lead to a proof of the fundamental theorem of calculus.

We start with the mean value theorem. In this context, the word "mean" is used as a technical term, not as a reference to anyone's experience or perception of this theorem. The term comes from its statistical usage, where "mean" indicates "average." The next question deals deals with a concept that motivates the idea of the mean value theorem (or average value theorem).

Question 4.7.2   If the average class grade on a midterm exam is 71%, did someone earn exactly a 71%? Give an example to support your answer.
■

From your experience with exams, you should realize that the answer to question 4.7.2 is "no." For example, perhaps two people took an exam and earned grades of 70% and 72%; the average score is 71%, but no one actually scored the average.

A startling fact, though, is that for continuous, differentiable functions, the answer to such a question is "yes"; for such functions the average rate of change must always occur. The mean value theorem claims that the average rate of change over an interval must actually be equal to the instantaneous rate of change at some specific point in the interval. An instantaneous rate of change is mathematically expressed as the value of the derivative at a particular point, while an average rate of change is expressed by the slope of a secant line from a point $(a, f(a))$ to a point $(b, f(b))$. The mean value theorem says that the slope of any secant line must always equal the value of the derivative at some point.

Theorem 4.7.1 Mean value theorem   *If a function f is continuous on $[a, b]$ and differentiable on (a, b), then there exists a value $c \in (a, b)$ such that*

$$f'(c) = \frac{f(b) - f(a)}{b - a}.$$

We outline a proof of the mean value theorem after developing a preliminary result. Notice that the mean value theorem is an existence result, only guaranteeing the existence of a particular $c \in \mathbb{R}$ with certain properties. One the other hand, the mean value theorem is not a uniqueness result; there are functions f and intervals $[a, b]$ over which many different real values c satisfy $f'(c) = [f(b) - f(a)]/(b - a)$. Furthermore, the theorem (and its proof) is nonconstructive in that it does not determine a value for c, but only guarantees that c exists in the right interval with the right properties.

Finally, the mean value theorem describes a relationship between a secant line and a tangent line as illustrated in figure 4.23. If a function f is continuous on $[a, b]$ and differentiable on (a, b), then a secant line through the two points $(a, f(a))$ and $(b, f(b))$ must be parallel to *some* line tangent to f at a point in (a, b).

Before proving the mean value theorem, we first isolate two important properties of the derivative and then prove a special case of the mean value theorem–for horizontal lines–known as Rolle's theorem. The section then finally considers a question that provides a detailed outline of a complete proof of the mean value theorem. We begin with the extreme value theorem for continuous functions.

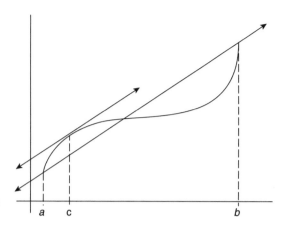

Figure 4.23 An illustration of the mean value theorem

Theorem 4.7.2 Extreme value theorem *If a function f is continuous on [a, b], then f has both an absolute maximum and an absolute minimum on [a, b].*

The extreme value theorem is traditionally studied in an introductory calculus course; the result guarantees the ability to find a continuous function's absolute extrema on any closed and bounded interval. The proof of the extreme value theorem relies on a notion of "compactness," which is beyond the scope of this text.

The next theorem is also a preliminary result needed to prove the mean value theorem; it says that a differentiable function's derivative is zero at a relative extrema. We state a partial proof of the result and leave further details for the exercises at the end of this section.

Theorem 4.7.3 *If a function f is continuous on [a, b] and differentiable on (a, b), and if f has a relative extrema at a point $c \in (a, b)$, then $f'(c) = 0$.*

Proof The proof proceeds under the assumption that $f(c)$ is a relative maximum. The proof is by way of contradiction; assume the function f has a relative maximum at $c \in (a, b)$ with $f'(c) \neq 0$. There are two cases to consider: $f'(c) > 0$ and $f'(c) < 0$.

First assume $f'(c) > 0$. Identify an interval around c such that the difference quotient from the alternative definition of the derivative is positive. Since

$$f'(c) = \lim_{x \to c} \frac{f(x) - f(c)}{x - c},$$

when we apply the definition of the limit with $\varepsilon = f'(c)$, there exists a value $\delta > 0$ such that $0 < |x - c| < \delta$ implies

$$\left| \frac{f(x) - f(c)}{x - c} - f'(c) \right| < \varepsilon = f'(c).$$

Hence

$$-f'(c) < \frac{f(x) - f(c)}{x - c} - f'(c) < f'(c) \quad \Rightarrow \quad 0 < \frac{f(x) - f(c)}{x - c} < 2f'(c).$$

If $(c - \delta, c + \delta)$ is not contained in (a, b), redefine δ as a sufficiently small positive value so that $(c - \delta, c + \delta) \subseteq (a, b)$. Then $[f(x) - f(c)]/(x - c) > 0$

whenever $x \in (c, c + \delta)$; for these x values, $f(x) - f(c) > 0$, and so $f(x) > f(c)$. But $f(c)$ is a relative maximum, and so $f(c) > f(x)$ must hold true for some open interval about c. This fact gives the desired contradiction; we conclude $f'(c) \not> 0$.

The proof that $f'(c) < 0$ leads to a similar contradiction; this case is left for you to answer in the next question. The proof of the case where $f(c)$ is a relative minimum is left for exercises 53–54 at the end of the section. Accepting these results, the theorem follows.

▪

Question 4.7.3 Using the following strategy, complete the proof of theorem 4.7.3 for the case where $f(c)$ is a relative maximum. Assume a function f is continuous on $[a, b]$ and differentiable on (a, b), and that f has a relative maximum at $c \in (a, b)$ with $f'(c) < 0$. Obtain a contradiction.

▪

We can now state and prove Rolle's theorem, which is a special case of the mean value theorem where the secant line in question is horizontal (and has a slope of zero). Rolle's theorem is named in honor of the English mathematician Michel Rolle who proved the result in 1691 using methods that the Danish mathematician Johann van Waveren Hudde developed. Ironically, Rolle personally regarded calculus as "a collection of ingenious fallacies" despite being best remembered for a result that plays a vital role in the modern proof of the fundamental theorem of calculus. Rolle also introduced the notation $\sqrt[n]{x}$ for the nth root of x.

Theorem 4.7.4 Rolle's theorem *If a function $f(x)$ is continuous on $[a, b]$ and differentiable on (a, b), and if $f(a) = f(b)$, then there exists $c \in [a, b]$ such that $f'(c) = 0$.*

Proof   If f is a constant function, then $f'(c) = 0$ for every $c \in [a, b]$; in this case any point in (a, b) may be chosen to satisfy the theorem's conclusion. Hence assume f is not constant. Because it is continuous on $[a, b]$, f has both an absolute maximum and an absolute minimum on $[a, b]$ by the extreme value theorem. Applying the assumptions that f is not constant and $f(a) = f(b)$, the endpoints can be at most one of these maximum and minimum values. Therefore, there exists $c \in (a, b)$ such that either f has its maximum at c or f has its minimum at c. Furthermore, because f is differentiable and $c \in (a, b)$, this extremum must be a relative maximum or minimum. Theorem 4.7.3 then implies $f'(c) = 0$.

▪

Rolle's theorem leads to the proof of the mean value theorem. The first proof of this result was given by Lagrange. The following sequence of questions outlines highlights of the proof, inviting you to fill in all necessary details.

Question 4.7.4   The following steps outline a proof of the mean value theorem. The proof's strategy is to define a new function g in terms of the given function f:

$$g(x) = f(x) - \left[\frac{f(b) - f(a)}{b - a}\right] x.$$

Prove that g satisfies the hypotheses of Rolle's theorem and then use the conclusion of Rolle's theorem to obtain the desired conclusion of the mean value theorem as follows.

(a) Prove that g is continuous on (a, b) using the continuity of f on $[a, b]$ and theorem 4.3.5 from section 4.3 (which details the properties of continuous functions). Extend the continuity of g to the closed interval $[a, b]$ by directly verifying that $\lim_{x \to a+} g(x) = g(a)$ and $\lim_{x \to b-} g(x) = g(b)$.

(b) Prove that g is differentiable on (a, b) by computing $g'(x)$ in terms of $f'(x)$.

(c) Finish the verification that g satisfies all conditions of Rolle's theorem: use the definition of g and the assumptions about f in the statement of the mean value theorem to prove $g(a) = g(b)$.

(d) Apply Rolle's theorem to g to obtain a value $c \in (a, b)$ such that $g'(c) = 0$. What does this fact about g say about f?

(e) Based on your answers to parts (a)–(d), write a proof of the mean value theorem, using complete sentences and appropriate supportive computations.

■

We will use the mean value theorem to prove the fundamental theorem of calculus. As a means of solidifying an understanding of the mean value theorem, we consider the process of actually finding the value c whose existence the theorem guarantees. In this case, the values for c may be found by solving the equation $f'(c) = [f(b) - f(a)]/(b - a)$.

Example 4.7.1 For the function $f(x) = x^3 + x^2$ on the interval $[0, 1.5]$, we identify $c \in (0, 1.5)$ whose existence the mean value theorem guarantees.

First verify that the hypotheses of the mean value theorem are satisfied. The function f is a polynomial and is therefore continuous and differentiable on all of $\mathbb{R}$ (and therefore on the particular interval $[0, 1.5]$). According to the mean value theorem, there exists a value $c \in (0, 1.5)$ such that $f'(c)$ is equal to

$$\frac{f(1.5) - f(0)}{1.5 - 0} = \frac{(1.5)^3 + (1.5)^2 - 0^3 - 0^2}{1.5} = 3.75.$$

We therefore seek $c \in (0, 1.5)$ such that $f'(c) = 3.75$. Since $f'(x) = 3x^2 + 2x$, we need to solve the quadratic $3c^2 + 2c = 3.75$. The quadratic equation implies

$$c = \frac{-2 \pm \sqrt{4 + 45}}{6},$$

and therefore $c = \frac{5}{6}$ or $c = -\frac{3}{2}$. Since $-\frac{3}{2} \notin (0, 1.5) = (a, b)$, there is only one value $c = \frac{5}{6} \in (0, 1.5)$ that satisfies the mean value theorem.

■

After proving a theorem, mathematicians often investigate if all the theorem's assumptions are necessary. For example, does the mean value theorem really need to assume f is continuous on $[a, b]$ in order for the conclusion to hold? What about differentiability? Sometimes hypotheses can be weakened to yield a more widely applicable result; in other cases, counterexamples are identified that demonstrate the necessity of a given hypothesis. The next example demonstrates that the differentiation hypothesis of the mean value theorem is necessary.

Example 4.7.2 We consider a function f that is continuous on an interval $[a, b]$ but not differentiable at every point in (a, b), and show that f fails to satisfy the conclusion of the mean value theorem.

Toward this end, define $f(x) = 5 - |x|$ on $[a, b] = [-2, 2]$. The function f is continuous on $[-2, 2]$ and differentiable on $(-2, 0) \cup (0, 2)$, but f is not differentiable at $x = 0$. The slope of the secant line from $x = -2$ to $x = 2$ is

$$\frac{f(2) - f(-2)}{2 - (-2)} = \frac{3 - 3}{4} = 0.$$

But f does not have any horizontal tangent lines, and so $f'(c)$ never equals 0 for any given $c \in [-2, 2]$. The conclusion of the mean value theorem therefore fails to hold. The assumption that f be differentiable on the entire interval is a vital one. ■

We now consider the process of antidifferentiation: given a function f, find F so that $F'(x) = f(x)$. In the language of section 4.2, antidifferentiation is (up to the addition of an arbitrary constant) the inverse of differentiation. The following simple example illustrates this notion.

Example 4.7.3 For $f(x) = 2x$, we determine those functions $F(x)$ such that $\frac{d}{dx} F(x) = 2x$.

Consider the following derivatives to motivate the solution.

$$\frac{d}{dx}[x^2] = 2x \qquad \frac{d}{dx}[x^2 + 3] = 2x \qquad \frac{d}{dx}[x^2 - 5] = 2x$$

As suggested by these examples and learned in any calculus course, the addition of an arbitrary constant to $F(x)$ does not affect its derivative. If we "undo" a differentiation process that yields $f(x) = 2x$ as a derivative $F'(x) = 2x$, then we obtain a function of the form $F(x) = x^2 + C$, where C is an arbitrary constant. ■

Antidifferentiation is studied and used in several areas of mathematics, and so there are many standard notations used for it.

Definition 4.7.1 *A function $F(x)$ is an* **antiderivative** *of a function $f(x)$ when*

$$\frac{d}{dx} F(x) = f(x).$$

Notationally, the following are equivalent:

- *$F(x)$ is an antiderivative of $f(x)$,*
- *$\int f(x)\, dx = F(x)$,*
- *$F(x)$ is the* **general solution** *of the differential equation $\frac{dy}{dx} = f(x)$.*

The function F is called the **indefinite integral** *of $f(x)$.*

We recall some important observations about this definition of an antiderivative. First of all, the article "an" is important in the definition, since a function f can have many antiderivatives. But we will soon prove that each antiderivative differs only by a constant, and so we may refer to the *general antiderivative*. Whenever a reference is given to *the* antiderivative of a function, the meaning should be understood as implicitly referring to the general antiderivative. For example, the antiderivative of $2x$ is $x^2 + C$, where C is an arbitrary real constant.

Second, every differentiation rule can be "reversed" to yield a corresponding antidifferentiation rule. The following familiar examples illustrate the relationship between the two types of rules.

Differentiation rule	*Antidifferentiation rule*
$\frac{d}{dx}[x^2] = 2x$	$\int 2x\, dx = x^2 + C$
$\frac{d}{dx}[x^4] = 4x^3$	$\int 4x^3\, dx = x^4 + C$
$\frac{d}{dx}[\cos(x)] = -\sin(x)$	$\int -\sin(x)\, dx = \cos(x) + C$
$\frac{d}{dx}[\ln(x)] = \frac{1}{x}$	$\int \frac{1}{x}\, dx = \ln(x) + C$

A fluency with derivative rules for functions therefore results in a corresponding fluency with antiderivatives. Not all functions have easily determined (or even closed-form) antiderivatives, but the many cases that do provide a strong incentive to apply the fundamental theorem of calculus—it links finding areas under curves to the antiderivative. The next question will remind you of computational aspects of antidifferentiation.

Question 4.7.5 Evaluate each indefinite integral and solve each differential equation. Some of these exercises use basic antidifferentiation rules, while others involve more sophisticated techniques of integration learned in calculus.

(a) $f(x) = 5x^4 + 2 + \dfrac{1}{2\sqrt{x}}$

(b) $f(x) = e^x + \cos x$

(c) $f(x) = \sec(x)\tan(x)$

(d) $f(x) = \dfrac{1}{x} + \dfrac{1}{x+1}$

(e) $y' = x\sqrt{x^2 + 1}$

(f) $y' = e^x(x^2 + 1)$

∎

Question 4.7.6 Recall from calculus that the antiderivative of a product of two functions is not equal to the product of the antiderivatives of the functions. Verify that $\int x^2\, dx = \int x \cdot x\, dx \neq \int x\, dx \cdot \int x\, dx$ provides a counterexample to such a supposed "rule" for antidifferentiation.

∎

How do we know that any antiderivative $F(x)$ is unique up to a constant? In the context of example 4.7.3, how can we be sure that every antiderivative of $f(x) = 2x$ is of the form $F(x) = x^2 + C$? Could some algebraic combination of trigonometric, logarithmic, or exponential functions also have a derivative of $2x$? We resolve such questions using theoretical mathematics. In this case we prove the uniqueness of the general antiderivative (up to a constant), not only for the function $f(x) = 2x$, but for all continuous functions simultaneously. The proof of this result follows after first proving an intuitive statement about the derivative: a function that has a slope of zero everywhere is a horizontal line.

Theorem 4.7.5   *If a function f is continuous on [a, b] and differentiable on (a, b), and if f'(x) = 0 for every x ∈ (a, b), then f is constant on [a, b].*

Proof   For any given x in the interval $(a, b]$, show f is constant by showing $f(x) = f(a)$. The function f satisfies the assumptions of the mean value theorem. Applying the mean value theorem to f on the interval $[a, x]$, there exists a value $c \in (a, x)$ such that

$$f'(c) = \frac{f(x) - f(a)}{x - a}.$$

But $f'(c) = 0$, hence $f(x) = f(a)$.

■

This theorem leads to a proof of the next result, which describes the uniqueness of the general antiderivative.

Theorem 4.7.6   *If the functions F(x) and G(x) are continuous on [a, b] and differentiable on (a, b), and if F(x) is an antiderivative of f(x) on (a, b), then the following are equivalent.*

(1) *G(x) = F(x) + C for some C ∈ ℝ*
(2) *G(x) is an antiderivative of f(x)*

Proof   First prove (1) implies (2). Assume $G(x) = F(x) + C$ for some $C \in \mathbb{R}$ and show that $G(x)$ is an antiderivative of $f(x)$. Differentiating $G(x)$ using the differentiation rules from theorem 4.4.1 in section 4.4,

$$\frac{d}{dx}[G(x)] = \frac{d}{dx}[F(x) + C] = \frac{d}{dx}[F(x)] + \frac{d}{dx}[C] = f(x) + 0 = f(x).$$

Therefore, by definition, $G(x)$ is an antiderivative of $f(x)$.

Now prove (2) implies (1). Assume $G(x)$ is an antiderivative of $f(x)$ and show that $G(x) = F(x) + C$ for some $C \in \mathbb{R}$. Differentiating the function $G(x) - F(x)$,

$$\frac{d}{dx}[G(x) - F(x)] = \frac{d}{dx}[G(x)] - \frac{d}{dx}[F(x)] = f(x) - f(x) = 0.$$

theorem 4.7.5 now implies that $G(x) - F(x)$ is constant on $[a, b]$; in other words, $G(x) - F(x) = C$ for some $C \in \mathbb{R}$. The result follows.

■

As a final step before presenting the fundamental theorem, we state a handful of properties of the definite integral that are familiar from calculus. Their formal proofs follow from the definition of the definite integral and are left for exercises 57–59 at the end of this section.

Theorem 4.7.7   *If f is Riemann integrable and a, b, c, r ∈ ℝ, then the following identities hold.*

(a) $\displaystyle\int_a^b f\, dx = \int_a^c f\, dx + \int_c^b f\, dx.$

(b) $\displaystyle\int_a^a f\, dx = 0.$

(c) $\displaystyle\int_a^b r\, dx = r \cdot (b - a).$

We are now ready to state and prove the fundamental theorem of calculus. Traditionally, the fundamental theorem is written in two parts. The first part asserts that differentiation is the inverse operation of taking a definite integral; if $F(x)$ equals the area under a continuous function f between values a and x, then $F'(x) = f(x)$. In the theorem's statement, the independent variable x will appear as a limit of integration, and a "dummy" variable t will be used in the integrand. The second part of the fundamental theorem evaluates the definite integral of a given function f in terms of its antiderivative.

Theorem 4.7.8   The fundamental theorem of calculus

(a) *If a function f is continuous on $[a, b]$, then for every $x \in (a, b)$*

$$\frac{d}{dx}\left[\int_a^x f(t)\,dt\right] = f(x).$$

(b) *If a function f is continuous on $[a, b]$ and $F(x)$ is any antiderivative of $f(x)$, then*

$$\int_a^b f(x)\,dx = F(b) - F(a).$$

Proof of (a)   Apply the definition of the derivative to the function $F(x) = \displaystyle\int_a^x f(t)\,dt$:

$$\frac{d}{dx}F(x) = \lim_{h \to 0} \frac{F(x+h) - F(x)}{h}$$

$$= \lim_{h \to 0} \frac{\int_a^{x+h} f(t)\,dt - \int_a^x f(t)\,dt}{h}$$

$$= \lim_{h \to 0} \frac{\int_x^{x+h} f(t)\,dt}{h}.$$

Now identify upper and lower bounds on the integration term in this limit—consider upper and lower Riemann sums and then apply the squeeze theorem (see theorem 4.3.4 from section 4.3). The following inequalities consider only the right hand limit with $h > 0$; the left hand limit with $h < 0$ is similar.

An upper bound is found by applying the partition with only one interval $P = \{x, x+h\}$ to the interval $[x, x+h]$; compute the corresponding upper and lower Riemann sum.

$$U(f, P) = \sup\{f(t) : t \in [x, x+h]\} \cdot [(x+h) - x] = M_1(f) \cdot h$$

$$L(f, P) = \inf\{f(t) : t \in [x, x+h]\} \cdot [(x+h) - x] = m_1(f) \cdot h$$

From the definition of the definite integral, the following inequalities hold.

$$
\begin{array}{ccccc}
L(f, P) & \leq & \int_x^{x+h} f(t)\,dt & \leq & U(f, P) \\
m_1(f) \cdot h & \leq & \int_x^{x+h} f(t)\,dt & \leq & M_1(f) \cdot h \\
m_1(f) & \leq & \dfrac{\int_x^{x+h} f(t)\,dt}{h} & \leq & M_1(f)
\end{array}
$$

As h approaches 0, $M_1(f) = m_1(f) = f(x)$. Therefore, by the squeeze theorem for limits,

$$\frac{d}{dx}\left[\int_a^x f(t)\,dt\right] = \lim_{h\to 0}\frac{\int_x^{x+h}f(t)\,dt}{h} = f(x)$$

▪

Proof of (b) Assume f is continuous on $[a, b]$. Let $G(x) = \int_a^x f(t)\,dt$ be the antiderivative for f given in part (a). If $F(x)$ is any antiderivative of f, then theorem 4.7.6 implies $F(x) = G(x) + C$ for some $C \in \mathbb{R}$. Substituting a and b into F as specified,

$$F(b) - F(a) = (G(b) + C) - (G(a) + C) = G(b) - G(a)$$

$$= \int_a^b f(t)\,dt - \int_a^a f(t)\,dt = \int_a^b f(t)\,dt - 0 = \int_a^b f(t)\,dt.$$

▪

The next several examples and questions use the fundamental theorem of calculus to evaluate derivatives and definite integrals. Many computations should be quite familiar from calculus; their ease illustrates the importance of the theorem. When using the fundamental theorem to differentiate a function $F(x) = \int_a^x f(t)\,dt$ whose domain variable x appears as an upper limit of integration, the variable t acts only as a placeholder. Using the chain rule, we may also differentiate a composition of the form $F(g(x)) = \int_a^{g(x)} f(t)\,dt$, obtaining

$$\frac{d}{dx}\int_a^{g(x)} f(t)\,dt = f(g(x))\cdot g'(x).$$

Example 4.7.4   We use the fundamental theorem of calculus to differentiate the following functions.

- $\dfrac{d}{dx}\left[\displaystyle\int_0^x t + 2\,dt\right] = x + 2$

- $\dfrac{d}{dx}\left[\displaystyle\int_2^x t^2 + \cos(t)\,dt\right] = x^2 + \cos(x)$

- $\dfrac{d}{dx}\left[\displaystyle\int_0^{x^4} \sec(t)\,dt\right] = \sec(x^4)\cdot 4x^3$

- $\dfrac{d}{dx}\left[\displaystyle\int_x^3 t + \frac{1}{t}\,dt\right] = \dfrac{d}{dx}\left[-\displaystyle\int_3^x t + \frac{1}{t}\,dt\right] = -(x + 1/x)$

▪

Question 4.7.7   Use the fundamental theorem to differentiate each of the following functions.

(a) $\displaystyle\int_1^x t + t^2\,dt$

(b) $\displaystyle\int_\pi^x \ln t\,dt$

(c) $\displaystyle\int_0^{x^3+e^x} \sin t\,dt$

(d) $\displaystyle\int_{x^2}^3 t + e^t\,dt$

▪

Example 4.7.5   We determine the area bounded by $f(x) = 6x^2$ and the x-axis on $[0, 4]$.

Apply the fundamental theorem of calculus. Since $2x^3$ is an antiderivative of $6x^2$, the bounded area is equal to

$$\int_0^4 6x^2 \, dx \;=\; 2x^3 \, \Big]_0^4 \;=\; 2(4)^3 - 2(0)^3 \;=\; 128.$$

∎

Example 4.7.6   We determine the area bounded by $f(x) = \sin(x)$ and the x-axis on $[0, \pi]$.

Applying the fundamental theorem of calculus, this bounded area is equal to

$$\int_0^\pi \sin(x) \, dx \;=\; -\cos(x) \, \Big]_0^\pi \;=\; -\cos(\pi) - (-\cos(0)) \;=\; -(-1) - (-1) \;=\; 2.$$

∎

Question 4.7.8   Use the fundamental theorem of calculus to determine the area bounded by $f(x)$ and the x-axis on the given interval.

(a) $f(x) = 4x^3 + 1$ on $[0, 1]$ (b) $f(x) = 4x^3 + 1$ on $[-1, 0]$

∎

A final example illustrates the necessity of continuity in the hypotheses of the fundamental theorem of calculus. The example shows that discontinuous functions may not satisfy the conclusion of the fundamental theorem.

Example 4.7.7   We consider the function $f(x) = 1/x^2$ on the interval $[-1, 1]$. If the fundamental theorem applied, then the corresponding definite integral of f would be

$$\int_{-1}^1 \frac{1}{x^2} \, dx \;=\; -\frac{1}{x} \, \Big]_{-1}^1 \;=\; -1 - [-(-1)] \;=\; -2.$$

But this value cannot be interpreted as the definite integral; in other words, it is *not* the area between $f(x) = 1/x^2$ and the x-axis over the interval $[-1, 1]$. This region is unbounded, and the area is infinite. Since $f(x) = 1/x^2$ is discontinuous at $x = 0 \in [-1, 1]$, the fundamental theorem of calculus cannot be applied.

∎

4.7.1   Reading Questions for Section 4.7

1. State the mean value theorem.
2. State the extreme value theorem and illustrate it with an example.
3. State theorem 4.7.3. How is this result useful when studying functions?
4. State Rolle's theorem.
5. What function is studied in proving the mean value theorem as outlined in question 4.7.4?
6. Define and give an example of an antiderivative.
7. Name three different ways in which mathematicians refer to antiderivatives.
8. Discuss the uniqueness of antiderivatives. What does theorem 4.7.6 tell us?

9. State theorem 4.7.5. Sketch an example and explain how this result is used in the proof of theorem 4.7.6.

10. State and give an example of the properties of the definite integral identified in theorem 4.7.7.

11. State the fundamental theorem of calculus. How is this result useful?

12. True or false: Any function $f(x)$ on $[a, b]$ satisfies $\int_a^b f(x)\, dx = F(b) - F(a)$, where $F'(x) = f(x)$ on $[a, b]$. Explain your answer.

4.7.2   Exercises for Section 4.7

In exercises 1–4, apply Rolle's theorem and the mean value theorem to the continuous and differentiable function $f(x) = (x - 2)(x + 1) = x^2 - x - 2$.

1. Compute the derivative of f.
2. Identify c between the zeros of f satisfying Rolle's theorem.
3. Identify $c \in [0, 3]$ satisfying the mean value theorem.
4. Identify $c \in [0, 5]$ satisfying the mean value theorem.

In exercises 5–10, identify a constant c guaranteed to exist by applying either Rolle's theorem or the mean value theorem to each continuous, differentiable function f. Indicate when you are using Rolle's theorem.

5. $f(x) = 3 - \dfrac{2}{x}$ on $[2, 4]$

6. $f(x) = x^3 - 4x$ on $[0, 1]$

7. $f(x) = x^3 - 4x$ on $[-2, 2]$

8. $f(x) = 2x^4 - 14x^2 + 20$ on $[-5, 5]$

9. $f(x) = 3x^5 - 50x^3 + 135x$ on $[0, 2]$

10. $f(x) = 3x^5 - 10x^3 + 15x$ on $[-1, 1]$

In exercises 11–12, apply the mean value theorem in each scenario.

11. Suppose two patrol cars are sitting along side the highway six miles apart when a red sporty car drives by. The first police officer clocks the car at 50 mph, then three minutes later the second police officer clocks the car at 45 mph. Prove the driver deserves a huge speeding ticket for exceeding the speed limit (of 50 mph) at some time during the three minutes.

12. Two toll booths are 25 miles apart on a turnpike. If a car stops at the first tollbooth at 12:42 PM and the second at 12:58 later that afternoon, then what can you say about the car's speed somewhere in between the tollbooths?

In exercises 13–16, find the general solution of each differential equation.

13. $f'(x) = 3x^2 + e^x$

14. $f'(x) = \sin(2x) + x^2 \cdot \ln(x^3)$

15. $\dfrac{dy}{dx} = x \cdot e^x$

16. $\dfrac{dy}{dx} = e^x \cdot \cos x$

Exercises 17–20 consider the "particular solution" of a differentiable equation $dy/dx = f'(x)$ with a given initial condition $f(x) = y$. A particular solution identifies a specific $C \in \mathbb{R}$ for the general antiderivative by substituting the given initial condition into the general antiderivative and solving for C.

For exercises 17–20, recall that the general solution of the differentiable equation $f'(x) = \cos(x) + e^x$ is $f(x) = \sin(x) + e^x + C$. Identify the particular solution satisfying each initial condition.

17. $f(0) = 6$

18. $f(\pi) = e^\pi + 7$

19. $f(\pi/2) = 2$

20. $f(\ln 2) = 0$

In exercises 21–24, determine the particular solution satisfying each differential equation and initial condition.

21. $f'(x) = 4x^3 + 2e^x$ and $f(0) = 2$

22. $f'(x) = 5\cos(2\pi x) + x^2 \cdot \ln(2x^3)$ and $f(1) = 2$

23. $dy/dx = (x + 2) \cdot e^x$ and $f(\ln 2) = 1$

24. $dy/dx = x^2 \cdot e^x$ and $f(0) = 1$

In exercises 25–32, state the antiderivative of each function.

25. $f(x) = 4x^3 + 5x^{2/3}$

26. $f(x) = (2x^2 + 1)^2$

27. $f(x) = \dfrac{1}{8x^3} + \dfrac{1}{2x}$

28. $f(x) = \dfrac{1}{x(x + 1)}$

29. $f(x) = \dfrac{1}{e^{2x}} + \cos x$

30. $f(x) = \tan(x) + \cot(x)$

31. $f(x) = e^x \cos(e^x) + \sin(2x)$

32. $f(x) = [\ln(x)]^2 \dfrac{1}{x}$

In exercises 33–40, differentiate each function using the fundamental theorem of calculus.

33. $\displaystyle\int_3^x t + \frac{1}{t}\, dt$

34. $\displaystyle\int_\pi^x \csc(t^2) + \ln(t)\, dt$

35. $\displaystyle\int_x^6 t^2 + \cos(t)\, dt$

36. $\displaystyle\int_4^{4x} \sec(t)\, dt$

37. $\displaystyle\int_{x^5}^3 \sqrt{t} + e^t\, dt$

38. $\displaystyle\int_1^{\cos(x)+e^x} t^3 + \ln(t)\, dt$

39. $\displaystyle\int_e^{x^2+\ln(x)} e^t + t^8\, dt$

40. $\displaystyle\int_x^{2x} \frac{3}{t}\, dt$

In exercises 41–48, evaluate each definite integral using the fundamental theorem of calculus.

41. $\displaystyle\int_0^1 24x^5 + e^x\, dx$

42. $\displaystyle\int_1^3 x^3 + 1/x\, dx$

43. $\displaystyle\int_0^2 (2 + t)\sqrt{t}\, dt$

44. $\displaystyle\int_{-\pi/2}^{\pi/2} 2t + \cos(t)\, dt$

45. $\displaystyle\int_1^4 4e^x + \frac{1}{x^3\sqrt{x}}\, dx$

46. $\displaystyle\int_{-\pi}^\pi \cos(x)\, dx$

47. $\displaystyle\int_{-\pi}^\pi \sin(x)\, dx$

48. $\displaystyle\int_0^{\pi/4} \sec(x)\tan(x)\, dx$

In exercises 49–52, explain why each definite integral *cannot* be evaluated using the fundamental theorem of calculus.

49. $\displaystyle\int_0^4 \frac{1}{x}\, dx$

50. $\displaystyle\int_2^6 \frac{1}{x^2 - 4x}\, dx$

51. $\displaystyle\int_{-1}^1 \sqrt{x}\, dx$

52. $\displaystyle\int_0^3 \frac{1}{\sqrt{x-2}}\, dx$

In exercises 53–64, prove each mathematical statement.

53. Complete half of the proof of theorem 4.7.3 for a relative minimum. Assume a function f is continuous on $[a, b]$, differentiable on (a, b), and $c \in (a, b)$ is a relative minimum of f on $[a, b]$ with $f'(c) < 0$ and obtain a contradiction.

54. Complete half of the proof of theorem 4.7.3 for a relative minimum. Assume a function f is continuous on $[a, b]$, differentiable on (a, b), and $c \in (a, b)$ is a relative minimum of f on $[a, b]$ with $f'(c) > 0$ and obtain a contradiction.

55. If f is not continuous on $[a, b]$, then f may not satisfy the extreme value theorem.

56. If f is not continuous on $[a, b]$, then f may not satisfy the mean value theorem.

57. If $f(x) = Ax^2 + Bx + C$, then for any interval $[a, b]$, the real number $c = (b + a)/2$ satistfies the mean value theorem for f on $[a, b]$.

58. If $F'(x) = G'(x)$ on $[a, b]$, then $F(b) - F(a) = G(b) - G(a)$.

59. Theorem 4.7.7(a): if f is Riemann integrable and $a, b, c \in \mathbb{R}$, then

$$\int_a^b f\, dx \;=\; \int_a^c f\, dx \;+\; \int_c^b f\, dx.$$

60. Theorem 4.7.7(b): if f is Riemann integrable and $a \in \mathbb{R}$, then

$$\int_a^a f\, dx \;=\; 0.$$

61. Theorem 4.7.7(c): if f is Riemann integrable and $a, b, r \in \mathbb{R}$, then

$$\int_a^b r\, dx \;=\; r \cdot (b - a).$$

62. If f and g are Riemann integrable and $a, b \in \mathbb{R}$, then

$$\int_a^b f + g\, dx \;=\; \int_a^b f\, dx + \int_a^b g\, dx.$$

63. *Integration by substitution:* If $g'(x)$ is positive and continuous on $[c, d]$, f is continuous on $[a, b]$, and $a = g(c)$ and $b = g(d)$, then

$$\int_a^b f(x)\, dx \;=\; \int_c^d f(g(t))g'(t)\, dt.$$

Hint: Apply the chain rule

$$\frac{d}{dx}\{F(g(x))\} = F'(g(x)) \cdot g'(x).$$

64. *Integration by parts:* If the functions F and G have continuous derivatives on $[a, b]$, then

$$\int_a^b F(x)G'(x)\,dx \; = \; F(x)G(x)\Big]_a^b \; - \; \int_a^b F'(x)G(x)\,dx.$$

Hint: Apply the product rule $\dfrac{d}{dx}\{F(x)G(x)\} = F(x)G'(x) + F'(x)G(x)$.

In exercises 65–69, *disprove* each *false* mathematical statement by giving an appropriate counterexample.

65. If $f'(x) = 0$ for every x in the domain of f, then f is a constant function.
66. For any function f with continuous derivative,

$$\int \frac{1}{f(x)}\,dx \; = \; \ln|f(x)| + C.$$

67. If the functions f and g have continuous derivatives on $[a, b]$, then

$$\int_a^b \frac{f(x)}{g(x)}\,dx \; = \; \frac{\int_a^b f(x)\,dx}{\int_a^b g(x)\,dx}.$$

68. If f is a bounded function on $[a, b]$ and $F'(x) = f(x)$, then

$$\int_a^b f\,dx \; = \; F(b) - F(a).$$

69. If f and g are bounded functions, then

$$\int f(x) + g(x)\,dx \; = \; \int f(x)\,dx + \int g(x)\,dx + C.$$

Exercise 70 outlines the proof that π is irrational that is most often studied by contemporary mathematicians. This proof was first given by Ivan Niven in 1947. Niven's proof that π is irrational proceeds by contradiction. Assume $\pi = a/b$, where $a, b \in \mathbb{N}$. Now define the following polynomials (where $n \in \mathbb{N}$ is a fixed, but at the moment unspecified positive integer).

$$f(x) \; = \; \frac{x^n \cdot (a - bx)^n}{n!} \qquad F(x) \; = \; f(x) - f^{(2)}(x) + \cdots + (-1)^n f^{(2n)}(x)$$

Prove each mathematical statement about these polynomials.

(a) If $0 \le j < n$, then $f^{(j)}(0) = 0$ and $f^{(j)}(\pi) = 0$.
(b) If $n \le j$, then $f^{(j)}(0) \in \mathbb{Z}$ and $f^{(j)}(\pi) \in \mathbb{Z}$. Hint: $n! \cdot f(x)$ has integer coefficients.
(c) $F(\pi) + F(0)$ is an integer. Hint: Use parts (a) and (b) and the definition of F.
(d) $F + F'' = f$. Hint: Differentiate F and algebraically combine terms.
(e) $\dfrac{d}{dx}\left[F'(x) \cdot \sin(x) - F(x) \cdot \cos(x)\right] \; = \; f(x) \cdot \sin(x)$. Hint: Use the difference rule, the product rule, and part (d).
(f) $\displaystyle\int_0^\pi f(x) \cdot \sin(x)\,dx$ is an integer. Hint: Use the fundamental theorem of calculus and parts (c) and (e).

(g) If $0 \leq x \leq \pi$, then $0 \leq f(x) \leq \dfrac{\pi^n \cdot a^n}{n!}$, and so $0 \leq f(x) \cdot \sin(x) \leq \dfrac{\pi^n \cdot a^n}{n!}$.

(h) Conclude from part (g) that $\displaystyle\int_0^\pi f(x) \cdot \sin(x)\, dx$ is positive but arbitrarily small for sufficiently large n. Obtain a contradiction from part (f).

4.8 Application: Differential Equations

One of the first and most important applications of calculus was the development of a mathematical model for planetary motion. Since ancient times humans have stared at the stars and sought to understand the motion of the planets. The ancient Greek mathematician and astronomer Ptolemy detailed the most successful and enduring of the ancient mathematical models in his treatise the *Almagest*. When subsequent astronomers' physical observations conflicted with Ptolemy's description of planetary motion, additions and adjustments were made to Ptolemy's model to account for the discrepancies. By the seventeenth century, the result was a complicated description—a tangled mess of adjustments and unexplainable fine-tunings that was of limited use and questionable accuracy. It also certainly lacked the elegance that mathematicians often seek when modeling the physical world.

In the 1660s Sir Isaac Newton applied the newly developed mathematics of derivatives and integrals to produce a model for planetary motion, including a mathematical formula describing the path of the Earth around the Sun. In general, an object's future physical position can often be expressed in terms of derivatives or integrals of functions for known present quantities; in this way, the techniques of calculus can provide information about an object's future position and behavior. Newton's model for planetary motion as detailed in the *Principia* is recognized as one of the greatest accomplishments of science and remains an enduring witness to the power and potential of calculus.

In only a short time, mathematical models were developed describing scores of different physical phenomena, including the flight of cannonballs, the drape of a tent roof, the flow of heat across a metal surface, and the vibration of violin strings. These successes profoundly influenced the philosophical perspectives of western Europe. Mathematicians and a host of others optimistically sought models and insights into diverse physical phenomena, confident that mathematics could answer any and all questions of science. Like a clock's mechanisms turning and ticking at a predictable rate, mathematics seemed to show that the various parts of our universe (the Moon, the planets, comets, and other celestial objects) all move in a highly predictable manner.

The equations that form the heart of such physical models are often *differential equations*—ones that explicitly incorporate derivatives as terms in the equations. In this section, we categorize basic types of differential equations and discuss the general algorithms that mathematicians have developed for solving them. This section begins with the definition of a differential equation and some associated terms.

Definition 4.8.1 *A **differential equation** expresses a relationship between a function and its derivatives. The order of a differential equation is the order of the highest derivative* *that appears in the equation. A* **linear** *differential equation can be written in the form*

$$F_n(x)y^{(n)} + \cdots + F_1(x)y' + F_0(x)y = G(x),$$

for some $n \in \mathbb{N}$ where $y^{(n)}$ denotes the nth derivative of a function y with respect to x. A **nonlinear** *differential equation is one that is not linear.*

This introduction to the study focuses on first-order and second-order differential equations for functions of a single-variable. Such equations are sometimes referred to as *ordinary* differential equations, in contrast to *partial* differential equations for multivariable functions involving partial derivatives. Definition 4.8.1 refers to ordinary differential equations, since the derivatives are of a single-variable function. The following example illustrates several common types of differential equations.

Example 4.8.1   We state several differential equations, their respective orders, and state whether or not they are linear or nonlinear.

- First-order, linear differential equation: $y' + \frac{1}{x}y = 5e^x$.
- First-order, nonlinear differential equation: $y' + \sin y = 6$
- Second-order, linear differential equation: $y'' + e^x y = x^3$.
- Second-order, nonlinear differential equation: $yy'' + e^y[y']^3 = x$.
- Fourth-order, linear differential equation: $y^{(4)} + y = 0$.

■

Question 4.8.1   Identify the order and linearity of each differential equation.

(a) $y' = \cos x + e^x$

(b) $y' = x^2 y^3$

(c) $y'' = y^{(3)} \cos x + e^x$

(d) $\sin(x + y)y' = y$

■

For most of this section, we focus our attention on the study of first-order differential equations. As we will see, the general study of first-order equations quickly becomes quite complicated, even for equations that appear relatively simple at first blush. For example, the nonlinear first-order differential equation $[y']^2 + 1 = 0$ has no real solutions, since the derivative of a real function cannot equal the imaginary number $i = \sqrt{-1}$. Such differential equations immediately bring to light challenging questions about characterizing when a solution to a first order equation might exist, as well as what might happen if we were to allow for solutions that are complex-valued. Answering such questions has been a major area of focus for the last two centuries, and progress has been made that has had an impact on applications as varied as electricity and magnetism to the study of the subatomic world.

The study of antiderivatives in section 4.7 distinguished between the general antiderivative and a particular antiderivative for a given function. The general antiderivative's form includes the addition of an arbitrary constant. In a similar fashion, a given differential equation has both a general solution and particular solutions, where a particular solution both satisfies the differential equation and passes through some given point on the plane. The point is often identified by means of an *initial condition*;

an exercise that seeks a particular solution to a differential equation is called an *initial condition problem* or an *initial value problem*. The following definition explains these notions.

Definition 4.8.2   *A function $y = f(x)$ is a* **solution** *to a differential equation when the equation is satisfied (or true) when $f(x)$ and its derivatives are substituted for y and its derivatives. A* **general** *solution to a first-order equation includes an arbitrary constant of integration that expresses every possible solution to the equation. A* **particular** *solution satisfies both the equation and an* **initial condition** *$y(a) = f(a) = b$ for $a, b \in \mathbb{R}$.*

In these settings we often distinguish between verifying a given function is a solution of a differential equation and the process of actually identifying a solution for a differential equation. We briefly consider the process of verifying a solution and spend the rest of the section working on finding solutions.

Example 4.8.2   We confirm that $y(x) = \cos x + 4$ is a solution of the initial condition problem $y'' + y = 4$ and $y(\pi) = 3$.

Substituting y into the above equation we find

$$y'' + y = \frac{d^2}{dx^2}[\cos x + 4] + \cos x + 4 = -\cos x + 0 + \cos x + 4 = 4.$$

Therefore, the given function satisfies the differential equation. Furthermore, $y(\pi) = \cos \pi + 4 = -1 + 4 = 3$, and the given function satisfies the initial condition.

■

Question 4.8.2   Verify that $y(x) = \cos x + 4$ is a solution of each initial condition problem.

(a) $y'' - y + 4 = -2\cos x$ and $y(0) = 5$
(b) $y^{(4)} - y = -4$ and $y(\pi/3) = 4.5$

■

Most problems do not provide a particular solution to a given differential equation—they instead require an analysis leading to a solution. Sometimes the analysis follows an algorithmic approach that has been shown to solve various types of differential equations. Simple first-order differential equations can sometimes be solved by applying basic antidifferentiation techniques, as the next example illustrates.

Example 4.8.3   We find the particular solution of the initial condition problem $y' = 2x$ and $y(3) = 17$.

Finding the general solution of $y' = 2x$ is equivalent to determining the general antiderivative of $f(x) = 2x$, which is of the form $y = x^2 + C$. In short,

$$\int 2x\, dx = 2\frac{x^2}{2} + C = x^2 + C.$$

The given initial condition computes C as

$$17 = y(3) = 3^2 + C \qquad \Rightarrow \qquad 17 = 9 + C \qquad \Rightarrow \qquad 8 = C.$$

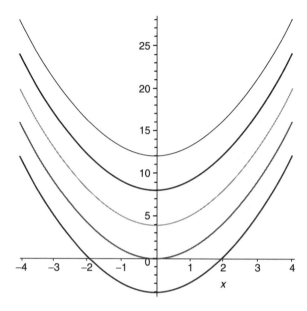

Figure 4.24 General and particular solutions for example 4.8.3

The particular solution is therefore $y = x^2 + 8$. As illustrated in figure 4.24, the addition of the arbitrary constant in the general solution may be graphically interpreted as translations of the parabola $y = x^2$ up and down the y-axis, and the particular solution $y = x^2 + 8$ is the parabola with vertex at $(0, 8)$.

∎

Question 4.8.3   Find the solution of the initial condition problem $y' = \cos x + e^x$ and $y(0) = 5$. ∎

Despite the relative ease of solving the differential equations given in example 4.8.3 and question 4.8.3, first-order (and higher-order) differential equations are often quite difficult to solve. Fortunately, mathematicians have successfully developed various techniques for solving many types of differential equations; a standard approach to finding solutions is to first determine if the given equation fits into a category with a known solution technique.

Definition 4.8.1 includes one such category—linear first-order differential equations. Every such equation is of the form $y' + F(x)y = G(x)$ and is often readily solved using a well-known, algorithmic process. The following theorem states the formula that gives the solution.

Theorem 4.8.1   *The general solution of a linear first-order differential equation $y' + F(x)y = G(x)$ is provided by the formula*

$$y = \left[e^{-\int F(x)\,dx} \right] \cdot \left[\int G(x) \cdot e^{\int F(x)\,dx}\,dx + C \right].$$

The derivation of the solution identified in theorem 4.8.1 is left for your later studies, perhaps in a course devoted exclusively to the study of differential equations. In order to apply the theorem practically, we need to be able to antidifferentiate $F(x)$ and the often more-complicated expression $\int G(x) \cdot e^{\int F(x)\,dx}\,dx$. But as seen in calculus courses, not every function $F(x)$ is elementary and antidifferentiated in closed form. The formula is useful when the integrals can be evaluated, and in these cases it identifies the solution. Finally, in cases where the integrals do not lend themselves to closed-form expressions, the formula can often be evaluated using numerical methods.

Example 4.8.4   We use theorem 4.8.1 to solve the initial condition problem

$$y' + (1/x)y = 5x \quad \text{and} \quad y(3) = 18.$$

Although unnecessary, we assume $x > 0$ to simplify the discussion.

The differential equation $y' + (1/x)y = 5x$ is a linear first-order differential equation with $F(x) = 1/x$ and $G(x) = 5x$. Applying the formula in theorem 4.8.1,

$$\int F(x)\,dx = \int \frac{1}{x}\,dx = \ln x.$$

In the end, any arbitrary constants in the solution are absorbed into the single constant value C that appears explicitly in the formula, and so this first integral calculation has chosen the arbitrary constant as 0 for simplicity. Now compute the right-hand expression from the formula in theorem 4.8.1:

$$\int G(x) \cdot e^{\int F(x)\,dx}\,dx = \int 5x \cdot e^{\ln x}\,dx = \int 5x^2\,dx = \frac{5}{3}x^3.$$

Putting the two pieces together, the general solution of $y' + (1/x)y = 5x$ is

$$y = e^{-\ln x} \cdot \left[\frac{5}{3}x^3 + C\right] = \frac{5}{3}x^2 + \frac{C}{x}.$$

Now use the given initial condition to compute C:

$$18 = y(3) = \frac{5}{3}3^2 + \frac{C}{3} \quad \Rightarrow \quad 18 = 15 + \frac{C}{3} \quad \Rightarrow \quad 9 = C.$$

The particular solution of the given initial value problem is therefore $y = \frac{5}{3}x^2 + \frac{9}{x}$. ∎

Question 4.8.4   Find the general solution of each linear first-order differential equation.

(a) $\dfrac{dy}{dx} + (1/x)y = 2x$

(b) $\dfrac{dy}{dx} - \dfrac{3}{x}y = 2x^3$ ∎

Another important category of first-order differential equations is referred to as *separable*. These equations can be algebraically manipulated so that the variables x and y are "separated" from one another, which enables a straightforward solution process. calculus courses often study separable differential equations after defining the natural logarithm and exponential functions; you may already be familiar with finding solutions to separable equations. The next definition and theorem details the solution to such an equation.

Definition 4.8.3 *A first-order differential equation is said to* **separable** *when the differential equation can be expressed in the form* $y' = F(x) \cdot G(y)$.

By writing the separable differential equations in terms of differentials as $dy/G(y) = F(x)dx$ and integrating both sides, the following theorem is immediately proven.

Theorem 4.8.2 *The general solution of a separable first-order differential equation* $y' = F(x) \cdot G(y)$ *is provided by the formula*

$$\int \frac{1}{G(y)}\, dy = \int F(x)\, dx.$$

Example 4.8.5 We use theorem 4.8.2 to solve the initial condition problem $y' = x^2 y^3$ and $y(3) = 0.5$.

The differential equation $y' = x^2 y^3$ is a separable first-order equation with $F(x) = x^2$ and $G(y) = y^3$. Hence

$$\frac{dy}{dx} = x^2 y^3 \qquad \Rightarrow \qquad \frac{dy}{y^3} = x^2 dx \qquad \Rightarrow \qquad \int y^{-3}\, dy = \int x^2\, dx.$$

Taking the antiderivative of both sides give the general solution as $-y^{-2}/2 = x^3/3 + C$. Now apply the given initial condition to compute C as

$$-\frac{1}{2 \cdot (0.5)^2} = \frac{27}{3} + C \qquad \Rightarrow \qquad -2 = 9 + C \qquad \Rightarrow \qquad -11 = C.$$

The particular solution of the given initial value problem is therefore

$$-\frac{1}{2y^2} = \frac{x^3}{3} - 11.$$

∎

Question 4.8.5 Find the general solution of each separable first-order differential equation.

(a) $\dfrac{dy}{dx} = x^3 y^3$

(b) $\dfrac{dy}{dx} = (x^2 + 1)(y^3 + y)$

∎

Not all first-order linear differential equations have closed-form solutions. Try finding a closed-form expression for the general solution to the separable first-order equation $dy/dx = e^{-x^2}$. The challenge presented by this fact only makes the study of differential equations more interesting—there are many different approaches to finding a solution, such as expressing the solution in the form of a power series, that are studied in a course dedicated to differential equations. Because the equations arise in many applied areas of mathematics and engineering, this study is not only interesting, it is useful.

First-order differential equations can be used to model many different physical and social phenomena, including the shape of hanging chains, the behavior of simple electric circuits, the decay of radioactive substances, the growth of money under various compounding schemes, and changing populations. The exercises at the end of section explore some of these applications, and many textbooks on differential equations describe these applications in great detail.

At the same time, many physical and social phenomena are properly described using higher-order equations. For example, the vibrations that occur in many physical systems are modeled by second-order differential equations. Newton recognized that a falling body is affected by gravity according to the second-order equation $d^2y/dt^2 = -g$, where $-g$ is the constant for gravitational acceleration, y is the body's vertical position, and t is time. In this setting, initial conditions are often expressed in terms of the body's initial position and initial velocity. When such an equation was solved, it provided a *deterministic* view of the physical world, since the equation determined the behavior of the object under study at any future time. The mathematics was seen as predicting the future—so long as any physical action (such as the orbit of a planet or a flight of a ball through the air) could be modeled in terms of a solvable differential equation, the mathematical solution would determine how the action would play out in any future time.

As mathematics has advanced into our present day understanding, the complexity of the universe has been shown to produce highly unpredictable, chaotic behavior of physical quantities, and the differential equations that model actions as simple as water dripping to those as complicated as weather turn out to lead to inherently chaotic solutions. Coupled with early twentieth century theory about physics on an atomic scale—primarily Niels Bohr's theory of quantum mechanics—this study of chaos indicates that many real-world features do not work in a deterministic framework. Instead they have chaotic properties that make future prediction—both on the micro and macro scale—difficult or impossible.

But the classical mathematics of the eighteenth century is useful in many controlled, nonchaotic settings. One such example is the motion of a clock pendulum. The mathematical model for the motion of a pendulum can be traced back to Galileo, who in the early 1600s first proved that the period of a pendulum's swing is independent of its amplitude. The Dutch mathematician Christian Huygens is often credited with the construction of the pendulum clock, which he described in a 1657 treatise titled *Horologium*. The Newtonian physics of the 1660s provided the right differential tools to mathematically model the motion of a pendulum.

Example 4.8.6   We develop a mathematical model for the motion of a clock pendulum. Assume the pendulum is swinging in a plane with a bob of mass m at the end of a pole (having insignificant mass) of length L (see figure 4.25). Furthermore, suppose the bob is drawn to one side so the pole makes an angle X with the vertical, and then released.

At any given point in time, the pendulum bob is therefore positioned along an arc at a distance D from the vertical, where the rod forms an angle x with the vertical. The relationship between these distances is expressed by the equation $D = L \cdot x$.

The law of physics known as the principle of conservation of energy ensures that the gain in kinetic energy attained by the pendulum moving along the arc is equal to its loss in potential energy. In this case, the kinetic energy is $\frac{1}{2}mv^2$, where v is the velocity of the pendulum bob along the arc. The loss in potential energy is $-gm(L\cos x - L\cos X)$, where $-g = -9.8$ m/s^2 is the gravitational constant.

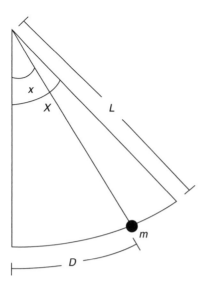

Figure 4.25 A clock pendulum with bob of mass m

The loss is calculated from the fact (which follows from unit circle analysis) that $L - L\cos x$ is the vertical distance of the pendulum bob above its lowest possible point.

Substituting $D = Lx$ and $v = dD/dt = L(dx/dt)$, the principle of conservation of energy implies the following equalities.

$$\frac{mv^2}{2} = -gm[L\cos x - L\cos X]$$

$$\frac{1}{2}\left[L\frac{dx}{dt}\right]^2 = -gL[\cos x - \cos X]$$

The last equality is thus a first-order, nonlinear differential equation that fully describes the motion of the pendulum.

∎

While any mathematician would feel pleased that a first-order differential equation will model the motion of a pendulum, the equation's nonlinearity makes the problem difficult to solve. The only way to proceed toward a solution is to manipulate the equation

$$\frac{1}{2}L\left[\frac{dx}{dt}\right]^2 = -g[\cos x - \cos X],$$

transforming it into a linear second-order differential equation. Despite the increase in order, such an equation is preferable since there exist well-known strategies for solving linear second-order equations. In fact, mathematicians often adopt this approach when working with nonlinear first-order equations. The next question outlines the steps to obtain the desired linear second-order equation and introduces an approximation technique that simplifies the resulting computations.

Question 4.8.6 Continuing the study of pendulum motion from example 4.8.6, we examine the nonlinear first-order differential equation

$$\frac{1}{2}L\left[\frac{dx}{dt}\right]^2 = -g[\cos x - \cos X].$$

(a) Implicitly differentiate both sides of the equation

$$\frac{1}{2}L\left[\frac{dx}{dt}\right]^2 = -g[\cos x - \cos X]$$

with respect to t to determine a second-order equation modeling pendulum motion. Note that in this equation x is a function of t while L, g, and X are constants. Simplify your result to obtain

$$L\frac{d^2x}{dt^2} = g \sin x.$$

(b) When the angle x is small, $\sin x$ is approximately equal to x. Substituting x for $\sin x$ in the linear second-order equation from part (a), demonstrate that an approximate model for pendulum motion is provided by

$$\frac{d^2x}{dt^2} - \frac{g}{L}x = 0.$$

(c) Verify that for all $C, D \in \mathbb{R}$, the following function is a solution of

$$\frac{d^2x}{dt^2} - \frac{g}{L}x = 0.$$

$$x = C \sin\left[\sqrt{\frac{-g}{L}}\, t\right] + D \cos\left[\sqrt{\frac{-g}{L}}\, t\right]$$

(d) When $t = 0$, both $x = X$ and $dx/dt = 0$. Using these initial conditions and the general solution from part (c), find an equation approximating the pendulum's position x at any time t.

■

Part (c) of question 4.8.6 asked you to verify a given general solution to the differential equation. It is likely unclear where that solution came from, but there exist standard techniques that solve linear second-order differential equations, and the general solution given in part (c) comes from those techniques. These methods are taught in any undergraduate course in differential equations; further details are left for your later studies.

We finish this section by studying a type of second-order equation that has become known as the *Hermite* differential equations. The nineteenth century French mathematician Charles Hermite made extremely significant contributions in many mathematical fields, even though as a young student he tested poorly and struggled with low grades. In 1873 he gave the first proof that e is transcendental. His work also included showing how to solve the general quintic polynomial using elliptic functions, and he determined properties of an important class of matrices now called Hermitian. The Hermite differential equations, a large class of equations that he investigated, turn out to have important applications in the study of physical processes called simple harmonic oscillators. We turn our attention to these equations here.

As defined below, the Hermite equations take the form $y'' - 2xy' + 2ny = 0$, where n is a nonnegative integer. These second-order differential equations play a key role in articulating a mathematical model for quantum mechanics. In addition, the Hermite equations will naturally lead us to discuss a central topic in advanced function theory—namely, the construction of a "space of functions." We will see that the equations give rise to a solution set of polynomials called Hermite polynomials. We first define these polynomials and then consider the corresponding Hermite differential equations.

Definition 4.8.4   *For every nonnegative integer $n = 0, 1, 2, \ldots$, the nth degree **Hermite polynomial** is defined by*

$$H_n(x) = (-1)^n e^{x^2} \frac{d^n}{dx^n}\left[e^{-x^2}\right].$$

Despite having exponential functions in its formula, every Hermite polynomial really *is* a polynomial; the nth derivative of e^{-x^2} includes itself as a factor, which then cancels with the formula's term e^{x^2}. The next example and question exhibit the first few Hermite polynomials.

Example 4.8.7   We compute the 0th and 1st Hermite polynomials.

The 0th derivative of a function is the function itself. The 0th Hermite polynomial is therefore a constant polynomial $H_0(x) = 1$:

$$H_0(x) = (-1)^0 e^{x^2} \frac{d^0}{dx^0}\left[e^{-x^2}\right] = 1 \cdot e^{x^2} \cdot e^{-x^2} = 1.$$

Similarly, a direct computation produces $H_1(x) = 2x$:

$$H_1(x) = (-1)^1 e^{x^2} \frac{d^1}{dx^1}\left[e^{-x^2}\right] = (-1) \cdot e^{x^2} \cdot (-2x)e^{-x^2} = 2x.$$

∎

Question 4.8.7   Show that the 2nd, 3rd, and 4th Hermite polynomials are as follows:

$$H_2(x) = 4x^2 - 2, \qquad H_3(x) = 8x^3 - 12x, \quad \text{and} \quad H_4(x) = 16x^4 - 48x^2 + 12.$$

∎

You can see that the formula for computing the nth Hermite polynomials can be applied at any finite value of n; the nth Hermite polynomial has degree n. The next theorem states one of the Hermite polynomials' more important features.

Theorem 4.8.3   *For every nonnegative integer $n = 0, 1, 2, \ldots$, the nth Hermite polynomial $H_n(x)$ is a solution of the **nth Hermite differential equation** $y'' - 2xy' + 2ny = 0$.*

A complete proof of theorem 4.8.3 is left for your later studies, but the next example and question verify a few special cases.

Example 4.8.8   We verify that the 2nd Hermite polynomial $H_2(x) = 4x^2 - 2$ is a solution of the 2nd Hermite differential equation $y'' - 2xy' + 4y = 0$.

We first compute the first and second derivatives of $H_2(x)$ to obtain $H_2'(x) = 8x$ and $H_2''(x) = 8$. Substituting into the 2nd Hermite differential equation we have

$$y'' - 2xy' + 4y = 8 - 2x \cdot 8x + 4 \cdot (4x^2 - 2) = 8 - 16x^2 + 16x^2 - 8 = 0.$$

■

Question 4.8.8 Verify that the 3rd Hermite polynomial $H_3(x) = 8x^3 - 12x$ is a solution of the 3rd Hermite differential equation $y'' - 2xy' + 6y = 0$.

■

The use of the article "a" in theorem 4.8.3 is important since the Hermite polynomials $H_n(x)$ are only one among many different solutions to the differential equations $y'' - 2xy' + 2ny = 0$. For example, $y = H_0(x) = 1$ is not the only solution to $y'' - 2xy' = 0$; for all $C, D \in \mathbb{R}$ the following functions are also solutions:

$$y = C + D \left[x + \sum_{j=1}^{\infty} \frac{(2)^j(1)(3) \cdots (2j-1)}{(2j+1)!} x^{2j+1} \right].$$

Similarly, the other Hermite differential equations have infinitely many solutions, but the Hermite polynomials are among the most important. First, they are the only polynomial solutions. In addition, they also satisfy an important integration property known as an *orthogonality relation*. In particular, when $n \neq m$ are nonnegative integers, then the following identity holds

$$\int_{-\infty}^{\infty} H_m(x) \cdot H_n(x) \, e^{-x^2} \, dx = 0.$$

Question 4.8.9 Verify that $H_0(x) = 1$ and $H_1(x) = 2x$ satisfy the orthogonality relation; that is, produce a direct computation proving that $\int_{-\infty}^{\infty} H_0(x) \cdot H_1(x) \, e^{-x^2} \, dx = 0$.

■

This orthogonality relation is intricately related to the set of all functions f for which the following integral is finite:

$$\int_{-\infty}^{\infty} |f(x)|^2 \, e^{-x^2} \, dx < \infty.$$

Such functions are said to belong to the "function space" denoted $L^2(-\infty, \infty)$. A rigorous definition of $L^2(-\infty, \infty)$ requires the Lebesgue integral and is beyond the scope of this text. However, it turns out that the Hermite polynomials are the basic building blocks for functions in that space, in the sense that for every function $f(x) \in L^2(-\infty, \infty)$, there exist $C_n \in \mathbb{R}$ such that

$$f(x) = \sum_{n=0}^{\infty} C_n H_n(x).$$

Formally, the Hermite polynomials are said to form a "basis" for the function space $L^2(-\infty, \infty)$. This fact is analogous to the identification of the ordered pairs $x_1 = (1, 0)$ and $x_2 = (0, 1)$ as forming a basis for $\mathbb{R}^2$, since every ordered pair $(a, b) \in \mathbb{R}^2$ can be represented as $(a, b) = a(1, 0) + b(0, 1)$. Just as the space $\mathbb{R}^2$ of two-dimensional points has a basis $\{x_1, x_2\}$, so does the space $L^2(-\infty, \infty)$ (defined by the finiteness of the above integral) have a basis consisting of the Hermite polynomials.

Since the early twentieth century, much important research in analysis has focused on the study of function spaces. These sets of functions can possess a great deal of inherent structure. The structure provides important insight into a function's behavior, as well as the relationship among these functions. For example, mathematicians speak of the "size" of a function in the same way they speak of the size of a number, and they refer to "angles between functions" in the same way they speak of "angles between vectors." In addition, there exist mathematical objects (known as "operators") that map functions to functions, in much the same way that functions map points to points. The quest to learn more about operators and function spaces is at the heart of research in function theory today.

4.8.1 Reading Questions for Section 4.8

1. Define and give an example of a differential equation.
2. Define the order of a differential equation and give examples of first-, second-, and fifth- order differential equations.
3. Define what is meant by a linear differential equation and give examples of first- and second- order linear differential equations.
4. Define the solution of a differential equation and give an example.
5. Explain the distinction between general and particular solutions of differential equations.
6. State theorem 4.8.1. How is this result helpful when studying differential equations?
7. Define and give an example of a separable differential equation.
8. State theorem 4.8.2. How is this result helpful when studying differential equations?
9. State both a first-order differential equation and a second-order differential equation modeling the motion of a pendulum.
10. Define the nth degree Hermite polynomial $H_n(x)$ and list the polynomials $H_1(x)$ and $H_2(x)$.
11. State theorem 4.8.3. How is this result helpful when studying differential equations?
12. State the orthogonality relation that exists between distinct Hermite polynomials.

4.8.2 Exercises for Section 4.8

In exercises 1–8, classify each differential equation based on order, linearity and separability. Assume all derivatives are taken with respect to x.

1. $y' = 0$
2. $y' + \sqrt{1+x}\, y = 4$
3. $y'' + y \sin y' = 2x$
4. $y'' = 3e^x \cos y$

5. $x/y' = 3y$
6. $y^{(3)} + 2xy = \ln x$
7. $\sin y^{(23)} + \sin y' = 0$
8. $y' + y = 2$

In exercises 9–16, use direct integration to find the general solution of each differential equation.

9. $y' = 0$

10. $xy' + 2x^2 - 1 = 0$

11. $y'' + \cos x = 2$

12. $y' = 3(1 + x)^{-1}$

13. $y' - xe^{3x} = 5x^2 - 2$

14. $y' = \cos(2x) + \sec(5x)\tan(5x) - x^3$

15. $y' = x\cos x + e^x \sin x$

16. $y'' + x = e^x$

In exercises 17–28, find the general solution of each linear or separable differential equation.

17. $y' = 0$

18. $y' + xy = 0$

19. $y' = 4e^x y$

20. $y' + xy + 2y^2 = (x + 2)y^2$

21. $y' - \frac{3}{x}y = 2x$

22. $y' - 2xy = e^{x^2}$

23. $y' = x^2 y^3 - x(y^3 + y) + x^2 y$

24. $y' + \sin xy = 0$

25. $y' - \sqrt{x}\, y = 0$

26. $y' + y\cos x = 0$

27. $y' - 2xy = x$

28. $y' + x^2 y = x^2$

In exercises 29–40, find the particular solution of each initial condition problem.

29. $y' - 2xy = x$ and $y(1) = 2$

30. $y' = \cos(2\pi x) - x^3$ and $y(1) = 3$

31. $y' + xy = 0$ and $y(0) = 2$

32. $y' + \frac{2}{x}y = 3x$ and $y(1) = 0$

33. $y' - \frac{2}{x}y = -x^3$ and $y(2) = 4$

34. $y' = 4e^{2x}y$ and $y(0) = 3$

35. $y' = 5xy^{-1}$ and $y(2) = 1$

36. $y' = x(1 + y)$ and $y(0) = 4$

37. $y' + \sqrt{1 + x^2} = 0$ and $y(0) = 5$

38. $y' + xy = 1 + x$ and $y(1) = 0$

39. $3x^{-1}y' = y\cos x$ and $y(0) = 1$

40. $\cos y(1 + x^2)y' = -x\sin y$ and $y(1) = \pi/2$

Exercises 41–50 consider solutions of certain second-order linear differential equations with constant coefficients. For a differential equation $ay'' + by' + cy = 0$, the solution is determined by the zeros of the characteristic equation $ar^2 + br + c = 0$ in the following fashion (where $C, D \in \mathbb{R}$ are arbitrary constants).

Types of zeros of $ar^2 + br + c = 0$	General solution of $ay'' + by' + cy = 0$
Real, distinct zeros p, q	$y = Ce^{px} + De^{qx}$
Real, repeated zero p	$y = Ce^{px} + Dxe^{px}$
Complex zeros $p \pm qi$	$y = Ce^{px}\cos(qx) + De^{px}\sin(qx)$

In exercises 41–50, find the general solution of each differential equation, or the particular solution of each initial condition problem.

41. $y'' + 2y' - 3y = 0$

42. $y'' + 5y' + 4y = 0$

43. $4y'' + 4y' + 5y = 0$

44. $y'' + y' + y = 0$

45. $y'' - 2y' + y = 0$

46. $9y'' + 6y' + y = 0$

47. $y'' + 5y' + 6y = 0$, $y'(0) = 3$, $y(0) = 2$

48. $y'' + 4y' - 2y = 0$, $y'(0) = 2$, $y(0) = 1$

49. $y'' + 2y' + 4y = 0$, $y'(0) = 1$,   50. $y'' + 4y' + 4y = 0$, $y'(0) = 3$,
    $y(0) = 1$ $y(0) = 2$

In exercises 51–58, prove each mathematical statement about Hermite polynomials.

51. The 0th Hermite polynomial $H_0(x) = 1$ is a solution of the 0th Hermite differential equation $y'' - 2xy' = 0$.

52. The 1st Hermite polynomial $H_1(x) = 2x$ is a solution of the 1st Hermite differential equation $y'' - 2xy' + 2y = 0$.

53. The 3rd Hermite polynomial $H_3(x) = 8x^3 - 12x$ is a solution of the 3rd Hermite differential equation $y'' - 2xy' + 6y = 0$.

54. The 4th Hermite polynomial $H_4(x) = 16x^4 - 48x^2 + 12$ is a solution of the 4th Hermite differential equation $y'' - 2xy' + 8y = 0$.

55. The Hermite polynomials $H_0(x) = 1$ and $H_2(x) = 4x^2 - 2$ satisfy the orthogonality relation; that is, produce a direct computation proving that
$$\int_{-\infty}^{\infty} H_0(x) \cdot H_2(x)\, e^{-x^2}\, dx = 0.$$

56. The Hermite polynomials $H_1(x) = 2x$ and $H_2(x) = 4x^2 - 2$ satisfy the orthogonality relation; that is, produce a direct computation proving that
$$\int_{-\infty}^{\infty} H_1(x) \cdot H_2(x) e^{-x^2} dx = 0.$$

57. The Hermite polynomial $H_0(x) = 1$ is an element of $L^2(-\infty, \infty)$; that is, prove that $H_0(x)$ satisfies the integral condition for membership:
$$\int_{-\infty}^{\infty} |H_0(x)|^2 e^{-x^2} dx < \infty.$$

Hint: Square this integral and convert to polar coordinates.

58. The Hermite polynomial $H_1(x) = 2x$ is an element of $L^2(-\infty, \infty)$; that is, prove that $H_1(x)$ satisfies the integral condition for membership:
$$\int_{-\infty}^{\infty} |H_1(x)|^2 e^{-x^2} dx < \infty.$$

Exercises 59–62 consider the Laguerre polynomials $L_n(x)$ which are defined by
$$L_n(x) = e^x \frac{d^n}{dx^n}(x^n e^{-x}).$$

These polynomials satisfy the Laguerre differential equation $xy'' + (1 - x)y' + \lambda x = 0$, where $\lambda \in \mathbb{R}$.

59. Compute the polynomial expression for the Laguerre polynomials $L_0(x)$ and $L_1(x)$.

60. Compute the polynomial expression for the Laguerre polynomials $L_2(x)$ and $L_3(x)$.

61. Prove the Laguerre polynomials $L_0(x)$ and $L_1(x)$ satisfy the orthogonality relation
$$\int_0^{\infty} L_0(x) \cdot L_1(x)\, e^{-x}\, dx = 0.$$

62. Prove that $L_1(x)$ satisfies the integral bound $\int_0^{\infty} |L_1(x)|^2\, e^{-x}\, dx < \infty$.

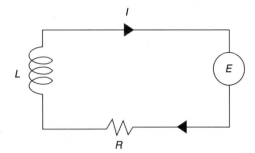

Figure 4.26 The electric circuit for
exercise 65

Exercises 63–66 consider physical settings modeled by first-order differential equations. Verify the model identified for each setting.

63. Radioactive decay satisfies the differential equation $dy/dt = ky$, where y is the amount of radioactive substance left at time t, and k is a constant of proportionality. Prove that $y = y_0 e^{kt}$, where y_0 is the amount y of the substance present at time $t = 0$.

64. The logistic model for population growth posits that the size of a population p at time t is determined by the differential equation

$$dp/dt = kp(1 - \frac{p}{L}),$$

where k is a proportionality constant and L is the carrying capacity of the local population. Using partial fractions, prove that in this setting we have

$$p = \frac{L}{1 + be^{-kt}}.$$

65. The flow of electricity around the electric circuit in figure 4.26 is governed by the differential equation $L(dI/dt) + RI = E$, where t is time, L is the constant inductance produced by the "inductor" in the circuit, I is the current, R is the constant resistance to the current generated by the "resistor," and E is a constant force produced by the battery. Find the general solution to the differential equation, expressing the current as a function of time.

66. Suppose that a hanging chain is modeled on the plane by the differential equation $y'' = c\sqrt{1 + y^2}$, where $c \in \mathbb{R}$ is a constant with initial conditions $y'(0) = 0$ and $y(0) = c^{-1}$. Prove that the shape of the chain (known as a catenary) is given by

$$y = \frac{1}{2c} \left(e^{cx} + e^{-cx} \right).$$

Exercises 67–70 apply the logistic model from exercise 64 to the population of humans on the Earth. For these exercises assume the carrying capacity of the Earth is approximately 10.76 billion.

67. If 2000 C.E. is taken as time $t = 0$ and the human population was approximately 5.96 billion, what is the value of b in this model?

68. Using the b from exercise 67, find the value of k in this model based on an actual approximate population of 6.47 billion in 2005.

69. Using the result from 67 and 68, predict the human population in the year 2020.

70. Using the result from 67 and 68, when does the human population reach the assumed carrying capacity of the Earth?

Notes

Real analysis has been a major area of mathematical research and application for the past four hundred years. In addition to the theoretical insights discussed in this chapter, the practical applications of calculus (for example, in engineering, the physical sciences, and the social sciences) have played a key role in the development of modern culture and civilization. Most mathematics students are first exposed to real analysis in a sequence of calculus courses. Many texts have been written to support the study of calculus at the basic undergraduate level, including Larson et al. [149], Stewart [234], and Thomas et al. [239]. Two texts that grew out of a 1990's "reform" movement in teaching calculus are Hughes-Hallet et al. [121] and Ostebee and Zorn [184]. An interesting text by Hahn [107] gives an historical perspective by discussing calculus in the context of questions studied and answered by Archimedes, Galileo, Newton, Leibniz, and others.

A rich and interesting theory supports the use of computational tools studied in initial calculus courses. Standard real analysis texts used in advanced undergraduate courses include those by Abbott [1], Goldberg [100], Kirkwood [140], and Rudin [202]. Standard graduate texts include those by Lang [148] and Rudin [203]. These books discuss proofs of the theorems we have studied in this chapter (including the Riemann–Lebesgue theorem) as well as many further insights into analysis. The Lebesgue integral plays a central role in a more sophisticated study of real analysis; it is described in texts by Goffman and Pedrick [101], Jones [133], and Royden [199]. Another interesting introduction to the Lebesgue integral is given by Weir in [251]; his novel approach allows for the computation of Lebesgue integrals with a minimal discussion of measure theory.

The historical development of calculus is discussed in any comprehensive treatment of the history of mathematics, including those by Boyer and Merzbach [28] and by Struik [235]. Studies focusing only on calculus include books by Boyer [27] and Dunham [62]. Two well-written and insightful books on the Age of Enlightenment (whose analysis is historical instead of mathematical) are by Will and Ariel Durant [66] and [67]. Part of a thorough eleven-volume set entitled *The story of civilization*, these works analyze "the Great Debate between faith and reason" (although many would agree that these two subjects are not mutually exclusive nor contradictory). The volumes on the Age of Enlightenment include passages on the contributions of Newton and Leibniz to mathematics as well as to science and philosophy; the authors clearly respect the importance of mathematical and scientific contributions to the development of history. One example of the detail the Durants offer is their description of Newton's 1704 work *Optiks*, where he listed 31 questions at the end of the book (reminiscent of Hilbert's 23 important unsolved problems offered in 1900). The Durants write, "Query I suggested prophetically: 'Do not bodies act upon light at a distance, and by their action bend its rays, and is not this action strongest at the least distance?' And Query XXX: 'Why may not Nature change bodies into light, and light into bodies?'" Written 200 years before Einstein's relativity theory introduced the fact that space is curved (and gave the formula for the equivalence between light energy and matter as $E = mc^2$), Newton's ability for conjecture and his instinctive understanding of the behavior of light seem remarkable (from only just these two "queries").

Many biographical reflections have been written about the geniuses who have been credited with the development of calculus. Christianson [40] and Westfall [255] are well-respected biographers of Sir Isaac Newton; both authors have written other books on his life and work. A translation of Newton's seminal treatise *Principia* has been prepared by Cohen and Whitman [180]. For a biography of Leibniz see Antognazza [7]. Thomson [240] is an excellent survey and overview of Leibniz's philosophical work. Scholars continue to study and discuss the priority dispute that arose between Newton and Leibniz. An excellent overview of this controversy (and others) is in *Acid tongues and tranquil dreamers: Eight scientific rivalries that changed the world* by White [256]. A more detailed discussion focusing on the controversy between Newton and Leibniz can be found in *Equivalence and priority: Newton versus Leibniz* by Meli [174].

A number of biographies have been written about René Descartes, including recent texts by Aczel [2] and Clarke [41]. In addition, Boyer [26] describes the history of analytic geometry, and Smith and Latham [58] is the definitive English translation of Descartes's *La géometrie*. For reflections on the life and work of Pierre de Fermat see Mahoney [163]; a similar book on the contributions of Isaac Barrow has been edited by Feingold [80]. After Newton and Leibniz proved the fundamental theorem of calculus, the most significant contributions to calculus were made by Augustin-Louis Cauchy and Bernhard Riemann. Belhoste [14] describes Cauchy's life; Grabiner [102] traces the development of Cauchy's rigorous treatment of calculus. In addition to the texts discussing the Riemann hypothesis mentioned in the notes for chapter 3, biographical works include *Bernhard Riemann 1826–1866: Turning points in the conception of mathematics* by Laugwitz and Shenitzer [150].

Many books describe the study of infinity. Dauben [53] is a biography of Georg Cantor, blending a description of his personal life with a detailed discussion of his professional contributions to mathematics. The first chapter is an illuminating reflection on eighteenth century analysis and the influences that led Cantor to explore the transfinite. Tiles [241] provides a historical introduction to Cantorian set theory. Contemporary translations of important essays by Georg Cantor [37] and by Richard Dedekind [56] are also available. A number of popular and accessible accounts of humanity's efforts to explore and understand infinity have been written by Maor [169], Rucker [201], Vicenkin [245], and Wallace [248]. In addition, Smullyan [224] is a fanciful and amusing discussion of the infinite. For a further discussion of Zeno's paradoxes, see Salmon [206]. A reference book that includes comments on countable vs. uncountable infinity is Gullberg and Hilton [104]. Sainsbury [205] provides a general discussion of mathematical paradoxes. Cantor's work is the starting point for continuing mathematical research in set theory. For an undergraduate text in set theory, see Halmos [108]; standard graduate level texts in set theory include those by Jech [130]. and by Kunen [146]

Differential equations are described briefly in most calculus courses, and undergraduate mathematics students typically study at least one course devoted to differential equations. Excellent undergraduate introductions to this field include Blanchard et al. [19], Boyce and DiPrima [25], Braun [29] Powers [189], and Simmons and Krantz [214]. In addition, Simmons [213] includes notes on the historical development of differential equations. A standard graduate text in this field is Walter [249]. While this chapter only briefly mentioned partial differential equations, a great deal of interesting mathematics has been developed in studying the derivatives of multivariate functions. Some popular introductions to this field include DuChateau and Zachmann [61], Farlow [79], and Logan [160].

5 Probability and Statistics

Probability and statistics are distinct fields of mathematics and yet are naturally intertwined with one another—sharing many common goals, questions, and techniques. Roughly speaking, probability is the mathematics of chance that seeks to measure the likelihood of an event occurring. Many people enjoy probability theory because its simplest problems reflect our natural intuition about real-life experience. For example, we expect a fair coin toss to show heads half the time, which matches the corresponding calculation from probability theory. As probability theorists grapple with more sophisticated questions, the required mathematics becomes more sophisticated and advanced. In this way, probability theory is an intricate field of mathematics with many surprises.

Probability theory draws on ideas from both discrete mathematics and real analysis and uses them in creative and useful ways. For example, the probability model for a coin toss is quite different from the probability model for the weight of a person; the first situation has only two outcomes (heads and tails) and is studied using discrete mathematics, while the second has a continuum of outcomes (since any weight greater than zero is theoretically possible) and is studied using the theory of functions with a positive domain. For even more sophisticated questions (such as the study of the price of stocks or the movement of microscopic particles in fluid), the Lebesgue integral may be required for a successful and productive analysis.

One important real-world application of probability theory is statistics. Roughly speaking, statistics is the mathematics of analyzing real-world data with the objective of identifying patterns, of better understanding the results, and of using inference to make decisions. Mathematicians typically distinguish between descriptive statistics, which focuses on the collection and organization of data, and inferential statistics, which focuses on drawing conclusions and making decisions. Our study of statistics develops inferential statistics. The mathematical procedure at the heart of basic statistical analysis is known as "hypothesis testing" or as "Fisher's procedure." Hypothesis testing enables us to interpret data (usually a sample of randomly collected numbers) in terms of known or assumed probabilities (the mathematics at work in the background). The procedure determines whether to reject or not reject a stated hypothesis based on an associated probability. Hypothesis testing was first developed in the 1920s by the British geneticist and mathematician Sir Ronald Fisher in terms of randomly collected data resulting from well-designed experiments; this procedure is now a standard method used in statistics.

In many settings, the computation of probabilities depends on knowing the number of objects in associated sets. We therefore first study combinatorics—the mathematics of counting. The chapter then considers an application of combinatorics in Pascal's triangle and the binomial theorem. These studies enable a development of basic probability theory in which the probability of an outcome is defined in terms of a mathematical formula. From this development we consider two important applications of statistics: hypothesis testing and linear regression. Hypothesis testing is a procedure for real-world decision-making. Linear regression is a systematic process for identifying linear models of relationships between variables. When a linear model is appropriate, linear regression determines the "best" such model.

The study of probability and statistics is relatively new. Scattered ancient historical sources reference probabilistic games. For example, the Christian Bible mentions Roman centurions gambling for Jesus Christ's garments during his crucifixion. The first, rudimentary formulations of probability theory took place around 1650 in a productive series of letters between the French mathematicians Blaise Pascal and Pierre de Fermat. The advances in probability theory for continuous settings occurred for the most part after the late 1800s. Statistics is mostly a twentieth-century-and-beyond sport, and has grown significantly with continuing development and use of more sophisticated computing devices. Probability and statistics both remain active areas of mathematical study and research—we hope that this brief introduction will motivate you to pursue further studies in these fields.

5.1　Combinatorics

In many settings, computing probabilities depends on knowing the number of objects in (that is, the cardinality of) related sets. Therefore, we begin a study of probability and statistics with *combinatorics*, or the mathematics of counting. We focus our attention on developing mathematical formulas and algorithmic methods for determining the number of elements in a given set or list, the two most common settings for counting objects. For simple scenarios, we are often able to reason through the process of identifying the variety of possible answers. However, for more complicated settings, manually counting elements in sets and lists is at best tedious and often impractical. Combinatorics provides efficient and elegant methods for determining the cardinality of sets and lists quickly and precisely.

In diverse settings, computing the number of lists and the number of sets formed from a given collection of objects plays an important role in the study of probability and statistics—these computations inform decision-making processes in real-life situations. Many important, real-life counting questions essentially boil down to finding the number of ways to select k objects from a given set of size n, with particular attention to two details:

- Is order relevant?
- Is repetition allowed?

For example, order is important when considering the number of different starting lineups of five players that a basketball coach can assign from a squad of 10 (playing center is different from playing guard), but not important when real estate agent is preparing to advertise 10 houses from a pool of 25 available for sale. Mathematicians typically express settings in which order is important in terms of lists, and settings in which order is not important in terms of sets. In addition, there are some settings in which repetition is allowed and others in which it is not. When forming three-letter words, letters may be repeated (for example, the word "BEE"); but if a set of refrigerator magnets has only one of each letter of the alphabet, then we cannot repeat letters. The following example and question consider further scenarios highlighting these distinctions.

Example 5.1.1   We state whether order is important and repetition is allowed in each possible way of selecting the executive committee of a student organization.

(a) The committee has offices that are designated, and a person can hold more than one office.
Since the offices are named, order is important so as to differentiate between, say, the president, vice-president, secretary, and treasurer. Since a person can hold more than one office, repetition is allowed.

(b) The committee has offices that are designated, and the officers must be distinct.
As in (a), order is important, but now repetition is not allowed because a single person cannot hold more than one office.

(c) The committee has offices that are equivalent to each other, and a person can hold more than one position.
Since the offices are equivalent, order of elevation to them is not important—it is a "set" rather than a "list" of officers. As in (a), repetition is allowed.

(d) The committee has offices that are equivalent to each other, and the officers must be distinct.
In this case order is not important and repetition is not allowed.

∎

Question 5.1.1   Determine if order is important and repetition is allowed when identifying each object. In part (d), a "multiset" may contain elements that appear more than once.

(a) Four-letter words
(b) Four-letter words with distinct letters
(c) Four-letter sets
(d) Four-letter multisets

∎

We will become quite adept at determining if order is important and if repetitions are allowed. The approach used to count objects is often determined by these two factors; knowing if order is important and if repetitions are allowed precisely identifies a consistent, algorithmic approach to counting the objects. You can see that there are four distinct pairings: (ordered) lists with repetition; lists without repetition; (unordered) sets with repetition (called "multisets"); and sets without repetition. We consider each of these combinations in turn, giving examples and developing mathematical formulas that make possible the immediate combinatoric solutions.

5.1.1 Lists with Repetition: The Multiplication Principle

An example of a simple action that is connected with probability might involve tossing coins or rolling dice multiple times. In these settings the order in which the outcomes occur is important, and repetitions are allowed—a (fair) coin tossed three times must repeat either a head or a tail! In any setting where order is important and repetition is allowed, we can count the number of objects having a given property by means of the *multiplication principle*.

We first consider this principle from a set-theoretic point of view. Recall from section 2.1 that the *Cartesian product $A \times B$* is the set of all ordered pairs with first-coordinate in A and second-coordinate in B; symbolically, $A \times B = \{(a, b) : a \in A \text{ and } b \in B\}$. The multiplication principle for sets expresses the Cartesian product in terms of the cardinality of the component sets.

Theorem 5.1.1 Multiplication principle for sets *If A and B are finite sets with A containing m elements and B containing n elements, then the set $A \times B$ contains $m \cdot n$ elements.*

Proof   For every fixed $a \in A$ in the first coordinate of an ordered pair $(a, b) \in A \times B$, there exist exactly n elements $b \in B$ that can be in the second coordinate. Since there are m elements from A that can be in the first coordinate of an ordered pair (a, b), there are $m \cdot n$ ordered pairs in the Cartesian product $A \times B$.

■

In addition to considering the Cartesian product of a pair of sets, we might also work generally with ordered n-tuples $A_1 \times \cdots \times A_n = \{(a_1, \ldots, a_n) : a_i \in A_i \text{ for } 1 \leq i \leq n\}$. The multiplication principle for sets extends in a natural way to these settings. The size of $A_1 \times \cdots \times A_n$ is the product of the (finite) sizes of the component sets; more formally, if set A_i contains m_i elements for $i = 1 \ldots n$, then the set $A_1 \times \cdots \times A_n$ contains $m_1 \cdot m_2 \cdots m_n$ elements.

Our interest in the multiplication principle is rooted in its application to combinatorics and probability, in which we focus primarily on finite processes. By properly interpreting this result in the context of step-by-step processes (rather than just in terms of Cartesian products of sets), the multiplication principle becomes quite useful and helps form the bedrock of a study of combinatorics. Consider the following expression of the multiplication principle.

Theorem 5.1.2 Multiplication principle for a two-step process *The number of outcomes from a two-step process is $m \cdot n$, when:*

- *The number of outcomes from Step 1 is m; and*
- *No matter what outcome results from Step 1, the number of outcomes from Step 2 is n.*

Example 5.1.2   We compute the total number of outcomes that can result from rolling two six-sided dice.

Viewed from the perspective of the multiplication principle, we can think of this activity as a two-step process, where Step 1 is the roll of the first die and Step 2 is the roll of the second. Observe that there are $m = 6$ possible outcomes from rolling the first die for Step 1 and also $n = 6$ possible outcomes from rolling the second die for Step 2 (no matter what is rolled on the first die). Therefore, by the multiplication principle, the total number of two-dice rolls is $6 \cdot 6 = 36$.

∎

Question 5.1.2   Using the multiplication principle, determine the total number of outcomes that can result from each activity.

    (a) Tossing a coin two times.
    (b) Rolling one die and tossing one coin.

∎

The multiplication principle extends to step-by-step processes with any (finite) number of steps. For example, we may want to count the number of outcomes from tossing a coin three, four, or five times, or the number of outcomes from rolling two dice and tossing two coins (these examples are simplistic, but they form good illustrations at an introductory level). In such settings, we use the multiplication principle for a multistep process as stated below.

Theorem 5.1.3 Multiplication principle for a multistep process *The total number of outcomes from a multistep process is the product of the number of outcomes from each step, provided that the number of outcomes in any one step is the same no matter what outcomes resulted in the previous steps.*

    Proof   The set of outcomes from an n-step process can be identified as a set of ordered n-tuples. Therefore, the total number of outcomes from the process is the same as the number of n-tuples in the corresponding set; according to the multiplication principle for sets, this number is the product of the number of outcomes from each step.

∎

We often refer to the multiplication principle for multistep processes as simply the multiplication principle. When forming a list of size k from a set of n objects with repetition allowed, the multiplication principle indicates the total number of such lists is n^k. Consider the following example and question.

Example 5.1.3   We apply the multiplication principle to determine the total number of outcomes that can result from each activity.

    • If a coin is tossed 10 times, the total number of possible outcomes is 2^{10}.
    • When rolling two dice and tossing two coins, the total number of possible outcomes is $6 \cdot 6 \cdot 2 \cdot 2 = 144$.
    • A 12-question multiple-choice test with four possible answers to each question (only one of which is correct) has 4^{12} different solutions that can be turned in by students.

∎

Question 5.1.3   Using the multiplication principle, compute the total number of outcomes that can result from each activity.

    (a) A student organization with 20 members elects a three-member executive committee with offices that are designated, allowing a person to hold more than one office.

    (b) A child lists all (possible) five-letter words.

■

5.1.2   Lists without Repetition: Permutations

When applying probability theory to model reality, mathematicians also need to be able to handle situations that involve lists without repetition. For example, most organizations only allow officers to hold one office—in the United States Government, the President and Vice-President must be different people. Similarly, batting orders, starting line-ups, the order of finish for a collection of racers, as well as some words and puzzles (such as Sudoku) can be interpreted as lists without repetition. Formally, mathematicians refer to such lists without repetition as *permutations*; these mathematical objects play important roles in the study of group theory, Galois theory, and graph theory. They are also important in combinatorics and probability theory.

Combinatorics is interested in determining the number of distinct permutations of a given set of n objects that have some prespecified length k. Sometimes $n = k$ (when every object available in the given set is listed), but we will also want to count situations with $k < n$ (when listing a proper subset of the objects available). We consider examples of permutations and then develop a formula for counting them.

Example 5.1.4   We determine the number of permutations of letters by manually listing every permutation with the specified features.

    • Form three-letter words without repeated letters using A, B, and G.
      Using each letter in the collection exactly once, the six possible words are: ABG, AGB, BAG, BGA, GAB, and GBA. These six lists are distinct; everyone knows that BAG is not the same as GAB (one holds groceries while the other describes trivial chatter).

    • Form three-letter words without repeated letters using A, B, G, and N.
      In this setting, there are many more possibilities; the following list describes the 24 distinct permutations satisfying this criteria.

      ABG, AGB, BAG, BGA, GAB, GBA, ABN, ANB, BAN, BNA, NAB, NBA,
      ANG, AGN, NAG, NGA, GAN, GNA, NBG, NGB, BNG, BGN, GNB, GBN.

■

Question 5.1.4   State the number of permutations of each type by manually listing every permutation with the given features.

    (a) Form two-letter words without repeated letters using T and O.

    (b) Form two-letter words with repeated letters using T and O (these are not permutations).

    (c) Form two-letter words without repeated letters using T, O, and P.

■

In general, we consider the number of lists without repetition of length k that can be formed from a collection of n objects; we use the notation $P(n, k)$ to denote the number of such permutations. Here order is important, and repetition is not allowed. In example 5.1.4, we found that $n = 3$ and $k = 3$ produce $P(3, 3) = 6$ permutations; similarly, $n = 4$ and $k = 3$ produce $P(4, 3) = 24$ permutations. As may be apparent, the total number of permutations quickly becomes quite large for even relatively small increases in the values of n and k. For example, when $n = 15$ and $k = 7$, there exist $P(15, 7) = 32, 432, 400$ permutations! Even in this relatively simple setting, writing out every permutation is clearly impractical. Instead, we find a general formula that counts permutations for any nonnegative integers n and k (with $k \leq n$).

The formula is written in terms of factorials. Recall that if $n \in \mathbb{N}$, then $n! = n \cdot (n - 1) \cdots 2 \cdot 1$ is the product of the positive integers less than or equal to n. For example, $3! = 3 \cdot 2 \cdot 1 = 6$ and $5! = 5 \cdot 4 \cdot 3 \cdot 2 \cdot 1 = 120$. By convention mathematicians define $0! = 1$. The next theorem states the formula for the number of permutations.

Theorem 5.1.4 *If n and k are nonnegative integers with $k \leq n$, then the number of permutations of length k that can be created from a given set of n objects is*

$$P(n, k) = \frac{n!}{(n - k)!}.$$

Proof   This result follows from an application of the multiplication principle. The process of creating a list of length k can be envisioned as a k-step process, where the first step fills in the first position in the list, the second step fills in the second position, and so on.

We now consider the number of objects available at each step in this process. For the first step, n different objects may be chosen. For the second step, the object chosen in the first step cannot be used again; only $(n - 1)$ objects are thus available (no matter what object is chosen in the first step). Continuing in this fashion, $(n - 2)$ objects are available in the third step, $(n - 3)$ in the fourth step, and so on. By the multiplication principle, there exist $P(n, k) = n \cdot (n - 1) \cdot (n - 2) \cdots [n - (k - 1)]$ distinct permutations of length k that can be created from a given set of n objects.

The final form of the solution (as given in the statement of this theorem) is obtained from algebra:

$$P(n, k) = n \cdot (n-1) \cdot (n-2) \cdots [n-(k-1)]$$

$$= n \cdot (n-1) \cdot (n-2) \cdots [n-(k-1)] \cdot \frac{(n-k) \cdot [n-(k+1)] \cdots 2 \cdot 1}{(n-k) \cdot [n-(k+1)] \cdots 2 \cdot 1} = \frac{n!}{(n-k)!}.$$

∎

Example 5.1.5   We apply the formula for $P(n, k)$ to the lists in example 5.1.4.

- From example 5.1.4, there are six three-letter words without repeated letters that can be formed using A, B, and G.
  We use the formula from theorem 5.1.4 to obtain the corresponding number of permutations of length $k = 3$.

$$P(3, 3) = \frac{3!}{(3 - 3)!} = \frac{3!}{0!} = \frac{6}{1} = 6.$$

- From example 5.1.4, there are 24 three-letter words without repeated letters that can be formed using A, B, G, and N.
  Using the formula from theorem 5.1.4,

$$P(4, 3) = \frac{4!}{(4 - 3)!} = \frac{4!}{1!} = \frac{24}{1} = 24.$$

- Using the formula from theorem 5.1.4, we verify the earlier assertion that

$$P(15, 7) = \frac{15!}{(15 - 7)!} = \frac{15!}{8!} = \frac{1307674368000}{40320} = 32,432,400.$$

■

Question 5.1.5 Using theorem 5.1.4, compute the value of each expression.

(a) $P(5, 1)$ (c) $P(15, 1)$

(b) $P(5, 3)$ (d) $P(15, 3)$

■

Example 5.1.6 We use theorem 5.1.4 to determine the number of starting line-ups for a basketball team with 10 women on the roster.

In basketball, the five starting positions are point guard, shooting guard, small forward, large forward, and center. Since there are $n = 10$ players available for the permutation of $k = 5$ players, there are $P(10, 5)$ distinct starting line-ups. From theorem 5.1.4, we have

$$P(10, 5) = \frac{10!}{(10 - 5)!} = \frac{10!}{5!} = \frac{10 \cdot 9 \cdots 1}{5 \cdot 4 \cdots 1} = 10 \cdot 9 \cdot 8 \cdot 7 \cdot 6 = 30,240.$$

■

Question 5.1.6 Using theorem 5.1.4, compute the total number of possible outcomes from each activity.

(a) A student organization with 20 members elects a three-member executive committee with named offices that must be held by different people.

(b) A child lists all (possible) five-letter words without repetition.

■

5.1.3 Sets without Repetition: Combinations

In many situations, the order in which objects are selected does not matter; instead, the important feature is simply the choice of which elements to include. For example, study groups, committees of equals, the players who make a sports team, and the "Choose Three Special" at your favorite restaurant, are just a few of the many scenarios in which the order of selection is not important. For those familiar with betting on horse races, the "unboxed trifecta" is another classic example; the bet is won by correctly identifying the collection of three horses that will win, place, or show, without needing to specify which horse will place in what position.

In these situations we are counting the number of sets that can be formed, rather than the number of lists that can be formed. This focus on sets implies that repetition is not allowed, since sets do not consider repeated elements; that is, we identify $\{A, A\}$ as $\{A\}$. Formally, mathematicians refer to such sets as *combinations*; they play important roles in many areas of mathematics, especially in combinatorics and probability theory.

Combinatorics is interested in determining the number of combinations (or groups) formed from a collection of n objects. Sometimes $n = k$ (where every object available is included in the given set), but we will also want to count situations with $k < n$ (when forming a proper subset of the objects available). We consider examples of combinations in the context of sets of letters and then develop a formula for counting them.

Example 5.1.7   We determine the number of combinations of letters by writing out every set with the given features.

- Three-letter sets that can be formed using A, B, G, and N.
  The four possible sets satisfying these conditions are $\{A, B, G\}$; $\{A, B, N\}$; $\{A, G, N\}$; and $\{B, G, N\}$.
- Three letter sets that can be formed using A, B, G, N, and P.
  The 10 distinct combinations satisfying this criteria are

$$\{A, B, G\}, \quad \{A, B, N\}, \quad \{A, B, P\}, \quad \{A, G, N\}, \quad \{A, G, P\},$$
$$\{A, N, P\}, \quad \{B, G, N\}, \quad \{B, G, P\}, \quad \{B, N, P\}, \quad \{G, N, P\}. \quad \blacksquare$$

Question 5.1.7   State the number of combinations of each type by writing out every combination with the given features.

(a) Two-letter sets formed using T, O, and P.
(b) Three-letter sets formed using T, O, and P.

$\blacksquare$

We find a general formula for the number of sets of size k that can be formed from a collection of n objects; we use the notation $C(n, k)$ to denote the number of such combinations. In example 5.1.7, we found that $n = 4$ and $k = 3$ produce $C(4, 3) = 4$ combinations; similarly, $n = 5$ and $k = 4$ produce $C(5, 4) = 5$. As with permutations, the total number of combinations quickly becomes quite large for even relatively small increases in the value of n and k. For example, $n = 35$ and $k = 7$ produce $C(35, 7) = 6{,}724{,}520$ combinations! Even in this relatively simple setting, writing out every combination is impractical. Instead, we identify a general formula that counts combinations for any nonnegative integers n and k (with $k \leq n$).

Theorem 5.1.5   *If n and k are nonnegative integers with $k \leq n$, then the number of combinations (or sets) of size k that can be formed from a given set of n objects is*

$$C(n, k) = \binom{n}{k} = \frac{n!}{k! \cdot (n - k)!}.$$

The notation $\binom{n}{k}$, which is read as "**n choose k**" is also commonly used for $C(n, k)$.

Proof   The proof uses the formula for the number of permutations from theorem 5.1.4, and then divides by $k!$ to factor out the permutations' attention to order (because combinations do not take order into account when forming sets). For each set of size k, there exist $k!$ different lists of length k of the elements in the set.

Think of forming a set as a two-step process, where the first step is to select k elements from the n available (there are $C(n, k)$ ways of choosing these k elements), and the second step is to arrange those k elements in a listed order (there are $k!$ ways to do this step). The multiplication principle then implies $P(n, k) = C(n, k) \cdot k!$. Solving for $C(n, k)$ and applying theorem 5.1.4,

$$C(n, k) = \frac{P(n, k)}{k!} = \frac{n!}{(n-k)!} \cdot \frac{1}{k!} = \frac{n!}{k! \cdot (n-k)!}.$$

■

Various notations are useful to denote the number of combinations of size k selected from n objects. We primarily use the notation $C(n, k)$ introduced in theorem 5.1.5. Equivalent notations are $C_{n,k}$ and C_k^n. The "n choose k" notation

$$\binom{n}{k}$$

is widely used in undergraduate and graduate mathematical studies; this symbolism was first introduced by the German mathematician Andreas von Ettinghausen in 1826.

Example 5.1.8   We apply the formula for $C(n, k)$ to the combinations in example 5.1.7.

- From example 5.1.7, there are four three-letter sets that can be formed using A, B, G, N.
  Using the formula from theorem 5.1.5,

$$C(4, 3) = \frac{4!}{3! \cdot (4-3)!} = \frac{4!}{3! \cdot 1!} = \frac{24}{6} = 4.$$

- From example 5.1.7, there are 10 three-letter sets that can be formed using A, B, G, N, P.
  Using the formula from theorem 5.1.5,

$$C(5, 3) = \frac{5!}{3! \cdot (5-3)!} = \frac{5!}{3! \cdot 2!} = \frac{120}{6 \cdot 2} = 10.$$

- Using the formula from theorem 5.1.5 and cancelling algebraically, we verify the earlier calculation of $C(35, 7)$:

$$C(35, 7) = \frac{35!}{7! \cdot (35-7)!} = \frac{35!}{7! \cdot 28!} = \frac{35 \cdot 34 \cdot 33 \cdot 32 \cdot 31 \cdot 30 \cdot 29}{7!}$$

$$= \frac{35 \cdot 34 \cdot 33 \cdot 32 \cdot 31 \cdot 30 \cdot 29}{7 \cdot 6 \cdot 5 \cdot 4 \cdot 3 \cdot 2 \cdot 1}.$$

Eliminating like factors, $C(35, 7) = 5 \cdot 17 \cdot 11 \cdot 8 \cdot 31 \cdot 29 = 6{,}724{,}520.$

■

Question 5.1.8 Using theorem 5.1.5, compute the value of each expression.

(a) $C(5, 1)$

(b) $C(5, 3)$

(c) $C(15, 1)$

(d) $C(15, 3)$

◾

Example 5.1.9 We use theorem 5.1.5 to determine the number of rosters available for a basketball squad of seven players when 12 women try out for the team. The number of different groups of women that could be used to form the team is a combination; a player wouldn't care about the order of selection, only that she was selected. In this case, we have $n = 12$ women trying out for $k = 7$ slots on the roster; from theorem 5.1.5, there are $C(12, 7) = 792$ rosters as computed by

$$C(12, 7) = \frac{12!}{7!(12 - 7)!} = \frac{479001600}{5040 \cdot 120} = 792.$$

◾

Question 5.1.9 Using theorem 5.1.5, compute the total number of possible outcomes from each activity.

(a) A student organization with 20 members elects a three-member executive committee of equals that must be held by different people.

(b) A child lists all (possible) five-letter sets without repeating a letter.

◾

5.1.4 Sets with Repetition

Finally, we consider the case in which we identify unordered collections of objects and allow repetition of elements in the collection. For example, a student organization may have an executive committee for which the same person may hold multiple positions; the collection of winners of the World Series and/or the Super Bowl might be presented with repeated team names in recognition of multiple national titles; or a mathematical model of stockholders in a company might incorporate a collection with repeated presentations of shareholders corresponding to the number of stocks owned by that individual. A simple model for such a setting is a game played with colored balls and a bag. The balls are drawn from the bag one at a time, their color noted, and then they are returned to the bag before the next draw—the results of the game are presented as a collection of colors with repetition allowed in light of returning each drawn ball to the bag before the next draw.

All of these settings involve sets of objects with repetition. As mentioned above, sets are defined so as to not allow repeated elements, but, as suggested here, mathematicians sometimes work with collections of objects and allow for repetitions in these collections. Formally, these mathematical objects are referred to as *multisets*. Combinatorics can determine the number of distinct multisets of size k formed from a given collection of n objects. This number turns out to be the combination $C(n + k - 1, k)$.

Example 5.1.10 We determine the number of two element multisets that can be formed using the letters A, B, and G. Here $n = 3$ and $k = 2$, and so the total number of such multisets is

$$C(3 + 2 - 1, 2) = C(4, 2) = \frac{4!}{2! \cdot (4 - 2)!} = \frac{4!}{2! \cdot 2!} = \frac{24}{4} = 6.$$

Alternatively, we can write out these six multisets.

$$\{A,A\}, \quad \{A,B\}, \quad \{A,G\}, \quad \{B,B\}, \quad \{B,G\}, \quad \{G,G\}.$$

■

Question 5.1.10 Using the formula $C(n + k - 1, k)$, compute the total number of possible outcomes from each activity.

(a) A student organization with 20 members elects a three-member executive committee of equals for which a person can hold more than one position.

(b) A child lists all (possible) four-letter multisets.

■

In summary, the following mathematical objects provide structure for a basic combinatoric study.

Definition 5.1.1 *A* **permutation** *is a finite ordered list without any repetitions. A* **combination** *is a finite set. A* **multiset** *is a finite collection of elements that may have repeated elements.*

In addition, the following formulas determine the number of objects that satisfy the corresponding conditions.

	Ordered list	Unordered set
Repetition	n^k	$C(n + k - 1, k)$
No repetition	$P(n, k)$	$C(n, k)$

The next several examples show how the multiplication principle, together with the formulas for combinations and permutations, can effectively count very complicated numbers of outcomes in various settings. We begin with a simple example that uses only the multiplication principle, and then combine the multiplication principle with the results of theorems 5.1.4 and 5.1.5 to handle more complicated situations.

Example 5.1.11 We use the multiplication principle to answer the question: In how many ways could a student answer a ten-question multiple choice test, if each question has four possible answers? We apply the multiplication principle by envisioning the test as a ten-step process, with the first step answering question one, the second step answering question two, and so on. Because there are four possible answers for each question, there are four possible outcomes in each step. Applying the multiplication principle, there are $4^{10} = 1,048,576$ possible ways to complete the test.

■

Example 5.1.12 In preparation for taking some pictures on vacation, a shopper selects two packs of film from a bin of 100 packs and three packs of batteries from a display of 40 packs. We determine how many distinct combinations of film and battery packages she could choose by viewing the selection as a two-step procedure: first, choose two packs of film; second, choose three battery packs. In both steps the order of selection is not important (we are simply choosing sets of film packs and battery packs) and repetition is not allowed (once a pack is selected it cannot be reselected); that is, we are interested in the number of combinations at each step. Applying theorem 5.1.5, there are $C(100, 2) = 4,950$ possible choices for film and $C(40, 3) = 9,880$ possible choices for batteries. Applying the multiplication principle, the total number of different selections is the product $C(100, 2) \cdot C(40, 3) = 48,906,000$.

∎

Example 5.1.13 A Midwest state charges $2.00 to play the game *LottoFun*, where a *LottoFun* ticketholder chooses five distinct numbers between one and 50, and wins if five, four, or three of the chosen numbers match the seven distinct numbers randomly drawn in the state's *LottoFun* draw. We use combinatorics to determine how many possible ways there are for a player to win. In analyzing this problem, the key is to recognize that a player can win in three different ways, and so there are three corresponding nonoverlapping cases that must be considered.

- **Case I:** Match five of the seven numbers drawn;
- **Case II:** Match four of the seven; and
- **Case III:** Match three of the seven.

Case I is the easiest to analyze—we determine how many ways a person can choose five out of seven numbers. In this case, order is irrelevant and repetition is not allowed, so there are $C(7, 5) = 21$ possibilities of winning in this way.

The other two cases are handled using the multiplication principle. For Case II, view matching four of the five chosen numbers as a two-step process. In the first step, choose four of the seven numbers drawn (to form a match); in the second step, choose one of the $50 - 7 = 43$ numbers not drawn (to complete the entire selection of five numbers by the player). Both steps form combinations, since order is not important and repetition is not allowed. The first step has $C(7, 4) = 35$ possible outcomes and the second step $C(43, 1) = 43$. Applying the multiplication principle, there are $35 \cdot 43 = 1,505$ ways to match four of the seven numbers.

Case III is similar to Case II; view matching three of the five numbers as a two-step process. For the first step, choose three of the seven numbers drawn (to form a match), which can be done in $C(7, 3) = 35$ different ways. For the second step, choose two of the 43 numbers not drawn (to complete the selection of five numbers by the player), which can be done in $C(43, 2) = 903$ ways. Applying the multiplication principle, there are $35 \cdot 903 = 31,605$ ways to match three of the seven numbers drawn.

The total number of ways to win is the sum of the possible ways to win from the three nonoverlapping cases: $21 + 1,505 + 31,605 = 33,131$.

∎

Example 5.1.14 A student is writing two term papers, both of which require five footnotes. Each footnote must refer to a distinct book selected from the school library; the library has 10 books on the first paper's topic and eight (other) books on the second. In light of this information, we determine the number of ways that the student can present the footnotes in these two papers.

We use combinatorics to analyze this situation. In this case, order does matter; a different organization of footnotes would result in a different presentation of the topic at hand (and so a different paper). In addition, repetition is not allowed because all 10 footnotes must refer to distinct texts. Thus, we can count permutations when finding the number of different presentations of footnotes.

Listing the footnotes is a two-step process. The first step lists $k = 5$ of the $n = 10$ relevant books available in the library; there are $P(10, 5) = 30,240$ possible outcomes for this step. The second step lists $k = 5$ of the $n = 8$ books; there are $P(8, 5) = 6,720$ possible outcomes for this step. Applying the multiplication principle, the total number of ways the student can present the footnotes is $P(10, 5) \cdot P(8, 5) = 30,240 \cdot 6,720 = 203,212,800$.

■

Without the benefit of combinatorics, it would have been nearly impossible to come up with the results in these last four examples—writing out every possible outcome is simply too large a task. These examples illustrate the power and scope of combinatorics, as broad classes of questions are managed by a handful of formulas. To conclude this section with a bit of practice using some of these formulas, the next question gives you the opportunity to imitate the techniques discussed in the preceding examples.

Question 5.1.11 Using the results and techniques presented in this section, compute the total number of possible outcomes from each activity.

(a) In how many ways can someone making a purchase form 73 cents, given that her purse holds 12 quarters, three dimes, and eight pennies, where all coins are distinguishable?

(b) A player immediately wins the casino game *Blackjack* if they are dealt an Ace and any other card worth 10 points. If Kings, Queens, Jacks, and tens are each worth 10 points, find the number of ways a player can win *Blackjack* on a two-card deal.

(c) A Northeastern state charges \$1.00 for the game *LottoNewEngland*, where a *LottoNewEngland* player chooses four numbers between one and 25, and wins if one, two, three, or four of the chosen numbers match the five numbers randomly chosen in the state's *LottoNewEngland* draw. How many possible ways can a player win?

(d) A governmental agency decides to select randomly four corporations for review. The agency decides to interview the top four officers in the first company selected, three of the top four officers in the second company selected, two of the top four officers in the third, and one of the top four officers in the fourth. In how many ways can the agency conduct the review process if 12 corporations are candidates for review?

■

5.1.5   Reading Questions for Section 5.1

1. State the two key questions that are considered when analyzing a counting question from the perspective of combinatorics.
2. State the multiplication principle for a multistep process.
3. What types of counting questions are answered using the multiplication principle?
4. Define and give an example of a permutation.
5. What does the mathematical symbol $P(n, k)$ represent?
6. Define $n!$ and compute the value of $1!$, $3!$, and $5!$.
7. What is a mathematical formula for calculating $P(n, k)$, where $n, k \in \mathbb{N}$ with $n \geq k$?
8. Define and give an example of a combination.
9. What does the mathematical symbol $C(n, k) =$ represent?
10. What is a mathematical formula for calculating $C(n, k)$, where $n, k \in \mathbb{N}$ with $n \geq k$?
11. Define and give an example of a multiset.
12. What formula do we use to compute the number of ways to form a multiset of k objects from a collection of n objects?

5.1.6   Exercises for Section 5.1

In exercises 1–12, compute the value of each expression without using a calculator.

1. $P(6, 2)$
2. $P(5, 4)$
3. $C(12, 3)$
4. $C(7, 5)$
5. $C(10, 1)$
6. $C(1,300, 1)$
7. $C(8, 5) \cdot P(5, 4)$
8. $C(10, 2) \cdot C(8, 2)$
9. $C(10, 3) + C(10, 2)$
10. $C(n, 1)$ for $n \in \mathbb{N}$
11. $P(n, 1)$ for $n \in \mathbb{N}$
12. $C(n, k) \cdot P(k, k - 1)$ for $k \geq 1$

In exercises 13–20, assume you are given the letters a, b, c, d and compute the total number of possible outcomes from forming each type of object. In answering this question, state whether order is important and whether repetition is allowed.

13. Three-letter words.
14. Three-letter words with distinct letters.
15. Three-letter sets.
16. Three-letter multisets.
17. Six-letter words.
18. Six-letter words with distinct letters.
19. Six-letter sets.
20. Six-letter multisets.

In exercises 21–39, answer each question using the results of this section. Explicitly state the role of order and repetition in your analysis of each situation.

21. How many lists of length nine can be formed from a set of 15 names?
22. How many different groups of size nine can be formed from a set of 15 names?
23. How many subsets of size six can be formed from the set {A, B, C, D, E, F, G, H, I, J}?

24. How many subsets of size nine can be formed from the set {A, B, C, D, E, F, G, H, I, J}? List every subset.

25. In how many ways can a mathematician arrange 12 objects in a list of size 12?

26. If order matters, how many ways can 13 objects be placed in an inventory of size 10?

27. If order doesn't matter, how many ways can 20 objects be placed in a catalog of size 12?

28. A mathematician counts the number of ways to complete a two-step process, where the first step arranges 10 names, chosen from a slate of 12, in a directory, and the second step chooses eight objects from a collection of 11. How many different outcomes are possible from this two-step procedure?

29. A three-step process consists of choosing five objects from a collection of 11, then listing 10 objects in an inventory, and finally arranging eight objects selected from 15 in a record. How many different ways can this three-step process be completed?

30. An airline with a hub in Chicago has five flights each morning from New York to Chicago and seven flights each afternoon and evening from Chicago to Dallas. Assuming any of the morning choices serve well as connecting flights for the afternoon alternatives, how many options are available to a traveler flying on the airline from New York to Dallas and eating lunch in Chicago?

31. A menu has six appetizers, five salad and soup options, 12 entrées, eight desserts, and nine beverage choices. How many different options are there for a customer who chooses one item from each of these food groups?

32. Suppose the customer looking at the menu described in the last exercise considers the possibility of not selecting from one or more of the food groups: if this patron decides to choose one item from at least three of the food groups (but not necessarily from all), how many options are available?

33. In how many ways can a vegetarian customer select a lunch of three distinct menu items, when the menu lists the following vegetables: salad, beans, peas, corn, potatoes, okra, spinach, and beets?

34. In a single elimination tournament, a team is eliminated if it loses one game (played between two teams). If the tournament starts with 16 teams, how many games must be played before the champion is declared? What about a tournament with 1,024 teams? What about with n teams, where n is of the form 2^k?

35. Whenever a game is played between two teams, there are two possible outcomes. After preliminary rounds, the NCAA basketball tournament starts with 64 teams. How many ways can a person fill out this tournament's brackets with winning teams? Hint: Use the result from exercise 34.

36. A conference organizer is to divide 120 participants into two rooms, each seating 60 people. In how many ways can he organize the participants into two groups?

37. A biologist plans to cross-pollinate two pea plants, selected from a crop of five different plants. In how many ways can she organize this pairing?

38. A poker player is dealt five cards from a 52-card deck. How many different poker hands can result? How many dealt hands result with the player receiving four cards that have the rank of Ace? How many result with the player receiving four cards of any single rank (this type of hand is called four of a kind)?

39. Which occurs more often in a deal of five poker cards—four of a kind or a straight flush (including a royal flush and where Aces are high or low)? A straight flush results when the five cards dealt are all in the same suit (spades, hearts, diamonds, or clubs), and are in adjacent rank order, such as nine, ten, Jack, Queen, and King. See the last exercise for interpretation of "four of a kind."

In exercises 40–49, suppose a jar contains 60 colored balls, with 40 green balls and 20 red balls, and we select three balls from the jar. In this setting, compute the total number of possible outcomes of each type, assuming that each ball is distinguishable from the others.

40. Total possible outcomes, if we do not replace balls.
41. Total possible outcomes, if we do replace balls.
42. An outcome with only green balls, if we do not replace balls.
43. An outcome with only green balls, if we do replace balls.
44. An outcome with two green balls and one red ball, if we do not replace balls.
45. An outcome with two green balls and one red ball, if we do replace balls.
46. An outcome with at least two green balls, if we do not replace balls.
47. An outcome with at least two green balls, if we do replace balls.
48. An outcome with a green ball chosen first, if we do not replace balls.
49. An outcome with a green ball chosen first, if we do replace balls.

In exercises 50–52, prove each mathematical statement about circular permutations; that is, ordered arrangements of objects in a circle.

50. There are two distinct ways that three people can seat themselves around a circular table. Hint: Draw a representation of each seating.

51. There are six distinct ways that four people can seat themselves around a circular table.

52. There are $(n - 1)!$ distinct ways to arrange n objects in a circle. Hint: Begin by placing one object anywhere in the circle, and then arrange the other objects.

Exercises 53–59 apply the pigeonhole principle, which asserts that if n "pigeons" reside in $n - 1$ "pigeonholes," then at least one pigeonhole holds more than one pigeon.

In exercises 53–59, prove each mathematical statement using the pigeonhole principle.

53. A dresser drawer contains four pairs of socks. The owner must remove at least five socks to guarantee that at least two of them match to form a pair. Why?

54. Suppose a dresser drawer contains m pairs of socks. The owner must remove at least $m + 1$ individual socks to guarantee that at least two of the socks match to form a pair. In your analysis, what objects did you identify as the pigeons? The pigeonholes?

55. Whenever eight women and 10 men are seated facing an audience, at least two men must sit next to each other.

56. A person needs to choose at least four numbers from the set $\{1,\ 2,\ 3,\ 4,\ 5\}$ to ensure that at least two of the values add to six.

    Hint: Think of the subsets $\{1,\ 5\}$, $\{2,\ 4\}$, and $\{3\}$ as pigeonholes.

57. If n is odd, then at least $(n + 1)/2 + 1$ numbers must be chosen from the set $\{1,\ 2,\ 3, \ldots, n\}$ to ensure that at least two of the values add to $n + 1$.

58. A person needs to choose at least four numbers from the set $\{1,\ 2,\ 3,\ 4,\ 5\}$ to ensure that at least two of the values have a difference of three.

59. If graduation requirements at a college mandate a student earn credit in seven courses from five disciplines of study, selecting at least one course from each discipline, then there must be at least one discipline from which the student takes at least two courses.

In exercises 60–65, prove each mathematical statement using the formulas for permutations and combinations given in theorems 5.1.4 and 5.1.5.

60. For every $n \in \mathbb{N}$, $P(n, 1) = n$.

61. For every $n \in \mathbb{N}$, $C(n, 0) = 1$.

62. For all nonnegative integers n and k with $k \le n$, $C(n, k) = C(n, n - k)$.

63. For all nonnegative integers n and k with $k \le n$, $C(n, k) + C(n, k - 1) = C(n + 1, k)$.

64. For all nonnegative integers n, j, k with $j \le n - k$, $P(n, k) \cdot P(n - k, j) = P(n, k + j)$.

65. Suppose n distinct objects are being placed into three different boxes, with n_1 in the first box, n_2 in the second, and n_3 in the third, where $n_1 + n_2 + n_3 = n$ and each n_i is a nonnegative integer. The number of different ways to accomplish this partitioning of the n objects is $n!/(n_1!n_2!n_3!)$.

    Hint: Use the multiplication principle to count the number of ways to first partition the n objects into three boxes as described above and then arrange each of the three groups of elements in a list within each box. Realize that procedure is exactly equivalent to forming a list of size n.

Exercises 66–70 consider situations in which n distinct objects are partitioned into k sets so that every object is placed into exactly one of set. If n_i is the number of objects in set i for $1 \le i \le k$, then $n_1 + n_2 + n_3 + \cdots + n_k = n$ (since each of the n objects is placed in some set). As proved in exercise 65 for the case $k = 3$, the number of ways to form such a partition is given by the formula (given with its standard notation):

$$\binom{n}{n_1, n_2, \ldots, n_k} = \frac{n!}{n_1! \cdot n_2! \cdots n_k!}.$$

For example, there are 12,600 ways to partition 10 objects into four sets, with three elements in the first set, two elements in the second set, one element in the third set, and four elements in the fourth set, because

$$\binom{10}{3, 2, 1, 4} = \frac{10!}{3! \cdot 2! \cdot 1! \cdot 4!} = 12{,}600.$$

66. In how many ways can we partition 12 objects into five sets, with two in the first three sets, five in the fourth, and one in the last?

67. In how many ways can we partition 25 objects into three sets, with 10 in the first set, 12 in the second, and three in the third?

68. In a test of quality of supply parts received, a manufacturer of lamp bulbs orders six filaments of identical specifications from six different suppliers. Two of the filaments received are assigned to each of three assembly lines, and the resulting produced bulbs are tested at the end of the manufacturing process. In how many different ways can the company assign the six distinct filaments to the three assembly lines?

69. A middle school men's basketball coach has 10 players on his squad, which he first divides into Team A and Team B. The coach thinks of these two teams as different: when holding a practice scrimmage drill between these two teams, the coach tells Team A to pass the ball without dribbling, but allows Team B to dribble the basketball as normal rules allow. After separating the squad into the two teams, the coach then assigns the five players on each team into the appropriate positions: one player is the center; two players are forwards; and two players are guards. In how many ways can the coach formulate the two squads with these lineups?

70. Prove that in the case where the n objects are partitioned into $k = 2$ sets, we have:

$$\binom{n}{n_1, n_2} = C(n, n_1) = \frac{n!}{n_1!(n - n_1)!}.$$

5.2   Pascal's Triangle and the Binomial Theorem

In this section we study a triangle of numbers named in honor of the French mathematician and philosopher Blaise Pascal. Pascal is counted among the most brilliant and insightful minds of the seventeenth century; he made many important contributions to both mathematics and philosophy that continue to be studied to this day. Among his many accomplishments is a noteworthy paper on projective geometry that he wrote when was just 16 years old. Projective geometry places a viewer's "eye" as a point on the plane of vision (called the projective plane), and then considers a geometric representation of objects in terms of this eye. For example, a ray that extends directly outward and away from the front of the eye is represented as a single point because the "rest" of the ray behind this point is hidden from the eye by the point. Amazingly enough, Pascal's work preceded the identifiable establishment of projective geometry as a distinctive mathematical discipline by almost 200 years!

Along with his compatriot Pierre de Fermat, Pascal is recognized as the founder of mathematical probability theory. These two great mathematicians developed their insights and expression of probability theory in a sequence of letters in the 1650s that fortunately has been preserved for study and admired by subsequent generations of mathematicians (they were first published in 1679). In this famous correspondence, Pascal and Fermat established the mathematical formulas used to compute probabilities of events that have at most a finite number of possible outcomes. In the next section, we develop this approach to studying probability and apply it to a study of hypothesis testing and linear regression.

You may be interested to know that Pascal and Fermat began their correspondence as a result of a question posed by the wealthy Antoine Gombaud (better known by his title of Chevalier de Méré), who frequently bet on games of chance and was searching for insights to increase his likelihood of winning. This led de Méré to ask how often two dice needed to be rolled before there is better than a 50–50 chance of rolling double sixes. And so probability theory was founded as the result of a question about gambling! Despite these very playful origins, probability theory has proven itself as an important and powerful tool for studying and understanding the world in which we live.

In addition to his mathematical genius, Pascal is also well-known for his many other talents and insights. After a mid-life religious experience and conversion, Pascal wrote several major treatises on theology and philosophy; in fact, he proved only one further mathematical result after this conversion experience—during a sleepless night when a painful toothache kept him awake, and Pascal turned to mathematics in search of diversion and relief. The book *Pensées* is his most famous work, especially in philosophical circles; this treatise describes his personal devotion to asceticism and contains the well-known "Pascal's Wager"—an attempt at developing a logical argument favoring belief in the divine over unbelief.

In mathematical circles, Pascal is best known in connection with the numerical *Pascal's triangle*, which we describe in this section as one of his main contributions to combinatorics. In preparation for presenting this triangle, we state and prove an important mathematical formula known as *Pascal's formula*. This formula explains the nature and structure of Pascal's triangle and indicates how Pascal's triangle produces numbers that are immediately applicable to combinatorial analysis. In fact, we stated Pascal's formula in the exercises of section 5.1; here we restate this important formula and present a complete proof.

Theorem 5.2.1 Pascal's formula *If $n, k \in \mathbb{N}$ with $1 \le k \le n - 1$, then*

$$C(n + 1, k + 1) = C(n, k) + C(n, k + 1).$$

Proof The proof is completely algebraic. We add the terms on the right-hand side of the formula by finding a common denominator to obtain the expression for $C(n + 1, k + 1)$ given by theorem 5.1.5.

$$
\begin{aligned}
C(n,k)+C(n,k+1) &= \frac{n!}{k!\cdot(n-k)!}+\frac{n!}{(k+1)!\cdot[n-(k+1)]!} \\
&= \frac{(k+1)\cdot n!}{(k+1)\cdot k!\cdot(n-k)!}+\frac{n!\cdot(n-k)}{(k+1)!\cdot[n-(k+1)]!\cdot(n-k)} \\
&= \frac{n!\cdot(k+1+n-k)}{(k+1)!(n-k)!}=\frac{(n+1)!}{(k+1)!\cdot(n-k)!}=C(n+1,k+1)
\end{aligned}
$$

■

Pascal's formula provides an alternative (and sometimes simpler) approach to calculate combinations. Instead of using the factorial formula from theorem 5.1.5, we can successively "build up" combinations for larger and larger values of n, by adding two combinations from preceding values of n. For example, knowing that $C(1, 0) = 1!/(0! \cdot 1!) = 1$ and that $C(1, 1) = 1!/(1! \cdot 0!) = 1$, we use Pascal's formula to find $C(2, 1)$ by means of the sum:

$$C(2, 1) = C(1, 0) + C(1, 1) = 1 + 1 = 2.$$

The other combinations with $n = 2$ are $C(2, 0) = 1$ and $C(2, 2) = 1$; recall that for every positive integer n, $C(n, 0) = 1$ and $C(n, n) = 1$. Summarizing, the combinations with $n = 2$ are

$$C(2, 0) = 1 \qquad C(2, 1) = 2 \qquad C(2, 2) = 1.$$

Now use the values for these $n = 2$ combinations to determine the number of combinations for $n = 3$ as follows:

- $C(3, 0) = 1$;
- $C(3, 1) = C(2, 0) + C(2, 1) = 1 + 2 = 3$;
- $C(3, 2) = C(2, 1) + C(2, 2) = 2 + 1 = 3$;
- $C(3, 3) = 1$.

These calculations are easy and fun! We always start and end such a list with $C(n + 1, 0) = 1$ and $C(n + 1, n + 1) = 1$ and, using Pascal's formula, sum the appropriate number of combinations for the preceding stage n to determine the middle values. The following question gives you practice.

Question 5.2.1   Using Pascal's formula, compute the value of $C(n, k)$ for $0 \le k \le n$ when

(a) $n = 4$ (c) $n = 6$

(b) $n = 5$ (d) $n = 7$

■

Definition 5.2.1   **Pascal's triangle** *consists of the values of $C(n, k)$ for $0 \le k \le n$ organized into rows based on the value of n, where the nth row contains the values for $C(n, k)$.*

$$C(0, 0)$$

$$C(1, 0) \quad C(1, 1)$$

$$C(2, 0) \quad C(2, 1) \quad C(2, 2)$$

$$C(3, 0) \quad C(3, 1) \quad C(3, 2) \quad C(3, 3)$$

$$C(4, 0) \quad C(4, 1) \quad C(4, 2) \quad C(4, 3) \quad C(4, 4)$$

$$\vdots$$

With the combinations calculated, Pascal's triangle is presented as either:

$$
\begin{array}{ccccccccc}
 & & & & 1 & & & & \\
 & & & 1 & & 1 & & & \\
 & & 1 & & 2 & & 1 & & \\
 & 1 & & 3 & & 3 & & 1 & \\
1 & & 4 & & 6 & & 4 & & 1
\end{array}
\qquad or \qquad
\begin{array}{ccccc}
1 & & & & \\
1 & 1 & & & \\
1 & 2 & 1 & & \\
1 & 3 & 3 & 1 & \\
1 & 4 & 6 & 4 & 1
\end{array}
$$

$$\vdots \qquad\qquad\qquad\qquad\qquad \vdots$$

The remainder of this section studies applications of Pascal's triangle; these combinations appear and provide helpful mathematical insight in many interesting and sometimes surprising ways. The first application is algebraic in nature: the binomial theorem. A *binomial* is an algebraic expression of the form $(a+b)^n$, where n is a positive integer. We are familiar with binomials from algebra and calculus. For example, we have all used the F.O.I.L. method to expand the polynomial $(x+3)^2 = x^2 + 6x + 9$ and the two-variable expression $(x+y)^2 = x^2 + 2xy + y^2$. However, for large exponents $n \in \mathbb{N}$, the time and space resources required to directly multiply out and simplify such expressions becomes impractical (and tedious if we are doing this work by hand). The binomial theorem is useful in these situations—providing a simpler approach to computing the coefficients of an expanded binomial in terms of combinations.

If we study a few binomial expansions and compare them with Pascal's triangle, we can intuit the binomial theorem's result. Consider the following binomial expansions and compare them with Pascal's triangle:

$$
\begin{aligned}
(x+y)^0 &= 1 \\
(x+y)^1 &= 1 \cdot x + 1 \cdot y \\
(x+y)^2 &= 1 \cdot x^2 + 2 \cdot xy + 1 \cdot y^2 \\
(x+y)^3 &= 1 \cdot x^3 + 3 \cdot x^2 y + 3 \cdot xy^2 + 1 \cdot y^3
\end{aligned}
$$

Can you see the pattern? The coefficients in the expansion of $(x+y)^1$ are the combinations identified in the $n = 1$ row of Pascal's triangle; the coefficients in the expansion of $(x+y)^2$ are the combinations identified in the $n = 2$ row of the triangle; and so on. As we might hope, this pattern continues, and the coefficients in the expansion of $(x+y)^n$ are the combinations identified in the nth row of Pascal's triangle; this includes identifying $(x+y)^0$ with the coefficient 1 in the $n = 0$ or the 0th row of Pascal's triangle. For this reason, combinations are sometimes referred to as *binomial coefficients*. The following famous theorem formalizes these observations and intuitions.

Theorem 5.2.2 The binomial theorem *If $n \in \mathbb{N}$, the coefficients in the expansion of $(a+b)^n$ are combinations of the form $C(n, k)$ for $0 \le k \le n$; symbolically, the expansion of $(a+b)^n$ is given by:*

$$(a+b)^n = \sum_{k=0}^{n} C(n, k) \cdot a^k \cdot b^{n-k}.$$

Proof   The coefficient of the term $a^k \cdot b^{n-k}$ is the number of ways to multiply a^k by b^{n-k} when expanding the product $(a+b)^n = (a+b) \cdot (a+b) \cdot (a+b) \cdots (a+b)$.

The term $a^k \cdot b^{n-k}$ is obtained in this expansion when we choose k of these n "$(a + b)$" terms to contribute a factor of a (the other $n - k$ "$(a + b)$" terms then automatically contribute a factor of b). In this situation, order is not important and repetition is not allowed. By definition, the total number of ways to obtain the term $a^k \cdot b^{n-k}$ is therefore $C(n, k)$.

∎

As mentioned above, the binomial theorem can be understood as an application of Pascal's triangle. The value $C(n, k)$ appears as the kth term in the nth row of the triangle, and so Pascal's triangle provides an alternative approach (different than the formula stated in theorem 5.1.5) to computing the value of $C(n, k)$. The case $n = 2$ of the binomial theorem was stated by Euclid as Proposition 4 in Book XI of *Elements* in the third century B.C.E. In addition, Indian, Chinese, and Islamic mathematicians are believed to have been familiar with various versions of this result. Pascal was the first European to give this rendition of the binomial theorem in his *Treatise on the Arithmetical Triangle* in 1665. By 1676, Sir Isaac Newton had extended this statement of the binomial theorem for positive integer exponents to arbitrary exponents, including negatives and fractional exponents. The next example and question provide further illustrations of the use of these results.

Example 5.2.1   We expand the binomial expressions $(a + b)^4$ and $(a + b)^8$ using the binomial theorem and Pascal's triangle.

The binomial theorem assures us that the fourth row provides the coefficients for the expansion of $(a + b)^4$. Consulting Pascal's triangle as given in definition 5.2.1, we see that the fourth row is $1, 4, 6, 4, 1$. These numbers are the desired coefficients, providing the expansion:

$$(a + b)^4 = \sum_{k=0}^{4} C(4, k) \cdot a^k \cdot b^{4-k} = 1a^4 + 4a^3b + 6a^2b^2 + 4ab^3 + 1b^4.$$

We also expand $(a + b)^8$, extending Pascal's triangle using Pascal's formula $C(n + 1, k + 1) = C(n, k) + C(n, k + 1)$ until we have determined its eighth row to be $1, 8, 28, 56, 70, 56, 28, 8, 1$. These numbers are the desired coefficients, providing us with the expansion:

$$(a+b)^8 = \sum_{k=0}^{8} C(8,k) \cdot a^k \cdot b^{8-k}$$

$$= 1a^8 + 8a^7b + 28a^6b^2 + 56a^5b^3 + 70a^4b^4 + 56a^3b^5 + 28a^2b^6 + 8ab^7 + 1b^8.$$

If you have ever had to manually multiply out such an expression as $(a + b)^8$, the relative ease of this approach is quite apparent!

∎

Question 5.2.2   Using the binomial theorem and Pascal's triangle expand each expression.

(a) $(a + b)^5$

(b) $(2x + b)^5$, setting $a = 2x$ in (a)

(c) $(2x - 3y)^5$, setting $b = -3y$ in (b)

(d) $(a + b)^6$

(e) $(3x + 2y)^6$

(f) $(3x - 2y)^6$

∎

The rest of this section continues to explore interrelationships among the numbers appearing in Pascal's triangle, as well as other diverse applications. We begin with an interesting and well-known result about the sum of the numbers appearing in any row of Pascal's triangle.

Theorem 5.2.3   *The sum of the numbers appearing in the nth row of Pascal's triangle is equal to* 2^n; *using the symbolism for combinations, we have:*

$$\sum_{k=0}^{n} C(n, k) = 2^n.$$

Proof   The strategy of this proof is to count the total number of outcomes from the same activity in two different ways, obtaining the desired equality of the two expressions. This approach is commonly used in combinatorics, and so it is helpful to develop the ability to recognize the different ways that an activity can occur. For this argument, consider the activity of tossing a (fair, two-sided) coin n times and count the total number of possible outcomes in two different ways.

First, consider the corresponding n-step process, with each step consisting of one toss of the coin. Each step has two possible outcomes (heads or tails), and the steps are independent of one another. Applying the multiplication principle, there are 2^n possible outcomes from tossing a coin n times.

Alternatively, count the number of outcomes by splitting them into distinct nonoverlapping sets based on the number of heads in each collection of n tosses. Define "Set 0" as those outcomes with 0 heads (and n tails), "Set 1" as those outcomes with 1 head (and $n - 1$ tails), and so on, until "Set n" consists of those outcomes with n head (and 0 tails). For these sets, the order in which the k heads are tossed is not important and repetitions are not allowed. Therefore, "Set k" has $C(n, k)$ elements. Since these sets are nonoverlapping, the total number of outcomes from tossing a coin n times is

$$C(n, 0) + C(n, 1) + \cdots + C(n, n) = \sum_{k=0}^{n} C(n, k).$$

Both counting methods identify the total number of outcomes of the same activity, and so the resulting numbers must be equal to one another. Thus $$\sum_{k=0}^{n} C(n, k) = 2^n.$$

            ■

As with other important mathematical results, this theorem can be proven in a variety of different ways. The following question indicates an algebraic proof.

Question 5.2.3   Use the binomial theorem with $a = 1$ and $b = 1$ to prove that $\sum_{k=0}^{n} C(n, k) = 2^n$.

            ■

Mathematicians have also recognized connections between numbers in Pascal's triangle and other important numbers that have be defined and explored in completely different settings. An example of this phenomenon occurs with *Catalan numbers*, which were defined in the mid-nineteenth century by the Belgian mathematician

Eugéne Charles Catalan. He was studying the *polygon division problem*, which asks: "In how many ways can a regular polygon with $n + 2$ sides be divided into n triangles using diagonals that connect vertices?" Here a different orientation of the same slicing is counted as a different division. This question was first solved in the eighteenth century by the Hungarian mathematician Johann Andreas van Segner (the first mathematics professor at Göttingen); both Leonhard Euler and the French mathematician Jacques Phillipe Marie Binet simplified Segner's solution. However, Catalan's approach is the most elegant and has endured. Consider the following formal definition.

Definition 5.2.2   *For $n \in \mathbb{N}$, the **nth Catalan number** C_n is the number of ways that a regular polygon with $n + 2$ sides can be divided into n triangular pieces using diagonals that connect vertices. Here a different orientation of the same slicing is counted as a different division. The nth Catalan number is computed using the formula*

$$C_n = \frac{C(2n, n)}{n + 1} = \frac{1}{n + 1} \cdot \binom{2n}{n} = \frac{(2n)!}{(n + 1)! \cdot n!}.$$

Example 5.2.2   We identify the first three Catalan numbers.

- $C_1 = 1$, since there is only one way to divide a regular polygon with three sides (a triangle) into triangles (it would be the triangle itself). Using the formula from definition 5.2.2,

$$C_1 = \frac{2!}{2! \cdot 1!} = \frac{2}{2} = 1.$$

- $C_2 = 2$, because there are two ways to divide a regular polygon with four sides (a square) into triangles. Here each diagonal of the square provides one such division. From definition 5.2.2,

$$C_2 = \frac{4!}{3! \cdot 2!} = \frac{24}{6 \cdot 2} = 2.$$

- $C_3 = 5$, since there are five ways to divide a regular pentagon into triangles; figure 5.1 illustrates these divisions. From definition 5.2.2,

$$C_3 = \frac{6!}{4! \cdot 3!} = \frac{720}{24 \cdot 6} = 5.$$

∎

Question 5.2.4   As in example 5.2.2, prove that $C_4 = 14$ both by sketching the 14 divisions of a regular hexagon into triangles and by using the formula from definition 5.2.2.   ∎

Catalan numbers can also be used to describe such results as the number of ways to place parentheses in a sequence of numbers to be multiplied two at a time, the number of "rooted trivalent trees with $n + 1$ nodes," and the number of paths of length $2n$

Figure 5.1   The five divisions of a pentagon into triangles

through an $n \times n$ grid that do not rise above the main diagonal. There is also an elegant relationship between Catalan numbers and Pascal's triangle: the Catalan number C_n is equal to the difference of two adjacent numbers on the $2n$th row of Pascal's triangle. We invite you to prove this identity in the following question.

Question 5.2.5   Using the formula for $C(n, k)$ given in theorem 5.1.5, prove the Catalan number identity $C_n = C(2n, n) - C(2n, n + 1)$.

                                                                                        ■

This section ends by considering a connection between Pascal's triangle and a famous fractal known as the Sierpinski triangle. Perhaps you have studied or seen pictures of such fractals as the Koch snowflake, the Mandelbrot set, or Julia sets (which we will study in the exercises of section 7.3). These self-similar, self-replicating sets have proven useful for modeling such diverse and interesting natural phenomena as weather patterns, the shape of coastlines, the branching of ferns, trees, and rivers, and blood and air flow in arterial and bronchial systems.

Sierpinski's triangle results from applying mod 2 arithmetic to the elements of Pascal's triangle. When we reduce each number mod 2, Pascal's triangle contains only 0's and 1's, as illustrated for the first 15 rows in figure 5.2.

Essentially, the mod 2 arithmetic distinguishes among the even and odd numbers appearing in Pascal's triangle, substituting 0 for the even numbers and 1 for the odd numbers. Can you see a triangular design in this mod 2 version of Pascal's triangle? Placing lines around the groups of 0's helps highlight the triangular shapes, as illustrated in figure 5.3.

Even though figure 5.3 only contains the first 15 rows of Pascal's triangle, a clear pattern is becoming apparent. As more and more rows are included, a surprising and pleasing complexity appears in these patterns; each of these images is a manifestation of a step in the construction of the Sierpinski triangle. The Sierpinski triangle is a *self-replicating set*; any such set is formed using a step-by-step process beginning with some given shape and repeatedly removing this same shape (of appropriately scaled size) from "middle regions" that remain from the previous step. The first steps in the construction of the Sierpinski triangle are illustrated in figure 5.4.

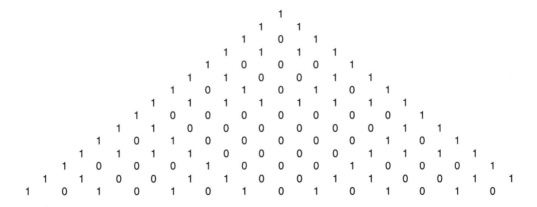

Figure 5.2   Pascal's triangle mod 2

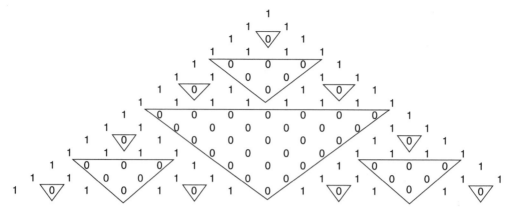

Figure 5.3   Pascal's Triangle mod 2

Figure 5.4   Constructing Sierpinski's triangle

As can be seen, an equilateral triangle is the shape used in the construction of the Sierpinski triangle. We begin with a solid equilateral triangle and remove the appropriately scaled equilateral triangle from the center of this given triangle. Three equilateral triangles remain, each one a smaller version of the original triangle. We now remove the appropriately scaled equilateral triangle from the center of each of these three triangles, leaving nine much smaller triangles. Continuing this process indefinitely produces the Sierpinski triangle.

The Sierpinski triangle is named in honor of the Polish mathematician Waclaw Sierpinski. Sierpinski is best known for his work in set theory, point set topology, and functions of real variables, publishing an impressive 724 papers and 50 books during his long and active mathematical career. In addition to defining the Sierpinski triangle, he also proved a theorem asserting that points in the plane can be specified with a single coordinate, gave the first example of an *absolutely normal* number whose digits occur with equal frequency in whatever base it is written, and produced the Sierpinski curve that has infinite length, visits every interior point of a given square, and bounds an area equal to 5/12 the area of the given square. Sierpinski's mathematical career and his support for the study of mathematics in Poland carried on through both World Wars despite the loss of close friends, colleagues, and students, and the loss of his personal library and papers.

Many other such self-replicating sets exist, and the same procedure applied to triangles can be implemented using other regular polygons. For example, the *Sierpinski carpet* is obtained by repeatedly removing the central square forming the middle-ninth of a larger square. In the mid-1800s, the German mathematician Georg Cantor defined the first of these self-replicating sets by removing the middle third of the interval [0, 1] and indefinitely repeating the removal of the middle third of each remaining interval.

This set is known as the *Cantor set* and is examined in further detail in the exercises for section 4.6. Self-replicating sets defined by such a process of repeated removals often turn out to have quite stunning geometrical properties and, in the late 1960s, *chaos theory* emerged as a distinct field of mathematics devoted to their study. More than just a fun study of patterns, Chaos theory has been used to model such complicated physical phenomena as weather patterns, oddly dripping water faucets and unevenly recurring ocean waves (more formally known as irregular dynamical systems), and behavior of some celestial objects.

This success with applying $\mod 2$ arithmetic to the elements of Pascal's triangle might lead us to wonder what patterns and results might appear if we consider $\mod n$ arithmetic using other, fixed values of $n \in \mathbb{N}$. The following questions invite you to explore other patterns within Pascal's triangle that are similar in nature to the $\mod 2$ results discussed above.

Question 5.2.6   List the first 27 rows of Pascal's triangle, reducing each element $\mod 3$. (Notice that you do not need to write out every element in the original Pascal's triangle; rather, the $\mod 3$ Pascal's triangle can be obtained by adding pairs of $\mod 3$ numbers on the previous row to obtain the next row's $\mod 3$ numbers.) Then draw triangles around each of grouping of 0's, 1's, and 2's and color the triangles for 0's red, for 1's black, and for 2's green. Compare and contrast the resulting picture with figures 5.3 and 5.4.   ■

Question 5.2.7   Repeat the construction outlined in question 5.2.6 for mod 4 arithmetic on the first 32 rows of Pascal's triangle, using four colors to distinguish the various triangles. Compare and contrast the resulting picture with figures 5.3 and 5.4.   ■

Many more interesting properties and mathematical results are associated with Pascal's triangle; we take up some of these ideas in the following exercises. In addition, section 5.3 uses combinations to calculate important probabilities that arise often in the real-world. Pascal's triangle frequently helps simplify calculations that involve combinations.

5.2.1   Reading Questions for Section 5.2

1. State Pascal's formula and give an example. How is this result helpful?
2. Give the first five rows of Pascal's triangle.
3. What are the first and last numbers in any row of Pascal's triangle? Why?
4. What is the connection between Pascal's triangle and expanding a binomial expression $(a + b)^n$?
5. State the binomial theorem. How is this result helpful?
6. What is the idea behind the proof of the binomial theorem?
7. According to the binomial theorem, what is the expansion of $(a + b)^3$? Using this result expand the binomial $(x + 2y)^3$.
8. Using the binomial theorem and the formula for computing combinations, determine the coefficient of $x^4 y^3$ in the expansion of $(x + y)^7$?
9. State theorem 5.2.3. What strategy is used in proving this result?

10. Define and give an example of a Catalan number.
11. State the relationship between the nth Catalan number and the elements in the $2n$th row of Pascal's triangle.
12. Describe the construction of Sierpinski's triangle.

5.2.2   Exercises for Section 5.2

In exercises 1–4, extend the portion of Pascal's triangle given in definition 5.2.1 up to the given row. By writing out these additional rows of Pascal's triangle, you are constructing a handy reference that is useful when working with combinatorics, and you are preparing to study this triangles for further patterns.

1. Extend up to row 8.
2. Extend up to row 12.
3. Extend up to row 16.
4. Extend up to row 20.

In exercises 5–10, identify the combination that answers each question and determine its value using Pascal's triangle.

5. An art dealer decides to buy four valuable paintings at auction. If 15 paintings are on the auction block, how many options are available to the dealer?
6. A genetic process involves randomly selecting two genomes from a collection of eight to form a health characteristic. In how many ways can the characteristic be expressed genetically from this part of the process alone?
7. A chef offers a menu consisting of 12 entrées. In how many ways can a group of six patrons ask the cook to prepare their meals if everyone in the group orders a different single entree? (Note: Who orders what doesn't matter.)
8. A neighborhood of 16 homes is awarded a tree-planting grant from the local community that directs the planting of 10 trees. If each tree is planted at a different community home, in how many ways can the plantings be done?
9. A group of seven teenagers arrive at an ice cream parlor that serves 21 flavors. How many different combinations of single scoop ice cream cones can the group ask for if each teenager gets a different flavor?
10. A college has 15 new computers, but 20 faculty members are in need of an upgrade. In how many ways can the college administration select 15 of the 20 to receive upgrades this academic year?

In exercises 11–30, use the extension of Pascal's triangle identified in exercises 1–4 to answer each question about combinations and patterns in Pascal's triangle.

11. Find the value of $C(7, 0)$ and $C(7, 7)$.
12. Find the value of $C(9, 0)$ and $C(9, 9)$.
13. Find the value of $C(11, 0)$ and $C(11, 11)$.
14. In light of the answers to exercises 11–13, make a conjecture about the value of $C(n, 0)$ and $C(n, n)$ for $n \in \mathbb{N}$. What property of Pascal's triangle supports your conjecture? Prove this result using the formula for combinations given in theorem 5.1.5.
15. Find the value of $C(8, 1)$ and $C(8, 7)$.
16. Find the value of $C(10, 1)$ and $C(10, 9)$.

17. Find the value of $C(12, 1)$ and $C(12, 11)$.

18. In light of the answers to exercises 15–17, make a conjecture about the value of $C(n, 1)$ and $C(n, n-1)$ for $n \in \mathbb{N}$. What property of Pascal's triangle supports your conjecture? Prove this result using the formula for combinations given in theorem 5.1.5.

19. Identify which numbers in the $n = 2$ row of Pascal's triangle are even.

20. Identify which numbers in the $n = 4$ row of Pascal's triangle are even.

21. Identify which numbers in the $n = 8$ row of Pascal's triangle are even.

22. In light of the answers to exercises 19–21, make a conjecture about the value of $C(n, k)$ when $n = 2^m$ is a power of two and k is not equal to 0 or n. What property of Pascal's triangle supports your conjecture? Prove this result using the formula for combinations given in theorem 5.1.5.

23. Show that 5 divides every non-one element in the 5th row of Pascal's triangle.

24. Show that 7 divides every non-one element in the 7th row of Pascal's triangle.

25. Show that 11 divides every non-one element in the 11th row of Pascal's triangle.

26. In light of the answers to exercises 23–25, make a conjecture about the divisibility of $C(n, k)$ when n is prime and k is not equal to 0 or n. What property of Pascal's triangle supports your conjecture? Prove this result using the formula for combinations given in theorem 5.1.5.

27. Find a counterexample *disproving* the claim that n divides every non-one element of the nth row of Pascal's triangle for every $n \in \mathbb{N}$.

28. The third main diagonal from right to left in Pascal's triangle consists of the numbers 1, 3, 6, 10, 15, ... Prove by direct computation that the sum of any two adjacent numbers in the first five terms of this sequence is a perfect square.

29. The third main diagonal from right to left in Pascal's triangle consists of the numbers 1, 3, 6, 10, 15, ... Prove by direct computation that the sum of any two adjacent numbers in the first 10 terms of this sequence is a perfect square.

30. In light of the answers to exercises 28–29, make a conjecture about the value of $C(n, 2) + C(n+1, 2)$ when n is an integer greater than or equal to two. Prove this result using the formula for combinations given in theorem 5.1.5.

In exercises 31–37, prove each mathematical statement about combinations and Pascal's triangle.

31. If n and k are integers with $0 \le k \le n$, then $C(n, k) = C(n, n-k)$. What property of Pascal's triangle follows from this equality?

32. If n is a nonnegative odd integer, then the alternating sum of the elements of the nth row of Pascal's triangle is equal to zero; that is,

$$C(n, 0) - C(n, 1) + C(n, 2) - \cdots + C(n, n-1) - C(n, n) = 0.$$

33. If n is a nonnegative even integer, then

$$C\left(n, \frac{n}{2}\right) = C\left(n-1, \frac{n}{2} - 1\right) + C\left(n-1, \frac{n}{2}\right).$$

34. For every positive integer n, $\sum\limits_{k=0}^{n} C(n, k)$ is even.

35. Using induction, prove that if n is a nonnegative integer, then $\sum\limits_{k=0}^{n} C(n, k) = 2^n$.

    Hint: This is theorem 5.2.3; use Pascal's formula in the inductive step.

36. Using induction, prove that if n is a positive integer, then $\sum\limits_{k=0}^{n-1} 2^k = 2^n - 1$.

37. The sum of the first n rows of Pascal's triangle (that is, the sum of all the numbers from the 0th row to the $n - 1$st row) is equal to $2^n - 1$.

    Hint: Use exercises 35 and 36.

In exercises 38–43, answer each question about the permutation triangle, a variation on Pascal's triangle obtained by recalling that $P(n, k) = C(n, k) \cdot k!$ and multiplying the elements in Pascal's triangle by $k!$ to obtain a triangle of permutations.

38. Determine the first five rows of the permutation triangle.
39. Determine the first 10 rows of the permutation triangle.
40. Using exercise 38, find the value of $P(4, 2)$ and $P(4, 3)$.
41. Using exercise 38, find the value of $P(5, 2)$ and $P(5, 3)$.
42. Using exercise 39, find the value of $P(8, 3)$ and $P(8, 5)$.
43. Using exercise 39, find the value of $P(9, 3)$ and $P(9, 5)$.

In exercises 44–57, expand each expression using the binomial theorem and Pascal's triangle.

44. $(x + y)^2 + (x - y)^4$
45. $(x^2 + y^2)^2 + (x - y)^4$
46. $(2x + y)^3 + (2x - y)^5$
47. $(x + y^2)^3 + (2x - y)^5$
48. $(x + y)^7$
49. $(5x + 2y)^7$
50. $(3x - y)^7$
51. $(xy + z)^7$
52. $(x + y)^9$
53. $(4t + 5s)^9$
54. $(4t - 5s)^9$
55. $(x^2 - y^2)^9$
56. $(x + y)^{10}$
57. $(x + y)^{11}$

In exercises 58–60, prove each mathematical statement, extending our understanding and results for the binomial theorem. For exercises 58–60, recall that the binomial power series was defined by Sir Isaac Newton as:

$$(1 + x)^n = 1 + nx + \frac{n(n - 1)}{2!} \cdot x^2 + \frac{n(n - 1)(n - 2)}{3!} \cdot x^3 + \cdots$$

58. If n is a nonnegative integer, then the coefficients in the above formula for the binomial power series are $C(n, k)$.
59. If n is a nonnegative integer, then the binomial power series is finite.
60. Using induction, prove the binomial theorem.

Exercises 61–65 consider connections between Pascal's triangle and both the Catalan numbers and the Fibonacci numbers.

61. Compute the first seven Catalan numbers.

62. In Pascal's triangle, identify the pairs of numbers $C(2n, n)$ and $C(2n, n + 1)$ for $n = 1, 2, 3, 4$.

63. Prove the Catalan number identity $C_n = C(2n, n) - C(2n, n + 1)$ for $n \geq 1$.

64. In section 3.6, the sequence of Fibonacci numbers was defined by $f_1 = 1$, $f_2 = 1$, and $f_{n+2} = f_n + f_{n+1}$; this sequence begins 1, 1, 2, 3, 5, 8, ... Prove by direct computation that the first 10 Fibonacci numbers can be obtained by summing the elements of Pascal's triangle along the diagonals indicated below.

```
              1
          ↗      1
    1        ↗      2
          ↗      3
    1     1    ↗      5
       ↗     ↗      8
    1     2     1    ↗
       ↗     ↗
    1     3     3     1
       ↗     ↗     ↗
    1     4     6     4     1
```

65. Following the model given in exercise 64, prove by direct computation that the first 20 Fibonacci numbers can be obtained by summing the elements of Pascal's triangle along the appropriate diagonals.

Exercises 66–67 examine Pascal's triangle using modular arithmetic with various integers.

66. State the first 24 rows of Pascal's triangle, reducing each element mod 3. Notice that you do not need to write out every element in the original Pascal's triangle; rather, the mod 3 Pascal's triangle can be obtained by adding the two mod 3 numbers on the previous row to obtain the next row's mod 3 numbers. Color the triangles for 0's red, for 1's black, and for 2's green. Compare and contrast the resulting picture with figure 5.4.

67. Repeat exercise 66 for 24 rows, reducing each element mod 4, and coloring the triangles for 0's red, for 1's black, for 2's green, and for 3's blue.

5.3	**Basic Probability Theory**

Probability is the mathematical calculation of chance. It measures the likelihood of a given event occurring. Throughout history, humans have been concerned with chance: What is the probability that a serious storm will develop? Can I predict if a tossed coin results in heads or tails? What is the chance that I will remain healthy? How does that chance change if I regularly smoke cigarettes? By adopting a systematic, mathematical approach to questions of chance, probabilists have developed a mathematics of inferential study: based on collected data, we can ask and answer questions about large populations and, as a result, better understand the world about us. Any application of probability begins with the description of an *experiment* (sometimes called a random experiment), which is an activity that can result in many different possible outcomes.

Definition 5.3.1   *An **experiment** is an activity that might result in many different outcomes, where an **outcome** is a technical term for the most fundamental of the results of an experiment. A **finite experiment** has only a finite number of possible outcomes. In addition, a finite experiment is **equiprobable** when each possible outcome is equally likely.*

As may be apparent, there are a variety of different types of experiments. The next example describes a few.

Example 5.3.1   In the experiment of rolling a standard die, the six possible outcomes are determined by the number of dots appearing on the die roll. This experiment is also equiprobable (when the die is fair) because each of the six rolls has an equal probability of occurring $1/6$ of the time.

When predicting the weather, a given day's outcomes might be broadly categorized as sunny, rainy, sleeting, or snowy. Based on these categories, predicting the weather is a finite experiment. However, it is rarely equiprobable; in some locales, the outcome of sunny is more common than rainy, while in others rainy is more common than sunny.

When measuring the distance between two objects, the outcomes are the possible distances. This experiment is not finite, since any nonnegative distance is possible.

■

Question 5.3.1   Provide additional real-life examples of:

(a) a finite, equiprobable experiment;
(b) a finite experiment that is not equiprobable;
(c) an infinite experiment.

■

In practical applications, probability theorists have found it useful to refer to various collections of outcomes. The largest such set is the collection of all possible outcomes; it is referred to as the *sample space* and labeled S. Subsets of the sample space are *events*. Simply put, an event is some result of the experiment (an outcome is a most basic type of event). For example, in the finite experiment of tossing a two-sided coin twice, the set of all possible outcomes (the sample space) is $S = \{HH, HT, TH, TT\}$, where H denotes heads and T denotes tails. An example of an event in this setting is $\{HH, HT, TH\}$, which can be described as "toss heads at least once." We can see that many different outcomes may result in the same event. Another event is $\{HH\}$, or "toss heads twice." We now give the formal mathematical definitions of these notions.

Definition 5.3.2   *The **sample space** of an experiment is the set S of all possible outcomes of the experiment. An **event** is a subset of the sample space, and so it is a collection of* *outcomes of the experiment. We use capital letters, such as A and B, to identify events.*

We consider further examples of sample spaces and events in the following example and question.

Example 5.3.2   We consider the experiment of rolling a standard six-sided die twice. In this setting, each element of the sample space may be conveniently represented as an ordered pair (x, y), where x is the roll on the first die and y is the roll on the second. Using this notation, the sample space of this experiment is the set S consisting of the following 36 ordered pairs:

$(1,1)$, $(1,2)$, $(1,3)$, $(1,4)$, $(1,5)$, $(1,6)$, $(2,1)$, $(2,2)$, $(2,3)$, $(2,4)$, $(2,5)$, $(2,6)$,

$(3,1)$, $(3,2)$, $(3,3)$, $(3,4)$, $(3,5)$, $(3,6)$, $(4,1)$, $(4,2)$, $(4,3)$, $(4,4)$, $(4,5)$, $(4,6)$,

$(5,1)$, $(5,2)$, $(5,3)$, $(5,4)$, $(5,5)$, $(5,6)$, $(6,1)$, $(6,2)$, $(6,3)$, $(6,4)$, $(6,5)$, $(6,6)$.

Many different events can be described in this setting, including:

- $A =$ "the first roll is 1" $= \{(1, 1), (1, 2), (1, 3), (1, 4), (1, 5), (1, 6)\}$;
- $B =$ "the sum of the dice is four" $= \{(1, 3), (2, 2), (3, 1)\}$;
- $C =$ "the sum of the dice is twelve" $= \{(6, 6)\}$.

■

Question 5.3.2   Continuing to work in the setting of example 5.3.2, state the elements in each event.

(a) $D =$ "the second roll is two or three"   (c) $F =$ "roll doubles"

(b) $E =$ "the sum of the digits is six"   (d) $G =$ "the first roll is prime"

■

We can now state Pascal and Fermat's definition of the probability of simple types of events, which they developed in an exchange of letters in 1654. Their formulation is widely recognized as the advent of probability theory. After Pascal and Fermat developed this theory, their ideas were taken up by other researchers and gradually became an important element of popular culture. You will hopefully find the formal definition of probability intuitive; it has become such a part of the natural discourse of society that we innately understand many of its basic notions. Recall that $|A|$ denotes the number of elements in a set A.

Definition 5.3.3   *For any finite, equiprobable experiment, the probability of an event A is*

$$P[A] = \frac{|A|}{|S|} = \frac{\text{the number of outcomes resulting in } A}{\text{the number of outcomes in the sample space } S}.$$

While this definition does not describe what happens for experiments that are not finite and equiprobable, it is an important start. Since Pascal's and Fermat's original work, mathematicians have developed mathematical models that can handle any type of experiment in probability theory; we will examine some of these models later in this section. But we first focus on finite, equiprobable experiments and study simple, easily managed examples.

Example 5.3.3   Example 5.3.2 considered the experiment of rolling a standard six-sided die twice and found that S is the set of 36 ordered pairs of dice rolls. Assuming the dice are fair, each of these outcomes is equally likely, and so the experiment is equiprobable. We can thus apply definition 5.3.3 to compute the probabilities of events; those discussed in example 5.3.2 follow.

- If $A =$ "the first roll is 1" $= \{(1, 1), (1, 2), (1, 3), (1, 4), (1, 5), (1, 6)\}$, then $P[A] = \dfrac{6}{36} = \dfrac{1}{6}$;
- If $B =$ "the sum of the dice is four" $= \{(1, 3), (2, 2), (3, 1)\}$, then $P[B] = \dfrac{3}{36} = \dfrac{1}{12}$;
- If $C =$ "the sum of the dice is twelve" $= \{(6, 6)\}$, then $P[C] = \dfrac{1}{36}$.

∎

Question 5.3.3   Continuing in the setting of example 5.3.3, state the probability of each event; your answers to question 5.3.2 may be helpful.

(a) $D =$ "the second roll is two or three"   (c) $F =$ "roll doubles"
(b) $E =$ "the sum of the digits is six"   (d) $G =$ "the first roll is prime"

∎

While considering such playful games as tossing games and rolling dice is a vital and helpful approach to first learning basic probability theory, it is also important to realize that the insights and understandings acquired from studying these basic scenarios extends far beyond such games. Many real-life situations involve finite, equiprobable experiments; the probabilities in these settings are computed using the techniques already outlined. The next example provides an easily understood query from an industrial setting.

Example 5.3.4   A quality-control engineer at a film manufacturing plant is testing for defective film cannisters. Assume that a batch of 50 cannisters contains two that are defective. If the engineer randomly tests five cannisters for defects, what is the probability that exactly one of them is defective?

We count the number of possible outcomes when randomly choosing five cannisters from the batch of fifty; since order is not important and repetition is not allowed, this number is $C(50, 5) = 2,118,760$. This experiment is therefore finite, and the selection being random ensures it is equiprobable. Defining the event $A =$ "a single defective cannister appears in the sample of five chosen," we think of A as being obtained by a two-step process. In the first step, one of the two defective cannisters in the batch of 50 is chosen for the sample; this step can be done in $C(2, 1) = 2$ ways. In the second step, four of the 48 good cannisters in the batch of 50 is chosen for the sample; this step can be done in $C(48, 4) = 194,580$ ways. Applying the multiplication principle and definition 5.3.3, $P[A] = 194,580 \cdot 2/2, 118, 760 = 9/49 \approx 0.18367$. Thus, there is approximately an 18 percent chance of finding exactly one defective film in a random selection of five cannisters.

∎

We have focused thus far on finite experiments. The discussion of experiments having infinitely many possible outcomes is more subtle, but turns out to be manageable using ideas from calculus. Mathematicians describe a given experiment in terms of a *random variable*, which assigns a numeric value to each outcome of the experiment. A random variable acts like a function; its domain value inputs are the outcomes of the experiment, and its range value outputs are the corresponding numbers that it assigns.

Mathematicians have naturally adopted a notation for random variables that closely mirrors the notation for a function. Just as f, g, and u are common names for functions, capital letters (such as X, Y, or Z) are common names for random variables. And just as $f(t)$ describes a functional output for a given input value t, the notation $X(\omega)$ describe a random variable's numeric output for a given input event ω. We make these ideas precise in the following definition.

Definition 5.3.4 *Given an experiment and resulting sample space S containing all possible outcomes, a **random variable** X assigns a number X(ω) to each element $\omega \in S$. In this way, the random variable X acts like a function on the sample space S; sometimes X is called a **random function**. In addition, for any value x that X might assign, the notation **X** $=$ **x** identifies the event that X assigns the number x. We write P[X = x] for the probability of this event.*

Example 5.3.5 We may define a simple (but useful) example of a random variable X on the experiment of tossing a coin, where $X(\omega) = 1$ if the experiment results in the event ω of "tossing heads," and $X(\omega) = 0$ if ω is "tossing tails." For a fair coin, the probability of tossing heads and the probability of tossing tails are both equal to $1/2$, and so $P[X = 0] = 1/2$ and $P[X = 1] = 1/2$.

■

Question 5.3.4 Define a random variable Y on the experiment of rolling a fair six-sided die, and determine the corresponding probabilities for each possible event of the form $Y = y$.

■

Some random variables are categorized as being *discrete*, which means that the random variable assigns only a countable number of possible outputs. For discrete X, a probability $P[X = x]$ is generally positive for any value x that X assigns. In contrast, a *continuous* random variable X assigns an uncountably infinite number of different values, and the probability $P[X = x]$ is generally zero for any value x. Other types of random variables are more complicated and more difficult to categorize; some are a combination of both discrete and continuous types, still others require advanced notions such as the Lebesgue integral to handle mathematically. In this chapter, we assume that any given random variable is either (purely) discrete or (purely) continuous, and we will develop the mathematical models for each of these types. As it turns out, almost every real-life probabilistic question is either discrete or continuous, and so these ideas are broadly applicable in many diverse settings.

5.3.1   Discrete Random Variables

We begin a study of discrete random variables by discussing the experiment that motivated the development of probability theory. Recall that the Chevalier de Méré asked Pascal and Fermat the question, "How many times must I roll two fair dice before there is better than a 50–50 chance of rolling double-sixes?" While this concern would not impact any of our lives, many profound insights and results have followed from its solution. The Chevalier's question asks about the experiment of "rolling the dice until double-sixes occurs." An important associated discrete random variable is $X = $ "the number of times the dice are rolled." Thus, if the Chevalier happens to roll double-sixes on the very first roll, then $X = 1$. But if it takes 30 rolls before he rolls double-sixes, then $X = 30$. Since it can only take on positive integer values, X is a discrete random variable. Furthermore, X is infinite because it can take on any positive integer value.

What are the probabilities associated with this random variable X? We might start by considering the probability that the Chevalier rolls double-sixes on the first roll, finding $P[X = 1]$. As we observed in example 5.3.3, the probability of rolling double-sixes on a single dice roll is $1/36$; we thus conclude that $P[X = 1] = 1/36$. In order to determine the probabilities of other values of this random variable X, we need to use three important results about probabilities, which we present in the next theorem.

Theorem 5.3.1   *Suppose A and B are given events.*

- *The probability of the* **complementary event** $A' = $ *"A does* **not** *happen" satisfies*

$$P[A'] = 1 - P[A].$$

- *If the occurrence of A does not affect the probability of B, then we say that A and B are* **independent events**; *in this case, the probability that both A and B occur is given by*

$$P[A \text{ and } B] = P[A] \cdot P[B].$$

- *If A and B are independent, so are A' and B.*

Comments on Proof   We consider the proof of the first statement only for finite, equiprobable experiments, though the results hold in general. In this case, the formula for probabilities of complementary events follows from the set-theoretic relationship between sizes of complementary sets. Since $A' = S \setminus A$, we have $|A'| = |S \setminus A| = |S| - |A|$. Substituting this fact into the formula in definition 5.3.3,

$$P[A'] = \frac{|A'|}{|S|} = \frac{|S| - |A|}{|S|} = 1 - \frac{|A|}{|S|} = 1 - P[A].$$

An alternative proof of this property can be given in terms of "mutually exclusive" events; we discuss this notion shortly and leave this proof for the exercises at the end of this section. The standard proof of the formula for independent events relies on the notion of "conditional" probability and is left for your later studies. ∎

Theorem 5.3.1 is useful in a continuing analysis of the Chevalier's question to Pascal and Fermat.

Example 5.3.6   Recall the Chevalier's question: "How many times must I roll two fair dice before there is better than a 50–50 chance of rolling double-sixes?" The corresponding discrete random variable $X =$ "the number of rolls until double-six occurs" has $P[X = 1] = 1/36$. We now determine the probabilities associated with the other possible values of the random variable X.

The event $X = 2$ happens when the Chevalier fails to roll double-sixes on the first roll, but succeeds in rolling double-sixes on the second. Defining events $A =$ "roll double sixes on the first roll" and $B =$ "roll double-sixes on the second roll," we see that A and B are independent because the chance that B occurs is unaffected by whether or not A has occurred (in either case, $P[B] = 1/36$). Applying theorem 5.3.1,

$$P[A'] = 1 - P[A] = 1 - \frac{1}{36} = \frac{35}{36}$$

and

$$P[X = 2] = P[A' \text{ and } B] = P[A'] \cdot P[B] = \frac{35}{36} \cdot \frac{1}{36}.$$

In the exact same way but adding one more layer of dice rolls,

$$P[X = 3] = \frac{35}{36} \cdot \frac{35}{36} \cdot \frac{1}{36} = \frac{1}{36} \cdot \left(\frac{35}{36}\right)^2.$$

For an arbitrary number of dice rolls n,

$$P[X = n] = \frac{1}{36} \cdot \left(\frac{35}{36}\right)^{n-1}.$$

■

Example 5.3.6 comes close to answering the Chevalier's question, as it provides the probability of getting double-sixes on any given number of dice rolls. But the Chevalier is asking for more: "How many times do I have to roll two fair dice before there is better than a 50–50 chance of rolling double-sixes?" Since $P[X = 1] = 1/36 \approx 0.02778$, there is less than a 3 percent chance of rolling double-sixes on the first roll. To determine the probability of rolling double-sixes on either of the first two rolls, we calculate $P[X = 1 \text{ or } X = 2]$. As you might expect, this probability will still be less than 50 percent, and so we turn to $P[X = 1 \text{ or } X = 2 \text{ or } X = 3]$, and $P[X = 1 \text{ or } X = 2 \text{ or } X = 3 \text{ or } X = 4]$, and so on until the probability reaches 50 percent. The next theorem provides an important mathematical tool to calculate these probabilities.

Theorem 5.3.2 *When events A and B can never happen at the same time, then they are called* **mutually exclusive**. *Mutually exclusive events satisfy*

$$P[A \text{ or } B] = P[A] + P[B].$$

Proof We consider a proof for finite, equiprobable experiments, though the results hold in general. In this case, the formula follows from the corresponding set-theoretic relationship between sizes of disjoint sets. Recall that an "or" relationship is expressed set-theoretically in terms of the union operation, denoted by "∪." Mutually exclusive events A and B have no outcome in common, and so they are disjoint sets that satisfy $|A \cup B| = |A| + |B|$. Combining this fact with the formula from definition 5.3.3,

$$P[A \text{ or } B] = \frac{|A \cup B|}{|S|} = \frac{|A| + |B|}{|S|} = \frac{|A|}{|S|} + \frac{|B|}{|S|} = P[A] + P[B].$$

■

We can now complete the analysis of the question posed by the Chevalier. This final part of that project utilizes the *geometric identity* from basic algebra:

$$1 + p + p^2 + \cdots + p^{n-1} = \frac{1 - p^n}{1 - p}.$$

This identity holds for every real number p and $n \in \mathbb{N}$ (except for $p = 1$ because of division by zero) and is easily verified by long division of the corresponding polynomial expressions.

Example 5.3.7   We determine the answer to the Chevalier's question, "How many times must I roll two fair dice before there is better than a 50–50 chance of rolling double-sixes?" From example 6, we know $P[X = n] = 1/36 \cdot (35/36)^{n-1}$, where $X = $ "the number of rolls until double-six occurs."

Events such as $X = 1$ and $X = 2$ are mutually exclusive, since the *first* occurrence of double-six cannot simultaneously happen on different numbers of dice rolls. By theorem 5.3.2, the probability of rolling double-sixes on one of the first two dice rolls is

$$P[X \le 2] = P[X = 1 \text{ or } X = 2] = P[X = 1] + P[X = 2] = \frac{1}{36} + \frac{1}{36} \cdot \left(\frac{35}{36}\right) \approx 0.05478.$$

This approach extends to the general case of rolling double-six on one of the first n rolls; using theorem 5.3.2 and the geometric identity,

$$
\begin{aligned}
P[X \le n] &= P[X = 1 \text{ or } X = 2 \text{ or } \ldots \text{ or } X = n] \\
&= P[X = 1] + P[X = 2] + \cdots + P[X = n] \\
&= \frac{1}{36} + \frac{1}{36} \cdot \left(\frac{35}{36}\right) + \cdots + \frac{1}{36} \cdot \left(\frac{35}{36}\right)^{n-1} \\
&= \frac{1}{36} \cdot \left[1 + \frac{35}{36} + \cdots + \left(\frac{35}{36}\right)^{n-1}\right] \\
&= \frac{1}{36} \cdot \frac{1 - \left(\frac{35}{36}\right)^n}{1 - \frac{35}{36}} = 1 - \left(\frac{35}{36}\right)^n.
\end{aligned}
$$

Substituting $n = 24$ and $n = 25$ into this equation gives

$$P[X \le 24] = 1 - \left(\frac{35}{36}\right)^{24} \approx 0.49140 \quad \text{and} \quad P[X \le 25] = 1 - \left(\frac{35}{36}\right)^{25} \approx 0.50553.$$

The Chevalier therefore needs to roll the dice 25 times to be assured of a better than 50–50 chance of rolling double-sixes.

■

For any discrete random variable, mathematicians refer to $P[X = x]$ as the *probability distribution function*, or simply the *probability distribution*. A probability distribution identifies all the probabilities associated with a given discrete random variable X in terms of the particular numbers x that X might assign. The probability distribution is therefore a complete description of the chance elements associated with the random variable. We formalize this notion in the following definition.

Definition 5.3.5 *If X is a discrete random variable, then every value x assigned by X has a corresponding probability P[X = x]. These values form the* **probability distribution function for** *X, or simply the* **probability distribution**.

Example 5.3.8 We consider the finite, equiprobable experiment of rolling a fair six-sided die once. An associated random variable is $X = $ "the numeric result of the die roll." The sample space is $S = \{1, 2, 3, 4, 5, 6\}$, and the probability distribution for X is $P[X = n] = \frac{1}{6}$ for $n = 1, \ldots, 6$. Sometimes probability distributions are illustrated graphically; in this example, the probability distribution for this experiment is given in figure 5.5.

■

Question 5.3.5 Consider the finite, equiprobable experiment of tossing a fair coin until heads is tossed, along with the associated random variable $X = $ "the number of tosses until heads is tossed for the first time." Find the formula for the probability distribution $P[X = n]$ for $n \in \mathbb{N}$ and sketch the (infinite) graph that illustrates it. In parallel with the Chevalier's question about dice, how many times must a fair coin be tossed until there is better than a 90 percent chance of tossing heads?

■

Mathematicians often categorize a discrete random variable based on the type of formula that describes its probability distribution. For example, a random variable with probability distribution $P[X = n] = p \cdot (1 - p)^{n-1}$ (as in example 5.3.6) is referred to as a *geometric random variable*. There are many other important categories of discrete random variables; we highlight two other examples in this text: the binomial and the hypergeometric. The accompanying discussion of these random variables includes examples and questions, which provide further illustration; they are based on real-world scenarios and provide some indication of the random variables' breadth of applicability and usefulness.

5.3.2 Geometric Discrete Random Variables

The analysis undertaken in examples 5.3.6 and 5.3.7 works for any experiment that involves a number of independent repetitions of some process (such as the repeated rolls

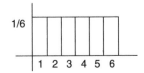

Figure 5.5 The Probability distribution for example 5.3.8

of two dice) and ends with the occurrence of some event (such as rolling double sixes). In such settings, the random variable counting how many repetitions have occurred at the end of the process (such as $X = $ "the number of rolls until double sixes") is sometimes referred to as the *stopping time*. In addition, because of the mathematical formula involved (the geometric formula identified just before example 5.3.7), this random variable is classified as a *geometric random variable*. If p is the probability that the event will occur on any given repetition (such as $p = 1/36$ in example 5.3.6), then the same analysis as above shows that for every $n \in \mathbb{N}$, we have

$$P[X = n] = p \cdot (1 - p)^{n-1}.$$

Example 5.3.9   A broker decides to sell a package of Dorsicom stock as soon as its value increases by at least five percent. Based on past performance, there is an approximately two percent chance of such an increase on any given day. What is the probability that the broker sells his package of Dorsicom stock by the end of the fifth business day?

The random variable $X = $ "the number of days the stock is held" is geometric with $p = 0.02$. Therefore, the corresponding geometric probability distribution is given by $P[X = n] = 0.02 \cdot (0.98)^{n-1}$. Proceeding as in example 5.3.7, we obtain

$$P[X \le 5] = P[X = 1] + P[X = 2] + \cdots + P[X = 5] = 1 - (.98)^4 \approx 0.09608.$$

∎

Question 5.3.6   An assembly line at Turbo Motor Corporation shuts down whenever three components malfunction at any one time. If there is approximately a one percent chance of a shutdown on any given day of operation, what is the probability that the assembly line will shut down by the end of the twentieth day of operation?

∎

5.3.3   Binomial Discrete Random Variables

The probability distribution of a *binomial random variable* Y is defined on $y = 0, 1, 2, \ldots, n$ for some fixed positive integer n by the formula

$$P[Y = y] = C(n, y) \cdot p^y \cdot (1 - p)^{n-y}.$$

In general, a binomial random variable Y counts the number of times some "good" outcome occurs when a process is repeated n times, where each independent repetition of the process results in exactly one of two outcomes (either a "good" or "bad" outcome). The repetitions are often called "trials." In the above formula, p is the probability of the good outcome occurring on any given trial, and so $1 - p$ is the probability of the bad outcome occurring. For example, for the experiment of tossing a fair coin n times, the random variable $Y = $ "the number of times heads is tossed" is binomial with $p = 1/2$. The formula for $P[Y = y]$ is obtained by first choosing the y trials out of n in which the good outcome occurs and then taking the product of the probabilities for the y good outcomes and the $n - y$ bad outcomes.

Example 5.3.10   Based on a previous analysis of its manufacturing process, an industrial manufacturer believes that an assembly line machine produces defective bolts three percent of the time, with no predictable indication of when a defective bolt will occur. During a follow-up test, the manufacturer counts the number of defective bolts appearing in a randomly selected group of 15. Assuming the previous analysis is correct and that there is an endless supply of bolts, what is the probability that one or fewer bolts in this group is defective?

The random variable Y = "the number of defective bolts in the group" is binomial with $n = 15$ members in the group and a probability $p = 0.03$ that a defective bolt is selected. A bolt is either defective or not (that is, considered "good" or "bad" if we swap the usual interpretation of these words). The trials are independent because the bolts are randomly selected, and therefore a future occurrence of a defective bolt is not predicted based on a previous selection. The corresponding binomial probability distribution is therefore given by $P[Y = y] = C(15, y) \cdot (0.03)^y \cdot (0.97)^{15-y}$, and the probability that Y is one or fewer is

$$P[Y \leq 1] = P[Y = 0] + P[Y = 1]$$

$$= C(15, 0) \cdot (0.03)^0 \cdot (0.97)^{15} + C(15, 1) \cdot (0.03)^1 \cdot (0.97)^{14} \approx 0.92703.$$

■

Question 5.3.7   A basketball player shoots 12 free throws in a contest. Assuming the player makes 85 percent of her free throws and that the outcomes for each free throw shot are independent, what is the probability that the player will make at least 10 out of the 12 shots?   ■

5.3.4   Hypergeometric Discrete Random Variables

The probability distribution of a *hypergeometric random variable* S is defined on $s = 0, 1, 2, \ldots, M + N$ for fixed positive integers M and N by the formula

$$P[S = s] = \frac{C(M, s) \cdot C(N, n - s)}{C(M + N, n)}.$$

In general, a hypergeometric random variable S counts the number of "good" elements randomly selected from a set containing M "good" elements and N "bad" elements, when a total of n elements are selected (without replacement). For example, the experiment of selecting $n = 10$ red and green balls from a jar containing $M = 15$ red balls and 16 green balls has an associated hypergeometric random variable S = "the number of red balls selected in 10 draws."

Example 5.3.11   Fifty people have applied for a position; 29 applicants are female. Suppose the company follows equal opportunity hiring guidelines and that the applicants are all essentially equally qualified for the position. What is the probability that a fair selection process for determining three final candidates does not result in a selection of any of the female applicants?

The random variable S = "the number of females in the group of three final candidates" is hypergeometric with $M = 29$ female applicants, $N = 21$ male applicants, and a group of $n = 3$ candidates that are selected. The corresponding hypergeometric probability distribution and the probability that no female applicant is selected for the group of three finalists are therefore

$$P[S=s] = \frac{C(29,s) \cdot C(21, 3-s)}{C(50,3)} \quad \text{and} \quad P[S=0] = \frac{C(29,0) \cdot C(21,3)}{C(50,3)} \approx 0.06786.$$

■

Question 5.3.8   Another example of a hypergeometric random variable is $S = $ "the number of Foodmerchant grocery stores that are profitable" when 20 Foodmerchant groceries are randomly selected from the total of 70 chain locations of which 52 are profitable.

(a) State the probability distribution formula for $P[S = s]$.
(b) What is the probability that 19 or 20 of the selected stores are profitable?
(c) If three Foodmerchant stores have already been selected and are known to be profitable, what is the probability that 17 stores randomly selected from the remaining 67 locations identifies 16 or 17 profitable stores?

■

5.3.5   Continuous Random Variables

Continuous random variables differ markedly from discrete random variables, both in their descriptions and in the mathematical structures used to model them. Discrete random variables are usually defined as the *number* of some object, such as "the number of cars traveling over the speed limit on a given stretch of highway," or "how many in a survey group were in favor of gun control," or "the number of diabetics who have properly controlled their weight." In contrast, continuous random variables are usually defined as the *amount* of some object, such as "the volume of paint remaining in a gallon bucket," or "the weight of a newborn baby," or "the speed of an automobile traveling a given stretch of highway."

For a discrete random variable X, the probability distribution function $P[X = x]$ is the mathematical structure best-suited to analyzing X and providing insight into its behavior. In contrast, probability distributions are useless for studying continuous random variables; it turns out that any continuous X has $P[X = x] = 0$ for every possible value x. Instead of studying them in terms of distribution functions, continuous random variables are mathematically modeled in terms of density functions, which we define below.

Definition 5.3.6   *A **probability density function** $f(x)$ is a real-valued, Riemann integrable function with domain $\mathbb{R}$ that satisfies the properties:*

- *$f(x) \geq 0$ for every $x \in \mathbb{R}$;*
- *$\displaystyle\int_{-\infty}^{\infty} f(x)\, dx = 1$.*

A continuous random variable X is mathematically defined as having a corresponding density function $f(x)$ such that for every $a, b \in \mathbb{R}$

$$P[a < X < b] = \int_a^b f(x)\, dx.$$

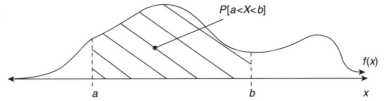

Figure 5.6 The correspondence between $P[a < X < b]$ and bounded area

For any value $a \in \mathbb{R}$ and any density function $f(x)$, $\int_a^a f(x)\,dx = 0$. We can therefore always replace "$<$" by "$\leq$" in the expression $P[a < X < b]$; in other words, for any continuous random variable X,

$$P[a < X < b] \;=\; P[a \leq X < b] \;=\; P[a < X \leq b] \;=\; P[a \leq X \leq b].$$

In addition, since the definite integral of f equals the area bounded by the x-axis and the integrand, Definition 5.3.6 links probabilities associated with continuous random variables to the corresponding areas bounded by the density function's curve. Figure 5.6 illustrates the relationship between probabilities and bounded areas.

In this way, the density function provides us with a mathematical framework upon which to build a suitable model for continuous random variables. The profound importance and usefulness of this model will become apparent as we explore its mathematical properties in the rest of the chapter.

Just as mathematicians have identified various categories for discrete random variables, there are also important categories of continuous random variables. Two of the most important types of random variable are the uniform and the normal. As usual, we define each of these random variables and develop insights into their behavior by considering examples and questions.

5.3.6 Uniform Continuous Random Variables

Uniform random variables are important because of their relative simplicity – the defining density functions for these random variables is constant over a given interval. The probability density function for a *uniform random variable X* is of the form

$$f(x) = \begin{cases} \frac{1}{b-a} & \text{if } x \in [a, b] \\ 0 & \text{otherwise.} \end{cases}$$

Equivalently, we say that a random variable X is *uniformly distributed* over the interval $[a, b]$ if the corresponding density function is of this form. The corresponding graphical presentation of the density function is illustrated in figure 5.7.

Example 5.3.12 In a proton accelerator, the projected distance between two protons in a "zone of impact" is uniformly distributed in the interval $[0, 90]$ (measured in microns). We determine the probability that a camera image of two protons (taken in the zone of impact) reveals protons between 20 and 40 microns apart.

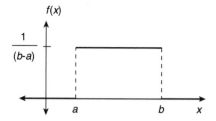

Figure 5.7 The graph of a uniform random variable

First, define the uniform random variable $X = $ "the distance in question," and then calculate

$$P[20 < X < 40] = \int_{20}^{40} \frac{1}{90 - 0} \, dx = \frac{x}{90} \Big]_{20}^{40} = \frac{40 - 20}{90} = \frac{2}{9} \approx 0.22222.$$

■

Question 5.3.9 The final horizontal position (measured in feet from the center point) of a projectile fired in a walled chamber 60 feet across is assumed to be uniformly distributed in the interval $[-30, 30]$. Find the probability that a projectile will have a final horizontal position that is within 10 feet of the target center point.

■

5.3.7   Normal Continuous Random Variables

Normal random variables are important because of their broad applicability—the density function for a normal random variable is a "bell-shaped curve." As a functional expression, a *normal random variable X* has a density function of the form

$$f(x) = \frac{1}{\sigma\sqrt{2\pi}} \cdot e^{-(x-\mu)^2/(2\sigma^2)}.$$

Here μ and $\sigma > 0$ are given constants, respectively called the *mean* and *standard deviation*. Equivalently, we say that a random variable X is *normally distributed with mean μ and standard deviation σ*. In addition, a normal density function is sometimes called *Gaussian* in honor of the contributions made by Carl Freidrich Gauss to the study of such functions. The normal density function f has a graphical representation that is illustrated as a bell-shaped curve in figure 5.8.

Example 5.3.13 The weight of full-grown labrador retrievers is believed to be approximately normally distributed with a mean of $\mu = 80$ pounds and a standard deviation of $\sigma = 30$. Suppose a labrador retriever is randomly selected from a registered

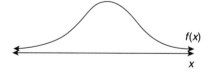

Figure 5.8 The graph of a normal (or Gaussian) random variable

list of newborn pups. We determine the probability that the selected retriever will eventually weigh between 75 and 90 pounds as an adult.

Set X = "the weight of an adult labrador retriever." The desired probability $P[75 < X < 90]$ matches the area under the normal density function (with $\mu = 80$ and $\sigma = 30$) and between $x = 75$ and $x = 90$. This area is found via the normal density function integral as

$$P[75 < X < 90] = \int_{75}^{90} \frac{1}{30\sqrt{2\pi}} \cdot e^{-(x-80)^2/(2 \cdot 30^2)} \, dx \approx 0.1967.$$

The approximate value of this integral was obtained by numerically estimating the integral using a computing device, which we discuss below.

■

We will want and need to use computing devices to obtain numerical approximations for the integrals of normal density functions. We can find these values using appropriate commands for the exponential function and for the numerical value of an integral. Alternatively, the use of the normal distribution as a statistical model in real-life settings has become so prevalent that many graphing calculators and computer algebra systems include special commands for numerically approximating the normal density function. We provide examples of several commands using the context provided by example 5.3.13, where we calculated $P[75 < X < 90]$ for $\mu = 80$ and $\sigma = 30$.

System	Command
TI-83	DISTR - normalcdf normalcdf(75,90,80,30)
Maple	[> with(stats): [> statevalf[cdf,normald[80,30]](90) - statevalf[cdf,normald[80,30]](75);
Mathematica	]:<<Statistics 'NormalDistribution']:CDF[NormalDistribution[80,30], 90] - CDF[NormalDistribution[80,30], 75]

When using computing devices to study the normal distribution, you should consult the manual for your calculator or computer algebra system to determine commands that return normal probabilities. Some systems return probabilities of the form $P[a \leq X \leq b]$ (as for the TI-83). Others return probabilities of the form $P[X \leq a]$ (as for Maple); in this case, a desired probability $P[a \leq X \leq b]$ is found by taking a difference

$$P[a \leq X \leq b] = P[X \leq b] - P[X \leq a].$$

Drawing a picture of the probability as an area under the density function can help you determine how to manipulate a probability algebraically; for example, since the total area under the normal density is 1, a picture as in figure 5.9 quickly indicates that $P[X \leq a] = 1 - P[X > a]$.

Question
5.3.10 Consider the normal random variable X = "the average quarterly profit" that is calculuted from 10 randomly selected quarterly profit/loss reports of Oilmark, a Canadian sand oil distributer. Based on historical performance, the mean of this

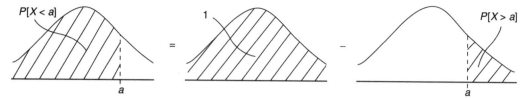

Figure 5.9   An illustration of $P[X \leq a] = 1 - P[X > a]$

average profit (in millions) is $\mu = 1.2$ with a standard deviation of $\sigma = 0.15$. Assuming these values remain constant over time, what is the probability that 10 randomly selected quarters over the next many years of operation will have an average between \$1.3 and \$1.5 million? In this setting, what is the probability of such an average quarterly profit being above \$1.5 million?

∎

Many other important types of continuous random variables play a prominent role in statistical models applied to real-life settings. Several important categories of continuous random variables are the Exponential and Weibull distributions (useful in engineering modeling problems), the Student's t- and F- distributions (useful in statistical analysis), and the Chi-square distribution (useful in analyzing risks associated with life factors). We present some of these distributions in the exercises at the end of this section.

5.3.8   Some Parameters for Statistics

Mathematicians approach a study of chance by means of experiments, defining random variables to isolate important properties of interest. *Statistics* is the study of random variables, analyzing the numbers assigned by random variables and seeking patterns in data sets for past events (often as a means of predicting future events). In statistics, mathematicians have isolated important values that help describe random variables. Referred to as *parameters*, the two most important of these values are the mean and standard deviation. We present the definitions of these parameters for both the discrete and continuous cases.

Definition   *If X is a random variable, then the "average" value of X is called the* **mean** *of*
5.3.7   *X and is denoted by μ; sometimes we refer to the mean as the* **expectation** *of X and denote it by E[X]. For a discrete random variable X, the mean is defined by a sum computed over each value x assigned by X:* $\mu = E[X] = \sum x \cdot P[X = x]$.

For a continuous random variable X with corresponding density function $f(x)$, the mean is defined by an integral: $\mu = E[X] = \displaystyle\int_{-\infty}^{\infty} x \cdot f(x)\, dx.$

The mean is the average value assigned by the random variable X. Many real-world problems depend on the calculation of an average or are concerned with the magnitude of an average. A simple example in the real estate business is the importance of an average home price in a neighborhood. The use of definition 5.3.7 to calculate $E[X]$ is straightforward and is the topic of the next two questions.

Figure 5.10 Different distributions
with $\mu = 3$

Question 5.3.11 Consider an experiment of tossing a fair coin one time. Define a discrete random variable X for this experiment by setting $X = 0$ if heads is tossed and $X = 1$ if tails is tossed. Using the definition for the mean of a discrete random variable, prove the mean of X is $\mu = 1/2$. ■

Question 5.3.12 Use the definition for the mean of a continuous random variable and the formula for a uniform density function to prove that a uniform random variable X on an interval $[a, b]$ has mean $E[X] = (a + b)/2$. ■

The measure of the "average" or "central" value of a random variable as provided by the mean gives us important insight into the behavior of the random variable. Another important feature of random variables is how the outcomes of an experiment are spread out around the mean. Figure 5.10 provides a visual presentation of two discrete random variables with identical means, but with very different distributions.

The statistical parameter used to describe the way in which a random variable X assigns values spread out around the mean is known as the standard deviation of X. The next definition provides formulas for each of the discrete and continuous cases.

Definition 5.3.8 *If X is a random variable, the **standard deviation of X**, which is denoted by σ, provides a measurement of how much the values assigned by X differ from the mean. For a discrete random variable X, the standard deviation is defined as the square root of a sum computed over each value x assigned:*

$\sigma = \sqrt{\sum (x - \mu)^2 \cdot P[X = x]}$. *For a continuous random variable X with corresponding density function $f(x)$, the standard deviation is defined as the square root of an integral:* $\sigma = \sqrt{\int_{-\infty}^{\infty} (x - \mu)^2 \cdot f(x)\, dx}$.

Question 5.3.13 Consider the experiment of tossing a fair coin one time. Define a discrete random variable X for this experiment by setting $X = 0$ if heads is tossed and $X = 1$ if tails is tossed. Using the definition for the standard deviation of a discrete random variable, prove the standard deviation of X is $\sigma = 1/2$. ■

Question 5.3.14 Using the definition for the standard deviation of a continuous random variable, prove that a uniform random variable X on an interval $[a, b]$ has standard deviation $\sigma = (b - a)/\sqrt{12}$. ■

The mean and standard deviation of a random variable are just two of the statistical parameters that are used to develop an insight and an understanding of random variables. We hope that this glance at these parameters will motivate you to continue an exploration of the mathematics of random variables in your later studies.

5.3.9   Reading Questions for Section 5.3

1. Define and give an example of an experiment.
2. When is a finite experiment equiprobable? Give an example of such an experiment.
3. Define and give an example of an event.
4. State the formula for the probability of an event A occurring as a result of a finite, equiprobable experiment.
5. What is a discrete random variable? Give two examples of discrete random variables—one finite and one infinite.
6. What is a probability distribution function for a discrete random variable?
7. What is a geometric random variable? Give an example of a real-life situation generating a geometric random variable.
8. Define and give an example of a continuous random variable.
9. What is a density function for a continuous random variable?
10. What is a normal random variable? Give an example of a real-life situation generating a normal random variable.
11. State the formulas for the mean and standard deviation of a discrete random variable X.
12. State the formulas for the mean and standard deviation of a continuous random variable X with density function f.

5.3.10   Exercises for Section 5.3

Exercises 1–10 consider the experiment of tossing three fair coins and recording the resulting sequence of heads and tails.

1. Without listing the sample space S, compute the number of elements in S using the appropriate combinatorial formula.
2. Identify the outcomes in the sample space S.
3. Identify the outcomes in the event $A =$ "heads is tossed on the first coin."
4. Determine the probability that heads is tossed on the first coin.
5. Identify the outcomes in the complementary event A' for event A from exercise 3.
6. Determine the probability of event A' from exercise 4 in two different ways.
7. Identify the outcomes in the event $B =$ "heads is never tossed."
8. Identify the outcomes in the event $C =$ "heads is tossed twice."
9. Determine the probability of events B from exercise 7 and event C from exercise 8.
10. Suppose this experiment is modified so that only the number of heads tossed is recorded. State the corresponding sample space for this modified experiment. Explain why this new experiment is no longer equiprobable.

Exercises 11–18 consider an experiment in which researchers randomly select participants in a medical study from a pool of 1,000 patients; in this pool 20 patients have some symptoms of cardiovascular disease.

11. Find the total number of outcomes if 15 patients are selected for the study.
12. If 15 are selected, what is the probability that exactly two have cardiovascular symptoms?
13. If 15 are selected, what is the probability that two or fewer have cardiovascular symptoms?
14. If 15 are selected, what is the probability that more than two have cardio-vascular symptoms?
15. Find the total number of outcomes if 30 patients are selected for the study.
16. If 30 are selected, what is the probability that exactly two have cardiovascular symptoms?
17. If 30 are selected, what is the probability that two or fewer have cardiovascular symptoms?
18. If 30 are selected, what is the probability that more than two have cardio-vascular symptoms?

Exercises 19–23 consider an experiment in which seven horses are running a race and two of the seven are fillies. Assume that each horse has an equal likelihood of winning the race.

19. If someone randomly chooses a list of three horses to finish in first, second, and third place, what is the probability the list is correct?
20. What is the probability that a filly finishes in first place?
21. If someone randomly chooses a set of three horses to finish in the top three, what is the probability the set is correct?
22. What is the probability that exactly one filly finishes in the top three?
23. What is the probability that no filly finishes in the top three?

Exercises 24–27 consider the situation in which local law enforcement officers have apprehended two suspects, Baker and Taylor, who are alleged to have robbed the First Union Town Bank and Trust. The police invite an eyewitness to pick out the two conspirators from a lineup of five people.

24. Determine the total number of ways that two people can be selected from a line-up of five people.
25. If the eyewitness does not recognize any of the subjects and randomly chooses two people from the lineup, what is the probability that both Baker and Taylor are selected?
26. If the eyewitness randomly chooses two people from the line-up, what is the probability that exactly one of Baker and Taylor is selected?
27. If the eyewitness randomly chooses two people from the line-up, what is the probability that neither Baker nor Taylor is selected?

In exercises 28–34, define clearly a random variable that satisfies the stated criteria. Verify that your definition identifies the value the random variable assigns to each outcome of the given experiment. Note that there are many different correct answers to each exercise. For example, when defining a random variable X on the experiment consisting of a person stepping on a scale, we can define $X =$ "the person's weight as determined by the scale" or we can define $X = 0$ if the person weighs at most 100 lbs and $X = 1$ if the person weighs more.

28. Define a random variable X on the experiment of rolling two dice. What value does your random variable assign when the first die rolls one and the second rolls four?

29. Define a random variable X on the experiment of rolling two dice, where X assigns only even numbers. What value does your random variable assign when the first die rolls one and the second rolls four?

30. Define a discrete random variable D on the experiment consisting of a person running a 100 yard dash.

31. Define a continuous random variable Y on the experiment consisting of a person running a 100 yard dash.

32. Define a discrete random variable C on the experiment consisting of a person purchasing groceries at a check-out register, where the number of different values that C could assign is countably infinite.

33. Define a finite random variable N on the sample space of drawing a card out of a standard 52-card deck, where N can assign one of 13 different values. What value does your random variable assign when the Ace of Spades is selected?

34. Define a random variable P on the sample space of an ornithologist surveying for different species of birds in a local recreation area.

In exercises 35–49, find the identified probabilities for each random variable X.

35. If X is geometric with $p = .65$, find $P[X = 10]$ and $P[X \neq 10]$.
36. If X is geometric with $p = .10$, find $P[X = 2]$ and $P[X \leq 2]$.
    Hint: If X is geometric, then $P[X \leq 2] = P[X = 1] + P[X = 2]$.
37. If X is geometric with $p = .25$, find $P[X = 1]$ and $P[X \leq 1]$.
38. If X is binomial with $n = 12$ and $p = .40$, find $P[X = 10]$ and $P[X \neq 10]$.
39. If X is binomial with $n = 10$ and $p = .20$, find $P[X = 1]$ and $P[X \leq 1]$.
    Hint: If X is binomial, then $P[X \leq 1] = P[X = 0] + P[X = 1]$.
40. If X is binomial with $n = 20$ and $p = .10$, find $P[X = 2]$ and $P[X \leq 2]$.
41. If X is hypergeometric with $n = 12$, $M = 20$ and $N = 10$, find $P[X = 10]$.
42. If X is hypergeometric with $n = 3$, $M = 10$ and $N = 2$, find $P[X = 2]$.
43. If X is uniform with $a = 0$ and $b = 4$, find $P[2 < X < 5]$ and $P[2 \leq X \leq 5]$.
44. If X is uniform with $a = 1$ and $b = 6$, find $P[X < 7]$ and $P[2 \leq X]$.
45. If X is uniform with $a = 2$ and $b = 6$, find $P[-1 \leq X < 5]$ and $P[X = 3]$.
46. If X is normal with $\mu = 3$ and $\sigma = 1$, find $P[2 < X < 4]$ and $P[2 < X < 5]$. State explicitly the integral for the corresponding normal density function and use a computing device to approximate these probabilities.
47. If Z is the standard normal random variable with $\mu = 0$ and $\sigma = 1$, find $P[1 < Z < 2]$ and $P[-2 < Z < -1]$. State explicitly the integral for the corresponding normal density function and use a computing device to approximate these probabilities. What does your answer suggest about the graph of the standard normal density function?

48. Let X be a continuous random variable with the following density function

$$f(x) = \begin{cases} x & \text{if } x \in [0, 1) \\ 2 - x & \text{if } x \in [1, 2] \\ 0 & \text{otherwise.} \end{cases}$$

Prove that $\int_{-\infty}^{\infty} f(x) \, dx = 1$ and find both $P[0 < X < 1]$ and $P[1/2 \leq X < 3/2]$.

49. Let X be a continuous random variable with density function

$$f(x) = \begin{cases} e^{-x} & \text{if } x \geq 0 \\ 0 & \text{if } x < 0. \end{cases}$$

Prove that $\int_{-\infty}^{\infty} f(x) \, dx = 1$ and find both $P[X > 1]$ and $P[-1 < X \leq 2]$.

In exercises 50–53, find the probabilities associated with each real-life situation.

50. When rolling two dice, what is the probability of rolling "lucky seven," where the number of dots appearing on the dice sum to seven?

51. A state lottery Prosperity Pick 4 asks players to choose four distinct numbers from 1 to 20, and then the state draws four such numbers from a random process. If all four of a player's numbers match the state's, then the player wins $10,000. If three numbers match, then the player wins $2,000. Let X be the discrete random variable $X =$"the amount of winnings" for a single play of the lottery. Find the values for $P[X = 10, 000]$, $P[X = 2, 000]$, and $E[X]$. Comparing the entrance fee against $E[X]$, which equals the expected winnings, would you be willing to play the lottery if it costs $5? What if it costs $2?

52. The accepted probability of Boris Stansky winning a game of chess against Peter Similov is 2/3. Assuming that each game of chess is independent of any previous game, what is the probability that Stansky will win at least three games out of four played against Similov? Use the binomial distribution to answer this question.

53. A machine that is put into service with the goal of filling paint cans to exactly one gallon has an error, measuring the overfill (a positive error) or an underfill (a negative error) for each can that is normally distributed with a mean of 1 gallon and a standard deviation of 0.001. Find the probability that the machine will fill a randomly selected can with an amount of paint between 0.9995 gallons and 1.002 gallons.

In exercises 54–61, compute the mean μ and the standard deviation σ of each random variable X; use the definitions given in this section.

54. A random variable X assigns only the values 1 and 2 with equal probability.

55. A discrete random variable X with $P[X = 0] = 0.2$, $P[X = 1] = 0.3$, $P[X = 2] = 0.1$, and $P[X = 3] = 0.4$.

56. A discrete random variable X with the following probability distribution table:

x	0	1	2	3	4
$P[X = x]$	0.2	0.1	0.2	0.35	0.15

57. A discrete random variable X with the following probability distribution table:

x	1/2	1	3/2	2	3	4
$P[X = x]$	0.3	0.2	0.1	0.1	0.1	0.2

58. A discrete random variable X with the following probability distribution table:

x	-1	0	1	2	π	8.5	$\sqrt{73}$
$P[X = x]$	0.15	0.25	0.1	0.2	0.15	0.05	0.1

59. The continuous random variable X with density function

$$f(x) = \begin{cases} x & \text{if } x \in [0, 1) \\ 2 - x & \text{if } x \in [1, 2] \\ 0 & \text{otherwise} \end{cases}$$

60. The exponential random variable X with density function:

$$f(x) = \begin{cases} e^{-x} & \text{if } x \geq 0 \\ 0 & \text{if } x < 0. \end{cases}$$

61. The continuous random variable X with density function:

$$f(x) = \begin{cases} 2/3 & \text{if } x \in [0, 1) \\ x^2/7 & \text{if } x \in [1, 2] \\ 0 & \text{otherwise .} \end{cases}$$

In exercises 62–66, prove each mathematical statement.

62. By the definition of mutually exclusive events, $P[A'] = 1 - P[A]$.

63. If $c \in \mathbb{R}$, then $E[cX] = c \cdot E[X]$. Hint: Prove this for both discrete and continuous random variables.

64. If X and Y are both discrete or both continuous random variables, then $E[X + Y] = E[X] + E[Y]$.

65. If $c \in \mathbb{R}$, then the standard deviation of cX is equal to the product of $|c|$ and the standard deviation of X; symbolically, we write $\sigma_{(cX)} = |c| \cdot \sigma_X$. Hint: Prove this for both discrete and continuous random variables.

66. If the *second moment* of a continuous random variable is defined as $E[X^2] = \int_{-\infty}^{\infty} x^2 \cdot f(x)\, dx$ and if the *variance* of a random variable is defined as $\text{Var}[X] = \sigma^2$ (that is, the square of the standard deviation), then $\text{Var}[X] = \int_{-\infty}^{\infty} (x - \mu)^2 \cdot f(x)\, dx = E[X^2] - (E[X])^2$.

Exercises 67–70 consider the gamma function defined for $x > 0$ by

$$\Gamma(x) = \int_0^\infty t^{x-1} e^{-t}\, dt.$$

The gamma function is often thought of as a generalization of factorials because when x is a nonnegative integer n, $\Gamma(n + 1) = n!$.

In exercises 67–70, evaluate the gamma function at each value; use a computing device to approximate the corresponding integrals.

67. $\Gamma(1)$

69. $\Gamma(3.5)$

68. $\Gamma(2.5)$

70. $\Gamma(10)$

Exercises 71–74 consider the Student's t-distribution. The random variable T_m with a Student's t-distribution has the following density function for $-\infty < t < \infty$:

$$f(t) = \frac{\Gamma((m + 1)/2)}{\sqrt{m\pi} \cdot \Gamma(m/2)} \cdot \left[1 + \frac{t^2}{m}\right]^{-(m+1)/2}$$

Note that the gamma function $\Gamma(x)$ is defined for exercises 67–70. The parameter m in this expression for T_m is called the random variable's degrees of freedom. The mean of T_m is $\mu = 0$, and the standard deviation is $\sigma = \sqrt{m/(m - 2)}$. The graph of this density function is similar to the normal density function with $\mu = 0$ and $\sigma = 1$, only the "bell-shape" is lower and wider. A probability $P[a < T_m < b]$ can be computed using a TI-83 with the command tcdf(a,b,m) accessed under the DISTR menu. Computer algebra systems have similar commands in their statistics libraries. Furthermore, since $P[T_m > 12] \approx 0 \approx P[T_m < -12]$, we can use the identities $P[a < T_m < 12] \approx P[T_m > a]$ and $P[-12 < T_m < b] \approx P[T_m < b]$ when computing probabilities.

In exercises 71–74, find each probability.

71. $P[-0.91 < T_{21} < 1.3]$

73. $P[T_7 < 0.21]$

72. $P[T_{11} > 0.21]$

74. $P[T_{21} < -2.03]$

5.4 Application: Statistical Inference and Hypothesis Testing

Many real-life situations require drawing valid conclusions and making decisions about large groups of people or objects. Researchers often want to answer a question about an entire group (or *population*) affected by an issue, but it is often impossible to have information on every element of the population.

- Sometimes the population may be too large for a timely gathering of information. For example, a national political consultant may want to understand how 150 million registered voters feel about the federal deficit vis-a-vis increased taxation; it is impossible to survey the entire group in a reasonable period of time at a reasonable cost.

- The population may be conceptually defined, rather then referring to an existing group of people or objects. For example, if a camera manufacturer is interested in determining the percentage of defective lenses produced in a particular factory, the population in question is the group of *all* lenses, including those that may be produced in the future. In this way, an examination of the entire group is again impossible, at least in time to contribute to a decision-making process.

- The population may be infinite, either considered as a conceptual notion or from a practical perspective. For example, an astronomer may develop a conjecture about the average apparent magnitude of the stars in our galaxy. The number of stars in this population is so vast that a practical study would essentially need to treat the population as infinite.

In these and many other real-world investigations, a researcher cannot collect data for every element of the population. In such settings with only partial information available, how might mathematics make a meaningful contribution?

In the 1940s, Sir Ronald Aylmer Fisher developed a decision-making process for these (and other) settings based on probability theory. Essentially, this process (widely known as *hypothesis testing*) is a model for inductive reasoning that uses data collected on only a sample to draw statistical conclusion about the entire population. Just as the mathematics of symbolic logic provided a framework to study rigorously the deductive reasoning process of chapter 1, so does the mathematics of probability theory provide a sound framework for inductive reasoning. This section describes how hypothesis testing lies at the heart of rigorous inductive thought and decision-making, sometimes referred to as *inferential mathematics* or *statistical inference*.

Fisher's influence on the practice of statistics was enormous. He began his professional career in 1919 as a lower-level statistician at a small agricultural station in Hertfordshire, England. His brilliant understanding of the mathematical framework behind statistics allowed him to develop guidelines for conducting experiments, and he promoted the idea of randomization. During the next decade, he created a process now known as analysis of variance and developed new statistical methods in small sample analysis and other new statistical methods, such as "maximum likelihood." In 1925, he wrote his first book, *Statistical Methods for Research Workers*, which almost instantly became the standard in the field. It was in this text that Fisher introduced a standard "level of significance" of $\alpha = 0.05$, which continues in common use today. Fisher was by far the most influential statistician in the twentieth century.

The process of statistical inference involves collecting *data* for only a *sample*—a small portion of the population – and then extrapolating these representative results to the population. This process has a number of specific and practical steps, beginning with a researcher's question. For example, a researcher might ask: What percentage of voters would favor a five percent increase in taxation to control the federal deficit? Should we expect a factory to produce an intolerable number of defective lenses? Is the average apparent magnitude less than six (that is, the magnitude of a star just visible to the human eye)? Not every question lends itself to an inferential study, but a well-directed initial question kick-starts the process. By its very nature, the question should focus attention on a population of interest, and therefore defines the population to be studied. Successful research answers the question by collecting and analyzing sample data.

Each step in the statistical inference process follows fixed guidelines, which provide a mathematical framework for inductively determining an answer to the question at hand. The next definition introduces this approach to inductive reasoning.

Definition 5.4.1 **Statistical inference** *consists of the following process.*

1. *Formulate a question that can ultimately be phrased in terms of a random variable.*
2. *Identify the corresponding population; this is the group of objects to which the random variable is applied.*
3. *Collect relevant data for a representative sample, using a random process to eliminate problems such as bias.*
4. *Analyze the random variable's statistical values for the sample, and use hypothesis testing (as described below) to extrapolate the sample results and answer the question about the entire population.*

The first step of statistical inference involves the formulation of a question in terms of a random variable. Since any random variable X assigns numbers to the outcomes of an experiment, the question must be phrased in terms of a numerical value or some other feature of X, such as its distribution. Sometimes this numerical value is a *parameter*, which is a quantity that describes the random variable on the population. Two examples of frequently studied parameters are the mean μ (the average value assigned by X) and the proportion p of the population that is assigned a particular value by X. Alternatively, the research question may ask about the distribution of X; for example, it might ask if X is normal. Although the question need not explicitly identify the random variable, the corresponding population, or the parameter in question, the process of statistical inference requires that they be identified clearly. The next example examines the first step in this process.

Example 5.4.1 We consider a researcher who asks, "What percentage of voters would favor a five percent increase in taxes to control the federal deficit?" This question can be phrased in terms of a random variable defined on an experiment: ask a registered voter whether or not (s)he is in favor of a five percent tax increase to control the deficit. Let $X = 1$ if the voter responds "Yes," and $X = 0$ if "No." The original question is then equivalent to a question about the parameter p, which is the proportion of time X assigns the value 1. For example, the researcher might ask, "Is $p > 0.50$?" in an effort to gauge the preference of the majority. In this way, the researcher's original query is formulated in terms of a parameter and identifies a random variable at the heart of the original research question.

■

The next question practices the formation of such inferential queries; there is a "natural" random variable that can be defined in each real-world setting.

Question 5.4.1 For each research question, define a random variable with its corresponding population and experiment, and then express the question in terms of the random variable and an associated parameter.

(a) A high-tech company runs an analysis of its disk drive factory, which has an acceptable tolerance level of fewer than $1/2$ of a percent of defective manufactured items. A quality control engineer asks, "Should the company expect a collection of 500 newly manufactured disk drives to satisfy this tolerance level?"

(b) A biologist studying tropical fish asks, "Does the visual presence of a male Mexican cichlid (*Herichthys cyanoguttatus*) increase the average food intake of female Mexican cichlid?"

(c) A certain pain reliever is known to bring relief in an average of 3.5 minutes. A pharmacist asks, "Is a new drug faster at providing relief from pain?"

∎

Once a research question is phrased in terms of a random variable, and once the corresponding population has been identified, the process of statistical inference directs us to collect relevant data for a representative *sample* (or subset) of the population. Samples must be chosen carefully. In the relatively simple settings we are studying (where the question can be phrased in terms of a single random variable), a sample of size n is often obtained by repeating the corresponding experiment n times. More sophisticated models of statistical inference consider questions about multiple random variables simultaneously. In such settings, the sampling procedures in these cases require more careful control. But when the question at hand involves only one random variable and is expressed in terms of a single parameter, the identification of a sample is relatively straightforward, as formalized in the following definition.

Definition 5.4.2   *A **sample of size** n is a subset of a random variable's population formed by repeating the associated experiment n times.*

Repeating an experiment n times and recording the numbers the random variable assigns produces a set of n numbers. This set of numbers is the desired representative data set. Once a sample is formed, statistical inference proceeds to its next step, where the random variable's statistical values are analyzed. Hypothesis testing can extrapolate the sample results and answer the question about the entire population. The following example illustrates the construction of a sample.

Example 5.4.2   We consider a gambler who wonders if a coin is fair. For the experiment of tossing the coin, she defines a random variable $X = 1$ if heads is tossed and $X = 0$ if tails is tossed. The population is the set of all coin tosses (past, present, and future), and a sample of any size n is formed by tossing the coin n times and recording the value $X = 0$ or $X = 1$ for each toss.

∎

Question 5.4.2   In parallel to the work in example 5.4.2, describe the first three steps in the statistical inference process for the Chevalier, who is interested in determining whether a pair of dice rolls double-sixes in a fair way.

∎

At times, the population exists as a real entity; in question 5.4.1 the populations under study are disk drives, fish, and patients in pain. In such settings, the sample must be representative, and we must be careful to avoid *lurking variables* – additional features that are not shared by the entire population and that influence the analysis. For example, if a researcher is studying the average weight of 12-year-olds, then lurking variables result from surveying only males (or only females), or only children living in urban areas, or only children with access to email, or only children attending a particular elementary school. A famous historical example of lurking variables affecting a survey occurred when the *Chicago Tribune* incorrectly forecast that Thomas Dewey would defeat Harry S. Truman in the 1944 presidential election. The forecast was based on a telephone survey at a time when telephones were a luxury, and apparently people with phones (and more disposable income) were inclined to vote for Dewey.

Question 5.4.3 Working in the context of example 5.4.1, a researcher asks registered voters whether or not they are in favor of a five percent tax increase to control the deficit. Describe a poorly designed selection process for survey participants that might introduce a lurking variable into the analysis.

▪

A *random selection process* is an important tool to avoid both lurking variables and unintended bias when choosing the sample. For example, suppose a student running for class president wonders how many of her 1,100 classmates agree that their university should freeze tuition hikes. Using the experiment of surveying classmates on this issue, she defines a random variable $X = 1$ if a classmate agrees and $X = 0$ if a classmate disagrees. The corresponding population is the set of all her classmates, but which classmates should be selected for a representative sample of survey participants?

To guaranteeing an unbiased sample in such settings, select the classmates by a random process. Many computing devices are programmed with "random number generators"; the commands for some are provided below. The candidate for class president can identify a sample by first assigning a number from 1 to 1,100 to each of her classmates. A computing device can then generate a random number between 0 and 1, which is then multiplied by 1,100 and rounded up. This method produces a random value between 1 and 1,100, and the classmate with this number is added to the sample. A classmate should only be chosen once (we wouldn't want to survey the same person multiple times); if the same random number is generated twice, simply toss it out the second time and use the next generated number. Continue in this fashion until a sample of the desired size n is identified. Some commands for generating random numbers are listed in the following table. The Maple command returns n random numbers.

System	Command
TI-83	MATH, PRB, rand
Maple	[>with(stats): [>stats[random, uniform](n);
Mathematica	]:= Random[ ]

Question 5.4.4 Using a computing device, randomly produce a list of five numbers.

▪

Question 5.4.5 Describe a procedure for constructing a representative sample for each scenario.

(a) A RISK player wonders if a game die really does "roll a one" one-sixth of the time.

(b) A medical doctor wonders if *existing* juvenile diabetics lost more than five percent of their bodyweight at the initial onset of the illness.

(c) A medical doctor wonders if a newly diagnosed juvenile diabetic under her care should be expected to lose more than five percent of his bodyweight at the initial onsent of the illness.

∎

We now focus on the last step of statistical inference, which uses hypothesis testing to answer researchers' questions. In this setting, possible answers are referred to as *hypotheses*. We test a first hypothesis against a second hypothesis, and decide whether or not to reject the first in favor of the second. As described below, hypothesis testing relies on the calculation of a probability, called the *P-value*. The *P*-value is the probability of an "observation" occurring. The observation is typically a statistical value obtained on a sample. We formally outline the process of Hypothesis Testing and then focus on developing a familiarity with it.

Hypothesis Testing
This process decides between two statements after making an observation. The statements are formulated as a *null-hypothesis* H_0 and an *alternative hypothesis* H_a (often as logical complements of one another).

- **Step 1.** Assume the null-hypothesis H_0 is true.
- **Step 2.** Based on this assumption, calculate the probability of the observation occurring. This probability is called the *P-value*; the observation is often a statistical value obtained on a representative sample.
- **Step 3.** If the *P*-value from Step 2 is low, reject the null hypothesis and conclude the alternative hypothesis H_a is true. If the *P*-value is high, do not reject H_0. The distinction between low and high *P*-values is based on a comparison against a predetermined *significance level* that is commonly denoted by α (the Greek letter "alpha").

∎

The decision made in hypothesis testing depends on the comparative sizes of the *P*-value and the significance level α. In statistics, there is no single standard for a numerical choice of α. Sir Ronald Fisher first articulated the hypothesis testing process, and he suggested that an appropriate benchmark is the five-percent level; in other words, reject the null hypothesis when the *P*-value is less than $\alpha = 0.05$.

In practice, the significance level used in a given setting depends on the repercussions of rejecting the null hypothesis. If researchers need to be very sure of a hypothesis test's conclusions, then they use a correspondingly smaller significance level (such as $\alpha = 0.01$ or $\alpha = 0.005$). For example, hypothesis tests studying the failure rate of gaskets would use a different level of significance for gaskets used in a lawn mowers versus gaskets used in a nuclear power plant. Unless otherwise stated, we will follow Fisher's convention and use $\alpha = 0.05$.

In addition, researchers should be careful to determine the significance level for a hypothesis test *before* implementing the test. Choosing it instead *after* the computation of the *P*-value can compromise its value as an independent benchmark level, which would leave the procedure open to criticism—the researcher could be accused of adjusting the selection of α to obtain a desired conclusion. In this way, a hypothesis test is conducted with integrity when a significance level is selected in advance. Many researchers follow a standard practice of comparing the P-value against $\alpha = 0.05$, $\alpha = 0.01$, or $\alpha = 0.001$; a small P-value in each case is respectively deemed significant, very significant, or highly significant.

The next example demonstrates the complete process of statistical inference.

Example 5.4.3   An accounting firm checks the accuracy of a company's records, which contains 13 inaccurate accounts out of a total of 50. Because of time constraints, the accounting firm can only audit eight of the 50 accounts. The company supplies the accounting firm with eight "randomly selected" accounts. However none of the eight accounts contain inaccuracies. In light of this information, an investigator asks, "Is it true that the company randomly selected the eight accounts to be audited, or did the company purposefully supply only accurate accounts?"

We use statistical inference to answer the investigator's question, commenting on each step of the process.

1. The experiment selects one of the company's 50 records to be audited; define a random variable $X = 1$ if the record is accurate, and $X = 0$ if not.
2. The population of this experiment is the set of all 50 records.
3. A sample of size $n = 8$ was taken (as described above), and the company claimed that the sample was randomly selected. The sample contained only accurate records, producing $X = 1$ for every element of the sample. The "observation" made about the random variable is thus $\{1, 1, 1, 1, 1, 1, 1, 1\}$.
4. Now perform a hypothesis test using the significance level $\alpha = 0.05$.

   - Step 1: State the null hypothesis H_0 and the alternative hypotheses H_a. The two choices for the hypotheses are: "the selection of accounts was random" and "the selection of accounts was not random," which are phrased as H_0 and H_a.

     H_0: The selection of the accounts was random.
     H_a: The selection of the accounts was not random.

     We'll say more below about the choice of "The selection of the accounts was random" as the null hypothesis. Following the guidelines in Step 1, assume that the null hypothesis is true; i.e., that the selection of the accounts was random.
   - Step 2: Based on this assumption, calculate the corresponding P-value (that is, probability of the observation occurring).

     Since there are 37 accurate accounts in the collection of 50 accounts, the probability that the company *randomly* selected eight accurate accounts is equal to

     $$P\text{-value} = \frac{C(37, 8)}{C(50, 8)} = 0.0719.$$

     You can see how the assumption and the observation both played a role in the calculation of the P-value: the assumption guarantees randomness, and the observation is the event of selecting eight accurate records.
   - Step 3: Base the decision to reject or not reject the null hypothesis H_0 on the relative size of the P-value and the significance level α.

In this case, the P-value is 0.0719, which is greater than $\alpha = 0.05$. We therefore do not reject the null hypothesis. Although there appears to be some evidence that the company selected only accurate files, this evidence is not strong enough at the five-percent significance level to reject the assumption that the company randomly selected the accounts to be audited.

■

Hypothesis testing depends on an ability to calculate the P-value, which must be done in terms of an observation and an assumption. The observation is based on the sample identified in the third step of the statistical inference process. The assumption is phrased as H_0. In example 5.4.3 the probability of the observation occurring (that is, the probability that every selected account is accurate) is calculated under the assumption as "the probability of randomly selecting a sample of size eight and obtaining $\{1,\ 1,\ 1,\ 1,\ 1,\ 1,\ 1,\ 1\}$ as the outcome." In practice, researchers must phrase both the null hypothesis and the observation in such a way as to allow for the computation of the P-value.

Question 5.4.6   Working in the context of example 5.4.3, state the outcome of the investigation under the following significance levels.

(a) $\alpha = 0.01$ (b) $\alpha = 0.10$

■

Question 5.4.6 demonstrates the importance of a researcher predetermining the significance level before computing the P-value: different significance levels can produce different results for the same observation.

Question 5.4.7   A judge wonders if a company followed Equal Opportunity Employer guidelines when narrowing a field of 20 female applicants and 35 male applicants to a pool of five final candidates, all of whom were male. Assuming the applicants were all equally qualified for the position, use statistical inference to conclude whether or not there was bias in the selection process. Use the significance level $\alpha = 0.05$.

■

The formation of the P-value depends upon the phrasing of the observation. In general, we frame the statistical evidence in the sample so that it includes *all* sample statistic values that provide evidence toward the alternative hypothesis.

The point is that we have some freedom in how we formulate the observation. For example, suppose that question 5.4.7 had discussed a scenario where the pool of five final candidates contained four males (rather than five). This observation can be expressed in terms of what happened on the sample of five in at least two different ways: as $Y = 4$ (where $Y =$ "the number of males in the pool of five candidates") or as $Y \geq 4$. The second formulation $Y \geq 4$ also includes the situation where all five of the final candidates are males, which would intuitively lend even more evidence toward the alternative hypothesis that "there was bias toward males in the selection process."

Succinctly put, use the following principle whenever forming the observation:

When framing the observation in terms of a sample statistic, include any value of the statistic that lends additional evidence toward the alternative hypothesis.

A researcher using a hypothesis test will sometimes make a mistake, either rejecting the null hypothesis when it is true, or deciding not to reject H_0 when it is false. In statistical circles, the first type of mistake is called a *Type I error*, and the second is called a *Type II error*. In practice, the researcher should always establish H_0 and H_a so that a Type I error is the more serious of the two. This structure is analogous to many important real-life decisions. For example, a jury choosing between guilt and innocence can make two types of mistakes: convicting an innocent person, or declaring innocent someone who committed a crime. Most people would agree that the first mistake is the more serious error; by formulating a corresponding null hypothesis as "H_0: the accused is innocent," a researcher would be establishing the test's Type I error as this more serious mistake.

5.4.1 The Central Limit Theorem

In many statistical tests, the calculation of the P-value is difficult without powerful probabilistic theorems. In most practical situations, the hypotheses are written in terms of a random variable's parameter (such as a mean or standard deviation) or in terms of a parameter for multiple random variables (such as a difference of means). But if the value of a parameter is unknown, then the random variable's distribution will surely also be unknown. And so it will be impossible to calculate a P-value if that calculation is based on this distribution.

Fortunately, mathematical theory exists to help us resolve this mathematical dead end. There are several powerful theorems that describe the probabilistic behavior of samples (regardless of the random variable's distribution) and enable a computation of the corresponding P-value. Each is suited to a particular type of hypothesis test; a full course in statistics would show how to employ several of these theorems. We focus on one of the most important and useful results—the central limit theorem. This theorem describes the probability distribution for a value called the sample mean, which is symbolically denoted $\overline{X}$ and defined as the average value of the data in the sample. The central limit theorem considers $\overline{X}$ as a random variable, defined on the experiment of collecting a sample of given size n, and states that when n is large, $\overline{X}$ is normally distributed. In this way, the central limit theorem enables the calculation of P-values involving an observation about a sample mean. We formally define the notion of a sample mean, consider an example, and then focus on the central limit theorem.

Definition 5.4.3 *If $X = \{X_1, X_2, \ldots, X_n\}$ is a sample of size n, then the **sample mean** $\overline{X}$ is the average of the sample values; symbolically, we define*

$$\overline{X} = \frac{1}{n} \cdot \sum_{i=1}^{n} X_i = \frac{X_1 + X_2 + \cdots + X_n}{n}.$$

Most people are familiar with the notion of a sample mean. It is simply the average of the numbers in a sample. Means such as the average exam grade for a class of 10 students, the average total rainfall for 10 months on record, and the average high temperature recorded for a collection of 10 days are all examples of a sample mean defined on samples of size $n = 10$. It's important to realize that a random variable's sample mean $\overline{X}$ is a completely different object than the population mean μ. The first is defined on a sample and can change depending on the choice of the sample, while the second is a fixed value defined probabilistically (as in definition 5.3.7 of section 5.3) on the population.

Question 5.4.8   A student at My University is interested in knowing the mean grade point average of graduating seniors. For confidentiality reasons, she does not have access to the grade point average of every senior, and so she conducts several surveys in an effort to estimate the population mean. Compute the sample mean for each survey.

(a) { 2.44, 2.98, 3.15 }   (c) { 2.22, 2.99, 3.14, 3.50, 3.75 }

(b) { 2.01, 2.33, 2.68, 3.01 }   (d) { 3.33, 3.89, 3.96 }

■

As you might expect, different samples of the same population often have different sample means; the sample means computed in question 5.4.8 illustrate this phenomenon. The central limit theorem provides important insight into the statistical behavior of the sample mean, as it considers the sample mean as a random variable. In particular, given any underlying random variable X, the experiment of selecting a random sample of size n from the population of outcomes for X determines the sample mean random variable $\overline{X}_n$. We define $\overline{X}_n =$ " the mean of a randomly selected sample of size n." The central limit theorem asserts that for large enough n, the random variable $\overline{X}_n$ is normally distributed, no matter what the probability distribution of the underlying random variable X. We present this important result as the following theorem.

Theorem 5.4.1 The central limit theorem   *Let X be a random variable with mean μ and standard deviation σ. For each $n \in \mathbb{N}$, define a random variable $\overline{X}_n$ as the sample mean (that corresponds to the experiment of selecting a random sample of size n for X). No matter what distribution X has, the probability distribution of $\overline{X}_n$ is normal in the limit:*

$$\lim_{n \to \infty} P[a \le \overline{X}_n \le b] = \int_a^b \frac{1}{(\sigma/\sqrt{n}) \cdot \sqrt{2\pi}} \cdot e^{-\frac{(\overline{x}-\mu)^2}{2(\sigma/\sqrt{n})^2}} \, d\overline{x}.$$

Furthermore, $\overline{X}_n$ has mean equal to μ (the mean of X) and standard deviation equal to $\sigma/\sqrt{n}$ (the standard deviation σ of X divided by the square root of the sample size).

The proof of this theorem is sophisticated and interesting, but it is also beyond the scope of this text and left for later studies. We instead focus on applying the central limit theorem in the statistical inference process.

The central limit theorem is especially useful in tests having hypotheses phrased in terms of an unknown population mean μ and when the observation is expressed in terms of a sample mean $\overline{X}$ on samples with large size n. Such a test is called a *large sample test on the mean*. In light of the central limit theorem, the P-value for the corresponding observation can be calculated; it is approximately a probability about a normal random variable $\overline{X}$ (assuming the sample size n is sufficiently large).

The central limit theorem implies that the approximation of the P-value becomes more accurate as the sample size n increases. The standard benchmark for a "sufficiently large" sample size is $n = 30$; unless the original random variable X is extraordinarily lopsided or has a very oddly shaped distribution or density, the sample mean for samples of size $n \geq 30$ is approximately normal.

Most standard tests phrase the null hypothesis H_0 in terms of given parameter equaling a specific value. A common example is a test on the mean, such as $H_0\colon \mu = -10$. When the null hypothesis has this form, there are three possible options for the alternative hypothesis:

- H_a: the parameter is less than a specified value (for example, $H_a\colon \mu < -10$); this format is commonly called a *left-tailed test*;
- H_a: the parameter is greater than the specified value (for example, $H_a\colon \mu > -10$); this format is commonly called a *right-tailed test*;
- H_a: the parameter is not equal to the specified value (for example, $H_a\colon \mu \neq -10$); this format is commonly called a *two-tailed test*.

The form of the alternative hypothesis plays a crucial role in determining the structure of the corresponding P-value. The general rule of thumb is that the inequality in the format of the P-value points the same way as that in the one-tailed alternative hypothesis, and the P-value is doubled for the two-tailed test. The next example illustrates this structure.

Example 5.4.4 Continuing to work with the example of a null hypothesis of the form $H_0\colon \mu = -10$, we consider scenarios that correspond to calculated values of the sample mean, and we identify the format of the P-value for each possible alternative hypothesis.

- Suppose that a random sample has a sample mean of $\overline{X} = -12$. For the left-tailed test with $H_a\colon \mu < -10$, the P-value is $P[\overline{X} \leq -12]$.
- Suppose that a random sample has a sample mean of $\overline{X} = -4$. For the right-tailed test with $H_a\colon \mu > -10$, the P-value is $P[\overline{X} \geq -4]$;
- Suppose that a random sample has a sample mean of $\overline{X} = 2$. For the two-tailed test with $H_a\colon \mu \neq -10$, the P-value is $2 \cdot P[\overline{X} \geq 2]$.

■

Any hypothesis test in terms of a parameter (such as μ) should be set up so that the alternative hypothesis has a left-tailed, right-tailed, or two-tailed format. The null hypothesis may first be structured as the parameter equaling a specified value, such as $H_0\colon \mu = -10$. A statistician should then always formulate the alternative hypothesis so that it is supported by the evidence provided in the sample. For example, it would make little sense to formulate a right-tailed alternative hypothesis as $H_a\colon \mu > -10$ when a large random sample had $\overline{X} = -150$; there is no way that a test would support a population mean larger than -10 when the sample mean is less than -10.

Besides requiring the sample evidence to support the result, the formation and choice of H_a should set up a Type I error as being more serious than a Type II.

The alternative hypothesis should also be of a form that makes good conclusive sense. For example, a researcher worried about global warming raising the average global temperature above 50 degrees Fahrenheit would construct the alternative hypothesis as $H_a : \mu > 50$ rather than the two-tailed $H_a : \mu \neq 50$. On the other hand, a quality controller studying if an assembly line machine has an average output quantity different from the standard average of 100 grams would typically formulate the alternative hypothesis in a two-tailed structure, using $H_a : \mu \neq 100$. As indicated by the next question, the P-value for a test on the mean follows from the sample mean and the formulation of the hypotheses.

Question 5.4.9   Suppose a hypothesis test on the mean has a null hypothesis $H_0: \mu = 2$ and that a random sample of size 40 has a sample mean of $\overline{X} = 3$ and standard deviation 7. Identify the P-value for the right-tailed and two-tailed test as in example 5.4.4.

∎

We now consider the application of the central limit theorem in a specific hypothesis test on the mean.

Example 5.4.5   A psychologist wonders if the mental states of clinically depressed patients would be significantly improved by an exercise regimen of at least 30 minutes of walking or jogging every day. She identifies significant improvement with an average decrease of more than 10 points in the patient's score on the Goldberg Depression Psychological Test. To test this theory, she randomly selects 45 clinically depressed patients with a Goldberg score of at least 36 (the cutoff level for a moderate chance of depression). Suppose that after four weeks of daily, controlled workouts, each patient is retested, indicating an average decrease in their Goldberg scores of 11 points with an estimated standard deviation of 3.1. Based on these results, can the psychologist extrapolate the average reduction of more than 10 points to *all* clinically depressed patients? As usual, use a significance level of $\alpha = 0.05$ for this hypothesis test.

The psychologist's expectation of an average decrease of more than 10 points in a patient's Goldberg score can be phrased in terms of a hypothesis test on the mean. For the experiment of determining a clinically depressed patient's Goldberg score before and after the four-week exercise regimen, define the random variable $X = $ "the change in a patient's Goldberg score." The psychologist's question can then be formulated as a choice between the two hypotheses $H_0: \mu = -10$ and $H_a: \mu < -10$.

In this setting, a negative mean indicates an average decrease in the score. The null hypothesis $H_0: \mu = -10$ has the standard format for a test on the mean, and the alternative hypothesis $H_a: \mu < -10$ is supported by the evidence provided on the sample; since the sample average $\overline{X} = -11$ is less than -10, it is reasonable to conjecture that the population mean is also less than -10. Notice the form of the alternative hypothesis $H_a: \mu < -10$ is for a left-tailed test.

Because the psychologist obtained a sample mean of $\overline{X} = -11$, the P-value is equal to $P[\overline{X} \leq -11]$. This form of the P-value matches that of a left-tailed test with alternative hypothesis $H_a: \mu < -10$. Since the sample size $n = 45$ is greater than 30, the central limit theorem applies with $n = 45$, $\mu = -10$, and $\sigma = 3.1$, obtaining

$$\text{P-value} = P[\overline{X} \le -11] \approx \int_{-\infty}^{-11} \frac{1}{(3.1/\sqrt{45}) \cdot \sqrt{2\pi}} \cdot e^{-\frac{(\overline{x}+10)^2}{2(3.1/\sqrt{45})^2}} \, d\overline{x} \approx 0.0152.$$

A computing device can provide the final value in the calculation; useful commands for such a calculation are listed in a chart at the end of section 5.3. If the computing device used cannot manage the lower limit of $-\infty$, then you may (by convention) substitute any value that is outside of four standard deviations from the mean. In this example, $\mu - 4(\sigma/\sqrt{n}) = -10 - 4 \cdot 3.1/\sqrt{45} \approx -10 - 1.85 \approx -12$, and so it is appropriate to substitute -12 for $-\infty$ when using a computing device.

Since the P-value is less than 0.05, we reject the null hypothesis and adopt the alternative hypothesis. And so (in this imaginary scenario), the psychologist's supposition was correct: a daily exercise regimen produces an average decrease of more than 10 points in the Goldberg score of clinically depressed patients.

■

Question 5.4.10   An automobile manufacturer conjectures that the mean mileage per gallon of one of its cars exceeds the mean EPA rating of 43 miles per gallon. A random sample of 40 cars produces a mean of 43.6 and a standard deviation of 1.3 miles per gallon. Is the company's conjecture correct?

■

Question 5.4.11   A machine is supposed to produce bolts with a mean length of 1 inch. A sample of 40 bolts has a mean of 1.02 inches and a standard deviation of 0.07 inches. Does this provide evidence to indicate that the machine is producing bolts with a mean length different from 1 inch?

■

The central limit theorem is one of several theorems that help calculate a P-value when a random variable's underlying probability distribution is unknown. Similar results exist for processes involving a small sample size ($n < 30$), or when many random variables' means are being compared against one another, or when the test studies the number of recurrences of categorized events. Such theorems and their applications are discussed in any statistics course, including introductory courses that require almost no mathematical prerequisites. Their proofs (including the proof of the central limit theorem) are described in both advanced undergraduate and graduate statistics courses. Theoretical statistics remains an active area of research, as mathematicians continue to develop new insights to handle inference procedures in increasingly advanced settings. We hope you might be interested in learning about this theory and studying more topics in mathematical statistics.

Because the process of statistical inference uses inductive reasoning (rather than deductive), its hypothesis test conclusions are never absolutely proven (independent of a corresponding deductive proof of the result). However, because it is rooted in sound mathematical principle, the statistical inference process is one of the most useful methods for exploring and understanding the variable nature of the world in which we live. Statistical inference allows us to extrapolate from a relatively small sample to an entire population, providing insight into how various aspects of our world influence us and one another. In this sense, the application of statistical inference and hypothesis testing is far reaching and important in our lives.

5.4.2   Reading Questions for Section 5.4

1. Define and give an example of a population and a sample. What motivates the study of samples rather than entire populations?
2. What are the four steps in the process of statistical inference?
3. Define and give an example of a parameter. What role do parameters play in the development of research questions?
4. Give an example of a sample that suffers from lurking variables and a sample that suffers from bias. How do we avoid such issues when selecting samples?
5. What is a hypothesis? Discuss the distinction between the null and alternative hypotheses in the hypothesis testing process.
6. What is a P-value?
7. State the three steps in the process of hypothesis testing.
8. Define the significance level α for a hypothesis test. What is the standard significance level used in most hypothesis tests?
9. Define Type I and Type II errors and give an example.
10. Define and give an example of a sample mean.
11. State the central limit theorem. How is this result helpful for statistical inference?
12. In a hypothesis test on the mean, how does the P-value depend on the form of the alternative hypothesis?

5.4.3   Exercises for Section 5.4

In exercises 1–8, define a random variable for each question with a corresponding population and experiment, and then express the question in terms of the random variable and an associated parameter.

1. A dietician is concerned about how much students are eating for lunch and asks, "Is the average student consuming more than 1,000 calories at lunch in the school cafeteria?"
2. An economist is studying the local economy and wonders, "What percentage of adults earn an annual income in excess of $100,000 per year?"
3. An archeological anthropologist is studying whether a severe drought may have caused the collapse of an ancient civilization and asks, "Does the soil in the region from the time of the collapse contain a severely small amount of moisture?"
4. A software developer has created a game-theoretic computer program to facilitate divorced couples' division of belongings and asks, "Does each party believe they have obtained at least 70 percent of the objects they originally wanted?"

5. The director of Information Technology wonders, "How much time does a typical student spend working on college-owned computers in the various labs around campus?"

6. A medical researcher asks, "Will a rare disease strike an average of 10 times per million people each year?"

7. A physicist wonders, "Does a mu-proton (a type of sub-atomic particle) pass through a particular chamber on average once every 10 days?"

8. An investor in the tourist industry asks, "Does a typical family take more than one family vacation each year to a destination at least 500 miles away from home?"

In exercises 9–13, identify the population whose properties are studied by each random variable.

9. $X =$ "the late arrival time (in minutes) for States Airline on a flight from New York to Chicago"; the researcher sets $X = 0$ if the plane arrives early or on time.

10. $X =$ "the amount of rainfall in one day in Washington, DC."

11. $X =$ "the amount of tar present in a Horse and Rider cigarette."

12. $X =$ "the miles per gallon consumed by a Sunburst automobile when the car is driven 500 miles over a controlled route."

13. $X =$ "the average grade point average of high school seniors."

In exercises 14–19 use a computing device to identify a random sample of the given size from the following data set identifying the number of books read by 24 graders in fulfilling a reading challenge. In addition, compute the sample mean $\overline{X}$ and compare $\overline{X}$ with the population mean $\mu = 49$.

Student	1	2	3	4	5	6	7	8	9	10	11	12
Books read	10	12	15	15	16	17	19	19	21	28	32	32

Student	13	14	15	16	17	18	19	20	21	22	23	24
Books read	32	34	35	38	38	52	64	75	80	160	162	170

14. A random sample of size $n = 4$.   17. A random sample of size $n = 10$.
15. A random sample of size $n = 6$.   18. A random sample of size $n = 12$.
16. A random sample of size $n = 8$.   19. A random sample of size $n = 14$.

In exercises 20–25, perform a hypothesis test on H_0: "the selection process is random" against H_a: "the selection process is not random" with a significance level of $\alpha = 0.05$ to determine if a company followed Equal Opportunity Employer guidelines requiring a random selection process to narrow each given collection of equally qualified candidates to a pool of four final candidates.

20. A field of 45 female and five male applicants to a pool of all male candidates.
21. A field of 45 female and five male applicants to a pool of all female candidates.
22. A field of 25 female and 25 male applicants to a pool of all male candidates.
23. A field of 25 female and 25 male applicants to a pool of all female candidates.

24. A field of five female and 45 male applicants to a pool of all male candidates.
25. A field of five female and 45 male applicants to a pool of all female candidates.

Exercises 26–35 use a hypothesis test to assess whether or not a selection process is random. A jar contains 50 colored balls, of which 30 are red and 20 are green. Perform the hypothesis test on H_0: "the selection process is random" against H_a: "the selection process is biased toward selecting red balls" with significance level $\alpha = 0.05$ for each selection (without replacement) of balls from the jar.

26. Five balls, five of which are red.
27. Six balls, five of which are red.
28. Three balls, three of which are red.
29. Eight balls, two of which are red.
30. 10 balls, five of which are red.
31. Seven balls, five of which are red.
32. 16 balls, one of which is red.
33. 15 balls, 13 of which are red.
34. 15 balls, 10 of which are red.
35. 20 balls, none of which are red.

In exercises 36–40, explicitly work through each of the four steps of the statistical inference process, including a hypothesis test in Step 4 with significance level $\alpha = 0.05$; example 5.4.3 may serve as a helpful model for these exercises.

36. An accounting firm checks the accuracy of a company's records, which contains 15 inaccurate accounts out of a total of 45. Because of time constraints, the accounting firm can only audit six of the 45 accounts. The company supplies the accounting firm with six "randomly selected" accounts. However, all six of the supplied accounts contain no inaccuracies. In light of this information, an investigator asks, "Is it true that the company randomly selected the six accounts to be audited, or did the company purposefully supply only accurate accounts?"

37. A travel agent assures his manager that, from his current pool of 60 potential customers, 40 will purchase an agency trip and 20 will make arrangements elsewhere or not go on vacation at all. In a subsequent review, the manager observes that out a randomly selected set of four from the pool of 60, none of four purchased an agency trip. In light of this information, was the travel agent's assumption reasonable, or should the estimate of 40 have been lower?

38. A travel agent assures her manager that, from her current pool of 50 potential customers, 34 will purchase an agency trip and 16 will make arrangements elsewhere or not go on vacation at all. In a subsequent review, the manager observes that out a randomly selected set of five from the pool of 50, two of five purchased an agency trip. In light of this information, was the travel agent's assumption reasonable, or should the estimate of 34 have been lower?

39. Responding to a tip that a large percentage of a company's specially manufactured machine bolts are defective, an inspector examines the company's inventory of 10, 000 bolts. Due to time constraints, the inspector initially examines only 200 bolts randomly selected by the company. The inspector finds that 199 of these bolts are perfect, and starts to wonder if they really were randomly selected. A subsequent test of all the bolts finds 530 defective bolts. Determine the validity of the company's claim that the original set of 200 bolts was randomly selected.

40. In a Pick 'em and Win game, a casino advertises that a basket contains 100 blue chips, 10 white chips, nine red chips, and nine green chips. A contestant selects five chips from the basket (without replacement) and wins if any chip is white or green. After playing once, a contestant selects five chips that are blue and red. Determine if the casino's advertisement about the contents of the basket is accurate.

In exercises 41–48, assume that a hypothesis test on the mean has a null hypothesis H_0: $\mu = 22$, that $\sigma = 1.5$, and that the sample size is $n = 40$. Identify the P-value for the given mean for the two-tailed test and the appropriate one-tailed test using a significance level of $\alpha = 0.01$.

41. $\overline{X} = 20.5$

42. $\overline{X} = 23.5$

43. $\overline{X} = 22.1$

44. $\overline{X} = 24$

45. $\overline{X} = 25$

46. $\overline{X} = 21.8$

47. $\overline{X} = 21.5$

48. $\overline{X} = 22.5$

In exercises 49–54, determine whether or not to reject the given null hypothesis in each hypothesis test on the mean using first a significance level of $\alpha = 0.05$ and then a significance level of $\alpha = 0.01$.

49. For H_0: $\mu = 0$ and H_a: $\mu > 0$, a random sample of size $n = 34$ from a population with a standard deviation of $\sigma = 15$ has a sample mean of $\overline{X} = 5$.

50. For H_0: $\mu = 2$ and H_a: $\mu > 2$, a random sample of size $n = 44$ from a population with a standard deviation of $\sigma = 1$ has a sample mean of $\overline{X} = 3$.

51. For H_0: $\mu = 5$ and H_a: $\mu < 5$, a random sample of size $n = 60$ from a population with a standard deviation of $\sigma = 2$ has a sample mean of $\overline{X} = 3$.

52. For H_0: $\mu = 80$ and H_a: $\mu < 80$, a random sample of size $n = 45$ from a population with a standard deviation of $\sigma = 11$ has a sample mean of $\overline{X} = 78$.

53. For H_0: $\mu = 5$ and H_a: $\mu \neq 5$, a random sample of size $n = 60$ from a population with a standard deviation of $\sigma = 9$ has a sample mean of $\overline{X} = 3$.

54. For H_0: $\mu = 50$ and H_a: $\mu \neq 50$, a random sample of size $n = 49$ from a population with a standard deviation of $\sigma = 15$ has a sample mean of $\overline{X} = 53$.

In exercises 55–60, determine whether or not to reject the given null hypothesis in each hypothesis test on the mean using a significance level of $\alpha = 0.05$

55. A consumer advocate group is concerned that a chain of restaurants is selling "quarter pounders" that weigh less than the advertised weight. In a test of H_0: $\mu = 4$ against H_a: $\mu < 4$, a random sample of 34 burgers is found to have an average weight of only 3.75 ounces with a standard deviation of 0.12 ounces. Should the advocate group conclude that the population average is less than 4 ounces?

56. A pediatrician randomly samples 95 twelve-year-old boys and finds that their average height is 57 inches with a standard deviation of 6.8 inches. The pediatrician suspects that the average height of the corresponding population is less than 62 inches. Perform a hypothesis test to investigate the validity of the pediatrician's belief.

57. A school counselor wonders if local high-income students generally score higher than the national average of 1,000 on the SAT test. In a test of $H_0: \mu = 1,000$ against $H_a: \mu > 1,000$, a random sample of 59 high-income students has an average scores of $1,053$ with a standard deviation of 60. Should the counselor conclude that the average of local high-income students is greater than the national average?

58. A random survey of 400 local high school boys between 65 and 70 inches tall found an average weight of 165 pounds with a standard deviation of 21 pounds. Based on this sample, should a researcher conclude that the corresponding population of boys has an average weight greater than a fitness upper bound of 157 pounds?

59. A realtor is curious about a neighborhood's average assessed home value and tests $H_0: \mu = \$275,000$ against $H_a: \mu \neq \$275,000$. A random sample of 30 homes in the neighborhood finds an average assessed value of $\$300,000$ with a standard deviation of $\$9,000$. What does the realtor conclude?

60. The average life of a standard fluorescent lightbulb is 900 hours. For a new type of lightbulb, a sample of 64 bulbs is found to have an average life of 920 hours with a standard deviation of 80 hours. Is the average life of the new type of fluorescent bulb different from the average life of the old standard fluorescent bulb?

5.5 Least Squares Regression

Many research efforts attempt to identify and describe a relationship that may exist between variables. Economists search for relationships between a good's price and its quantity demanded. Medical teams work to find causal relationships between a patient's blood pressure and the body's cholesterol level. Golfers wonder about increasing clubhead speed in order to gain yardage on tee shots. Environmentalists study the impact of a high consumption of fossil fuels on the average global temperature—the "Greenhouse Effect." These examples are just a handful of the many relationships among real-world quantities. Often a change in one causes a change in the other, but sometimes the apparent relationship is due to some third factor, or may just be coincidental.

Mathematics can provide some clarity and direction in an effort to understand the often complicated and subtle relationships that exist between random variables. An appropriate hypothesis test can assess whether certain types of relationships (such as a linear relationship) exist. If a linear relationship is determined to hold, then an appropriate mathematical model can be applied. Data sets and mathematical models are often viewed from a graphical perspective, and so the use of such a model is sometimes referred to as "curve-fitting." In this section, we seek the "optimal" linear model matching a set of data by means of a process called *least squares regression*.

A first step in this direction is to develop the basic terminology used by statisticians who study relationships between random variables, say X and Y. We start with the data upon which the model is based; they are called *bivariate data* and are often presented as ordered pairs (x, y), where the value x corresponds to the random variable X and y corresponds to Y.

Definition 5.5.1 *For given random variables X and Y, a determination about a possible relationship between X and Y is based on **bivariate data**, which are ordered pairs of numerical values of the form (x, y), where we observe simultaneously $X = x$ and $Y = y$. A bivariate sample of size n consists of n such ordered pairs, and the graph of these points on an X-Y plane is called a **scatterplot**.*

Example 5.5.1 Medical researchers study the relationship between an adult male's height X as compared to his weight Y. A random sample of size n is obtained by randomly selecting n adult males from the population and recording each man's height $X = x$ and weight $Y = y$ as an ordered pair (x, y). Consider the following sample of size three presented as a set of ordered pairs, as a table, and as a scatterplot in Figure 5.11.

$\{(69, 168), (69, 155), (72, 184)\}$

height	69	69	72
weight	168	155	184

Real-world data is often not functional in nature; for this example, the height of 69 inches corresponds to two different weights of 168 pounds and 155 pounds. Even so, functions often serve as reasonably accurate models for such data sets.

■

Question 5.5.1 Extend the sample of size three from example 5.5.1 to a sample of size six by asking three more adult male's their height and weight. Present your data as a set of ordered pairs, as a table, and as a scatterplot.

■

When using a statistical model to relate bivariate data, researchers often begin with the simplest of all curves—the line—where two variables X and Y are related by the equation $Y = mX + b$. Instead of the notations m and b for the slope and Y-intercept, statisticians traditionally use the Greek letter "beta" for both, with subscripts: β_1 is the slope and β_0 is the intercept. The graph of the equation $Y = \beta_1 X + \beta_0$ is a line in the $X - Y$ plane. The slope β_1 of that line predicts the change in one variable (the dependent variable Y) per unit change in the other, X. Whenever a mathematical model is used, a researcher should ask if the form of the model is an appropriate choice. In this setting, the researcher should confirm if it is reasonable

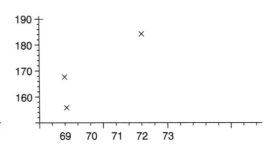

Figure 5.11 A scatterplot for example 5.5.1

to assume that the variables are related linearly, or is there an absence of a linear relationship? Approaching this question from the perspective of statistical inference requires that the researcher base the answer on an analysis of the corresponding bivariate data.

The full statistical model also introduces an error term. The random fluctuations that occur in nature, the accuracy of equipment used to gather data, and the inevitable mistakes by human observers can all introduce variance and error into data sets. The error term in a linear model indicates that the relationship is theoretical and that real-world values for (x, y) often do not lie directly on the theoretical line. Using ε to denote the error term, a linear statistical model describing the (assumed linear) relationship between two random variables X and Y is expressed as

$$Y = \beta_0 + \beta_1 X + \varepsilon.$$

A complete mathematical model establishing a linear relationship between two variables depends on a number of standard assumptions:

- The values of Y are independent of each other, so that one observed value of Y does not influence the value of another.
- At any given value for X, an expected value $E[Y|X]$ for Y exists. When different values of $E[Y|X]$ are plotted on an X–Y plane, they will lie on a straight line.
- The standard deviation of the error random variable ε is the same for each value of X.
- The error random variable ε is normally distributed.

We take these assumptions for granted throughout this section's introduction. Their role in the analysis may not be readily apparent. Rest assured that there exist statistical inference procedures that test each of these assumptions as it applies to a given bivariate data set. Further discussion of these details is left for your later studies.

An important statistical process is to determine the line that "best" runs through the data collected, as visualized in the scatterplot. This line is often referred to as the *regression line* or the *line of best fit* for the given set of bivariate data. The regression line is not the actual, theoretical relationship $Y = \beta_0 + \beta_1 X + \varepsilon$ between X and Y, but is instead an estimate of that relationship based on given bivariate data. In fact, different bivariate data will generally produce a different regression line. Regression lines are therefore typically denoted by $\widehat{Y} = \widehat{\beta}_0 + \widehat{\beta}_1 X$, where the "hat" notation distinguishes the regression line's estimation of the linear relationship between X and Y from the actual linear relationship between X and Y.

Statisticians have identified a precise formulation of what is meant by the "best-fitting line"—it is one that minimizes the sum of squares of distances between the data points and the line. Using subscripts to identify the elements of a bivariate data set as $\{(x_i, y_i) : i = 1, \ldots, n\}$, figure 5.12 presents a graphical illustration of this overarching idea. The vertical distances between the regression line and a given data point (x_i, y_i) is denoted v_i. As we can see, data points lie both above and below the regression line, resulting in some v_i being positive and others negative. Squaring each distance makes each term v_i^2 positive and eliminates the possibility of negative and positive values cancelling. The regression line minimizes the sum of these squared values.

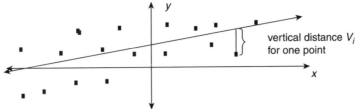

Figure 5.12 The regression line minimizes $\sum\limits_{i=1}^{n} v_i^2$

Definition 5.5.2 *If $\{(x_i, y_i) : i = 1, \ldots, n\}$ is a bivariate data set for random variables X and Y, then the corresponding **regression line** minimizes the sum of the squares of the vertical distances from the data points to the line. Symbolically, the regression line is the line $\widehat{Y} = \widehat{\beta}_0 + \widehat{\beta}_1 X$ that minimizes the sum*

$$\sum_{i=1}^{n} v_i^2 = \sum_{i=1}^{n} [\widehat{Y}_i - y_i]^2 = \sum_{i=1}^{n} [(\widehat{\beta}_0 + \widehat{\beta}_1 x_i) - y_i]^2.$$

The regression line's slope $\widehat{\beta}_1$ approximates the theoretical line's slope β_1, and the intercept $\widehat{\beta}_0$ approximates β_0.

Statisticians have determined the formulas for the slope $\widehat{\beta}_1$ and y-intercept $\widehat{\beta}_0$ of the regression line. They are expressed in terms of the sample mean for each random variable X and Y on the data set $\{(x_i, y_i) : i = 1, \ldots, n\}$. As in definition 5.4.3 of section 5.4,

$$\overline{X} = \frac{1}{n} \cdot \sum_{i=1}^{n} x_i \quad \text{and} \quad \overline{Y} = \frac{1}{n} \cdot \sum_{i=1}^{n} y_i.$$

The next theorem incorporates these values in the formulas for $\widehat{\beta}_0$ and $\widehat{\beta}_1$.

Theorem 5.5.1 *Let $\{(x_i, y_i) : i = 1, \ldots, n\}$ be a bivariate data set for random variables X and Y. The regression line $\widehat{Y} = \widehat{\beta}_0 + \widehat{\beta}_1 X$ is defined by the **least squares estimators***

$$\widehat{\beta}_1 = \frac{\sum_{i=1}^{n} (x_i - \overline{X}) \cdot (y_i - \overline{Y})}{\sum_{i=1}^{n} (x_i - \overline{X})^2} \quad \text{and} \quad \widehat{\beta}_0 = \overline{Y} - \widehat{\beta}_1 \cdot \overline{X}.$$

We apply theorem 5.5.1 in examples and questions, and then discuss its proof.

Example 5.5.2 An economist studies the relationship between the price P of Shine toothpaste and the quantity Q demanded by consumers. She collects the following data from three randomly selected local markets:

P = price	1.95	2.32	1.85
Q = quantity sold	20	12	16

To find the regression line, first compute the sample means

$$\overline{P} = \frac{1.95 + 2.32 + 1.85}{3} = \frac{6.12}{3} = 2.04$$

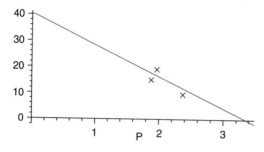

Figure 5.13 The regression line for example 5.5.2

and

$$\overline{Q} = \frac{20 + 12 + 16}{3} = \frac{48}{3} = 16.$$

Using these values, find the regression line's slope and intercept:

$$\widehat{\beta_1} = \frac{(1.95 - 2.04) \cdot (20 - 16) + (2.32 - 2.04) \cdot (12 - 16) + (1.85 - 2.04) \cdot (16 - 16)}{(1.95 - 2.04)^2 + (2.32 - 2.04)^2 + (1.85 - 2.04)^2}$$

$$= \frac{(-0.09) \cdot 4 + (0.28) \cdot (-4) + (-0.19) \cdot 0}{(-0.09)^2 + (-0.28)^2 + (-0.19)^2} = \frac{-1.48}{0.1226} \approx -12.072;$$

$$\widehat{\beta_0} = 16 - (-12.072) \cdot 2.04 \approx 40.627.$$

According to theorem 5.5.1, the regression line is

$$\widehat{Q} = 40.627 - 12.082P.$$

Assuming a linear relationship between P and Q, this regression line provides an estimate of the demand function for Shine toothpaste in the local area studied by the economist. Figure 5.13 illustrates this curve with the corresponding data points.

∎

Question 5.5.2   The economist considers using the line $Q^* = 41 - 12P$, which has rounded off slope and intercept values, as a model for the demand function. Verify that the regression line $\widehat{Q} = 40.627 - 12.082P$ is a "better-fitting" curve for the data given in example 5.5.2 by finding the sum of squares of vertical distances for each line; that is, compute both

$$\sum_{i=1}^{3} [\widehat{Q}_i - Q_i]^2 = \sum_{i=1}^{3} [(40.627 - 12.082P_i) - Q_i]^2$$

and

$$\sum_{i=1}^{3} [Q_i^* - Q_i]^2 = \sum_{i=1}^{3} [(41 - 12P_i) - Q_i]^2.$$

∎

In the relatively simple settings of example 5.5.2 and question 5.5.2, the usefulness of computing devices is readily apparent. Whenever data sets are sufficiently large, computing devices are needed to calculate the regression line in a reasonable amount of time. The commands listed in the following table describe how to input bivariate data into a given computing device.

System	Command
TI-83	STAT - EDIT - Type X data into L1 and Y data into L2
Maple	[> with(stats): [> datax:=$[x_1, x_2, x_3, \ldots, x_n]$; datay:=$[y_1, y_2, y_3, \ldots, y_n]$;
Mathematica	]:< <Statistics'LinearRegression']:=data=$\{\{x_1, y_1\}, \{x_2, y_2\}, \ldots \{x_n, y_n\}\}$;

Once the bivariate data are stored, a computing device will calculate the regression line, using the following commands.

System	Command
TI-83	STAT - CALC - LinReg(ax+b)
Maple	[> with(stats): [> fit[leastsquare[[x,y]]([datax,datay]);
Mathematica	]:< <Statistics'LinearRegression']:=func = Fit[data, $\{1, x\}$, x]

Example 5.5.3 We use a computing device to identify the least squares regression line for the bivariate data introduced in example 5.5.1:

Height	69	69	72
Weight	168	155	184

A computing device finds the rounded values as $\widehat{\beta}_1 = 7.5$ and $\widehat{\beta}_0 = -356$. The least squares regression line is therefore $\widehat{Y} = -356 + 7.5 \cdot X$.

■

Question 5.5.3 A college believes that the sum of the algebra and trigonometry subscores X of a standardized test is an indicator of a student's grade Y in the fall term calculus course. Determine the regression line for the following randomly selected data:

$X = $ subscore	31	27	22	20	15
$Y = $ grade	4.0	3.5	3.0	2.5	2.0

■

The next question considers the proof of theorem 5.5.1. Its argument is very algebraic in nature; the sum of squares of vertical distances turns out to be quadratic in the variable β_0. Therefore the minimum of the sum of squares corresponds to the minimum of the quadratic, which occurs at its vertex.

Question 5.5.4 This question outlines the proof that the regression line formula given in theorem 5.5.1 minimizes the sum of the squares of the vertical distances. As in the context of the theorem, assume $\{(x_i, y_i) : i = 1, \ldots, n\}$ is a bivariate data set for X and Y.

(a) For a general line $Y = \beta_0 + \beta_1 X$, express the sum of the squares of the vertical distances given below as a quadratic in the variable β_0. Explicitly determine the coefficients a, b, and c. You may wish to use the fact that $\sum\limits_{i=1}^{n} \beta_0^2 = n \cdot \beta_0^2$.

$$\sum_{i=1}^{n}[Y_i - y_i]^2 = \sum_{i=1}^{n}[\beta_0 + (\beta_1 x_i - y_i)]^2 = a \cdot \beta_0^2 + b \cdot \beta_0 + c.$$

(b) The minimum of a quadratic $a \cdot \beta_0^2 + b \cdot \beta_0 + c$ occurs at the vertex for which $\beta_0 = -b/2a$. Using the result from part (a), show that the vertex of the above quadratic occurs when $\widehat{\beta_0} = \overline{Y} - \beta_1 \cdot \overline{X}$.

(c) Note that part (b) identifies the value of $\widehat{\beta_0}$ that minimizes the sum of the squares independent of any specific value of β_1. Therefore, we can use this expression to find the value of β_1 that minimizes the sum of the squares. To this end, algebraically "complete the square" in the quadratic expression for β_1 as follows:

$$\sum_{i=1}^{n}(Y_i - y_i)^2 = \sum_{i=1}^{n}[(\widehat{\beta_0} + \beta_1 x_i) - y_i]^2 = \sum_{i=1}^{n}[\beta_1 x_i + (\widehat{\beta_0} - y_i)]^2$$

$$= \sum_{i=1}^{n}[\beta_1^2 x_i^2 + 2(\widehat{\beta_0} - y_i)\beta_1 x_i + (\widehat{\beta_0} - y_i)^2]$$

$$= \sum_{i=1}^{n} x_i^2 \cdot \left[\beta_1^2 + \frac{2\beta_1 \sum_{i=1}^{n} x_i(\widehat{\beta_0} - y_i)}{\sum_{i=1}^{n} x_i^2} + \left(\frac{\sum_{i=1}^{n} x_i(\widehat{\beta_0} - y_i)}{\sum_{i=1}^{n} x_i^2}\right)^2\right]$$

$$+ \left(\sum_{i=1}^{n}(\widehat{\beta_0} - y_i)^2 - \frac{(\sum_{i=1}^{n} x_i(\widehat{\beta_0} - y_i))^2}{\sum_{i=1}^{n} x_i^2}\right)$$

$$= \sum_{i=1}^{n} x_i^2 \cdot \left[\beta_1 + \frac{\sum_{i=1}^{n} x_i(\widehat{\beta_0} - y_i)}{\sum_{i=1}^{n} x_i^2}\right]^2 + \{\text{terms that do not involve } \beta_1\}.$$

Based on the bracketed squared term at the beginning of the last line of this string of equalities, what choice of β_1 minimizes this expression?

(d) Part (c) shows that the sum of squares is minimized by

$$\widehat{\beta_1} = -\frac{\sum_{i=1}^{n} x_i(\widehat{\beta_0} - y_i)}{\sum_{i=1}^{n} x_i^2}.$$

Substituting the expression $\widehat{\beta}_0 = \overline{Y} - \beta_1 \overline{X}$,

$$\widehat{\beta}_1 = \frac{\sum_{i=1}^{n} x_i \cdot (y_i - \overline{Y} + \widehat{\beta}_1 \overline{X})}{\sum_{i=1}^{n} x_i^2}.$$

Algebraically manipulate this equation by solving for $\widehat{\beta}_1$ and appropriately substituting the algebraic identity

$$\sum_{i=1}^{n}(a_i - \overline{A})(b_i - \overline{B}) = \sum_{i=1}^{n} a_i b_i - \frac{1}{n}\sum_{i=1}^{n} a_i \sum_{i=1}^{n} b_i$$

twice (once for $a_i = x_i$ and $b_i = y_i$, and once for $a_i = b_i = x_i$) to prove the final desired expression for $\widehat{\beta}_1$:

$$\widehat{\beta}_1 = \frac{\sum_{i=1}^{n} x_i \cdot (y_i - \overline{Y} + \widehat{\beta}_1 \overline{X})}{\sum_{i=1}^{n} x_i^2} \qquad \Rightarrow \qquad \widehat{\beta}_1 = \frac{\sum_{i=1}^{n}(x_i - \overline{X})(y_i - \overline{Y})}{\sum_{i=1}^{n}(x_i - \overline{X})^2}.$$

■

5.5.1 Hypothesis Test for Linearity

We now consider the question: How useful is a linear model in predicting the relationship between X and Y? Sometimes bivariate data points indicate that a linear pattern may exist, but sometimes they do not; the random variables may not be related at all, or they may have a nonlinear relationship. The scatterplots in figure 5.14 illustrate bivariate data that seem to follow linear, quadratic, and exponential patterns.

How might we inferentially test a set of given data to determine if a linear model is useful in describing the relationship?

To determine statistically if two random variables X and Y share a linear relationship, first collect a sample of bivariate data, and then carry out a straightforward hypothesis test to determine the usefulness of a linear model. This test is formulated in terms of the slope β_1 of the theoretical line. As it turns out, if the scatterplot of X and Y is filled with points that show no discernable pattern (and do not indicate any relationship between X and Y), then the regression model results in $\widehat{\beta}_1 = 0$. This value estimates that the true line's slope would also equal 0. To check for linearity between X and Y, we therefore test $H_0: \beta_1 = 0$ vs. $H_a: \beta_1 \neq 0$. Rejecting the null hypothesis H_0 is equivalent to concluding that a linear model is useful. We call this procedure a hypothesis test for linearity.

The P-value for this two-tailed test can be phrased in terms of the Student's t-distribution T_m that was introduced in exercises 67–70 of section 5.3. The random

Figure 5.14 Scatterplots with linear, quadratic, and exponential patterns

variable T_m having m degrees of freedom is defined using the *gamma function*

$$\Gamma(x) = \int_0^\infty t^{x-1} e^{-t} \, dt, \text{ where } x > 0.$$

The density function for T_m is then

$$f(t) = \frac{\Gamma((m+1)/2)}{\sqrt{m\pi} \cdot \Gamma(m/2)} \cdot \left[1 + \frac{t^2}{m} \right]^{-(m+1)/2}, \text{ where } -\infty < t < \infty.$$

The population mean of T_m is $\mu = 0$, and the standard deviation is $\sigma = \sqrt{m/(m-2)}$. The density function's graph is similar to that of the normal density having mean 0 and standard deviation 1, except the "bell-shape" for $f(t)$ is lower and wider. The following commands enable the use of computing devices to calculate probabilities $P[a < T_m < b]$:

System	Command
TI-83	DISTR - tcdf(a,b,m)
Maple	[> with(stats): [>int(GAMMA((m+1)/2)/GAMMA(m/2)/sqrt(m*Pi)/(1+t^2/m)^((m+1)/2), t=a..b);
Mathematica	]:< <Statistics'ContinuousDistribution']:CDF[StudentTDistribution[m],b] - CDF[StudentTDistribution[m],a]

Furthermore, since $P[T_m > 12] \approx 0 \approx P[T_m < -12]$, we can use the identities $P[a < T_m < 12] \approx P[T_m > a]$ and $P[-12 < T_m < b] \approx P[T_m < b]$ when computing probabilities.

The following theorem describes how to compute the P-value for the test of linearity. The proof of the theorem is given in advanced statistics courses and is left for your later studies.

Theorem 5.5.2 **Hypothesis test for linearity** *Based on bivariate data $\{(x_i, y_i) : i = 1, 2, \ldots, n\}$, a test for the usefulness of a linear model describing the relationship between random variables X and Y is conducted using H_0: $\beta_1 = 0$ vs. H_a: $\beta_1 \neq 0$. The corresponding P-value is*

$$P\text{-value} = 2 \cdot P\left[T_{n-2} > \frac{|\widehat{\beta_1}|}{(s/\sqrt{SS_X})} \right],$$

which uses $SS_X = \sum_{i=1}^n (x_i - \overline{X})^2$, $SS_Y = \sum_{i=1}^n (y_i - \overline{Y})^2$, *and* $s = \sqrt{\dfrac{SS_Y - \widehat{\beta_1}^2 \cdot SS_X}{n-2}}$.

The values SS_X and SS_Y are often called the sums of squares for the x-values and y-values, respectively. These sums of squares are related to the sample standard deviations s_x and s_y, which a computing device will calculate, by the equations $SS_X = (n-1) \cdot (s_x)^2$ and $SS_Y = (n-1) \cdot (s_y)^2$.

Example 5.5.4 A real estate agent studies the relationship between home size X (in thousands of square feet) in a local neighborhood and the corresponding sale price Y (in tens of thousands of dollars). She is interested in determining the usefulness of a linear model in predicting home prices and collects the following data for seven randomly selected homes:

X = size	1	2	3	4	5	6	7
Y = price	10	13	17	18	21	26	28

To test for linearity between X and Y, we test H_0: $\beta_1 = 0$ vs. H_a: $\beta_1 \neq 0$. To calculate the P-value, we first compute the necessary statistics.

$$\overline{X} = \frac{1+2+3+4+5+6+7}{7} = 4 \qquad \overline{Y} = \frac{10+13+17+18+21+26+28}{7} = 19$$

$$SS_X = \sum_{i=1}^{n}(x_i - \overline{X})^2 = 28 \qquad SS_Y = \sum_{i=1}^{n}(y_i - \overline{Y})^2 = 256$$

$$\widehat{\beta_1} = \frac{\sum_{i=1}^{n}(x_i - \overline{X})(y_i - \overline{Y})}{SS_X} = \frac{84}{28} = 3 \qquad s = \sqrt{\frac{SS_Y - \widehat{\beta_1}^2 \cdot SS_X}{n-2}} = \sqrt{\frac{256 - 3^2 \cdot 28}{7-2}}$$

$$= 0.894427$$

Applying the formula from theorem 5.5.2,

$$P\text{-value} = 2 \cdot P\left[T_{n-2} > \frac{\widehat{\beta_1}}{(s/\sqrt{SS_X})}\right] = 2 \cdot P\left[T_5 > \frac{3}{(.894427/\sqrt{28})}\right] =$$
$$2 \cdot P[T_5 > 17.74] \approx 2 \cdot 0 = 0.$$

Since the P-value is less than $\alpha = 0.05$, the real estate agent should reject the null hypothesis in favor of the alternative H_a: $\beta_1 \neq 0$. She should conclude that the linear model is useful in predicting the sale price Y in terms of the size of a home X, where $Y = \beta_0 + \beta_1 X + \varepsilon$. Using the value for $\widehat{\beta_1}$ calculated above, along with $\widehat{\beta_0} = \overline{Y} - \widehat{\beta_1} \cdot \overline{X} = 19 - 3 \cdot 4 = 7$, the regression line is $\widehat{Y} = 3 \cdot X + 7$.

■

Question 5.5.5   A medical researcher studies the relationship between the age X (in years) of a female child and the child's weight Y (in pounds). Wondering if it is reasonable to conclude that X and Y are related linearly, she collects the following data for eight randomly selected children:

X = age	1	2	3	5	7	9	10	12
Y = weight	14	26	30	41	54	60	81	87

Using theorem 5.5.2, conduct a hypothesis test to determine the usefulness of a linear model for the relationship between X and Y. If so, determine the corresponding regression line using theorem 5.5.1.

■

5.5.2   Predicting Response Values

We now consider using the regression line to predict an output value $\widehat{Y}$ that corresponds to an input $X = x_0$. These output values are often called *response values*. When a test for linearity concludes that a linear model appropriately describes the relationship between X and Y, then it is appropriate to substitute the value x_0 into the regression line to obtain the response: $\widehat{Y} = \widehat{\beta}_0 + \widehat{\beta}_1 \cdot x_0$. The only restriction that applies to this substitution is that the value $X = x_0$ should lie inside the range of X values in the bivariate data. This restriction makes sense: the bivariate data set that generated a positive conclusion about linearity does not contribute information about any value outside of its range; the relationship might be linear in the range of bivariate data, but could become curvilinear or degenerate into no relationship outside of that range.

In example 5.5.4, all of the values for X lie between 1 and 7, and the data indicate the usefulness of a linear model for X and Y. The corresponding regression line $\widehat{Y} = 3 \cdot X + 7$ may therefore be used as a predictor of output values $\widehat{Y}$ corresponding to any input that falls between 1 and 7. The next example illustrates the use of the regression line as a predictive model.

Example 5.5.5   In example 5.5.4, the regression line is identified as $\widehat{Y} = 3 \cdot X + 7$, where the home size X is measured in thousands of square feet, and the sales price Y is measured in tens of thousands of dollars. We use this regression line to predict the value of a home in the neighborhood with 1,800 square feet of living space.

The corresponding input value is $x_0 = 1.8 \in [1, 7]$. Substituting into the regression line produces the response value $\widehat{Y} = 3 \cdot 1.8 + 7 = 12.4$. Therefore, we predict that home in this neighborhood with 1,800 square feet can sell for $\$124,000$.

■

Question 5.5.6   Working in the context of example 5.5.4, find the predicted value of homes in the neighborhood that have each of the following square footage.

(a) 2,200   (c) 3,100

(b) 1,250   (d) 4,000

Explain why might it not be appropriate to use the regression line $\widehat{Y} = 3 \cdot X + 7$ to predict the sale value of a home with 900 square feet.

■

Many statisticians and mathematicians have invested a great deal of research effort into the development of statistical inference processes for regression analysis. This brief introduction has touched on only a few of the most basic ideas; many other interesting topics await your study in later courses. The notes at the end of the section indicate several sources to help with further investigation.

We finish this section by briefly touching on some of the more advanced topics. A number of hypothesis tests exist for other regression parameters. For example, to test $H_0: \beta_1 = C$ vs. $H_a: \beta_1 \neq C$ with $C \neq 0$ (theorem 5.5.2 dealt with the choice of

$C = 0$), compute the corresponding P-value:

$$P\text{-value} = 2 \cdot P\left[T_{n-2} > \frac{|\widehat{\beta}_1 - C|}{(s/\sqrt{SS_X})}\right].$$

Hypothesis tests also exist for the Y-intercept β_0 and the response value Y; these statistical inference processes are discussed in most standard texts devoted exclusively to statistics.

Models also exist for situations in which the bivariate data indicate a nonlinear relationship between random variables X and Y. If the nonlinear pattern matches a well-known function such as an exponential, logarithmic, or sinusoidal, then a transformation can be applied to the data, producing new bivariate data having a linear pattern. A test for linearity would then indicate the usefulness of a linear model relating the transformed data. An inverse transformation can then convert the regression line for the transformed data into a regression curve that predicts the response $\widehat{Y}$ in terms of the original random variable X.

Further statistical processes exist for models that involve multiple variables. The standard "multiple linear regression model" assumes an underlying relationship between an output variable Y that depends on n input variables $X_1, X_2, \ldots, X_n$ of the form $Y = \beta_0 + \beta_1 X_1 + \beta_2 X_2 + \cdots + \beta_n X_n + \varepsilon$. Such models are more sophisticated than the simple linear regression model outlined in this section and involve matrix theory in the construction of "best-fitting" lines. But they follow an approach similar to the process outlined in this section. Further details and discussion are left for your later studies in statistics.

5.5.3 Reading Questions for Section 5.5

1. Define and give an example of bivariate data.
2. Give an example of a bivariate data set that is functional and one that is not.
3. What is a scatterplot? Sketch an example.
4. What value does a regression line estimate? Include a discussion of the distinction between $Y = \beta_0 + \beta_1 X + \varepsilon$ and $\widehat{Y} = \widehat{\beta}_0 + \widehat{\beta}_1 X$.
5. What sum is minimized by the regression line?
6. Define and give an example of a sample mean.
7. State theorem 5.5.1. How is this result helpful when studying linear regression?
8. Sketch scatterplots illustrating a linear and a nonlinear relationship between random variables.
9. State the null and alternative hypotheses in a hypothesis test for linearity. Identify which hypothesis indicates the usefulness of a linear model.
10. What is the relationship between Student's t-distribution and the standard normal distribution?
11. State theorem 5.5.2. How is this result helpful when studying linear regression?
12. What input values may be substituted into a regression line to obtain a good predictor for a corresponding response value?

5.5.4   Exercises for Section 5.5

Exercises 1–6 study the following bivariate data set.

X	1	2	4 .	5
Y	2	6	8	12

1. Using theorem 5.5.2, confirm the usefulness of a linear model for X and Y.
2. Using theorem 5.5.1, determine the linear regression line for this data set.
3. Sketch the scatterplot for this data set and the linear regression line on the Cartesian plane.
4. Compute the sums of squares of vertical distances for the linear regression line.
5. Confirm that $y = 2x + 3$ is not the "best-fitting" line for this data set by computing the sums of squares of vertical distances for this line at each data point and comparing the result with the corresponding sum for the regression line.
6. Confirm that $y = 3x + 2$ is not the "best-fitting" line for this data set by computing the sums of squares of vertical distances for this line at each data point and comparing the result with the corresponding sum for the regression line.

Exercises 7–12 consider a bivariate data set that is extended one observation at a time. Perform a hypothesis test to determine if a linear model is useful for the corresponding random variables. If so, compute the regression line for the data set; if not, sketch the scatterplot and make a conjecture about the nature of the relationship between the random variables.

7.
X	1	2
Y	8	6

8.
X	1	2	3
Y	8	6	4

9.
X	1	2	3	4
Y	8	6	4	8

10.
X	0	1	2	3	4
Y	4	8	6	4	8

11.
X	0	1	2	3	4	5
Y	4	8	6	4	8	24

12.
X	−1	0	1	2	3	4	5
Y	−12	4	8	6	4	8	24

Exercises 13–18 consider the analysis of the bivariate data set from exercise 12; use the corresponding regression line to answer the following questions.

13. Compute the response for $X = 1.5$.
14. Compute the response for $X = 2.5$.
15. Compute the response for $X = 2$. Explain the difference between this response and the data set in exercise 12.
16. Compute the response for $X = 3$. Explain the difference between this response and the data set in exercise 12.
17. Compute the response for $X = 4$. Explain the difference between this response and the data set in exercise 12.

18. Compute the response for $X = 5$. Explain the difference between this response and the data set in Exercise 12.

In exercises 19–28, perform a hypothesis test to determine if the corresponding random variables are linearly related and, if so, compute the linear regression line for the data set.

19. To study the difference in average weights of brothers, based on age, a researcher randomly selects five sets of brothers, all over 30 years old, and compares their weights, with the following results (in pounds):

X = older brother's weight	178	150	197	168	200
Y = younger brother's weight	173	139	178	170	200

20. To determine the relationship between family income and IQ of nine-year-old children, a school psychologist randomly collected the following bivariate data, where the family income is measured in thousands of dollars.

X = IQ	178	150	197	168	200	90	200
Y = income	173	139	178	170	200	300	200

21. An environmentalist collects total rainfall amounts on her hometown Caribbean island, with the following results:

X = month	1	3	4	6	8	10	12
Y = rainfall in inches	73	62	59	50	112	104	74

22. A physician looks at the effect of pH levels on the number of bacterial cells, measured in tens of thousands, appearing in a culture at the midpoint of a controlled experiment.

X = pH	2	3	3	5	6	6
Y = number of bacteria	103	139	146	140	196	212

23. The advising office at a major university is curious about the relationship between student study times and success on a sociology test. Seven students in a large introductory course were randomly selected, and their grades on the final test were recorded, with the following results:

X = hours studying	3	7	11	9	18	20	6
Y = test grade	58	87	89	89	93	90	77

24. A medical research team performs a study on patients not taking a beta-blocker medication. The results linking a patient's age to a corresponding systolic blood pressure follow:

X = age	17	25	55	68	76
Y = systolic blood pressure	119	121	158	136	168

25. An academic research group studies the number of Ph.D.s awarded by American universities in the given year. The results follow.

X = year	1980	1985	1990	1995	2000
Y = Ph.D.s in physics	873	739	778	670	720

26. The white-tailed deer population, given in thousands, in a midwest state was examined by a governmental agency, with the results below:

X = year	1949	1970	1980	1990	1995	2000	2005
Y = deer population	2.3	50	82	135	436	476	487

27. A study is performed on the effect of a nitrate fertilizer on plant growth follows, where the fertilizer amount is given in pounds, and the crop yield is given in bushels of wheat.

X = fertilizer amount	3	7	15	17	21	25	25
Y = crop yield	55	66	79	88	90	90	95

28. The Environmental Group Agency studied American-built automobiles' highway miles per gallon rating against the amount of horsepower produced by the cars' engines. The results follow.

X = miles per gallon	15	18	21	23	25	29	31
Y = horsepower	189	167	158	159	145	144	138

In exercises 29–33, determine the regression line using summation formulas; you may wish to use the formulas:

$$\sum_{i=1}^{n}(x_i-\overline{X})(y_i-\overline{Y})=\sum_{i=1}^{n}x_iy_i-(\sum_{i=1}^{n}x_i)\cdot(\sum_{i=1}^{n}y_i)/n \text{ and}$$

$$\sum_{i=1}^{n}(x_i-\overline{X})^2=\sum_{i=1}^{n}x_i^2-(\sum_{i=1}^{n}x_i)^2/n.$$

29. A researcher collects data in the form (x_i, y_i) to find the regression line and, from a data set of $n = 15$ points, computes the following sums:

$$\sum_{i=1}^{n}x_i = 3; \quad \sum_{i=1}^{n}x_iy_i = 12; \quad \sum_{i=1}^{n}y_i = 2; \quad \sum_{i=1}^{n}x_i^2 = 15.$$

30. An engineer collects data on X = "the horsepower produced by an automobile engine" as compared to Y = "the size of the engine as measured in liters." From a data set of $n = 21$ points, the engineer computes the following sums:

$$\sum_{i=1}^{n}(x_i-\overline{X})(y_i-\overline{Y})=4,419; \quad \sum_{i=1}^{n}x_iy_i=8,475; \quad \sum_{i=1}^{n}y_i=44.5; \quad \sum_{i=1}^{n}x_i^2=865,380.$$

31. A researcher collects data in the form (x_i, y_i) to find the regression line and, from a data set of $n = 15$ points, computes the following sums:

$$\sum_{i=1}^{n}(x_i-\overline{X})(y_i-\overline{Y})=1,160; \quad \sum_{i=1}^{n}(x_i-\overline{X})^2=482; \quad \sum_{i=1}^{n}x_i=29; \quad \sum_{i=1}^{n}y_i=25.$$

32. To study the growth in federal expenditures on urban development, an economist collects data on X = "the year (measured in numbers of years after 1995)" and Y = "percent of total outlays devoted to development."

The data set of $n = 12$ points yields the following sums:

$$\sum_{i=1}^{n}(x_i - \overline{X})(y_i - \overline{Y}) = -75; \quad \sum_{i=1}^{n}(x_i - \overline{X})^2 = 275; \quad \sum_{i=1}^{n} x_i = 7.2; \quad \sum_{i=1}^{n} y_i = 11.8.$$

33. A researcher collects data in the form (x_i, y_i) and, from the $n = 20$ data points, computes the following sums. Based on these results, is it appropriate to conclude that there exists a linear relationship between the variables X and Y?

$$\sum_{i=1}^{n}(x_i - \overline{X})(y_i - \overline{Y}) = 115; \quad \sum_{i=1}^{n}(x_i - \overline{X})^2 = 480; \quad \sum_{i=1}^{n}(y_i - \overline{Y})^2 = 610;$$

$$\sum_{i=1}^{n} x_i = 30; \quad \sum_{i=1}^{n} y_i = 34.$$

We say that X and Y are positively correlated when an increase in one causes an increase in another, negatively correlated when an increase in one causes a decrease in another, and have no correlation when a change in one does not affect the other.

In exercises 34–45, sketch a scatter plot with the given features.

34. A low positive correlation.
35. A high positive correlation.
36. A perfect positive correlation.
37. A low negative correlation.
38. A high negative correlation.
39. A perfect negative correlation.

40. No correlation.
41. A quadratic correlation.
42. A cubic correlation.
43. An exponential correlation.
44. A logarithmic correlation.
45. A sinusoidal correlation.

Exercises 46–55 consider an alternative approach to the hypothesis test for linearity. Define

$$r = \hat{\beta}_1 \sqrt{\frac{SS_X}{SS_Y}}$$

and use the P-value equal to

$$2 \cdot P\left[T_{n-2} > |r|\sqrt{\frac{n-2}{1-r^2}}\right].$$

When the P-value is less than α, conclude that a linear model is useful in describing the relationship between X and Y.

In exercises 46–55, conduct a hypothesis test using the alternative approach on each bivariate data set to determine if the corresponding random variables are linearly related.

46. The data set from exercise 19.
47. The data set from exercise 20.
48. The data set from exercise 21.
49. The data set from exercise 22.
50. The data set from exercise 23.

51. The data set from exercise 24.
52. The data set from exercise 25.
53. The data set from exercise 26.
54. The data set from exercise 27.
55. The data set from exercise 28.

Exercises 56–60 consider an alternative derivation of the formulas for the least squares estimators $\widehat{\beta_0}$ and $\widehat{\beta_1}$ than the one provided by question 5.5.4. These same formulas can be derived using the techniques of calculus by minimizing the sum of squares of vertical distances (also called the sum of squares due to error, or SSE). In addition, we consider further properties of these least squares estimators.

56. Part (a) of question 5.5.4 identifies the sum of vertical distances as $SSE =$ $\sum_{i=1}^{n}(\beta_0 + \beta_1 \cdot x_i - y_i)^2$. This expression may be thought of either as a function of β_0 treating β_1 as a constant, or as a function of β_1 treating β_0 as a constant. For this exercise we adopt the first perspective, writing $f(\beta_0) = \sum_{i=1}^{n}(\beta_0 + \beta_1 \cdot x_i - y_i)^2$. Differentiate this function and set the derivative equal to zero to obtain an equation in the terms of β_1 and β_0.

57. Following the approach begun in exercise 56, adopt the second perspective suggested above; think of SSE as a function $g(\beta_1)$ treating β_0 as a constant and write $g(\beta_1) = \sum_{i=1}^{n}(\beta_0 + \beta_1 \cdot x_i - y_i)^2$. Differentiate this function and set the derivative equal to zero to obtain an equation in the terms of β_1 and β_0.

58. The solutions obtained in exercises 56 and 57 are two equations in terms of β_1 and β_0. Based on multivariable calculus, the minimum value for SSE is found by solving these two equations for the unknowns β_1 and β_0. Solve the equation from exercise 56 for the unknown β_0 and the equation from exercise 57 for the unknown β_1 (using the formulas from the instructions for exercises 29–33) to show that SSE is minimized at

$$\widehat{\beta_0} = \overline{Y} - \widehat{\beta_1}\overline{X} \qquad \text{and} \qquad \widehat{\beta_1} = \frac{\sum_{i=1}^{n}(x_i - \overline{X})(y_i - \overline{Y})}{\sum_{i=1}^{n}(x_i - \overline{X})^2}.$$

59. Assume that the relationship between random variables X and Y follows a linear model with a y-intercept equal to 0; that is, $\beta_0 = 0$ and so $Y = \beta_1 \cdot X + \varepsilon$. Find the value of β_1 that minimizes the sum of squares $\sum_{i=1}^{n}(Y_i - y_i)^2 = \sum_{i=1}^{n}(\beta_1 x_i - y_i)^2$.
    Hint: As in exercises 57 and 58, differentiate with respect to β_1, set the resulting derivative equal to 0, and solve for β_1.

60. Treating $\widehat{\beta_1}$ as a random variable defined on a sample of bivariate data, we may calculate its expected value $E[\widehat{\beta_1}]$. From algebraic computations and properties of the expected value:

$$E[\widehat{\beta_1}] = E\left[\sum_{i=1}^{n}\frac{(x_i - \overline{X})(y_i - \overline{Y})}{SS_X}\right] = E\left[\sum_{i=1}^{n}\frac{(x_i - \overline{X})y_i}{SS_X}\right]$$

$$= \sum_{i=1}^{n}\frac{(x_i - \overline{X})E[y_i]}{SS_X} = \sum_{i=1}^{n}\frac{(x_i - \overline{X})(\beta_0 + \beta_1 x_i)}{SS_X}.$$

Expanding this final expression, we have:

$$E[\widehat{\beta_1}] = \beta_0\sum_{i=1}^{n}\frac{x_i - \overline{X}}{SS_X} + \beta_1\sum_{i=1}^{n}\frac{(x_i - \overline{X}) \cdot x_i}{SS_X}.$$

Use this result and the fact $\sum_{i=1}^{n}(x_i - \overline{X}) = 0$ and $SS_X = \sum_{i=1}^{n}(x_i - \overline{X}) \cdot x_i$ to prove that $E[\widehat{\beta_1}] = \beta_1$.

Notes

Combinatorics is an intriguing and active area of mathematical study; there are many well-written books describing its approach to answering mathematical questions. Basic combinatorics is often studied in "Discrete Mathematics" courses, which are supported by such texts as those by Epp [72], Richmond and Richmond [193], and Scheinerman [209]. Standard texts devoted exclusively to the study and development of combinatorics at the undergraduate level include those by Andreescu and Feng [5], Brualdi [33], Cameron [36], and van Lint and Wilson [157]. In addition, Tucker [243] approaches combinatorics from an applied perspective. Benjamin and Quinn [16] adopt an especially insightful view of combinatorial questions by developing a visual analysis; much of the work they present was a result of undergraduate research projects.

Section 5.2 is only an introduction to Pascal's triangle. Conway and Guy [45] provide an enjoyable discussion of many different types of numbers and sequences of numbers that are defined using Pascal's triangle. Colledge [42] is a resource developed for teachers that explores various aspects of this triangle. The history of Pascal's triangle is traced in Edwards [69]. An interesting biography of Blaise Pascal has been written by O'Connell [183]; this work describes the physical and intellectual climate of seventeenth century France and its impact on this insightful mathematician. Krailsheimer [185] is a good translation of Pascal's *Pensées*, the most important of his philosophical works.

Section 5.2 included a basic introduction to fractals. You may have seen some of the beautiful fractal images that have become popular with the general public. In the 1990s, television shows describing fractals and their applications to modeling occurred on such series as NOVA on PBS. Benoit M. Mandelbrot [165] drew attention to the mathematical study of fractals; his work received wide acclaim. Briggs [30] provides another well-written introduction to fractals. Several interesting and accessible books about fractals have recently been published, including an anthology of original papers creating an historical context of the study in Mandelbrot [164], a collection of essays by leading researchers that describes recent developments in Stewart et al. [232], and a connection to economic applications in Mandelbrot and Hudson [166]. To learn how to create some of those beautiful and intriguing fractal images, see Stevens [227].

Probability theory is often studied at an introductory level and can be found in the discrete mathematics textbooks cited above. For a more thorough and complete development of probability theory at an undergraduate level, see Bertsekas and Tsitsiklis [17], Jaynes [129], Ross [198], and Rozanov [200]. The complete correspondence on probability theory between Pascal and Fermat is presented in Smith [217]. Two textbooks that focus on the development of probability theory with a focus on applications include Feller [81] and Scheaffer [208]. Anyone interested in working on problems in probability will find Mosteller [177] fun and intriguing.

The study of probability is naturally intertwined with the study of statistics. Statistics is taught at the undergraduate level both from an introductory applied standpoint and in a way that illuminates the underlying mathematical theory. Textbooks serving an advanced, calculus-based undergraduate course include such classics as Wackerly et al. [246] and Hogg and Tanis [119]. A more computational approach that describes the methods (but does not generally prove the mathematical results) is taken in Fleiss et al. [86], Johnston [132], Peck et al. [186], and Watkins et al. [250]. Advanced discussions of hypothesis testing are in Casella and Berger [38] and Lehmann and Romano [154]. Linear regression is developed more fully in Montgomery et al. [176], Seber and Lee [210], and Weisberg [252].

For an overview of the development of statistics in the twentieth century, *The Lady Tasting Tea: How Statistics Revolutionized Science in the Twentieth Century* by Salsburg [207] is an excellent reference. As discussed in section 5.4, Sir Ronald Fisher played a pivotal role in the development of statistical inference. Fisher's seminal treatise *The Genetical Theory of Natural Selection* [83] details this process in the context of genetics, developing an important synthesis of Darwinian evolution and Mendelian genetics. *Philosophical Problems of Statistical Inference: Learning from R. A. Fisher* by Seidenfeld [211] is an insightful reflection on Fisher's continuing influence in this area. Finally, the book *R. A. Fisher, the Life of a Scientist* by Fisher's daughter Joan Fisher Box [24] blends the stories of the personal life and the professional accomplishments of this scientist, who contributed so much to mathematical statistics.

6 Graph Theory

Graph theory is the area of mathematics that grapples with the notion of connectedness among objects. Viewed from the proper perspective, our world is filled with many different relationships from the large-scale to the microscopic, from the concrete to the more abstract. Airline routes and interstates connect cities, bridges connect land masses, and wires connect computer circuits. There are relationships among family members, friends, business associates, countries, and corporations. Many aspects of our society are improved by analyzing and optimizing such connections. On a more personal level, navigating our way through daily life often involves understanding social relationships.

Many of the questions that challenge and motivate mathematicians who study graph theory are rooted in visual representations of our world. For example, we might wonder, "What is the best route from Cincinnati, Ohio to Nashville, Tennessee?" The road map in figure 6.1 can help us answer this question. When considering this map, we should keep in mind that "best" can have many different meanings, including least distance, least time, or most picturesque. Perhaps you can think of some other criterion; for example, having a close friend in Lexington, Kentucky might influence your choice of routes.

Graph theory enables us to analyze such questions by representing entities such as cities on the map as points called "vertices." We then represent relationships between entities such as roads between cities as lines or curves called "edges" connecting the corresponding vertices. These abstract representations cull away unimportant details, allowing us to focus on the essential features of and to obtain precise answers to a variety of important questions. As we develop a theory of graphs, we will begin to reintroduce some of these details, expanding the scope and power of associated mathematical tools. This more advanced work in graph theory has led mathematicians to include weights on edges to represent such details as distance or time, directions on edges to represent such features as one-way streets, and colors on both edges and vertices to represent levels of connection and distinction between measurable quantities in relationships.

The abstract study of relationships via graph theory is an active area of mathematical research that grapples with many issues relevant to our lives at personal, professional, and societal levels. Many find the perspectives and algorithms of graph theory rather intuitive, as well as both interesting and elegant. In addition, a large number of the open questions of graph theory are readily explored and studied by both professional and amateur mathematicians. We hope that you will

Figure 6.1   Possible major routes from Cincinnati to Nashville

want to contribute to the way mathematicians articulate and understand the ideas of graph theory.

This chapter begins with an introduction to the basic terminology of graph theory. As in any field of mathematics, a precise vocabulary of words with unique meanings has gradually been developed in graph theory as mathematicians have explored the intuitive ideas that motivate its study. The chapter identifies the fundamental notions and defines the corresponding terms that elucidate these notions. The remainder of this chapter studies specific graph-theoretic questions, all of which have a very practical flavor. We study Euler's solution of the *explorer's problem* of visiting every edge of a graph exactly once as well as the similar (but far more subtle) *traveling salesman problem* of visiting every vertex exactly once. We then search for optimal paths through both standard graphs and weighted graphs. All of these ideas and questions are of significant importance. Mathematicians have succeeded in several different ways in making progress on understanding graph theory.

6.1   An Introduction to Graph Theory

We begin with the basic ideas and terminology of graph theory. Graph theory approaches the notion of connectedness from a visual perspective, using graphical representations to express relationships among entities. As mathematicians, we are interested in culling away unimportant details and identifying the essential features of a given situation by expressing seemingly complicated scenarios with simple pictures. Working with such graphs enables us to analyze and (in many cases) solve many different and important questions.

The first question that led to the development of graph theory was the *Königsberg bridge problem*. The town of Königsberg (now known as Kaliningrad in central Russia) is divided into four parts by the Pregel River. After flowing around the central island

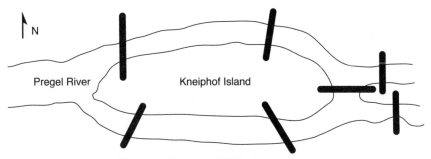

Figure 6.2 A model of the overall layout of Königsberg

of Kneiphof, the Pregel splits into two branches on the downstream side of the island. As Königsberg grew, bridges were built over various stretches of the river until there were a total of seven bridges connecting the four parts of the town. The general layout of the river and the bridges is illustrated in figure 6.2.

To relax in the evenings, the people of Königsberg would often go for leisurely walks, crossing the various bridges during their strolls through town. During one of these walks, some curious and clever person asked: "What route allows someone to cross every bridge exactly once and end the walk at its starting place?" As news of this challenge spread, many townspeople walked through the town in search of a solution. In the 1730s, the study of this Königsberg bridge problem was taken up by the eminent Swiss mathematician Leonhard Euler (pronounced "Oiler" because it a Germanic name like "Freud" rather than a Greek name like "Euclid"). In the process of answering this question, the study of graph theory was born.

In thinking about the Königsberg bridge problem, Euler chose to identify the four land masses of the town (as partitioned by the Pregel) as four points (or *vertices*) and to connect these vertices with seven curves (or *edges*) representing the seven bridges. In this way, Euler obtained the *graph* in figure 6.3 representing the city of Königsberg with its seven bridges. For example, the center vertex in the left-side column of three represents Kneiphof Island.

This simplified picture facilitated Euler's analysis of the Königsberg bridge problem and enabled Euler not only to answer this particular question, but also to identify general criteria that an arbitrary graph must satisfy for one to be able to traverse every edge exactly once and return to the starting point. We study Euler's solution in section 6.2. But first, we define the basic vocabulary and explore the corresponding ideas of graph theory for the remainder of this section. For now, you might be interested

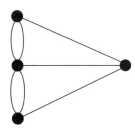

Figure 6.3 Euler's representation of Königsberg as a graph

in working with the graph in figure 6.3 to see if you can identify Euler's response to the Königsberg bridge problem.

As we have mentioned before, Archimedes, Newton, and Gauss are widely regarded as three of the most preeminent mathematicians of recorded human history. Leonhard Euler could certainly be added as a fourth. Euler was born and educated in Basel, Switzerland in the early 1700s. His early academic studies were in theology, but his first love was mathematics, which he began studying and researching under the tutelage of Johann Bernoulli. By the age of 19, Euler had published his first paper, and he went on to become the most prolific writer of mathematics of all time; in fact, his papers continued to be published for almost 50 years after his death. For most of his professional life, Euler worked at the St. Petersburg Academy in Russia and at the Berlin Academy in Germany. He made contributions in all areas of mathematics, despite gradually going blind and raising and supporting a large family. Euler's important influence on mathematics is reflected in some of the mathematical notation attributed to him: $f(x)$ for a function from 1734; e for the base of the natural logarithm from 1722; i for $\sqrt{-1}$ from 1777; π for pi; Σ for summations from 1755; and Δy for finite differences.

What are the basic notions and terminology of graph theory? As we have seen, the choice of words to denote mathematical concepts plays an important role in our understanding of the ideas at hand. Since graph theory is such a visually motivated subject, we rely on related English terms, being careful to distinguish between formal definitions and informal meanings when the situation warrants. We begin with the definition of a graph.

Definition 6.1.1   *A **graph** is a set of points called **vertices** and a set of curves called **edges**. Every edge **joins** exactly two vertices (these vertices may not be distinct). The two vertices are called the **endpoints** of an edge joining them and are said to be **adjacent** to one another. An edge is said to be **incident** to its two endpoints.*

Example 6.1.1   Three graphs are given in figure 6.4.

∎

Question 6.1.1   Mathematicians distinguish among the three graphs given in figure 6.4 for example 6.1.1 by categorizing their different features. For example, graph (a) has three edges, while graph (b) has five. Identify as many differences as you can among these graphs. Developing the ideas and vocabulary for graph theory allows us to express more carefully these distinctions.

∎

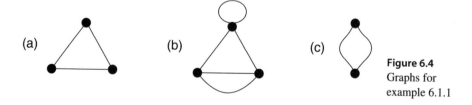

Figure 6.4
Graphs for
example 6.1.1

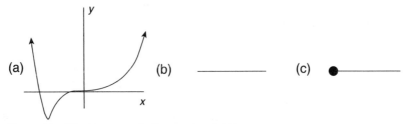

Figure 6.5 Three non-graphs for question 6.1.2

Figure 6.6 Three more graphs

Question 6.1.2   Explain why the pictures given in figure 6.5 are not graphs in the sense of definition 6.1.1.

■

We will consider only finite graphs (that is, graphs with a finite number of vertices and edges), unless explicitly stated otherwise. Finite graphs are sufficient to address most questions of interest. As you might expect, there are many different ways to present graphs. The pictorial representation illustrated in example 6.1.1 is often the most intuitive and useful. However, some questions and ideas are best explored from other perspectives, and so we introduce alternative ways to represent graphs in the exercises at the end of sections 6.1 and 6.2.

For now, we focus on enriching our language for describing important features of graphs. In question 6.1.1 we began developing an intuition for distinguishing among graphs. To help sharpen these insights further, we consider the three graphs given in figure 6.6.

There are some important differences among these three graphs. Some edges share endpoints while others do not, and other edges have a unique endpoint. Motivated by these observations, we define several terms that describe such features.

Definition 6.1.2   *Two edges are* **parallel** *if they have the same endpoints. An edge is a* **loop** *if its endpoints are identical. A graph is* **simple** *if the graph has no loops or parallel edges. We sometimes refer to a graph that is not simple as a* **multigraph**.

Example 6.1.2   We consider the graphs given in figure 6.6 in light of definition 6.1.2. Graph (a) is simple, while graphs (b) and (c) are not. In particular, graph (b) has two parallel edges and graph (c) has a loop.

■

Question 6.1.3   Sketch a graph with the following features, or explain why such a graph does not exist.

(a) A graph (or multigraph) with two vertices, two loops, and two parallel edges.

(b) A graph (or multigraph) with one vertex and two loops.

(c) A simple graph with two loops.

■

As suggested by definition 6.1.2 and these examples, the number of edges connecting a pair of vertices is of interest in our study of graph theory. Mathematicians have also found that focusing on the total number of edges incident to a single vertex is helpful in the study of graph theory; this total is known as the degree of a vertex, as defined below.

Definition 6.1.3   *The **degree of a vertex** is the total number of times the vertex is an endpoint of an edge; alternatively, the degree is the total number of edges incident to the vertex, provided we count incident edges that are loops twice. If a vertex is labeled V, then **d(V)** denotes the degree of the vertex. The **total degree of a graph** is the sum of the degrees of all vertices in the graph.*

In some applications of graph theory, the degree of a vertex is referred to as the *valence* of a vertex. This terminology arose from the British mathematician Arthur Caley's application of graph theory to the study of molecules in chemistry. In chemistry, the "valence" of an atom in a molecule is important and corresponds to the degree of a vertex in an appropriate representative graph.

Example 6.1.3   We identify the degree of each vertex and the total degree of the graph in figure 6.7. As customary in graph theory, vertices are labeled with letters for ease of reference.

- $d(A) = 4$
- $d(B) = 3$
- $d(C) = 0$

- $d(D) = 2$
- $d(E) = 1$
- total degree $= 4 + 3 + 0 + 2 + 1 = 10$

■

Question 6.1.4   Specify the degree of each vertex and the total degree of the graphs given in figure 6.8.

■

Question 6.1.5   Based on example 6.1.3 and question 6.1.4, what conjectures can you make about the total degree of a graph? What conjecture can you make about the relationship between the total degree of a graph and the number of edges in the graph? Can you see any other patterns? Try to answer these questions before reading further.

■

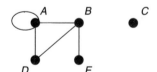

Figure 6.7   Graph for example 6.1.3

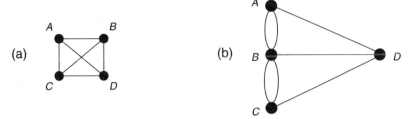

Figure 6.8　Graphs for question 6.1.4

Your conjectures in response to question 6.1.5 are part of a search for general facts about graphs. We have learned that mathematicians often begin the quest for theorems and mathematical truths by forming conjectures based on repeated patterns observed in multiple examples. In many cases, a conjecture is an important part of the process of mathematics—a process in which a mathematician seeks the reasons behind observed patterns, refines an understanding of the underlying ideas, and ultimately crafts a proof expressing these insights as a theorem.

In this context, we are considering a potential relationship between the total degree of a graph and the number of edges in the graph. If we study the graph in figure 6.7 for example 6.1.3, we observe that the total degree is 10 and that there are five edges in the graph. Similarly, for graph (a) in figure 6.8 for question 6.1.4, the total degree is 12 and there are six edges. Based on these observations, we might start to wonder if this doubling pattern holds for all graphs (or perhaps it is particular to these examples, and other graphs do not satisfy this pattern). At this point, we might notice that every edge has exactly two endpoints and contributes to the degree of exactly two vertices. This observation is the key to an argument proving that a doubling pattern is indeed universal. In light of these reflections, we state and prove a first theorem of graph theory; this result first appeared in Euler's 1736 paper that presented a solution of the Königsberg bridge problem and began the study of graphs.

Theorem 6.1.1　*The total degree of a graph is twice the number of edges in the graph.*

Proof　Recall from definition 6.1.3 that the total degree of a graph is the sum of the degrees of all vertices in the graph. According to the definition of the degree of a vertex, we are summing the number of times a vertex is an endpoint of an edge. Since every edge of a graph has exactly two endpoints, the total degree must be equal to twice the number of edges.

■

Question 6.1.6　Prove that the total degree of a graph is even. Use definition 1.7.1 in section 1.7.

■

As suggested by the Königsberg bridge problem and the map examples mentioned in the introduction to this chapter, many important questions in graph theory address the goal of traversing graphs under certain assumptions or restrictions. In graph-theoretic models of these physical settings, traversing a graph corresponds to crossing the bridges or traveling the roads on a map. Some of the most common restrictions on such traversals include minimizing time or distance, minimizing or avoiding repetition, and beginning and ending traversals at the same vertex. The following definitions facilitate this study; we consider several examples after stating these several definitions to help clarify the distinctions among these terms.

Definition 6.1.4 • *For a graph with labeled vertices and edges, a* **walk** *from a vertex V to another vertex W is a sequence of edges such that the first edge has V as an endpoint, the last edge has W as an endpoint, and any two adjacent edges in the sequence are incident to a common vertex. Vertex V is the* **initial** *endpoint of the walk and vertex W is the* **final** *endpoint of the walk.*
   • *A* **path** *is a walk with no repeated edges.*
   • *A* **simple path** *has no repeated vertices.*
   • *A* **closed walk** *has the same initial and final endpoint. We say that a closed walk is* **based** *at this common endpoint.*
   • *A* **circuit** *is a closed path; that is, a walk with a common initial and final endpoint that does not repeat an edge.*
   • *A* **cycle** *is a circuit that does not repeat any vertex besides the base vertex.*

The following table summarizes definition 6.1.4 and may help with understanding and recalling the distinctions among these terms.

Traversal	Repeated edges	Repeated vertices	Initial = final vertex
Walk	maybe	maybe	maybe
Path	no	maybe	maybe
Simple path	no	no	maybe
Closed walk	maybe	at least initial	yes
Closed path	no	at least initial	yes
Circuit	no	maybe	yes
Cycle	no	exactly initial	yes

When only the vertices in a graph are labeled, we present walks and paths as a sequence of vertices (rather than as a sequence of edges). Similarly, when neither the vertices nor the edges of a graph are labeled, walks and paths are presented in some other appropriate fashion. For example, a pictorial representation of a graph may have a sequence of edges highlighted to denote a path. In such settings, particular care may be required to uniquely express a walk or path through a graph. In particular, when multiple traversed edges are incident to the same vertex and the order of traversal is important, we must somehow indicate the intended ordering of edges in the path.

Example 6.1.4   The graph in figure 6.9 represents a selection of possible routes from Cincinnati, Ohio to Nashville, Tennessee. We label the graph's vertices (with uppercase letters) and edges (with lowercase letters) to facilitate the identification of various walks, paths, circuits, and cycles.

   We identify several of the many different walks from Cincinnati (denoted by vertex C) to Nashville (denoted by vertex N) in figure 6.9.

(a) u, u, x, y (c) u, v, y

(b) u, v, x, z (d) z

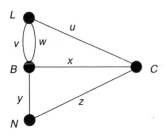

Figure 6.9 A graph representing some routes from
Cincinnati (C) to Nashville (N)

The walk (a) is not a path because it repeats edge u, while the last three walks are
paths. The walks (a) and (b) are not simple because they repeat vertex C, while
walks (c) and (d) are simple paths because no edge or vertex is repeated. There
are other simple paths from C to N; perhaps you can find one of them?

We also consider circuits and cycles in figure 6.9 based at Louisville, Kentucky
(denoted by vertex L). Directly from the definitions, we observe that every cycle
is a circuit. Therefore, the cycles u, x, v and v, y, z, u are also examples of circuits.
On the other hand, not every circuit is a cycle; for example, the path v, y, z, x, w
is a circuit but not a cycle (because vertex B is visited twice).   ■

As mentioned above, we sometimes present paths by boldfacing the appropriate
edges in the graph. For example, the path u, v, y in the graph given in figure 6.9 can
be represented by boldfacing edges as illustrated in figure 6.10 below.

Question 6.1.7 Working with the graph in figure 6.11, identify paths, circuits, and cycles with the
following properties.

  (a) Identify a simple path and a non-simple path with initial vertex A and final
      vertex D.
  (b) Identify two distinct circuits with base vertex B that are not cycles; that is,
      circuits with some repeated vertex in addition to the common base.
  (c) Identify two distinct cycles with base vertex C.   ■

With this understanding of traversals in hand, we carefully consider various types
of connectedness among the vertices of a graph. Reflecting on the graphs we have
studied thus far, you may recognize at least three different levels of connectedness

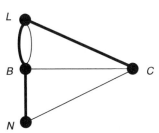

Figure 6.10 The path u, v, y highlighted from figure 6.9

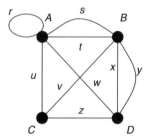

Figure 6.11   Graph for question 6.1.7

among pairs of vertices: some pairs of vertices are adjacent and directly connected by an edge; other pairs are only connected by a path using edges through other vertices; and still other pairs of vertices are not even connected by a path. We have also seen graphs in which some vertices are adjacent to every other vertex in a graph, while other vertices are adjacent to only a proper subset of vertices. We can readily visualize a graph with a vertex that is not adjacent to any other vertex in the graph. Motivated by such examples and reflections, we define some terms to help us articulate such distinctions among graphs and vertices.

Definition 6.1.5   • *A graph is* **connected** *if for every pair of distinct vertices V and W in the graph there exists a path with V and W as the the initial and final vertex of the path.*
• *A* **complete** *graph is a graph in which every vertex is joined to every other vertex in the graph by exactly one edge; that is, if every pair of distinct vertices are the endpoints of exactly one edge. We denote the unique complete graph with n vertices by* K_n.
• *A* **cycle** *graph is a graph consisting of a single cycle. We denote the unique cycle graph with n vertices by* C_n.
• *A* **null** *graph is a graph with no edges. We denote the unique null graph with n vertices by* N_n.
• *A vertex is* **isolated** *if the vertex is not adjacent to any other vertex in the graph.*

In definition 6.1.5, notice that "connected," "complete," "cycle," and "null" are properties of graphs, while "isolated" is a property of vertices. We illustrate these terms in the following example.

Example 6.1.5   We discuss the connectedness of the graphs in figure 6.12.
Graph (a) is both the complete graph K_3 with three vertices and the cycle graph C_3 with three vertices; notice that every complete graph and every cycle graph is connected. Graph (b) is a connected graph, but is not complete since there

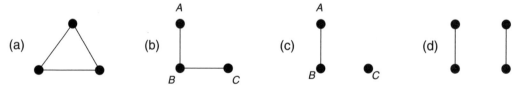

Figure 6.12   Graphs for example 6.1.5

is no edge between vertices A and C. In contrast, graph (c) is not connected since there is no edge from either A or B to vertex C; for this same reason, graph (c) has the single isolated vertex C. Similarly, graph (d) is not connected, but graph (d) does not have any isolated vertices.

■

We note that graph (d) in figure 6.12 is an example that motivates the definition of the "connected components" of a graph; this graph has two connected components, each consisting of one edge with its two incident vertices. In this context, we can readily conceive of a *connected component* of a graph as a portion of a graph whose vertices are all connected by paths. Further details are left for your later studies.

Question 6.1.8 Sketch graphs with the following properties.

    (a) The complete graph K_4 with four vertices.
    (b) The cycle graph C_4 with four vertices.
    (c) The null graph N_4 with four vertices.
    (d) A graph with four vertices that is not connected.
    (e) A graph with four vertices exactly two of which are isolated.

■

Question 6.1.9 Prove the following mathematical statements for graphs with at least two vertices.

    (a) Every complete graph is connected.
    (b) Every cycle graph is connected.
    (c) Every null graph is not connected.

■

We end this section by describing four important questions of graph theory; we study some of these questions later in this chapter.

6.1.1 The Explorer's Problem

The most famous example of the explorer's problem is the first question that motivated the study of graph theory—the Königsberg bridge problem. Extended to a more general setting, we consider an explorer who is interested in traveling every possible route and eventually returning home. As we discuss in section 6.2, Euler isolated a criterion that determines when a graph has a solution to the explorer's problem and when a graph does not. Once a solution is known to exist, we can find these solutions by implementing an algorithm, by the method of exhaustion, or by clever insight. In addition, for graphs that do not have solutions, we can identify an "eulerization" of the graph by "duplicating" edges in the given graph (without a solution to the explorer's problem) to obtain a graph that does have a solution.

6.1.2 The Traveling Salesman Problem

This variation of the explorer's problem focuses on visiting vertices rather than edges. Contemporary mathematicians describe this problem in terms of a traveling salesman leaving the home office, visiting each of a predetermined set of locations exactly once, and finally returning home. The visited locations are represented as vertices on a graph and the available routes between these locations are represented as edges. The goal is to determine the route of the salesman that is most efficient, where efficiency might be measured in terms of the total distance, time, or money (or some other measure) required to complete the route.

Specific instances of the traveling salesman problem were studied by Euler in 1759, by the French mathematician Alexandre-Théophile Vandermonde in 1771 in the context of a knight visiting every square of a chess board exactly once, and by the Irish mathematician Sir William Rowan Hamilton in the nineteenth century. Hamilton designed a puzzle called the "Icosian game" based on solving the traveling salesman problem on the planar graph of a regular dodecahedron. The first general study of the traveling salesman problem appears to have been undertaken by the English mathematician Thomas Penyngton Kirkman in his 1856 paper *On the Representation of Polyhedra*. Interestingly enough, the first actual use of the phrase "traveling salesman problem" appears to have occurred during conversations among Hassler Whitney, Albert William Tucker, and Merrill Flood at Princeton University in the 1930s; the first refereed publication explicitly using this phrase was published in 1949 by the logician Julia Robinson.

Many practical questions of how best to do business or provide services can be understood as traveling salesman problems, including work in such diverse industries as transportation (airline routes), public service (mail routes), energy (power grid networks), communications (phone, cable, and computer networks), and hardware development (computer circuit design). Consequently there has been a strong and steady interest in developing solutions to such questions. As you might expect, researchers are interested in a general algorithm for solving any traveling salesman problem. However, at this point, no such general algorithm is known; in fact, such questions are known to be "NP-hard" requiring nonpolynomial computations of significant complexity. And so the traveling salesman problem remains an active area of ongoing research in graph theory.

6.1.3 The Four-Color Problem

Do you remember the large political maps of the world and the United States that adorned the walls of your elementary classroom? Regions on such maps were colored various shades of pink, yellow, green, and orange (or other colors) to distinguish among countries and states. In 1852, mathematics student Francis Guthrie (later a mathematics professor at the University of Cape Town in South Africa) conjectured that it is possible to color the regions of any map on a plane with at most four colors so that no two adjacent regions have the same color. Guthrie shared this conjecture with his brother Frederick, who was studying at University College London in England, and Frederick in turn passed the claim along to his mathematics professor Augustus De Morgan. De Morgan discussed this question with others (most notably Sir William Rowan Hamilton and Arthur Cayley), and soon the four-color problem became one of the most famous open questions in mathematics.

Many talented mathematicians invested more than a century's worth of effort in working to prove (or disprove) Guthrie's conjecture. Progress was made in special cases.

For example, if a map can be drawn so that a single, continuous, closed curve determines the boundaries of its various regions, then three colors are sufficient. Mathematicians overcame the difficulty of grappling with regions of any shape arranged in any configuration by means of graph theory. Distinct regions of a map are represented by vertices that are joined by an edge whenever the corresponding regions are adjacent. We then seek to color the vertices (using just four colors) so that no two adjacent vertices have the same color.

Finally, in 1976 (more than 120 years after Guthrie's original conjecture), Kenneth Appel and Wolfgang Haken of the University of Illinois at Urbana-Champaign proved the four-color theorem. Their argument consists of a proof by contradiction in which they show that if a map cannot be four-colored, then it must contain a "smaller" map that cannot be four-colored. The contradiction is obtained by considering a map that is "minimal" in the sense that every submap can be four-colored. The Appel–Haken proof uses computers; they wrote a program to exhaustively rule out the need for more than four colors in the more than 1000 special map configurations to prove the general result. From a positive perspective, we know that the four-color theorem is true. But, at the same time, the Appel–Haken proof does not provide any insight into why the theorem is true. In this sense, mathematicians continue to be interested in a more theoretically based and more elegant proof of the four-color theorem.

Variations of this problem have also been studied by mathematicians. The Appel–Haken proof of the four-color theorem ensures that any map on a plane or a sphere can be four-colored. In 1890, the English mathematician Percy John Heawood studied a donut-shaped surface (formally referred to as a *torus*) and proved that maps on such surfaces require at most seven colors. The following is an open question in the study of graph colorings.

If we color every point in the plane, how many colors are needed if we require any two points that are exactly one unit apart to have different colors?

First posed in 1954, this question is known as the *Chromatic plane number problem*, where *chromatic number* refers to the required number of colors. While it is known that either four, five, six, or seven colors is necessary, the exact number required remains unknown.

6.1.4   The Crossing Number Problem

*The complete bipartite graph $K(n, n)$ of size n is formed by placing n vertices along a top row in the plane and another n vertices along a bottom row, and then joining every vertex in the top row with every vertex in the bottom row by a single edge. A depiction of the complete bipartite graph $K(3, 3)$ of size $n = 3$ is given in figure 6.13.

Notice that one of the edges crosses another edge in this presentation of the graph. In fact, every presentation of $K(3, 3)$ must have at least one such crossing; mathematicians identify the *crossing number* of $K(3, 3)$ as 1.

The crossing numbers of many other complete bipartite graphs are known. For example, the crossing number of $K(4, 4)$ is 4, the crossing number of $K(5, 5)$ is 16, and the crossing number of $K(6, 6)$ is 36. However, as we increase the size n of the

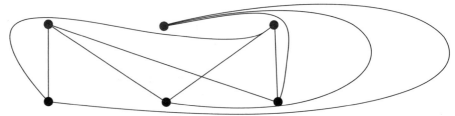

Figure 6.13   The complete bipartite graph $K(3, 3)$

complete bipartite graph, there is a corresponding significant increase in the difficulty of determining the crossing number (because $K(n, n)$ can be presented in so many more ways). In fact, for the relatively small integer $n = 9$, the crossing number of $K(9, 9)$ is unknown. In 1954, the Polish mathematician Kazimierz Zarankiewicz conjectured a formula for crossing numbers that predicts the crossing number of $K(9, 9)$ is 256, but no definitive proof of this fact is known.

6.1.5   Reading Questions for Section 6.1

1. Define and give an example of a graph, parallel edges, a loop, and a simple graph.
2. Define and give an example of the degree of a vertex and the total degree of a graph.
3. State and explain theorem 6.1.1. What is the main idea behind the proof of this theorem?
4. Define and give an example of a walk, a path, a closed walk, a circuit, and a cycle.
5. Discuss the relationship between a walk and a path, a path and a cycle, and a circuit and a cycle.
6. Define and give an example of a connected graph and an isolated vertex.
7. Can a connected graph have an isolated vertex? Explain your answer.
8. Can a complete graph have an isolated vertex? Explain your answer.
9. Can a cycle graph have an isolated vertex? Explain your answer.
10. Can a null graph have an isolated vertex? Explain your answer.

6.1.6   Exercises for Section 6.1

In exercises 1–6, explain why each picture does (or does not) represent a graph.

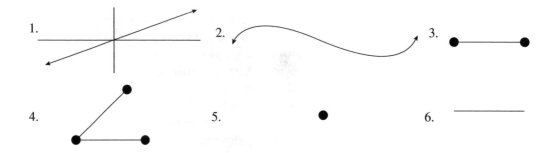

In exercises 7–12, explain why each graph is (or is not) simple, connected, complete, and/or a cycle graph.

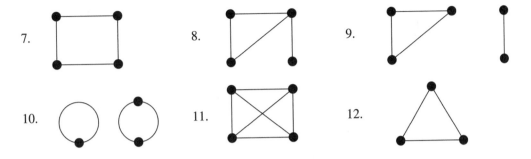

7. 8. 9.

10. 11. 12.

In exercises 13–20, sketch a graph with the following properties, or explain why such a graph does not exist.

13. A connected graph with four vertices that is not complete.
14. A connected, complete graph with four vertices.
15. A connected, simple graph with four vertices.
16. A simple graph with four vertices that is not connected.
17. A connected, cycle graph with four vertices.
18. A null graph with five vertices.
19. A connected, null graph with five vertices.
20. A connected, simple graph with one isolated vertex.

In exercises 21–26, identify the following six objects in the labeled graph, or explain why such an object does not exist.

(a) a walk from vertex A to vertex E
(b) a path from vertex A to vertex E
(c) a circuit based at vertex B
(d) a cycle based at vertex E
(e) the degree of each vertex in the graph
(f) the total degree of the graph

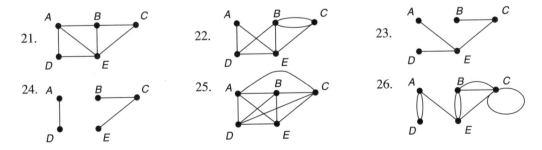

21. 22. 23.

24. 25. 26.

In exercises 27–32, prove each mathematical statement about graphs.

27. Every graph has an even number of vertices of odd degree.
28. A walk with a repeated edge must have a repeated vertex.

29. If vertices V and W are connected by a walk, then V and W are connected by a path.

30. If vertices V and W are in a circuit of a connected graph G, then a graph obtained by removing any one edge from G still contains a path from V to W.

31. If a graph has exactly two vertices V and W of odd degree, then V and W are connected by a path.

32. The complete bipartite graph $K(n, n)$ has $2n$ vertices and n^2 edges.

Exercises 33–38 consider the complete graph K_n with n vertices and the cycle graph C_n with n vertices.

33. Sketch the complete graphs K_1, K_2, and K_3 and specify the number of edges in each graph.

34. Sketch the complete graphs K_4 and K_5 and specify the number of edges in each graph.

35. Prove by induction that K_n has $n(n - 1)/2$ edges, where $n \in \mathbb{N}$.

36. Sketch the cycle graph C_3 and specify its the number of edges.

37. Sketch the cycle graphs C_4 and C_5 and specify the number of edges in each graph.

38. Prove by induction that C_n has n edges, where $n \geq 3$.

Exercises 39–42 consider the complement $\overline{G}$ of a simple graph G, where the set of vertices of $\overline{G}$ is the same as the vertices of G, but vertex V is adjacent to vertex W in $\overline{G}$ iff V is not adjacent to W in G.

In exercises 39–42, identify the complement of each simple graph.

 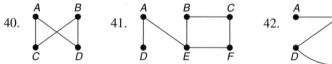

39.   40.   41.   42.

Exercises 43–49 consider an adjacency list representation of graphs that lists all the vertices adjacent to each vertex in the graph. For example, we represent the graph in figure 6.14 with either of the given adjacency lists.

Glenn: Highland, Bigtown, Glenn   Glenn: Highland (w), Bigtown (x), Glenn (v)
Highland: Glenn, Bigtown   Highland: Glenn (w), Bigtown(y)
Bigtown: Glenn, Highland, Union   Bigtown: Glenn (x), Highland (y), Union (z)
Union: Bigtown   Union: Bigtown(z)

These adjacency lists indicate that Glenn is joined by an edge to Highland, Bigtown, and Glenn (perhaps by a bypass), as identified in the graph. Similarly, Highland is adjacent to both Glenn and Bigtown, and so on. Any graph with labeled vertices can be represented by an adjacency list. In the rightmost adjacency list, the labels for the edges are identified by writing the edge label next to each incident vertex. Parallel edges are denoted by repeating vertices in the adjacency matrix.

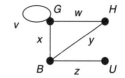

Figure 6.14 Routes between Glenn (G), Highland (H), Bigtown (B), and Union (U)

In exercises 43–45, sketch the graph represented by each adjacency list.

43. Capital: Mall (v), Natural History Museum (q), Smithsonian (r),
    White House: Mall (w)
    Mall: White House (w), Washington Monument (x), Smithsonian (u),
        Capital (v), Natural History Museum (t)
    Washington Monument: Mall (x)
    Natural History Museum: Mall (t), Capital (q), Smithsonian (s)
    Smithsonian: Natural History Museum (s), Capital (r), Mall (u)

44. $A : B, D, D, E$ 45. $A : B, C, D, E$
    $B : A$ $B : A, C, D, E$
    $C :$ $C : A, B, D, E$
    $D : A, A$ $D : A, B, C, E$
    $E : A$ $E : A, B, C, D$

In exercises 46–49, state the adjacency list representation of each graph.

46. The graph from exercise 39.   48. The graph from exercise 41.

47. The graph from exercise 40.   49. The graph from exercise 42.

Exercises 50–58 consider an adjacency relation on vertices defined by graphs. Given a graph G and two vertices V and W in G, we write $V \sim_G W$ iff V is adjacent to W in G. Every graph induces an adjacency relation on its set of vertices.

In exercises 50–53, use the notation $V \sim_G W$ to identify the vertices adjacent to vertex A and vertex B in each graph.

50. The graph from exercise 39.   52. The graph from exercise 41.

51. The graph from exercise 40.   53. The graph from exercise 42.

In exercises 54–58, prove each property of a relation holds for the adjacency relation $\sim_G$ for an arbitrary graph G, or sketch a counterexample demonstrating otherwise. If a given property does not hold, state a condition on graphs that ensures the property holds for every graph satisfying the condition.

54. Reflexivity: For every vertex V, $V \sim_G V$.

55. Irreflexivity: For every vertex V, $V \not\sim_G V$.

56. Symmetry: For all vertices V and W, $V \sim_G W$ implies $W \sim_G V$.

57. Transitivity: For all vertices V, W, and X, $V \sim_G W$ and $W \sim_G X$ implies $V \sim_G X$.

58. Comparability: For all vertices V and W, either $V \sim_G W$ or $V = W$ or $W \sim_G V$.

Exercises 59–70 consider a matrix representation of graphs. A matrix represents a graph when every vertex in the graph is numbered and the ith vertex is identified with both ith row and the ith column of the matrix. The $a_{i,j}$ entry of the matrix (the entry in the ith row and jth column) is a nonnegative integer identifying the number of edges connecting the graph's ith vertex with the jth vertex. For example, the following two matrices represent the given graphs.

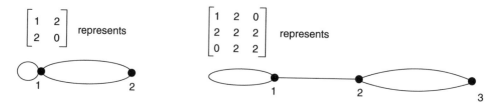

For the left graph and matrix, matrix entry $a_{11} = 1$ identifies the loop based at vertex 1, matrix entries $a_{12} = a_{21} = 2$ identify the two edges adjoining vertex 1 and vertex 2, and matrix entry $a_{22} = 0$ identifies that there are zero loops based at vertex 2. The right graph and matrix are similar.

In exercises 59–62, sketch the graph represented by each matrix.

59. $\begin{bmatrix} 1 & 1 & 1 \\ 1 & 0 & 1 \\ 1 & 1 & 0 \end{bmatrix}$

61. $\begin{bmatrix} 2 & 1 & 1 & 2 \\ 1 & 2 & 1 & 1 \\ 1 & 1 & 2 & 1 \\ 2 & 1 & 1 & 2 \end{bmatrix}$

60. $\begin{bmatrix} 1 & 2 & 0 \\ 2 & 2 & 0 \\ 0 & 0 & 1 \end{bmatrix}$

62. $\begin{bmatrix} 1 & 1 & 0 & 0 \\ 1 & 1 & 0 & 0 \\ 0 & 0 & 1 & 1 \\ 0 & 0 & 1 & 1 \end{bmatrix}$

In exercises 63–66, state the matrix representation of each graph.

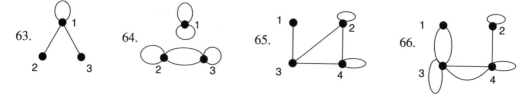

Exercises 67–70 consider general properties of matrix representations of graphs.

67. Describe the graph represented by the zero matrix Z_n with 0 in every entry.

68. Describe the graph represented by the identity matrix I_n with 1's down the main diagonal and 0's in all other entries.

69. How does the matrix representation of a graph indicate that a vertex is isolated?
70. What is the relationship between a_{ij} and a_{ji} in the matrix representation of a graph? Matrices with this property are said to be *symmetric* and (among other things) can be helpful when solving systems of linear equations.

6.2   The Explorer and the Traveling Salesman

In this section we consider two of the questions that led to the study and development of graph theory: the traveling salesman problem and the explorer's problem. You may recall the brief outline and discussion of these questions at the end of section 6.1. The *explorer's problem* is often presented by imagining an explorer who wants to develop a thorough understanding of some area and is interested in traveling every possible road in the given region. Adopting a graph-theoretic perspective on this scenario, we consider a graph whose edges represent all possible routes that are available; the explorer's mission can be translated into the goal of visiting every edge of the corresponding graph. Furthermore, motivated by a sense of efficiency, we seek paths and circuits through graphs that visit every edge exactly once. Perhaps this task reminds you of the Königsberg bridge problem described in section 6.1; in fact, the Königsberg bridge problem is the first rendition of the explorer's problem studied from such a graph-theoretic point of view.

A solution of the explorer's problem is relevant not just to entertaining puzzles and amusing distractions such as walking across bridges. Although puzzles are important for raising and expressing this and similar questions, a solution to the explorer's problem has significant practical applications. Such everyday tasks as delivering mail, picking up trash, and plowing roads during snow storms can be accomplished in an optimal fashion by solving the corresponding explorer's problem. The resulting increase in efficiency has saved cities, states, and countries millions of dollars and thousands of hours of labor. Similarly, companies have realized significant savings in manufacturing processes and business transactions by modeling processes with graphs and implementing solutions of the explorer's problem. As mathematicians, we have the good fortune of working with relatively simple renditions of such questions and finding optimal solutions in the idealized world of graph theory—solutions that (more often than not) are relevant to the more complicated real-world settings we have described.

We also study the *traveling salesman problem* in which our traveler seeks to visit every vertex of a given graph (rather than every edge). Euler, Vandermonde, Hamilton, and Kirkman studied this question in the eighteenth and nineteenth centuries, and contemporary mathematicians and computer scientists continue to grapple with this subtle and delicate question. As described in section 6.1, the traveling salesman problem is often presented by imagining a traveling salesman who leaves the home office to visit a predetermined set of locations before returning home. In this section, we seek the simplest possible "efficiency" in the salesman's journey by requiring the route to begin and end at home and to visit the other locations exactly once.

Viewed from a graph-theoretic perspective, we represent the locations to be visited as vertices and possible routes as edges; the traveling salesman problem asks for a cycle visiting every vertex of this representative graph exactly once. Many important, practical questions can be phrased in terms of this problem. Telecommunications companies, computer manufacturers, and transportation industries must regularly resolve networking and connection questions that are traveling salesman problems. In fact, whenever you leave your home or residence hall to visit friends, to attend classes, to go shopping, or to run other errands, you are essentially solving such a problem. As we learn in this section, insights into the traveling salesman problem are more difficult to develop than for the explorer's problem. However, there are many interesting partial results, and mathematicians, computer scientists, industrial engineers, and others continue to work on optimal solutions to a given traveling salesman problem.

6.2.1   The Explorer's Problem and Eulerian Circuits

We begin our study of the explorer's problem by describing Euler's approach to the Königsberg bridge problem. Recall that the Pregel River divides the town of Königsberg into four parts that were interconnected by seven bridges in the 1730s. Is there a path that starts in one part of the town, crosses every bridge exactly once, and returns to the place where the path began? As we mentioned in section 6.1, Euler answered this question by modeling the physical setting of Königsberg with a graph G in which the vertices represented the four parts of the town and the edges represented the bridges; figure 6.15 presents Euler's model of Königsberg.

Working with this graphical representation, Euler translated the question of traversing every bridge exactly once into the graph-theoretic question: "Does there exist a circuit traversing every edge of G?" In other words, does there exist a closed path in G that traverses every edge of G exactly once?

As we can see, this question is an example of the explorer's problem. Perhaps you tried to find a solution to this specific problem when it was introduced in section 6.1—and hopefully you didn't succeed! In Euler's 1736 paper *Solutio problematic ad geometriam situs pertinentis* (that is, *The Solution of a Problem Relating to the Geometry of Position*), he proved that such a circuit of G does not exist; either we come up one bridge short or we have to cross at least one bridge twice. In fact, Euler proved a much more general result characterizing precisely those graphs containing a circuit traversing every edge exactly once. Motivated by Euler's success, we define a

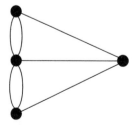

Figure 6.15   Euler's representation of Königsberg as a graph G

solution of the explorer's problem in terms of an Eulerian circuit along with the closely related notion of an Eulerian path.

Definition 6.2.1 *If G is a graph then, an* **Eulerian circuit of G** *is a circuit of G that includes every edge of G exactly once. Recall that every circuit is closed, and so the initial endpoint and the final endpoint of an Eulerian circuit are the same. An* **Eulerian path in G** *is a path in G that includes every edge of G exactly once.*

Often a graph G is said to be *Eulerian* if there is an Eulerian circuit of G, and *semi-Eulerian* if there is an Eulerian path in G (that is not an Eulerian circuit). From these definitions, we see that every Eulerian circuit is an Eulerian path. On the other hand, not every Eulerian path is an Eulerian circuit, because the initial and final endpoints may differ in an Eulerian path. For example, the complete graph K_2 with two vertices does not have an Eulerian circuit (you can verify this fact with a sketch of the graph); after traversing the one available edge, it is impossible to return to the initial endpoint. However, K_2 does have an Eulerian path that starts at either vertex and traverses the single edge to the other vertex. The next example identifies further Eulerian circuits and paths.

Example 6.2.1 We discuss Eulerian circuits and paths in the graphs given in figure 6.16.

In graph (a), the path given by the sequence of edges s, t, u, v, w, x, y, z is an Eulerian circuit of graph (a) based at the top left vertex. The reverse of this path is also an Eulerian circuit of graph (a). Similarly, we can obtain other Eulerian circuits of graph (a) based at other vertices by beginning with an edge adjacent to the given vertex and otherwise preserving the ordering of the above sequence of edges. For example, v, w, x, y, z, s, t, u is an Eulerian circuit based at the top right vertex of graph (a). Since every Eulerian circuit is an Eulerian path, this graph also contains multiple Eulerian paths.

Graph (b) contains many different Eulerian paths including those given by the sequence of edges v, w, x, y, z and by v, w, z, y, x. Can you identify other distinct Eulerian paths in this graph? Having identified Eulerian paths, we naturally search for Eulerian circuits of graph (b). However, there does not exist an Eulerian circuit of graph (b); can you articulate why?

■

Question 6.2.1 If possible, identify an Eulerian circuit of each graph given in figure 6.17; if such a circuit does not exist, identify an Eulerian path.

■

The four graphs we considered in example 6.2.1 and question 6.2.1 have either an Eulerian circuit or an Eulerian path. As you might expect, not every graph has an Eulerian circuit or path. For example, the complete graph K_4 on four vertices has

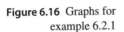

Figure 6.16 Graphs for example 6.2.1

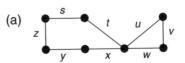

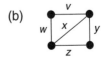

(a) 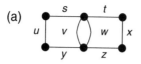   (b)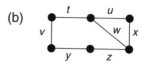

Figure 6.17 Graphs for question 6.2.1

neither an Eulerian circuit nor an Eulerian path. If you sketch K_4 and try to identify Eulerian circuits and Eulerian paths in K_4, you may come to recognize why this graph does not have such circuits or paths.

Such an example highlights important questions concerning the existence of Eulerian circuits and Eulerian paths. When does a given graph have an Eulerian circuit or path? Is there a criterion that a graph must satisfy to contain such a circuit or path? Are any criteria on a graph equivalent to the existence of an Eulerian circuit or path? Answers to these questions are of interest not only from an abstract, theoretical perspective, but also because of the corresponding real-world applications of solutions to this question. Fortunately, rather than blindly (or even cleverly) searching for Eulerian circuits or paths in a graph, we can first determine the existence of such a circuit or path based on the degrees of vertices and the connectedness of the graph. Once we know that the conditions guaranteeing the existence of Eulerian circuit or path are satisfied, then we can search for such a circuit or path confident that we will eventually find at least one (if not more). When these conditions fail to hold, we know that such a search would be fruitless, and we are free to turn our attention to other interesting questions.

Recall that a graph is *connected* if there exists a path between any two vertices in the graph, and that the *degree* of a vertex is the total number of times the vertex is the endpoint of an edge (alternatively, we can compute the total number of edges incident to the vertex, provided loops are counted twice). As Euler recognized, these two features of a graph G are all that is needed to determine the existence of an Eulerian circuit or path in G. Consider the following theorem.

Theorem 6.2.1 *If G is a connected graph, then the following hold:*

    (a) There exists an Eulerian circuit of G iff the degree of every vertex is even.

    *(b) There exists an Eulerian path in G (which is **not** an Eulerian circuit) between distinct vertices V and W iff V and W are the only vertices in G of odd degree.*

Proof of (a) We prove this "if and only if" statement in two parts, one for each half of the biconditional. We first assume that there exists an Eulerian circuit of G and prove that an arbitrary vertex V in G has even degree. The Eulerian circuit of G traverses every edge of G exactly once, and (most importantly for this argument) must therefore traverse every edge incident to V exactly once. We determine the parity of the degree of V by following the Eulerian circuit around G. Whenever the given Eulerian circuit enters V along an edge, the circuit exits V along some other distinct edge, and so contributes 2 to $d(V)$. Because every edge of G is traversed exactly once, $d(V)$ is a sum of 2's and so even.

We now consider the other half of the biconditional. We assume G is a connected graph such that every vertex has even degree. We prove that there exists an Eulerian circuit of G by describing an algorithm for constructing the desired Eulerian circuit. If G consists of a single vertex V with no edges, the trivial walk from V to V is an Eulerian circuit. If G contains more than one vertex, we choose an arbitrary vertex V in G to serve as the base for the Eulerian circuit under construction. We then traverse the edges of G in some arbitrary fashion (under the one restriction that we never repeat an edge) until we return to V. Since every vertex of G has even degree, we can exit every vertex we enter along some other edge; this fact, together with the assumption of finitely many edges in G, ensures that this process must eventually close the path under construction by returning to V. For ease of reference, we refer to this circuit as C.

If C contains every edge of G, then C is the desired Eulerian circuit of G. If not, the algorithm extends C to include additional edges of G. We first identify another circuit C^* in G that shares at least one vertex, but no edge, in common with C. Let G^* denote the subgraph of G consisting of every edge in G that is not in C along with the corresponding set of vertices incident to an edge not in C; note that every vertex of G^* is also of even degree. Since C does not contain every edge of G and G is connected, the original circuit C and this subgraph G^* share some common vertex V^*. As before, we identify a circuit in G^* based at V^*. With C^* in hand, we extend the original circuit C by "patching together" C and C^*. We start at the orginal base V, traverse C until V^*, hop off C and traverse C^* from start to finish, and finally complete the traversal of C from V^* back to V.

If this new, extended circuit contains every edge of G, this is the desired Eulerian circuit of G. If not, we repeat the extension process outlined above. Since G is finite, this algorithm must eventually terminate, providing the desired Eulerian circuit of the given graph G. Notice the constructive flavor of this proof—we are provided an algorithm for identifying an Eulerian circuit in any connected graph with every vertex of even degree.

■

Comments on the proof of (b) We assume G is a connected graph with exactly two distinct vertices V and W of odd degree. The strategy in this setting is to add an extra edge from V to W, apply the first part (1) of theorem 6.2.1, and then remove the added edge to obtain the desired conclusion. As indicated in the statement of the theorem, an Eulerian path in such a graph G must have the two vertices of odd degree as its initial and final vertex. Further details are left for the exercises at the end of this section.

■

Example 6.2.2 We use theorem 6.2.1 to determine if the graphs in figure 6.18 have an Eulerian path or Eulerian circuit.

Graph (a) is the representation of Königsberg studied by Euler. The graph is connected, and all four vertices have odd degree: the center vertex representing the island has degree five and the other vertices have degree three. Since more than two vertices have odd degree, the graph has neither an Eulerian circuit nor

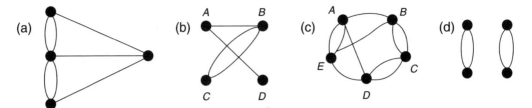

Figure 6.18   Graphs for example 6.2.2

an Eulerian path. In this way, Euler proved that the Königsberg bridge problem has no solution.

Graph (b) is also connected. Vertices A and C have degree two, while vertices B and D have odd degrees of three and one, respectively. Since this graph has vertices of odd degree, there does not exist an Eulerian circuit of graph (b). However, because this graph has exactly two vertices of odd degree, there does exist an Eulerian path in graph (b). Recall from theorem 6.2.1 that such an Eulerian path must begin and end at the vertices of odd degree. One example of an Eulerian path in graph (b) is B, C, B, A, D. Can you identify another Eulerian path in graph (b)?

Graph (c) is connected and every vertex has degree four, so this graph has an Eulerian circuit. One Eulerian circuit of graph (c) is: A, B, C, D, E, A, D, C, B, E, A. Can you identify another Eulerian circuit of graph (c)?

Finally, in graph (d), every vertex has an even degree of two. However, the graph is not connected and theorem 6.2.1 does not apply. As we can see, the existence of an Eulerian circuit or path is not guaranteed by every vertex having even degree. In particular, there is no way to move from one "connected component" of graph (d) to the other, and so it is impossible to traverse every edge of the graph with a single path or circuit.

∎

Example 6.2.2 demonstrates the usefulness of theorem 6.2.1. We do not need (or want) to engage in a trial- and- error search through the graphs in figure 6.18 for example 6.2.2 in a perhaps vain quest for an Eulerian circuit or path. Instead, we simply observe connectedness and check the degrees of vertices (both relatively straightforward processes) to determine the existence of Eulerian circuits and Eulerian paths. Once we are assured of the existence of such circuits or paths we begin our search for them—if necessary, using the algorithmic processes outlined in the proof of theorem 6.2.1.

Finally, note that graph (c) in figure 6.18 for example 6.2.2 demonstrates the necessity of searching for Eulerian circuits rather than Eulerian cycles in most graphs. Recall that a cycle does not repeat edges or vertices. For most graphs, we must revisit some vertices in order to traverse every edge of the graph.

Question 6.2.2   Using theorem 6.2.1, identify either an Eulerian circuit or path in each graph in figure 6.19, or explain why such a circuit or path does not exist.

∎

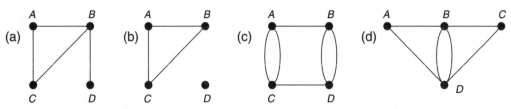

Figure 6.19 Graphs for question 6.2.2

6.2.2 The Traveling Salesman Problem and Hamiltonian Cycles

We now turn our attention to the explorer problem's close cousin: the traveling salesman problem. For this question, we focus on visiting every vertex exactly once (instead of every edge exactly once). This slight shift in the type of "efficiency" we seek from our paths yields a profoundly different mathematical question that is far more subtle than the explorer's problem. We have already presented this question in the context of a salesman visiting some collection of locations without repetition and finishing the journey where he began. The existence of cycles solving specific traveling salesman problems is significant for many practical applications.

Unfortunately, the only known general method for finding such cycles employs the method of exhaustion; as we have seen in other settings, exhaustion is not feasible in any sufficiently complex setting. In addition, there is no theorem similar to Theorem 6.2.1 that would allow us to determine if a solution to the traveling salesman problem exists for a given graph by examining some collection of relatively apparent features of the graph. Mathematicians (and others) continue to seek such a theorem, and the study of the traveling salesman problem remains an important and active area of research in graph theory.

The study of cycles and paths that visit every vertex exactly once was popularized by the Irish mathematician Sir William Rowan Hamilton in the nineteenth century. Hamilton was born in Dublin in 1805 and showed an early skill for languages, mastering Latin, Greek, and Hebrew by the age of five under the tutelage of his uncle. He entered Trinity College in Dublin at the age of 18 and, while he was still an undergraduate, he (amazingly) was appointed Professor of Astronomy and Royal Astronomer of Ireland. However, most of Hamilton's research efforts were devoted to the mathematical study of optics, dynamics, and abstract algebra. He is best known for his definition of the quaternions, the first noncommutative algebra studied by mathematicians and an important tool in the study of physics. After struggling to define an appropriate multiplication of triples of real numbers for almost a decade, Hamilton had the flash of insight defining the quaternions via the equation $i^2 = j^2 = k^2 = ijk = -1$ (see the exercises at the end of section 3.5) while on a walk with his wife on the Brougham Bridge over Dublin's Royal Canal on Monday, October 16, 1843. Hamilton was so delighted that he carved this equation on the bridge!

Hamilton's association with graph theory is traced to his description of an "Icosian calculus" in 1856 that can be interpreted in terms of paths on the surface of a regular dodecahedron (a solid figure with twelve identical pentagonal faces). In 1858, Hamilton developed the Icosian game based on this calculus. He labeled the vertices of the planar rendition of a regular dodecahedron with the names of cities from around the world, including Brussels, Delhi, Canton, ..., and Zanzibar. The players of this game were challenged to start in a city and tour the world, visiting all the other cities exactly once until returning to the starting city; that is, they were asked to solve a specific instance of the traveling salesman problem. In light of this connection, we name such cycles and paths that visit every vertex of a given graph in honor of Hamilton.

Definition 6.2.2 *If G is a graph, then a* **Hamiltonian cycle in G** *is a cycle in G that includes every vertex of G and a* **Hamiltonian path** *is a simple path in G that includes every vertex of G.*

Often a graph G is said to be *Hamiltonian* if there is a Hamiltonian cycle in G, and *semi-Hamiltonian* if there is a Hamiltonian path in G (that is not a Hamiltonian cycle). From the definition of cycle and simple path, every Hamiltonian cycle and path contains every vertex of the given graph exactly once. Every Hamiltonian cycle is a Hamiltonian path; on the other hand, not every Hamiltonian path is a Hamiltonian cycle because the initial and final endpoints may differ in a Hamiltonian path. As in the Eulerian setting, the complete graph K_2 on two vertices illustrates this distinction. The next example considers further examples of such cycles and paths.

Example 6.2.3   We discuss Hamiltonian cycles and paths in the graphs given in figure 6.20.

In graph (a), the cycle A, B, C, D, A is a Hamiltonian cycle. This path is based at A and visits every vertex of the graph exactly once. In fact, this path also traverses every edge of the graph exactly once and so is an Eulerian circuit. Can you identify the other Hamiltonian cycle in graph (a) that is based at vertex A?

Graph (b) is the complete graph K_2 on two vertices. This graph has a Hamiltonian path but no Hamiltonian cycle. For example, the path beginning at vertex A and following the one edge to vertex B is a Hamiltonian path. However, after traversing the one available edge, it is impossible to return to the initial vertex; thus, K_2 does not have a Hamiltonian cycle.

Finally, graph (c) has neither a Hamiltonian cycle nor a Hamiltonian path. The key issue is vertex C. A path exiting vertex C must choose one of the three edges to follow and, once it has done so, the path cannot reach the other branches of the graph without violating the Hamiltonian condition of visiting vertices exactly once. Thus, any path in graph (c) entering C from (at most) one of the three vertices A, B, D can exit C to visit one of A, B, D. Therefore, at least just one of A, B, D is not visited.

■

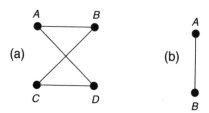
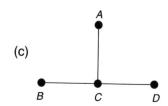

Figure 6.20   Graphs for example 6.2.3

Reflecting on the graphs studied in figure 6.20 for example 6.2.3, mathematicians are particularly interested in graph (c) that contains neither a Hamiltonian cycle nor a Hamiltonian path. Such a relatively simple example can often yield insights into the essential features of graphs that determine the existence and nonexistence of Hamiltonian cycles and paths. Perhaps you can develop a conjecture as to what general feature of graph (c) led to the nonexistence of a Hamiltonian cycle or path, and suggest some corresponding sufficient condition(s) that a graph must satisfy to (possibly) contain a Hamiltonian cycle or path.

Question 6.2.3 For each graph given in figure 6.21, either identify a Hamiltonian cycle or path in the graph, or explain why such a cycle or path does not exist.
■

As mentioned above, mathematicians have not yet identified a general result characterizing graphs with Hamiltonian cycles or paths. While Euler's success with the explorer's problem has not been achieved with the traveling salesman problem, various partial results have been proven that provide sufficient conditions for the existence of Hamiltonian cycles and paths. We present one of these results, first proved by the Norwegian mathematician Øystein Ore in 1960.

Theorem 6.2.2 *If G is a simple, connected path with n vertices such that $n \geq 3$ and for every pair of distinct non-adjacent vertices V and W in G, $d(V) + d(W) \geq n$, then there exists a Hamiltonian cycle in G.*

The proof of this result is left for your later studies of mathematics. Instead, we focus on applying theorem 6.2.2 to specific graphs.

Example 6.2.4 We discuss the application of theorem 6.2.2 to each graph given in figure 6.22.
In graph (a) with $n = 4$ vertices, there are two pairs of distinct, nonadjacent vertices: A, D and B, C. We observe that both $d(A) + d(D) = 2 + 2 = 4 \geq n$ and $d(B) + d(C) = 2 + 2 = 4 \geq n$. Therefore, since graph (a) is simple and

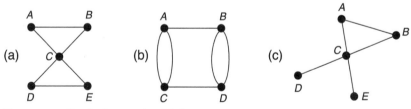

Figure 6.21 Graphs for question 6.2.3

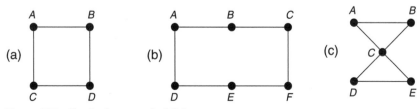

Figure 6.22 Graphs for example 6.2.4

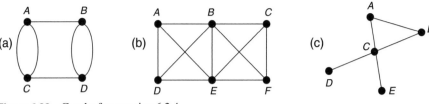

Figure 6.23   Graphs for question 6.2.4

connected, theorem 6.2.2 ensures that there exists a Hamiltonian cycle in G. The path A, B, C, D is one such cycle.

In graph (b) with $n = 6$ vertices, no pair of distinct, nonadjacent vertices V and W satisfies the required condition that $d(V) + d(W) \geq n = 6$. For example, $d(A) + d(E) = 2 + 2 = 4 < n$. As we have seen, when the hypotheses of a theorem are false, the conclusion may be either true or false. In this case, there does exists a Hamiltonian cycle in graph (b); the path A, B, C, F, E, D, A is one such cycle. This graph demonstrates that the hypotheses of Ore's theorem 6.2.2 are not necessary for the existence of a Hamiltonian cycle in a given graph (even though they are sufficient).

Graph (c) also does not satisfy the hypotheses of theorem 6.2.2, but in this case there is no Hamiltonian cycle in the graph. However, there does exist a Hamiltonian path in graph (c); can you identify such a path?

■

Question 6.2.4   Determine if theorem 6.2.2 applies to each graph given in figure 6.23. If so, identify a Hamiltonian cycle in the graph.

■

In section 6.4, we study variations of the traveling salesman problem that extend the goal of efficiency beyond visiting every vertex exactly once. We consider "weighted" graphs that have numeric labels assigned to their edges. These labels may represent distance, time, money, or some other measurable quantity of interest. In this context, we seek "minimum weight" Hamiltonian cycles and paths that visit every vertex exactly once at the least possible expense. Again, we will not practically be able to obtain ideal solutions for every such graph, but we can obtain some measure of optimization by means of various "heuristic" algorithms.

6.2.3   Reading Questions for Section 6.2

1. State the Königsberg bridge problem, the explorer's problem, and the traveling salesman problem.
2. Define and give an example of a connected graph.
3. Define and give an example of a vertex of even degree and a vertex of odd degree.
4. Define and give an example of an Eulerian circuit and an Eulerian path.
5. State theorem 6.2.1. How is this result helpful when studying graphs?

6. Discuss the possibility of defining an Eulerian cycle. What are the potential difficulties with such a notion?

7. Define and give an example of a Hamiltonian cycle and a Hamiltonian path.

8. Identify at least two ways in which an Eulerian circuit and a Hamiltonian cycle differ.

9. Discuss the possibility of defining a Hamiltonian circuit. What are the potential difficulties with such a notion?

10. Explain why a Hamiltonian walk and a Hamiltonian path are identical; that is, why must a walk visiting every vertex exactly once necessarily be a path?

11. State theorem 6.2.2. How is this result helpful when studying graphs?

12. In light of the application of theorem 6.2.2 to the graph (b) in example 6.2.4, discuss the distinction between "necessary" and "sufficient" conditions for mathematical properties.

6.2.4 Exercises for Section 6.2

In exercises 1–8, identify an Eulerian circuit or an Eulerian path in each graph, or explain why such a circuit or path does not exist.

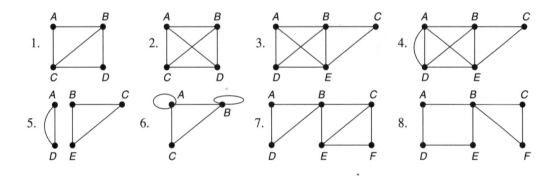

In exercises 9–16, identify a Hamiltonian cycle or a Hamiltonian path in each graph, or explain why such a cycle or path does not exist.

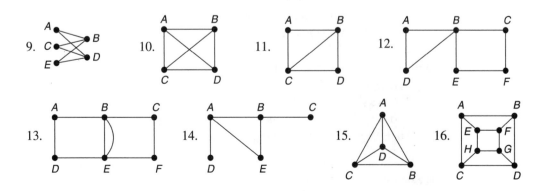

In exercises 17–24, sketch a graph containing the given object or explain why such a graph does not exist. While many different answers may be possible, try to find an example using as few vertices and edges as possible.

17. A graph with an Eulerian circuit and a Hamiltonian cycle.
18. A graph with an Eulerian circuit but no Hamiltonian cycle.
19. A graph with no Eulerian circuit but a Hamiltonian cycle.
20. A graph with no Eulerian circuit and no Hamiltonian cycle.
21. A graph with an Eulerian path and a Hamiltonian path.
22. A graph with an Eulerian path but no Hamiltonian path.
23. A graph with no Eulerian path but a Hamiltonian path.
24. A graph with no Eulerian path and no Hamiltonian path.

In exercises 25–34, prove each mathematical statement about graphs.

25. Theorem 6.2.1, part(2): If there exists an Eulerian path in a connected graph G between distinct vertices V and W, then V and W are the only vertices in G of odd degree.
26. Theorem 6.2.1, part(2): If V and W are distinct vertices in a connected graph G and are the only vertices in G of odd degree, then there exists an Eulerian path in G between V and W.
27. If every vertex of a graph G has even degree, then G can be partitioned into distinct cycles so that no two cycles share an edge in common.
    Hint: See the proof of theorem 6.2.1 part (a).
28. If a graph has a vertex of odd degree, then the graph does not have an Eulerian circuit.
29. If a graph G has a Eulerian circuit and no isolated vertices, then G is connected. Using the contrapositive of this result, sketch a graph that does not have an Eulerian circuit.
30. If a graph G has a Hamiltonian cycle, then G is connected. Using the contrapositive of this result, sketch a graph that does not have a Hamiltonian cycle.
31. If a graph G has a cycle that is both an Eulerian circuit and a Hamiltonian cycle, then the number of vertices in G is the same as the number of edges in G. Using the contrapositive of this result, sketch a graph that does not have a cycle that is both Eulerian and Hamiltonian.
32. If a graph G has a cycle that is both an Eulerian circuit and a Hamiltonian cycle, then every vertex of G has degree two. Using the contrapositive of this result, sketch a graph that does not have a cycle that is both Eulerian and Hamiltonian.
33. There exists a Hamiltonian cycle in every complete graph with at least three vertices.
34. If G is a simple, connected graph with n vertices such that $n \geq 3$ and for every vertex in G, $d(V) \geq n/2$, then there exists a Hamiltonian cycle in G.
    Hint: Use theorem 6.2.2 to prove this result.

In exercises 35–41, sketch a graph (or graphs) representing the following modifications of the Königsberg bridge problem. In addition, either identify an Eulerian circuit or an

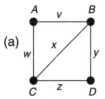
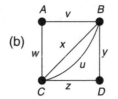
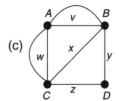

Figure 6.24 Eulerizing graphs

Eulerian path in the resulting graph(s), or explain why such a circuit or path does not exist. You may be interested in referring to figure 6.2 and figure 6.15.

35. An eighth Königsberg bridge is built downriver on the north-eastern branch of the Pregel River.
36. An eighth Königsberg bridge is built downriver on the south-eastern branch of the Pregel River.
37. An eighth Königsberg bridge is built over the Pregel River that provides an Eulerian path in the corresponding graph; identify one such solution.
38. An eighth Königsberg bridge is built over the Pregel River; we do not specify which part of the river is spanned, so there are several representative graphs.
39. An eighth and a ninth Königsberg bridge are built over the Pregel River that provide an Eulerian circuit in the corresponding graph; identify one such solution.
40. The north-eastern Königsberg bridge is removed.
41. The Königsberg bridge running from east to west is removed.

Exercises 42–47 consider an eulerization of a graph. Even if a connected graph does not contain an Eulerian circuit or path, we can obtain a close approximation by eulerizing the graph. A connected graph does not contain an Eulerian circuit or path if the graph contains the wrong number of vertices of odd degree. We eulerize a graph by duplicating existing edges until we have the correct number of vertices of odd degree. In example 6.2.1, we observed that graph (a) (shown in figure 6.24) does not has an Eulerian circuit because vertices B and C have odd degree.

However, if we duplicate edge x in graph (a), we obtain graph (b); now every vertex has even degree, and graph (b) contains several Eulerian circuits, including v, x, u, y, z, w. In this case, graph (b) is said to be an eulerization of the original graph (a). Eulerizations of graphs are not unique; for example, graph (c) in figure 6.24 is another distinct eulerization of graph (a). Usually we seek eulerizations that are optimal in the sense of duplicating the fewest number of edges possible.

In exercises 42–47, explain why each graph does not have an Eulerian circuit and provide an (optimal) eulerization of the graph.

42.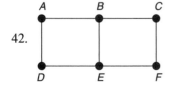

43.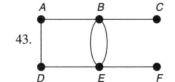

44.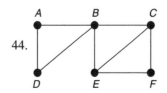

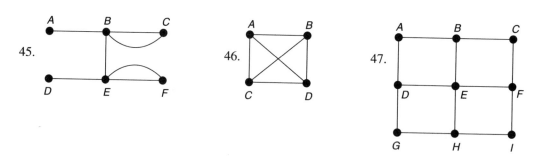

Exercises 48–53 consider the notion of a subgraph. Intuitively, a graph H is a subgraph of a graph G if H can be obtained from G by (possibly) deleting some vertices and edges of G. More formally, H is a subgraph of a graph G if every vertex and edge of H is in G and every edge of H has the same endpoints as in G.

In exercises 48–53, specify which of the given graphs (a) and (b) is a subgraph of the other, or explain why neither is a subgraph of the other.

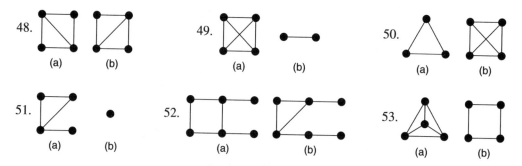

Exercises 54–61 study set-theoretic representations of graphs through a collection of two-element multisets. The two elements in each multiset are vertices that are adjacent to one another. For example, $\{ \{A, A\}, \{A, B\}, \{B, C\} \}$ represents graph (a) in figure 6.25. Similarly, we can represent the complete graph K_3 on three vertices given in graph (b) in figure 6.25 with the set: $\{ \{A, B\}, \{B, C\}, \{A, C\} \}$.

The equality of edges AB and BA is expressed by the equality of the corresponding multisets $\{A, B\}$ and $\{B, A\}$. In addition, this representation uses multisets in the same way (allowing repeated elements in the two element sets) in order to represent loops with their identical endpoints.

In exercises 54–57, sketch the graph represented by each set, or explain why the set does not represent a graph.

54. $\{ \{A, A\}, \{B, B\}, \{A, B\} \}$
55. $\{ \{A, A\}, \{A, B\}, \{A, C\}, \{B, B\}, \{C, C\} \}$

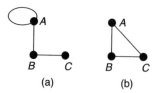

Figure 6.25   Set-theoretic representations of graphs

56. { {A, B}, A}
57. { {A, A}, {A, C}, {B, D}, {C, D}, {D, D} }

In exercises 58–61, state the set-theoretic representation of each graph.

58. The graph from exercise 2. 60. The graph from exercise 6.
59. The graph from exercise 4. 61. The graph from exercise 8.

Exercises 62–70 consider the connected to relation on vertices defined via graphs: given a graph G and two vertices V and W in G, we define $V \sim_G W$ iff V is connected to W via some path in graph G. Every graph induces a connected to relation on its set of vertices.

In exercises 62–65, use the notation $V \sim_G W$ to identify the vertices connected to vertex A and connected to vertex B in each graph.

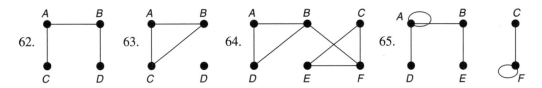

In exercises 66–70, prove each property of a relation holds for the connected to relation $\sim_G$ for an arbitrary graph G, or sketch a counterexample demonstrating otherwise. If a given property does not hold, state a condition on graphs that ensures the property holds for every graph satisfying the condition.

1. Reflexivity: For every vertex V, $V \sim_G V$.
2. Irreflexivity: For every vertex V, $V \not\sim_G V$.
3. Symmetry: For all vertices V and W, $V \sim_G W$ implies $W \sim_G V$.
4. Transitivity: For all vertices V, W, and X, $V \sim_G W$ and $W \sim_G X$ implies $V \sim_G X$.
5. Comparability: For all vertices V and W, either $V \sim_G W$ or $V = W$ or $W \sim_G V$.

6.3 Shortest Paths and Spanning Trees

We continue a study of graph-theoretic questions with an optimization theme. In section 6.2, we sought efficiency with respect to visiting edges by means of Eulerian circuits and paths, and with respect to visiting vertices by means of Hamiltonian cycles and paths. In this section, we are interested in identifying optimal paths that traverse a minimal number of edges. Such a shortest path always exists between any two connected vertices. Connected vertices are the endpoints of some path, and there exist only finitely many paths joining such vertices because graphs are finite; one of these paths is of minimum length. We consider a *shortest path algorithm* that enables the identification of one of the shortest paths between any two given vertices. This algorithm employs a variation on the method of exhaustion and introduces a labeling system for vertices that is useful in our continuing study of graph-theoretic algorithms.

In addition to optimizing paths with respect to length, we are interested in efficiently connecting vertices. Recall the complete graph K_n on n vertices joins every pair of distinct vertices with an edge. In some settings such an explicit connection between vertices is essential, but in many applications simply having a path between distinct vertices is sufficient. For example, airline companies commonly operate with a few designated "hub" airports. Rather than taking direct flights between two small, regional airports, passengers fly from the closest regional airport to their airline's hub, and then catch a connecting flight to their final destination. Expressed graph-theoretically, there is a path between any two distinct vertices (or airports), even when there is not an edge between them. Similar choices are often made in establishing computer networks, running gas and electric lines, routing mass transportation systems, and a host of other settings in which people are interested in the (sometimes) competing goals of connectedness and efficiency in creating the connectedness. In the language of graph theory, these applications search for a *spanning tree* that includes every vertex of a given connected graph. We will study spanning trees in the second half of this section and will continue this work in the context of weighted graphs in section 6.4.

For each algorithm studied in this chapter, we will first work carefully through a detailed example and then state the general algorithm. Along the way, we also discuss proofs that these algorithms always achieve their stated goals.

6.3.1   A Shortest Path Algorithm

This algorithm identifies a path between two connected vertices that traverses the minimum possible number of edges. In essence, the shortest path algorithm uses the method of exhaustion, considering every possible initial segment of a path based at one of the two given vertices. Since the given vertices are assumed to be connected, the algorithm eventually identifies a path to the other vertex; the first such path identified by the algorithm is a shortest path. As we study this algorithm, we should keep in mind that such an exhaustive search is sufficient and appropriate for relatively small graphs. However, as graphs increase in size and complexity, the corresponding (significant) increase in the resources required to implement this algorithm can render its use infeasible.

Example 6.3.1   We implement the shortest path algorithm to identify a shortest path between vertices A and J in the graph S given in figure 6.26.

The algorithm first selects one of the two given vertices to serve as the initial endpoint of the path; for this example, we select A. The algorithm now assigns labels to the vertices in S while constructing a path to the other given vertex (in this case, vertex J). First, the label 1:A is assigned to every vertex adjacent to A; in this graph, both B and E are labeled 1:A. Proceeding alphabetically on B and E, the algorithm labels every previously unlabeled vertex adjacent to B with the notation 2:B; for graph S, both C and F are labeled 2:B. Notice that even though E is adjacent to B, the algorithm does not "relabel" the previously labeled vertex E. The algorithm completes this second step by considering E and

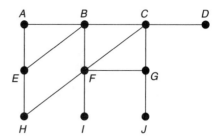

Figure 6.26 Graph S for example 6.3.1

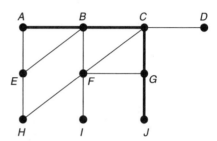

Figure 6.27 A shortest path in graph S for
example 6.3.1

labeling the adjacent unlabeled vertex H with 2:E. Proceeding alphabetically on C, F, and H, the above process is repeated so that D and G are labeled 3:C and I is labeled 3:F. Finally, J is labeled 4:G and the labeling process is stopped because J (the other given vertex) is now labeled.

The algorithm identifies a shortest path by working backwards from J to A based on the assigned labels. Since J is labeled 4:G, the algorithm includes the edge JG from vertex J to vertex G in the path under construction. Continuing in this fashion, the shortest path from J to A also includes edge GC (since G is labeled 3:C), edge CB (since C is labeled 2:B), and edge BA (since B is labeled 1:A). Thus, a shortest path from A to J follows the four edges connecting the vertices A, B, C, G, J as highlighted in figure 6.27.

■

The graph given in figure 6.26 for example 6.3.1 actually has two distinct paths of minimum length four from vertex A to vertex J; can you find the other path of length four? This other shortest path is identical to the path identified by the shortest path algorithm, except for visiting vertex F instead of vertex C. As in other such settings, we must therefore refer to the path identified by this algorithm as "a" shortest path between given vertices, rather than "the" shortest path (which would incorrectly indicate a unique solution). We now present a general description of the shortest path algorithm.

A shortest path algorithm This algorithm identifies a shortest path between two connected vertices. For ease of reference, we assume the given vertices are labeled A and B and let A denote the initial endpoint of the path under construction.

Step 1. Label every vertex adjacent to A with the notation 1:A.

Step 2. Consider each vertex V labeled with the notation 1:A. If more than one vertex is labeled 1:A, use any ordering of these vertices. Label every vertex adjacent to V that is previously unlabeled (either in Step 1 or in a previous iteration of Step 2) with the notation 2:V.

Step 3. Repeat Step 2 until vertex B is labeled. Throughout this process, label vertices using the notation n:V, where n is the number of iterations of Step 2 and V is a previously labeled vertex.

Step 4. Construct a shortest path by working backwards from B to A using the assigned labels. If vertex B is labeled m:V (for some adjacent vertex V and integer m), include the edge BV from B to V in the path under construction. Continuing in this fashion, include the edge from V to the vertex designated by its label, and so on back to vertex A. The length of this shortest path is m.

∎

If any step in this shortest path algorithm requires an ordering of vertices, then any ordering may be used. In fact, the particular ordering is not important from one step to the next, although we often fix some ordering at the beginning and consistently use this ordering throughout the implementation of the algorithm. We now discuss a proof that this algorithm always achieves its goal of identifying a "shortest" path between any two connected vertices.

Theorem 6.3.1   *If V and W are distinct, connected vertices in a graph G and a path in G identified by the shortest path algorithm has length m, then every path from V to W has length greater than or equal to m.*

Sketch of proof We outline a proof by contradiction. Such a proof assumes that p is a path of length m from vertex V to W given by the shortest path algorithm and that q is a distinct path from V to W of length n with $n < m$. We obtain a contradiction by considering the labeling process utilized by the shortest path algorithm. In particular, vertex W must be labeled n:X for some vertex X in G based on path q, and so cannot later be labeled m:Y for a vertex Y in G based on path p. This contradicts the construction of p by means of the shortest path algorithm.

∎

Question 6.3.1   Using the shortest path algorithm, identify a shortest path from vertex A to vertex H in the graph given in figure 6.28.

∎

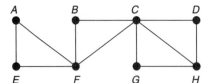

Figure 6.28   Graph for question 6.3.1

6.3.2   Spanning Trees

For the remainder of this section, we develop and explore the notion of a "spanning tree" of a graph. As discussed above, many real-life problems are solved using graph-theoretic models to examine connections between vertices in a representative graph. The strongest response to a need for connectivity is to supply edges that make every vertex adjacent to every other vertex; the complete graph on n vertices models this approach. However, in many practical settings, such a use of resources may be wasteful and (more often) impossible. In such cases, an effective, alternative approach is to require every pair of vertices simply to be connected by a path. Airline routes, bus routes, and computer networks are just a few of the many interrelationships implemented in this fashion; and all of these connectivity problems are readily solved by finding a *spanning tree* of an appropriate representative graph.

The first mathematical treatise on spanning trees was written by Gustav Robert Kirchoff in the mid-1800s. Kirchoff was born and raised in Königsberg (of graph-theoretic bridge fame) and studied mathematical physics at Albertus University of Königsberg in the early 1840s. While still an undergraduate, Kirchoff made his first significant research contribution, extending Ohm's laws to allow for the calculation of currents, voltages, and resistances in electrical circuits with multiple loops; his interest in spanning trees arose naturally in this study of electrical networks. Soon after, Kirchoff moved to Berlin, Germany (eventually becoming chair of mathematical physics at the University of Berlin), where he played a prominent role in the analysis and teaching of physics and influenced the generation of physicists that followed.

Requiring only connectedness among vertices (rather than adjacency) enables us (like Kirchoff) to optimize potential networks of relationships among vertices; we can eliminate repeated or redundant paths linking two vertices, thereby obtaining a greater level of efficiency. For example, after implementing the shortest path algorithm in example 6.3.1, we observed that there exist two shortest paths connecting vertices A and J in figure 6.26; in fact, many different paths connect these two vertices. In this setting, we can pare down the graph in figure 6.26 so that there exists exactly one path connecting vertices A and J, and any other pair of distinct vertices.

In graph-theoretic terms, we seek a subgraph that includes every vertex of the original graph G and preserves all connectedness relations in G, but also contains as few edges of G as possible. Formally, we refer to this type of subgraph as a *spanning tree*. The key to obtaining such an ideal subgraph is to avoid cycles; recall from definition 6.1.4 in section 6.1 that a cycle is a closed path that does not repeat any vertices or edges. If we identify a subgraph of a given graph not containing any cycles, then the resulting subgraph avoids any redundant paths and contains exactly one path connecting any distinct pair of vertices. With these reflections in mind, we state the related definitions.

Definition 6.3.1   *A* **tree** *is a connected graph that does not contain any cycles. A* **spanning tree** *of a connected graph G is a subgraph of G that is a tree and that contains every vertex of G.*

As you might expect, every graph that is a tree is a spanning tree of itself. However, most often, we consider graphs that are not trees and seek spanning trees of these more complicated graphs. Consider the following example.

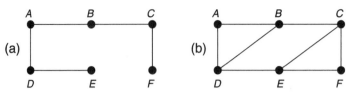

Figure 6.29   Graphs for example 6.3.2

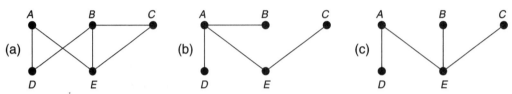

Figure 6.30   Graphs for question 6.3.2

Example 6.3.2   We examine the graphs given in figure 6.29 in light of the notions of a tree and a spanning tree from definition 6.3.1.

Graph (a) is a tree because the graph is connected and does not contain any cycles. In contrast, graph (b) is not a tree. Although this graph is connected, there are many different cycles in graph (b), including A, B, D; can you identify another cycle in graph (b)? In addition, graph (a) is a subgraph of graph (b) because every vertex and every edge in (a) is also in (b). Since (a) is a subgraph of (b), and graph (a) is a tree, and every vertex of (b) appears in (a), we conclude that graph (a) is a spanning tree of graph (b).

∎

As illustrated in example 6.3.2, the following three properties must be verified to prove that a graph T is a spanning tree of a graph G:

- T is a subgraph of G; that is, every vertex and every edge of T is in G;
- T is a tree; that is, T is a connected graph not containing any cycles;
- every vertex of G is contained in T.

If even one of these three properties fails, then T is not a spanning tree. We continue to refine our understanding of these ideas in the following question.

Question 6.3.2   Explain why each graph given in figure 6.30 is (or is not) a tree. Also, prove that one of these three graphs is a spanning tree of one of the other graphs.

∎

When given a visual representation of a graph, we are often readily able to determine if the graph is a tree; the human eye and mind almost instinctively identify the presence of cycles in graphs. In addition to such a geometric approach to identifying trees, the following theorem provides an arithmetic test for trees based on a numeric relationship between the number of vertices and the number of edges in a graph; consider the following theorem.

Theorem 6.3.2   *If n is a nonnegative integer and G is a connected graph with $n + 1$ vertices, then G is a tree iff G has exactly n edges.*

Comments on proof Throughout this discussion, we assume that n is a nonnegative integer and that G is a connected graph with $n + 1$ vertices. Induction is often used to prove "If G is a tree, then G has exactly n edges." This portion of the proof uses the fact that every tree with more than one vertex contains at least one vertex of degree one. Often a proof by contradiction is given to prove "If G has n edges, then G is a tree." This portion of the proof uses the fact that connected graphs remain connected when a single edge is removed from a nontrivial circuit in the graph.

■

Theorem 6.3.2 may bring to mind theorem 6.3.2 from section 6.1, in which the relatively straightforward properties of connectedness and degrees of vertices characterize the apparently much stronger assertion that Eulerian circuits or paths exist. In this setting, counting vertices and edges in a given graph is a routine process; comparing the resulting numbers enables the determination of cycles existing (and whether the given graph is a tree). We now reconsider the graphs given in the previous example and question in light of this result.

Example 6.3.3 We use theorem 6.3.2 to determine if the graphs given in figure 6.31 are trees.

Graph (a) is connected with six vertices and five edges; thus, theorem 6.3.2 implies that (a) is a tree. In contrast, graph (b) is not a tree, because (b) has six vertices but eight edges (and six is not equal to eight plus one).

■

Question 6.3.3 Using theorem 6.3.2, determine if the graphs given in figure 6.32 are trees.

■

Question 6.3.4 Sketch a graph with the following properties, or explain why such a graph does not exist.

(a) A tree with eight vertices and nine edges.

(b) A connected graph with eight vertices and nine edges.

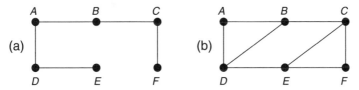

Figure 6.31 Graphs for example 6.3.3

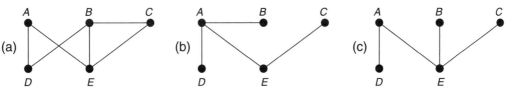

Figure 6.32 Graphs for question 6.3.3

(c) A tree with eight vertices and seven edges.

(d) A graph with eight vertices and seven edges that is not a tree.

■

Two algorithms identify a spanning tree of a given graph: the breadth-first spanning tree algorithm and the depth-first spanning tree algorithm. As with the shortest path algorithm, we carefully work through a detailed example and then give a general description of each algorithm.

6.3.3   The Breadth-First Spanning Tree Algorithm

The main idea of this algorithm, which identifies a spanning tree of a given connected graph, is to move from one vertex to the next, adjoining as many edges as we can at each vertex. As you might expect, placing a priority on "breadth" results in spanning trees that tend to be "thicker" around a few vertices having many adjacent vertices. Consider the following example.

Example 6.3.4   We implement the breadth-first spanning tree algorithm to identify a spanning tree of the graph given in figure 6.33.

The algorithm first selects any two adjacent vertices, labels them V_1 and V_2, and includes these vertices along with the incident edge $V_1 V_2$ in the spanning tree. For this graph, vertex A is labeled V_1 and vertex B is labeled V_2; vertices $V_1 = A$ and $V_2 = B$ and edge $V_1 V_2 = AB$ are then included in the spanning tree under construction. The algorithm now focuses on vertex $V_1 = A$ and considers all unlabeled vertices adjacent to V_1. In this graph, vertex E is the only such vertex; therefore, E is labeled V_3, and vertex $V_3 = E$ and edge $V_1 V_3 = AE$ are included in the spanning tree under construction.

Because no further unlabeled vertices are adjacent to V_1, the algorithm moves onto $V_2 = B$ and considers all unlabeled vertices adjacent to V_2. In this graph, C and F are the two vertices satisfying this criteria. Using an alphabetical ordering, C is labeled V_4 and F is labeled V_5, and we includes vertices $V_4 = C$ and $V_5 = F$ and the edges $V_2 V_4 = BC$ and $V_2 V_5 = BF$ in the spanning tree under construction.

The algorithm then considers $V_3 = E$, labeling H as V_6 and including vertex $V_6 = H$ and edge $V_3 V_6 = EH$ in the spanning tree. Continuing in this fashion until all the vertices are labeled, we obtain the following results.

- For $V_4 = C$: vertex D is labeled V_7, vertex G is labeled V_8, and vertices $V_7 = D$ and $V_8 = G$ and edges $V_4 V_7 = CD$ and $V_4 V_8 = CG$ are included in the spanning tree;

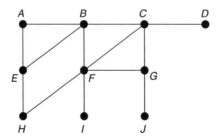

Figure 6.33   Graph for example 6.3.4

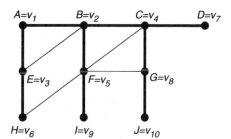

Figure 6.34 A breadth-first spanning tree for example 6.3.4

- For $V_5 = F$: vertex I is labeled V_9, and vertex $V_9 = I$ and edge $V_5 V_9 = FI$ are included in the spanning tree;
- For $V_6 = H$: there are no unlabeled adjacent vertices;
- For $V_7 = D$: there are no unlabeled adjacent vertices;
- For $V_8 = G$: vertex J is labeled V_{10}, and vertex $V_{10} = J$ and edge $V_8 V_{10} = GJ$ are included in the spanning tree.

At this point, every vertex in the original graph has been labeled and included in the spanning tree under construction. The set of vertices and edges identified by the algorithm is a subset of the original graph; that is, this set is a subgraph. Furthermore, this subgraph is a tree because only edges incident to unlabeled vertices are added to the graph, ensuring that no cycles are created during the construction. Thus, the breadth-first spanning tree algorithm has successfully identified a spanning tree of the original graph as illustrated in figure 6.34.

∎

The breadth-first spanning tree algorithm identifies just one of (typically) many possible spanning trees for a given connected graph. As may be apparent, different choices for the initial vertices V_1 and V_2 often led to the breadth-first algorithm identifying different spanning trees of the same graph. In addition, the depth-first spanning tree algorithm (discussed below) provides an alternative approach to finding a spanning tree of a given graph and often identifies a different spanning tree than the breadth-first algorithm. We now present the general description of the breadth-first spanning tree algorithm.

The breadth-first spanning tree algorithm This algorithm identifies a spanning tree of any given connected graph.

Step 1. Choose any two adjacent vertices in the graph, label them V_1 and V_2, and include the vertices V_1 and V_2 and the edge $V_1 V_2$ in the spanning tree under construction. In addition, two counters are defined to facilitate the construction. First, let m denote the vertex for which the algorithm is currently seeking adjacent unlabeled vertices and set $m = 1$. Second, let n denote a "label counter" that identifies the index of the label V_n to be assigned to the next vertex and set $n = 3$.

Step 2. Consider vertex V_m and identify an unlabeled vertex adjacent to V_m. If such a vertex does not exist, increment the counter m (setting $m = m + 1$) and consider the next vertex. If there exists more than one unlabeled vertex adjacent to V_m, select one using any ordering of the adjacent unlabeled vertices. The selected vertex is labeled V_n, and both the vertex V_n and the edge $V_m V_n$ are included in the spanning tree; also, increment the label counter n. We continue this process with unlabeled vertices adjacent to V_m until every vertex adjacent to V_m has been labeled and included in the spanning tree under construction (along with the appropriate corresponding edge). At this point, increment the counter m and consider the next vertex.

Step 3. Repeat Step 2 until every vertex of the given graph has been labeled. Once n is one more than the number of vertices in the given graph, the algorithm is complete and the resulting tree is a spanning tree of the original graph.

■

Theorem 6.3.3 *The breadth-first spanning tree algorithm identifies a spanning tree of any given connected graph.*

Sketch of proof The graph identified by this algorithm includes only vertices and edges from the original graph, ensuring the constructed graph is a subgraph of the original. In addition, this algorithm always identifies a tree because only edges from a labeled vertex to a new, previously unlabeled vertex are included in the graph under construction. A cycle could only be created by including edges from a labeled vertex to a labeled vertex, and the algorithm does not allow this possibility. Finally, the algorithm continues adding vertices and edges until every vertex in the original graph is included in the tree under construction. Thus, the graph identified by the breadth-first spanning tree algorithm is a spanning tree of the original graph.

■

In this proof sketch of theorem 6.3.3, notice the three strands of the argument that must be developed: subgraph, tree, and spanning. Every proof of the validity of such an algorithm must address these three criteria to ensure that a spanning tree of the original graph has been identified.

Question 6.3.5 Using the breadth-first spanning tree algorithm, identify a spanning tree of the graph given in figure 6.35; for the first step, use the labels $V_1 = A$ and $V_2 = E$.

■

6.3.4 The Depth-First Spanning Tree Algorithm

This algorithm is an alternative approach to identifying a spanning tree of a given connected graph. The main idea of the depth-first spanning tree algorithm is to move from one vertex to the next, adjoining just one edge incident to each vertex, unless

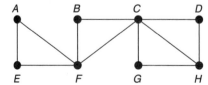

Figure 6.35 Graph for question 6.3.5

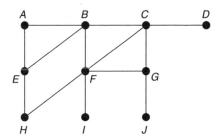

Figure 6.36 Graph for example 6.3.5

forced to backtrack and add more edges incident to a previously included vertex (because of the particular structure of the given graph). As you might expect, placing a priority on "depth" results in spanning trees that tend to be "thinner" with fewer edges incident to any one vertex. Consider the following example.

Example 6.3.5 We implement the depth-first spanning tree algorithm to identify a spanning tree for the (now familiar) graph given in figure 6.36.

The algorithm first selects an arbitrary vertex to serve as the starting point and proceeds through the original graph assigning labels to vertices in a fashion similar to that of the shortest path algorithm. In this case, we begin with vertex A, labeling A as vertex "1", and including A in the spanning tree under construction. The remaining vertex labels are of the form $n{:}V$ where n refers to the number of vertices already included in the spanning tree, and V refers to the "originating" adjacent vertex.

Beginning with vertex A, the algorithm identifies adjacent vertices B and E. Proceeding alphabetically, B is labeled with the notation $2{:}A$ and both vertex B and edge AB are included in the spanning tree. We now consider vertex $2{:}A = B$ and identify the adjacent vertices A, C, F. Continuing to work with the alphabetic ordering of vertices, the first vertex (vertex A) is already labeled, so the next (unlabeled) vertex C is labeled $3{:}B$ and both vertex C and edge BC are included in the spanning tree. Considering vertex $3{:}B = C$, the algorithm identifies adjacent vertices B, D, F, G. Proceeding alphabetically, vertex $B = 2{:}A$ is already labeled, so the next (unlabeled) vertex D is labeled $4{:}C$ and both vertex D and edge CD are included in the spanning tree under construction.

Moving on to the next vertex, we see that $4{:}C = D$ has no unlabeled adjacent vertices, even though some vertices in the original graph remain unlabeled; consulting the label counter, the number of labeled vertices is $n = 4$ which is less than 10 (the total number of vertices in the graph). When such a vertex is reached, the algorithm backtracks along the path of the most recently labeled vertices until reaching a vertex with an adjacent unlabeled vertex. For this graph, we backtrack to $3{:}B = C$ with adjacent unlabeled vertices F and G. Using the alphabetic ordering, vertex F is labeled $5{:}C$ and both vertex F and edge CF are included in the spanning tree. Continuing in this fashion until all vertices are labeled, we obtain the following results.

- Vertex G is labeled $6{:}F$ and both G and FG are included in the spanning tree.
- Vertex J is labeled $7{:}G$ and both J and GJ are included in the spanning tree.

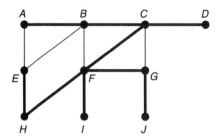

Figure 6.37 A depth-first spanning tree for example 6.3.5

- Vertex H is labeled $8{:}F$ and both H and FH are included in the spanning tree.
- Vertex E is labeled $9{:}H$ and both E and HE are included in the spanning tree.
- Vertex I is labeled $10{:}F$ and both I and FI are included in the spanning tree.

At this point, every vertex in the original graph has been labeled and included in the spanning tree under construction. Based on an argument similar to that for the breadth-first spanning tree algorithm, the depth-first spanning tree algorithm has successfully identified a spanning tree of the original graph as illustrated in figure 6.37.

∎

The depth-first spanning tree algorithm identifies just one of (typically) many possible spanning trees of a given connected graph. As may be apparent, different choices for the initial vertex often lead to the depth-first spanning tree algorithm identifying different spanning trees of the same graph. In addition, the depth-first spanning tree algorithm and breadth-first spanning tree algorithm often identify different spanning trees; compare figure 6.37 with figure 6.34 for an example of different spanning trees produced by these algorithms for the same graph. We now present a general description of the depth-first spanning tree algorithm.

The depth-first spanning tree algorithm This algorithm identifies a spanning tree of any given connected graph.

Step 1. Choose a fixed ordering of the vertices in the given graph; often an alphabetic or numeric ordering is chosen. This description of the algorithm uses an alphabetic ordering and labeling of vertices, but is easily modified to handle any ordering of vertices.

Step 2. The first vertex A is labeled "1" and included in the spanning tree under construction. We now consider the vertex adjacent to A that appears first in the fixed ordering from Step 1. This vertex is labeled $2{:}A$ and both vertex $2{:}A$ and edge with endpoints A and $2{:}A$ are included in the spanning tree under construction.

Step 3. Let $V = n{:}W$ denote the most recently labeled vertex, where n is the total number of vertices already labeled and W is the vertex adjacent to V that added V to the spanning tree. Based on the fixed ordering of vertices, identify the next unlabeled vertex adjacent to $V = n{:}W$. This vertex is

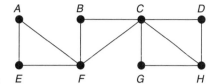

Figure 6.38   Graph for question 6.3.6

labeled $(n + 1){:}V$ and both the vertex $(n + 1){:}V$ and the edge with endpoints V and $(n + 1){:}V$ are included in the spanning tree.

Step 4. Repeat Step 3 until every vertex in the graph is labeled. If Step 3 cannot be implemented, but some vertex in the original graph remains unlabeled, then there must not be any unlabeled vertices adjacent to the most recently labeled vertex $V = n{:}W$. In this case, we backtrack along the path of most recently labeled vertices until we reach a labeled vertex U with at least one adjacent unlabeled vertex; such a vertex U exists because the given graph is connected. Now apply Step 3 to U, assigning $(n + 1){:}U$ to the (first) unlabeled vertex adjacent to U and include both vertex $(n + 1){:}U$ and the edge with endpoints U and $(n + 1){:}U$ in the spanning tree under construction. Repeat Step 3 for this newly labeled vertex $(n + 1){:}U$.

Step 5. Repeat Step 3 (and if necessary Step 4) until every vertex of the original graph is labeled. Once n is equal to the total number of vertices in the original graph, the algorithm is complete and the resulting tree is a spanning tree of the original graph.

∎

Theorem 6.3.4   *The depth-first spanning tree algorithm identifies a spanning tree of any given connected graph.*

Comments on proof   The proof is essentially identical to that of theorem 6.3.3 for the breadth-first spanning tree algorithm; further details are left to the reader.

∎

Question 6.3.6   Using the depth-first spanning tree algorithm, identify a spanning tree for the graph given in figure 6.38; start at vertex A and use alphabetic ordering of vertices.

∎

Question 6.3.7   Compare and contrast the spanning trees produced using the breadth-first spanning tree algorithm in question 6.3.5 and the depth-first spanning tree algorithm in question 6.3.6. What insights does this comparison provide into these algorithms?

∎

Among other things, your response to question 6.3.7 might include the observation that the breadth-first spanning tree algorithm produces trees that tend to be "thicker" around a few vertices, where these vertices have more adjacent vertices. In contrast, the depth-first spanning tree algorithm produces "thinner" trees with fewer incident edges at any one vertex.

6.3.5   Reading Questions for Section 6.3

1. Describe the strategy implemented by the shortest path algorithm.
2. State theorem 6.3.1. In what sense is the path identified by the shortest path algorithm the shortest path in a given graph?
3. Why do we say that the shortest path algorithm identifies "a" shortest path, rather than "the" shortest path between two vertices in a connected graph?
4. Define and give an example of a tree.
5. State theorem 6.3.2. How is this result helpful when studying graphs?
6. Using theorem 6.3.2, give an example of a tree and a graph that is not a tree.
7. Define and give an example of a spanning tree.
8. Discuss some real-world applications of spanning trees.
9. Describe the strategy implemented by the breadth-first spanning tree algorithm.
10. Describe the strategy implemented by the depth-first spanning tree algorithm.
11. Compare and contrast the breadth-first and depth-first spanning tree algorithms.
12. State theorem 6.3.3 and theorem 6.3.4. Why are these results both helpful and necessary?

6.3.6   Exercises for Section 6.3

In exercises 1–12, sketch a graph with the following properties, or explain why such a graph does not exist.

1. A tree with seven vertices and eight edges.
2. A connected graph with seven vertices and eight edges.
3. A tree with seven vertices and six edges.
4. A graph with seven vertices and six edges that is not a tree.
5. A connected graph with seven vertices, eight edges, and no cycles.
6. A graph with seven vertices, eight edges, and no cycles.
7. A tree with eight vertices and a total degree of 14.
8. A tree with eight vertices and a total degree of 15.
9. A graph with four vertices, five edges, and no cycles.
10. A connected graph with four vertices, three edges, and one cycle.
11. A graph with four vertices, three edges, and one cycle.
12. A tree with one vertex.

In exercises 13–20, use the shortest path algorithm to identify a shortest path from vertex A to vertex H in each graph.

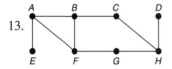   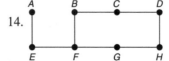   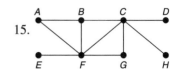

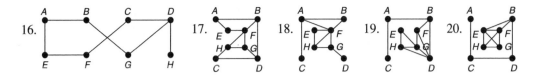

16. | 17. | 18. | 19. | 20.

In exercises 21–28, use the breadth-first spanning tree algorithm to identify a spanning tree of each graph; for the first step, let $V_1 = A$ and $V_2 = B$.

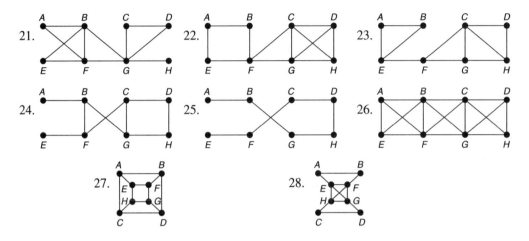

21. | 22. | 23.

24. | 25. | 26.

27. | 28.

In exercises 29–36, use the depth-first spanning tree algorithm to identify a spanning tree of each graph; start at vertex A and use an alphabetic ordering of vertices.

29. The graph from exercise 21.
30. The graph from exercise 22.
31. The graph from exercise 23.
32. The graph from exercise 24.

33. The graph from exercise 25.
34. The graph from exercise 26.
35. The graph from exercise 27.
36. The graph from exercise 28.

Exercises 37–47 consider the leaves of a tree; a vertex in a tree with degree one is called a leaf of the given tree. For example, vertices D, E, and F are leaves of the tree in figure 6.39.

In exercises 37–40, identify the leaves in each graph. Also, add a minimal number of edges to the given graph to obtain an extension of the given graph with no leaves.

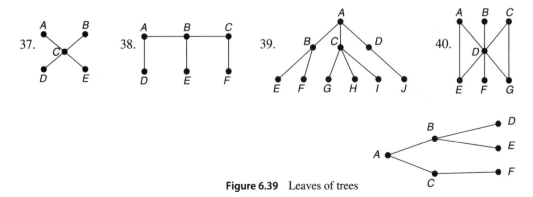

37. | 38. | 39. | 40.

Figure 6.39 Leaves of trees

In exercises 41–47, prove each mathematical statement about trees and leaves of trees. As defined before exercise 37, a vertex of degree one is called a leaf of a tree.

41. A tree with two or more vertices has at least one leaf.
    Hint: Develop a proof by construction, describing an algorithm that selects an arbitrary vertex in the tree and searches outward along a path until a vertex of degree one is found. The fact that a graph has finitely many vertices is key.

42. A tree with two or more vertices has at least two leaves.
    Hint: Develop a proof by contradiction on the length of the longest path in the tree; the two endpoints of this longest path are the desired leaves.

43. There exist exactly two trees with no leaves.
    Hint: Use exercises 41 and 42 to identify and sketch the two examples.

44. If T is a tree and vertex V is a leaf of T, then $T \setminus \{V\}$ is a tree.

45. If G is a tree with $n + 1$ vertices, then G has exactly n edges, where $n \in \mathbb{N}$.
    Hint: Develop a proof by induction on the number of vertices in the tree. In the inductive step, use the results stated in exercises 41–44.

46. The average degree of a vertex in a tree with two or more vertices is less than two.
    Hint: Use exercise 45.

47. Using exercise 46, prove that every tree with two or more vertices has at least one leaf.
    Note: This exercise calls for a statistical argument, rather than the constructive argument suggested for exercise 41.

Exercises 48–63 consider the notion of a binary tree. Working in this direction, a rooted tree is a tree with one distinguished vertex identified as the root of the tree. Also, the level of a given vertex V in a rooted tree is the number of edges in the (unique) path from the root to V. A binary tree is a rooted tree with the property that each vertex is adjacent to at most two vertices on the next level; these two vertices are called "children" as motivated by diagrams representing family trees. For example, consider the graphs given in figure 6.40.

Graph (a) is a binary tree with root A. Although vertex C has only one child F, graph (a) is a binary tree because each vertex has at most two children on the next level. In contrast, graph (b) is not a binary tree, because vertex B has three children (even though every other vertex has at most two children). Finally, a full binary tree is a binary tree in which every vertex (except the leaves) has exactly two children.

In exercises 48–53, sketch a rooted tree with the following properties or explain why such a graph does not exist.

48. A binary tree with six vertices and three leaves.
49. A binary tree with six vertices and five leaves.

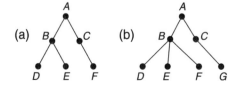

Figure 6.40 A tree that is binary and a tree that is not

50. A binary tree with four vertices and three leaves.
51. A binary tree with four vertices and two leaves.
52. A full binary tree with five vertices.
53. A full binary tree with 12 vertices and seven leaves.

Exercises 54–57 consider the height of a tree. The maximum length of a path in a tree with the root as one endpoint is called the height of the tree.

In exercises 54–57, identify upper and lower bounds on the height of rooted trees with the following properties.

54. A full binary tree with eight leaves.

55. A binary tree with eight leaves.

56. A full binary tree with nine vertices.

57. A binary tree with nine vertices.

Exercises 58–63 consider complete binary trees, The complete tree B_n is a full binary tree of height n with the additional property that every vertex at any level less than n has exactly two children; the only leaves in the complete binary tree B_n occur at level n. For example, consider the graphs in figure 6.41.

Both graphs in figure 6.41 are full binary trees of height two. Since graph (a) has a leaf at level one, this graph is not complete; in contrast, graph (b) is the complete binary tree B_2 of height two.

In exercises 58–63, explain why each rooted tree is (or is not) binary, full, and/or complete.

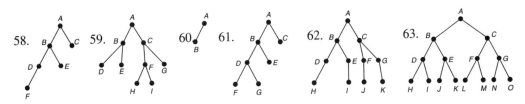

Exercises 64–69 consider connected graphs with a vertex at every intersection of two edges. Such a graph has no vertices with degree one and subdivides the plane into distinct regions. For example, the graph in figure 6.42 yields four distinct planar regions that are labeled I, II, III, and IV.

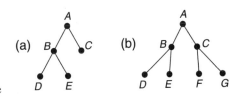

Figure 6.41 A full and a complete binary tree

Figure 6.42 A graph producing four planar regions

In exercises 64–67, specify the number of planar regions r, the number of edges e, the number of vertices v, and the value of $r - e + v$ for each graph.

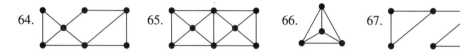

64.   65.   66.   67.

Exercises 68–69 consider properties of connected graphs with a vertex at every intersection of two edges.

68. In light of the results from exercises 64–67, make a conjecture about the value of $r - e + v$.

The correct conjecture is often referred to as *Euler's formula*.

69. Sketch a disconnected graph with a vertex at every intersection of two edges that also fails to satisfy your conjecture for the value of $r - e + v$. What is the value of $r - e + v$ for this graph?

Exercise 70 considers the notion of isomorphic graphs. The term isomorphism is used in many mathematical fields and is derived from the Greek/Latin words "iso" (meaning "same") and "morph" (meaning "form"). In mathematics, we say that two objects are isomorphic if they are identical with respect to some mathematical property under study (even if other details, such as names, are different). We often identify isomorphic objects by defining a one-to-one, onto function between the two objects that also satisfies some further condition(s).

70. In graph theory, we prove that two graphs are isomorphic by defining a one-to-one correspondence from the vertices of one graph to the vertices of the other, and from the edges of one graph to the edges of the other that also satisfy an additional graph-theoretic property. What properties of vertices and edges must be preserved by such a one-to-one correspondence to ensure that the graphs are identical with respect to their graph-theoretic properties?

6.4   Application: Weighted Graphs

In this section we add another layer of sophistication to our study of graphs by including "weights" on graphs. In many applications, we need more information than just the connectedness of two objects as modeled by vertices joined by edges or paths. Often, important aspects of the relationships among objects can be expressed by assigning numeric values to the edges of a graph. These numeric values are called *weights* and a graph with such labels is called a *weighted graph*. Working with weighted graphs as representative models of real-life settings enables us to successfully address such sophisticated questions as the weighted renditions of the spanning tree problem and the traveling salesman problem.

For example, suppose we are deciding whether to fly or drive from New York City to Los Angeles. After doing some research, we find that flying will take 6 hours and 12 minutes at a cost of $383, while driving will take 41 hours and 29 minutes at a

Figure 6.43 Two weighted graphs

cost of $277. These comparative relationships can be represented in the two weighted graphs (one for time and one for cost) given in figure 6.43.

In deciding whether to fly or drive, the traveler needs to assess the relative importance of the weights presented in each graph. Some other criteria may also influence the decision-making process; for example, enjoying the scenery while driving or visiting Aunt Gertrude in Saint Louis on the way out west may be more important than either time or money. In short, the information presented by weighted graphs can greatly assist decision-making processes in a variety of real-life settings.

Similar decisions are routinely made by trucking companies when setting up routes, by phone and electric companies when running lines, by oil and chemical companies when laying pipe, by computer manufacturers when designing chips and hardwiring computers, and by a host of other service and manufacturing industries. A common goal in many of these settings (as when making travel plans) is to minimize the use of some resource, whether it be time, or distance, or money, or physical objects. This goal of optimizing solutions has played an influential role in shaping the study of algorithms for working with weighted graphs. While sometimes real-world problem solvers are able to get by with rough guesses and estimates, more often they need a systematic approach that produces a a desired, precise, optimal solution. Just as calculus successfully addresses such questions in the continuous context, graph theory has proven marvelously applicable in many discrete settings. Since the 1930s, graph theorists have developed a variety of algorithms that identify optimal (or at least nearly optimal) solutions to many questions about weighted graphs. We begin with two definitions that set the stage for this study.

Definition 6.4.1 *A* **weighted graph** *is a graph whose edges are labeled with numbers that are called the* **weights** *of the edges. The* **total weight** *of a weighted graph is the sum of the weights of all the edges of the graph.*

We have already given two simple examples of weighted graphs for a traveler from New York City to Los Angeles; additional examples are given in the following example and question.

Example 6.4.1 The graphs in figure 6.44 are weighted because each edge is labeled with a numerical value or weight.

■

Question 6.4.1 State two destinations in the world that you would like to visit and sketch two weighted graphs with your hometown (or college town) as one vertex and the destinations as the other two vertices. As above, determine the time and cost of

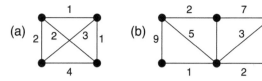

Figure 6.44 Weighted graphs for
example 6.4.1

the journey represented by each edge and assign these values as weights to the
corresponding edges. This graph is a weighted version of the complete graph K_3
on three vertices.

■

Question 6.4.2   Describe another scenario (besides traveling) for which a weighted graph could
serve as a helpful model and sketch an example of a corresponding weighted
graph.

■

Most applications of weighted graphs seek an optimal (often minimal) solution to
some question, and often these questions are solved by identifying spanning trees
or Hamiltonian cycles (which are already familiar from our work in sections 6.2
and 6.3). Therefore, we focus our study in this section on the very practical task
of identifying *minimum weight spanning trees* that solve the weighted spanning tree
problem and *minimum weight Hamiltonian cycles* that solve the weighted traveling
salesman problem.

6.4.1   The Weighted Spanning Tree Problem

Recall from definition 6.3.1 in section 6.3 that a spanning tree of a connected graph
is a subgraph of that is a tree and that contains every vertex of the original graph. For
a weighted graph, we compute the (total) weight of a spanning tree by summing the
weights of the edges included in the spanning tree. Many connected graphs have several
distinct spanning trees with different weights and (as suggested by the discussion
above) we are interested in the spanning tree with the least possible weight. As you
might expect from our work in section 6.3, various algorithms have been developed for
identifying a minimum weight spanning tree of a given connected, weighted graph; we
study two such algorithms in this section: Kruskal's algorithm and Prim's algorithm.
We begin with the formal definition of a minimum weight spanning tree and then
develop these algorithms.

Definition 6.4.2   *A **minimum weight spanning tree** of a connected, weighted graph G is a spanning
tree of G whose total weight is less than or equal to the total weight of any other
spanning tree of G.*

Every connected, weighted graph has a minimum weight spanning tree—the
challenge is identifying this tree! For sufficiently small graphs, we can use the method
of exhaustion to find the minimum weight spanning tree; consider the following
example and question.

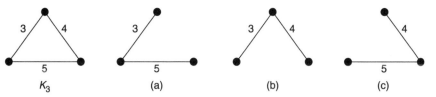

Figure 6.45   A weighted K_3 graph and its spanning trees for example 6.4.2

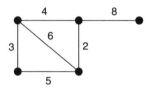

Figure 6.46   A weighted graph for question 6.4.3

Example 6.4.2   We identify a minimum weight spanning tree of the weighted version of K_3 given in figure 6.45; the three spanning trees of this graph are also given as graphs (a), (b), and (c).

Summing the weights of the edges, we see that graph (a) has total weight 8, graph (b) has total weight 7, and graph (c) has total weight 9. Thus, graph (b) is the minimum weight spanning tree of this weighted version of K_3.

■

Question 6.4.3   Using the method of exhaustion (as in example 6.4.2) identify a minimum weight spanning tree of the graph given in figure 6.46; this graph has seven distinct spanning trees.

■

For the small graphs given in example 6.4.2 and question 6.4.3, we are able to find minimum weight spanning trees with relative ease by using the method of exhaustion to identify all possible spanning trees and comparing their total weights. As we have come to recognize, such exhaustive searches are infeasible for graphs of sufficient complexity. Just the relatively slight increase in the complexity of the graphs from example 6.4.2 to question 6.4.3 results in a significant increase in the number of spanning trees that must be considered. Another example that provides numerical insight is the complete graph K_n on n vertices; this graph has n^{n-2} spanning trees, leading quickly to an impractical number of spanning trees for increasing values of n. For large values of n we therefore need some other approach to identifying minimum weight spanning trees, and so we turn our attention to studying Kruskal's algorithm and Prim's algorithm.

6.4.2   Kruskal's Algorithm

This algorithm for identifying minimum weight spanning trees was developed by the American mathematician and statistician Joseph Bernard Kruskal while he has working at Bell Laboratories in the 1950s. At that time, researchers at Bell Labs were grappling with a key problem in computer network design that is solved by finding a minimum weight spanning tree. Kruskal first outlined this algorithm that now bears his name in a 1956 paper entitled on the shortest spanning tree and the traveling salesman problem [144]. In addition, Kruskal made important contributions to statistics in his study of "multidimensional scaling"; his two older brothers were also accomplished researchers in mathematics and statistics.

Kruskal's algorithm is a classical example of a "greedy" algorithm that constructs a minimum weight spanning tree by selecting edges from the given connected, weighted graph that are of overall minimum weight; the algorithm is "greedy" in the sense that it focuses on including what appear to be the best possible edges in the spanning tree under construction. Computer scientists have determined that Kruskal's algorithm can be implemented (using an appropriate data structure) in "the order of $m \log n$ time," where m is the number of edges and n is the number of vertices in the given graph; this is considered a relatively "short" time and indicates the algorithm is pretty efficient. We carefully work through an example; a general description of Kruskal's algorithm follows.

Example 6.4.3   We implement Kruskal's algorithm to identify a minimum weight spanning tree of the connected, weighted graph G given in figure 6.47.

The overall strategy of Kruskal's algorithm is to include one edge at a time in the spanning tree under construction from least to greatest weight. Edges are added until every vertex of the given graph is included in the tree, at which point we have the desired spanning tree. Since the goal is to identify a tree, the algorithm cannot include any edge that would create a cycle. This is the one restriction in our choice of edges: if including some edge would result in a cycle, then the algorithm skips that edge and considers another edge of the same or of the next largest weight. When multiple edges have the same weight and are simultaneously eligible for inclusion in the spanning tree, we consider one edge at a time using any ordering of edges. With this discussion in mind, we apply Kruskal's algorithm to the graph G given in figure 6.47.

The algorithm first considers all edges in G with minimum weight. In this graph, the minimum weight is two and BE is the unique edge of weight two; the algorithm includes edge BE and its endpoints, the vertices B and E, in the spanning tree under construction.

The algorithm now considers all edges in G with the next largest weight; in this graph, both edges AE and CF have weight three. Proceeding in this listed order, we check if adding edge AE would create a cycle in the tree under construction; since it does not, edge AE and vertices A and E are included in the tree. Similarly, adding

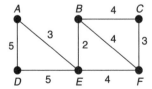

Figure 6.47   A weighted graph G for example 6.4.3

edge CF would not create a cycle in the tree under construction, and the algorithm includes edge CF and vertices C and F in the tree. Since not every vertex of the original graph is included in the tree under construction, the algorithm proceeds to the next largest weight.

In this graph, edges BC, BF, and EF all have the next largest weight of four; we consider whether these edges can be added to the tree under construction in the order listed. Adding BC does not create a cycle and so edge BC is included in the tree under construction (the vertices B and C are already in the tree). At this point, we observe that adding either BF or EF to this tree would create a cycle: including BF creates cycle BC, CF, BF, and including EF creates cycle BC, CF, FE, EB. Therefore, the algorithm does not include any additional edges of weight four.

Since vertex D is still not in the tree under construction, the algorithm considers adding edges of the next largest weight; in this graph, edges AD and DE both have weight five, and we consider including these edges in the listed order. Adding edge AD does not create a cycle and so edge AD and vertex D are included in the tree.

Every vertex of the original graph has now been included in the spanning tree under construction. The set of vertices and edges identified by this algorithm is a subset of the original graph; that is, the construction has produced a subgraph. Furthermore, this subgraph is a tree because edges that would create a cycle are not included by construction, and so the subgraph does not contain any cycles. Kruskal's algorithm has thus identified a spanning tree of the original graph as illustrated in figure 6.48.

The total weight of this spanning tree is $2 + 3 + 3 + 4 + 5 = 17$. As discussed below, Kruskal's algorithm always identifies a minimum weight spanning tree of a given connected, weighted graph, and so we are guaranteed that every spanning tree of G has total weight greater than or equal to 17.

■

As may be apparent, Kruskal's algorithm produces "a" minimum weight spanning tree for a given connected, weighted graph G, rather than "the" minimum weight spanning tree, because G may have several different spanning trees with the same minimum weight. In example 6.4.3, we could have listed the last two edges in the order DE and AD, and instead included DE in the spanning tree under construction; this different spanning tree has the same minimum weight of 17 as the spanning tree identified in figure 6.48.

Question 6.4.4   Working with the graph given in figure 6.48 for example 6.4.3, identify another distinct spanning tree with minimum weight 17.

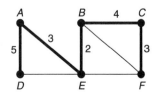

Figure 6.48 Kruskal's minimum weight spanning tree for example 6.4.3

Hint: Consider another ordering of the edges having weight four from that used in example 6.4.3.

∎

We now present a general description of Kruskal's Algorithm.

Kruskal's algorithm for minimum weight spanning trees This algorithm identifies a minimum weight spanning tree of any given connected, weighted graph.

Step 1. Choose any edge of minimum weight in the given graph and include this edge and its two endpoints in the spanning tree under construction.

Step 2. Identify an edge of minimum weight that has not been included in or been previously rejected for inclusion in the spanning tree under construction. If more than one edge satisfies this criteria, then choose one based on any ordering of the edges. Determine if adding this selected edge would create a cycle. If not, include the edge and its endpoints in the tree under construction; if so, identify another such edge of the same or the next largest weight and consider this new edge for inclusion in the tree as above.

Step 3. Repeat Step 2 until the graph under construction is a tree containing every vertex of the given graph. At this point, the algorithm is complete and the resulting tree is a minimum weight spanning tree of the original graph.

∎

Theorem 6.4.1 *Kruskal's algorithm identifies a minimum weight spanning tree of any given connected, weighted graph.*

Comments on proof As discussed at the end of example 6.4.3, Kruskal's algorithm always constructs a subgraph that is a tree and that includes every vertex of the original graph; that is, this algorithm identifies a spanning tree of the given graph. The proof that this spanning tree obtained has minimum weight is more subtle. The standard proof proceeds by contradiction, assuming that there is some other spanning tree with total weight less than the spanning tree identified by Kruskal's algorithm. By comparing the edges and their corresponding weights in this assumed tree with the edges and their corresponding weights in Kruskal's tree, a contradiction follows from the choices made in the construction of Kruskal's tree. Further details are left to the reader.

∎

Question 6.4.5 Using Kruskal's algorithm, identify a minimum weight spanning tree of the graph given in figure 6.49.

∎

6.4.3 Prim's Algorithm

We describe a second algorithm that identifies a minimum weight spanning tree of a given connected, weighted graph. Widely known as *Prim's algorithm*, this algorithm

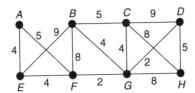

Figure 6.49 A weighted graph for question 6.4.5

was developed by several different researchers working independently of one another and is sometimes referred to as the *Jarnik algorithm* or the *DJP algorithm* because of this history. Prim's algorithm was articulated by the Czech mathematician Vojtech Jarnik in 1930, by the American mathematician Robert Prim [190] in 1957, and by the Dutch computer scientist Edsger Dijkstra in 1959. Jarnik is best known for his contributions to number theory, functions of real variables, and the support of mathematical education and research in Czechoslovakia through much of the twentieth century. Prim was a colleague of Kruskal at Bell Laboratories in the 1950s and served as director of mathematics research at Bell Labs from 1958 to 1961. Dijkstra is best known for his contributions to computer science in the area of programming languages and in 1972 was honored with the Turing Award for this work. For different reasons, each of them became interested in minimum weight spanning trees, and they all isolated this same algorithm.

This situation is just one example of the many times in mathematical history when different individuals have independently developed the same result at approximately the same time in history. Newton's and Leibniz's independent articulation of the fundamental theorem of calculus is another well-known example. At such times in humanity's intellectual history, the confluence of ideas and culture seem to have primed simultaneous creative insights from multiple individuals. In addition, slow means of communication, language barriers, and reluctance to publish results have sometimes delayed the dissemination of mathematical knowledge and provided the opportunity for such independent work. At their best, mathematicians have graciously acknowledged and celebrated such independent, identical contributions to humanity's collective knowledge; at their worst, bitter rivalries have erupted that negatively impacted other mathematicians and the ongoing development of new mathematical insights.

Prim's algorithm identifies a minimum weight spanning tree by focusing on vertices (rather than edges), constructing a spanning tree in connected links from a fixed initial vertex. Also classified as a "greedy" algorithm, Prim's algorithm is essentially equivalent to Kruskal's in terms of efficiency; using an appropriate data structure, this algorithm can be implemented in the "order of $(m + n) \log n$" time, where m is the number of edges and n is the number of vertices in the given graph. We carefully work through an example; a general description of Prim's algorithm follows.

Example 6.4.4 We implement Prim's algorithm to identify a minimum weight spanning tree of the (by now familiar) connected, weighted graph G given in figure 6.50.

Prim's algorithm constructs a minimum weight spanning tree by adding edges to a tree built up from a fixed initial vertex. In this example, we select A as the initial vertex and include A in the spanning tree under construction. The algorithm now considers all edges incident to vertex A and adds the edge of minimum weight.

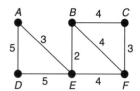

Figure 6.50   A weighted graph G for example 6.4.4

In this graph, edge AD has weight five and edge AE has weight three, so the algorithm includes edge AE and vertex E in the tree.

Since the tree under construction does not contain every vertex of G, the algorithm continues to add vertices. We consider every edge that is incident to some vertex in the current tree under construction but is not already included in the tree. At this step, the available edge incident to vertex A is AD with weight five, and the available edges incident to vertex E are DE with weight five, BE with weight two, and EF with weight four. The algorithm selects the edge of minimum weight from this collection of available edges and includes edge BE and vertex B in the tree under construction.

The current tree still does not span G, so the algorithm again considers all available (unincluded) edges incident to the vertices in the current tree. At this step, the available edges are: AD with weight five; DE with weight five; EF with weight four; BF with weight four; and BC with weight four. Four is the minimum weight available and we (arbitrarily) choose to include edge EF and vertex F in the tree under construction.

The current tree does not span G (vertices C and D are still not included), so the algorithm adds another edge in this same manner. At this step, every edge of graph G is either in or incident to a vertex in the tree under construction; that is, every unincluded edge is available for inclusion in the tree. Among the available edges, CF has minimum weight three, and the algorithm includes edge CF and vertex C in the tree under construction.

Finally, only vertex D is not included in the tree. The minimum weight edges still available are the two edges BF and BC with weight four, but neither can be added because doing so would create a cycle in the tree under construction. Therefore, the algorithm considers the two edges AD and DE of weight five. Either of these edges can be added and we (arbitrarily) choose to include edge AD and vertex D in the tree.

At this point, every vertex of the original graph is included in the spanning tree under construction. Thus, Prim's algorithm is complete and has identified a spanning tree of the original graph G as illustrated in figure 6.51.

The total weight of this spanning tree is $2 + 3 + 3 + 4 + 5 = 17$. As discussed below, Prim's algorithm always identifies a minimum weight spanning tree of a given connected, weighted graph, and so we are guaranteed that every spanning tree of G has total weight greater than or equal to 17.

■

As with Kruskal's algorithm, Prim's algorithm always produces "a" minimum weight spanning tree. Notice that the spanning tree of graph G produced by

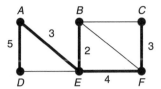

Figure 6.51 Prim's minimum weight spanning tree for example 6.4.4

Prim's algorithm (given in figure 6.51 for example 6.4.4) is different from the spanning tree of G produced by Kruskal's algorithm (given in figure 6.48 for example 6.4.3). Furthermore, both of these graphs have the same minimum weight of 17. Thus, a connected, weighted graph does not necessarily have a unique minimum weight spanning tree. We now present a general description of Prim's algorithm.

> **Prim's algorithm for minimum weight spanning trees** This algorithm identifies a minimum weight spanning tree of any given connected, weighted graph.
>
> **Step 1.** Choose any vertex V in the given graph and include V in the tree under construction.
>
> **Step 2.** Identify every edge of the original graph that is incident to a vertex in the tree under construction, that is not already in this tree, and that would not create a cycle if added to this tree. From this collection of available edges, select one of minimum weight, and include both this edge and the (new) endpoint in the tree under construction. If more than one available edge has minimum weight, then choose one based on any ordering of the edges.
>
> **Step 3.** Repeat Step 2 until the tree under construction contains every vertex of the given graph. At this point, the algorithm is complete and the resulting tree is a minimum weight spanning tree of the original graph.
>
> ■

Theorem 6.4.2 *Prim's algorithm identifies a minimum weight spanning tree of any given connected, weighted graph.*

Comments on proof The proof is essentially identical to that for Kruskal's Algorithm; further details are left to the reader.

■

Question 6.4.6 Using Prim's algorithm, identify a minimum weight spanning tree of the graph given in figure 6.52; use vertex A as the initial vertex.

■

6.4.4 The Weighted Traveling Salesman Problem

In the traveling salesman problem, we are given a connected graph and we seek to find a Hamiltonian cycle in this graph that visits every vertex in exactly once along some closed path. In real-life applications, additional information is often relevant

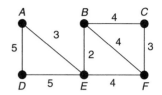

Figure 6.52   A weighted graph for question 6.4.6

to identifying a useful solution of the traveling salesman problem; this information is incorporated into a corresponding graph-theoretic model by means of weighted graphs. In such settings, real-life problem solvers seek minimum weight solutions of the traveling salesman problem that are naturally referred to as minimum weight Hamiltonian cycles. In this way, we not only identify a solution of the (unweighted) traversal question, but also conserve the resource represented by the weights on the graph (perhaps money, time, or some physical objects). We begin with the formal definition of a minimum weight Hamiltonian cycle and then develop two algorithms that identify approximate solutions to the weighted traveling salesman problem.

Definition 6.4.3   *A **minimum weight Hamiltonian cycle** in a graph G is a Hamiltonian cycle in G whose total weight is less than or equal to the total weight of any other Hamiltonian cycle in G.*

Recall from section 6.2 that some connected graphs do not have Hamiltonian cycles; naturally, not every connected, weighted graph has a minimum weight Hamiltonian cycle. When a graph does have such a cycle, identifying one can be a difficult process. For sufficiently small weighted graphs, we can use the method of exhaustion to find minimum weight Hamiltonian cycles; consider the following example and question.

Example 6.4.5   We identify a minimum weight Hamiltonian cycle in the weighted version of K_4 given in figure 6.53; the three Hamiltonian cycles in this graph are also given as graphs (a), (b), and (c).

Summing the weights of the edges, we see that graph (a) has total weight 11, graph (b) has total weight 10, and graph (c) has total weight nine. Thus, graph (c) is the minimum weight Hamiltonian cycle in this weighted version of K_4.

∎

Question 6.4.7   Using the method of exhaustion (as in example 6.4.5) identify a minimum weight Hamiltonian cycle in the weighted version of K_4 given in figure 6.54.

∎

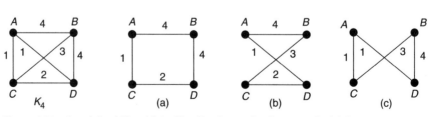

Figure 6.53   A weighted K_4 with its Hamiltonian cycles for example 6.4.5

Figure 6.54 A weighted graph for question 6.4.7

For the small graphs studied in example 6.4.5 and question 6.4.7, we are able to find minimum weight Hamiltonian cycles with relative ease by using the method of exhaustion to identify all possible Hamiltonian cycles and then comparing their total weights. However, for graphs of sufficient complexity, using exhaustion to identify a minimum weight Hamiltonian cycle is simply not feasible. For example, the complete graph K_{10} with 10 vertices has $1,814,400$ distinct Hamiltonian cycles and the complete graph K_{20} with twenty vertices has 1.216×10^{18} distinct Hamiltonian cycles! Even the fastest of today's supercomputers cannot generate and compare the weights of 1.216×10^{18} Hamiltonian cycles in a person's lifetime. Thus, real-life problem solvers need some other approach to answering practical questions expressed as weighted traveling salesman problems.

Unfortunately, as we learned in section 6.2, there is no known general algorithm (besides exhaustion) for determining if there exists a Hamiltonian cycle in a connected graph, let alone a general algorithm for identifying such a Hamiltonian cycle when one exists. Adding a minimum weight requirement to the desired solution only increases the question's complexity. Thus there is no known general algorithm (besides exhaustion) answering the weighted traveling salesman problem for an arbitrary connected, weighted graph.

At the same time, we often need an answer to weighted traveling salesman problems: roads need to be snowplowed, mail delivered, electric and phone lines run, and computers assembled—and all (hopefully) by the most efficient means possible. Fortunately, graph-theorists have isolated several "heuristic" algorithms employing minimization strategies that quickly and efficiently identify a Hamiltonian cycle (if one exists). Sometimes these algorithms stumble across a minimum weight Hamiltonian cycle, but all they really promise to achieve is some measure of optimization; in this sense, these heuristic algorithms typically identify an "approximately" minimum weight Hamiltonian cycle. Despite this shortcoming, the increase in speed, coupled with their ability to achieve some optimization, have proven the importance and utility of these algorithms.

We study two of these heuristic algorithms in this section: the sorted edges algorithm and the nearest neighbor algorithm. The goal of these algorithms is to identify a Hamiltonian cycle, and so we must state the resulting restrictions on adding edges to the cycle under construction. Both of these algorithms:

(1) do not add an edge to the cycle under construction that would result in three edges incident to a single vertex (because a Hamiltonian cycle can visit each vertex exactly once); and

(2) do not add an edge to the cycle under construction that would create a cycle until every vertex has been included in the cycle (at which point a last edge is included to close the cycle).

Abiding by theses two restrictions at each step in the process ensures that these algorithms identify a Hamiltonian cycle in the given graph (if one exists). With these reflections in mind, we carefully work through an example and then provide a general description of each heuristic algorithm.

6.4.5 The Sorted Edges Algorithm

This algorithm resembles Kruskal's algorithm for identifying minimum weight spanning trees, only adapted to finding minimum weight Hamiltonian cycles. The strategy of Kruskal's algorithm is to focus on edges and include edges one at a time in the tree under construction from least to greatest weight under the appropriate restrictions that ensure the end result is a spanning tree. The sorted edges algorithm also includes edges from least to greatest weight under the appropriate restrictions (discussed above) for identifying a cycle visiting every vertex exactly once. Since not every connected graph has a Hamiltonian cycle, this algorithm does not always obtain the desired solution. However, when the algorithm succeeds, the Hamiltonian cycle identified is "approximately" minimal. We demonstrate the sorted edges algorithm in the relatively simple setting provided by the weighted version of K_4 studied in example 6.4.5.

Example 6.4.6   We implement the sorted edges algorithm to identify an approximately minimum weight Hamiltonian cycle in the graph given in figure 6.55.

The sorted edges algorithm adds edges to the cycle under construction from least to greatest weight. Recall that in order to obtain a Hamiltonian cycle, the algorithm cannot add an edge that would result in either three edges incident to a single vertex or a cycle (until the last edge is added). At each step in the following process, the algorithm verifies that including an edge in the cycle under construction would not violate either of these restrictions.

The first step is to list every edge of the given graph in order from least to greatest weight; multiple edges with the same weight may be listed in any order. In this example, we use the following ordering of edges (identified by weight to facilitate the discussion):

$$1 : AC, AD \qquad 2 : CD \qquad 3 : BC \qquad 4 : AB, BD.$$

Working with this ordered list of edges, the algorithm adds edges to the cycle under construction from least to greatest weight (under the two restrictions). Both edges of weight one satisfy these conditions, and so the edges AC and AD and the vertices $A, C,$ and D are included in the cycle under construction. In fact, the sorted

Figure 6.55   A weighted graph for example 6.4.6

Figure 6.56 The sorted edges Hamiltonian cycle for example 6.4.6

edges algorithm always includes the first two edges from such a weight-ordered list; can you articulate why?

The algorithm now considers the next edge in this list: edge CD with weight two. Notice that adding CD to the current cycle would create the cycle AC, CD, DA, and so CD cannot be included in the (Hamiltonian) cycle under construction. Moving on to the next edge in the ordered list, BC can be added; thus, edge BC and vertex B are included in the cycle under construction. This cycle now contains every vertex of the original graph and the last edge can now be added to close the cycle.

The algorithm considers the two edges of weight four in the order listed above. Edge AB cannot be added to the cycle under construction because doing so would violate both restrictions on adding edges. In particular, vertex A already has two incident edges AC and AD, and edge AB would be a third incident edge; furthermore, the edges AC, CB, AB would form a cycle that does not contain vertex D. In contrast, edge BD can be added to the cycle under construction, completing the construction of the subgraph illustrated in figure 6.56.

As we can see, this subgraph is a closed path visiting every vertex of the given graph exactly once; that is, the algorithm has identified a Hamiltonian cycle. The total weight of this Hamiltonian cycle is $1 + 1 + 3 + 4 = 9$. As mentioned above, the sorted edges algorithm finds an approximately minimum weight Hamiltonian cycle; as determined by the method of exhaustion in example 6.4.5, this subgraph does happen to be the minimum weight Hamiltonian cycle in the original graph.

■

In many settings, the approximately optimal solution provided by the sorted edges algorithm enables real-life problem solvers to make important and timely decisions. The strength of the sorted edges algorithm (and the nearest neighbor algorithm) lies in quickly and systematically identifying a Hamiltonian cycle in a given connected, weighted graph (if one exists). Even if the minimum weight Hamiltonian cycle is not identified by applying these algorithms, some measure of optimization has been achieved and, in this sense, the solution is approximately minimal. We now present a general description of the sorted edges algorithm.

The sorted edges algorithm for minimum weight Hamiltonian cycles

This algorithm identifies an approximately minimum weight Hamiltonian cycle in a given connected weighted graph (if such a cycle exists).

Step 1. Sort the edges of the given graph in order from least to greatest weight. Multiple edges with the same weight can be listed in any order relative to one another.

Step 2. Based on the ordering of edges from least to greatest weight from Step 1, add one edge at a time to the cycle under construction, provided including an edge does not produce a cycle or result in either three edges incident to a single vertex.

Step 3. Repeat Step 2 until every vertex is included in the graph and then include a last edge to close the cycle under construction, producing the desired Hamiltonian cycle.

Since not every graph has a Hamiltonian cycle, Step 2 and Step 3 cannot always be implemented. In such cases, the sorted edges algorithm (indeed any algorithm) cannot identify a Hamiltonian cycle in the given graph. ∎

Question 6.4.8 Using the sorted edges algorithm, identify an approximately minimum weight Hamiltonian cycle in the graph given in figure 6.57. This graph is a weighted version of K_5, and the sorted edges algorithm determines a Hamiltonian cycle of weight 14. ∎

Question 6.4.9 Compare the weight of the Hamiltonian cycle identified in question 6.4.8 with the weight of the Hamiltonian cycle obtained by traversing exactly the outer edges of the graph given in figure 6.57. What does this comparison tell you about the sorted edges algorithm? ∎

6.4.6 The Nearest Neighbor Algorithm

We develop a second heuristic algorithm for approximating a minimum weight Hamiltonian cycle in a given connected, weighted graph. This nearest neighbor algorithm resembles Prim's algorithm for identifying minimum weight spanning trees, only adapted to finding minimum weight Hamiltonian cycles. The strategy of Prim's algorithm is to focus on vertices, and include edges in the tree under construction only if they are incident to a vertex in the current tree. Similarly, the nearest neighbor algorithm works from vertex to vertex in the given graph, including the minimum weight edge incident to the current vertex in the cycle under construction (under the standard restrictions needed to ensure a Hamiltonian cycle). In this way, the new included edge

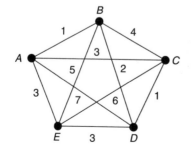

Figure 6.57 A weighted graph for question 6.4.8; for edges in the interior of the graph, each numerical weight goes with the edge next to it and toward the center of the figure. For example, 6 goes with edge CE and 7 to AD

Figure 6.58 A weighted graph for example 6.4.7

identifies the "nearest" vertex (or the nearest neighbor) of the current vertex. Therefore, the nearest neighbor algorithm is a "greedy" algorithm because the algorithm always makes the best possible choice available at any one vertex. This locally good choice often turns out to be the best choice (or at least an approximately best choice) for the entire graph. We demonstrate the nearest neighbor algorithm in the relatively simple setting provided by the weighted version of K_4 studied in examples 6.4.5 and 6.4.6.

Example 6.4.7 We implement the nearest neighbor algorithm to identify an approximately minimum weight Hamiltonian cycle in the graph given in figure 6.58.

The nearest neighbor algorithm arbitrarily identifies some vertex in the given graph to serve as the base vertex for the cycle under construction. Although different choices for this base vertex may result in different Hamiltonian cycles, there is no particular strategy for selecting the starting point of the construction. The algorithm then includes the minimum weight edge incident to the base vertex and the other endpoint of this edge in the cycle under construction. The other endpoint of this included edge is the "nearest neighbor" of the base vertex and is considered next. The algorithm continues to add available (that is, unincluded) minimum weight edges incident to the most recently added vertex until every vertex of the original graph is included in the cycle under construction. Finally, an edge is included to close the Hamiltonian cycle. As with the sorted edges algorithm, the algorithm does not add an edge that would result in either three edges incident to a single vertex or a cycle (until the last edge is added).

For this example, we choose A as the base vertex. Vertex A has three incident edges: AC with weight one; AD with weight one; and AB with weight four. Either edge of minimum weight one can be added; based on the order they are listed, we include edge AC and vertex C in the cycle under construction.

The algorithm now considers vertex C. The edges incident to C not already in the current cycle are: BC with weight three and CD with weight two. Edge CD has the minimum weight and both edge CD and vertex D are included in the cycle under construction.

The available edges incident to vertex D are: AD with weight one and BD with weight four. Edge AD has the minimum weight of one, but adding this edge would create the cycle AC, CD, AD, and so AD cannot be included in the (Hamiltonian) cycle under construction. Therefore, the algorithm includes edge BD and vertex D in the cycle under construction.

Finally, the algorithm considers the available edges incident to vertex B: AB with weight one and BD with weight three. Adding minimum weight edge BC

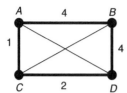

Figure 6.59 Nearest neighbor Hamiltonian cycle for example 6.4.7

would violate both restrictions on adding edges. In particular, vertex C already has two incident edges AC and CD, and edge BC would be a third incident edge; furthermore, the edges BC, CD, BD would form a cycle that does not contain vertex A. Instead, the algorithm includes edge AB, completing the construction of the desired Hamiltonian cycle as illustrated in figure 6.59.

The total weight of this cycle is $1 + 2 + 4 + 4 = 11$, which is an approximately minimum weight Hamiltonian cycle in the graph given in figure 6.58. ■

In example 6.4.5 we used the method of exhaustion to identify the minimum weight Hamiltonian cycle (with a total weight of nine) in the graph given in figure 6.58; therefore, the nearest neighbor algorithm does not necessarily identify a minimum weight Hamiltonian cycle. However, the nearest neighbor algorithm does efficiently find an approximately minimum weight Hamiltonian cycle and, in some cases, this algorithm actually does identify the minimum weight Hamiltonian cycle (if one exists). The choice of the base vertex in the first step of this algorithm has an important influence on the Hamiltonian cycle that is identified; if we had chosen B as the base vertex in example 6.4.7, the nearest neighbor algorithm would have identified the same minimum weight Hamiltonian cycle as the method of exhaustion (and the sorted edges algorithm). We now present a general description of the nearest neighbor algorithm.

The nearest neighbor algorithm for minimum weight Hamiltonian cycles   This algorithm identifies an approximately minimum weight Hamiltonian cycle in a given connected weighted graph (if such a cycle exists).

Step 1. Choose any vertex V in the given graph to serve as a base vertex for the cycle under construction and include V in the cycle under construction.

Step 2. Consider all available (that is, unincluded) edges incident to the vertex most recently added to the cycle under construction. From this collection of edges, include the minimum weight edge and its other endpoint in the cycle under construction, provided that including an edge does not produce a cycle or result in either three edges incident to a single vertex.

Step 3. Repeat Step 2 until every vertex is included in the graph and then include a last edge to close the cycle under construction, producing the desired Hamiltonian cycle.

Since not every graph has a Hamiltonian cycle, Step 2 and Step 3 cannot always be implemented. In such cases, the nearest neighbor algorithm

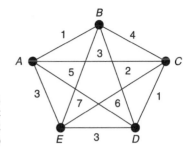

Figure 6.60 A weighted graph for question 6.4.10; each numerical weight goes with the edge next to it and toward the center of the figure. For example, 2 goes with edge BD

(indeed any algorithm) cannot identify a Hamiltonian cycle in the given graph.

■

Question 6.4.10 Using the nearest neighbor algorithm, identify an approximately minimum weight Hamiltonian cycle in the graph given in figure 6.60. Use vertex A as base vertex; with this base vertex, the nearest neighbor algorithm determines a Hamiltonian cycle of weight 13.

■

Question 6.4.11 Compare the weight of the Hamiltonian cycle identified in question 6.4.10 with the weight of the Hamiltonian cycle obtained by traversing exactly the outer edges of the graph given in figure 6.60. What does this comparison tell you about the nearest neighbor algorithm?

■

Question 6.4.12 (a) Using vertex B as the base vertex in the nearest neighbor algorithm, identify a an approximately minimum weight Hamiltonian cycle in the graph given in figure 6.60 for question 6.4.10.
(b) Compare the solutions obtained from using vertices A, B, and E as the base vertex in the nearest neighbor algorithm for the graph given in figure 6.60.

■

6.4.7 Reading Questions for Section 6.4

1. Define and give an example of a weighted graph.
2. Discuss the real-world questions that motivate the study of weighted graphs.
3. Define and give an example of a minimum weight spanning tree.
4. Discuss the distinction between "a" minimum weight spanning tree and "the" minimum weight spanning tree of a graph.
5. Describe the strategy implemented by Kruskal's algorithm.
6. Describe the strategy implemented by Prim's algorithm.
7. Define and give an example of a minimum weight Hamiltonian cycle.
8. Discuss the necessity and nature of an "approximately" minimum weight Hamiltonian cycle.

9. State the weighted traveling salesman problem.
10. Describe the strategy implemented by the sorted edges algorithm.
11. Describe the strategy implemented by the nearest neighbor algorithm.
12. Discuss the impact of the base vertex on the solution identified by the nearest neighbor algorithm.

6.4.8   Exercises for Section 6.4

In exercises 1–8, use Kruskal's algorithm to identify a minimum weight spanning tree of each graph.

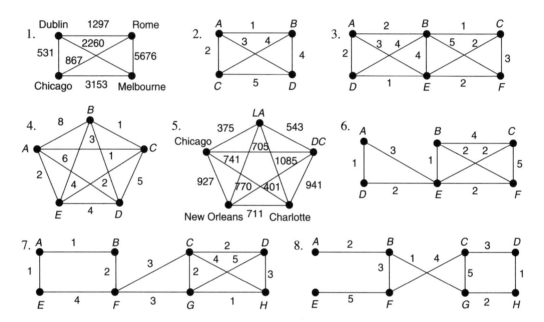

In exercises 9–16, use Prim's algorithm to identify a minimum weight spanning tree of each graph.

9. The graph from exercise 1.   13. The graph from exercise 5.
10. The graph from exercise 2.   14. The graph from exercise 6.
11. The graph from exercise 3.   15. The graph from exercise 7.
12. The graph from exercise 4.   16. The graph from exercise 8.

In exercises 17–24, use the sorted edges algorithm to identify an approximately minimum weight Hamiltonian cycle in each graph (if such a cycle exists).

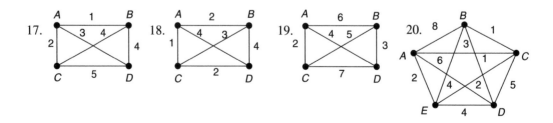

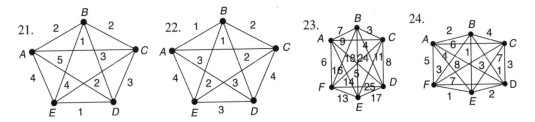

In exercises 25–32, use the nearest neighbor algorithm to identify an approximately minimum weight Hamiltonian cycle in each graph (if such a cycle exists).

25. The graph from exercise 17.
26. The graph from exercise 18.
27. The graph from exercise 19.
28. The graph from exercise 20.

29. The graph from exercise 21.
30. The graph from exercise 22.
31. The graph from exercise 23.
32. The graph from exercise 24.

Exercises 33–40 consider the *extended nearest neighbor algorithm* for specifying an approximately minimum weight Hamiltonian cycle in a given connected, weighted graph. As discussed in question 6.4.12, implementing the nearest neighbor algorithm with different base vertices can result in different Hamiltonian cycles with different weights. The extended nearest neighbor algorithm applies the nearest neighbor algorithm to every vertex in a given graph and identifies the Hamiltonian cycle from this collection with minimum weight as the solution. For example, applying the nearest neighbor algorithm to every vertex of the weighted version of K_4 in figure 6.61 produces the Hamiltonian cycles identified in the following table.

Base vertex	Hamiltonian cycle	Cycle weight
A	A, C, D, B, A	15
B	B, C, A, D, B	13
C	C, A, D, B, C	13
D	D, A, C, B, D	13

The minimum cycle weight is 13 and the extended nearest neighbor algorithm identifies any of the last three cycles as an approximately minimum weight Hamiltonian cycle in the given graph. In this particular example, the three vertices B, C, D all happen to identify the same Hamiltonian cycle; they only appear to differ because of the different base vertices and the direction of traversal of the cycle. This pattern occurs often, but is not necessary. This algorithm requires more resources than the original nearest neighbor algorithm because of the repeated use of the nearest neighbor algorithm. However,

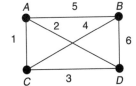

Figure 6.61 A weighted graph for the extended nearest neighbor algorithm

for most graphs, far fewer resources are used by the extended nearest neighbor algorithm than by the method of exhaustion, and a greater measure of optimization is achieved than just using the nearest neighbor algorithm based at one vertex in a given graph.

In exercises 33–40, use the extended nearest neighbor algorithm to identify an approximately minimum weight Hamiltonian cycle in each graph (if such a cycle exists).

33. The graph from exercise 17.   37. The graph from exercise 21.
34. The graph from exercise 18.   38. The graph from exercise 22.
35. The graph from exercise 19.   39. The graph from exercise 23.
36. The graph from exercise 20.   40. The graph from exercise 24.

Exercises 41–52 consider regular graphs. A graph is regular if every vertex of the graph has the same degree; sometimes we say that a regular graph is r-regular if the degree of every vertex is r.

In exercises 41–46, determine if each graph is regular; if not, identify two vertices of different degree.

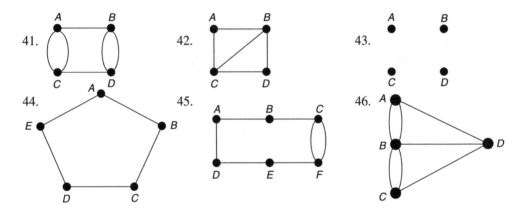

In exercises 47–50, sketch an r-regular graph with the indicated number of vertices.

47. A three-regular graph with four vertices.

48. A three-regular graph with six vertices.

49. A four-regular graph with two vertices.

50. A four-regular graph with five vertices.

In exercises 51–53, prove each mathematical statement about regular graphs.

51. Every complete graph K_n is $(n-1)$-regular.
52. An r-regular graph with n vertices has $nr/2$ edges.
53. If a graph G with more than two vertices has an Eulerian path between distinct vertices, then G is not regular.

Exercises 54–60 consider isomorphic graphs. We say that graphs G and G^* are isomorphic if there exist one-to-one correspondences $f : V(G) \to V(G^*)$ on the set of vertices and $g : E(G) \to E(G^*)$ on the set of edges that preserve the edge–endpoint

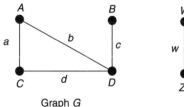

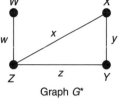

Figure 6.62 Isomorphic graphs

Graph G

Graph G^*

relations of G and G^*; that is, vertex $V \in V(G)$ is an endpoint of edge $e \in E(G)$ iff vertex $f(V) \in V(G^*)$ is an endpoint of edge $g(e) \in E(G^*)$. Thus, two graphs are isomorphic if they have the same form in the sense that the vertices and edges identified by the maps share the same edge–endpoint relationships. For example, the two graphs given in figure 6.62 are isomorphic under the mappings f (from vertices to vertices) and g (from edges to edges) defined by

$$f: \quad A \to X \quad B \to W \quad C \to Y \quad D \to Z$$
$$g: \quad a \to y \quad b \to x \quad \quad c \to w \quad d \to z.$$

In Exercises 54–59, state an isomorphism for each pair of graphs, or explain why such a mapping does not exist.

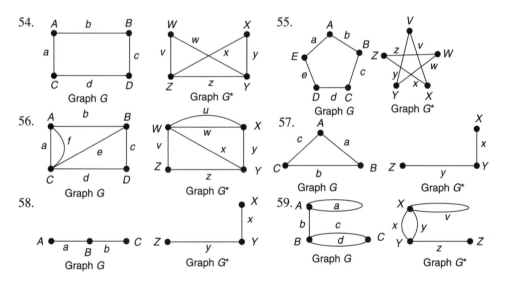

In exercise 60, prove one of many isomorphism theorems that hold for graphs.

60. If two simple graphs G and H are isomorphic, then the complements of the two graphs $\overline{G}$ and $\overline{H}$ are also isomorphic.

Exercises 61–62 consider the infinite complete binary tree, which is denoted by either $2^{<\omega}$ or $B_{<\omega}$. As illustrated in figure 6.63, the graph $2^{<\omega}$ contains every complete graph of height n as a subgraph, where we identify the root of each graph B_n with the root of $2^{<\omega}$.

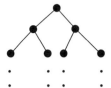

Figure 6.63 The infinite complete binary tree $2^{<\omega}$

In exercises 61–62, prove each mathematical statement about the infinite complete binary tree $2^{<\omega}$.

61. The infinite complete binary tree $2^{<\omega}$ has countably many paths of finite length with one endpoint at the root.

    Hint: Define a one-to-one correspondence between such paths and the natural numbers by means of a one-to-one correspondence between the vertices of $2^{<\omega}$ and the natural numbers $\mathbb{N}$. State why identifying this second mapping is sufficient to prove the statement.

62. There exist uncountably many paths of "infinite" length (i.e., paths passing through every level) in the infinite complete binary tree $2^{<\omega}$.

    Hint: One approach is to give a proof by contradiction modeled on the proof of the uncountability of $\mathbb{R}$ in section 4.5. Alternatively, construct a one-to-one correspondence between infinite binary numbers and infinite paths of $2^{<\omega}$ by assigning the binary digits 0 and 1 to right and left branches in the tree; then prove the uncountability of the set of infinite binary numbers.

Exercises 63–70 consider parse trees, which play an important role in the analysis of languages (a fundamental task relevant to many different areas of computer science). Parse trees for algebraic expressions in a single variable x are generated by context-free grammars. We work with such a grammar defined by the set of rules:

$$S \to (S) \qquad S \to -S \qquad S \to S+S \qquad S \to S-S$$
$$S \to S*S \qquad S \to S/S \qquad S \to x.$$

For example, this grammar generates the string $-(x * x)$ by the following derivation; the corresponding parse tree is given in figure 6.64.

$$S \quad \to \quad -S \quad \to \quad -(S) \quad \to \quad -(S*S) \quad \to \quad -(x*S) \quad \to \quad -(x*x)$$

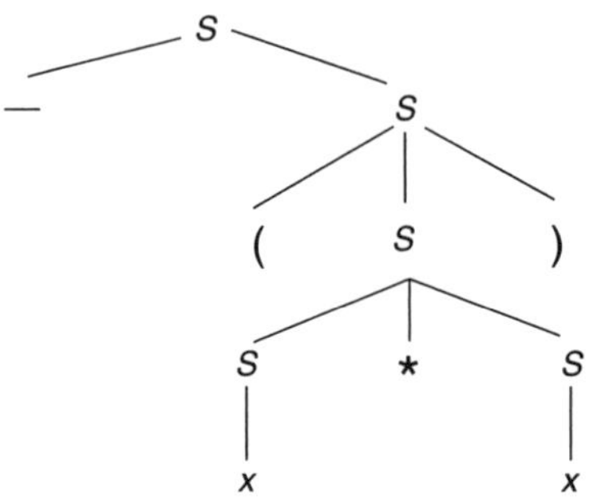

Figure 6.64 The parse tree for a derivation of $-(x * x)$

In exercises 63–70, use the grammar given above to identify a derivation of each algebraic expression in x and sketch the corresponding parse tree for each derivation.

63. x

64. $x + x$

65. $-(x + x)$

66. $(-x) + x$

67. $x - (-x)$

68. $-(x - x)$

69. $(x + x) * x$

70. $(x * x)/(x + x)$

Notes

Graph theory is an exciting and active area of mathematical study with lots of open questions. Many of these questions are easily stated and understood, and undergraduate mathematics students have engaged research-level questions in this field. A number of books surveying graph theory have been written in recent years. Among the excellent undergraduate introductions to graph theory are those by Alduous and Wilson [4], Chartrand [39], Trudeau [242], and West [254]; both Bollobas [21] and Diestal [59] are standard graduate level texts in graph theory. Another interesting text that is accessible to advanced undergraduates has been written by Foulds [87] and focuses on the practical application of graph theory to a striking number of different areas.

In addition to books focused exclusively on surveying graph theory, many of the ideas presented in this chapter are studied in both discrete mathematics and "liberal arts" mathematics courses. Some standard textbooks used in discrete mathematics courses include Epp [72], Richmond and Richmond [193], and Scheinerman [209]; supporting texts for liberal arts mathematics courses have been written by Burger and Starbird [34] and the Consortium for Mathematics and Its Applications [43].

Some of the questions we have introduced in this chapter are (by themselves) the focus of entire books. Recall our discussion of the four-color theorem in section 6.1. Recently, Wilson [258] has written an interesting book exploring the history and proof of the four-color theorem; Fristch et al. [92] is another good exploration of mathematicians' study and solution of this same result. In addition, we have extensively discussed the traveling salesman problem in this chapter. An even more thorough study of this question can be found in Lawler et al. [151] and in *The Traveling Salesman Problem and its Variations* edited by Gutin and Punnen [106]. As we mentioned in sections 6.2 and 6.4, there is no known general algorithm for solving the traveling salesman problem, and mathematicians, computer scientists, and many others continue to actively search for such a solution.

The most prominent and widely acclaimed mathematician associated with the study of graph theory is Leonhard Euler. Dunham's *Euler: The Master of Us All* [63] details the life and the significant contributions of Euler to the ongoing study of many areas of mathematics. Any general survey of mathematical history, such as Boyer and Merzbach [28], will discuss Euler's work. Anthologies of biographies of mathematicians with dedicated, extensive essays on Euler include *Remarkable Mathematicians: From Euler to von Neumann* by James [128] and *Men of Mathematics* by Bell [15]. Recently, two of Euler's most important mathematical works were translated by Blanton: *Introduction to Analysis of the Infinite* [74] and *Foundations of Differential Calculus* [75].

Finally, we mention Hankin's engaging biography [111] of Sir William Rowan Hamilton, blending the story of his personal life with a discussion of his professional accomplishments. In addition to his contributions to graph theory, Hamilton is best known for his study and development of the quaternions—a noncommutative extension of the complex numbers. Both Kuipers [145] and Smith and Conway [218] are good introductions to significant aspects of this interesting and important number system.

7 Complex Analysis

This chapter introduces the elegant and useful mathematics of complex-valued functions. The underlying characterization of a single-valued complex function is the same as for a real-valued function—every (complex number) input is mapped to a unique (complex number) output. In the same way as for real functions, the set of all possible inputs is called the domain of the function and the set of all possible outputs is the range. As we defined in chapter 3, the field of complex numbers contains the reals as a proper subset; we will see that $\mathbb{C}$ is two-dimensional, whereas the set of reals is one-dimensional. Because of this increase in dimensionality, the resulting functional behavior is much more intricate for complex functions than for real-valued ones. As a result, the study of complex functions is a rich, interesting field of endeavor, containing many beautiful and sometimes surprising mathematical phenomena.

Mathematicians have historically long been conscious of the algebraic issues involved with taking square roots of negative numbers. In ancient recorded human history, the Babylonians understood the quadratic formula from a procedural perspective; they recognized the need to take the square root of the discriminant $b^2 - 4ac$ when solving for x in the quadratic equation $ax^2 + bx + c = 0$. Chapter 3 has already discussed the fact that the discriminant is often negative, such as when solving $x^2 + 2x + 3 = 0$. In this case, $b^2 - 4ac = -8$.

Until relatively recent times, mathematicians concluded that the appearance of square roots of negative numbers implied the nonexistence of solutions; they thereby would have asserted that $x^2 + 2x + 3$ has no roots. The belief that square roots of negative numbers formed an impasse to finding roots continued until the Age of Enlightenment. Even the many insightful Italian mathematicians of the Renaissance who advanced techniques to find zeros of polynomials perpetuated this perspective. Recall from section 3.5 that Sciopione del Ferro (who lived until 1526) was able to factor the cubic equation $x^3 + mx + n = 0$ into a linear term and a quadratic. But as it turns out, when m is positive and n is negative, the resulting quadratic factor of del Ferro's cubic always has a negative discriminant. What was del Ferro's conclusion in this case? It was the same as any other mathematician's of his day: the corresponding cubic had only one root!

It was not until the late 1700s that the idea of a complex number $a + bi$ began to gain widespread understanding and acceptance by mathematicians. The Swiss mathematician Leonhard Euler made early discoveries about the algebraic properties of complex numbers. Euler, who developed the notion of the natural logarithm and described the importance of the irrational number e (whose notational symbol 'e' he first introduced) discovered such important formulas as

$$e^{it} = \cos t + i \sin t.$$

Here the symbol i stands for $\sqrt{-1}$ and the angle t is a given real value. Euler demonstrated this formula early—in the mid-1700s. We will use this important formula over and over again in this chapter, and we will see that Euler's insights resulted from what he considered a natural extension of facts about power series of real valued functions. Beginning in the 1770s, Euler was the first mathematician to use the symbol i to denote $\sqrt{-1}$, and he went on to derive many curious and often surprising algebraic truths about complex numbers. For example, we will see that i^i is real-valued (interestingly enough, i^i takes on an infinite number of values, but all of them are real!).

The study of complex numbers grew out of mathematicians' efforts to explain nonreal algebraic solutions of equations. But there are many other aspects to complex analysis. The subject is extremely useful in many mathematical fields such as complex number theory, ordinary and partial differential equations, physical chemistry, homotopy theory, mathematical physics, and operator theory. Because of its broad applicability, complex analysis is often recognized as a field that opens doors to many advanced studies of mathematics. This chapter develops a theory of complex functions, describing many ideas and applications that result from working with the derivative of complex-valued functions. As you might expect, there is a parallel theory of the integral as it applies to complex functions, one that can be traced back to Augustin-Louis Cauchy in the late 1700s. That material is left for your later studies in mathematics.

We will find that a study of just the differential properties of complex functions provides many interesting facts and insights. After developing several basic algebraic and geometric properties of complex numbers and functions, the chapter will examine what it means for a complex function to be differentiable. This investigation will include a description of the "partial derivatives" of associated functions, which will lead to a study of the "Cauchy–Riemann" equations that characterize differentiability. Section 7.3 then develops power series representations of differentiable functions. Later sections study a type of function known as "harmonic," which can be used to model many real-life phenomena in the study of fluid flow and other physical processes.

7.1   Complex Numbers and Complex Functions

By the 1730s a number of mathematicians (most notably Euler) were working with complex numbers and had identified various algebraic facts for this number system. But it was not until a geometric understanding of these numbers appeared in the late 1700s and early 1800s that complex analysis really became a fruitful mathematical field.

These early geometric insights involved representing complex numbers as points and are now attributed to the Norwegian mathematician Caspar Wessel. In 1797 Wessel published these notions shortly after a presentation to the Royal Danish Academy of Sciences. Unfortunately, Wessel was an impoverished surveyor and mapmaker and only an amateur mathematician. His lowly professional position explains in part why his geometric insight was known only to a few. Like a spark that may flicker but fails to light a fire, important geometric properties associated with complex numbers did not catch on with most mathematicians around the world.

In 1806, the French mathematician Jean-Robert Argand independently developed results equivalent to Wessel's work. Argand published his ideas at his own expense in a small book that did not even list his name as the author. The treatise was passed around the mathematical community sparingly and read by a handful of mathematicians, including Adrien-Marie Legendre. But little came of it, and the flame of complex analysis continued to sputter unsteadily.

Around this same time, Carl Friedrich Gauss began making important contributions to mathematics that incorporated the field of complex numbers in an essential way. As part of his dissertation (written in 1799 when he was 22), Gauss formulated and proved the famous fundamental theorem of algebra that was presented in section 3.5, describing the algebraic factorization of an arbitrary polynomial into linear and quadratic factors. The zero of any linear factor $ax + b$ was understood to be $x = -b/a$, but (before Gauss' work) the zeros of a quadratic $ax^2 + bx + c$ were not always considered to be meaningful numbers; a negative discriminant $b^2 - 4ac$ results in the quadratic formula having a square root of a negative number. Such terms were said to be "imaginary," reflecting the distrust mathematicians had for the existence of such numbers.

Gauss's statement and proof of the fundamental theorem of algebra showed the world why complex numbers were important. Because the complex field $\mathbb{C}$ is crucial to factoring all polynomials into linear terms, complex numbers were finally understood to be of essential value to understanding the basic underlying structure of polynomials. By 1831 Gauss had also independently reproduced Wessel's and Argand's geometric depiction of complex numbers as points in the plane, and he had used these results to prove significant mathematical theorems in complex analysis. Faced with the fundamental theorem of algebra, mathematicians everywhere were persuaded that complex analysis deserved their engaged attention. The sputtering flame finally caught fire.

How did the fundamental theorem work in producing complex numbers as important quantities when studying polynomials? We can answer this question by looking at an example. We know that the quadratic $x^2 + 2x + 5$ has zeros (according to the quadratic formula) equal to $x = -1 + \sqrt{-4}$ and $-1 - \sqrt{-4}$, which do not make sense as real numbers. Gauss declared these values to exist in a very important way, as they provided the complete set of zeros for the polynomial (the fundamental theorem says there are two, since the polynomial has degree two). Gauss used the jargon already established by that time: he called the values complex and used Euler's symbol $i = \sqrt{-1}$ to provide a coherent interpretation of their nature; the terms become $x = -1 + \sqrt{4}\sqrt{-1} = -1 + 2i$ and $x = -1 - 2i$, respectively, and the quadratic factors as $x^2 + 2x + 5 = (x - (-1 + 2i))(x - (-1 - 2i))$.

Question 7.1.1   Determine the zeros of each quadratic polynomial.

(a) $x^2 - 1$ (c) $x^2 + 4x + 4$

(b) $x^2 + 1$ (d) $x^2 - 4x + 5$

■

By expressing a complex number $x + iy$ in terms of its "real part" x and its "imaginary part" y, Gauss realized (as did Wessel and Argand) that complex numbers $x + iy$ need to be understood in terms of two components. Since these two pieces can assume any real value, the complex number $x + iy$ corresponds to a point (x, y) on a two-dimensional coordinate plane. Gauss called this plane the *complex plane* $\mathbb{C}$; it is sometimes now called the *Argand plane* after Jean-Robert Argand. The following definition expresses these ideas.

Definition 7.1.1   *The set of* **complex numbers** *is* $\mathbb{C} = \{a + bi : a, b \in \mathbb{R}, \text{ where } i = \sqrt{-1}\}$. *A complex number* $z = a + bi$ *has* **real part** $a = Re(z)$ *and* **imaginary part** $b = Im(z)$; *the number* $z = a + ib$ *is graphically represented as the point* (a, b) *on the two-dimensional plane having horizontal "real" axis labeled* $\mathbb{R}$ *and vertical "imaginary" axis labeled* i.

A graphical illustration of the complex plane is provided in figure 7.1. Note, for example, that the number i equals $0 + 1i$, and so it is identified on the complex plane as the point $(0, 1)$, which sits on the imaginary axis.

Example 7.1.1   We identify the real and imaginary parts of several complex numbers and then graph these numbers on the complex plane in figure 7.2.

(a) The complex number $z = -1 + 2i$ has $Re(z) = -1$ and $Im(z) = 2$.

(b) The number $z = -2i$ has $Re(z) = 0$ and $Im(z) = -2$.

(c) The number $z = 4$ has $Re(z) = 4$ and $Im(z) = 0$.

■

Figure 7.1 A point $a + bi$ on the complex plane

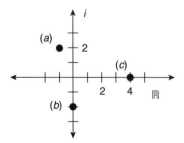

Figure 7.2 Graph of complex numbers for example 7.1.1

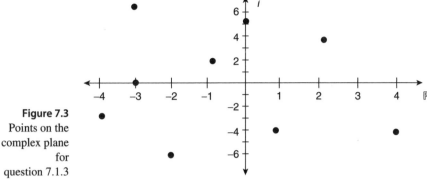

Figure 7.3
Points on the
complex plane
for
question 7.1.3

Question 7.1.2 Identify the real and imaginary parts of each complex number and then graph each one as a point on the complex plane.

(a) $z = 2 + i$

(b) $z = -5 + 3i$

(c) $z = 2i$

(d) $z = 4 - 7i$

■

Question 7.1.3 State the complex number that corresponds with each point on the complex plane identified in figure 7.3.

■

Geometric representations of complex numbers provide important insights into their algebraic representations and properties. In addition to the rectangular coordinate representation described in definition 7.1.1, each complex number also has a polar coordinate representation, one that turns out to be extremely important in many algebraic calculations. Rather than describing the two dimensions in terms of real and imaginary parts, a polar representation thinks of a complex number in terms of: (i) the distance from the point to the origin; and (ii) the angle that is formed between the positive real axis $\mathbb{R}$ and the ray emanating from the origin through the point. The following definition establishes a complex number's polar representation.

Definition 7.1.2 *A complex number z is represented by its two **polar coordinates**: the **modulus** $|z|$ of z equal to the distance from z to the origin; and the **polar angle** θ equal to the angle formed between the positive real axis $\mathbb{R}$ and the ray emanating from the origin through the point.*

Figure 7.4 provides a graphical illustration of a number's polar representation; notice that every point on the dashed circle has the same modulus.

The polar representation of a value z is not unique. If z has polar angle θ, then any 2π multiple can be added to θ to obtain the same polar angle; $\theta + 2n\pi$ for $n \in \mathbb{Z}$ are also polar angles for z. A unique polar representation can be assigned to a complex number z by requiring, for example, that $-\pi < \theta \leq \pi$, or in some other such predetermined 2π range of values for the polar angle. This unique polar representation is often useful when making algebraic calculations or when representing a set of values such as the

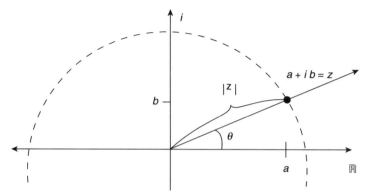

Figure 7.4   The polar representation of a complex number z

range outputs from a given function. In these settings, such a chosen range of values is often called the *principal range*, and the unique representation value for the polar angle is called the *principal value*.

It is a straightforward process to move from the rectangular coordinate representation of a complex number z to the polar representation. The next theorem lists the relationships between the variables involved. The proof uses the fact, first defined by Euler and verified by him as reasonable (in terms of properties of an exponential), that $e^{i\theta} = \cos\theta + i\sin\theta$, where $\theta \in \mathbb{R}$. We'll discuss this fact further in section 7.3.

Theorem 7.1.1   *For a complex number $z = a + bi$ having modulus $|z|$ and polar angle θ, the following identities hold:*

$$a = |z|\cos\theta, \quad b = |z|\sin\theta, \quad |z| = \sqrt{a^2 + b^2}, \text{ and } z = |z|e^{i\theta}.$$

When $-\pi/2 < \theta < \pi/2$, the polar angle is $\theta = \tan^{-1}\left(\dfrac{b}{a}\right)$; otherwise, $\tan^{-1}\left(\dfrac{b}{a}\right)$ will serve to indicate a reference angle for θ.

Proof   The right triangle with hypotenuse from (a, b) to the origin and base on the real axis has a base with length $|a|$, height with length $|b|$, and hypotenuse with length $|z|$. The first two identities follow from basic definitions of the cosine and sine ratios. Since $\tan\theta = (b/a)$, it follows that $\theta = \tan^{-1}(b/a)$ when $-\pi/2 < \theta < \pi/2$. The Pythagorean theorem implies $a^2 + b^2 = z^2$, and so $|z| = \sqrt{a^2 + b^2}$. Since $e^{i\theta} = \cos\theta + i\sin\theta$,

$$z = a + bi = |z|\cos\theta + i|z|\sin\theta = |z|(\cos\theta + i\sin\theta) = |z|e^{i\theta}.$$

∎

In some settings the rectangular representation of a complex number is more helpful, while in other settings the polar representation is better used. It is helpful to become adept at working with both representations and to develop an ability to move freely back and forth between them. The next example considers the polar representation of several complex numbers.

Example 7.1.2   We consider the complex number $z = 1 + i$ using a principal range of $-\pi < \theta \leq \pi$. Both the real and imaginary part of z are $a = b = 1$, and its modulus is $|z| = \sqrt{1^2 + 1^2} = \sqrt{2}$. The graph of the point z on the complex plane shows that its principal value polar angle is $\theta = \pi/4$. Allowing for multiple-value polar angles, $\theta = \tan^{-1} 1 + 2n\pi = (\pi/4) + 2n\pi$, for $n \in \mathbb{Z}$. The point $1 + i$ therefore has polar representations $\sqrt{2}e^{i(9\pi/4)} = \sqrt{2}e^{-i(7\pi/4)}$, and so on. Using the principal range $-\pi < \theta \le \pi$, the polar representation is $1 + i = \sqrt{2}e^{i(\pi/4)}$.

■

Example 7.1.3   We consider the complex number $z = \sqrt{6} + \sqrt{2}i$ using a principal range of $-\pi < \theta \le \pi$. The modulus is $|z| = \sqrt{6 + 2} = 2\sqrt{2}$ and the (multi-valued) polar angle is $\theta = \tan^{-1}(\sqrt{2}/\sqrt{6}) = \arctan(1/\sqrt{3}) = (\pi/6) + 2n\pi$, for $n \in \mathbb{Z}$. Using the principal range $-\pi < \theta \le \pi$, the polar representation is found using the principal value $\theta = \pi/6$ as $\sqrt{6} + \sqrt{2}i = 2\sqrt{2}e^{i(\pi/6)}$.

■

Question 7.1.4   Identify the modulus and the (multi-valued) polar angle of each complex number and then graph each one as a point on the complex plane. Then use the principal range of $-\pi < \theta \le \pi$ to identify the corresponding single-valued polar representation.

(a) $z = 2 + 2i$ (d) $z = -4$

(b) $z = 2\sqrt{3} + 2i$ (e) $z = 3e^{i(\pi/6)}$

(c) $z = 4i$ (f) $z = 2e^{i(11\pi/6)}$

■

7.1.1   The Arithmetic of Complex Numbers

The field of complex numbers has well-defined addition and multiplication operations, along with their associated inverse operations of subtraction and division. We describe these operations in terms of the rectangular representation of complex numbers and then in terms of the polar representation.

Definition 7.1.3   *Suppose $z = a + bi$ and $w = c + di$ are complex numbers with $a, b, c, d \in \mathbb{R}$. Algebraic operations are:*

- **addition:** $z + w = (a + bi) + (c + di) = (a + c) + (b + d)i;$
- **subtraction:** $z - w = (a + bi) - (c + di) = (a - c) + (b - d)i;$
- **multiplication:** $z \cdot w = (a + bi) \cdot (c + di) = (ac - bd) + (ad + bc)i;$
- **complex conjugate:** $\bar{z} = a - bi;$
- **division:** $\dfrac{z}{w} = \dfrac{z \cdot \bar{w}}{w \cdot \bar{w}} = \dfrac{(a + bi) \cdot (c - di)}{(c + di) \cdot (c - di)} = \dfrac{ac + bd}{c^2 + d^2} + \dfrac{bc - ad}{c^2 + d^2} i.$

In a descriptive sense, addition and subtraction are defined componentwise in terms of real and imaginary parts. Multiplication uses the familiar F.O.I.L. method of multiplying First, Outer, Inner, and Last terms and the fact that $i^2 = \sqrt{-1}\sqrt{-1} = -1$. Division is defined by multiplying both the numerator and denominator by the complex conjugate of the denominator, and then simplifying the resulting expression. The next example shows how each operation works.

Example 7.1.4   We apply each arithmetic operation to the complex numbers $z = -1 + 2i$ and $w = -1 - 2i$.

- $z + w = (-1 + 2i) + (-1 - 2i) = (-1 + (-1)) + (2 + (-2))i = -2 + 0i = -2$
- $z - w = (-1 + 2i) - (-1 - 2i) = (-1 - (-1)) + (2 - (-2))i = 4i$
- $z \cdot w = (-1 + 2i) \cdot (-1 - 2i) = (-1)(-1) + (-1)(-2i) + (2i)(-1) + (2i)(-2i) = 1 + 2i - 2i + 4 = 5$
- $\overline{w} = -1 - (-2i) = -1 + 2i$ and $\overline{z} = -1 - (2i) = -1 - 2i$
- $\dfrac{z}{w} = \dfrac{z \cdot \overline{w}}{w \cdot \overline{w}} = \dfrac{(-1 + 2i)(-1 + 2i)}{(-1 - 2i)(-1 + 2i)} = \dfrac{1 - 2i - 2i - 4}{5} = -\dfrac{3}{5} - \dfrac{4}{5}i$

∎

The result of arithmetic operations applied to complex numbers in rectangular form should always generate an answer that is also expressed in rectangular form as $a + bi$ with $a, b \in \mathbb{R}$. The next question practices working with the operations.

Question 7.1.5   For $t = 4 + 7i$, $u = \sqrt{2} - 0.5i$, $v = 10 - 5i$, and $w = -\pi + ln(2)i$, express each of the following terms in rectangular form.

(a) $z = t + v$ (g) $z = t \cdot v$

(b) $z = v - t$ (h) $z = t^2 + u \cdot v$

(c) $z = u + w$ (i) $z = t/v$

(d) $z = t - u$ (j) $z = u/w$

(e) $z = \overline{v}$ (k) $z = |t|$

(f) $z = \overline{v} + v$ (l) $z = |u|$

∎

The polar representation is especially useful when multiplying and dividing complex numbers, as well as for the more advanced operations of taking integer powers or roots. Consider the following definition.

Definition 7.1.4   *Suppose $z = re^{it}$ and $w = pe^{is}$ are complex numbers written in polar format with $r, t, p, s \in \mathbb{R}$. If $n \in \mathbb{Z}$, then:*

- **multiplication:** $z \cdot w = re^{it} \cdot pe^{is} = (r \cdot p)e^{i(s+t)}$;
- **division:** $\dfrac{z}{w} = \dfrac{re^{it}}{pe^{is}} = \dfrac{r}{p}e^{i(t-s)}$;
- **integer powers:** $z^n = (re^{it})^n = r^n e^{int}$.

It can be helpful to visualize these operations in terms of actions on polar angles. From the definition, we see that multiplication adds the two polar angles to get the new polar angle, division subtracts one from the other, and taking an integer power corresponds to multiplying the polar angle by that power. The next example shows how each operation works.

Example 7.1.5   The complex numbers $z = 4 + 4i$ and $w = -4 - 4i$ have polar representation using the principal range $-\pi < \theta \leq \pi$ as $z = 4\sqrt{2}e^{i(\pi/4)}$ and $w = 4\sqrt{2}e^{i(-3\pi/4)}$. We apply each of the operations from definition 7.1.4 to z and w. The final answer is given in terms of the rectangular coordinate system using the fact that $e^{it} = \cos t + i \sin t$.

- $z \cdot w = 4\sqrt{2}e^{i(\pi/4)} \cdot 4\sqrt{2}e^{i(-3\pi/4)} = (4\sqrt{2} \cdot 4\sqrt{2})e^{i(\pi/4 - 3\pi/4)} = 32e^{i(-\pi/2)} = -32i$

- $\dfrac{z}{w} = \dfrac{4\sqrt{2}e^{i(\pi/4)}}{4\sqrt{2}e^{i(-3\pi/4)}} = \dfrac{4\sqrt{2}}{4\sqrt{2}}e^{i[\pi/4 - (-3\pi/4)]} = e^{i\pi} = -1$

- $z^5 = (4\sqrt{2}e^{i(\pi/4)})^5 = (4\sqrt{2})^5 e^{i(5\pi/4)} = 4096\sqrt{2}e^{i(-3\pi/4)}$
  $= 4096\sqrt{2}(\cos(-3\pi/4) + i \sin(-3\pi/4)) = -4096 - 4096i$

■

The calculation of z^5 in terms of its polar angle in example 7.1.5 was finalized only when it was given in terms of the principal value for the polar angle. Whenever a calculation is made and a principal value is mandated, such an adjustment is necessary.

Question 7.1.6 Use the polar representation with principal range $-\pi < \theta \le \pi$ to find the value of each complex number. Then write your final answer in rectangular coordinates.

(a) $z = (3 + 3i) \cdot (2 - 2i)$

(b) $z = (1 + i) \cdot (-3i)$

(c) $z = \dfrac{3 + 3i}{2 - 2i}$

(d) $z = \dfrac{1 + i}{-3i}$

(e) $z = (3 + 3i)^8$

(f) $z = (1 + i)^{-3}$

■

The polar representation of complex numbers is especially useful when computing roots (fractional powers) of complex numbers. Because a complex number can have multiple roots (in the same way that a positive real value x has two square roots $\pm\sqrt{x}$), the computations for fractional powers require greater care. Though there are some real numbers, such as $x = -1$, that do not have real square roots, the square root of any arbitrary complex number can be found using a polar representation and has two values. Similarly, the cube root of any complex number has three values, the fourth root has four values, and so on. The next example presents a standard algebraic method for determining a cube root; the approach for an arbitrary nth root is similar.

Example 7.1.6 We find the cube root of $1 + i$ using the polar representation with principal range $-\pi < \theta \le \pi$. We first consider the polar representation of $1 + i$ and allow for polar angles outside the principal range: $1 + i = \sqrt{2}e^{i(\pi/4)} = \sqrt{2}e^{i(\pi/4 + 2k\pi)}$ for any integer $k \in \mathbb{Z}$. Expressing the cube root in terms of the 1/3 power, $(1 + i)^{1/3} = (\sqrt{2})^{1/3}e^{(i/3)(\pi/4 + 2k\pi)} = \sqrt[6]{2}e^{i(\pi/12 + 2k\pi/3)}$. Three of these polar angles are in the principal range: $\pi/12$; $\pi/12 + 2\pi/3 = 3\pi/4$; and $\pi/12 - 2\pi/3 = -7\pi/12$. Pictured geometrically on the complex plane, the three cube roots are spaced evenly around the circle with modulus $\sqrt[6]{2}$ as illustrated in figure 7.5.

■

The algebraic process demonstrated in example 7.1.6 can be implemented to obtain the values for the nth root $\sqrt[n]{z}$ of any complex number z. This process proves there exist n distinct nth roots of any given nonzero complex value z, each of these roots have modulus $\sqrt[n]{|z|}$, and they are equally spaced at $2\pi/n$ intervals around the circle having radius $\sqrt[n]{|z|}$. We summarize these results in the following theorem.

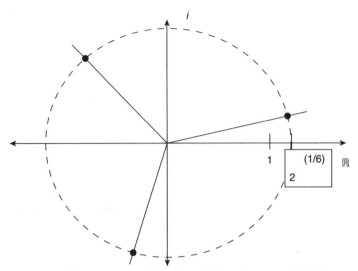

Figure 7.5   The cube roots of $(1 + i)$ as calculated in example 7.1.6

Theorem 7.1.2 *There are n distinct complex values for the nth root of a nonzero complex number z. These values are equally spaced at $2\pi/n$ intervals around the circle of points in the complex plane with modulus $\sqrt[n]{|z|}$.*

Applying theorem 7.1.2 to $z = 1$, there must exist n distinct *nth roots of unity*; that is, the equation $z^n = 1$ has n distinct solutions. The nth roots of unity play an important role in the study of solutions of complex equations and the study of complex functions. This section's exercises provide an opportunity to study these numbers in greater detail.

In addition, theorem 7.1.2 indicates how to calculate a complex number raised to any fractional exponent. To calculate $z^{m/n}$, where $m, n \in \mathbb{Z}$ and $n \neq 0$, express m/n as $m(1/n)$, find z^m, and then take this value's nth root.

Question 7.1.7   Determine the n distinct complex nth roots as indicated below, using the principal range $-\pi < \theta \leq \pi$. Graph these roots on the complex plane.

(a) $\sqrt{5e^{i(\pi/7)}}$

(b) $\sqrt{10 + 10i}$

(c) $\sqrt[3]{-5\sqrt{3} + 5i}$

(d) $\sqrt[4]{-5\sqrt{3} + 5i}$

(e) $\sqrt[6]{2 - 2i}$

(f) $\sqrt[5]{-4 - 4\sqrt{3}i}$

■

7.1.2   Complex Functions

We now turn our attention to the study of variable expressions in $z \in \mathbb{C}$, focusing on the expressions that can be interpreted as functions. Not surprisingly, a complex function is a type of relation between elements in a domain set D of complex numbers and those in a range set R of complex numbers. But in complex analysis, mathematicians often extend the study functions to those that produce multiple outputs for a given input (rather than just those that produce a unique output). At first this choice may seem odd, as functions have been defined previously so that every input value maps to a unique output. But we have already seen, in the calculation of the nth root of a complex number, a situation where a study of multiple outputs is valuable. In the same way, the study of "multiple-valued functions" will be of great benefit.

Our study of functions begins with complex polynomials, whose definition is identical to that of real polynomials, but whose coefficients and input variables are allowed to be complex-valued.

Definition 7.1.5 *A **polynomial over** $\mathbb{C}$ with degree $n \in \mathbb{Z}$ and coefficients $a_0, \ldots, a_n \in \mathbb{C}$ is an expression of the form $p(z) = a_n z^n + a_{n-1} z^{n-1} + \cdots + a_1 z + a_0$, where $z \in \mathbb{C}$.*

Any real polynomial $p(x)$ generalizes to a complex polynomial $p(z)$ by appropriately substituting the complex variable. For example, $p(x) = x^2$ yields the complex quadratic $p(z) = z^2$. At the same time, many polynomials are unique to the complex numbers; for example, $q(z) = z^2 + i$ is not a real polynomial. As with complex numbers, complex polynomials are often presented in "rectangular" form $p(x + iy) = u(x, y) + iv(x, y)$ with real part $u(x, y) = Re[p(x + iy)]$ and imaginary part $v(x, y) = Im[p(x + iy)]$. For example, the real and imaginary parts of $p(z) = z^2$ are calculated as $p(z) = z^2 = (x + iy)^2 = (x^2 - y^2) + i(2xy)$, and so $u(x, y) = x^2 - y^2$ and $v(x, y) = 2xy$. The real and imaginary parts are seen to be real valued functions of the two input variables x and y, where $z = x + iy$.

For any polynomial $p(z)$, each input z is related (or mapped) to a unique output element $p(z)$. In this way polynomials act as a function with a single output; mathematicians call such relationships "single-valued functions." But complex analysts also consider relationships that have many output values for each single input z, interpreting these relationships as "multiple-valued functions." The following definition makes these notions precise.

Definition 7.1.6 *If to each value z in a **domain** set D there corresponds, via a relation, one or more values w in a **range** set R, then we call the relation a **complex function from D to R** and describe the function using such notation as $f(z) = w$. When only one value w is in relation with each value $z \in D$, then $f(z) = w$ is a **single-valued function** of z. When more than one value w is in relation with each value $z \in D$, then $f(z) = w$ is a **multiple-valued function** of z.*

We have already studied many relations among complex numbers that are functions. The next example highlights several.

Example 7.1.7 Every complex polynomial is a single-valued function, since applying the algebraic operations of taking nonnegative integer powers, multiplying by a complex coefficient, and adding or subtracting complex numbers always produces a single outcome.

In contrast, $f(z) = \sqrt{z}$ is a multiple-valued function with two output values for each nonzero input $z \in \mathbb{C}$. For this reason it is sometimes called a two-valued function. For any nonzero $z = re^{i\theta}$ represented using the principal range of $-\pi < \theta < \pi$, the two square roots of z are $w_1(z) = \sqrt{r}e^{i\theta/2}$ and $w_2(z) = \sqrt{r}e^{i(\theta/2+\pi)}$.

Therefore $f(z) = \sqrt{z}$ is a multiple-valued function. Notice that $f(0) = \sqrt{0} = 0$ is unique; definition 7.1.6 only requires *some* inputs z to have more than one output $\sqrt{z}$ for the function to be multiple-valued.

■

Question 7.1.8 Identify each complex function as single-valued or multiple-valued; explain your answer.

(a) $f(z) = 5z^3 - 9z + 102$

(b) $g(z) = \sqrt[3]{4z + 15}$

(c) $h(z) = 15z + 4\bar{z}$

(d) $j(z) = \sqrt[4]{5z^2 - (2 + 15i)}$

■

The polar angle for a given complex number generates both a single-valued function and a multiple-valued function. We know that a complex number z can be represented as $re^{i\theta}$ where $r = |z|$ and θ is the polar angle. For any given polar angle θ, the values $\theta \pm 2n\pi$ for $n \in \mathbb{Z}$ serve to define the same polar angle. Defining the function $f(z) = \arg(z)$, where $\arg(z)$ is any angle θ satisfying $z = re^{i\theta}$, we see that $\arg(z)$ is multiple-valued. In contrast, we can also consider the polar angle in terms of the principal range $-\pi < \theta \le \pi$. Defining the function $f(z) = \mathrm{Arg}(z)$, where $\mathrm{Arg}(z)$ is the (unique) polar angle θ for z in the principal range $(-\pi, \pi]$, we see that $\mathrm{Arg}(z)$ is single-valued with domain $D = \mathbb{C}$. (Note: for $z = 0$ we simply choose $\theta = 0$.) The next definition formalizes the meaning of these functions.

Definition 7.1.7 *If $z \in \mathbb{C}$ is a complex number, then*

- $\arg(z)$ *is any polar angle θ that satisfies $z = re^{i\theta}$, and*
- $\mathrm{Arg}(z) = \theta$ *is the unique angle θ that satisfies $z = re^{i\theta}$ with*
  $-\pi < \theta = \mathrm{Arg}(z) \le \pi.$

Definition 7.1.7 purposefully indicates the notational difference that distinguishes the two functions: the multiple-valued function $\arg(z)$ has a lower-case lettering while the single-valued function $\mathrm{Arg}(z)$ has a capital letter; this notation has become standard in complex analysis. Both functions are referred to as the *argument function*, and the polar angle is often called the *argument* of the complex number. Definition 7.1.7 shows that the functions are related to each other according to $\arg(z) = \mathrm{Arg}(z) + 2k\pi$, where $k \in \mathbb{Z}$; mathematicians sometimes call $\mathrm{Arg}(z)$ a *branch* of the function $\arg(z)$ because it generates a single choice for an output value from the collection of outputs for $\arg(z)$. The principal range $(-\pi, \pi]$ is considered so standard that the function $\mathrm{Arg}(z)$ is sometimes called the *principal branch* of the argument function. The argument functions will be very useful in the definition of other important functions, such as the logarithm function described in the next section.

Question 7.1.9 For the following complex numbers from question 7.1.4, determine the value of $\arg(z)$ and $\mathrm{Arg}(z)$ for each complex z.

(a) $z = 2 + 2i$

(b) $z = 2\sqrt{3} + 2i$

(c) $z = 4i$

(d) $z = -4$

(e) $z = 3e^{i(\pi/6)}$

(f) $z = 2e^{i(11\pi/6)}$

■

7.1.3 Partial Derivatives

The rest of this section deals with "multivariate real-valued functions." We have seen that single-valued complex functions are often separated into real and imaginary parts as $f(x + iy) = u(x, y) + iv(x, y)$, and so $u(x, y)$ and $v(x, y)$ are real-valued functions defined on two real independent variables x and y. Mathematicians refer to functions such as $u(x, y)$ and $v(x, y)$ as *multivariate* functions; these functions are studied in an undergraduate multivariate calculus course. They arise in the study of complex functions because they appear as a function's real and imaginary parts.

In an analysis of real functions in the single variable setting, the derivative (when it exists) generates the slope of the line tangent to a given function at a point; the slope is the instantaneous rate of change in the dependent variable with respect to the independent variable. In the multivariable setting, the rate of change for the given function can be calculated with respect to each of the independent variables—these rates of change are called *partial derivatives*. They turn out to be very helpful in analyzing the properties for the real and imaginary parts of many complex functions.

A real-valued function $u(x, y)$ can have two partial derivatives—one with respect to x and the other with respect to y. These partial derivatives are described in terms of difference quotients, as the next definition points out.

Definition 7.1.8 *The **partial derivatives** of $u(x, y)$ with respect to the variable x and with respect to the variable y are*

$$u_x = \frac{\partial u}{\partial x} = \lim_{\Delta x \to 0} \frac{u(x + \Delta x, y) - u(x, y)}{\Delta x} \quad and$$

$$u_y = \frac{\partial u}{\partial y} = \lim_{\Delta y \to 0} \frac{u(x, y + \Delta y) - u(x, y)}{\Delta y}.$$

*When the first limit exists, we say that $u(x, y)$ is **differentiable with respect to** x and, when the second limit exists, we say that $u(x, y)$ is **differentiable with respect to** y.*

Using the definition to calculate a partial derivative can sometimes be quite involved. But other partial derivatives have very straightforward calculations, as in the following example.

Example 7.1.8 We use the definition of the partial derivative to compute u_x and u_y for the function $u(x, y) = x \cdot y$.

$$u_x = \frac{\partial u(x, y)}{\partial x} = \frac{\partial [xy]}{\partial x} = \lim_{\Delta x \to 0} \frac{(x + \Delta x)y - xy}{\Delta x} = \lim_{\Delta x \to 0} \frac{\Delta x \cdot y}{\Delta x} = \lim_{\Delta x \to 0} y = y$$

$$u_y = \frac{\partial u(x, y)}{\partial y} = \frac{\partial [xy]}{\partial y} = \lim_{\Delta y \to 0} \frac{x(y + \Delta y) - xy}{\Delta y} = \lim_{\Delta y \to 0} \frac{x \cdot \Delta y}{\Delta y} = \lim_{\Delta y \to 0} x = x$$

Both of these limits exist everywhere, and so at any point (x, y), $u(x, y) = x \cdot y$ is differentiable with respect to both x and y.

■

Question 7.1.10   Using the definition of a partial derivative, find both partial derivatives of each function.

(a) $u(x, y) = 2x - y$

(b) $v(x, y) = x + 2y$

(c) $f(x, y) = x^2 y$

(d) $g(x, y) = \dfrac{x}{y}$

■

From a practical standpoint, we will not want to use the definition of partial derivatives to compute the derivative too often. Instead, the familiar rules for single-variable differentiation extend to the calculation of a partial derivative! When differentiating $u(x, y)$ with respect to x, we may simply employ standard differentiation rules for x while treating the variable y as a "constant." The notion of treating y as a constant makes sense, since y remains unchanged when x changes. Similarly, when differentiating $u(x, y)$ with respect to y, simply use standard differentiation rules for y while treating x as a "constant." The next example illustrates these calculations.

Example 7.1.9   We use differentiation rules to compute u_x and u_y for the function $u(x, y) = y^3 x^5$.

$$u_x \;=\; \frac{\partial}{\partial x}[y^3 x^5] \;=\; y^3 \frac{\partial}{\partial x}[x^5] \;=\; y^3 5x^4 = 5y^3 x^4,$$

$$u_y \;=\; \frac{\partial}{\partial y}[y^3 x^5] \;=\; x^5 \frac{\partial}{\partial x}[y^3] \;=\; x^5 3y^2 = 3x^5 y^2.$$

Both of these computations use first a scalar multiple rule to factor out the "constant" variable, and then the standard power rule for differentiation.

■

Example 7.1.10   We use differentiation rules to compute the partial derivatives v_x and v_y for the function $v(x, y) = (y^3 + \cos x)(x^5 + 2y)$.

The function $v(x, y)$ is a product of two functions $f(x, y)$ and $g(x, y)$. Using the product rule, we have $v_x = \partial/\partial x(f \cdot g) = f_x \cdot g + f \cdot g_x$, and similarly for v_y. First compute the partial derivatives of each component of the product.

$f_x = (\partial/\partial x)[y^3 + \cos x] = 0 - \sin x = -\sin x$ $g_x = (\partial/\partial x)[x^5 + 2y] = 5x^4 + 0 = 5x^4$

$f_y = (\partial/\partial y)[y^3 + \cos x] = 3y^2 - 0 = 3y^2$ $g_y = (\partial/\partial y)[x^5 + 2y] = 0 + 2 = 2$

Now use the product rule to compute v_x and v_y.

$$v_x \;=\; \frac{\partial}{\partial x}[(y^3 + \cos x)(x^5 + 2y)] \;=\; (-\sin x)(x^5 + 2y) + (y^3 + \cos x)(5x^4),$$

$$v_y \;=\; \frac{\partial}{\partial y}[(y^3 + \cos x)(x^5 + 2y)] \;=\; 3y^2(x^5 + 2y) + (y^3 + \cos x)2.$$

■

The examples show that standard differentiation rules enable straightforward calculations of the partial derivatives of many multivariate functions.

Question 7.1.11   Find both partial derivatives with respect to each variable x and y for the following functions.

(a) $f(x, y) = \sin(x^2 + y^2)$

(b) $g(x, y) = 10x^2 e^{2y}$

(c) $h(x, y) = (x + y) \sin(x^2 + y^2)$

(d) $j(x, y) = \dfrac{x^2}{y^3}$

(e) $u(x, y) = x^4 - 6x^2y^2 + y^4$

(f) $v(x, y) = 4x^3y - 4xy^3$

■

We have mentioned that real-valued multivariate functions arise in the study of single-valued complex functions $w = f(z) = f(x + iy)$ because the real and imaginary parts of the output are themselves functions of the form $u(x, y)$ and $v(x, y)$. For example, the quadratic $p(z) = z^2$ has $p(x + iy) = (x + iy)^2 = (x^2 - y^2) + i(2xy)$, and so $u(x, y) = Re[p(z)] = x^2 - y^2$ and $v(x, y) = Im[p(z)] = 2xy$. For many single-valued complex functions, including all polynomials, the real and imaginary parts might appear unrelated, but they happen to share a rather remarkable association that is expressed in terms of their partial derivatives. The next question begins an exploration of this relationship; section 7.2 will develop and explain it fully.

Question 7.1.12 Consider the partial derivatives of $p(z) = z^2 = (x + iy)^2 = (x^2 - y^2) + i(2xy)$.

(a) Evaluate the partial derivatives u_x and u_y for $u(x, y) = x^2 - y^2$.

(b) Evaluate the partial derivatives v_x and v_y for $v(x, y) = 2xy$.

(c) Based on a comparison of u_x and v_y for $p(z) = z^2$, formulate a conjecture about the relationship between u_x and v_y for an arbitrary polynomial $p(x + iy) = u(x, y) + iv(x, y)$.

(d) Based on a comparison of v_x and u_y for $p(z) = z^2$, formulate a conjecture about the relationship between v_x and u_y for an arbitrary polynomial $p(x + iy) = u(x, y) + iv(x, y)$.

■

As we have discussed, the partial derivative $\partial u/\partial x$ describes the rate of change in a given function $u(x, y)$ in the x-direction; similarly, $\partial u/\partial y$ describes the rate of change in $u(x, y)$ in the y-direction. These changes in $u(x, y)$ are well-defined (and often readily computed) when the corresponding partial derivatives exist. But when can we say that the rate of change in $u(x, y)$ is well-defined in *any* direction? Mathematicians say that a multivariate function is *differentiable* at a point when the rate of change exists in every direction; the concept of being differentiable is discussed in any multivariate calculus course. It turns out that a multivariate function is differentiable whenever each of its partial derivatives with respect to either independent variable are continuous. The concept of continuity for a multivariate function is detailed in the following definition.

Definition 7.1.9 *Let (x_0, y_0) be a point in a disk $S = \{z : |z - c| < R\}$ having fixed center $c \in \mathbb{C}$ and radius $R \in \mathbb{C}$. A function $u(x, y)$ whose domain contains S is* **continuous** *at (x_0, y_0) when $\lim\limits_{(x,y) \to (x_0,y_0)} u(x, y) = u(x_0, y_0)$. By this limit we mean: Given $\varepsilon > 0$, there exists a value $\delta > 0$ such that $|u(x, y) - u(x_0, y_0)| < \varepsilon$ whenever $\sqrt{(x - x_0)^2 + (y - y_0)^2} < \delta$.*

This definition of a continuous function of two variables may appear a bit complicated, as it involves algebra associated with both variables x and y. Intuitively, though, it is quite familiar. It says that the value of $u(x, y)$ is close to $u(x_0, y_0)$ whenever the point (x, y) is close to (x_0, y_0); the square root term in the definition is simply the two-dimensional distance formula applied to these last two points. Nonetheless, we will restrict the introductory remarks here to fairly simple functions, as the algebra quickly becomes complicated for more intricate situations.

Example 7.1.11   We consider the continuity of $u(x, y) = x - y^2$ at the point $(0, 0)$, showing that $\lim_{(x,y)\to(0,0)} u(x, y) = u(0, 0) = 0$. Given any $\varepsilon > 0$, choose $\delta = (-1 + \sqrt{1 + 4\varepsilon})/2$.

A straightforward calculation shows $\delta + \delta^2 = \varepsilon$. Whenever $\sqrt{(x - 0)^2 + (y - 0)^2} < \delta$, we have $|x| = \sqrt{x^2} \leq \sqrt{x^2 + y^2} < \delta$, and similarly $|y| < \delta$. Therefore by the triangle inequality,

$$|u(x, y) - u(0, 0)| = |(x - y^2) - 0| = |x - y^2| \leq |x| + |y|^2 < \delta + \delta^2 = \varepsilon.$$

The function $u(x, y)$ is therefore continuous at $(0, 0)$.

∎

The function $u(x, y) = x - y^2$ is an example of a multivariate polynomial, which is any function consisting of a sum of terms that are products of real-valued coefficients and powers of x and/or y. It turns out, not surprisingly, that all multivariate polynomials are continuous at any given point. Hence the function $u(x, y) = x - y^2$ is not only continuous at the origin, but also at any given point in the two-dimensional plane.

As in any instance of applying the definition of limit, the choice of δ may certainly not be apparent at the onset of calculations involved with the definition. The strategy is to find an upper bound for $|u(x, y) - u(x_0, y_0)|$ that involves only constants and δ, assuming $\sqrt{(x - x_0)^2 + (y - y_0)^2} < \delta$ (whatever δ may turn out to be). Then set that upper bound equal to ε and solve for δ. In the last example, when $\delta + \delta^2$ is set equal to ε, the quadratic $1\delta^2 + 1\delta - \varepsilon = 0$ has (using the quadratic formula with $a = 1$, $b = 1$, and $c = -\varepsilon$) the positive root $\delta = (-1 + \sqrt{1 + 4\varepsilon})/2$. A similar strategy may be applied to the functions in the next question.

Question 7.1.13   Verify that the real part $u(x, y) = x^2 - y^2$ and the imaginary part $v(x, y) = 2xy$ of the polynomial $f(z) = z^2 = u(x, y) + iv(x, y)$ are continuous at $(0, 0)$. Also verify that the partial derivatives $u_x(x, y) = 2y$ and $u_y(x, y) = 2x$ are continuous at any point (x_0, y_0).

∎

This brief study of partial derivatives and continuity for multivariable functions is a first step in the direction of developing the elegant and interesting theory of differentiability for complex functions. As we will see in the next section, for any complex function that is differentiable, the real and imaginary parts of the function will always be related to one another. These relationships are called the Cauchy–Riemann equations. For example, we will see that every complex polynomial has real and imaginary parts that follow the Cauchy–Riemann equations. These facts will lead us to interesting theory about real-valued multivariate functions, which will in turn lead to many important real-world applications and advanced mathematical analysis.

7.1.4 Reading Questions for Section 7.1

1. Define and give an example of the real part and the imaginary part of a complex number.
2. Define and give an example of the modulus and the polar angle of a complex number.
3. Define the principal range and the principal value of a complex number.
4. State theorem 7.1.1. What does this theorem accomplish?
5. Define and give an example of addition and subtraction for two complex numbers $z = a + ib$ and $w = c + id$ in rectangular form.
6. Define and give an example of multiplication and division for two complex numbers $z = a + ib$ and $w = c + id$ in rectangular form.
7. Define and give an example of multiplication and division for two complex numbers $z = re^{i\theta}$ and $w = se^{i\alpha}$ in polar form.
8. State theorem 7.1.2. How many cube roots does an arbitrary complex number have?
9. Discuss the distinction between single-valued and multiple-valued complex functions.
10. Define the two argument functions defined on the complex numbers and give an example.
11. Define the partial derivatives $u_x = \partial u/\partial x$ and $u_y = \partial u/\partial y$ for a multivariate real-valued function $u(x, y)$. Give an example using the rules for differentiation.
12. State the definition of a function $u(x, y)$ continuous at a point (x_0, y_0).

7.1.5 Exercises for Section 7.1

In exercises 1–6, graph each complex number $a + bi$ on the complex plane and determine its modulus and polar angle in radians.

1. $-3\sqrt{3} - 3i$
2. $\sqrt{2} - \sqrt{2}i$
3. $8 + 8i$
4. $-4 + 4\sqrt{3}i$
5. 2
6. $-8i$

In exercises 7–12, graph each complex number on the complex plane and express each in rectangular form $a + ib$.

7. $e^{i3\pi/2}$
8. $6e^{i\pi/4}$
9. $\sqrt{2}e^{i\pi/3}$
10. $5e^{i13\pi/6}$
11. $e^{i\pi/2}e^{i\pi/4}$
12. $7e^{i2\pi/3}e^{i9\pi/4}$

In exercises 13–24, evaluate each expression in rectangular form.

13. $(7 + 3i) - (2 - 4i)$
14. $(8 + 21i) + (14 - 3i) - (3 + 5i)$
15. $5i - (8 + 3i) + 16$
16. $(-2 + i) \cdot (-2 - i)$
17. $(4 + 2i) \cdot (8 + 3i)$
18. $(9 - 2i) \cdot (-2 + 3i)$
19. $\overline{8 - 3i}$
20. $\overline{-2 \cdot i}$
21. $\dfrac{4 + 2i}{8 - 3i}$
22. $\dfrac{9 - 2i}{-2 - 3i}$

23. $(1 + 3i) + \dfrac{3 - 2i}{8 + 3i}$

24. $(1 + 2i) \cdot \dfrac{8 + 2i}{2 - 3i}$

In exercises 25–30, evaluate each expression using a polar representation with principal range $-\pi < \theta \le \pi$.

25. $(4 + 4i)^4$

26. $(2 - 2\sqrt{3}i)^{10}$

27. $(-3 - 3i)^2$

28. $(-3 - 3i)^{-1}$

29. $(8i)^7$

30. $(-4i)^5$

In exercises 31–36, find the n distinct complex nth roots using the principal range $-\pi < \theta \le \pi$. Graph these roots on the complex plane.

31. $\sqrt{-3 + 3i}$

32. $\sqrt{9e^{i(\pi/12)}}$

33. $\sqrt[3]{2 - 2\sqrt{3}i}$

34. $\sqrt[3]{2\sqrt{3} - 2i}$

35. $\sqrt[4]{-8 - 8i}$

36. $\sqrt[5]{16\sqrt{3} + 16i}$

In exercises 37–50, prove each of the theorems about complex numbers. Use the fact that the complex conjugate of $z = a + ib$ is $\bar{z} = a - ib$.

37. For every $r \in \mathbb{R}, \bar{r} = r$.

38. For every $z, w \in \mathbb{C}, \overline{z + w} = \bar{z} + \bar{w}$.

39. For every $z, w \in \mathbb{C}, \overline{z \cdot w} = \bar{z} \cdot \bar{w}$.

40. For every $z, w \in \mathbb{C}, \overline{z/w} = \bar{z}/\bar{w}$.

41. For every $z \in \mathbb{C}, Re(z) = \frac{1}{2}(z + \bar{z})$.

42. For every $z \in \mathbb{C}, Im(z) = \frac{1}{2}(z - \bar{z})$.

43. For every $\theta \in \mathbb{R}, \overline{(e^{i\theta})} = e^{-i\theta}$, where $e^{i\theta} = \cos\theta + i\sin\theta$.
    Hint: Use the trignometric identities for $\cos(-)$ and $\sin(-\theta)$.

44. For every $z = re^{i\theta}, \bar{z} = re^{-i\theta}$.

45. For every $z \in \mathbb{C}, |z|^2 = z \cdot \bar{z}$.

46. For every $z \in \mathbb{C}, |z| = |\bar{z}|$.

47. For every $z = a + ib, |z| \le |a| + |b|$.

48. If $z = a + ib$ and $w = c + id$, then $ac + bd = Re[z \cdot \bar{w}]$ and $ac + bd \le |z \cdot \bar{w}|$.

49. For every $z, w \in \mathbb{C}, |z + w| \le |z| + |w|$.
    Note: This relationship is known as the *triangle inequality* and is important for our study of limits later in this chapter.

50. For every $z, w \in \mathbb{C}, |z - w| \ge |z| - |w|$.
    Hint: use the triangle inequality from exercise 49.

In exercises 51–54, use the definition of the partial derivative to find both partial derivatives for each function.

51. $f(x, y) = 2x - 4y + 8$

52. $g(x, y) = x(y + 2)$

53. $u(x, y) = (x + 1)y^2$

54. $v(x, y) = 2x\sqrt{y + 3}$

In exercises 55–60, use differentiation rules and the fundamental theorem of calculus to find both partial derivatives for each function.

55. $u(x, y) = (x + 1)y^2$

56. $v(x, y) = 2x\sqrt{x^2 + y^2}$

57. $u(x, y) = (x + y)^2 + 2x - y$

58. $v(x, y) = y\sqrt[3]{x - 2y}$

59. $u(x, y) = \int_0^x \sqrt{1 + t^2}\, dt + \int_0^y x + t^3\, dt$

60. $v(x, y) = xy + \int_y^x \sqrt{t + t^2}\, dt$

In exercises 61–64 prove that each polynomial function is continuous at the point $(0, 0)$ by using the definition of continuity.

61. $u(x, y) = 3x^2 + y$

62. $u(x, y) = 2x^2 y$

63. $u(x, y) = x^2 - 2x + 3y^2$

64. $u(x, y) = x + 2y^2 + 5$

In exercises 65–70, prove each function is continuous at the given point using the formal definition of continuity.

65. $u(x, y) = x - 2y$ at $(a, b) = (1, 2)$

66. $u(x, y) = 3x + 2y + 7$ at $(a, b) = (3, 4)$

67. $u(x, y) = \dfrac{x}{y}$ at $(a, b) = (0, 1)$

68. $u(x, y) = \dfrac{x + i}{y}$ at $z = 2 + i$

In exercises 69–75, prove each mathematical statement about real-valued continuous functions.

69. Constant functions are continuous: if $c \in \mathbb{R}$ and $u(x, y) = c$, then $u(x, y)$ is continuous.

70. Scalar multiplication preserves continuity: if $c \in \mathbb{R}$ and $u(x, y)$ is continuous at (a, b), then $c \cdot u(x, y)$ is continuous at (a, b).

71. Addition preserves continuity: if $u(x, y)$ and $v(x, y)$ are continuous at (a, b), then $u(x, y) + v(x, y)$ is continuous at (a, b).

72. Subtraction preserves continuity: if $u(x, y)$ and $v(x, y)$ are continuous at (a, b), then $u(x, y) - v(x, y)$ is continuous at (a, b).

73. Squaring preserves continuity: if $u(x, y)$ is continuous at (a, b), then $[u(x, y)]^2$ is continuous at (a, b).

74. Multiplication preserves continuity: if $u(x, y)$ and $v(x, y)$ are continuous at (a, b), then $u(x, y) \cdot v(x, y)$ is continuous at (a, b).

75. Reciprocals preserve continuity (when defined): if $u(x, y)$ is continuous and nonzero at (a, b), then $1/u(x, y) = 1/u(x, y)$ is continuous at (a, b).

In exercises 76–79 show that each nonpolynomial function is continuous at $(0, 0)$ by applying the definition of continuity.

76. $u(x, y) = xe^y$

77. $v(x, y) = e^x \sin y$

78. $g(x, y) = e^x \cos y$

79. $f(x, y) = (x + 1) \cdot \sin y$

Exercises 80–81 consider the Laplacian equation, which asserts that the sum of the second partial derivatives u_{xx} and u_{yy} is zero; symbolically,

$$\frac{\partial^2 u}{\partial x^2} + \frac{\partial^2 u}{\partial y^2} = u_{xx} + u_{yy} = 0,$$

where $u_{xx} = (u_x)_x$ and $u_{yy} = (u_y)_y$. The Laplacian is named in honor of the French mathematician Pierre-Simon Laplace who made many important contributions to the development of complex analysis and differential equations in the nineteeth century. In exercises 80–81, verify that both the real and the imaginary parts of each complex function satisfy the Laplacian equation.

80. $f(z) = z^2 = (x^2 - y^2) + i(2xy)$ 81. $g(z) = z^3$

7.2   Analytic Functions and the Cauchy–Riemann Equations

This section defines and develops basic properties of differentiable single-valued complex functions. Such functions $f(z)$ are also called analytic or holomorphic. We will see that they have a number of important properties and applications, many of which are expressible in terms of their real and imaginary parts u and v, where $f(z) = u(x, y) + iv(x, y)$. Definition 7.1.9 in section 7.1 described how to determine the continuity of u and v at any given point in the domain: verify that the limit as the function approaches the point is equal to the function's value at that point.

But many important questions remain: How do we determine the differentiability of the single-valued complex function $f(z)$? Is the derivative of $f(x + iy) = u(x, y) + iv(x, y)$ related to the partial derivatives of u and v? Does the differentiability of f force the partial derivatives of u and v to satisfy certain differential equations? These questions have attracted the interest and attention of mathematicians since the study of complex analysis first blossomed after the publication of Gauss's insights, and they have been answered by many of the great functional analysts in history, including Augustin-Louis Cauchy, Pierre-Simon Laplace, and the German mathematician Bernhard Riemann. This section's study of analytic functions will examine their insightful answers to these and other questions.

As in a study of real functions, differentiability of complex functions is defined in terms of a limit of a difference quotient. We will therefore need to establish what it means for a single-valued complex function $f(z)$ to approach a limit L as the complex number z approaches the origin 0 in the two-dimensional complex plane $\mathbb{C}$. Symbolically, we will describe what is meant by $\lim_{z \to 0} f(z) = L$.

The fact that z approaches 0 means that the distance gets very small between the complex number $z = x + iy$ and the origin $(0, 0)$ in the complex plane. This distance is measured by the two-dimensional distance formula $\sqrt{(x - 0)^2 + (y - 0)^2} = \sqrt{x^2 + y^2} = |z|$. In the same way, the distance between $f(z)$ and L may be calculated as $|f(z) - L|$. Since we are working in the two-dimensional plane, z may approach 0 along any path; figure 7.6 illustrates two paths that approach the origin—one along a radial path, and the second along a spiral path. These are just two of the many different paths approaching the origin in the complex plane; a definition of the limit must allow for any path of approach. With these reflections in mind, the definition follows.

Definition 7.2.1   *Suppose $f : D \to \mathbb{C}$ is a given single-valued complex function with $0 \in D \subseteq \mathbb{C}$ and L is a complex value. The notation $\lim_{z \to 0} f(z) = L$ means: given any $\varepsilon > 0$, there exists $\delta > 0$ such that $0 < |z| < \delta$ and $z \in D$ implies $|f(z) - L| < \varepsilon$. In this*

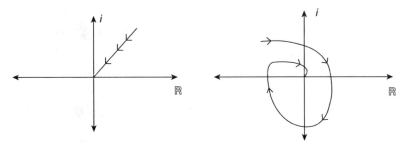

Figure 7.6 Two paths approaching the origin in the complex plane

case, we say that the limit of f as z approaches 0 exists and is equal to L. The term δ may depend on ε but not on z.

This definition of the limit of a complex function closely mirrors the definition of a limit for a real function and historically follows from the work of Augustin-Louis Cauchy. As you work with this definition, keep in mind that the term $|z|$ is calculated as $|x + iy| = \sqrt{x^2 + y^2}$. The next example applies this definition to verify a general limit statement.

Example 7.2.1 We use the formal definition of the limit to prove that for $m, b \in \mathbb{C}$, $\lim\limits_{z \to 0} m \cdot z + b = b$.

Proof Let $\varepsilon > 0$ be a given real number and define $\delta = \varepsilon/|m|$. Then whenever $0 < |z| < \delta$, we have

$$|f(z) - L| \;=\; |(m \cdot z + b) - b| \;=\; |m \cdot z| \;=\; |m| \cdot |z| \;<\; |m| \cdot \delta \;=\; |m| \cdot \frac{\varepsilon}{|m|} \;=\; \varepsilon.$$

The definition of limit is therefore satisfied, and so $\lim\limits_{z \to 0} m \cdot z + b = b$. ∎

Question 7.2.1 Identify a formula for δ expressed in terms of an arbitrary $\varepsilon > 0$ that satisfies the formal definition of the limit. (Hint: the proof given in example 7.2.1 indicates that $\delta = \varepsilon/|m|$, where m is the coefficient of z.)

(a) $\lim\limits_{z \to 0} -12z + (7 - 8i) \;=\; 7 - 8i$ (b) $\lim\limits_{z \to 0} (2 - 3i)z + 5 \;=\; 5$ ∎

The result given in example 7.2.1 is just a first step in the full development of the theory of limits (and continuity) for complex functions. Since this section's main goal is to study derivatives and analytic functions, we simply state the primary results that parallel our study of limits in real analysis. The proofs are similar to those from real analysis; further details are left for the exercises at the end of this section.

Theorem 7.2.1 *Let $c, L, M \in \mathbb{C}$ and let both f and g be single-valued complex functions with $\lim\limits_{z \to 0} f = L$ and $\lim\limits_{z \to 0} g = M$. Then the following equalities hold.*

- **Limits are unique:** *If $\lim\limits_{z \to 0} f(z) = L$ exists, then L is unique.*

- **Limit of a constant:** $\lim\limits_{z \to 0} c \;=\; c$

- **Limit of a scalar multiple:** $\lim_{z \to 0} c \cdot f(z) = c \cdot L$
- **Limit of a sum:** $\lim_{z \to 0} f + g = L + M$
- **Limit of a difference:** $\lim_{z \to 0} f - g = L - M$
- **Limit of a product:** $\lim_{z \to 0} f \cdot g = L \cdot M$
- **Limit of a quotient:** $\lim_{z \to 0} \dfrac{f}{g} = \dfrac{L}{M}$, *provided that $M \neq 0$*

We now focus on the concept of the derivative. For complex functions $f(z)$, differentiability is defined in terms of a limit of a difference quotient, which closely mimics the definition of the derivative for real-valued functions.

Definition 7.2.2   *If $f(z)$ is a single-valued complex function, then the **derivative** of $f(z)$ is*

$$\frac{df}{dz} = f'(z) = \lim_{\Delta z \to 0} \frac{f(z + \Delta z) - f(z)}{\Delta z},$$

*provided this limit exists. Given a point $z_0 \in \mathbb{C}$, we say that $f(z)$ is **differentiable** at z_0 when $f'(z_0)$ exists. Given a disk $S = \{z : |z - c| < R\}$ having fixed center $c \in \mathbb{C}$ and radius $R \in \mathbb{C}$, we say that $f(z)$ is **analytic on** S when $f'(z)$ exists for all $z \in S$. If $f'(z)$ exists for every $z \in \mathbb{C}$, then $f(z)$ is simply said to be **differentiable** or **analytic**. Such functions are also called **entire**.*

Many algebraic calculations used to compute derivatives of complex functions are identical to those used for real functions. The following question and example apply the definition of the derivative. We will later describe differentiation rules that will greatly simplify the computation of derivatives.

Example 7.2.2   We use the definition of the derivative to differentiate $f(z) = z^2$.

Applying the definition, expanding the resulting quadratic, and simplifying yields the following result.

$$\frac{d}{dz}[z^2] = \lim_{\Delta z \to 0} \frac{(z + \Delta z)^2 - z^2}{\Delta z} = \lim_{\Delta z \to 0} \frac{[z^2 + 2z \cdot \Delta z + (\Delta z)^2] - z^2}{\Delta z}$$

$$= \lim_{\Delta z \to 0} \frac{2z \cdot \Delta z + (\Delta z)^2}{\Delta z} = \lim_{\Delta z \to 0} 2z + \Delta z = 2z.$$

∎

Question 7.2.2   Using the definition, find the derivative of the following complex functions.

(a) $f(z) = (1 + i)z^3$ (b) $g(z) = 2z^2 + iz + 1 - i$

∎

The complex functions in both example 7.2.2 and question 7.2.2 are defined and differentiable on the complex plane, and so we say that they are analytic functions. In contrast, the function $f(z) = \bar{z}$ is not differentiable at every point on the complex plane, and so is not analytic. In fact, $f(z) = \bar{z}$ is nowhere differentiable, as the following example shows.

Example 7.2.3   We prove that $f(z) = \bar{z}$ is not differentiable at the origin.

Proof Consider two paths approaching the origin and compute the corresponding limits of the difference quotient. Each path will result in a different limit, and so the limit defining the derivative does not exist. The computations use the fact that $\overline{z+w} = \overline{z} + \overline{w}$ for $z, w \in \mathbb{C}$.

The first path approaches 0 along the real axis (here, $\Delta z = x$):

$$\lim_{\Delta z \to 0} \frac{\overline{0 + \Delta z} - \overline{0}}{\Delta z} = \lim_{\Delta z \to 0} \frac{\overline{0} + \overline{\Delta z} - \overline{0}}{\Delta z} = \lim_{\Delta z \to 0} \frac{\overline{\Delta z}}{\Delta z} = \lim_{x \to 0} \frac{x}{x} = 1.$$

The second path approaches 0 along the imaginary axis (here, $\Delta z = iy$):

$$\lim_{\Delta z \to 0} \frac{\overline{0 + \Delta z} - \overline{0}}{\Delta z} = \lim_{\Delta z \to 0} \frac{\overline{0} + \overline{\Delta z} - \overline{0}}{\Delta z} = \lim_{\Delta z \to 0} \frac{\overline{\Delta z}}{\Delta z} = \lim_{x \to 0} \frac{-iy}{iy} = -1.$$

As we have seen, two different paths of approach to the origin result in two different values for the limit of the difference quotient. Thus, the derivative does not exist. We note that an appropriate extension of this argument demonstrates that $f(z) = \overline{z}$ is not differentiable at any point on the complex plane.

■

As in real analysis, we rely on the definition of the derivative when proving that certain specific rules of differentiation are valid. However, this definition with its reliance on the limit is much too cumbersome to use for typical derivative computations, and so we state a number of rules of differentiation. You can see that the close parallel between the the definitions of the derivative for complex functions and for real functions makes the familiar differentiation rules from real analysis carry over to the complex setting. We summarize the most important of these rules in the following theorem.

Theorem 7.2.2 *If $c \in \mathbb{C}$ and both $f(z)$ and $g(z)$ are differentiable functions, then the following hold.*

- **The constant rule:** $\dfrac{d}{dz}[\,c\,] = 0$

- **The scalar multiple rule:** $\dfrac{d}{dz}[\,c \cdot f(z)\,] = c \cdot f'(z)$

- **The sum rule:** $\dfrac{d}{dz}[f + g] = f' + g'$

- **The difference rule:** $\dfrac{d}{dz}[f - g] = f' - g'$

- **The power rule:** $\dfrac{d}{dz}[\,z^n\,] = n \cdot z^{n-1}, \text{ for } n \in \mathbb{N}$

- **The product rule:** $\dfrac{d}{dz}[f \cdot g] = g \cdot f' + f \cdot g'$

- **The quotient rule:** $\dfrac{d}{dz}\left[\dfrac{f}{g}\right] = \dfrac{g \cdot f' - f \cdot g'}{g^2}$,

  provided that $g(z) \neq 0$

- **The chain rule:** $\dfrac{d}{dz}[f(g(z))] = f'(g(z)) \cdot g'(z)$

The rules greatly simplify the process of differentiating complex functions. Before considering the proofs of various parts of this theorem, the next question gives you the opportunity to practice them.

Question 7.2.3 Using theorem 7.2.2, compute the derivative of each function.

(a) $f(z) = 10z^3 - 7iz^2 + 5 - i$

(b) $g(z) = (iz + 2)^3$

(c) $h(z) = \dfrac{iz + 2}{5z^2 + 3i}$

(d) $j(z) = (2z + 9i)^5 \cdot (4iz^3 + 7)^4$

∎

The proofs of the rules in theorem 7.2.2 are identical in algebraic structure to those from real analysis. The next question and examples prove a couple of these results; further details are left for the exercises at the end of this section and for later studies.

Example 7.2.4 We prove the power rule for positive integers from theorem 7.2.2: If $n \in \mathbb{N}$ is a positive integer, then $d/dz[z^n] = n \cdot z^{n-1}$.

Proof The proof closely parallels the corresponding one for real functions given in example 4.4.5 in section 4.4. From the definition of the derivative and the binomial theorem from section 5.2,

$$
\begin{aligned}
\frac{d}{dz}[z^n] &= \lim_{\Delta z \to 0} \frac{(z + \Delta z)^n - z^n}{\Delta z} \\
&= \lim_{\Delta z \to 0} \frac{[z^n + nz^{n-1}\Delta z + \cdots + nz(\Delta z)^{n-1} + (\Delta z)^n] - z^n}{\Delta z} \\
&= \lim_{\Delta z \to 0} \frac{nz^{n-1}\Delta z + \cdots + nz(\Delta z)^{n-1} + (\Delta z)^n}{\Delta z} \\
&= \lim_{\Delta z \to 0} nz^{n-1} + \cdots + nz(\Delta z)^{n-2} + (\Delta z)^{n-1} = nz^{n-1}.
\end{aligned}
$$

∎

Question 7.2.4 Using the definition of the derivative, prove the scalar multiple rule from theorem 7.2.2; prove that if $c \in \mathbb{C}$ and $f(z)$ is a differentiable function, then

$$
\frac{d}{dz}[c \cdot f(z)] = c \cdot f'(z).
$$

∎

When calculating the derivative of a complex function, we are free to apply either the definition of the derivative or the differentiation rules stated in theorem 7.2.2. As you would expect, it is generally easier to apply the differentiation rules.

Example 7.2.5 We compare the calculations that the derivative of $f(z) = 3z + 1$ is $f'(z) = 3$, using first the definition of the derivative and then the differentiation rules.

Applying the definition,

$$
f'(z) = \lim_{\Delta z \to 0} \frac{3(z + \Delta z) + 1 - (3z + 1)}{\Delta z} = \lim_{\Delta z \to 0} \frac{3z + 3\Delta z + 1 - 3z - 1}{\Delta z}
$$

$$
= \lim_{\Delta z \to 0} \frac{3\Delta z}{\Delta z} = 3.
$$

Alternatively, we apply four differentiation rules from theorem 7.2.2: the sum rule, the scalar multiple rule, the power rule, and the constant rule. Justifying each step in the calculation,

$$
\begin{aligned}
\frac{d}{dz}[3z + 1] &= \frac{d}{dz}[3z] + \frac{d}{dz}[1] && \text{Sum rule} \\
&= 3 \cdot \frac{d}{dz}[z] + 0 && \text{Scalar multiple and constant rule} \\
&= 3 \cdot 1 = 3 && \text{Power rule (and simplification)}
\end{aligned}
$$

■

The differentiation rules are also used effectively to obtain general results about analytic complex functions. For example, an application to a general polynomial of the sum rule, the scalar multiple rule, the power rule, and the constant rule as in example 7.2.5 shows that every complex polynomial $p(z)$ is differentiable at any value $z \in \mathbb{C}$. The many important results that can be proven using the differentiation rules include the following:

- Every complex polynomial is analytic;
- Every sum, difference, and product of analytic functions is analytic;
- Every quotient of analytic functions is analytic (except for when the denominator is zero);
- A composition of analytic functions is analytic.

And these are only a handful of the many significant and interesting corollaries of the differentiation rules.

We are now ready to explore the famous Cauchy–Riemann equations, which begin to describe the powerful relationship between the real and imaginary parts of analytic functions and which Augustin-Louis Cauchy and Bernhard Riemann recognized. When $f(z)$ has real and imaginary parts $u(x, y)$ and $v(x, y)$ that are continuous and have continuous first partial derivatives on some disk $S = \{z : |z - c| < R\}$, then f is differentiable at any given point in S exactly when the following equations are satisfied at the point: $u_x = v_y$ and $u_y = -v_x$. Known as the Cauchy–Riemann equations, this condition therefore exactly determines when a complex function $f(z)$ is differentiable at a given point in the complex plane. The following theorem states this result.

Theorem 7.2.3 The Cauchy–Riemann theorem *Suppose the multivariate functions $u(x, y)$ and $v(x, y)$ and their first partial derivatives with respect to x and y are all continuous in some disk in $\mathbb{C}$. Then the complex function $f(z) = f(x + iy) = u(x, y) + iv(x, y)$ is differentiable at any point (x_0, y_0) in that disk iff $u(x, y)$ and $v(x, y)$ satisfy the* **Cauchy–Riemann equations** *at the point:*

$$
u_x = v_y \qquad and \qquad u_y = -v_x.
$$

Proof   We first prove that if the function f is differentiable at (x_0, y_0) in the given disk, then the Cauchy–Riemann equations are satisfied by $f = u + iv$ at (x_0, y_0). Assume $f(z) = f(x + iy) = u(x, y) + iv(x, y)$ is differentiable at $z_0 = x_0 + iy_0$ and calculate $f'(z_0)$ in two ways, examining the definition of the derivative along two different path approaches to the origin. Since f is differentiable at z_0, the resulting limits must agree; the Cauchy–Riemann equations will follow.

The first path approaches the origin along the real axis. Let $\Delta z = h$, where h is real:

$$
\begin{aligned}
f'(z_0) &= \lim_{h\to 0} \frac{f(z_0 + h) - f(z_0)}{h} = \lim_{h\to 0} \frac{f((x_0 + h) + iy_0) - f(x_0 + iy_0)}{h} \\
&= \lim_{h\to 0} \frac{u(x_0 + h, y_0) + iv(x_0 + h, y_0) - u(x_0, y_0) - iv(x_0, y_0)}{h} \\
&= \lim_{h\to 0} \left[\frac{u(x_0 + h, y_0) - u(x_0, y_0)}{h} + i\frac{v(x_0 + h, y_0) - v(x_0, y_0)}{h} \right] \\
&= u_x(x_0, y_0) + iv_x(x_0, y_0).
\end{aligned}
$$

The second path approaches the origin along the imaginary axis. Let $\Delta z = i \cdot h$ for h real:

$$
\begin{aligned}
f'(z_0) &= \lim_{ih\to 0} \frac{f(z_0 + ih) - f(z_0)}{ih} = \lim_{h\to 0} \frac{f(x_0 + i(y_0 + h)) - f(x_0 + iy_0)}{ih} \\
&= \lim_{h\to 0} -i\frac{u(x_0, y_0 + h) + iv(x_0, y_0 + h) - u(x_0, y_0) - iv(x_0, y_0)}{h} \\
&= \lim_{h\to 0} \left[\frac{v(x_0, y_0 + h) - v(x_0, y_0)}{h} - i\frac{u(x_0, y_0 + h) - u(x_0, y_0)}{h} \right] \\
&= v_y(x_0, y_0) - iu_y(x_0, y_0).
\end{aligned}
$$

Because $f'(x_0 + iy_0)$ exists, these two expressions for $f'(z_0)$ must be equal, which means

$$
f'(z_0) = u_x(x_0, y_0) + iv_x(x_0, y_0) = v_y(x_0, y_0) - iu_y(x_0, y_0).
$$

Equating the real and imaginary parts of these expressions proves that the Cauchy–Riemann equations are satisfied; that is, $u_x = v_y$ and $u_y = -v_x$ at (x_0, y_0).

We now indicate the proof of the converse. Assume the functions $u(x, y)$ and $v(x, y)$ satisfy the Cauchy–Riemann equations in the disk. We show f is differentiable at any point in the disk by showing it satisfies the definition of the derivative. We do so by working with the Taylor series (with remainder) expansion of $u(x, y)$ and $v(x, y)$ near any point (x_0, y_0) of the disk. This multivariate expansion can be written in terms of the function's first partial derivatives, just as the one-variable Taylor series expansion can be written in terms of the function's first derivative. The details are messy, but when the Cauchy–Riemann equations are applied to the Taylor series expansions, we obtain the following expression for any point $z_0 = x_0 + iy_0$ in the disk.

$$
f(z_0 + h) = f(z_0) + h[u_x(x_0, y_0) + iv_x(x_0, y_0)] + \delta_1(A + iB) + \delta_2(C + iD):
$$

where $h = \delta_1 + i\delta_2$ has an arbitrarily small modulus and $A, B, C,$ and D approach 0 as h approaches 0. Subtracting $f(z_0)$ from both sides of the above expression and dividing by h, we obtain the function's difference quotient at the point z_0. Taking the limit as h approaches 0 and applying the fact that the partial derivatives are continuous, $f'(z) = u_x(x_0, y_0) + iv_x(x_0, y_0)$. Thus, $f(z)$ is analytic in the disk.

■

The history of the Cauchy–Riemann equations is long in development. Using a technique that he pioneered of integrating complex functions along a closed curve, Augustin-Louis Cauchy published the Cauchy–Riemann equations in an involved treatise on definite integrals; he initially presented this result in 1814 and published it as a book in 1827. As part of his 1851 dissertation, Bernhard Riemann gave the first general, formal proof of the Cauchy–Riemann theorem; his proof of sufficiency was essentially the same as that presented above. Much previous to both Riemann and Cauchy, the French mathematician Jean le Rond d'Alembert had stated a form of the equations in a 1752 essay on fluid dynamics.

The Cauchy–Riemann equations turn out to have far-reaching implications about the relationships between an analytic function and its real and imaginary parts. Such relationships make the study of analytic functions useful in physics, especially in the field of harmonic analysis; section 7.5 will illustrate a small portion of this real-world application. For the moment, we simply practice verifying that the Cauchy–Riemann equations are satisfied for given examples of analytic functions.

Example 7.2.6   We verify the conclusion of the Cauchy–Riemann theorem at every point in $\mathbb{C}$ for the analytic function $f(z) = z^4 = (x + iy)^4$.

First note that $f(z) = z^4$ is analytic (that is, differentiable at every point in $\mathbb{C}$), and $f'(z) = 4z^3$ by the power rule from theorem 7.2.2. Expanding the expression $f(x + iy)^4$, the function $f(z)$ has real part $u(x, y) = x^4 - 6x^2y^2 + y^4$ and imaginary part $v(x, y) = 4x^3y - 4xy^3$. Now compute the partial derivatives of u and v at an arbitrary point z in $\mathbb{C}$:

$$u_x = 4x^3 - 12xy^2 \qquad\qquad v_y = 4x^3 - 12xy^2$$
$$u_y = -12x^2y + 4y^3 \qquad\qquad v_x = 12x^2y - 4y^3.$$

Notice that both $u(x, y)$ and $v(x, y)$ are continuous, as are their partial derivatives because they are polynomials in x and y. Also, $u_x = v_y$ and $u_y = -v_x$, and so the Cauchy–Riemann equations are satisfied.

■

Question 7.2.5   Following the technique used in example 7.2.6, verify the conclusion of the Cauchy–Riemann theorem for the following analytic functions.

(a) $f(z) = z^5$ (b) $g(z) = 2z^5 - iz^3$

■

The Cauchy–Riemann theorem can also be used to confirm that a given complex function is not differentiable. In particular, $f(z)$ cannot be differentiable whenever the partial derivatives of the real and imaginary parts of f do not satisfy the Cauchy–Riemann equations. Consider the following example.

Example 7.2.7   We use the Cauchy–Riemann theorem to prove that the conjugate function $f(z) = \overline{z}$ is not analytic.

Proof   The function $f(z) = \overline{z}$ is $f(x + iy) = x - iy$, and therefore has real part $u(x, y) = x$ and imaginary part $v(x, y) = -y$. The corresponding partial derivatives are $u_x = 1$, $u_y = 0$, $v_x = 0$, and $v_y = -1$. Even though $u_y = -v_x = 0$, the Cauchy–Riemann equations are not satisfied at any point $z \in \mathbb{C}$ because $u_x = 1 \neq -1 = v_y$. Therefore $f(z) = \bar{z}$ cannot be analytic at any $z \in \mathbb{C}$.

∎

The analytic functions we have thus far studied have all been complex polynomials (such as the functions $f(z) = z^5$ and $g(z) = 2z^5 - iz^3$ in question 7.2.5). As you might expect, there are many other types of analytic functions that play a prominent role in the study of complex analysis. Among the most important is the complex exponential $f(z) = e^z$. This function arises quite naturally in many applications of complex analysis, much as its real counterpart e^x often arises in applications of real analysis. The following definition of the exponential function is based on *Euler's formula* $e^{iy} = \cos y + i \sin y$.

Definition 7.2.3   *For every $z \in \mathbb{C}$, we define $e^z = e^{x+iy} = e^x e^{iy} = e^x(\cos y + i \sin y)$. Therefore, the real part of e^z is $u(x, y) = e^x \cos y$ and the imaginary part is $v(x, y) = e^x \sin y$.*

When considering a new complex function, mathematicians often ask if it satisfies certain desirable properties. In this context, we are immediately led to ask if $f(z) = e^z$ is analytic on its domain $\mathbb{C}$. The Cauchy–Riemann theorem will confirm the fact that the exponential function is analytic, as stated in the next theorem.

Theorem 7.2.4   *The exponential function $f(z) = e^z$ is analytic.*

Proof   Definition 7.2.3 identifies the real and imaginary parts of $f(z) = e^z$ as $u(x, y) = e^x \cos y$ and $v(x, y) = e^x \sin y$. These functions are continuous and turn out to have continuous partial derivatives at any point (x, y) in the sense of definition 7.1.9 in section 7.1. The derivatives are $u_x = e^x \cos y$, $u_y = -e^x \sin y$, $v_y = e^x \cos y$, and $v_x = e^x \sin y$. Because $u_x = v_y$ and $u_y = -v_x$, the Cauchy–Riemann equations are satisfied. Therefore the complex exponential function $f(z) = e^z$ is differentiable for every $z \in \mathbb{C}$; in other words $f(z) = e^z$ is entire.

∎

Question 7.2.6   Prove that $f(z) = e^z$ has derivative $f'(z) = e^z$ using the definition of the derivative along with the facts that $e^{z+\Delta z} = e^z e^{\Delta z}$ and

$$\lim_{\Delta z \to 0} \frac{e^{\Delta z} - 1}{\Delta z} = 1.$$

∎

With this understanding of the complex exponential function in hand, we consider the definition of its inverse function. Just as for real-valued functions discussed in section 4.2, a complex function g is an *inverse function* of a complex function f if $w = f(z)$ whenever $z = g(w)$. Equivalently, these conditions mean $g(f(z)) = z$ for all z in the domain of f and $w = f(g(w))$ for all w in the range of f. Any complete study of calculus includes the fact that the inverse of the real exponential function $f(x) = e^x$ is the natural logarithm function $g(x) = \ln(x)$.

The complex exponential function $f(z) = e^z$ has an inverse function $g(z)$ satisfying $z = g(w)$ whenever $w = e^z$. As in the real-valued situation, the function g is called the logarithm function. If we allow the logarithm function to be multiple-valued, then its definition is in terms of the multiple-valued argument function $\arg(z) = \theta$, where θ is any polar angle satisfying $z = re^{i\theta}$. If, however, we insist on the logarithm function being single-valued, then its definition must be chosen to be in terms of a single-valued argument function, such as the principal branch given in definition 7.1.7 of section 7.1: $\text{Arg}(z) = \theta$ is the unique polar angle satisfying $z = re^{i\theta}$ with $-\pi < \theta = \text{Arg}(z) \leq \pi$. The next definition describes both of these complex logarithm functions $\log(z)$ and $\text{Log}(z)$.

Definition 7.2.4   *If $z \neq 0$ is a complex number, then*

- $\log(z) = \ln|z| + i\arg(z)$ *is the multiple-valued complex logarithm function;*
- $\text{Log}(z) = \ln|z| + i\text{Arg}(z)$ *is the single-valued complex logarithm function.*

The next example helps clarify the definition through several computations.

Example 7.2.8   We determine the value of the complex logarithm(s) of $2 + 2i$ and $-4i$.
For $2 + 2i$, we observe that $|2 + 2i| = \sqrt{8} = 2\sqrt{2}$, that $\text{Arg}(2 + 2i) = \pi/4$, and that $\arg(2 + 2i) = \pi/4 + 2n\pi$ for $n \in \mathbb{Z}$. Thus, we have:

$$\text{Log}(2 + 2i) = \ln(2\sqrt{2}) + i\frac{\pi}{4} \quad \text{and} \quad \log(2 + 2i) = \ln(2\sqrt{2}) + i\left[\frac{\pi}{4} + 2n\pi\right] \text{ for } n \in \mathbb{Z}.$$

For $-4i$, we observe that $|-4i| = 4$, that $\text{Arg}(-4i) = -\pi/2$, and that $\arg(-4i) = -\pi/2 + 2n\pi$ for $n \in \mathbb{Z}$. Thus, we have:

$$\text{Log}(-4i) = \ln 4 - i\frac{\pi}{2} \quad \text{and} \quad \log(-4i) = \ln 4 + i\left[-\frac{\pi}{2} + 2n\pi\right] \text{ for } n \in \mathbb{Z}.$$

■

Question 7.2.7   Compute the value of $\text{Log}(z)$ and $\log(z)$ for each complex number.

(a) $z = -4$

(b) $z = 2\sqrt{3} + 2i$

(c) $z = 3e^{i(\pi/6)}$

(d) $z = 2e^{i(11\pi/6)}$

■

In example 7.2.8 and question 7.2.7, the value of $\text{Log}(z)$ for $z \neq 0$ is just one of the values identified by the multiple-valued function $\log z$, as $\text{Log}(z)$ has an imaginary part in the range $(-\pi, \pi]$. To describe the fact that the function $\text{Log}(z)$ is one choice for the output value of the multiple-valued function $\log z$, mathematicians refer to $\text{Log}(z)$ as the *principal branch* of the logarithm function. This label distinguishes it as the single-valued logarithm function that arises from the principal range $(-\pi, \pi]$ of the polar angle $\text{Arg}(z)$. In part, our interest in the logarithm function is based on its relationship with the complex exponential function. The following theorem confirms that $\text{Log}(z)$ serves as an inverse function for e^z.

Theorem 7.2.5   *The complex logarithm $\text{Log}(z)$ is an inverse function of e^z.*

Proof   The theorem follows by showing that $e^{\text{Log}(z)}$ and $\text{Log}(e^z)$ both equal z. For the first composition,

$$e^{\text{Log}(z)} = e^{\ln|z| + i\text{Arg}(z)} = e^{\ln|z|}e^{i\text{Arg}(z)} = |z|e^{i\text{Arg}(z)} = z.$$

The second calculation uses two identities: $\ln|e^z| = Re(z)$ and $\text{Arg}(e^z) = Im(z)$. The proofs of these identities are left for the exercises at the end of this section (see exercises 65 and 66). Then the second calculation follows:

$$\text{Log}(e^z) = \ln|e^z| + i\text{Arg}(e^z) = Re(z) + iIm(z) = z.$$

The single-valued function $g(z) = \text{Log}(z)$ therefore serves as the inverse function of $f(z) = e^z$.

■

The single-valued function $\text{Log}(z)$ is not a continuous function; two points close to each other but on opposite sides of the negative real axis will not have logarithm values that are close (the point with positive imaginary part will have a corresponding functional imaginary part near π, while the one with negative imaginary part will have a functional imaginary part near $-\pi$). For examples of complex points close to $z = -1$, if ε is a small positive number, then $\text{Log}(-1 + i\varepsilon) = \sqrt{1 + \varepsilon^2} + i\text{Arg}(-1 + i\varepsilon)$, but $\text{Log}(-1 - i\varepsilon) = \sqrt{1 + \varepsilon^2} + i\text{Arg}(-1 - i\varepsilon)$, and these functional imaginary parts are far apart. To work with a logarithm function that is continuous, complex analysts often restrict the domain of $f(z) = \text{Log}(z)$ to $\{z : -\pi < \text{Arg}(z) < \pi\}$. This function will not only be continuous in the sense that nearby complex points will have nearby function values, but this restricted function also turns out to be analytic across its domain; several exercises at the end of this section discuss the differentiability of this restricted function.

The function $f(z) = \text{Log}(z)$ having restricted domain $\{z : -\pi < \text{Arg}(z) < \pi\}$ is sometimes said to have a "branch cut," a reference to the fact that points in $\mathbb{C}$ along the negative real axis have been "cut out of" the domain. Whenever complex analysts talk about such functions as differentiable, it is always assumed that a branch cut has been taken. The negative real axis is often called the "principal branch cut" for the logarithm function, since it corresponds to the single-valued branch of $\log z$ that is sometimes called the principal branch.

As in real analysis, a well-defined natural logarithm function enables us to rigorously define the expression z^w when z and w are *any* complex numbers. Thus far, we have only considered rational powers, so this is a significant step forward in developing an understanding of exponentiation for complex numbers. Consider the following definition.

Definition 7.2.5   *For $z, w \in \mathbb{C}$, define $z^w = e^{w \log z}$.*

As we should expect and demand, definition 7.2.5 for z^w agrees with our previous definition of z^w when $w = n \in \mathbb{Z}$ is an integer power or when $w = 1/n$ for $n \in \mathbb{Z}$ is a fractional power. The following example is a specific illustration of this claim.

Example 7.2.9   We determine the three cube roots of $1 + i$.

The use of definition 7.2.5 in this case depends on knowing the value of $\log(1 + i)$. Here, $|1 + i| = \sqrt{2}$ and $\arg(1 + i) = \pi/4 + 2k\pi$ for $k \in \mathbb{Z}$. Applying the definition of exponentiation for complex numbers,

$$(1 + i)^{1/3} = e^{(1/3)\cdot\log(1+i)} = e^{(1/3)[\ln(\sqrt{2})+i(\pi/4+2k\pi)]}$$

$$= e^{\ln(\sqrt[6]{2})}e^{i(\pi/12+2k\pi/3)} = \sqrt[6]{2}e^{i(\pi/12+2k\pi/3)}, \text{ where } k \in \mathbb{Z}.$$

Three of these polar angles are in the principal range: $\pi/12$; $\pi/12 + 2\pi/3 = 3\pi/4$; and $\pi/12 - 2\pi/3 = -7\pi/12$. We note that this approach results in the same three values obtained in example 7.1.6 from section 7.1, which worked with the polar representation of $1 + i$.

■

In addition to agreeing with known exponentiation results for rational powers, definition 7.2.5 can be used to calculate new and interesting results for complex exponentials. As stated in the introduction to this chapter, Euler was the first mathematician to observe that the expression i^i is real-valued and, in fact, assumes infinitely many distinct real numbers. Calculations using definition 7.2.5 confirm Euler's insightful observation.

Example 7.2.10 We evaluate the expression i^i in rectangular form.

To use definition 7.2.5, first find $\log(i)$, which is calculated from the facts that $|i| = 1$ and $\arg(i) = \pi/2 + 2k\pi$ for $k \in \mathbb{Z}$. Then

$$i^i = e^{i\log i} = e^{i[\ln|i| + i\arg(i)]} = e^{i[0 + i(\pi/2 + 2k\pi)]} = e^{-(\pi/2 + 2k\pi)}, \text{ where } k \in \mathbb{Z}.$$

Therefore i^i is equal to (countably) infinite distinct values. Furthermore, because there are no imaginary components in this final expression, these values are always real!

■

Question 7.2.8 Evaluate each expression in rectangular form using the appropriate definition.

(a) $e^{3 - i\pi}$

(b) $\log(1 + i\sqrt{3})$

(c) $\mathrm{Log}(1 + i\sqrt{3})$

(d) $(-2)^i$

■

We end this section with a final look at an arbitrary monomial $f(z) = z^n$. As stated in theorem 7.2.2 and proven in example 7.2.4, any monomial $f(z) = z^n$ with a positive integer power $n \in \mathbb{N}$ is differentiable at every $z \in \mathbb{C}$ and has derivative $f'(z) = nz^{n-1}$. Therefore, according to the Cauchy–Riemann theorem, the real and imaginary parts of $f(z) = z^n$ (which are continuous along with the first partial derivatives) satisfy the Cauchy–Riemann equations. We verify this last observation in the following example.

Example 7.2.11 Through direct calculation, we verify that the real and imaginary parts of the analytic function $f(z) = z^n$ satisfy the Cauchy–Riemann equations for every $n \in \mathbb{N}$.

The calculations use the greatest integer function (denoted by $\lfloor x \rfloor$) to index the sums that result from the expansion of $f(x + iy) = (x + iy)^n$. Expand $f(x + iy) = (x + iy)^n = u(x, y) + iv(x, y)$ using the binomial theorem from section 5.2:

$$u(x, y) = \sum_{k=0}^{\lfloor n/2 \rfloor} \frac{(-1)^k n! x^{n-2k} y^{2k}}{(2k)!(n - 2k)!} \quad \text{and} \quad v(x, y) = \sum_{k=0}^{\lfloor (n+1)/2 \rfloor - 1} \frac{(-1)^k n! x^{n-2k-1} y^{2k+1}}{(2k + 1)!(n - 2k - 1)!}.$$

Now compute the partial derivative of $u_x(x, y)$:

$$u_x = \sum_{k=0}^{\lfloor(n+1)/2\rfloor-1} \frac{(-1)^k n!(n-2k)x^{n-2k-1}y^{2k}}{(2k)!(n-2k)!} = \sum_{k=0}^{\lfloor(n+1)/2\rfloor-1} \frac{(-1)^k n!x^{n-2k-1}y^{2k}}{(2k)!(n-2k-1)!}.$$

Similarly, the partial derivative with respect to y is

$$v_y = \sum_{k=0}^{\lfloor(n+1)/2\rfloor-1} \frac{(-1)^k n!x^{n-2k-1}(2k+1)y^{2k}}{(2k+1)!(n-2k-1)!} = \sum_{k=0}^{\lfloor(n+1)/2\rfloor-1} \frac{(-1)^k n!x^{n-2k-1}y^{2k}}{(2k)!(n-2k-1)!}.$$

As we can see, $u_x = v_y$ for every (x, y). Similar computations yield the equality $u_y = -v_x$ for every (x, y), and so the Cauchy–Riemann equations hold for the analytic function $f(z) = z^n$, where $n \in \mathbb{N}$.

$\blacksquare$

7.2.1   Reading Questions for Section 7.2

1. Define and give an example of the limit $\lim_{z \to 0} f(z) = L$.
2. Discuss the role that different path approaches play in computing the limit of a complex function.
3. State theorem 7.2.1. Why is this result interesting?
4. In this section, what motivates a consideration of only those limits that have z approaching the origin?
5. Define the derivative $f'(z)$ for a complex function $f(z)$.
6. Define and give an example of a function $f(z)$ that is analytic at a given point $z_0 \in \mathbb{C}$.
7. Give an example of a function that is not analytic at a point $z_0 \in \mathbb{C}$.
8. State theorem 7.2.2 and differentiate the complex function $f(z) = (4 + 3i)e^{(z^2+i)(3iz^3+2z)}$.
9. State the Cauchy–Riemann theorem. What are the Cauchy–Riemann equations?
10. Define e^z, $\mathrm{Log}(z)$, and $\log(z)$. What is the relationship among these three functions?
11. Express e^{3+4i}, $\mathrm{Log}(3 + 4i)$, and $\log(3 + 4i)$ in rectangular form.
12. Define z^w for complex numbers $z, w \in \mathbb{C}$. Give an example of an expression z^w in rectangular form, where w is not real.

7.2.2   Exercises for Section 7.2

In exercises 1–6, evaluate each limit expressing the solution in rectangular form.

1. $\lim_{z \to 0} (3 + i)(z - i) + 2 + 2i$
2. $\lim_{z \to 0} 2z^3 + 4z^2 + 8z + 16$
3. $\lim_{z \to 0} (4 - i)e^z$
4. $\lim_{z \to 0} i(z - 7e^{3z-i})$
5. $\lim_{z \to 0} \mathrm{Log}(z + i)$
6. $\lim_{z \to 0} z^z$

In exercises 7–12, prove each limit using the formal definition.

7. $\lim\limits_{z \to 0} iz + i = i$

8. $\lim\limits_{z \to 0} (3 + i)(z - i) = 1 - 3i$

9. $\lim\limits_{z \to 0} (1 + i)z + (3 + 2i) = 3 + 2i$

10. $\lim\limits_{z \to 0} z^2 + 1 - i = 1 - i$

11. $\lim\limits_{z \to 0} \dfrac{1}{z + 1} = 1$

12. $\lim\limits_{z \to 0} \dfrac{x^2}{z} = 0$

Hint: $0 \le x^2 \le x^2 + y^2$.

In exercises 13–20, prove each mathematical statement about limits.

13. The property that limits are unique from theorem 7.2.1.
    Hint: Assume there exist two limits L and M and prove that $L = M$.

14. The limit of a constant rule from theorem 7.2.1.
    Hint: Use the definition of the limit with any $\delta > 0$.

15. The limit of a scalar multiple rule from theorem 7.2.1.
    Hint: Use the definition of the limit with $\delta = \varepsilon/|c|$ for a given $\varepsilon > 0$.

16. The limit of a sum rule from theorem 7.2.1.
    Hint: Given $\varepsilon > 0$, apply the definition of the limit to each of $f(z)$ and $g(z)$ for $\varepsilon/2$ and let $\delta = \min\{\delta_f\left(\frac{\varepsilon}{2}\right), \delta_g\left(\frac{\varepsilon}{2}\right)\}$.

17. The limit of a difference rule from theorem 7.2.1.

18. The limit of a square rule: if $\lim_{z \to 0} f = L$, then $\lim_{z \to 0} f^2 = L^2$.

19. The limit of a product rule from theorem 7.2.1.

20. The limit of a quotient rule from theorem 7.2.1.

In exercises 21–26, use the definition to compute the derivative of each function.

21. $f(z) = 3z + 4i$

22. $g(z) = 5z^2 + 1$

23. $h(z) = (3z + i)^2$

24. $p(z) = \dfrac{1}{z}$

25. $q(z) = \dfrac{i}{z + 2i}$

26. $r(z) = \sqrt{z}$

In exercises 27–36, use the differentiation rules from theorem 7.2.2 to compute the derivative $f'(z)$ of the following functions.

27. $f(z) = 5z^2 + 1 + i$

28. $f(z) = 4iz^4 - 3z^2 + 5iz - 10$

29. $f(z) = (3z + i)^2$

30. $f(z) = (2 - i)(z^4 - iz^2)^3$

31. $f(z) = \dfrac{iz^2 + 2}{z^3 + 2i}$

32. $f(z) = \sqrt{4z^4 + iz^3}$

33. $f(z) = e^{z^2}$

34. $f(z) = 3e^{5iz}(3z^4 - 2z)$

35. $f(z) = 2i^z$

Hint: $i^z = e^{z \log i}$

36. $f(z) = i^z \sqrt{e^{z^2}}$

In exercises 37–44, prove each mathematical statement about derivatives.

37. The derivative of a (single-valued) analytic function is unique.

38. The derivative of a constant rule from theorem 7.2.2.

39. The derivative of a scalar multiple rule from theorem 7.2.2.

40. The derivative of a sum rule from theorem 7.2.2.

41. The derivative of a difference rule from theorem 7.2.2.

42. For any constant $w \in \mathbb{C}$, the derivative of $f(z) = z^w$ is $f'(z) = wz^{w-1}$.
43. The derivative of $f(z) = e^{g(z)}$ is $f'(z) = g'(z)e^{g(z)}$ for any (single-valued) analytic function $g(z)$.
44. If $f(z)$ is analytic and $f(z) = f(-z)$, then $f'(z) = -f'(-z)$.

In exercises 45–50, verify that the Cauchy–Riemann equations are satisfied for each analytic function $f(z)$.

45. $f(z) = z + i$
46. $f(z) = z^2 + z$
47. $f(z) = (3z + i)^2$

48. $f(z) = e^{2z}$
49. $f(z) = z + e^z$
50. $f(z) = ze^z$

In exercises 51–64, evaluate each expression in rectangular form using the appropriate definition.

51. $\left(e^{2-3i}\right)^2$
52. $e^{(2-3i)^2}$
53. $\text{Log}(2\sqrt{3} - 2i)$
54. $\log(2\sqrt{3} - 2i)$
55. $e^{2+i}\log(-4 + 4i)$
56. $e^{2+i}\text{Log}(-4 + 4i)$
57. $e^{\text{Log}(-4+4i)}$

58. $\text{Log}(e^{-4+4i})$
59. $(5 - 2i)e^{i \cdot \text{Log}(i)}$
60. $e^{i \cdot \text{Log}(-4+4i)}$
61. 2^i
62. $(-1)^{1+i}$
63. $(1 + i)^{1+i}$
64. $\sqrt[3]{-3i}$

In exercises 65–67, prove each mathematical statement about the complex exponential function.

65. For every $z \in \mathbb{C}$, $|e^z| = e^{\text{Re}(z)}$.
    Note: This result was used in the proof of theorem 7.2.5.
66. For every $z \in \mathbb{C}$, $\text{Arg}(e^z) = \text{Im}(z)$ (because of the polar representation of e^z).
    Note: This result was used in the proof of theorem 7.2.5.
67. For every $z \in \mathbb{C}$, $\overline{(e^z)} = e^{(\bar{z})}$.

In exercises 68–70, prove each mathematical statement, and so develop a proof that the derivative of $\text{Log}(z)$ is $1/z$. In these problems, we are working with the function $f(z) = \text{Log}(z)$ having domain restricted to the set $\{z : -\pi < \text{Arg}(z) < \pi\}$ (where the principal branch cut of the nonpositive real axis has been removed).

68. For any nonzero $z = x + iy \in \mathbb{C}$ with $-\pi < \text{Arg}(z) < \pi$, we have

$$\text{Arg}(z) = \begin{cases} \arctan\left(\frac{y}{x}\right) & \text{if } x > 0 \\ \pi + \arctan\left(\frac{y}{x}\right) & \text{if } x < 0 \text{ and } y > 0 \\ -\pi + \arctan\left(\frac{y}{x}\right) & \text{if } x < 0 \text{ and } y < 0 \\ \pi/2 & \text{if } x = 0 \text{ and } y > 0 \\ -\pi/2 & \text{if } x = 0 \text{ and } y < 0 \end{cases}$$

69. The partial derivatives of $v(x, y) = \text{Arg}(x + iy)$ are $v_x = -y/(x^2 + y^2)$ and $v_y = x/(x^2 + y^2)$.
    Hint: Use the piecewise expression for $\text{Arg}(z)$ given in exercise 68.
70. The derivative of $f(z) = \text{Log}(z)$ is $f'(z) = \frac{1}{z}$.
    Hint: Use the partial derivatives from exercise 69 and the fact that for analytic functions $f(z) = u(x, y) + iv(x, y)$ the derivative is $f'(z) = u_x + iv_x$.

7.3 Power Series Representations of Analytic Functions

The last section defined complex analytic functions as being differentiable at every point $z \in \mathbb{C}$. We learned that all complex polynomials are analytic and that the complex exponential function $f(z) = e^z$ is also analytic. This section studies power series representations of functions analytic on a disk $S = \{z : |z - c| < R\}$ having fixed center $c \in \mathbb{C}$ and radius $R \in \mathbb{C}$. These representations are important mathematical objects that, among other things, show how to define complex versions of trigonometric functions.

We will see that the power series for a function $f(z)$ analytic on a disk is often expressable in the same algebraic form as the power series for the corresponding real-valued function $f(x)$. For example, the power series for the real-valued exponential function is $e^x = \sum_{n=0}^{\infty} x^n/n!$; this section will soon prove that the power series for the complex exponential function $f(z) = e^z$ is $e^z = \sum_{n=0}^{\infty} z^n/n!$. Such natural extensions from real power series to complex ones provide important guidance and motivation in deciding how to *define* many complex functions, and these extensions often result in complex functions that share many well-understood properties of their real-valued counterparts.

Complex power series are important for other reasons as well. For example, they provide a type of representation that is valid for all analytic functions. Mathematicians have proven a variety of "representation theorems," which characterize a class of functions in terms of some mathematical object such as an infinite power series. Representation theorems are extremely useful in advanced complex analysis and allow mathematicians to understand powerful and general characteristics of all the functions in the given category. Mathematicians have identified representations for many types of complex functions, including "meromorphic," "harmonic," and "subharmonic" functions, but a general theory of these representations is beyond the scope of this book (they generally require an understanding of complex integration). Rather, this setting provides only a taste of the general theory as it studies complex functions with convergent power series representations and considers several important examples of such functions.

Any discussion of representing analytic functions as complex power series must consider the issues of convergent and divergent power series. It would be foolhardy to try to identify a well-defined complex function as an infinite series that has no sensible values to which it converges. Roughly speaking, a power series converges if the infinite sum at a given domain value can be identified with a finite number, and a power series diverges if it cannot. While this rough description is sufficient for developing an intuitive understanding of infinite sums, it does not develop a precise mathematical analysis; this section articulates a careful rigorous definition of power series convergence. As for real power series, convergence follows from a corresponding convergence of the sequence of partial sums. We must therefore describe what it means for a complex sequence to converge before defining what it means for a complex power series to converge.

7.3.1 Complex Sequences

A *sequence* is an infinite list of numbers; in a sense that emphasizes structure, a sequence is a function whose domain is the set of natural numbers $\mathbb{N}$. For a *complex sequence*, every element of the sequence (that is, every element in the range) is a complex number. Complex sequences are often denoted by $\{z_n\}_{n=1}^{\infty}$. The indices are often omitted for brevity's sake.

Example 7.3.1   We state two examples of complex sequences.

- $\{(1+i)/n\}_{n=1}^{\infty} = \{1+i, \frac{1}{2}+\frac{1}{2}i, \frac{1}{3}+\frac{1}{3}i, \ldots\}$
- $\{i^n\} = \{i, -1, -i, 1, i, -1, \ldots\}$

    ■

These two simple examples already indicate that complex sequences can exhibit very different behaviors when compared. We intuitively recognize that the first sequence in example 7.3.1 is approaching 0 and should be identified as a convergent sequence. In contrast, the second sequence keeps hopping around the complex plane on the unit circle and should thus be identified as a divergent sequence. The following definition makes these notions precise.

Definition 7.3.1   *A sequence of complex numbers* $\{z_n\}_{n=1}^{\infty} = \{z_1, z_2, z_3, \ldots\}$ **converges** *to a limit* $L \in \mathbb{C}$ *if for every* $\varepsilon > 0$, *there exists* $N > 0$ *such that* $|z_n - L| < \varepsilon$ *whenever* $n \geq N$. *In this case, we write* $\lim_{n \to \infty} z_n = L$. *We note that* N *may depend on* ε. *A sequence that does not converge is called* **divergent**.

This definition for complex sequences closely mirrors the definition for a convergent sequence of real numbers. It is thereby important to keep in mind that the values (for z_n and L) and the operations (taking the difference and the absolute value) are complex numbers and operations, not real ones. Intuitively, this definition says that a sequence of complex numbers converges to a limit L when every term far enough along the sequence (that is, every term past some Nth term) is close to L (that is, within an arbitrary small distance ε). The following examples illustrate how the definition works in practice.

Example 7.3.2   We use the definition of a convergent sequence to prove the sequence of complex numbers

$$\left\{ \frac{1}{(3+4i)^n} \right\}_{n=1}^{\infty}$$

converges to 0; that is, we prove

$$\lim_{n \to \infty} \frac{1}{(3+4i)^n} = 0.$$

Proof   Let $\varepsilon > 0$ be a given small value, and choose

$$N = \frac{\ln(1/\varepsilon)}{\ln(5)}.$$

Then, whenever $n > N$, we have

$$|z_n - L| = \left| \frac{1}{(3+4i)^n} - 0 \right| = \left| \frac{1}{3+4i} \right|^n = \left(\frac{1}{\sqrt{3^2+4^2}} \right)^n = \left(\frac{1}{5} \right)^n < \left(\frac{1}{5} \right)^N = \varepsilon.$$

The last equality follows from the choice of N, as $N \ln(5) = ln(1/\varepsilon)$ implies $\ln(5^N) = \ln(1/\varepsilon)$, and so $5^N = 1/\varepsilon$. We have therefore proven that the limit is zero, since we have shown $|z_n - L| < \varepsilon$ whenever $n > N$.

■

An important strategy to employ when using the definition is to find a choice for the positive number N that satisfies the right conditions. For most sequences, such as the one in example 7.3.2, no readily apparent value for N will exist, and so the choice for N may seem quite mysterious. As mentioned above, an effective strategy for identifying N is to work with the expression $|z_n - L|$, simplifying it as much as possible and comparing it to the same term with the smaller (though yet undetermined) N replacing n. Set this expression in N equal to ε and solve for N to obtain an appropriate choice of N. The next example more fully illustrates this strategy.

Example 7.3.3 We use the definition of a convergent sequence to prove that

$$\lim_{n \to \infty} \frac{n + i}{n - i} = 1.$$

Proof Let $\varepsilon > 0$ be given and employ the strategy just described to identify a choice for N. We have

$$|z_n - L| = \left| \frac{n + i}{n - i} - 1 \right| = \left| \frac{n + i - (n - i)}{n - i} \right| = \frac{|2i|}{|n - i|} = \frac{2}{\sqrt{n^2 + 1}} < \frac{2}{\sqrt{n^2 + 0}}.$$

The last inequality follows from the simple fact that replacing a denominator with a smaller value results in a larger fraction. Therefore,

$$|z_n - L| < \frac{2}{\sqrt{n^2 + 0}} = \frac{2}{n} < \frac{2}{N}$$

whenever $n > N$ (though N is as yet undetermined).

Now set $2/N = \varepsilon$ and solve to obtain $N = 2/\varepsilon$. With this choice of N, $|z_n - L| < \varepsilon$ whenever $n > N$, and so the limit is proven.

■

You can employ the same strategy to find an appropriate choice of positive number N to many other examples. The next question allows you to practice this technique in a straightforward way.

Question 7.3.1 Use the definition of a convergent sequence to prove each mathematical statement.

(a) $\displaystyle\lim_{n \to \infty} \frac{1}{(9 - 2i)^n} = 0$ 　　　　(b) $\displaystyle\lim_{n \to \infty} \frac{n}{(2i)^n} = 0$

■

Using the definition can be a slow and tedious approach to computing the limit of a sequence. Thankfully, the standard computational limit theorems so useful in real analysis are also at our disposal for complex sequences. These results can be used to find many limit values without reverting to the definition. The next theorem combines many helpful statements into a single list.

Theorem 7.3.1 *Let $c, L, M \in \mathbb{C}$ and let both $\{z_n\}$ and $\{w_n\}$ be complex sequences that converge to L and M respectively. Then the following hold.*

- **Limits are unique:**   *If* $\lim\limits_{n \to \infty} z_n = L$ *exists, then L is unique.*

- **Limit of a constant:** $\lim\limits_{n \to \infty} c = c$

- **Limit of a scalar multiple:** $\lim\limits_{n \to \infty} c \cdot z_n = c \cdot L$

- **Limit of a sum:** $\lim\limits_{n \to \infty} z_n + w_n = L + M$

- **Limit of a difference:** $\lim\limits_{n \to \infty} z_n - w_n = L - M$

- **Limit of a product:** $\lim\limits_{n \to \infty} z_n \cdot w_n = L \cdot M$

- **Limit of a quotient:** $\lim\limits_{n \to \infty} \dfrac{z_n}{w_n} = \dfrac{L}{M}$, *provided that* $M \neq 0$

- **Limit of a modulus:** $\lim\limits_{n \to \infty} |z_n| = |L|$

- **Limit of a conjugate:** $\lim\limits_{n \to \infty} \overline{z_n} = \overline{L}$

The proofs for each of the statements in theorem 7.3.1 are essentially the same as for real-valued sequences. Several of these proofs make essential use of the triangle inequality for the complex absolute value; we recall from exercise 49 in section 7.1 that for every $z, w \in \mathbb{C}$, we have $|z + w| \leq |z| + |w|$. To provide a flavor of the proofs of these types of statements, we prove in the next example the limit of a sum rule from theorem 7.3.1. The proofs of the others are similar (although some are a bit more algebraically complicated); we leave the rest for the exercises and for your further studies of mathematics.

Example 7.3.4 We use the definition of a convergent sequence to prove the limit of a sum rule; that is, we prove if $\{z_n\}$ and $\{w_n\}$ are complex sequences that converge respectively to L and M, then the sequence $\{z_n + w_n\}$ converges to $L + M$.

Proof Let $\varepsilon > 0$ be given. Since $\{z_n\}$ converges to L, for the given positive value $\varepsilon/2$ there exists a positive value N_z such that $|w_n - M| < \varepsilon/2$ whenever $n > N_z$. Similarly, there exists a positive value N_w such that $|z_n - L| < \varepsilon/2$ whenever $n > N_w$.

Now show $\{z_n + w_n\}$ satisfies the definition of a convergent sequence. For the given $\varepsilon > 0$, define $N = \max\{N_z, N_w\}$. Then whenever $n > N$, we have both $n > N_z$ and $N > N_w$. Applying the triangle inequality,

$$|(z_n + w_n) - (L + M)| = |(z_n - L) + (w_n - M)| \leq |z_n - L| + |w_n - M| < \frac{\varepsilon}{2} + \frac{\varepsilon}{2} = \varepsilon.$$

The result follows from the definition of limit.

∎

As we might hope and expect, theorem 7.3.1 can greatly simplify the process of evaluating the limits of convergent sequences. Consider the following example and question.

Example 7.3.5 We use theorem 7.3.1 to evaluate $\lim\limits_{n \to \infty} \dfrac{(n + 1) + (n + 1)i}{n - i}$.

Recall from example 7.3.3 that $\lim_{n\to\infty}(n+i)/(n-i) = 1$. Using this result and theorem 7.3.1,

$$\lim_{n\to\infty}\frac{(n+1)+(n+1)i}{n-i} = \lim_{n\to\infty}\frac{(n-i^2)+(ni+i)}{n-i} = \lim_{n\to\infty}\frac{ni-i^2+n+i}{n-i}$$

$$= \lim_{n\to\infty}\frac{ni-i^2}{n-i} + \frac{n+i}{n-i} = \lim_{n\to\infty} i + \frac{n+i}{n-i} = i+1.$$

■

Question 7.3.2 Determine the limit of each convergent sequence.

(a) $\left\{\dfrac{(7-3i)(n+1)(1+i)}{n-1}\right\}$

(b) $\left\{\dfrac{4n}{5n+i}\right\}$

(c) $\left\{\dfrac{in^2+3i}{(2-3i)n^2}\right\}$

(d) $\left\{\dfrac{(1-i)n}{n+1} + \dfrac{i}{n+1}\right\}$

■

There are many other theorems that help evaluate limits for complex sequences. We highlight one of these results before moving on to a discussion of power series.

Theorem 7.3.2 *If* $\lim\limits_{n\to\infty}|z_n| = 0$, *then* $\lim\limits_{n\to\infty} z_n = 0$.

The proof of this theorem uses the definition of the limit and follows quickly from the fact that $|z_n - 0| = ||z_n| - 0|$. We leave further details to the reader. The next example illustrates a typical application of this result.

Example 7.3.6 We use theorem 7.3.2 to evaluate $\lim\limits_{n\to\infty}\dfrac{1}{(in)}$.
First compute the modulus of the terms of this sequence as

$$\left|\frac{1}{in}\right| = \sqrt{0^2 + \left(\frac{-1}{n}\right)^2} = \frac{1}{n}.$$

Recall that $\lim\limits_{n\to\infty} 1/n = 0$. Therefore, by theorem 7.3.2, $\lim\limits_{n\to\infty} 1/(in) = 0$.

■

The next question provides valuable experience applying theorem 7.3.2; it is an often-applied and helpful result.

Question 7.3.3 Applying theorems 7.3.1 and 7.3.2 whenever necessary, determine the limit of each convergent sequence.

(a) $(1+i)^{-n}$

(b) $\left\{\dfrac{1}{(4+i)^n}\right\}$

(c) $\left\{\dfrac{n}{(2-3i)^n}\right\}$

(d) $\left\{(1+i)^{-n} + \dfrac{n+i}{n}\right\}$

■

With this understanding of complex sequences, we are now ready to discuss convergence aspects of complex power series.

7.3.2 Complex Power Series

A *complex series* is an infinite sum of complex-valued terms. Complex series are often denoted by $\sum_{n=0}^{\infty} w_n$ or simply $\sum w_n$ (where the indexing implicitly begins at $n = 0$); here each of terms w_n is in $\mathbb{C}$. To determine if the series $\sum w_n$ converges, we form the *sequence of partial sums* $\{S_k\}_{k=0}^{\infty} = \{w_0 + w_1 + w_2 + \cdots + w_k\}_{k=0}^{\infty}$. We say that the series $\sum w_n$ *converges* to a value L when the associated sequence of partial sums $\{S_k\}_{k=0}^{\infty}$ converges to L, and in this case we write $\sum w_n = L$. If a series does not converge, then we say that the series *diverges*. The next example illustrates these notions.

Example 7.3.7   We determine the convergence of the complex series $\sum_{n=0}^{\infty} i/2^n$.

The sequence of partial sums for this series is

$$S_0 = i, \quad S_1 = i + \frac{i}{2} = \frac{3i}{2}, \quad S_2 = i + \frac{i}{2} + \frac{i}{4} = \frac{7i}{4}, \quad S_3 = i + \frac{i}{2} + \frac{i}{4} + \frac{i}{8} = \frac{15i}{8}, \dots$$

From the pattern apparent from these first few terms, the general partial sum term and the associated limit are

$$S_n = i + \frac{i}{2} + \frac{i}{4} + \cdots + \frac{i}{2^n} = \frac{(2^{n+1} - 1)i}{2^n} \quad \text{and} \quad \lim_{n \to \infty} S_n = \lim_{n \to \infty} \left[2 - \frac{1}{2^n} \right]$$

$$i = (2 - 0)i = 2i.$$

Therefore the series converges to $2i$; we write $\sum i/2^n = 2i$.

∎

Question 7.3.4   List the first eight terms of the sequence of partial sums for the complex series $\sum_{n=0}^{\infty} i$. Using your result, find a general pattern for the terms in the sequence of partial sums, and show that the series $\sum_{n=0}^{\infty} i$ diverges.

∎

The rest of this section focuses on power series, which is a type of complex series that contain variable expressions and can be used to represent analytic functions. You can intuitively think of a power series as an infinite series extension of complex polynomials (which are finite series since they have finite degree). The next definition shows how complex power series have the same algebraic structure as real power series studied in calculus.

Definition 7.3.2   A **power series** is an infinite series of the form

$$\sum_{n=0}^{\infty} a_n(z - c)^n = a_0 + a_1(z - c) + a_2(z - c)^2 + \cdots + a_n(z - c)^n + \cdots .$$

For a complex power series, $a_n \in \mathbb{C}$ is the **nth coefficient** of the power series, z is a complex variable, and $c \in \mathbb{C}$ is a fixed constant. Such a series is said to be **expanded about the point** $z = c$.

A power series can be intuitively thought of as an infinite polynomial in the sense that the corresponding sequence of partial sums consists of complex polynomials. Every complex polynomial forms a power series in its own right, having coefficients $a_n = 0$ for every n greater than the degree of the polynomial.

When we substitute a given complex number into the variable z of a power series, the resulting series expression may converge or it may diverge. The set of values $z \in \mathbb{C}$ for which a given power series converges (as a limit of partial sums) is called the *region of convergence*. In general, any given power series expanded about $c \in \mathbb{C}$ turns out to converge for every complex number z in a disk $S = \{z : |z - c| < R\}$, where the real number R is known as the *radius of convergence*. The series also turns out to diverge for z outside of the disk and in the set $\{z : |z - c| > R\}$.

Theorem 7.3.3 *For a given power series $\sum a_n(z - c)^n$, one of the following scenarios characterizes the radius of convergence R:*

- *$R = 0$ and $\sum a_n(z - c)^n$ converges at the single point $z = c$;*
- *R is a positive real number so that $\sum a_n(z - c)^n$ converges for every complex point in the disk $S = \{z : |z - c| < R\}$, and additionally **may** converge for a value in the disk's boundary $\{z : |z - c| = R\}$;*
- *$R = \infty$ so that $\sum a_n(z - c)^n$ converges for every $z \in \mathbb{C}$.*

The proof of theorem 7.3.3 is left for your later studies in complex analysis. The result provides an elegant and simple characterization of the possible regions of convergence for complex power series. Figure 7.7 illustrates the geometry for a region of convergence when the radius of convergence R is a positive real number. The series converges inside the disk, diverges outside the disk, and converges for all, some, or none of the points on the boundary of the disk.

It might seem surprising that more-complicated sets in the complex plane cannot serve as regions of convergence; and yet theorem 7.3.3 guarantees that only a point, a disk, or the entire plane can act as regions of convergence. For each $z_0 \in \mathbb{C}$ for which the power series $\sum a_n(z - c)^n$ converges, we denote the corresponding series value by $f(z_0) = \sum a_n(z_0 - c)^n$. In this way a power series defines a function whose domain is the set of complex numbers for which the series converges. Those power series having an infinite radius of convergence $R = +\infty$ are called *entire*. An entire power series has a domain that consists of the entire complex plane $\mathbb{C}$.

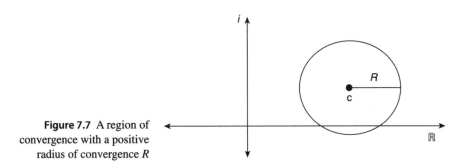

Figure 7.7 A region of convergence with a positive radius of convergence R

Example 7.3.8   Every complex polynomial is a power series with only a finite number of terms and converges for every $z \in \mathbb{C}$. Thus, every complex polynomial is entire.

In contrast, the power series $\sum_{n=0}^{\infty} z^n/2^n$ converges for some, but not every, complex number $z \in \mathbb{C}$. For example, it converges for $z = 1$, since $\sum_{n=0}^{\infty} 1/2^n = 2$. But it diverges for $z = 2$, since $\sum_{n=0}^{\infty} 2^n/2^n = \sum_{n=0}^{\infty} 1$. Thus, this series is not entire.

■

Theorem 7.3.3 has characterized the possibilities for the radius of convergence R; we now turn our attention to the practical calculation of the value of R for a given series. As was the case in calculus for the corresponding real-valued discussion, an important concept leading to this calculation is the absolute convergence of a power series.

Theorem 7.3.4   *If $\sum z_n$ is a complex series and the real-valued series $\sum |z_n|$ converges, then $\sum z_n$ also converges. When $\sum |z_n|$ converges, we say that $\sum z_n$ converges absolutely.*

Proof   To prove the result, separate $\sum z_n$ into its real and imaginary parts and then apply the corresponding version of this result for real-valued power series. Since a complex series converges exactly when the two series forming its real and imaginary parts both converge (that is, $\sum z_n = \sum (x_n + iy_n)$ converges when both $\sum x_n$ and $\sum y_n$ converge), absolute convergence of $\sum z_n$ can be analyzed in terms of absolute convergence of its real and imaginary parts.

Assume that $\sum |z_n|$ converges. Since $|z_n| \geq |x_n|$ and $|z_n| \geq |y_n|$, both $\sum |x_n|$ and $\sum |y_n|$ converge by the (real-valued) comparison test. Since real-valued series that converge absolutely always converge, both $\sum x_n$ and $\sum y_n$ are convergent series. But then $\sum z_n = \sum x_n + i\sum y_n$ converges.

■

Theorem 7.3.4 shows that the convergence of the real-valued series $\sum |a_n(z - c)^n|$ guarantees the convergence of the corresponding complex power series $\sum a_n(z - c)^n$. Therefore any applicable real-valued series convergence test familiar from calculus may help determine absolute convergence of complex power series. The ratio test, as described in the following theorem, is one such test that applies perfectly to the complex power series setting.

Theorem 7.3.5   The ratio test   *For a given complex power series $\sum a_n(z - c)^n$ having nonzero radius of convergence R,*

$$\frac{1}{R} = \lim_{n \to \infty} \left| \frac{a_{n+1}}{a_n} \right|$$

whenever the limit exists. When the limit is equal to 0, then $R = \infty$ and the power series is entire.

The proof of the ratio test can be found in any standard calculus text; these details are left to the reader. We focus on using the ratio test to determine the radius of convergence for a given power series.

Example 7.3.9 We employ the ratio test to prove that $f(z) = \sum_{n=0}^{\infty} z^n/n!$ is entire.

For this series, $a_n = 1/n!$. Applying the ratio test, we calculate the limit to determine the value for $1/R$:

$$\frac{1}{R} = \lim_{n\to\infty} \left| \frac{a_{n+1}}{a_n} \right| = \lim_{n\to\infty} \left| \frac{1}{(n+1)!} \cdot \frac{n!}{1} \right| = \lim_{n\to\infty} \left| \frac{1}{n+1} \right| = \lim_{n\to\infty} \frac{1}{n+1} = 0.$$

Therefore, $R = \infty$ and the series $\sum z^n/n!$ is entire.

■

Question 7.3.5 For each power series, explicitly identify the coefficient a_n for the nth term and determine the radius of convergence R.

(a) $\displaystyle\sum_{n=0}^{\infty} \frac{z^n}{n+3i}$

(b) $\displaystyle\sum_{n=0}^{\infty} \frac{n!z^n}{(3n+i)(n+1)!}$

(c) $\displaystyle\sum_{n=0}^{\infty} \frac{in(z-i)^n}{(n+2)!}$

(d) $\displaystyle\sum_{n=0}^{\infty} (z-2i)^n$

■

The real version of the power series considered in example 7.3.9 may be familiar from previous studies in mathematics. The (real) series $\sum x^n/n!$ is the Maclaurin series for the function $f(x) = e^x$, where $x \in \mathbb{R}$; we have $f(x) = e^x = \sum x^n/n!$. Mathematicians are naturally interested in the complex power series $\sum z^n/n!$ to determine its connection to the exponential function e^z. We will soon see that $e^z = \sum z^n/n!$ for every $z \in \mathbb{C}$. Example 7.3.9 shows that $\sum z^n/n!$ is entire, and so the function $f(z) = \sum z^n/n!$ is defined for all $z \in \mathbb{C}$. The following theorem characterizes the derivative of any complex power series, taking an important first step in understanding the analytic properties of a given power series.

Theorem 7.3.6 *A power series $f(z) = \sum_{n=0}^{\infty} a_n(z-c)^n$ with radius of convergence R is analytic in the disk $S = \{z : |z| < R\}$. The derivative of this series is $f'(z) = \sum_{n=1}^{\infty} na_n(z-c)^{n-1}$, and this derivative $f'(z)$ is also analytic with radius of convergence R. In this way, the power series $f(z)$ is differentiable any number of times, and each derivative has the same radius of convergence.*

Proof We prove that the derivative of $f(z) = \sum_{n=0}^{\infty} a_n(z-c)^n$ is $f'(z) = \sum_{n=1}^{\infty} na_n(z-c)^{n-1}$. The proof that the radius of convergence for f' is the same as the radius of convergence for f is developed in question 7.3.6 immediately following this proof.

We first assume that $c = 0$ (the proof for other values of c would then follow from a translation of the function to $g(z) = f(z+c)$). We are therefore assuming that z is a given point inside the disk of radius R centered at the origin, so that $|z| < R$. The strategy of this proof is to examine the distance between the series expression for $f'(z)$ given above and the difference quotient from the definition of the derivative. We will show that this distance approaches 0 as Δz approaches 0, which will imply that the derivative equals the series $f'(z) = \sum_{n=1}^{\infty} na_n z^{n-1}$.

Substituting the series expansion for $f(z)$ into the difference quotient and expanding, we obtain the following equality.

$$\frac{f(z+\Delta z) - f(z)}{\Delta z} - \sum_{n=1}^{\infty} na_n z^{n-1} = \sum_{n=1}^{\infty} a_n \left(\frac{(z+\Delta z)^n - z^n}{\Delta z} - nz^{n-1} \right).$$

For the moment we focus our attention on the nth term of the right-hand series; it satisfies the following sequence of equalities and inequalities, where justification for each algebraic step is listed.

$$\left| \frac{(z+\Delta z)^n - z^n}{\Delta z} - nz^{n-1} \right| = \left| \sum_{k=2}^{n} C(n,k) z^{n-k} \Delta z^{k-1} \right| \quad \text{Binomial theorem.}$$

$$\leq \sum_{k=2}^{n} C(n,k) |z|^{n-k} |\Delta z|^{k-1} \quad \text{Triangle inequality.}$$

$$< \sum_{k=2}^{n} C(n,k) |z|^{n-k} |\Delta z| \left[\frac{R-|z|}{2} \right]^{k-2} \quad \begin{array}{l} \text{As } \Delta z \to 0, \\ \text{we may assume } |\Delta z| < \dfrac{R-|z|}{2}. \end{array}$$

$$= |\Delta z| \left[\frac{R-|z|}{2} \right]^{-2} \sum_{k=2}^{n} C(n,k) |z|^{n-k} \left[\frac{R-|z|}{2} \right]^{k} \quad \text{Algebra.}$$

$$< |\Delta z| \left[\frac{R-|z|}{2} \right]^{-2} \sum_{k=0}^{n} C(n,k) |z|^{n-k} \left[\frac{R-|z|}{2} \right]^{k}$$

$$\text{Adding more terms.}$$

The final sum in this sequence is an expansion of a binomial; applying the Binomial theorem (but in the opposite direction) and simplifying implies

$$\left| \frac{(z+\Delta z)^n - z^n}{\Delta z} - z^{n-1} \right| < |\Delta z| \left[\frac{R-|z|}{2} \right]^{-2} \left[|z| + \frac{R-|z|}{2} \right]^{n}$$

$$= |\Delta z| \left[\frac{R-|z|}{2} \right]^{-2} \left[\frac{R+|z|}{2} \right]^{n}.$$

Putting it all together,

$$\left| \frac{f(z+\Delta z) - f(z)}{\Delta z} - \sum_{n=1}^{\infty} na_n z^{n-1} \right| = \left| \sum_{n=1}^{\infty} a_n \left(\frac{(z+\Delta z)^n - z^n}{\Delta z} - nz^{n-1} \right) \right|$$

$$< |\Delta z| \left[\frac{R-|z|}{2} \right]^{-2} \sum_{n=1}^{\infty} |a_n| \cdot \left[\frac{R+|z|}{2} \right]^{n}.$$

As Δz goes to 0, this last expression approaches 0, since z is fixed in the disk S and the series on the right converges by the ratio test. As discussed above, this fact proves $f'(z) = \sum_{n=1}^{\infty} n a_n z^{n-1}$.

■

Question 7.3.6 Complete the proof of theorem 7.3.6 for $c = 0$, answering the following questions and proving that if $f(z) = \sum_{n=0}^{\infty} a_n z^n$ has a radius of convergence R, then $f'(z) = \sum_{n=1}^{\infty} n a_n z^{n-1}$ also has radius of convergence R. The proof for power series with $c \neq 0$ is essentially identical to the proof outlined here.

(a) Prove that if $K \in \mathbb{C}$ with $|K| < 1$, then $\lim_{n \to \infty} n K^n = 0$. Hint: Use theorem 7.3.2.

(b) Let $r \in \mathbb{R}$ be a positive real number such that $|z| < r < R$. Use part (a) to prove that there exists an $N > 0$ such that $n|z|^{n-1} \leq r^n$ for all integers $n > N$. Conclude that $n|a_n||z|^{n-1} \leq |a_n|r^n$ for all $n > N$. Furthermore, since $r < R$, we know that the series $f(r) = \sum_{n=0}^{\infty} a_n r^n$ converges absolutely by the ratio test.

■

According to theorem 7.3.6, every complex power series is analytic in the region determined by its radius of convergence. As it turns out, the converse result is also true; that is, every analytic function can be expressed as a complex power series. This powerful and important result is obtained by investigating the *Taylor series* representation of analytic functions. Consider the following result.

Theorem 7.3.7 Taylor's theorem *Every function $f(z)$ analytic in the disk $S = \{z : |z - c| < R\}$ can be represented in terms of its Taylor series*

$$f(z) = f(c) + f'(c)(z - c) + \frac{f''(c)}{2!}(z - c)^2 + \cdots + \frac{f^{(n)}(c)}{n!}(z - c)^n + \cdots ,$$

which converges to $f(z)$ for every $z \in S$. In this formula, the notation $f^{(n)}(c)$ denotes the nth derivative of f evaluated at c.

The statement of this theorem is identical in structure to the real analysis version of Taylor's theorem that can be found in just about any calculus text. However the standard proof of theorem 7.3.7 relies on the "Cauchy integral formula" and is therefore beyond the scope of our investigation; we leave the proof for your later studies of complex analysis. Taylor's theorem "closes the loop" on this section's discussion of power series representations of complex functions. Every function analytic in a disk of radius R can be represented by a convergent power series, and every convergent power series is analytic inside a disk with radius of convergence R. These facts combine to form a *representation theory* for analytic functions. The next example begins an exploration of complex functions represented or defined in terms of their Taylor series.

Example 7.3.10   We determine the power series representation of the complex function $f(z) = e^z$.

Recall from theorem 7.2.4 in section 7.2 that the complex exponential function is analytic (for every $z \in \mathbb{C}$). Furthermore, in question 7.2.6 in section 7.2, you showed that the derivative of $f(z) = e^z$ is $f'(z) = e^z$, and so $f^{(n)}(z) = e^z$ for all $n \in \mathbb{N}$. Applying Taylor's theorem to $f(z) = e^z$ to obtain the power series $\sum a_n z^n$ about $c = 0$,

$$a_n = \frac{f^{(n)}(0)}{n!} = \frac{e^0}{n!} = \frac{1}{n!}, \quad \text{and so} \quad e^z = \sum_{n=0}^{\infty} \frac{z^n}{n!}.$$

Finally, example 7.3.9 indicates that the $\sum z^n/n!$ is entire, showing that its representation for $f(z) = e^z$ is valid on all of $\mathbb{C}$. Because, as in this case, entire power series represent functions that are always analytic (for all $z \in \mathbb{C}$), we often call analytic functions entire.   ■

Question 7.3.7   Use Taylor's theorem to find the power series representation for the function $1/(1 - z)$, expanded about $c = 0$. Then determine the radius of convergence for the resulting power series expression.   ■

We now consider complex trigonometric functions; recall that the real-valued functions $\cos x$ and $\sin x$ have Taylor series

$$\cos x = \sum_{n=0}^{\infty} (-1)^n \frac{x^{2n}}{(2n)!} \quad \text{and} \quad \sin x = \sum_{n=0}^{\infty} (-1)^n \frac{x^{2n+1}}{(2n + 1)!}.$$

Motivated by the fact that the complex exponential function agreed in algebraic structure with its real-valued counterpart, we define the complex cosine and sine functions as complex-valued generalizations of these familiar real power series. In addition, the other four complex trigonometric functions are defined as the familiar ratios of these two functions.

Definition 7.3.3   *For $z \in \mathbb{C}$,*

$$\cos z = \sum_{n=0}^{\infty} (-1)^n \frac{z^{2n}}{(2n)!} \quad \text{and} \quad \sin z = \sum_{n=0}^{\infty} (-1)^n \frac{z^{2n+1}}{(2n + 1)!}.$$

Furthermore, whenever the denominator is nonzero, we define

$$\tan z = \frac{\sin z}{\cos z}, \quad \cot z = \frac{\cos z}{\sin z}, \quad \sec z = \frac{1}{\cos z}, \quad \text{and } \csc z = \frac{1}{\sin z}.$$

As you may recall from calculus, the other four real trigonometric functions can also be represented as real Taylor series (in addition to this familiar definition in terms of ratios). At this point you won't be surprised to learn that the Taylor series representation of the complex trigonometric functions is the same as for the real power series written with a complex variable z. For example, the real and complex tangent functions are:

$$\tan x = x + \frac{x^3}{3} + \frac{2x^5}{15} + \frac{17x^7}{315} + \cdots \quad \text{and} \quad \tan z = z + \frac{z^3}{3} + \frac{2z^5}{15} + \frac{17z^7}{315} + \cdots.$$

Whenever defining a new function, mathematicians seek to ascertain its basic properties. When defined in terms of a power series, an initial question of interest is to determine the function's radius of convergence, and hence its domain. In this case, the following question asks you to prove that the complex cosine and sine functions are defined (converge) for every complex number.

Question 7.3.8   Use the ratio test to prove that both $\sin z$ and $\cos z$ are entire.   ■

In addition to determining the radius of convergence, we might wonder about the relationships among functions in light of their respective power series representations. Indeed, many important algebraic properties of functions can be readily proven by examining the features of corresponding power series representations. The following example provides one such instance.

Example 7.3.11   We show that the exponential power series representation is consistent with Euler's formula: if $z = it$ with $t \in \mathbb{R}$, then $e^{it} = \cos t + i \sin t$.
Example 7.3.10 showed that $e^z = \sum z^n/n!$. Setting $z = it$,

$$e^z = e^{it} = \sum_{n=0}^{\infty} \frac{(it)^n}{n!} = \sum_{n=0}^{\infty} \frac{i^n t^n}{n!}.$$

Since $i^n = i$ if $n = 1 \bmod 4$, $i^n = -1$ if $n = 2 \bmod 4$, $i^n = -i$ if $n = 3 \bmod 4$, and $i^n = 1$ if $n = 0 \bmod 4$, this summation splits into even and odd terms based on n. Reindexing, we obtain the desired formula

$$e^{it} = \sum_{n=0}^{\infty} \frac{i^n t^n}{n!} = \sum_{n=0}^{\infty} (-1)^n \frac{t^{2n}}{(2n)!} + i \sum_{n=0}^{\infty} (-1)^n \frac{t^{2n+1}}{(2n+1)!} = \cos t + i \sin t.$$

■

Question 7.3.9   Following the strategy illustrated in example 7.3.11, prove that Euler's formula holds for every $z \in \mathbb{C}$; that is, prove that if $z \in \mathbb{C}$, then $e^{iz} = \cos z + i \sin z$.   ■

The verification of Euler's formula given in example 7.3.11 is based on the composition of a complex power series (in particular, the power series $e^z = \sum z^n/n!$) with an analytic function (in this case, the function $f(z) = it$) to obtain a new power series. After algebraic manipulation, we obtained Euler's formula. We can employ the same composition strategy when examining other complex power series and functions to obtain further power series representations and identities. The following example illustrates this point.

Example 7.3.12   We determine a power series representation for e^{-z^2}.
We may compose the known power series representation for e^z with the term $-z^2$, obtaining:

$$e^z = \sum_{n=0}^{\infty} \frac{z^n}{n!} \qquad \Rightarrow \qquad e^{(-z^2)} = \sum_{n=0}^{\infty} \frac{(-z^2)^n}{n!} = \sum_{n=0}^{\infty} \frac{(-1)^n z^{2n}}{n!}.$$

■

Question Prove the following algebraic identities by composing the power series for e^z with
7.3.10 iz and $-iz$ and performing the indicated algebraic operations.

$$\cos z = \frac{e^{iz} + e^{-iz}}{2} \qquad \text{and} \qquad \sin z = \frac{e^{iz} - e^{-iz}}{2i}.$$

∎

We note that many complex analysis textbooks *define* the complex trigonometric functions $\cos z$ and $\sin z$ using the formulas in question 7.3.10. The formulas are easily applied to evaluate the trigonometric functions at specific values of $z \in \mathbb{C}$; for example,

$$\cos i = \frac{e^{i^2} + e^{-i^2}}{2} = \frac{e^{-1} + e}{2} \qquad \text{and} \quad \sin i = \frac{e^{i^2} - e^{-i^2}}{2i} = \frac{i(e - e^{-1})}{2}.$$

Hence $\cos i$ is real and $\sin i$ is purely imaginary. Values for the other trigonometric functions follow similarly; for example,

$$\tan i = \frac{\sin i}{\cos i} = i\frac{e - e^{-1}}{e^{-1} + e} \approx -0.7616i.$$

There are many more enjoyable aspects to the study of complex power series. For example, a remarkable theorem of Cauchy's expresses the coefficients of an analytic function's power series in terms of a complex-valued integral (instead of the nth derivative). Additionally, the results in this section may be applied to important complex functions such as $f(z) = \text{Log}(z)$ or the complex inverse trigonometric functions. Hopefully this introduction to analytic functions has motivated you to continue investigating them in your later studies.

7.3.3   Reading Questions for Section 7.3

1. Define and give an example of a complex sequence.
2. Define what it means for a sequence to converge or diverge and give an example of each.
3. State theorem 7.3.1. How is this result helpful to a study of complex sequences?
4. Give an example of a complex series and display its associated sequence of partial sums.
5. Define what it means for a series $\sum z_n$ to converge absolutely, converge, or diverge.
6. Define and give an example of a power series. What is the corresponding sequence of partial sums?
7. Define the radius of convergence for a given power series.
8. Define and give an example of an entire function.
9. State the ratio test. How is this result helpful for a study of complex power series?
10. State theorem 7.3.6 and Taylor's theorem. What does this pair of theorems say about the relationship between analytic functions and complex power series with positive radius of convergence?
11. State Euler's formula.
12. Define the complex cosine and sine functions.

7.3.4　Exercises for Section 7.3

In exercises 1–8, use the definition of a convergent sequence to prove each mathematical statement.

1. $\lim\limits_{n\to\infty} \dfrac{1}{(4-3i)^n} = 0$

2. $\lim\limits_{n\to\infty} \dfrac{9+2i}{(4-3i)n} = 0$

3. $\lim\limits_{n\to\infty} 5i + \dfrac{1}{in} = 5i$

4. $\lim\limits_{n\to\infty} \dfrac{2in}{3n+1} = \dfrac{2i}{3}$

5. $\lim\limits_{n\to\infty} |e^{in}| = 1$

6. $\lim\limits_{n\to\infty} \dfrac{n+i^n}{n} = 1$

7. $\lim\limits_{n\to\infty} Re\, Log\left[\dfrac{n+i}{n-i}\right] = 0$

8. $\lim\limits_{n\to\infty} Arg\left[\dfrac{1}{n}+i\right] = \dfrac{\pi}{2}$

    Hint: $Arg(x+iy) = \arctan y/x$
    when $-\pi/2 < Arg(z) < \pi/2$

In exercises 9–18, determine the limit of each complex sequence, expressing your answer in rectangular form.

9. $\left\{\dfrac{2n+i}{n}\right\}_{n=1}^{\infty}$

10. $\left\{\dfrac{1}{(1+3i)^n}\right\}_{n=1}^{\infty}$

11. $\left\{\dfrac{(4-i)(2n+1)}{n-i}\right\}_{n=1}^{\infty}$

12. $\left\{\dfrac{1-[\frac{1}{3}(1+i)]^{n+1}}{1-\frac{1}{3}(1+i)}\right\}_{n=1}^{\infty}$

13. $\left\{\dfrac{i\sqrt{2n}+4}{n^2+2n}\right\}_{n=1}^{\infty}$

14. $\left\{\dfrac{in^2-2in+i}{n^2-1}\right\}_{n=0}^{\infty}$

15. $\{(x+iy)^n\}_{n=0}^{\infty}$　for $|x+iy| < 1$

16. $\{(3+2i)z^n + (5-2i)\}_{n=0}^{\infty}$　for $|z| < 1$

17. $\left\{\dfrac{1-z^{n+1}}{1-z}\right\}_{n=0}^{\infty}$　for $|z| < 1$

18. Prove $\lim\limits_{n\to\infty} Log\left[i+\dfrac{1}{n}\right] = i\dfrac{\pi}{2}$.

    Hint: Use exercise 8.

In exercises 19–28, prove each mathematical statement about complex sequences.

19. The limit of a complex sequence is unique; from theorem 7.3.1.

20. The limit of a constant sequence rule from theorem 7.3.1.

21. The limit of a scalar multiple of a sequence rule from theorem 7.3.1.

22. The limit of a difference of sequences rule from theorem 7.3.1.

23. The limit of a product of sequences rule from theorem 7.3.1.

24. The limit of a modulus rule from theorem 7.3.1.

25. There exists a sequence $\{z_n\}$ such that $\lim\limits_{n\to\infty} |z_n|$ exists, but for which $\lim\limits_{n\to\infty} z_n$ does not exist.

26. The limit of a conjugate rule from theorem 7.3.1.

27. Theorem 7.3.2; if $\lim\limits_{n\to\infty} |z_n| = 0$, then $\lim\limits_{n\to\infty} z_n = 0$.

28. If $\{x_n\}$ is a real sequence with $\lim\limits_{n\to\infty} x_n = L$, then $\lim\limits_{n\to\infty} e^{i\cdot x_n} = e^{i\cdot L}$.

In exercises 29–36, explicitly identify the coefficient a_n for the nth term of the complex power series and use the ratio test to determine the radius of convergence R.

29. $\sum\limits_{n=0}^{\infty} \dfrac{z^n}{2n-i}$

30. $\sum\limits_{n=0}^{\infty} \dfrac{n^2 z^n}{2n(3n+i)}$

31. $\displaystyle\sum_{n=0}^{\infty} \frac{(2i)^n z^n}{n!}$

32. $\displaystyle\sum_{n=0}^{\infty} \frac{n! z^n}{(2i)^n}$

33. $\displaystyle\sum_{n=0}^{\infty} \frac{(z-2i)^n}{(6i)^n}$

34. $\displaystyle\sum_{n=0}^{\infty} \frac{n(z-2i)^n}{n+i}$

35. $\displaystyle\sum_{n=0}^{\infty} \frac{(z+i)^n}{n!(5+i)^n}$

36. $\displaystyle\sum_{n=0}^{\infty} \frac{(z+i)^n}{(5+i)^n}$

In exercises 37–42, use Taylor's theorem to determine a power series representation for each complex function expanded about the given complex point $c \in \mathbb{C}$. Note that

$$\frac{d}{dz}\text{Log } z = \frac{1}{z}$$

and that

$$f'(z) = \frac{1}{1+z^2}$$

for $f(z) = \arctan(z)$, the inverse function for $\tan(z)$.

37. $f(z) = \text{Log}(z)$ about $c = 1$
38. $f(z) = \text{Log}(z)$ about $c = i$
39. $f(z) = \frac{1}{z}$ about $c = 1$

40. $f(z) = \frac{1}{z}$ about $c = i$
41. $f(z) = \arctan(z)$ about $c = 0$
42. $f(z) = \sqrt{z}$ about $c = -1$

In exercises 43–50, use known power series representations for analytic functions to determine power series representations expanded about $c = 0$ for each complex function.

43. $f(z) = ie^{-z}$
44. $f(z) = ze^{z^2}$
45. $f(z) = e^z + e^{-z}$
46. $f(z) = 4 + 3i - \cos z$

47. $f(z) = 5i \cos(iz)$
48. $f(z) = \cos\left[\frac{(3+i)z}{1+z}\right]$
49. $f(z) = \sin(iz/5)$
50. $f(z) = \sin(iz/5) + i \cos z$

In exercises 51–54, evaluate each expression, writing the answer in rectangular form.

52. $\sin(i\pi)$
53. $\cos(\pi + i)$

54. $\cot(i)$
55. $\sec(10i)$

Exercises 55–58 consider the derivatives of the complex trigonometric functions.

55. Prove that the derivative of $f(z) = \cos z$ is $f'(z) = -\sin z$ using the power series representations from definition 7.3.3.
56. Prove that the derivative of $g(z) = \sin z$ is $g'(z) = \cos z$ using the power series representations from definition 7.3.3.
57. Prove that the derivative of $f(z) = \cos z$ is $f'(z) = -\sin z$ and the derivative of $g(z) = \sin z$ is $g'(z) = \cos z$ using the ratio from question 7.3.10.
58. Differentiate the complex trigonometric functions $\tan z$, $\cot z$, $\sec z$, and $\csc z$ using the quotient rule and the derivatives from exercises 55–57.

In exercises 59–62, differentiate each function.

59. $f(z) = \cos(z^2 + 1)\sin^2(5z^3 + iz)$
60. $f(z) = e^{iz}\tan^2(2z)$

61. $f(z) = \sec(z^3 + i)\csc^2(2iz^2 + iz)$
62. $f(z) = \cot^6(iz)$

Exercises 63–64 consider the relationship between the complex cosine and sine functions and the hyperbolic cosine and sine functions. Recall that for every real value t,

$$\cosh t = \frac{e^t + e^{-t}}{2} \qquad \text{and} \qquad \sinh t = \frac{e^t - e^{-t}}{2}.$$

In exercises 63–64, prove each algebraic identity for $z = x + iy \in \mathbb{C}$.

63. $\cos(x + iy) = \cos x \cosh y - i \sin x \sinh y$
64. $\sin(x + iy) = \sin x \cosh y + i \cos x \sinh y$

In exercises 65–66, prove each mathematical statement about series.

65. If $n \in \mathbb{N}$ and $z_k \in \mathbb{C}$, then $\left| \sum_{k=1}^{n} z_k \right| \leq \sum_{k=1}^{n} |z_k|$. Hint: Use induction on $n \in \mathbb{N}$.

66. If $\sum_{k=1}^{\infty} z_k = L < \infty$, then $\left| \sum_{k=1}^{\infty} z_k \right| \leq \sum_{k=1}^{\infty} |z_k|$.

Exercises 67–70 consider sequences that are defined based on iterations of a selected complex function. Given a function $f(z)$ and a complex point z_0, a resulting iterated sequence is $\{z_n\} = \{f(z_{n-1})\}$, where $n = 1, 2, 3, \ldots$. In 1918 the French mathematicians Gaston Julia and Pierre Fatou determined that many of these sequences produce fractal images. Following Julia and Fatou's lead, we study iterations resulting from quadratic complex functions of the form $f_c(z) = z^2 + c$ where $c \in \mathbb{C}$ is constant.
In exercises 67–70, answer the following questions about $f_c(z)$.

67. Prove each statement about $f_0(z) = z^2$.

   (a) If $z_0 \in \mathbb{C}$ with $|z_0| < 1$, then the sequence $\{z_n\} = \{f_0(z_{n-1})\}$ converges to $L = 0$.
   (b) If $z_0 \in \mathbb{C}$ with $|z_0| > 1$, then the sequence $\{z_n\} = \{f_0(z_{n-1})\}$ diverges to $L = +\infty$; that is, given any real value $M > 0$, there exists $N \in \mathbb{N}$ such that $n > N$ implies $|z_n| > M$.
   (c) If $z_0 \in \mathbb{C}$ with $|z_0| = 1$, then the sequence $\{z_n\} = \{f_0(z_{n-1})\}$ either oscillates around the unit circle in the complex plane or converges to $L = 1$. Hint: Examine the polar representation of z_n based on the polar representation of z_0.

68. Working with $f_0(z) = z^2$ and the results from exercise 67, prove that the set of $z_0 \in \mathbb{C}$ for which the sequence $\{z_n\} = \{f_0(z_{n-1})\}$ does not diverge to infinity is the closed unit disk $\overline{D} = \{z : |z| \leq 1\}$.

69. Consider $f_{-2}(z) = z^2 - 2$. Prove that the sequence $\{z_n\} = \{f_{-2}(z_{n-1})\}$ does not diverge to infinity for any $z_0 \in S$, where $S = \{z : z \in \mathbb{R} \text{ and } -2 \leq z \leq 2\}$.

70. The functions f_0 and f_{-2} considered in exercises 67–69 are the only functions of the form $f_c(z) = z^2 + c$ with a simple set of values z_0 for which the sequence $\{z_n\} = \{f_0(z_{n-1})\}$ does not diverge to infinity. For every other $c \in \mathbb{C}$, this set is a "fractal," which is a self-replicating set as defined for the Sierbinski triangle in section 5.2. The determination of every element of these sets and

their corresponding illustrations is impractical without the aid of a computer. For this exercise, consider $f_i(z) = z^2 + i$ and prove that $z_0 = 0$, $z_0 = i$, and $z_0 = -i$ are in the set of values S for which the the sequence $\{z_n\} = \{f_i(z_{n-1})\}$ does not diverge to infinity.

Hint: Determine the sequence $\{z_n\}$ by direct computation.

7.4   Harmonic Functions

Real and complex differentiable functions have proven themselves important in mathematicians' efforts to model and understand the real world. Many applications involve differentiable functions, since an application studying a functional quantity's change with respect to an independent variable will often involve the derivative. In real-life situations involving multiple variables (and hence multivariate functions), mathematicians have come to recognize that other categories of functions share a similar importance, including those that are harmonic.

Harmonic functions arise naturally in many physical applications. Their study began in the 1800s with Joseph Fourier's investigations of temperatures and temperature change. In 1807, Fourier published his seminal article *On the Propagation of Heat in Solid Bodies*, which developed a mathematical model for the physical behavior of heat—a model in which harmonic functions serve as the fundamental entities for describing temperature change. In addition, harmonic functions are useful in the study of electromagnetism, aerodynamics, and fluid flow, in which they are used to reduce two—and three—dimensional vector fields to single-variable functions; we explore this type of application in section 7.5. In these settings, physicists and mathematicians often refer to harmonic functions as "potential functions" and harmonic analysis (that is, the study of harmonic functions) as "potential theory."

Though their labels involve the same term, a harmonic function arises in a *different* physical setting than an object that follows the familiar term *harmonic motion*—an object in "harmonic motion" is behaving in an oscillating movement common in nature. The pendulum problem in section 4.8 is an example of *simple harmonic motion*; the differential equation describing the pendulum's position u is $d^2u/dt^2 + c^2u = 0$ for a constant c. In contrast, harmonic functions satisfy Laplace's equation $u_{xx} + u_{yy} = 0$. They can also be described from a physical perspective as satisfying the *maximum principle*. This principle asserts that a harmonic function $u(x, y)$ defined on a two-dimensional disk cannot have a maximum or minimum *inside* the disk unless $u(x, y)$ is a constant function. Therefore, the maximum and minimum must always occur on the boundary.

The maximum principle makes harmonic functions useful when mathematically modeling real-life settings. For example, suppose we are studying the temperature $u(x, y, t)$ of a point (x, y) on a circular griddle at time t. The griddle may cool or warm in various ways, but as time goes on (that is, as time approaches infinity), the griddle's temperature becomes stable and approaches a limit $u(x, y)$. This limit $u(x, y)$ is called the *steady-state* temperature. The steady-state temperature does not change over time (mathematically, there is no dependence on time t as a result of the limiting process).

In addition, the steady-state temperature at one point in the interior of the griddle cannot be greater than any nearby point; if the temperature was greater at some point, then (in the limiting process over time) the heat at that point would flow out to the other nearby points resulting in a different steady-state distribution. Therefore the maximum of the steady-state limit $u(x, y)$ cannot occur in the interior of the griddle—exactly the physical manifestation of the maximum principle for $u(x, y)$. The function $u(x, y)$ must be harmonic!

Other physical quantities have harmonic steady-state values satisfying the maximum principle. Another example is the vertical position of a drumhead membrane stretched inside a circular frame whose edge may be bent out of the horizontal plane. The steady-state position of such a membrane cannot be highest at any point on the interior of the membrane; if the highest point were on the interior, then the lower position of the surrounding points would collectively "pull" down the membrane from this high point. The study of harmonic functions does not lead to only simple trivia that might inform drum-design in some way. Instead, it has contributed to a better understanding of a number of important elements of our modern lifestyle, such as electromagnetic radiation and the corresponding development of radio, television, and cell-phone technologies.

From these physical descriptions you might conjecture that a harmonic function's value on a circular boundary would completely determine the function's behavior at all the interior points of the corresponding disk. This conjecture is known as the *Dirichlet problem* and is correct, provided the function $u(x, y)$ satisfies certain continuity properties. In short, a continuous function on the circular boundary will have exactly one continuous extension into the interior that is harmonic. The proof of this claim is an important element in the advanced study of harmonic analysis, but is beyond the scope of this text and is left for later studies that include complex integration.

Perhaps at this point you are convinced that the study of harmonic functions is both important and mathematically interesting. We might now begin to ask several of the many corresponding questions: How are these functions defined? What are some examples of harmonic functions? What properties are satisfied by harmonic functions? We devote the rest of this section to developing and exploring the answers to these questions, and we begin with the definition of a harmonic function.

Definition 7.4.1 *A function $u(x, y)$ is **harmonic in a disk** $S = \{z : |z - c| < R\}$ with a fixed center $c \in \mathbb{C}$ and a positive radius R if both second partial derivatives of $u(x, y)$ are continuous in S and if $u(x, y)$ satisfies **Laplace's equation** at every point in S:*

$$\frac{\partial^2 u}{\partial x^2} + \frac{\partial^2 u}{\partial y^2} = 0.$$

*If $S = \mathbb{C}$ (that is, if $R = \infty$), then we say that $u(x, y)$ is harmonic on the entire plane $\mathbb{C}$, or simply that the function is **harmonic**.*

Laplace's equation $u_{xx} + u_{yy} = 0$ was first isolated by the French mathematician Pierre Simon Laplace while he was studying gravity and its relation to planetary motion. As we can see from the definition, proving that a given function $u(x, y)$ is harmonic over a disk involves two main steps: verifying that u satisfies Laplace's equation and verifying that both second partial derivatives u_{xx} and u_{yy} are continuous. The following example illustrates this process.

Example 7.4.1   We prove that $u(x, y) = 3x^2 + 5x - 3y^2 + 2$ is harmonic (on the entire complex plane).

We first compute the second partials for $u(x, y)$. Differentiating $u(x, y)$ with respect to x, we obtain $u_x = 6x + 5$ and so $u_{xx} = 6$; similarly, differentiating $u(x, y)$ with respect to y, we obtain $u_y = -6y$ and so $u_{yy} = -6$. Since all constant functions are continuous at every point on the plane, both partial derivatives $u_{xx} = 6$ and $u_{yy} = -6$ are continuous. In addition, Laplace's equation is satisfied at every point on the plane since $u_{xx} + u_{yy} = 6 + (-6) = 0$. Therefore $u(x, y) = 3x^2 + 5x - 3y^2 + 2$ is harmonic.

∎

If we extend the definition of continuity of a multivariate function $f(x, y)$ to include possibly the case that f is complex-valued, then a function does not have to be real-valued to be harmonic. For example, if $u(x, y)$ and $v(x, y)$ are both harmonic, then so is $f(x, y) = u(x, y) + iv(x, y)$, since the sum of two continuous functions is continuous. Among the more important results of this section is the theorem and proof that the real and imaginary parts of every analytic function are harmonic, and so any complex analytic function (such as a complex polynomial) is harmonic. The next example illustrates the way that a complex polynomial satisfies the Laplacian.

Example 7.4.2   We prove that $f(z) = z^3$ is harmonic.

First write f as a function $f(x, y)$ of two variables x and y by setting $z = x + iy$. Applying the binomial theorem from section 5.2 (or using a direct computation),

$$f(x, y) = = f(x + iy) = (x + iy)^3 = (x^3 - 3xy^2) + i(3x^2y - y^3).$$

Computing both second partial derivatives,

$$f_x = 3x^2 - 3y^2 + i6xy \qquad \Rightarrow \quad f_{xx} = 6x + i6y,$$
$$f_y = -6xy + i(3x^2 - 3y^2) \qquad \Rightarrow \quad f_{yy} = -6x - i6y.$$

Every multivariate polynomial is continuous at every point on the plane, and so both f_{xx} and f_{yy} are continuous everywhere. In addition, $f(x, y)$ satisfies Laplace's equation because $f_{xx} + f_{yy} = 6x + i6y + (-6x - i6y) = 0$. Therefore the analytic function $f(z) = z^3$ is harmonic on all of $\mathbb{C}$.

∎

The same strategy can be used to determine if any complex function $f(z)$ is harmonic at a point or on the plane: substitute $z = x + iy$ to write $f(z)$ as the multivariate function $f(x, y)$, and then verify that f satisfies the properties of a harmonic function. The next questions gives practice is working with both real and complex harmonic functions.

Question 7.4.1   Prove each function is harmonic on the given set.

   (a) $u(x, y) = x^4 - 6x^2y^2 + y^4$ on $\mathbb{C}$.
   (b) $u(x, y) = \arctan\left[\frac{y}{x}\right]$ having domain equal to the disk with radius 1 and center
       $c = (1, 0)$. Hint: Recall that the derivative of $f(t) = \arctan t$ is $f'(t) = 1/(1 + t^2)$.
   (c) $f(z) = z^4$ on $\mathbb{C}$.
   (d) $f(z) = \text{Log}(z)$ defined on the disk with radius 1 and center $c = 1$.

   ■

The real and imaginary parts of every analytic function are harmonic. As we will
see in the proof of the following theorem, the Cauchy–Riemann equations play an
important role in verifying that every analytic function has this property.

Theorem 7.4.1   *If $f(z) = u(x, y) + iv(x, y)$ is a complex function that is analytic on a disk S, then
the real part $u(x, y)$ and the imaginary part $v(x, y)$ are harmonic in S.*

   Proof   Consider the second partial derivatives u_{xx}, u_{yy}, v_{xx}, and v_{yy}. It turns out
   that complex differentiability implies continuity of these functions; since f is
   differentiable at any point in S, it and the second partial derivatives are continuous
   there.

   Verify that $u(x, y)$ satisfies Laplace's equation. Since f is analytic in S, u
   and v satisfy the Cauchy–Riemann equations in S: $u_x = v_y$ and $u_y = -v_x$.
   Differentiating each side of the first equation with respect to x gives $u_{xx} = v_{yx}$.
   Similarly, differentiating each side of the second equation with respect to y gives
   $u_{yy} = -v_{xy}$. Now apply an important result from multivariable calculus: if the
   second partial derivatives of any multivariate function g are continuous, then
   $g_{yx} = g_{xy}$ (that is, mixed partial derivatives are equal to one another). Applying
   this result to v at any point in the disk S,

   $$u_{xx} + u_{yy} \;=\; v_{yx} - v_{xy} \;=\; v_{yx} - v_{yx} \;=\; 0.$$

   The proof that $v(x, y)$ satisfies Laplace's equation is similar. Thus, both the real
   part $u(x, y)$ and the imaginary part $v(x, y)$ are harmonic in S.

   ■

The proof of theorem 7.4.1 used in a key way the fact that u and v satisfy
the Cauchy–Riemann equations. It turns out that any pair of real-valued functions
satisfying the Cauchy–Riemann equations and having continuous second partials are
harmonic; we will next show that any single real-valued harmonic function can also
be thought of as the real part and/or the imaginary part of an analytic function. This
result might seem extraordinarily "natural"; since any given analytic function can be
"broken apart" to obtain harmonic functions, we might wonder if a given pair of real-
valued harmonic functions can be pieced together to obtain an analytic function. The
delightful fact is that they always can be, so long as the two harmonic functions are
paired correctly—one is called the "harmonic conjugate" of the other. In addition, up
to an arbitrary real constant, each given real-valued harmonic function has exactly
one harmonic conjugate; it's a bit like a marriage made in heaven. When u is a given
real-valued harmonic function, we often denote its harmonic conjugate by $v = u^*$.

The marriage of the two functions (adding them together as real and imaginary parts) always forms an analytic function. This result is extremely important in applications; it means that an analysis of real-valued harmonic functions (which arise in many real-world situations) can be examined through an investigation of complex analytic functions. The next theorem describes the details of this fact, which you can think of as a partial converse to theorem 7.4.1.

Theorem 7.4.2 *If the function $u(x, y)$ is real-valued and harmonic in a disk S centered at $z_0 = x_0 + iy_0$, then there exists a function $v(x, y)$ harmonic in S such that $f(z) = u(x, y) + iv(x, y)$ is analytic in S. Such a function $v(x, y)$ is called the **harmonic conjugate of $u(x,y)$** and is uniquely determined by $u(x, y)$ up to an arbitrary real constant; we write $v(x, y) = u^*(x, y)$.*

The proof of theorem 7.4.2 is constructive; it provides an algorithm for producing the harmonic conjugate $v = u^*$ for a given harmonic function u. The algorithm is based on the process of "partial integration." We first discuss and illustrate this process and then give the proof of theorem 7.4.2.

A real two-variable function $f(x, y)$ can be integrated to obtain an antiderivative with respect to either variable x or y. If we integrate $f(x, y)$ with respect to x, then the integral $\int f(x, y)\, dx$ is calculated by treating the variable y as a "constant" and antidifferentiating f with respect to x. The result is a real two-variable function $F(x, y) = \int f(x, y)\, dx$ that satisfies the partial differential equation

$$\frac{\partial F}{\partial x} = \frac{\partial}{\partial x}\left[\int f(x, y)\, dx\right] = f(x, y).$$

A parallel result holds when integrating $f(x, y)$ with respect to y.

The function $F(x, y)$ obtained from the partial integration process will contain an addition of an arbitrary constant. Because the partial integration was with respect to only one variable (and the other variable was held constant), this arbitrary constant is actually a function in the variable treated as a constant. When integrating $f(x, y)$ with respect to x, the arbitrary constant is therefore a function of y, say $C(y)$. Similarly, integrating $f(x, y)$ with respect to y results in an addition of a function $C(x)$. The next example illustrates this notion.

Example 7.4.3 We integrate $f(x, y) = 5y\cos(x + y^2)$ with respect to each variable x and y.
First integrate with respect to x, obtaining $\int 5y\cos(x + y^2)\, dx = 5y\sin(x + y^2) + C(y)$. The constant of integration is a function of y. Now integrate with respect to y, obtaining $\int 5y\cos(x + y^2)\, dy = 5/2\sin(x + y^2) + C(x)$.

∎

Question 7.4.2 Integrate each function with respect to the variables x and y.
(a) $x^2 - y^2$ (b) $x^2 e^y$

∎

The partial integration process discussed in example 7.4.3 produces an indefinite integral; the corresponding "definite partial integral" evaluates the indefinite integral at the given limits of integration. If $F(x, y) = \int f(x, y) \, dx$, then we define $\int_a^b f(x, y) \, dx = F(x, y)\big|_{x=a}^{x=b} = F(b, y) - F(a, y)$. The next example illustrates the process.

Example 7.4.4   We evaluate $\int_0^\pi 5y \cos(x + y^2) \, dx$.

The corresponding indefinite partial integral is $\int 5y \cos(x + y^2) \, dx = 5y \sin(x + y^2)$. Hence,

$$\int_0^\pi 5y \cos(x + y^2) \, dx = 5y \sin(x + y^2)\big|_0^{x=\pi} = 5y \sin(\pi + y^2) - 5y \sin(y^2).$$

■

Question 7.4.3   Evaluate each definite partial integral.

(a) $\displaystyle\int_0^3 x^2 - y^2 \, dx$

(c) $\displaystyle u(x, y) = \int_0^x s^2 - y^2 \, ds$

(b) $\displaystyle\int_2^4 x^2 - y^2 \, dy$

(d) $\displaystyle v(x, y) = \int_0^y x^2 - t^2 \, dt$

■

We are now ready to prove theorem 7.4.2; the proof gives a constructive algorithm to produce the desired harmonic conjugate $v(x, y)$ for a given harmonic function $u(x, y)$.

Proof of theorem 7.4.2   Assume that $u(x, y)$ is harmonic in a disk S centered at $z_0 = x_0 + iy_0$. Define $f(z) = f(x + iy) = u(x, y) + iv(x, y)$, where $v = u^*$ is defined as

$$v(x, y) = \int_{x_0}^x -\frac{\partial u(s, y)}{\partial y} \, ds + \int_{y_0}^y \frac{\partial u(x_0, t)}{\partial x} \, dt = \int_{x_0}^x -u_y(s, y) \, ds + \int_{y_0}^y u_x(x_0, t) \, dt.$$

The fact that u is harmonic (and hence that its second partial derivatives are continuous) and properties of the integral turn out to make v and its second partial derivatives continuous. We claim $v(x, y)$ is a harmonic function in S for which f is analytic on S. This statement would follow completely from the fact that u and v together satisfy the Cauchy–Riemann equations in S—the Cauchy–Riemann theorem would then guarantee that f is analytic on S, and theorem 7.4.1 would prove the rest. We therefore need only verify that $f = u + iv$ satisfies the Cauchy–Riemann equations.

First compute the partial derivative v_x, using the fundamental theorem of calculus to simplify the resulting expression as follows

$$v_x(x, y) = \frac{\partial}{\partial x}\left[\int_{x_0}^x -u_y(s, y) \, ds + \int_{y_0}^y u_x(x_0, t) \, dt\right] = -u_y(x, y) + 0 = -u_y(x, y).$$

Therefore $v_x(x, y) = -u_y(x, y)$ for any $(x, y) \in S$, and one of the Cauchy–Riemann equations is satisfied. For the other equation, calculate and simplify f_y as follows:

$$f_y = u_y + iv_y$$

$$= u_y + i \int_{x_0}^{x} -u_{yy}(s, y) \, ds + iu_x(x_0, y) \qquad \text{Fundamental theorem on } v(x, y).$$

$$= u_y + i \int_{x_0}^{x} u_{xx}(s, y) \, ds + iu_x(x_0, y) \qquad u(x, y) \text{ is harmonic, so } u_{xx} = -u_{yy}.$$

$$= u_y + i\big[u_x(x, y) - u_x(x_0, y)\big] + iu_x(x_0, y) \quad \text{Fundamental theorem on } u_x(x, y).$$

$$= u_y + iu_x \qquad \text{Algebra cancellation.}$$

Since the imaginary parts of $f_y = u_y + iv_y = u_y + iu_x$ must be equal, we obtain the other Cauchy–Riemann equation $u_x = v_y$. The result follows.

■

Every real-valued harmonic function u has a harmonic conjugate v. We can implement the algorithm outlined in the proof to identify the harmonic conjugate $v = u^*(x, y)$ for a given real-valued harmonic function $u(x, y)$. Consider the following example.

Example 7.4.5   We determine the harmonic conjugate of $u(x, y) = x^2 - y^2 + 6x + 2y$.

We first verify that $u(x, y)$ is harmonic on the complex plane. Computing the partial derivatives, $u_x = 2x + 6$ and so $u_{xx} = 2$. Similarly $u_y = -2y + 2$, and so $u_{yy} = -2$. Since the second partial derivatives are constants, they are continuous at every point on the plane. Finally, direct substitution verifies that Laplace's equation is satisfied at every point on the plane since $u_{xx} + u_{yy} = 2 + (-2) = 0$. By theorem 7.4.2, the harmonic conjugate $v = u^*$ is $v(x, y) = \int_{x_0}^{x} -u_y(s, y) \, ds +$ $\int_{y_0}^{y} u_x(x_0, t) \, dt$. We are free to choose any base point (x_0, y_0) on the plane when using this formula (different choices result in different arbitrary constants being added to the conjugate function); we choose $x_0 = y_0 = 0$ and obtain the harmonic conjugate

$$v(x, y) = \int_0^x 2y - 2 \, dt + \int_0^y 2 \cdot 0 + 6 \, ds = 2yt - 2t\big|_0^x + 6s\big|_0^y = 2xy - 2x + 6y.$$

Instead of defining the harmonic conjugate in terms of an integral *formula*, a straightforward *process* leads to the harmonic conjugate function $u^*(x, y)$. The process uses the Cauchy–Riemann equations $u_x = v_y$ and $u_y = -v_x$ For any given real-valued harmonic function u, compute u_x and u_y. Apply the Cauchy–Riemann equations to realize these functions as v_y and $-v_x$. Using partial integration, integrate the first function v_y with respect to y obtaining $v(x, y) + C_1(x)$, where $C_1(x)$ is a function of x. Also integrate with respect to x the negation v_x of the second function, obtaining $v(x, y) + C_2(y)$, where $C_2(y)$ is a function of y. A comparison of the two resulting expressions will determine the conjugate function v.

To illustrate the process, examine $u(x, y) = x^2 - y^2 + 6x + 2y$ as in the last example; we have $v_y = u_x = 2x + 6$ and $= -v_x = u_y = -2y + 2$ (hence $v_x = 2y - 2$). Integrate both sides of these equations with respect to x and y, respectively, to obtain

$$2xy + 6y + C_1(x) = v(x, y) \quad \text{and} \quad 2xy - 2x + C_2(y) = v(x, y).$$

Comparing the two equations, we set $C_1(x) = -2x$ and $C_2(y) = 6y$, and conclude that the harmonic conjugate, determined up to a real-valued constant, is $v(x, y) = u^*(x, y) = 2xy + 6y - 2x$.

■

Question 7.4.4 Find the analytic function $f(z)$ with real part $Re(f) = e^x \cos y$ by the following steps.

(a) Prove that $u(x, y) = e^x \cos y$ is harmonic on $\mathbb{C}$.
(b) Find a harmonic conjugate $v(x, y)$ of $u(x, y) = e^x \cos y$.
(c) Prove that $f(z) = u(x, y) + iv(x, y)$ is entire by computing its derivative.

■

Question 7.4.5 Example 7.4.2 verified that the complex monomial $f(z) = z^3$ is harmonic. This question considers an arbitrary monomial in z of any degree $n \in \mathbb{N}$.

(a) Substitute $z = x + iy$ into z^n and use the binomial theorem to obtain

$$u(x, y) = \sum_{k=0}^{\lfloor n/2 \rfloor} (-1)^k \frac{n!}{(2k)!(n - 2k)!} x^{n-2k} y^{2k},$$

where $\lfloor \ \rfloor$ denotes the greatest integer function. Prove that $u(x, y)$ is harmonic by verifying $u(x, y)$ satisfies Laplace's equation.
(b) Find the harmonic conjugate $u^*(x, y)$ for $u(x, y)$ given in (a) and explicitly state the corresponding analytic function $f(z) = u(x, y) + iu^*(x, y)$.

■

There are many important and unexpected properties of harmonic functions that we have not had a chance to explore in this brief introduction. Many of these properties involve integration over paths in the complex plane—topics discussed in any complete course in complex analysis. We've already highlighted the important property known as the maximum principle. Another is the famous *mean value property*—a startling fact about a harmonic function's value at a given point being dependent on the function's value at surrounding points. The harmonic function's value $u(x_0, y_0)$ at a given point $z_0 = x_0 + iy_0$ turns out to equal the average value of u taken over *any* circle of points in S that is centered at z_0. You might expect such a rigid property to only be satisfied by very simple functions (such as constant functions), but harmonic functions can be quite sophisticated and intricate (as *any* analytic function's real and imaginary parts are harmonic). This dichotomy of facts about harmonic functions between a sophisticated analytic structure and relatively simple configuration properties characterizes harmonic functions as useful and elegant mathematical objects.

The next section considers one of many important applications of harmonic functions. Many people are surprised to learn that complex numbers can be applied to the real-world in any meaningful way, perhaps believing the term "imaginary number" suggests that complex analysis must be abstract mathematical game-playing. Instead, the powerful analytic tools of complex analysis have made many complicated real-world models manageable.

7.4.1 Reading Questions for Section 7.4

1. Name two physical quantities whose study motivates the definition of harmonic functions.
2. Describe the maximum principle for harmonic functions.
3. Define and give an example of a continuous function $u(x, y)$ in two variables.
4. What property of continuous functions is helpful for evaluating limits?
5. State the definition of a harmonic function, using Laplace's equation.
6. Give an example of a real-valued function that is harmonic and one that is not.
7. If $f(z)$ is entire, what are two associated real-valued harmonic functions that are harmonic conjugates?
8. Identify two real-valued harmonic functions associated with $f(z) = e^z$.
9. Define and give an example of a harmonic conjugate of a harmonic function $u(x, y)$.
10. What formula does the proof of theorem 7.4.2 give to produce the harmonic conjugate of a given harmonic function $u(x, y)$?
11. What does the mean value property say about a function u harmonic in a disk S?

7.4.2 Exercises for Section 7.4

In exercises 1–10, show that each function is harmonic in the plane by calculating the Laplacian.

1. $u(x, y) = 4x^3 + 12x^2 - 12y^2 - 12xy^2$
2. $u(x, y) = x^5 - 10x^3y^2 + 5xy^4$
3. $u(x, y) = xy^3 - x^3y$
4. $u(x, y) = (3 - i)(x + iy)^3 + x$
5. $u(x, y) = e^{2i(x+iy)}$
6. $u(x, y) = 2ix + (3 - i)y - e^{2i(x+iy)}$
7. $u(x, y) = \sin x \cos(iy)$
8. $u(x, y) = e^{x+1} \cos(y + 1)$
9. $u(x, y) = e^{ix} \cos(iy)$
10. $u(x, y) = e^{(1+i)x} \sin((1 + i)y)$

In exercises 11–16, show that each function is harmonic inside the disk S centered at $z = x + iy = 1$ with radius 1.

11. $u(x, y) = \dfrac{x}{x^2 + y^2}$
12. $u(x, y) = y - \dfrac{y}{x^2 + y^2}$
13. $u(x, y) = \ln(x^2 + y^2)$
14. $u(x, y) = \ln\left(|3(x + iy)|^4\right)$
15. $u(x, y) = \text{Arg}(x + 1 + iy)$
16. $u(x, y) = \dfrac{x - iy}{x^2 + y^2}$

In exercises 17–28, determine whether each function is harmonic on the plane. For exercises 27 and 28, recall that $\cosh y = (e^y + e^{-y})/2$ and $\sinh y = (e^y - e^{-y})/2$.

17. $u(x, y) = 4x^3 + 12x^2 - 4y^2 - 4xy^2$

18. $u(x, y) = x^5 - 10x^3y^2 + 5xy^4$

19. $u(x, y) = x^3 - y^3$

20. $u(x, y) = x^n - y^n$ for $n \in \mathbb{N}$

21. $u(x, y) = xy^3 - x^3y$

22. $u(x, y) = 3(x + iy)^2$

23. $u(x, y) = \sin x \cos y$

24. $u(x, y) = \sin(x^2 + 1) \cos(y^2 + 1)$

25. $u(x, y) = e^{5x+3} \cos(5y + 3)$

26. $u(x, y) = e^{x^2} \cos y^2$

27. $u(x, y) = \cos x \cosh y$

28. $u(x, y) = \sin x \sinh y$

In exercises 29–38, directly prove the real and imaginary parts of each analytic function are harmonic.

29. $f(z) = 2z + 4$

30. $f(z) = mx + b$ for $m, b \in \mathbb{C}$

31. $f(z) = z^2 + z$

32. $f(z) = z^3 + z^2$

33. $f(z) = e^z$

34. $f(z) = \text{Log } z$

35. $f(z) = \cos z$

36. $f(z) = \sin z$

37. $f(z) = a_n z^n + \cdots + a_1 z + a_0$ for $a_k \in \mathbb{C}$

38. $f(z) = \sum\limits_{k=0}^{\infty} a_k z^k$ for $a_k \in \mathbb{C}$ and the series convergent on $\mathbb{C}$

In exercises 39–49, prove each mathematical statement about harmonic functions.

39. If $m, b \in \mathbb{C}$, then the linear function $u(x, y) = m(x + iy) + b$ is harmonic.

40. A scalar multiple of a harmonic function is harmonic; that is, if $c \in \mathbb{C}$ and $u(x, y)$ is harmonic, then $cu(x, y)$ is harmonic.

41. The sum of two harmonic functions is harmonic.

42. The difference of two harmonic functions is harmonic.

43. Linear combinations of harmonic functions are harmonic; that is, if $a, b \in \mathbb{C}$ and both $u(x, y)$ and $v(x, y)$ are harmonic, then $au(x, y) + bv(x, y)$ is harmonic.

44. The square of a harmonic function is sometimes not harmonic.

45. The product of two harmonic functions is sometimes not harmonic.

46. The composition of a harmonic function $u(x, y)$ with two real-valued differentiable functions $g(x)$ and $h(y)$ to form $u(g(x), h(y))$ is sometimes not harmonic.

47. If $a, b, c \in \mathbb{R}$ and $u(x, y)$ is harmonic, then $u(ax + b, ay + c)$ is harmonic.

48. If $f(z)$ is entire, then $g(z) = e^{f(z)}$ is harmonic.

49. If $c \in \mathbb{C}$, then $u(x, y) = cx^2 - cy^2$ is harmonic.

In exercises 50–59, find the harmonic conjugate for each harmonic function and state the corresponding analytic function.

50. $u(x, y) = 3x(1 - y)$

51. $u(x, y) = 2x^2 - 2y^2 + 5$

52. $u(x, y) = 13x^4 - 78x^2y^2 + 13y^4$

53. $u(x, y) = 3x^2 - x - 3y^2$

54. $u(x, y) = 3x^2y - y^3$

55. $u(x, y) = e^y \sin x$

56. $u(x, y) = e^{x^2-y^2} \sin(2xy)$

57. $u(x, y) = \ln(x^2 + y^2)$

58. $u(x, y) = \sin x \cosh y$

59. $u(x, y) = \sin x \sinh y$

In exercises 60–62, prove each mathematical statement about harmonic conjugates.

60. If $v(x, y)$ is a harmonic conjugate of $u(x, y)$, then $-u(x, y)$ is a harmonic conjugate of $v(x, y)$

61. If $v(x, y)$ is a harmonic conjugate of $u(x, y)$, then $h(x, y) = u^2 - v^2$ is a harmonic function.

62. If $u(x, y)$ is harmonic, then the harmonic conjugate $v(x, y)$ is unique up to a complex constant.

Hint: Assume v_1 is a different harmonic conjugate, and apply the Cauchy–Riemann equations to calculate

$$\frac{\partial}{\partial x}(v - v_1) \quad \text{and} \quad \frac{\partial}{\partial x}(v - v_1).$$

7.5   Application: Streamlines and Equipotentials

The study of harmonic functions is important in many applications, such as fluid dynamics or aerodynamics—two situations where real-life phenomenon can be modeled using a mathematical object known as a "vector field." The complexity of studying a vector field (which typically has a high-dimensional structure) can often be reduced to an analytical discussion of real-valued harmonic functions. As we have learned in section 7.4, every harmonic function can be paired with a harmonic conjugate to obtain a complex-valued analytic function; the analytic structure of this function can in turn provide information about the original vector field. The history of the study of harmonic functions can be traced to applied settings, and so the real and imaginary parts of an analytic function are often referred to using terms from these fields: the real part $u(x, y)$ is sometimes called the *potential function*, and its harmonic conjugate $v(x, y)$ is called the *stream function*. This section describes the relationships that exist between harmonic functions and vector fields and explains how these notions are used to model physical phenomena in the world around us.

We start with the concept of vectors, which are often first studied in either a multivariable calculus or a linear algebra course; they also play a prominent role in the study of many aspects of physics and engineering. A vector $\vec{v}$ is a mathematical object having both direction and magnitude; if $\vec{v}$ is a two-dimensional vector, then it is geometrically represented by an arrow in a two-dimensional plane (such arrows are often called *directed line segments*). Of course, two vectors that have the same direction and length are equal (or equivalent), and so the directed line segment representing $\vec{v}$ can be placed anywhere in the plane. We represent two-dimensional vectors analytically in "component form" as $\vec{v} = \langle a, b \rangle$. The next definition makes these ideas precise.

Definition 7.5.1   *A **vector** is a mathematical object expressing both magnitude and direction. Geometrically, a vector is a **directed line segment** in the plane. If a vector $\vec{v}$ can be placed on the plane so that the initial end of $\vec{v}$ is at the origin and the*

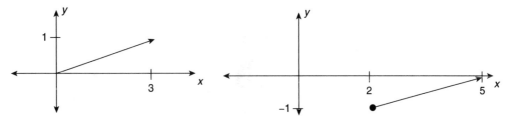

Figure 7.8 The vector ⟨3, 1⟩ drawn in two positions on the complex plan

terminal end of $\vec{v}$ is at the point (a, b), then the **component form** *of the vector is* $\vec{v} = \langle a, b \rangle$. *The real numbers a and b in this expression are called the* **components** *of $\vec{v}$, and the* **zero vector** *is denoted by* $\vec{0} = \langle 0, 0 \rangle$.

The definition describes vectors as having magnitude and direction, and not position. Placing the initial end of a vector at the origin is helpful for designating its component form, and this location is referred to as the vector's *standard position*. But a vector can be placed anywhere in the plane; any two vectors in the plane with the same direction and magnitude are said to be *equivalent* (or equal) regardless of where they are located. Figure 7.8 provides a graphical illustration of the vector ⟨3, 1⟩ in two different locations: in standard position (on the left) and emanating from the point $(2, -1)$ (on the right).

Question 7.5.1 Sketch each vector, both in standard position and emanating from the point $(-2, 3)$.

(a) $\langle 0, 2 \rangle$ (c) $\langle 0, 0 \rangle$

(b) $\langle -1, -2 \rangle$ (d) $\langle -3, 0 \rangle$

■

Operations can be defined on vectors, such as addition, subtraction, multiplication by scalars and two distinct vector products. In this section we need only define one operation on vectors—the length (or norm) operation, which gives the vector's magnitude. In terms of the component form (which identifies a vector in terms of its x and y components), the formula for the vector length is the familiar distance formula.

Definition 7.5.2 *The* **length** *or* **norm** *of a vector* $\langle a, b \rangle$ *is* $\| \langle a, b \rangle \| = \sqrt{a^2 + b^2}$.

Example 7.5.1 We compute the length of three vectors.

- The length of $\langle 2, 2 \rangle$ is $\| \langle 2, 2 \rangle \| = \sqrt{2^2 + 2^2} = 2\sqrt{2}$.
- The length of $\langle 0, 1 \rangle$ is $\| \langle 0, 1 \rangle \| = \sqrt{0^2 + 1^2} = 1$.
- The length of the zero vector $\vec{0} = \langle 0, 0 \rangle$ is $\| \langle 0, 0 \rangle \| = \sqrt{0^2 + 0^2} = 0$.

■

Vectors with length one are called *unit vectors*. The zero vector $\vec{0}$ is the unique vector of length zero.

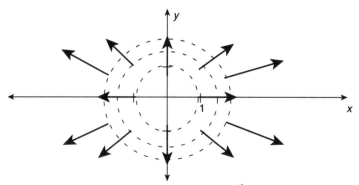

Figure 7.9   A representative plot of the vector field $\vec{F}(x, y) = \langle x, y \rangle$ for example 7.5.2

Question 7.5.2   Compute the length of each vector from question 7.5.1.

(a) $\langle 0, 2 \rangle$ (c) $\langle 0, 0 \rangle$

(b) $\langle -1, -2 \rangle$ (d) $\langle -3, 0 \rangle$ ∎

We now consider mappings known as *vector fields*, which are mathematical objects having the structure of a function—an input and an output. The input is a point in the plane, and the output is a vector. Vector fields are commonly used in physics to represent force fields and velocity fields. The following definition expresses this notion.

Definition 7.5.3   *A two-dimensional **vector field** is a function F whose domain is the set of ordered pairs of real numbers and whose target space is the set of two-dimensional vectors. For every such vector field F, there exist multivariate real-valued functions $M(x, y)$ and $N(x, y)$ such that $\vec{F}(x, y) = \langle M(x, y), N(x, y) \rangle$.*

A representative plot of a vector field $\vec{F}$ is formed by selecting a finite number of points (a, b) in the plane and sketching each vector $\vec{F}(a, b)$ so that it emanates from the point. When sketching a representative plot of a vector field, mathematicians sometimes find it helpful to group the vectors based on their respective lengths. A given vector field $\vec{F}(x, y) = \langle M(x, y), N(x, y) \rangle$ has length at the point (a, b) equal to $\|\vec{F}(a, b)\| = \sqrt{[M(a, b)]^2 + [N(a, b)]^2}$. The next example provides a representative plot of a vector field.

Example 7.5.2   We sketch the plot of $\vec{F}(x, y) = \langle x, y \rangle$.

The vector field $\vec{F}$ has $M(x, y) = x$ and $N(x, y) = y$. The length of any vector $\vec{F}(x, y) = \langle x, y \rangle$ is $\|\vec{F}(x, y)\| = \|\langle x, y \rangle\| = \sqrt{x^2 + y^2}$. Therefore vectors $\vec{F}$ with equal length $r = \sqrt{x^2 + y^2}$ all emanate from points (x, y) in a circular group centered at the origin with radius r.

Figure 7.9 provides a representative plot of $\vec{F}(x, y) = \langle x, y \rangle$, using 12 vectors to indicate the general pattern of the vectors in the vector field $\vec{F}$. The figure organizes vectors in circular groups that share the same length. For example, the four vectors drawn closest to the origin are $\langle 0, 1 \rangle$, $\langle 1, 0 \rangle$, $\langle -1, 0 \rangle$, and $\langle 0, -1 \rangle$, and all four have length one.

∎

Question 7.5.3   Determine the length of an arbitrary vector in each vector field and sketch a representative plot of each vector field using at least 12 vectors. Based on this plot, provide a written description of the vector field.

(a) $\vec{F}(x,y) = \langle 0, y \rangle$

(b) $\vec{F}(x,y) = \left\langle \dfrac{-y}{\sqrt{x^2+y^2}}, \dfrac{x}{\sqrt{x^2+y^2}} \right\rangle$

■

What do harmonic functions have to do with vector fields? Many vector fields that show up in applications are of a special structure; they are the "gradient" operation applied to a multivariate function $f(x, y)$. The gradient is sometimes referred to as a vector differential "operator"; it maps $f(x, y)$ to vector fields whose components are the partial derivatives of f. The following definition describes the gradient.

Definition 7.5.4   *Applied to multivariate functions, the* **gradient** *operation is denoted by*

$$\vec{\nabla} = \langle \frac{\partial}{\partial x}, \frac{\partial}{\partial y} \rangle,$$

and the gradient of a function $u(x, y)$ is the vector field

$$\vec{\nabla} u = \left(\frac{\partial u}{\partial x}, \frac{\partial u}{\partial y} \right) = \langle u_x, u_y \rangle.$$

Any vector field that can be expressed as the gradient $\vec{\nabla} u$ of a function is called a **conservative** *vector field, and the corresponding function u is called the* **potential** *function for $\vec{\nabla} u$. If $u(x, y)$ is a real-valued harmonic function, then the harmonic conjugate $u^*(x, y)$ is called the* **stream** *function.*

The next example describes a conservative vector field in terms of the gradient operator, focusing the discussion around analytic and harmonic functions.

Example 7.5.3   We apply the gradient operator to produce a conservative vector field that corresponds to the real part of the complex analytic function $f(z) = z^2$.

First expand $f(z) = z^2 = (x + iy)^2 = x^2 + 2xyi - y^2$ to determine the real and imaginary parts, which are harmonic. These functions are $u(x, y) = x^2 - y^2$ together with its harmonic conjugate $v(x, y) = 2xy$. Therefore u is the potential function and v is the stream function for the conservative vector field

$$\vec{\nabla} u = \langle u_x, u_y \rangle = \langle 2x, -2y \rangle.$$

■

Example 7.5.3 points out that an analytic function $f(z)$ can play a prominent role in identifying the potential function for a conservative vector field. In light of this connection, such functions $f(z)$ are sometimes referred to as the *complex potential*.

Question 7.5.4   Use the gradient operator to determine the conservative vector field for each given complex potential $f(z)$.

(a) $f(z) = z^2 - z$ (c) $f(z) = e^{2z}$

(b) $f(z) = z^3$ (d) $f(z) = \cos z$

■

7.5.1   Fluid Flow

Vector fields are often used to model various types of force fields and velocity fields—including the forces and velocities associated with fluid flow. The choice for the mathematical model of the flow is in terms of real-valued functions $M(x, y)$ and $N(x, y)$ forming a vector field $\vec{F}(x, y) = \langle M(x, y), N(x, y) \rangle$. The vector field provides information about the velocity of the fluid at any given point (x, y); $\vec{F}$ is often referred to as a *velocity field*. The velocity field has an output that is two-dimensional, and so the model is therefore assuming that the fluid is determined (by its velocity) in a given two-dimensional x–y plane, which is equivalent to $\mathbb{C}$. In short, the vectors $\vec{F}(a, b)$ describe the velocity of the fluid at each point $a + ib \in \mathbb{C}$. The model assumes that there is no time dependence; that is, a time variable t does not appear in the equations defining $\vec{F}(x, y)$, and the fluid's velocities are unchanging at each point. Such time-independent fluid flow is commonly said to be *stationary* (with respect to time).

This model focuses on fluid flows that are "incompressible" and "irrotational." An *incompressible* fluid flow occurs when the density of the fluid is constant throughout the fluid. An *irrotational* fluid flow occurs when the fluid is "circulation free." As it turns out, a fluid flow that is incompressible and irrotational is a conservative vector field with a harmonic potential function. In addition, the potential and stream function determine the two most important characteristics of the flow, as the following theorem describes.

Theorem 7.5.1   *Let $\vec{F}$ be an incompressible and irrotational fluid flow.*

(a) *The velocity field for the flow is a vector field of the form $\vec{\nabla}u$, where $u(x, y)$ is a harmonic function. In this setting, the function u is called the* **harmonic potential**, *the* **velocity potential**, *or the* **scalar potential**. *A representative plot of $\vec{\nabla}u$ provides an image of the velocity of the fluid at given points. Curves of the form $u(x, y) = C$ for real constants C are called* **equipotentials**.

(b) *The harmonic conjugate $v(x, y) = u^*(x, y)$ determines the path of an object caught in the flow as $v(x, y) = C$, where C is a real constant. The graphs of the equations $v(x, y) = C$ are called* **streamlines**.

(c) *For real constants C, the streamlines $v(x, y) = C$ and equipotentials $u(x, y) = C$ form an orthogonal system; that is, the curves intersect at right angles.*

The graphical presentation is helpful when considering theorem 7.5.1. The next examples illustrate its ideas through a depiction of different velocity fields.

Example 7.5.4   We apply theorem 7.5.1 to the incompressible and irrotational fluid flow having complex potential $f(z) = z^2/4 = (x + iy)^2/4$.

Example 7.5.3 used the fact that z^2 has real part $x^2 - y^2$ and imaginary part $2xy$. Therefore $f(z) = z^2/4$ has real part $u(x, y) = (x^2 - y^2)/4$ and imaginary part $v(x, y) = xy/2$. Since f is analytic, u and v are harmonic. We determine the velocity field for the flow by taking the gradient of $u(x, y)$, obtaining $\vec{\nabla}u = \langle x/2, -y/2 \rangle$. A representative plot of $\vec{\nabla}u$ provides a graphical illustration of the velocity of the fluid at each point on the plane; figure 7.10 indicates this flow in the first quadrant.

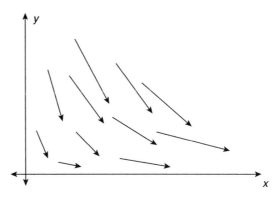

Figure 7.10 A first-quadrant plot of
$\vec{\nabla}u = \langle x/2, -y/2 \rangle$ for
example 7.5.4

Theorem 7.5.1(b) describes the streamlines for this velocity field; each one corresponds to the path followed by an object trapped in the fluid flow. For any real constant C,

$$v(x, y) = C \quad \Rightarrow \quad \frac{xy}{2} = C \quad \Rightarrow \quad y = \frac{2C}{x}.$$

To find the path of an object dropped into the flow at any given point (a, b), substitute $x = a$ and $y = b$ into the streamline equation, solve for the constant C, and conclude that the object must travel along the streamline path having that constant.

For example, if an object caught in the flow passes through the point $(3, 5)$, then (substituting $x = 3$ and $y = 5$) we have $C = xy/2 = 3 \cdot 5/2 = 15/2$. The corresponding streamline is therefore $y = 2 \cdot [15/2] \cdot 1/x = 15/x$. As illustrated in figure 7.11, the graph of this streamline may be superimposed on the velocity vector field to provide a geometric description for the object's path.

Finally, we consider an illustrative graph for theorem 7.5.1 (c) when $\langle x/2, -y/2 \rangle$. Since the streamlines and the equipotentials form an orthogonal system, the corresponding curves intersect at right angles. The streamlines just calculated are of the form $y = 2C/x$, where $C \in \mathbb{R}$. The harmonic potential is

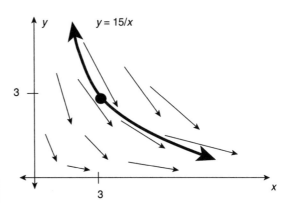

Figure 7.11 Streamline through
$(3, 5)$ for example 7.5.4

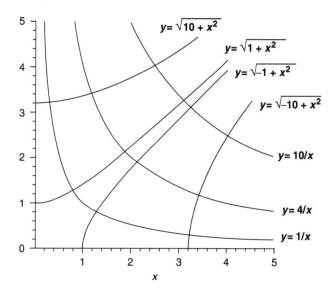

Figure 7.12 The orthogonal system of streamlines and equipotentials for example 7.5.4

$u(x, y) = (x^2 - y^2)/4$, and so the equipotentials $u(x, y) = C$ are of the form

$$\frac{x^2 - y^2}{4} = C \quad \Rightarrow \quad y = \sqrt{4C + x^2}.$$

Figure 7.12 provides an illustration of the orthogonal system in the first quadrant with streamlines chosen using $C = 0.5, 2$, and 5, and equipotentials using $C = -2.5, -0.25, 0.25$, and 2.5.

∎

The next question gives practice in analyzing a comprehensive model of a fluid flow.

Question 7.5.5 Answer each question about a fluid flow with complex potential $f(z) = z$. Assume the fluid is incompressible and irrotational and apply the fluid flow model from theorem 7.5.1.

(a) State the harmonic potential $u(x, y)$ and its harmonic conjugate $v(x, y)$ for this fluid flow.

(b) Exhibit the corresponding expression for the velocity field $\vec{\nabla} u$. Sketch a representative plot of the velocity field by graphing the vectors at the following points:

$(0, 0), \ (-1, 2), \ (1, 2), \ (-1, -1), \ (1, -1), \ (-1, 2), \ (1, 2), \ (-2, 0), \ (2, 0).$

Based on this plot, provide a written description of the fluid flow.

(c) State the general equation for the streamlines and the equipotentials, and plot the streamlines and equipotentials for $C = 1, 3, 5$ on the same graph (as modeled in figure 7.12).

(d) Determine the path followed by an object dropped in the fluid at $(3, 5)$.

∎

Many other applications besides fluid flow follow a similar vector field model that is generated from a complex potential. For example in the study of electrostatics, a famous theorem by Gauss implies that the two-dimensional electric field (which describes the force, or distribution and intensity of electric charge) is a conservative vector field with a harmonic potential. Simply put, such an electric field has the form $\vec{\nabla}u$ where $u(x, y)$ is harmonic. Gauss's theorem provided a mathematical description of Benjamin Franklin's famous explanation of electricity as a fluid (which he called "electric fluid"). While best known to contemporary Americans as one of the Founding Fathers, Franklin was a famous, highly regarded scientist in his day. Franklin was the first to chart the Gulf Stream current during his voyages across the Atlantic, and he was the first person to adopt a scientific model for electricity—an insight that soon brought him worldwide fame as the inventor of the lightning rod.

When applying the vector field model developed here to electric fields, $u(x, y)$ is called the *electrostatic potential*, and the streamlines $v(x, y) = C$ are called the *flux lines*. Despite having different labels for the mathematical elements (due to the history of their study and application), the electric field model is mathematically identical to the fluid flow model; theorem 7.5.1 can be applied to electric fields as well.

In a similar fashion, heat flow in a two-dimensional structure follows this section's vector field model, provided there is no net buildup of heat in the system. In this application, the vector field is called the *heat flux*, the streamlines are called *flux lines*, and the equipotentials are called *isothermals*. In this setting every point on a given isothermal has the same temperature. The following example applies the fluid flow model to the case of heat flux.

Example 7.5.5   We study the vector field representing a heat flux across a planar surface and with complex potential $f(z) = -i \, \text{Log}(z - (1 + i))$. We assume there is no net buildup of heat in the system and apply the vector field model described in theorem 7.5.1.

Since the heat flux has a complex potential function $f(z) = -i \, \text{Log}(z - (1+i))$, the potential function for this vector field is $u(x, y) = Re[f(z)] = \text{Arg}(z - (1+i))$. Exercise 68 in section 7.2 shows how to express $\text{Arg}(x + iy)$ in terms of $\arctan(y/x)$ when the argument is between $-\pi$ and π. Here $z - (1 + i) = (x - 1) + i(y - 1)$, and so $\text{Arg}(z - (1 + i)) = \arctan(y - 1)/(x - 1)$. Applying the gradient operator to $u(x, y) = \arctan(y - 1)/(x - 1)$, the heat flux is expressed as

$$\vec{\nabla}u = \langle u_x, u_y \rangle = \left\langle -\frac{y - 1}{(x - 1)^2 + (y - 1)^2}, \frac{x - 1}{(x - 1)^2 + (y - 1)^2} \right\rangle.$$

Question 7.5.3 considered a vector field very similar to this heat flux. As illustrated in figure 7.13, the vectors exhibit a pattern of counterclockwise rotation around a center of $1 + i$, which is plotted at the origin of the plane in figure 7.13.

In addition to a display of the heat flux, figure 7.13 provides a graph of the vector field's flux lines (the streamlines $v(x, y) = C$) and the isothermals (the equipotentials $u(x, y) = C$). The flux lines are equations of the form $v(x, y) = -\ln[(x - 1)^2 + (y - 1)^2] = C$, where $C \in \mathbb{R}$. Negating both sides of this equation and raising each as a power of e, the resulting expression is equivalent to the equation of a circle $(x - 1)^2 + (y - 1)^2 = e^{-C}$. The flux lines are therefore circles with center $(1, 1) = 1 + i$. Similarly, the isothermals are $u(x, y) = \text{Arg}[z - (1 + i)] = C$; for any given C, this equation describes a ray emanating from the point $1 + i$ (since for all points on this ray, the difference with $1 + i$ generates the same constant polar angle). Figure 7.13 indicates the characteristic orthogonal relationship between flux lines and isothermals.

∎

Question 7.5.6   Answer each question about a heat flux with complex potential $f(z) = -iz$. Assume there is no net buildup of heat in the system and apply the fluid flow model from theorem 7.5.1.

  (a) State the isothermal potential $u(x, y)$ and flux line function $v(x, y)$ for this heat flux.

  (b) State the analytic vector formula $\vec{\nabla}u$ for this heat flux. Sketch a representative plot of this vector field graphing the vectors at the points

$$(0, 0), \ (-1, 2), \ (1, 2), \ (-1, -1), \ (1, -1), \ (-1, 2), \ (1, 2), \ (-2, 0), \ (2, 0).$$

Based on this plot, provide a written description of the heat flux.

  (c) State the general equation for the flux lines and the isothermals, and plot the flux lines and isothermals for $C = 1, 3, 5$ on the same graph (as modeled in figure 7.13).

∎

As a final application, we consider the study of fluid flow around a stationary object. Such models play an important role in aerodynamics when studying the motion of air around an airplane wing, in hydrodynamics when studying the movement of water around the hull of a ship or submarine, and in engineering when developing industrial processes for working with fluids (including fossil fuels, chemicals, glass, and steel). Under the assumption that the fluid is incompressible and irrotational, the corresponding velocity field satisfies the fluid flow model from theorem 7.5.1; that is, the velocity field is conservative (of the form $\vec{\nabla}u$) with a harmonic potential $u(x, y)$.

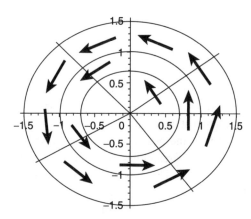

Figure 7.13 Heat flux, flux lines, and isothermals for example 7.5.5

The corresponding harmonic conjugate $v(x, y) = u^*(x, y)$ generates the streamlines, one of which describes the boundary of the stationary obstacle. The following example illustrates the model.

Example 7.5.6   We study the fluid flow around a circular obstacle described by the unit circle, where the fluid flow has complex potential function $f(z) = z + (1/z)$ for $|z| > 0$. We assume the fluid is incompressible and irrotational and apply the fluid flow model from theorem 7.5.1. This example provides a thorough analysis of this fluid flow, identifying the vector field $\vec{\nabla} u$, the harmonic potential $u(x, y)$, the harmonic conjugate $v(x, y)$, and the general equations for the streamlines $v(x, y) = C$ and equipotentials $u(x, y) = C$. We also verify the fact that $x^2 + y^2 = 1$ is a streamline, confirming that the fluid flows around the circular obstacle described by the unit circle.

First determine the analytic vector formula $\vec{\nabla} u = \langle u_x, u_y \rangle$ for the velocity field. The complex potential $f(z) = z + (1/z)$ is analytic over any disk S that does not contain the origin $z = 0$, and the derivative is expressible in terms of the partial derivatives of u: $f'(x + iy) = u_x(x, y) + iu_y(x, y)$ for $x + iy \neq 0$. We can therefore determine $\vec{\nabla} u$ by computing $f'(z)$ and expressing it in rectangular form. Since $f(z) = z + z^{-1}$, the derivative of $f(z)$ is $f'(z) = 1 - z^{-2}$; substituting $z = x + iy$ and multiplying by the denominator's complex conjugate,

$$
\begin{aligned}
f'(x + iy) &= 1 - \frac{1}{(x + iy)^2} = 1 - \frac{1}{x^2 - y^2 + i2xy} = 1 - \frac{x^2 - y^2 - i2xy}{(x^2 - y^2)^2 + 4x^2y^2} \\
&= \left[\frac{(x^2 + y^2)^2 - x^2 + y^2}{(x^2 + y^2)^2} \right] + i \left[\frac{2xy}{(x^2 + y^2)^2} \right].
\end{aligned}
$$

Therefore, the velocity vector field is

$$
\vec{\nabla} u(x, y) = \langle u_x, u_y \rangle = \left(\frac{(x^2 + y^2)^2 - x^2 + y^2}{(x^2 + y^2)^2}, \frac{2xy}{(x^2 + y^2)^2} \right).
$$

Now determine the harmonic potential $u(x, y)$ and the stream function $v(x, y)$ (the harmonic conjugate of u). Because the obstacle is circular, the analysis is most readily performed using a polar representation for z; as described in section 7.1, $z = re^{i\theta}$. Substituting this term into the complex potential,

$$
\begin{aligned}
f(z) = f(r, \theta) &= re^{i\theta} + r^{-1}e^{-i\theta} = r(\cos\theta + i\sin\theta) + r^{-1}[\cos(-\theta) + i\sin(-\theta)] \\
&= r\cos\theta + ri\sin\theta + r^{-1}\cos\theta - r^{-1}i\sin\theta \\
&= (r + r^{-1})\cos\theta + i[(r - r^{-1})\sin\theta].
\end{aligned}
$$

The real and imaginary parts of the complex potential are therefore $u(r, \theta) = (r + r^{-1})\cos\theta$ and $v(r, \theta) = (r - r^{-1})\sin\theta$.

The streamlines are found by setting the stream function equal to a constant; they are $(r - r^{-1})\sin\theta = C$ for $C \in \mathbb{R}$. The streamline generated by $C = 0$ must have either $r - r^{-1} = 0$ or $\sin\theta = 0$. The equation $r - r^{-1} = 0$ results in $r = 1$ or $r = -1$, both of which produce the circular obstacle having the unit circle as a boundary. This streamline path with $C = 0$ is extended to the region exterior to the circle by the solutions to $\sin\theta = 0$, which results in $\theta = n\pi$ for $n \in \mathbb{Z}$. We see

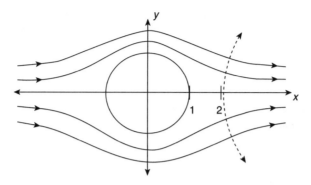

Figure 7.14 Streamlines and one (dashed) equipotential for example 7.5.6

that these curves are equal to the x-axis. Thus, the streamline path determined by $C = 0$ includes the unit circle and the entire x-axis exterior to the circle. There cannot be any flow across a streamline; the circular boundary of the obstacle fits this description. The streamline for $C = 0$ and the streamlines for other values of $C \in \mathbb{R}$ are graphically illustrated in figure 7.14.

Determine the equipotentials to complete the analysis. They are of the form $u(r, t) = (r + r^{-1}) \cos t = C$. The equipotential corresponding to $C = 0$ is the portion of the y-axis exterior to the unit circle. To illustrate the fact that the equipotentials are orthogonal to the streamlines, the equipotential for $C = 1$ is graphed as a dashed curve with the streamlines in figure 7.14.

■

Question 7.5.7   As in example 7.5.6, analyze the streamlines of the fluid flow with complex potential function

$$f(z) = z + \frac{1}{z} + i\frac{\text{Log } z}{2\pi}.$$

Verify that one streamline is the circle $x^2 + y^2 = 1$. The streamlines for this fluid flow are graphed in figure 7.15; they exhibit a flow around the unit circle obstacle that is slightly different from the one in example 7.5.6.

■

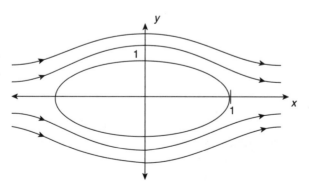

Figure 7.15 Streamlines for question 7.5.7

This chapter has considered just a few of the many applications of harmonic functions and differential complex analysis. A full course in complex analysis will provide a more complete development of the ideas touched on in this chapter, as well as integration theory, geometric issues in the complex plane, and a host of other interesting ideas and applications.

7.5.2   Reading Questions for Section 7.5

1. What is a vector? Describe the relationship between a geometric and a component form depiction of a vector.
2. What does it mean for a vector to be in standard position? Sketch an example of a vector in standard position and an equivalent vector emanating from $(1, 1)$.
3. Define the length of a vector and give an example of a unit vector.
4. What is a vector field?
5. Define and give an example of the gradient operator applied to a multivariate function u.
6. Define and give an example of a conservative vector field. Identify the potential function and the stream function for the example.
7. Define and give an example of a complex potential. How do we determine the corresponding (real-valued) potential function?
8. State theorem 7.5.1. How is this result helpful to a study of fluid flow?
9. State the mathematical form of a streamline. What information is provided by a graph of the streamlines for a fluid flow?
10. State the mathematical form of an equipotential. What information is provided by a graph of the equipotentials for a fluid flow?
11. What is the geometric relationship between the streamlines and the equipotentials of an incompressible, irrotational fluid flow?
12. When studying heat flow, what names are traditionally associated with the corresponding vector field, streamlines, and equipotentials?

7.5.3   Exercises for Section 7.5

In exercises 1–10, sketch each vector in standard position and emanating from $(1, 0)$ and from $(-1, 0)$. Also, determine the length of each vector.

1. $\langle 1, 1 \rangle$	6. $\langle 0, 0 \rangle$
2. $\langle 2, -1 \rangle$	7. $\langle 4, 0 \rangle$
3. $\langle -1, 1 \rangle$	8. $\langle -2, 0 \rangle$
4. $\langle -1, -2 \rangle$	9. $\langle 0, 3 \rangle$
5. $\langle -2, 1 \rangle$	10. $\langle 0, -1 \rangle$

In exercises 11–24, determine the length of an arbitrary vector in each vector field and sketch a representative plot of each vector field using at least eight vectors.

11. $\vec{F}(x, y) = \langle 1, 1 \rangle$ 12. $\vec{F}(x, y) = \langle -1, 2 \rangle$

13. $\vec{F}(x, y) = \langle x, 0 \rangle$

14. $\vec{F}(x, y) = \langle 0, x \rangle$

15. $\vec{F}(x, y) = \langle -x, -y \rangle$

16. $\vec{F}(x, y) = \langle y, -x \rangle$

17. $\vec{F}(x, y) = \langle x + y, 0 \rangle$

18. $\vec{F}(x, y) = \langle x - y, 0 \rangle$

19. $\vec{F}(x, y) = \langle 1, e^x \rangle$

20. $\vec{F}(x, y) = \langle e^y, y \rangle$

21. $\vec{F}(x, y) = \langle 1, \cos x \rangle$

22. $\vec{F}(x, y) = \langle x, \sin x \rangle$

23. $\vec{F}(x, y) = \langle e^y \cos x, e^y \sin x \rangle$

24. $\vec{F}(x, y) = \langle e^x \cos x, e^y \sin y \rangle$

In exercises 25–40, compute the conservative vector field determined by each harmonic function $u(x,y)$. Also, identify the corresponding stream function $v(x,y)$ (that is, the harmonic conjugate) and complex potential function $f(z)$.

25. $u(x, y) = 2$

26. $u(x, y) = 8x(1 - y)$

27. $u(x, y) = (x - a)(y - b)$ for $a, b \in \mathbb{R}$

28. $u(x, y) = 4x^2 - 4y^2 + 5$

29. $u(x, y) = 3x^2 - x - 3y^2 + 2y$

30. $u(x, y) = cx^2 - cy^2$, for $c \in \mathbb{R}$

31. $u(x, y) = x^3 - 3xy^2$

32. $u(x, y) = 3x^2y - y^3$

33. $u(x, y) = e^y \sin x$

34. $u(x, y) = e^x \cos y$

35. $u(x, y) = e^{x^2-y^2} \sin(2xy)$

36. $u(x, y) = e^{x^2-y^2} \cos(2xy)$

37. $u(x, y) = \sin y \sinh x$

38. $u(x, y) = \sin x \sinh y$

39. $u(x, y) = \sin x \cosh y$

40. $u(x, y) = \ln[x^2 + y^2]$ for $x^2 + y^2 > 0$

In exercises 41–44, answer each question about an incompressible and irrotational fluid that has a fluid flow with complex potential function $f(z) = -iz^2$.

41. State the harmonic potential $u(x, y)$ and streamline function $v(x, y)$ for this fluid flow.

42. State the analytic vector formula $\vec{\nabla}u$ for the velocity field of this fluid flow. Sketch a representative plot and provide a written description of the fluid flow.

43. State the general equation for the streamlines and the equipotentials and graph at least two streamlines and at least two equipotentials on the same axes.

44. Describe the path followed by an object dropped in the fluid at $(3, 5)$.

In exercises 45–48, answer each question about an incompressible and irrotational fluid that has a fluid flow with complex potential function $f(z) = e^z$. Use a computing device as appropriate.

45. State the harmonic potential $u(x, y)$ and streamline function $v(x, y)$ for this fluid flow.

46. State the analytic vector formula $\vec{\nabla}u$ for the velocity field of this fluid flow. Sketch a representative plot and provide a written description of the fluid flow.

47. State the general equation for the streamlines and the equipotentials and graph at least two streamlines and at least two equipotentials on the same axes.

48. Describe the path followed by an object dropped in the fluid at $(2, \pi/4)$.

In exercises 49–58, find the streamlines and equipotentials for a fluid flow satisfying the given criteria. On the same axes sketch a representative plot of the corresponding fluid flow, at least two streamlines, and at least two equipotentials. Assume the fluid is incompressible and irrotational, and apply the fluid flow model from theorem 7.5.1.

49. Potential $u(x, y) = 2$

50. Potential $u(x, y) = K$, where $K \in \mathbb{R}$

51. Potential $u(x, y) = 4x + 5y$

52. Potential $u(x, y) = 3x^2 - 2x - 2y^2$

53. Potential $u(x, y) = e^x \cos y$

54. Potential $u(x, y) = \text{Arg}(x + iy)$

55. Stream function $v(x, y) = 3$

56. Stream function $v(x, y) = 3x + 7y$

57. Complex potential $f(z) = z^2 + z$

58. Complex potential $f(z) = 2z^3$

In exercises 59–62, find the flux lines and isothermals for a heat flux satisfying the given criteria. On the same axes sketch a representative plot of the corresponding heat flux, at least two flux lines, and at least two isothermals. Assume there is no net buildup of heat in the system and apply the fluid flow model from theorem 7.5.1.

59. Isothermal $u(x, y) = 2x$

60. Flux $v(x, y) = K$, where $K \in \mathbb{R}$

61. Isothermal $u(x, y) = \text{Log}(x^2 + y^2)$

62. Complex isothermal $f(z) = \text{Log} z$

In exercises 63–64, answer each question about the velocity vector field for fluid flow of an incompressible, irrotational fluid.

63. What is the relationship between the vectors in the velocity fields generated by the potential functions $u(x, y)$ and $u(x, y) + C$, where $C \in \mathbb{C}$?

64. What is the relationship between the vectors in the velocity fields generated by the potential functions $u(x, y)$ and $C \cdot u(x, y)$, where $C \in \mathbb{C}$?

Exercises 65–70 consider the algebraic properties of two-dimensional vectors under various operations. For $a,b,c,d,r \in \mathbb{R}$, we define

Scalar multiplication	$r\langle a, b \rangle = \langle ra, rb \rangle$
Vector addition	$\langle a, b \rangle + \langle c, d \rangle = \langle a + c, b + d \rangle$
Vector dot product	$\langle a, b \rangle \cdot \langle c, d \rangle = \langle ac, bd \rangle$

In exercises 65–70, give an example and prove each statement about these operations on vectors using the field-theoretic properties of the real numbers.

65. Scalar multiplication distributes over scalar addition; that is, $(r + s)\langle a, b \rangle = r\langle a, b \rangle + s\langle a, b \rangle$.

66. Scalar multiplication distributes over vector addition; that is, $r [\langle a, b \rangle + \langle c, d \rangle] = r\langle a, b \rangle + r\langle c, d \rangle$.

67. Scalar multiplication does *not* distribute over vector dot product; that is, there exist $a, b, c, d, r \in \mathbb{R}$ such that

$$r [\langle a, b \rangle \cdot \langle c, d \rangle] \neq [r\langle a, b \rangle] \cdot [r\langle c, d \rangle].$$

68. Vector dot product distributes over vector addition; that is,

$$\langle a, b \rangle \cdot [\langle c, d \rangle + \langle e, f \rangle] = \langle a, b \rangle \cdot \langle c, d \rangle + \langle a, b \rangle \cdot \langle e, f \rangle.$$

69. Vector addition is a commutative operation. What is the identify for vector addition?

70. Vector dot product is a commutative operation. What is the identity for vector dot product?

Notes

The history of the development of complex analysis is reported in many popular books with an expository flavor, including [23] by Bottazini and the more recent [221] by Smithies. Nahin's book [179] traces humanity's efforts to understand $i = \sqrt{-1}$ and the complex numbers, detailing both the corresponding history and the mathematics. Among many technical books in complex analysis that address the history of the subject are the undergraduate text by Mathews and Howell [171] and the graduate text by Remmert [191].

Remmert [191], promotes the idea that our current understanding of complex analysis evolved from three distinct lines of development, arising from the work of Bernhard Riemann, Augustin-Louis Cauchy, and Karl Weierstrass. Each were aware of and used the others' breakthroughs and insights, but they also each had a unique approach to complex analysis. Riemann characterized analytic functions from a geometric standpoint as mappings between domains (now called Riemann surfaces) in the complex plane, creating "conformal equivalences" between these domains. Cauchy characterized analytic functions in terms of integrals: every analytic function can be represented as an integral in a set format. In contrast, Weierstrass characterized analytic functions via power series: every analytic function can be represented as a power series with a radius of convergence describing a domain over which the representation is guaranteed valid. For Weierstrass, power series representations turned analytic issues into algebraic ones.

The notes at the end of chapter 4 provide references on the life and work of Riemann, Cauchy, and Gauss (who also made many important contributions to the study of real analysis considered in chapter 4). A description of the life of French mathematician Pierre-Simon Laplace is given in [98]. Laplace was one of the most influential mathematicians of his time, and his work remains a staple in mathematics and engineering courses to this day. This well-written biography blends together the engaging story of Laplace's personal life with his professional work on some of the most challenging questions of the eighteenth and nineteenth centuries. Those interested in an historical description of Benjamin Franklin's development of a scientific theory of electricity would enjoy reading Isaacson [125]. Among his many accomplishments, Franklin's keen interpretations of observations led him to state in lay terms what has become known as the law of conservation of charge, a fundamental axiom in physics today. In addition, Franklin wrote an engaging autobiography [89] that may be of interest.

A full course in complex analysis includes a more thorough development of the ideas we have touched on in this chapter, as well as the complex integral, conformal mappings, transform methods, and further aspects of analytic and harmonic functions. Standard undergraduate texts in complex analysis include those by Brown and Churchill [31], Fisher [84], Gamelin [94], Marsden and Hoffman [170], and Mathews and Howell [171]. Many students find the complex variables workbook by Speigel [226] helpful. Standard graduate texts in complex analysis include those by Lang [147] and Remmert [191]. A recent book by Krantz [143] addresses advanced topics in complex analysis, touching on recent research developments in the field.

Much work has been done with harmonic functions; Axler et al. [9] is a good introduction to the field. Another classic text is Zygmund [259], originally published in Warsaw in 1935, which describes trigonometric series and Fourier series from the viewpoint of complex functions. Beerends et al. [13] and Dyke [68] introduce Laplace and Fourier transforms and are accessible to undergraduates.

Ian Stewart's popular book [231] discusses the success of various models of mathematics for understanding the real-life physical world in which we live. In chapter 5 (From Violins to Videos) Stewart traces the development of several mathematical ideas from Euler's first study of violins to modern electronic applications—with the mathematical study of drumskins playing an important role. Stewart has written a number of engaging and enjoyable expository mathematical books, including [228] and [230]. This last book of Stewart's was inspired by G. H. Hardy's classic book *A Mathematician's Apology* [112].

Answers to Questions

1.1 **The Formal Language of Sentential Logic**

Question 1.1.1

(a) I am going to [bike and run] or swim.
I am going to bike and [run or swim.]

(b) I am going to both bike and run, or swim.
I am going to bike and either run or swim.

Question 1.1.2

(a) or
(b) not
(c) if and only if
(d) both–and
(e) if
(f) when
(g) if–then; not

Question 1.1.3

(a) $P \wedge Z$
(b) $Q \leftrightarrow S$

(c) $(S \wedge Z) \rightarrow P$
(d) $S \wedge (Z \rightarrow P)$
(e) $S \rightarrow Q$
(f) $(P \wedge Z) \vee Q$
(g) $P \wedge (Z \vee Q)$
(h) The number n is prime or rational.
(i) If the number n is rational, then n is both not prime and square.
(j) The number n is prime exactly when n is not rational.
(k) The number n is not prime or both an integer and not rational.

Question 1.1.4

(a) sentence, outer parentheses may be dropped
(b) nonsentence, missing parentheses
(c) nonsentence, adjacent connectives
(d) sentence
(e) sentence
(f) nonsentence, $\wedge$ is the conjunction (or "and") symbol, not &

1.2 **Truth and Sentential Logic**

Question 1.2.1

(a)

p	$(\sim p)$	$(\sim p) \vee p$
T	F	T
F	T	T

(b)

p	q	$\sim p$	$\sim q$	$(\sim p) \wedge (\sim q)$
T	T	F	F	F
T	F	F	T	F
F	T	T	F	F
F	F	T	T	T

Question 1.2.2

(a) contradiction

p	$(\sim p)$	$p \leftrightarrow (\sim p)$
T	F	F
F	T	F

(b) tautology

p	$p \leftrightarrow p$
T	T
F	T

(c) contingency

p	q	$p \vee q$	$p \leftrightarrow (p \vee q)$
T	T	T	T
T	F	T	T
F	T	T	F
F	F	F	T

(d) contingency

p	q	$p \wedge q$	$p \leftrightarrow (p \wedge q)$
T	T	T	T
T	F	F	F
F	T	F	T
F	F	F	T

Question 1.2.3

(a)

p	q	$p \wedge q$	$\sim(p \wedge q)$
T	T	T	F
T	F	F	T
F	T	F	T
F	F	F	T

p	q	$\sim p$	$\sim q$	$(\sim p) \vee (\sim q)$
T	T	F	F	F
T	F	F	T	T
F	T	T	F	T
F	F	T	T	T

(b)

p	q	$p \vee q$	$\sim(p \vee q)$
T	T	T	F
T	F	T	F
F	T	T	F
F	F	F	T

p	q	$\sim p$	$\sim q$	$(\sim p) \wedge (\sim q)$
T	T	F	F	F
T	F	F	T	F
F	T	T	F	F
F	F	T	T	T

1.3 An Algebra for Sentential Logic

Question 1.3.1

(a) $p \vee q \equiv \sim [\sim (p \vee q)]$
$\equiv \sim [(\sim p) \wedge (\sim q)]$

(b) $p \rightarrow q \equiv (\sim p) \vee q$
$\equiv (\sim p) \vee (\sim \sim q) \equiv \sim [p \wedge (\sim q)]$

(c) $p \leftrightarrow q \equiv (p \rightarrow q) \wedge (q \rightarrow p)$
$\equiv [\sim [p \wedge (\sim q)]] \wedge [\sim [q \wedge (\sim p)]]$

Question 1.3.2

(a) $[p \wedge (\sim q)] \vee [(\sim p) \wedge q]$

(b) $[p \wedge q \wedge (\sim r)] \vee [(\sim p) \wedge q \wedge$
$(\sim r)] \vee [(\sim p) \wedge (\sim q) \wedge (\sim r)]$

Question 1.3.3

Given adequate $\{\sim, \wedge, \vee\}$		Proving adequate $\{\sim, \vee\}$
$\sim p$	$\equiv$	$\sim p$
$p \wedge q$	$\equiv$	$\sim [(\sim p) \vee (\sim q)]$
$p \vee q$	$\equiv$	$p \vee q$

Application: Designing Computer Circuits

Question 1.4.1

(a) top $= \sim p = \sim 1 = 0$
middle $= p \wedge q = 1 \wedge 1 = 1$
bottom $= q \vee r = 1 \vee 1 = 1$
$0 \vee 1 \vee 1 = 1$

(b) top $= \sim p = \sim 0 = 1$
middle $= p \wedge q = 0 \wedge 1 = 0$
bottom $= q \vee r = 1 \vee 0 = 1$
$1 \vee 0 \vee 1 = 1$

Question 1.4.2

(a) $[p \wedge (\sim q)] \vee [(\sim p) \wedge q]$
(b) $[p \wedge q \wedge (\sim r)] \vee [(\sim p) \wedge q \wedge (\sim r)] \vee [(\sim p) \wedge (\sim q) \wedge (\sim r)]$

Question 1.4.3

$\sim q$

Question 1.4.4

$q \vee [(\sim p) \wedge (\sim r)]$

Natural Deductive Reasoning

Question 1.5.1

(a)

p	q	$p \wedge q$	$(p \wedge q) \to p$
T	T	T	T
T	F	F	T
F	T	F	T
F	F	F	T

(b)

p	q	$p \to q$	$\sim q$	$\sim p$	$[(p \to q) \wedge (\sim q)] \to (\sim p)$
T	T	T	F	F	T
T	F	F	T	F	T
F	T	T	F	T	T
F	F	T	T	T	T

(c)

p	q	$p \vee q$	$\sim p$	$[(p \vee q) \wedge (\sim p)] \to q$
T	T	T	F	T
T	F	T	F	T
F	T	T	T	T
F	F	F	T	T

Question 1.5.2

(a) row 3

p	q	$p \to q$	$\sim p$	$\sim q$	$[(p \to q) \wedge (\sim p)] \to (\sim q)$
T	T	T	F	F	T
T	F	F	F	T	T
F	T	T	T	F	F
F	F	T	T	T	T

(b) row 1

p	q	$p \vee q$	$\sim q$	$[(p \vee q) \wedge p] \rightarrow (\sim q)$
T	T	T	F	F
T	F	T	T	T
F	T	T	F	T
F	F	F	T	T

Question 1.5.3

1. premise
2. premise
3. 2,3—modus tollens
4. premise
5. 3,4—modus tollens
6. 5—double negation
7. premise
8. 6,7—modus ponens

Question 1.5.4

1. premise
2. premise

3. 2—DeMorgan's laws
4. 3—conjunctive simplification
5. 1,4—modus tollens
6. 5—DeMorgan's laws
7. 6—double negation
8. premise
9. 7,8—disjunctive syllogism

Question 1.5.5

$$p = F; \quad q = T$$

Question 1.5.6

$$p = T; \quad q = T$$

1.6 The Formal Language of Predicate Logic

Question 1.6.1

(a) $B \vee A$

(b) $A \rightarrow (\sim D)$

(c) $(\sim C) \wedge (\sim B)$

(d) $P \wedge T$

(e) $O \rightarrow G$

(f) $(\sim O) \wedge T$

Question 1.6.2

$Z(x)$: x is an integer

$A(x, y)$: x is the antiderivative of y

3: three

t: three

$a(3, 3)$: sum of 3 and 3

Question 1.6.3

(a) $[L(c, b) \vee L(c, d)]$

(b) $L(d, c) \rightarrow \sim L(d, m)$

(c) $[\sim L(b, c)] \wedge [\sim L(c, b)]$ or $\sim [L(b, c) \vee L(c, b)]$

(d) $P(2) \wedge E(2)$

(e) $O(5) \rightarrow E[a(5, 5)]$

(f) $[\sim E(5)] \wedge E(2)$

Question 1.6.4

(a) $\exists x L(b, x)$

(b) $\exists x[L(x, x) \wedge L(x, d)]$

(c) $\forall x[L(x, c) \rightarrow L(x, m)]$

(d) $\exists x[P(x) \wedge O(x)]$

(e) $\forall n[E(n) \rightarrow E(s(n))]$

(f) $\forall n[E(n) \rightarrow E(s(n))]$

Question 1.6.5

(a) $\forall x \exists y\, L(x, y)$

(b) $\exists x \forall y\, L(x, y)$

(c) $\forall x[\exists y L(x, y) \to \exists z L(z, x)]$ or
    $\forall x[\exists y L(x, y) \to \exists y L(y, x)]$

(d) $\forall x \forall y\{[E(x) \wedge E(y)] \to {\sim} P(x+y)\}$

(e) ${\sim}\exists x \exists y[E(x) \wedge O(y) \wedge E(x+y)]$

(f) $\exists x \exists y[(x+y)^2 = x^2 + y^2]$

1.7   Fundamentals of Mathematical Proofs

Question 1.7.1

Let m and n be even integers. Then there exist $i \in \mathbb{Z}$ such that $m = 2i$. Then $m \cdot n = 2i \cdot n = 2(in)$, so $m \cdot n$ is even.

Contrapositive: If n is even, then n^2 is even.

Let n be even. Then $n = 2k$ for some $k \in \mathbb{Z}$. Thus, $n^2 = n \cdot n = 2k \cdot n = 2(kn)$, which is even.

Question 1.7.2

Let $x \in \mathbb{Q}^* = \mathbb{Q} \setminus \{0\}$, let $y \in \mathbb{R} \setminus \mathbb{Q}$, and suppose $x \cdot y \in \mathbb{Q}$. Then there exist $p, q, r, s \in \mathbb{Z}$ with $p, q, s \neq 0$ such that $x = p/q$ and $x \cdot y = r/s$. Therefore, we have

$$\frac{p}{q} y = \frac{r}{s} \quad \Rightarrow \quad y = \frac{q}{p}\frac{r}{s} = \frac{qr}{ps}$$

Since $p, s \neq 0, ps \neq 0$ by the zero product property, and so $y \in \mathbb{Q}$. This contradicts the assumption that $y \in \mathbb{R} \setminus \mathbb{Q}$.
If $x = 0$, then $x \cdot y = 0 \cdot y = 0 \in \mathbb{Q}$ for all $y \in \mathbb{R}$.

Question 1.7.3

Contrapositive: If n is not odd, then n^2 is not odd.   or

Question 1.7.4

(a) 3 is odd and prime.

(b) $\sqrt{2}$ is irrational, so is π, and many other reals.

Question 1.7.5

(a) Every odd prime provides a counterexample; consider the odd prime $3 = 2 \cdot 1 + 1$.

(b) Both $n = 3$ and $n = 4$ provide counterexamples. When $n = 3 > 2$, then $3^2 = 9 \not\geq 25$. When $n = 4 > 2$, then $4^2 = 16 \not\geq 25$.

Question 1.7.6

$\sqrt{2}$ is irrational, and so not every square root is rational.

2.1   The Algebra of Sets

Question 2.1.1

$A = \{2, 3\}$

$B = \{1, 3, 5, 7, \ldots\}$ = odd positive integers

Question 2.1.2

Every element of X is odd and so in Y; in particular, $1 = 2 \cdot 0 + 1, 3 = 2 \cdot 1 + 1$, and $5 = 2 \cdot 2 + 1$. Therefore, $X \subseteq Y$.

On the other hand, $7 \in Y$ since $7 = 2 \cdot 3 + 1$ is odd, but $7 \notin X$. Thus, X is a proper subset of Y.

Question 2.1.3

$a \in A$ iff $\{a\} \subseteq A$

($\Rightarrow$) Assume $a \in A$. The only element of $\{a\}$ is a. By assumption $a \in A$, and so every element of $\{a\}$ is in A. Thus, $\{a\} \subseteq A$ by definition.

($\Leftarrow$) Assume $\{a\} \subseteq A$. Then every element of $\{a\}$ is in A. Since $a \in \{a\}$, we have $a \in A$.

Question 2.1.4

$X^C = \{\ldots, -2, -1, 0, 2, 4, 6, 7, 8, \ldots\}$

$W \cap Y = \{1\}$

$W \cup Y = \{n : n \text{ is odd }\} \cup \{2\}$

$W \setminus Y = \{2\}$

$Y \setminus W = \{n : n \text{ is odd and not } 1\}$

$X \times W = \{(1, 1), (1, 2), (3, 1), (3, 2),$
$\qquad\qquad (5, 1), (5, 2)\}$

$W \times W = \{(1, 1), (1, 2), (2, 1), (2, 2)\}$

$W \times Y = \{(1, n), (2, n) : n \text{ is odd }\}$

$\mathbb{P}(X) = \{\emptyset, \{1\}, \{3\}, \{5\}, \{1, 3\}, \{1, 5\},$
$\qquad\qquad \{3, 5\}, \{1, 3, 5\}\}$

$\mathbb{P}(Y)$ contains $\{1\}$, $\{3\}$, $\{5\}$, $\{7\}$, $\{1, 3\}$, $\{1, 5\}$, and many other sets.

Question 2.1.5

If $x \in B^C$, then $x \notin B$. Since $x \in A$ implies $x \in B$, then $x \notin B$ implies $x \notin A$. Then $x \notin A$, and so $x \in A^C$.

Question 2.1.6

$a \in (A \cup B)^C$ iff $\quad a \notin A \cup B$

iff $\quad a$ not in A or B

iff $\quad a$ not in A and
$\qquad\quad a$ not in B

iff $\quad a \notin A$ and $a \notin B$

iff $\quad a \in A^C$ and $a \in B^C$

iff $\quad a \in A^C \cap B^C$

Question 2.1.7

Following the second approach suggested by example 2.1.9, define $A = \{1\}$, $B = \{2\}$, and $C = \{2\}$. This gives us $A \cap (B \cup C) = \{1\} \cap \{2\} = \emptyset$ and $(A \cap B) \cup C = \emptyset \cup \{2\} = \{2\}$, which are not equal to each other.

Question 2.1.8

(a)

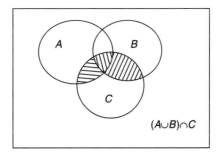

$(A \cup B) \cap C$

(b)

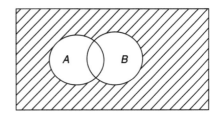

Question 2.1.9

We provide counterexamples disproving the supposed "equalities."

(a) Let $A = \{1\}$, $B = \{2\}$, and $C = \{2\}$. Then $(A \cup B) \cap C = \{2\}$, while $A \cup (B \cap C) = \{1, 2\}$.

(b) Let $A = \{2\}$, $B = \{1\}$, and $U = \{1, 2, 3\}$. Then $A^C \setminus B = \{3\}$, while $(A \setminus B)^C = \{1, 3\}$.

2.2 The Division Algorithm and Modular Addition

Question 2.2.1

(a) $q = 6$ and $r = 3$
(b) $q = -2$ and $r = 3$
(c) $q = 32$ and $r = 3$
(d) $q = -101$ and $r = 3$

Question 2.2.2

(a) $1, 7, -5, -11$
(b) $5, 11, -1, -7$

Question 2.2.3

(a) $0, 0, 0, 0$
(b) $9, 5, 1, 7$
(c) $1, 1, 1, 0$

Question 2.2.4

(a) for $n = 3$, $\{0, 1, 2\}$
(b) for $n = 6$, $\{0, 1, 2, 3, 4, 5\}$
(c) for $n = 9$, $\{0, 1, 2, 3, 4, 5, 6, 7, 8\}$
(d) for n, $\{0, 1, 2, \ldots, n - 1\}$

Question 2.2.5

$\mathbb{Z}_3 = \{0, 1, 2\}$
$\mathbb{Z}_6 = \{0, 1, 2, 3, 4, 5\}$
$\mathbb{Z}_{10} = \{0, 1, 2, 3, 4, 5, 6, 7, 8, 9\}$

Question 2.2.6

(a) $0 + 3 = 3 \in \mathbb{Z}_6$

(b) $3 + 3 = 6 \notin \mathbb{Z}_6$
(c) $0 + 4 = 4 \in \mathbb{Z}_6$
(d) $3 + 4 = 7 \notin \mathbb{Z}_6$

Question 2.2.7

(a) $1 \oplus 4 = (1 + 4) \bmod 6 = 5 \bmod 6 = 5$
(b) $4 \oplus 5 = (4 + 5) \bmod 6 = 9 \bmod 6 = 3$
(c) $2 \oplus 4 = (2 + 4) \bmod 6 = 6 \bmod 6 = 0$
(d) $3 \oplus 4 = (3 + 4) \bmod 6 = 7 \bmod 6 = 1$

Question 2.2.8

$0 \oplus 1 = (0 + 1) \bmod 6 = 1 \bmod 6 = 1$;
$1 \oplus 0 = (1 + 0) \bmod 6 = 1 \bmod 6 = 1$;
$0 \oplus 2 = (0 + 2) \bmod 6 = 2 \bmod 6 = 2$;
$2 \oplus 0 = (2 + 0) \bmod 6 = 2 \bmod 6 = 2$;
and similarly for $a = 3, 4, 5, 0$.

Question 2.2.9

(a) $2 \oplus 4 = (2 + 4) \bmod 6 = 6 \bmod 6 = 0$
$4 \oplus 2 = (4 + 2) \bmod 6 = 6 \bmod 6 = 0$
(b) $b = 3$ since $3 \oplus 3 = (3 + 3) \bmod 6 = 6 \bmod 6 = 0$
(c) Since $5 \oplus 1 = 6 \bmod 6 = 0$ and $1 \oplus 5 = 6 \bmod 6 = 0$, we know that 1 is the inverse of 5 under addition mod 6.

2.3 Modular Multiplication and Equivalence Relations

Question 2.3.1

(a) $1 \odot 4 = (1 \cdot 4) \bmod 7 = 4 \bmod 7 = 4$
(b) $3 \odot 5 = (3 \cdot 5) \bmod 7 = 15 \bmod 7 = 1$

(c) $4 \odot 5 = (4 \cdot 5) \bmod 7 = 20 \bmod 7 = 6$
(d) $4 \odot 6 = (4 \cdot 6) \bmod 7 = 24 \bmod 7 = 3$

Question 2.3.2

$1 \odot 0 = (1 \cdot 0) \bmod 7 = 0 \bmod 7 = 0$;
$0 \odot 1 = (0 \cdot 1) \bmod 7 = 0 \bmod 7 = 0$;
$1 \odot 1 = (1 \cdot 1) \bmod 7 = 1 \bmod 7 = 1$;
$1 \odot 1 = (1 \cdot 1) \bmod 7 = 1 \bmod 7 = 1$;
$1 \odot 2 = (1 \cdot 2) \bmod 7 = 2 \bmod 7 = 2$;
$2 \odot 1 = (2 \cdot 1) \bmod 7 = 2 \bmod 7 = 2$;
and similarly for the rest of $\mathbb{Z}_7$, namely
$\quad a = 3, 4, 5, 6.$

Question 2.3.3

(a) if $a = 1$, then $a^{-1} = 1$ in $\mathbb{Z}_7$ since
$\quad 1 \odot 1 = 1$
(b) if $a = 2$, then $a^{-1} = 4$ in $\mathbb{Z}_7$ since
$\quad 2 \odot 4 = 1$ and $4 \odot 2 = 1$
(c) if $a = 3$, then $a^{-1} = 5$ in $\mathbb{Z}_7$ since
$\quad 3 \odot 5 = 1$ and $5 \odot 3 = 1$
(d) if $a = 6$, then $a^{-1} = 6$ in $\mathbb{Z}_7$ since
$\quad 6 \odot 6 = 1$

Question 2.3.4

$U(5) = \{1, 2, 3, 4\}$ and $U(6) = \{1, 5\}$

Question 2.3.5

(a)

$\oplus$	0	1	2	3
0	0	1	2	3
1	1	2	3	0
2	2	3	0	1
3	3	0	1	2

(b)

$\oplus$	0	1	2	3	4	5
0	0	1	2	3	4	5
1	1	2	3	4	5	0
2	2	3	4	5	0	1
3	3	4	5	0	1	2
4	4	5	0	1	2	3
5	5	0	1	2	3	4

(c)

$\odot$	1	5
1	1	5
5	5	1

(d)

$\odot$	1	2	3	4	5	6
1	1	2	3	4	5	6
2	2	4	6	1	3	5
3	3	6	2	5	1	4
4	4	1	5	2	6	3
5	5	3	1	6	4	2
6	6	5	4	3	2	1

Question 2.3.6

(a) $m = 7$ and $r = 1$; $m = 4$ and $r = 1$;
$\quad m = -5$ and $r = 1$
(b) $\{3k + 1 : k \in \mathbb{Z}\} = \{\ldots, -5,$
$\quad -2, 1, 4, 7, 10, \ldots\}$
(c) for $r = 0$,
$\quad \{3k : k \in \mathbb{Z}\} = \{\ldots, -6, -3, 0, 3,$
$\quad 6, 9, \ldots\}$

$\quad$ for $r = 2$,
$\quad \{3k + 2 : k \in \mathbb{Z}\} = \{\ldots, -4,$
$\quad -1, 2, 5, 8, 11, \ldots\}$

Question 2.3.7

$[0] = \{\ldots, -8, -4, 0, 4, 8, \ldots\} =$
$\quad \{4k : k \in \mathbb{Z}\}$
$[2] = \{\ldots, -6, -2, 2, 6, 10, \ldots\} =$
$\quad \{4k + 2 : k \in \mathbb{Z}\}$

Question 2.3.8

(a) $7/3 \sim 28/12$ since $7 \cdot 12 = 94 =$
$\quad 3 \cdot 28$
(b) $7/3 \not\sim 28/3$ since $7 \cdot 3 = 21 \neq$
$\quad 84 = 28 \cdot 3$
(c) $3/4 \sim 6/8$ since $3 \cdot 8 = 24 = 4 \cdot 6$
(d) $5/4 \not\sim -10/8$ since $5 \cdot 8 = 40 \neq$
$\quad -40 = -4 \cdot 10$

Question 2.3.9

(a) By the reflexivity of equality of
integers, $m \cdot n = m \cdot n$ and so
$m/n \sim m/n.$
(b) If $m/n \sim s/t$, then $m \cdot t = n \cdot s.$
By the symmetry of equality of
integers, $n \cdot s = m \cdot t$ and so
$s/t \sim m/n.$

(c) If $m/n \sim s/t$ and $s/t \sim u/v$, then $m \cdot t = n \cdot s$ and $s \cdot v = t \cdot u$. Multiplying provides the equality $(m \cdot t)(s \cdot v) = (n \cdot s)(t \cdot u)$. By commutativity of integer multiplication, $(st)(mv) = (st)(nu)$, and so $m \cdot v = n \cdot u$ (note that either $m, s, u \neq 0$, or else $m = s = u = 0$). Thus, $m/n \sim u/v$.

2.4 An Introduction to Groups

Question 2.4.1

(a) No. $\mathbb{Z}$ is closed under addition.

(b) $1 + (2 + 3) = 1 + 5 = 6$ and $(1 + 2) + 3 = 3 + 3 = 6$. These sums are equal and such an equality holds for every triple of integers; that is, for every $a, b, c \in \mathbb{Z}$, we have $a + (b + c) = (a + b) + c$.

(c) $0 + a = a = a + 0$ for every $a \in \mathbb{Z}$

(d) For $a = 3$, add -3, and for n, add $-n$. Thus, $n + (-n) = 0 = (-n) + n$.

Property (a) is closure, (b) is associativity, (c) is identity, and (d) is inverses.

Question 2.4.2

(a) $\dfrac{p}{q} \cdot \dfrac{r}{s} = \dfrac{pr}{qs}$. Since $q, s \neq 0$, the zero product property implies that $qs \neq 0$.

(b) $\dfrac{1}{2} \cdot \left[\dfrac{3}{5} \cdot \dfrac{8}{7} \right] = \dfrac{1}{2} \cdot \dfrac{24}{35} = \dfrac{12}{35}$;

$\left[\dfrac{1}{2} \cdot \dfrac{3}{5} \right] \cdot \dfrac{8}{7} = \dfrac{3}{10} \cdot \dfrac{8}{7} = \dfrac{12}{35}$;

$\dfrac{m}{n} \cdot \left[\dfrac{p}{q} \cdot \dfrac{r}{s} \right] = \dfrac{m}{n} \cdot \dfrac{pr}{qs} = \dfrac{m(pr)}{n(qs)} = $

$\dfrac{(mp)r}{(nq)s} = \dfrac{mp}{nq} \cdot \dfrac{r}{s} = \left[\dfrac{m}{n} \cdot \dfrac{p}{q} \right] \cdot \dfrac{r}{s}$.

(c) The identity is $1 = \frac{1}{1}$.

Question 2.3.10

(a) $\left[\dfrac{2}{4} \right] = \left\{ \dfrac{2n}{4n} : n \in \mathbb{Z} \right\}$

$= \left\{ \dfrac{n}{2n} : n \in \mathbb{Z} \right\}$

(b) $\left[\dfrac{5}{3} \right] = \left\{ \dfrac{5n}{3n} : n \in \mathbb{Z} \right\}$

(d) $\left[\dfrac{3}{2} \right]^{-1} = \dfrac{2}{3}$; and $\left[\dfrac{m}{n} \right]^{-1} = \dfrac{n}{m}$ for $\dfrac{m}{n} \neq 0$.

(e) The inverse axiom fails for zero.

Question 2.4.3

(a) By the division algorithm $r \in \{0, \ldots, n - 1\}$, and so the possible values of $a \oplus b$ are the elements of $\mathbb{Z}_n$; that is, $a \oplus b \in \mathbb{Z}_n$ for every $a, b \in \mathbb{Z}_n$.

(b) $(4 \oplus 5) \oplus 3 = 3 \oplus 3 = 0$ and $4 \oplus (5 \oplus 3) = 4 \oplus 2 = 0$

$(2 \oplus 5) \oplus 1 = 1 \oplus 1 = 2$ and $2 \oplus (5 \oplus 1) = 2 \oplus 0 = 2$

(c) 0 is the identity since for every $a \in \mathbb{Z}_n$, both $a \oplus 0 = a$ and $0 \oplus a = a$.

(d) $0 \oplus 0 = (0 + 0) \bmod n = 0 \bmod n = 0$

$(n - a) \oplus a = (n - a + a) \bmod n = n \bmod n = 0$ and

$a \oplus (n - a) = (a + n - a) \bmod n = n \bmod n = 0$

Question 2.4.4

(a) $h = 5$ and $k = -2$

(b) $h = 2$ and $k = -3$

2.5 Dihedral Groups

Question 2.5.1

(a) R_{240}
(b) R_{240}
(c) R_{120}
(d) R_0
(e) R_{240}
(f) R_{120}

Question 2.5.2

$\circ$	R_0	R_{120}	R_{240}	F_T	F_R	F_L
R_0	R_0	R_{120}	R_{240}	F_T	F_R	F_L
R_{120}	R_{120}	R_{240}	R_0	F_R	F_L	F_T
R_{240}	R_{240}	R_0	R_{120}	F_L	F_T	F_R
F_T	F_T	F_L	F_R	R_0	R_{240}	R_{120}
F_R	F_R	F_T	F_L	R_{120}	R_0	R_{240}
F_L	F_L	F_R	F_T	R_{240}	R_{120}	R_0

Question 2.5.3

(a) $R_{120} \circ [F_T \circ R_{240}]$
$\quad = R_{120} \circ F_R = F_L$
$\quad [R_{120} \circ F_T] \circ R_{240}$
$\quad = F_R \circ R_{240} = F_L$
(b) $F_T \circ [R_{120} \circ F_R]$
$\quad = F_T \circ F_L = R_{120}$
$\quad [F_T \circ R_{120}] \circ F_R$
$\quad = F_L \circ F_R = R_{120}$

Question 2.5.4

The identity is R_0 and the inverses are given by
$$R_0^{-1} = R_0, \ R_{120}^{-1} = R_{240}, \ R_{240}^{-1} = R_{120},$$
$$F_T^{-1} = F_T, F_L^{-1} = F_L, \text{ and } F_R^{-1} = F_R.$$

Question 2.5.5

(a) True
(b) False
(c) False
(d) True
(e) False
(f) True
(g) True
(h) False

Question 2.5.6

(a) $C(R_{240}) = \{R_0, \ R_{120}, \ R_{240}\}$
(b) $C(F_R) = \{R_0, \ F_R\}$
(c) $C(F_L) = \{R_0, \ F_L\}$

Question 2.5.7

Let $A = R_{240}$.

2.6 Application: Check Digit Schemes

Question 2.6.1

(a) 23455
(b) 46755
(c) 3455
(d) 3456547655

Question 2.6.2

(a) Valid since $15 \bmod 10 = 5$.
(b) Invalid since $1253 \bmod 10 = 3 \neq 2$.

Question 2.6.3

(a) Incorrect last digit, so 18988 would be correct; incorrect next to last digit, so 18933 would be correct; or both of the last digits could be incorrect, so that one of 18911, 18922, 18944, 18955, 18966, 18977, or 18999 would be correct.

(b) The mod 10 check digit scheme does not detect this error, since the scheme only checks if 1234 mod 10 = 4, and does not check the third position.

(c) The mod 10 check digit scheme does not detect this error—the scheme only checks if the last two digits are equal.

(d) Time for a new check digit scheme!

Question 2.6.4

(a) 23455
(b) 46754
(c) 3453
(d) 3456547650

Question 2.6.5

(a) Invalid. If an error occurred in the last position, the correct record number is 156, or perhaps some other error occurred.
(b) Valid.

Question 2.6.6

(a) Since $1284 \bmod 9 = 6 \neq 1$, the mod 9 scheme detects this error.
(b) Since $1888 \bmod 9 = 7 \neq 8$, the mod 9 scheme detects this error.
(c) Since $1808 \bmod 9 = 8$, the mod 9 scheme does not detect this error.
(d) Since $1294 \bmod 9 = 7$, the mod 9 scheme does not detect this error.

In (c), the digit 0 is substituted for 9; and in case (d), the digit 9 is substituted for 0.

Question 2.6.7

(a) $3124 \bmod 9 = 1$; $4123 \bmod 9 = 1$; $2314 \bmod 9 = 1$; and there are others.
(b) Integer addition is commutative, so transpositions (switching positions of digits) are not detected.
(c) $a_1 \cdots a_n \bmod 9$

$$= (a_1 + \cdots + a_n) \bmod 9$$

$$= (a_{k_1} + \cdots + a_{k_n}) \bmod 9$$

$$= a_{k_1} \cdots a_{k_n} \bmod 9$$

Question 2.6.8

$8479 - 2642 - 1937 - 847^\wedge 5$

Question 2.6.9

Invalid—the check digit should be 2 rather than 4. These examples indicate that the Codabar check digit scheme detects all single-digit errors.

Question 2.6.10

(a) $1234^\wedge 0$
(b) $1235^\wedge 4$
(c) $1284^\wedge 7$
(d) $2134^\wedge 9$

Question 2.6.11

(a) The check digit should be $c = 0$.
(b) The check digit should be $c = 2$.
(c) The check digit should be $c = 8$.
(d) The check digit should be $c = 5$.

3.1   Prime Numbers

Question 3.1.1

(a) True
(b) False—$r = 4$
(c) False—$r = 7$
(d) True

Question 3.1.2

The positive integer divisors of 98 are 1, 2, 7, 14, 49, 98.
The positive integer divisors of 120 are 1, 2, 3, 4, 5, 6, 8, 10, 12, 15, 20, 24, 30, 40, 60, 120.

Question 3.1.3

(a) 11 is prime.
(b) 34 is nonprime with positive integer divisors 1, 2, 17, 34.
(c) -3 is nonprime (since $-3 \not> 2$) with positive integer divisors 1 and 3.
(d) 1 is nonprime (since $1 \not> 2$) with positive integer divisor 1.
(e) 83 is prime.
(f) 6 is nonprime with positive integer divisors 1, 2, 3, and 6.

Question 3.1.4

Only one prime is even—two.
If n is even, then $n = 2j$; if $n \neq 2$, then n is divisible by the distinct integers $1, 2, j$, and $2j$.

Question 3.1.5

(a) The first 10 primes: $2, 3, 5, 7, 11,$ $13, 17, 19, 23, 29.$
(b) The first 10 positive nonprimes: $1, 4, 6, 8, 9, 10, 12, 14, 15, 16.$

Question 3.1.6

(a) $30 = 2 \cdot 3 \cdot 5$
(b) $5 = 5$
(c) $12 = 2^2 \cdot 3$
(d) $27 = 3^3$

Question 3.1.7

(a) 2 and 3
(b) 2 and 3

Question 3.1.8

$28,17,1962,000 = 2^4 \cdot 3^5 \cdot 5^3 \cdot 7^3 \cdot 13^2$

Question 3.1.9

(a) $2 + 1 = 3$ is prime;
  $2 \cdot 3 + 1 = 7$ is prime;
  $2 \cdot 3 \cdot 5 + 1 = 31$ is prime; and
  $2 \cdot 3 \cdot 5 \cdot 7 + 1 = 211$ is prime.
(b) It appears that $p_1 \cdots p_n + 1$ is prime.
(c) $2 \cdot 3 \cdot 5 \cdot 7 \cdot 9 \cdot 11 \cdot 13 + 1 = 30031 =$ $59 \cdot 509$
(d) The number $p_1 \cdots p_n + 1$ is not divisible by any of the primes $p_1, \ldots, p_n$.

3.2 Application: Introduction to Coding Theory and Cryptography

Question 3.2.1

(a) TWO IS PRIME
(b) 13 | 01 | 20 | 08 | 09 | 19 | 06 | 21 | 14

Question 3.2.2

(a) 16 | 08 | 15 | 14 | 05 | 08 | 15 | 13 | 05
(b) 11 | 08 | 15 | 14 | 10 | 08 | 15 | 13 | 10
(c) FERMAT

Question 3.2.3

(a) 0979 | 0079 | 1391 | 0850 | 0098 | 1282 | 0001 | 1188 | 1101
(b) TWO IS EVEN

Question 3.2.4

(a) $\bullet \begin{bmatrix} 2 & -10 \end{bmatrix} \cdot \begin{bmatrix} 13 \\ 1 \end{bmatrix} = 26 - 10 = 16$

  $\bullet \begin{bmatrix} 2 & 8 & 10 \end{bmatrix} \cdot \begin{bmatrix} 3 \\ 5 \\ 1 \end{bmatrix} = 6 + 40 + 10$

  $= 56$

(b) The product is only defined if the length of the row vector is the same as the length of the column vector.

Question 3.2.5

(a) • $\begin{bmatrix} 123 & 46 \\ 50 & 18 \end{bmatrix}$

   • $\begin{bmatrix} -11 & 2 & -30 \end{bmatrix}$

(b) The product AB is only defined if the number of columns of A is the same as the number of rows of B, but $2 \neq 3$.

Question 3.2.6

$\begin{bmatrix} 0 & 1 & 1 & 1 & 0 & 0 & 0 \end{bmatrix}$
$\begin{bmatrix} 0 & 0 & 1 & 1 & 1 & 1 & 0 \end{bmatrix}$
$\begin{bmatrix} 1 & 0 & 1 & 0 & 0 & 1 & 0 \end{bmatrix}$

Question 3.2.7

(a) $\begin{bmatrix} 1 & 1 & 1 & 1 & 1 & 0 & 0 \end{bmatrix}$ has a single digit error in the fourth position and corrects as
   $\begin{bmatrix} 1 & 1 & 1 & 0 & 1 & 0 & 0 \end{bmatrix}$

(b) $\begin{bmatrix} 0 & 1 & 1 & 1 & 0 & 0 & 1 \end{bmatrix}$ has a single digit error in the seventh position and corrects as
   $\begin{bmatrix} 0 & 1 & 1 & 1 & 0 & 0 & 0 \end{bmatrix}$

3.3 From the Pythagorean Theorem to Fermat's Last Theorem

Question 3.3.1

(a) $L = a + b$ and
   $A_E = L^2 = (a+b)^2$

(b) $L = c$ and $A_I = L^2 = c^2$

(c) $A_T = \frac{1}{2}ab$

(d) $A_E = A_I + 4A_T = c^2 + 4\frac{1}{2}ab$

(e) $(a+b)^2 = c^2 + 2ab$

   $a^2 + 2ab + b^2 = c^2 + 2ab$

   $a^2 + b^2 = c^2$

Question 3.3.2

(a) $c = 5$
(b) $c = 7.623\ldots$
(c) $b = 15$
(d) $b = 12.961\ldots$

Question 3.3.3

(a) $5^2 + 12^2 = 25 + 144$
   $= 169 = 13^2$

(b) $10^2 + 24^2 = 26^2$
(c) $15^2 + 36^2 = 39^2$
(d) $(5n)^2 + (12n)^2 = (13n)^2$

Question 3.3.4

The Diophantine equation $x^2 - y = 0$ has solutions $x = 1, y = 1$; and $x = -1$, $y = 1$; one nonDiophantine equation is $x - y = \sqrt{2}$. There are many other examples.

Question 3.3.5

$x = 4, y = 2$; $x = 9, y = 3$; and $x = 24$, $y = 4$.

Question 3.3.6

If b and c are both even, they have a common prime divisor of two—contradicting theorem 3.3.4.

3.4 Irrational Numbers and Fields

Question 3.4.1

(a) $2 = 2/1$ and $-3 = -3/1$

(b) $k = k/1$

(c) $1/2 \notin \mathbb{Z}$

(d) Infinitely many.

Question 3.4.2

(a) $\dfrac{2,965}{10,000} = \dfrac{593}{2,000}$

(b) $\dfrac{10,505}{100} = \dfrac{2,101}{20}$

(c) $\dfrac{22,965}{10,000} = \dfrac{4,593}{2,000}$

(d) $\dfrac{10,505}{10,000} = \dfrac{2,101}{20,000}$

Question 3.4.3

(a) $\sqrt{3}n = m$, so $m^2 = 3n^2$.

(b) If 3 divides n^2, then 3 divides m and $m = 3k$.

(c) If $m^2 = 3n^2$, then $3^2 k^2 = 3n^2$ and $3k^2 = n^2$. In this case, 3 divides m^2, so 3 divides m and $m = 3j$.

(d) 3 is a common factor of m and n.

(e) This contradicts the assumption that m and n have no common factors, and so $\sqrt{3}$ is irrational.

(f) The proof is the same as that for theorem 3.4.1 with 3 substituted for 2 throughout the proof.

Question 3.4.4

The proof is the same as that for theorem 3.4.1 with p substituted for 2 throughout the proof.

Question 3.4.5

(a) 0

(b) $-r$

(c) $r = 0$

(d) 1

(e) $r^{-1} = \dfrac{1}{r}$

Question 3.4.6

(a) $(1 + i) + (3 + 5i) = 4 + 6i$
$(1 + i) \cdot (3 + 5i) = -2 + 8i$

(b) $2 + i$ and $2i$

(c) $(2 - i) + (-4 + 3i) = -2 + 2i$
$(2 - i) \cdot (-4 + 3i) = -5 + 10i$

(d) $i + (3 + 5i) = 3 + 6i$
$i \cdot (3 + 5i) = -5 + 3i$

Question 3.4.7

(a) $(a + bi) + (0 + 0i) = (a + 0) + (b + 0)i = a + bi$
$(0 + 0i) + (a + bi) = (0 + a) + (0 + b)i = a + bi$

(b) $-a - bi$

(c) $(a + bi) + (c + di)$
$= (a + c) + (b + d)i$
$= (c + a) + (d + b)i$
$= (c + di) + (a + bi)$

(d) $(a + bi) \cdot (1 + 0i) = a + 0bi + 0ai + bi = a + bi$
$(1 + 0i) \cdot (a + bi) = a + 0bi + 0ai + bi = a + bi$

(e) $\dfrac{1}{a + bi} = \dfrac{a - bi}{(a + bi)(a - bi)}$
$= \dfrac{a}{a^2 + b^2} + \dfrac{-b}{a^2 + b^2}i$

(f) $(a + bi) \cdot (c + di)$
$= (ac - bd) + (ad + bc)i$
$= (ca - db) + (da + cb)i$
$= (c + di) \cdot (a + bi)$

(g) $(a + bi) \cdot [(c + di) + (e + fi)]$

$= (a + bi) \cdot [(c + e) + (d + f)i]$

$= [a(c + e) - b(d + f)]$

$\quad + [b(c + e) + a(d + f)]i$

$= [ac + ae - bd - bf]$

$\quad + [bc + be + ad + af]i$

$= [ac - bd] + [bc + ad]i$

$\quad + [ac - bf] + [be + af]i$

$= (a+bi) \cdot (c+di) + (a+bi) \cdot (e+fi)$

(f)

$\odot$	1	2	3	4
1	1	2	3	4
2	2	4	1	3
3	3	1	4	2
4	4	3	2	1

(g) Since only elements of $U(5)$ appear in the Cayley table, $U(5)$ is closed under $\odot$, multiplication mod 5.

(h) 1

(i) $1^{-1} = 1$; $2^{-1} = 3$; $3^{-1} = 2$; $4^{-1} = 4$

(j) $2 \odot (3 \oplus 4) = 2 \odot 2 = 4$

$(2 \odot 3) \oplus (2 \odot 4) = 1 \oplus 3 = 4$

Question 3.4.8

(a) $\mathbb{Z}_5 = \{0, 1, 2, 3, 4\}$

(b)

$\oplus$	0	1	2	3	4
0	0	1	2	3	4
1	1	2	3	4	0
2	2	3	4	0	1
3	3	4	0	1	2
4	4	0	1	2	3

(c) Since only elements of $\mathbb{Z}_5$ appear in the Cayley table, $\mathbb{Z}_5$ is closed under $\oplus$, addition mod 5.

(d) 0

(e) $-0 = 0$; $-1 = 4$; $-2 = 3$; $-3 = 2$; $-4 = 1$

Question 3.4.9

(a) $\mathbb{Z}_4 = \{0, 1, 2, 3\}$

(b)

$\oplus$	0	1	2	3
0	0	1	2	3
1	1	2	3	0
2	2	3	0	1
3	3	0	1	2

(c)

$\odot$	1	2	3
1	1	2	3
2	2	0	2
3	3	2	1

(d) 2 has no inverse under multiplication mod 4.

3.5   Polynomials and Transcendental Numbers

Question 3.5.1

(a) Degree 4 with $a_4 = 3$, $a_3 = 2$, $a_2 = -7$, $a_1 = 5$, and $a_0 = -1$. One possible finite field is $F = \mathbb{Z}_{11}$.

(b) Degree 5 with $a_5 = 2$, $a_4 = 0$, $a_3 = 0$, $a_2 = 4$, $a_1 = 1$, and $a_0 = 0$. One possible finite field is $F = \mathbb{Z}_5$.

(c) Degree 3 with $a_3 = 1 + i$, $a_2 = 0$, $a_1 = 2i$, and $a_0 = -4$.

(d) Degree 7 with $a_7 = 2i$, $a_6 = a_5 = a_4 = a_3 = a_2 = a_1 = 0$, and $a_0 = 1 + i$.

Question 3.5.2

There are many possible answers, including x^{17}, x^2, x, and 1.

Question 3.5.3

There are many possible answers, including $\cos(x^2 + 1)$, $\ln(x^2 + 1)$, and $\arctan(x^2 + 1)$.

Question 3.5.4

(a) $2 \cdot 3^3 - 5 \cdot 3^2 - 9 \cdot 3 + 18 = 0$
(b) Using mod 6 arithmetic, $2 \cdot 3^3 + 3^2 + 3 \cdot 3 = 72 = 0(\bmod 6)$
(c) $2 \cdot 2^3 - 5 \cdot 2^2 - 9 \cdot 2 + 18 = -4$
(d) $2 \cdot i^3 + 3 \cdot i^2 + 2 \cdot i + 3 = 0$

Question 3.5.5

(a) There are many such polynomials, including $(2 + i)x^2 - 4x + 3$.
(b) There are many such polynomials, including $x^2 + 1$.
(c) There are many such polynomials, including $x^2 - 6x + 25$.
(d) There are many such polynomials, including $x - 1$.

Question 3.5.6

(a) $x = 4$ has multiplicity 1.
(b) $x = 1$ has multiplicity 2.
(c) $x = 1$, $x = -\dfrac{1}{2} + \dfrac{\sqrt{3}}{2}i$, and

$x = -\dfrac{1}{2} - \dfrac{\sqrt{3}}{2}i$ each have
multiplicity 1.

(d) $x = 1$, $x = -1$, $x = i$, and $x = -i$ each have multiplicity 1.

Question 3.5.7

(a) $x = 2$
(b) $x = 4\sqrt{2}$
(c) $x = \dfrac{-(\pi + 1)}{e}$
(d) $x = \dfrac{24}{13} - \dfrac{36}{13}i$

Question 3.5.8

(a) $x = -2$ has multiplicity 2.
(b) $x = -\dfrac{1}{3}$ and $x = 2$.
(c) $x = \dfrac{1}{4} \pm \dfrac{\sqrt{159}}{12}i$
(d) $x = \dfrac{1}{2} \pm \dfrac{\sqrt{2}}{4}$

Question 3.5.9

The del Ferro–Tartaglia solution is $x_1 = -4$.

Question 3.5.10

(a) $a = 1, b = -6, c = 11, d = -6$
(b) $m = -1$ and $n = 0$
(c) $y_1 = 0$
(d) $x_1 = 2$
(e) $x_2 = 1$ and $x_3 = 3$

3.6 Mathematical Induction

Question 3.6.1

Base case. For $n = 1$, $\displaystyle\sum_{i=1}^{2} 2 = 2 = 2 \cdot 1$.

Inductive step. Assume $\displaystyle\sum_{i=1}^{n} 2 = 2n$ and

prove that $\displaystyle\sum_{i=1}^{n+1} 2 = 2(n + 1)$. Then the

following equalities hold

$$\sum_{i=1}^{n+1} 2 = \left[\sum_{i=1}^{n} 2\right] + 2$$

$$= 2n + 2$$

$$= 2(n + 1).$$

Question 3.6.2

Base case. For $n = 5, 5^2 = 25 < 32 = 2^5$.
Inductive step. Assume $n^2 < 2n$ and prove that $(n + 1)^2 < 2^{n+1}$. Then the following relations hold

$$(n + 1)^2 = n^2 + 2n + 1$$
$$< 2^n + 2n + 1$$
$$< 2^n + 2^n$$
$$= 2 \cdot 2^n$$
$$= 2^{n+1}$$

Question 3.6.3

$a_1 = 1$, $a_2 = 3$, $a_3 = 1 + 2 \cdot 3 = 7$, $a_4 = 3 + 2 \cdot 7 = 17$, $a_5 = 7 + 2 \cdot 17 = 41$, and $a_6 = 17 + 2 \cdot 41 = 99$.

Question 3.6.4

Base case. $b_1 = 9/10 < 1$ and $b_2 = 10/11 < 1$.
Inductive step. Assume $b_n = r < 1$ and $b_{n+1} = q < 1$ and prove that $b_{n+2} < 1$. Since the product of two positive numbers less than 1 is also less than 1, we have

$$b_{n+2} = b_n \cdot b_{n+1} = r \cdot q < 1.$$

Question 3.6.5

$(B \vee C)$ has $1 + m + n$ left parentheses and $m + n + 1$ right parentheses; these numbers are the same, and so $(B \vee C)$ has the same number of left and right parentheses. The proofs for $(B \rightarrow C)$ and $(B \leftrightarrow C)$ are identical.

4.1   Analytic Geometry

Question 4.1.1

There are many such points, including $(0, 5)$, $(1, 7)$, and $(-1, 3)$; we call such a curve a line.

Question 4.1.2

(a) $D = 2\sqrt{2}$
(b) $D = \sqrt{35}$

Question 4.1.3

(a) $(x - 1)^2 + (y - 1)^2 = 1$
(b) $(x + 2)^2 + (y - 5)^2 = 64$
(c) $(x - 2)^2 + (y + 4)^2 = 1$
(d) $(x + 3)^2 + (y + 4)^2 = 25$

Question 4.1.4

(a) $x^2 + y^2 = 1$
(b) $(\sqrt{3}/2, 1/2)$ and $(\sqrt{3}/2, -1/2)$

(c) There are many such points, including $(1, 0)$, $(0, 1)$, $(0, -1)$, and $(0, -1)$.

Question 4.1.5

(a)

$x = \cos\theta$	1	$\sqrt{3}/2$	$\sqrt{2}/2$	1/2	0
$y = \sin\theta$	0	1/2	$\sqrt{2}/2$	$\sqrt{3}/2$	1

(b) $\cos(45°) = \sqrt{2}/2 = \sin(45°)$
    $\cos(\pi/3) = 1/2$ and $\sin(\pi/3) = \sqrt{3}/2$

(c) $\tan(0) = 0$
    $\tan(\pi/6) = 1/\sqrt{3}$
    $\tan(\pi/4) = 1$
    $\tan(\pi/3) = \sqrt{3}$
    $\tan(\pi/2) = $ undefined

Question 4.1.6

(a) Algebraic manipulations similar to those in example 4.1.3 enable us to begin with

$$\sqrt{(x+c)^2 + (y-0)^2}$$

$$- \sqrt{(x-c)^2 + (y-0)^2} = 2K$$

and obtain

$$\frac{x^2}{K^2} + \frac{y^2}{K^2 - c^2} = 1.$$

Since $K < c$, we have $K^2 - c^2 < 1$ and so $c^2 - K^2 > 1$. Defining $a = K$ and $b = \sqrt{c^2 - K^2}$ provides the desired equation

$$\frac{x^2}{a^2} + \frac{y^2}{-b^2} = 1.$$

(b) For foci $(-4, 0)$ and $(4, 0)$, we have $c = 4$ and $2K = 6$ implies $K = 3$. Substituting into the equation from part (a) yields

$$\frac{x^2}{3^2} + \frac{y^2}{-(4^2 - 3^2)} = 1$$

$$\Rightarrow \frac{x^2}{9} + \frac{y^2}{-7} = 1.$$

Question 4.1.7

(b) A linear pattern.

(c) $x = 2y$

Question 4.1.8

(a) $(-4, 4)$, $(-3, 3)$, $(-2, 2)$, $(-1, 1)$, $(0, 0)$, $(1, 1)$, $(2, 2)$, $(3, 3)$, $(4, 4)$

(b) The line $y = x$ when $x \geq 0$ and the line $y = -x$ when $x < 0$.

(c) $y = |x|$

Question 4.1.9

Many different graphs satisfy the stated conditions.

Question 4.1.10

(a) $y = -2x + 15$

(b) $d = \dfrac{(-2) \cdot 1 + 15 - 2}{\sqrt{(-2)^2 + 1}} = \dfrac{11}{\sqrt{5}}$.

(c) base $= 2\sqrt{5}$ and area $= 1/2 \cdot 2\sqrt{5} \cdot 11/\sqrt{5} = 11$

Question 4.1.11

(a) $4x + 4y + 4z = 4 + 4 + 4 \Rightarrow x + y + z = 3$

(b) $(3, 0, 0)$, $(0, 3, 0)$, and $(0, 0, 3)$.

Question 4.1.12

Squaring both sides of $\sqrt{(x-h)^2 + (y-j)^2 + (z-k)^2} = r$ results in $(x-h)^2 + (y-j)^2 + (z-k)^2 = r^2$.

4.2 Functions and Inverse Functions

Question 4.2.1

(a) Not a function since 3 maps to both 6 and 25.

(b) A function since every input maps to a unique output. The domain is $D = \{0, 1, 2, 3, 4\}$ and the range is $R = \{0, 1, 4, 9, 16\}$.

(c) A function since every input maps to a unique output. The domain is $D = \mathbb{N}$ and the range is $R = \mathbb{N}$.

(d) A function since every input maps to a unique output. The domain is $D = \mathbb{N}$ and the range is $R = \{2\}$.

(e) Not a function since 2 maps to every natural number.

Question 4.2.2

(a) For one-to-one:

$$f(a) = f(b)$$

$$\Rightarrow 12a - 10 = 12b - 10$$

$$\Rightarrow 12a = 12b$$

$$\Rightarrow a = b.$$

For onto: Let $a \in \mathbb{R}$. Then $(a + 10)/12 \in \mathbb{R}$ and

$$f\left(\frac{a + 10}{12}\right) = 12\left[\frac{a + 10}{12}\right] - 10$$

$$= [a + 10] - 10 = a.$$

(b) The function $f(x) = \sin(x)$ is not one-to-one since $\sin(0) = 0$ and $\sin(\pi) = 0$, but $0 \neq \pi$. Also, this function is not onto since $2 \in \mathbb{R}$ (the target space), but $\sin(x) \neq 2$ for every $x \in \mathbb{R}$. However, if the domain is restricted to $D = [-\pi/2, \pi/2]$ and the range is restricted to $R = [-1, 1]$, then the resulting function is both one-to-one and onto.

Question 4.2.3

(a) $f^{-1} : \{x \in \mathbb{R} : x \geq -5\}$ is algebraically defined by $f^{-1}(x) = \sqrt{\dfrac{x + 5}{3}}$.

(b) $(f^{-1} \circ f)(x) = f^{-1}(f(x))$

$$= f^{-1}(3x^2 - 5)$$

$$= \sqrt{\frac{(3x^2 - 5) + 5}{3}}$$

$$= \sqrt{\frac{3x^2}{3}}$$

$$= \sqrt{x^2}$$

$$= x$$

(c) $(f \circ f^{-1})(x) = f(f^{-1}(x))$

$$= f\left(\sqrt{\frac{x + 5}{3}}\right)$$

$$= 3\left(\sqrt{\frac{x + 5}{3}}\right)^2 - 5$$

$$= 3\left(\frac{x + 5}{3}\right) - 5$$

$$= (x + 5) - 5$$

$$= x$$

Question 4.2.4

$$y = \frac{\ln[(x - 3)^4]}{12} \Rightarrow 12y = \ln[(x - 3)^4]$$

$$\Rightarrow e^{12y} = (x - 3)^4$$

$$\Rightarrow e^{3y} = x - 3$$

$$\Rightarrow e^{3y} + 3 = x$$

Question 4.2.5

(a) The six points are: $\left(-2, \frac{1}{4}\right)$, $\left(-1, \frac{1}{2}\right)$, $(0, 1)$, $(1, 2)$, $(3, 8)$, $(5, 32)$.

4.3   Limits and Continuity

Question 4.3.1

(a) $\lim\limits_{x \to 0} f(x)$ does not exist

(b) $\lim\limits_{x \to 1} f(x) = 1$

(c) $\lim\limits_{x \to 2} f(x)$ does not exist

(d) $\lim_{x \to 2.5} f(x) = 2$

(e) $\lim_{x \to 3} f(x) = 2$

(f) $\lim_{x \to 4} f(x) = 1$

Question 4.3.2

(a) Let $\varepsilon > 0$ be a real number and define $\delta = \varepsilon/4$. Assuming $0 < |x - 3| < \delta$,

$$|f(x) - L| = |4x - 10 - 2|$$

$$= |4x - 12|$$

$$= 4|x - 3|$$

$$< 4 \cdot \delta = 4 \cdot \frac{\varepsilon}{4} = \varepsilon.$$

(b) Let $\varepsilon > 0$ be a real number and define $\delta = \varepsilon/2$. Assuming $0 < |x - 1| < \delta$,

$$|f(x) - L| = |-2x + 5 - 3|$$

$$= |-2x + 2|$$

$$= 2|x - 1|$$

$$< 2 \cdot \delta = 2 \cdot \frac{\varepsilon}{2} = \varepsilon.$$

(c) Let $\varepsilon > 0$ be a real number and define $\delta = \varepsilon/4$. Assuming $0 < |x - 5| < \delta$,

$$|f(x) - L| = |4x + 15 - 35|$$

$$= |4x - 20|$$

$$= 4|x - 5|$$

$$< 4 \cdot \delta = 4 \cdot \frac{\varepsilon}{4} = \varepsilon.$$

Question 4.3.3

(a) Let $\varepsilon > 0$ be a real number and define $\delta = \min\{1, \varepsilon/11\}$.

Assuming $0 < |x - 5| < \delta$,

$$|f(x) - L| = |x^2 - 25|$$

$$= |x + 5| \cdot |x - 5|$$

$$< 11 \cdot \delta \leq 11 \cdot \frac{\varepsilon}{11} = \varepsilon.$$

(b) Let $\varepsilon > 0$ be a real number and define $\delta = \min\{1, \varepsilon/3\}$. Assuming $0 < |x - 3| < \delta$,

$$|f(x) - L| = |(x - 2)^2 - 1|$$

$$= |x^2 - 4x + 3|$$

$$= |x - 1| \cdot |x - 3|$$

$$< 3 \cdot \delta \leq 3 \cdot \frac{\varepsilon}{3} = \varepsilon.$$

(c) Let $\varepsilon > 0$ be a real number and define $\delta = \min\{1, 2\varepsilon\}$. Assuming $0 < |x - 3| < \delta$,

$$|f(x) - L| = \left| \frac{1}{x - 1} - \frac{1}{2} \right|$$

$$= \left| \frac{2 - (x - 1)}{2(x - 1)} \right|$$

$$= \frac{1}{2|x - 1|} \cdot |x - 3|$$

$$< \frac{1}{2} \cdot \delta \leq \frac{1}{2} \cdot 2\varepsilon = \varepsilon.$$

(d) Let $\varepsilon > 0$ be a real number and define $\delta = \min\{1, 20\varepsilon\}$. Assuming $0 < |x - 2| < \delta$,

$$|f(x) - L| = \left| \frac{1}{x + 3} - \frac{1}{5} \right|$$

$$= \left| \frac{5 - (x + 3)}{5(x + 3)} \right|$$

$$= \frac{1}{5|x + 3|} \cdot |x - 2|$$

$$< \frac{1}{20} \cdot \delta \leq \frac{1}{20} \cdot 20\varepsilon = \varepsilon.$$

Question 4.3.4

(a) Assume $\lim_{x \to a} f(x) = L$ and $\lim_{x \to a} f(x) = M$ with $L \neq M$.

(b) Let $\varepsilon = |L - M|/2$, then there exists $\delta_L > 0$ such that $0 < |x - a| < \delta_L$ implies $|f(x) - L| < \varepsilon$ and there exists $\delta_M > 0$ such that $0 < |x - a| < \delta_M$ implies $|f(x) - M| < \varepsilon$.

(c) Let $\delta = \min\{\delta_L, \delta_M\}$. Then if $0 < |x - a| < \delta$, both $|f(x) - L| < \varepsilon$ and $|f(x) - M| < \varepsilon$.

(d) The contradiction follows from the fact that ε is half the distance between L and M and so $f(x)$ cannot be both closer to L than M and closer to M than L.

Question 4.3.5

Let $\varepsilon > 0$. Since $\lim\limits_{x \to a} f = L$, there exists $\delta_L > 0$ such that $0 < |x - a| < \delta_L$ implies $|f(x) - L| < \varepsilon/2$. Similarly, since $\lim\limits_{x \to a} g = M$, there exists $\delta_M > 0$ such that $0 < |x - a| < \delta_M$ implies $|g(x) - M| < \varepsilon/2$. Choose $\delta = \min\{\delta_L, \delta_M\}$ so that both inequalities involving $\varepsilon/2$ are true when $0 < |x - a| < \delta$. Therefore, when $0 < |x - a| < \delta$,

$$|f(x) + g(x) - (L + M)|$$

$$= |f(x) - L + g(x) - M|$$

$$\leq |f(x) - L| + |g(x) - M|$$

$$< \frac{\varepsilon}{2} + \frac{\varepsilon}{2} = \varepsilon.$$

Question 4.3.6

(a) Let $\varepsilon > 0$ be a real number and define $\delta = \varepsilon/2$. Assuming $0 < |x - 5| < \delta$,

$$|f(x) - f(a)| = |2x - 3 - 7|$$

$$= |2x - 10|$$

$$= 2|x - 5|$$

$$< 2 \cdot \delta = 2 \cdot \frac{\varepsilon}{2} = \varepsilon.$$

(b) Let $a \in \mathbb{R}$ and $\varepsilon > 0$ be a real number. Define $\delta = \varepsilon/2$ and, assuming $0 < |x - a| < \delta$,

$$|f(x) - f(a)| = |(2x - 3) - (2a - 3)|$$

$$= |2x - 2a|$$

$$= 2|x - a|$$

$$< 2 \cdot \delta = 2 \cdot \frac{\varepsilon}{2} = \varepsilon.$$

Question 4.3.7

(a) $\lim\limits_{x \to 3} 4x - 10 = 12 - 10 = 2$

(b) $\lim\limits_{x \to 2} x^2 = 2^2 = 4$

(c) $\lim\limits_{x \to 4} \dfrac{1}{x - 2} = \dfrac{1}{4 - 2} = \dfrac{1}{2}$

(d) $\lim\limits_{x \to 0} \dfrac{1}{(x - 3)^2} = \dfrac{1}{9}$

Question 4.3.8

(a) $\lim\limits_{x \to 3} \dfrac{x - 3}{x^2 + x - 12} = \dfrac{1}{7}$.

The following redefined function is continuous at $x = 3$:

$$g(x) = \begin{cases} \dfrac{x - 3}{x^2 + x - 12} & \text{if } x \neq 3 \\[2mm] \dfrac{1}{7} & \text{if } x = 3. \end{cases}$$

(b) $\lim\limits_{x \to 3} \dfrac{5x - 2}{5x^2 - 32x + 12} = -\dfrac{1}{3}$.

The following redefined function is continuous at $x = 3$:

$$g(x) = \begin{cases} \dfrac{5x - 2}{5x^2 - 32x + 12} & \text{if } x \neq 3 \\[2mm] -\dfrac{1}{3} & \text{if } x = 3. \end{cases}$$

(c) $\lim\limits_{x\to 3} \dfrac{x-3}{x^2-9} = \dfrac{1}{6}$.

The following redefined function is continuous at $x = 3$:

$$g(x) = \begin{cases} \dfrac{x-3}{x^2-9} & \text{if } x \neq 3 \\ \dfrac{1}{6} & \text{if } x = 3. \end{cases}$$

(d) $\lim\limits_{x\to 3} \dfrac{x-3}{x^3-27} = \dfrac{1}{27}$.

The following redefined function is continuous at $x = 3$:

$$g(x) = \begin{cases} \dfrac{x-3}{x^3-27} & \text{if } x \neq 3 \\ \dfrac{1}{27} & \text{if } x = 3. \end{cases}$$

Question 4.3.9

(a) Let $M > 0$ be a real number and choose $\delta = 1/\sqrt[4]{M}$.

Assuming $0 < |x - 3| < \delta$,

$$|x - 3| < \delta = \frac{1}{\sqrt[4]{M}}$$

$$(x - 3)^4 < \frac{1}{M}$$

$$\frac{1}{(x-3)^4} > M.$$

Therefore, $\lim\limits_{x\to 3} \dfrac{1}{(x-3)^4} = \infty$.

(b) Let $M > 0$ be a real number, choose $\delta = \min\left\{\dfrac{1}{2}, \sqrt{\dfrac{2}{3M}}\right\}$,

and assume $0 < |x - 1| < \delta$; note that under these assumptions,

$$\frac{1}{x} > \frac{1}{\delta + 1} \geq \frac{2}{3}.$$

$$\left|\frac{1}{x(x-1)^2}\right| > \frac{2}{3\delta^2} \geq M.$$

Therefore, $\lim\limits_{x\to 1} \dfrac{1}{x(x-1)^2} = \infty$.

4.4   **The Derivative**

Question 4.4.1

(a) For $f(x) = 2x + 1$,

$$\begin{aligned} f'(x) &= \lim_{h\to 0} \frac{2(x+h)+1-(2x+1)}{h} \\ &= \lim_{h\to 0} \frac{2x+2h+1-2x-1}{h} \\ &= \lim_{h\to 0} \frac{2h}{h} = \lim_{h\to 0} 2 = 2 \end{aligned}$$

(b) For $g(x) = 7x^3$,

$$\begin{aligned} g'(x) &= \lim_{h\to 0} \frac{7(x+h)^3 - 7x^3}{h} \\ &= \lim_{h\to 0} \frac{7x^3 + 21x^2h + 21xh^2 + 7h^3 - 7x^3}{h} \\ &= \lim_{h\to 0} \frac{21x^2h + 21xh^2 + 7h^3}{h} \\ &= \lim_{h\to 0} 21x^2 + 21xh + 7h^2 = 21x^2 \end{aligned}$$

(c) For $s(x) = \dfrac{1}{x+5}$,

$$s'(x) = \lim_{h \to 0} \frac{1/[(x+h)+5] - 1/(x+5)}{h}$$

$$= \lim_{h \to 0} \frac{x+5 - (x+h+5)}{h(x+h+5)(x+5)}$$

$$= \lim_{h \to 0} \frac{-h}{h(x+h+5)(x+5)} = \frac{-1}{(x+5)^2}$$

(d) For $t(x) = \dfrac{1}{3\sqrt{x}}$,

$$t'(x) = \lim_{h \to 0} \frac{\frac{1}{3\sqrt{x+h}} - \frac{1}{3\sqrt{x}}}{h}$$

$$= \lim_{h \to 0} \frac{\sqrt{x} - \sqrt{x+h}}{3h\sqrt{x+h}\sqrt{x}} \cdot \frac{\sqrt{x} + \sqrt{x+h}}{\sqrt{x} + \sqrt{x+h}}$$

$$= \lim_{h \to 0} \frac{x - x - h}{3h\sqrt{x+h}\sqrt{x} \cdot (\sqrt{x} + \sqrt{x+h})}$$

$$= \frac{-1}{6\sqrt{x^3}}$$

Question 4.4.2

(a) $f'(x) = 30x^2 - 14x$

(b) $g'(x) = \frac{1}{2} \cdot (5x + 2)^{-1/2} \cdot 5$

(c) $h'(x) = \dfrac{(5x^2 + 2) \cdot [-3\sin(3x + 1)]}{(5x^2 + 2)^2}$
$\qquad - \dfrac{\cos(3x + 1) \cdot 10x}{(5x^2 + 2)^2}$

(d) $p'(x) = \tan(2x) \cdot (5x^4 + 1) + (x^5 + x) \cdot 2\sec^2(2x)$

(e) $q'(x) = \sin^2(5x + 3) \cdot \frac{|8x|}{4x^2 + 1} + \ln(4x^2 + 1) \cdot 2\sin(5x + 3) \cdot \cos(5x + 3) \cdot 5$

(f) $r'(x) = \sqrt[3]{4e^x + 6x} \cdot 10x^4 + (2x^5 + 3) \cdot \frac{1}{3} \cdot (4e^x + 6x)^{-2/3} \cdot (4e^x + 6)$

Question 4.4.3

(a) $\dfrac{f(x+h)/g(x+h) - f(x)/g(x)}{h}$

(b) $\dfrac{f(x+h) \cdot g(x) - f(x) \cdot g(x+h)}{h \cdot g(x) \cdot g(x+h)}$

(c) $\dfrac{f(x+h)g(x) - g(x)f(x) + g(x)f(x) - f(x)g(x+h)}{h \cdot g(x) \cdot g(x+h)}$

$$= \frac{g(x) \cdot [f(x+h) - f(x)]}{h \cdot g(x) \cdot g(x+h)} - \frac{f(x) \cdot [g(x+h) - g(x)]}{h \cdot g(x) \cdot g(x+h)}$$

(d) $\dfrac{d}{dx}\left[\dfrac{f}{g}\right] = \dfrac{g \cdot f' - f \cdot g'}{g^2}$

Question 4.4.4

(a) $\dfrac{h(t) - h(x)}{t - x} = \dfrac{f(g(t)) - f(g(x))}{t - x}$

(b) $\dfrac{f(g(t)) - f(g(x))}{t - x} \cdot \dfrac{g(t) - g(x)}{g(t) - g(x)}$

   $\dfrac{f(g(t)) - f(g(x))}{g(t) - g(x)} \cdot \dfrac{g(t) - g(x)}{t - x}$

(c) $\displaystyle\lim_{t \to x} \dfrac{f(g(t)) - f(g(x))}{g(t) - g(x)} \cdot \dfrac{g(t) - g(x)}{t - x} = f'(g(x)) \cdot g'(x)$

Question 4.4.5

(a) $|x - 1|$

(b) $\dfrac{1}{x^2 - 1}$

(c) $\cot(x)$

(d) $\cot\left(\frac{\pi x}{2}\right)$

4.5   **Understanding Infinity**

Question 4.5.1

(a) The function $f : \mathbb{Z} \to \mathbb{Z}$ defined by $f(x) = x + 1$ is one-to-one since $f(a) = f(b)$ implies that $a + 1 = b + 1$; subtracting one from both sides yields $a = b$. The function f is also onto. For $n \in \mathbb{Z}$, consider $n - 1 \in \mathbb{Z}$: $f(n - 1) = (n - 1) + 1 = n$.

(b) The function $g : \mathbb{Z} \to \mathbb{Z}$ defined by $g(x) = 2x$ is one-to-one since $g(a) = g(b)$ implies that $2a = 2b$; dividing both sides by two yields $a = b$. The function g is *not* onto. Consider $1 \in \mathbb{Z}$; for all $n \in \mathbb{Z}$, $g(n) = 2n \neq 1$, and so 1 is not in the range of g even though 1 is in the target space.

(c) The function $h : \mathbb{R} \to \mathbb{R}$ defined by $h(x) = x^2$ is *not* one-to-one since $h(-2) = (-2)^2 = 4$ and $h(2) = 2^2 = 4$, but $-2 \neq 2$. The function h is *not* onto. Consider $-1 \in \mathbb{R}$; for all $r \in \mathbb{R}$, $h(r) = r^2 \neq -1$, and so -1 is not in the range of h even though -1 is in the target space.

Question 4.5.2

The function $f : \mathbb{N} \to E$ defined by $f(x) = 2x$ is one-to-one since $f(a) = f(b)$ implies that $2a = 2b$; dividing both sides by two yields $a = b$. The function f is also onto. For $n \in E$, there exists $j \in \mathbb{N}$ such that $n = 2j$. Then $f(j) = 2j = n$.

Question 4.5.3

(a) $2, 4, 6, 8, \ldots, 2j, \ldots$

(b) $0, 2, -2, 4, -4, 6, -6, \ldots, 2j, -2j, \ldots$

(c) $2, 3, 5, 7, 11, 13, 17, 19, 23, \ldots$

(d) Adding one to each element in the last sequence given in example 4.5.5,
$1, 1\frac{1}{2}, 1\frac{1}{3}, 1\frac{1}{4}, 1\frac{2}{4}, 1\frac{3}{4}, \ldots.$

Question 4.5.4

(a) For every $m \in \mathbb{N}$, the sets
$$A_m = \{(m, n) : n \in \mathbb{N}\}$$
are countable. Thus, $\mathbb{N} \times \mathbb{N} = \bigcup_{m \in \mathbb{N}} A_m$ is a countable union of countable sets and so countable.

(b) $\mathbb{Q}^n = \bigcup_{q_1 \in \mathbb{Q}} \bigcup_{q_2 \in \mathbb{Q}} \cdots \bigcup_{q_n \in \mathbb{Q}} \{(q_1, q_2, \ldots, q_n)\}$

(c) Define the following sequence of countable sets:
$Z_1 = \{(n_1, 0, 0, 0, \ldots) : n \in \mathbb{Z}\}$,
$Z_2 = \{(n_1, n_2, 0, 0, \ldots) : n, m \in \mathbb{Z}\}$,
$Z_3 = \{(n_1, n_2, n_3, 0, \ldots) : n, m \in \mathbb{Z}\}$,
and so on $\ldots$
Since the given set is equal to $\bigcup_{n \in \mathbb{N}} Z_n$, it is a countable union of countable sets and so countable.

Question 4.5.5

(a) $\mathbb{P}(A) = \{\emptyset, \{0\}\}$

(b) $\mathbb{P}(A) = \{\emptyset, \{0\}, \{1\}, \{0, 1\}\}$

Question 4.5.6

(a) For $x \in \mathbb{N}$, map x to $x + 2$, a to 1, and b to 2. Formally, let $A = \{a, b\} \cup \mathbb{N}$ represent

$2 + \omega$ and let $\mathbb{N}$ represent ω and define a one-to-one correspondence $f : A \to \mathbb{N}$ by
$$f(x) = \begin{cases} 1 & \text{if } x = a \\ 2 & \text{if } x = b \\ x + 2 & \text{if } x \in \mathbb{N}. \end{cases}$$

(b) For $x \in \mathbb{N}$, map x to $x + 3$, a to 1, b to 2, and c to 3. Formally, let $A = \{a, b, c\} \cup \mathbb{N}$ represent $3 + \omega$ and let $\mathbb{N}$ represent ω and define a one-to-one correspondence $f : A \to \mathbb{N}$ by
$$f(x) = \begin{cases} 1 & \text{if } x = a \\ 2 & \text{if } x = b \\ 3 & \text{if } x = c \\ x + 3 & \text{if } x \in \mathbb{N}. \end{cases}$$

(c) Let $A = \{a_1, \ldots, a_7\} \cup \mathbb{N}$ represent $7 + \omega$ and let $\mathbb{N}$ represent ω and define a one-to-one correspondence $f : A \to \mathbb{N}$ by
$$f(x) = \begin{cases} k & \text{if } x = a_k \text{ for } k = 1, \ldots, 7 \\ x + 7 & \text{if } x \in \mathbb{N}. \end{cases}$$

(d) Let $n \in \mathbb{N}$ be some *fixed* natural number. Let $A = \{a_1, \ldots, a_n\} \cup \mathbb{N}$ represent $n + \omega$ and let $\mathbb{N}$ represent ω and define a one-to-one correspondence $f : A \to \mathbb{N}$ by
$$f(x) = \begin{cases} k & \text{if } x = a_k \text{ for } k = 1, \ldots, n \\ x + n & \text{if } x \in \mathbb{N}. \end{cases}$$

Question 4.5.7

(a) Let $A = \{a_n : n \in \mathbb{N}\}$ and $B = \{b_n : n \in \mathbb{N}\}$. Define a one-to-one correspondence $f : A \cup B \to \mathbb{N}$ by
$$f(x) = \begin{cases} 2n - 1 & \text{if } x = a_n \\ 2n & \text{if } x = b_n. \end{cases}$$

(b) Let $A = \{a_n : n \in \mathbb{N}\}$, $B = \{b_n : n \in \mathbb{N}\}$, and $C = \{c_n : n \in \mathbb{N}$. Define a one-to-one correspondence $f : A \cup B \cup C \to \mathbb{N}$ by

$$f(x) = \begin{cases} 3n - 2 & \text{if } x = a_n \\ 3n - 1 & \text{if } x = b_n \\ 3n & \text{if } x = c_n. \end{cases}$$

A Hilbert Hotel illustrating $\omega + \omega + \omega = \omega$ would begin by listing the first element a_1 from the first ω, then the first element b_1 from the second ω, and then the first element c_1 from the third copy of omega. The list would then wrap back around

picking up the second element from each set, then the third, and so on. In short, we list: $a_1, b_1, c_1, a_2, b_2, c_2, a_3, b_3, \ldots$.

(c) For $1 \le m \le n$, let $A_m = \{a_{m,k} : k \in \mathbb{N}\}$ and define a one-to-one correspondence $f : \bigcup_{1 \le m \le n} A_m \to \mathbb{N}$ by

$$f(x) = \begin{cases} km - (m - 1) & \text{if } x = a_{1,k} \\ km - (m - 2) & \text{if } x = a_{2,k} \\ \vdots & \\ km & \text{if } x = a_{n,k}. \end{cases}$$

4.6 The Riemann Integral

Question 4.6.1

(a) $A = \pi \cdot \left(\dfrac{d}{2}\right)^2$

(b) $A = a \cdot b$

(c) $A = \dfrac{1}{2} \cdot (a + b) \cdot h$

(d) $A = \dfrac{5}{4} \cot\left(\dfrac{\pi}{5}\right) \cdot a^2$

(e) $A = \dfrac{1}{2} \cdot 3 \cdot 3 = \dfrac{9}{2}$

(f) $A = \displaystyle\int_0^4 100 - 6x^2 \, dx$

Question 4.6.2

(a) For base $[0, 1]$, the area is $A = 94$;
For base $[1, 2], A = 1 \cdot (100 - 6 \cdot 2^2) = 76$;
For base $[2, 3], A = 1 \cdot (100 - 6 \cdot 3^2) = 46$;
For base $[3, 4], A = 1 \cdot (100 - 6 \cdot 4^2) = 4$.

(b) The total enclosed area is the sum of the four areas computed in

part (a); that is, $A = 94 + 76 + 46 + 4 = 220$.

(c) The eight right-rectangle area is given by
$A = \frac{1}{2} \cdot (100 - 6 \cdot (\frac{1}{2})^2) + \frac{1}{2} \cdot 94 + \frac{1}{2} \cdot (100 - 6 \cdot (\frac{3}{2})^2) + \frac{1}{2} \cdot 74 + \frac{1}{2} \cdot (100 - 6 \cdot (\frac{5}{2})^2) + \frac{1}{2} \cdot 46 + \frac{1}{2} \cdot (100 - 6 \cdot (\frac{7}{2})^2) + \frac{1}{2} \cdot 4 = 256.$

Question 4.6.3

(a) $\mathbb{Z}$

(b) The set of reals in the interval $[0, 4]$.

(c) The set of reals in the interval $[0, 5)$.

(d) $\{6, 5\frac{1}{2}, 5\frac{1}{3}, 5\frac{1}{4}, \ldots\}$

(e) $\{6, 6\frac{1}{2}, 6\frac{2}{3}, 6\frac{3}{4}, \ldots\}$

Question 4.6.4

(a) $\mathbb{Z}$

(b) The set of reals in the interval $[5, 6]$.

(c) The set of reals in the interval $(4, 6]$.

(d) $\{3, 3\frac{1}{2}, 3\frac{2}{3}, 3\frac{3}{4}, \ldots\}$ or $\{3, 4, 5, 6, \ldots\}$

(e) $\{3, 2\frac{1}{2}, 2\frac{1}{3}, 2\frac{1}{4}, \ldots\}$

Question 4.6.5

If $S \subseteq \mathbb{R}$ is bounded below, then $M = \inf S$ iff both

- M is a lower bound of S, and
- for every $\varepsilon > 0$, there exists $s \in S$ such that $s < M + \varepsilon$.

Question 4.6.6

$P = \{0, 2, 3, 5, \}$
$Q = \{0, \sqrt{2}, 2, 3, 5\}$
$R = \{0, 1, 5\}$

Question 4.6.7

Since $f(x) = 3x^2 - 2x$ is increasing on $[1, 7]$ the suprema $M_i(f)$ occur at the right endpoint of each subinterval and the infima $m_i(f)$ occur at the left endpoint.
For $[x_0, x_1] = [1, 2]$, $M_1(f) = 8$ and $m_1(f) = 1$.
For $[x_1, x_2] = [2, 3]$, $M_2(f) = 21$ and $m_2(f) = 8$.
For $[x_2, x_3] = [3, 4]$, $M_3(f) = 40$ and $m_3(f) = 21$.
For $[x_3, x_4] = [4, 5]$, $M_4(f) = 65$ and $m_4(f) = 40$.
For $[x_4, x_5] = [5, 6]$, $M_5(f) = 96$ and $m_5(f) = 65$.
For $[x_5, x_6] = [6, 7]$, $M_6(f) = 133$ and $m_6(f) = 96$.

Question 4.6.8

(a) Note that $f(x) = 4x^2 - 6$ is increasing on $[0, 5]$; thus, $U(f, P)$

$$= \sum_{i=1}^{4} M_i(f) \cdot (x_i - x_{i-1})$$

$$= (-2)(1 - 0) + 10(2 - 1)$$

$$+ 30(3 - 2) + 94(5 - 3)$$

$$= -2 + 10 + 30 + 188 = 226.$$

(b) Note that $f(x) = |x + 2|$ is increasing on $[0, 5]$; thus, $U(f, P)$

$$= \sum_{i=1}^{4} M_i(f) \cdot (x_i - x_{i-1})$$

$$= 3(1 - 0) + 4(2 - 1)$$

$$+ 5(3 - 2) + 7(5 - 3)$$

$$= 3 + 4 + 5 + 14 = 26.$$

Question 4.6.9

First, find an algebraic expression for $L(f, P_n)$ as follows:

$$L(f, P_n) = \sum_{i=1}^{n} m_i(f) \cdot (x_i - x_{i-1})$$

$$= \sum_{i=1}^{n} x_{i-1} \cdot (x_i - x_{i-1})$$

$$= \sum_{i=1}^{n} \frac{i-1}{n} \cdot \frac{1}{n}$$

$$= \frac{1}{n^2} \cdot \sum_{i=1}^{n} i - 1$$

$$= \frac{1}{n^2} \cdot \left[\frac{n(n+1)}{2} - n \right]$$

$$= \frac{n-1}{2n}.$$

Applying Darboux's theorem,

$$L(f) = \lim_{n \to \infty} L(f, P_n)$$

$$= \lim_{n \to \infty} \frac{n-1}{2n} = \frac{1}{2}.$$

Question 4.6.10

Since the function $f(x) = x^2 + 1$ is Riemann integrable on $[0, 2]$, we can compute either $U(f)$ or $L(f)$ to obtain the value of $\int_0^2 x^2 + 1 \, dx$. The solution presented here evaluates $U(f)$. First, define a sequence of partitions (that are refinements as n increases)

$$P_n = \left\{0, \frac{2}{n}, \frac{4}{n}, \dots, 2\right\}$$

$$= \left\{\frac{2i}{n} : 0 \le i \le n\right\}.$$

Now find an algebraic expression for $U(f, P_n)$ as follows:

$$U(f, P_n) = \sum_{i=1}^{n} M_i(f) \cdot (x_i - x_{i-1})$$

$$= \sum_{i=1}^{n} [x_i^2 + 1] \cdot (x_i - x_{i-1})$$

$$= \sum_{i=1}^{n} \left[\left(\frac{2i}{n}\right)^2 + 1\right] \cdot \frac{2}{n}$$

$$= \sum_{i=1}^{n} \frac{8i^2}{n^3} + \frac{2}{n}$$

$$= \frac{1}{n^3} \cdot 8n(n+1)(2n+1)6n^3 + \frac{2n}{n}$$

$$= \frac{2(n+1)(2n+1)}{3n^2} + 2$$

Using the Riemann integrability and applying Darboux's theorem (for the sequence of partitions $\{P_{2n}\}$,

$$\int_0^2 x^2 + 1 \, dx = U(f)$$

$$= \lim_{n \to \infty} U(f, P_n)$$

$$= \lim_{n \to \infty} \frac{2(n+1)(2n+1)}{3n^2}$$

$$+ 2 = \frac{14}{3}.$$

Question 4.6.11

(a) finite: $\{(0, 3), (3, 6)\}$
   infinite: $\{(0, n) : n \in \mathbb{Z}\}$
(b) finite: $\{(1, 2), (16, 20)\}$
   none
   infinite: $\{(0, n) : n \in \mathbb{Z}\}$

(c) finite: none
   infinite: $\left\{\left(\frac{2n+1}{2}, \frac{2n+3}{2}\right) : n \in \mathbb{Z}\right\}$
(d) finite: none
   infinite: $\{(n, n+1) : n \in \mathbb{Z}\}$

Question 4.6.12

Let $\varepsilon > 0$ and consider the interval open cover consisting of $I_n = \left(x_n - \frac{\varepsilon}{2 \cdot 2^n}, x_n + \frac{\varepsilon}{2 \cdot 2^n}\right)$. Then the following equalities hold

$$\sum_{n=1}^{\infty} m(I_n) = \sum_{n=1}^{\infty} \frac{\varepsilon}{2^n}$$

$$= \varepsilon \cdot \sum_{n=1}^{\infty} \frac{1}{2^n}$$

$$= \varepsilon \cdot \frac{1/2}{1 - 1/2} = \varepsilon.$$

Question 4.6.13

(a) Since $f(x)$ has no discontinuities and the empty set has measure zero, the given function $f(x)$ is Riemann integrable by the Riemann–Lebesgue theorem.
(b) Since $f(x)$ has one discontinuity at $x = 0$, a set with one element is countable, and every countable has measure zero, the given function $f(x)$ is Riemann integrable by the Riemann–Lebesgue theorem.
(c) Since $f(x)$ has one discontinuity at $x = 0$, a set with one element is countable, and every countable has measure zero, the given function $f(x)$ is Riemann integrable by the Riemann–Lebesgue theorem.
(d) Since $f(x)$ has countably infinitely many discontinuities at $x = 1/2^k$ for $k \in \mathbb{N}$ and every countable has measure zero, the given function $f(x)$ is Riemann integrable by the Riemann–Lebesgue theorem.

4.7   The Fundamental Theorem of Calculus

Question 4.7.1

(a) Since $f(x) = 1/(x + 1)$ is decreasing on $[0, 1]$ the suprema $M_i(f)$ occur at the left endpoint of each subinterval. First, define a sequence of partitions (that are refinements as n increases)

$$P_n = \left\{ 0, \frac{1}{n}, \frac{2}{n}, \dots, 1 \right\}$$

$$= \left\{ \frac{i}{n} : 0 \leq i \leq n \right\}.$$

Now find an algebraic expression for $U(f, P_n)$ as follows:

$$U(f, P_n) = \sum_{i=1}^{n} M_i(f) \cdot (x_i - x_{i-1})$$

$$= \sum_{i=1}^{n} \frac{1}{x_{i-1} + 1} \cdot (x_i - x_{i-1})$$

$$= \sum_{i=1}^{n} \frac{1}{[(i-1)/n] + 1} \cdot \frac{1}{n}$$

$$= \sum_{i=1}^{n} \frac{n}{i - 1 + n} \cdot \frac{1}{n}$$

$$= \sum_{i=1}^{n} \frac{1}{i - 1 + n}.$$

Thus, $U(f) = \lim_{n \to \infty} \sum_{i=1}^{n} \frac{1}{i - 1 + n}.$

(b) Since $f(x) = \sin(x)$ is increasing on $[0, 1]$ the suprema $M_i(f)$ occur at the right endpoint of each subinterval. First, define a sequence of partitions (that are refinements as n increases)

$$P_n = \left\{ 0, \frac{1}{n}, \frac{2}{n}, \dots, 1 \right\}$$

$$= \left\{ \frac{i}{n} : 0 \leq i \leq n \right\}.$$

Now find an algebraic expression for $U(f, P_n)$ as follows:

$$U(f, P_n) = \sum_{i=1}^{n} M_i(f) \cdot (x_i - x_{i-1})$$

$$= \sum_{i=1}^{n} \sin(x_i) \cdot (x_i - x_{i-1})$$

$$= \sum_{i=1}^{n} \sin\left(\frac{i}{n}\right) \cdot \frac{1}{n}.$$

Thus, $U(f) = \lim_{n \to \infty} \sum_{i=1}^{n} \sin\left(\frac{i}{n}\right) \cdot \frac{1}{n}.$

Question 4.7.2

Not necessarily—the text gives an example of a two-person class with earned grades of 70 percent and 72 percent, which results in a mean of 71 percent and no one earning the mean.

Question 4.7.3

Assume $f'(c) < 0$. Identify an interval around c such that the difference quotient from the alternative definition of the derivative is positive. Since

$$f'(c) = \lim_{x \to c} \frac{f(x) - f(c)}{x - c},$$

when we apply the definition of the limit with $\varepsilon = |f'(c)| = -f'(c)$, there exists a value $\delta > 0$ such that $0 < |x - c| < \delta$ implies

$$\left| \frac{f(x) - f(c)}{x - c} - f'(c) \right| < \varepsilon = |f'(c)|.$$

Hence

$$f'(c) < \frac{f(x) - f(c)}{x - c} - f'(c) < -f'(c)$$

which implies that

$$2f'(c) < \frac{f(x) - f(c)}{x - c} < 0.$$

If $(c - \delta, c + \delta)$ is not contained in (a, b), redefine δ as a sufficiently small positive value so that $(c - \delta, c + \delta) \subseteq (a, b)$. Then

$$\frac{f(x) - f(c)}{x - c} > 0$$

whenever $x \in (c - \delta, c + \delta)$; for $x \in (c - \delta, c)$, $x - c < 0$ and so $f(x) - f(c) > 0$, and so $f(x) > f(c)$. But $f(c)$ is a relative maximum, and so $f(c) > f(x)$ must hold true for some open interval about c. This fact gives the desired contradiction; we conclude $f'(c) \neq 0$.

Question 4.7.4

(a) The linear expression

$$\left[\frac{f(b) - f(a)}{b - a}\right] \cdot x$$

is continuous on $\mathbb{R}$. Since $f(x)$ is continuous on $[a, b]$ and a difference of continuous functions is continuous

$$g(x) = f(x) - \left[\frac{f(b) - f(a)}{b - a}\right] \cdot x$$

is continuous on $[a, b]$. Note that the right continuity of g at $x = a$ and the left continuity of g at $x = b$ follow from the corresponding continuity of f at these points and the continuity of the linear expression on all of $\mathbb{R}$.

(b) $g'(x) = f'(x) - \left[\dfrac{f(b) - f(a)}{b - a}\right]$

(c) The following equalities hold by direct substitution and the assumption that $f(a) = f(b)$.

$$g(a) = f(a) - \left[\frac{f(b) - f(a)}{b - a}\right] \cdot a$$

$$= f(a) - 0 \cdot a = f(a)$$
$$= f(b) = f(b) - 0 \cdot b$$
$$= f(b) - \left[\frac{f(b) - f(a)}{b - a}\right] \cdot b$$
$$= g(b)$$

(d) By Rolle's theorem, there exists $c \in (a, b)$ such that $g'(c) = 0$. Substituting into the expression from part (b) produces

$$f'(c) - \left[\frac{f(b) - f(a)}{b - a}\right] = 0,$$

and so

$$f'(c) = \left[\frac{f(b) - f(a)}{b - a}\right].$$

Question 4.7.5

(a) $\int f \, dx = x^5 + 2x + \sqrt{x} + C$

(b) $\int f \, dx = e^x + \sin(x) + C$

(c) $\int f \, dx = \sec(x) + C$

(d) $\int f \, dx = \ln|x| + \ln|x + 1| + C$

(e) $y = \frac{1}{3}(x^2 + 1)^{3/2} + C$

(f) $y = x^2 e^x - 2xe^x + 3e^x + C$

Question 4.7.6

$$\int x^2 \, dx = \frac{x^3}{3} + C, \text{ but}$$

$$\int x \, dx \cdot \int x \, dx = \left[\frac{x^2}{2} + C\right] \cdot \left[\frac{x^2}{2} + C\right]$$

$$= \frac{x^4}{4} + Cx^2 + D.$$

Question 4.7.7

(a) $x + x^2$

(b) $\ln|x|$

(c) $[\sin(x^3 + e^x)] \cdot (3x^2 + e^x)$

(d) $(-1) \cdot (x^2 + e^x) \cdot 2x$

Question 4.7.8

(a) $\displaystyle\int_0^1 4x^3 + 1 \, dx = x^4 + x \, \Big]_0^1 = 2$

(b) $\displaystyle\int_{-1}^0 4x^3 + 1 \, dx = x^4 + x \, \Big]_{-1}^0 = 0$

4.8 Application: Differential Equations

Question 4.8.1

(a) First-order, linear differential equation.

(b) First-order, nonlinear differential equation.

(c) Third-order, linear differential equation.

(d) First-order, nonlinear differential equation.

Question 4.8.2

(a) $y'' - y + 4 = [\cos x + 4]'' - [\cos x + 4] + 4 = -\cos x + 0 - \cos x - 4 + 4 = -2\cos x$
$y(0) = \cos 0 + 4 = 1 + 4 = 5$

(b) $y^{(4)} - y = [\cos x + 4]^{(4)} - [\cos x + 4] = \cos x - \cos x - 4 = -4$
$y(\pi/3) = \cos(\pi/3) + 4 = \frac{1}{2} + 4 = 4.5$

Question 4.8.3

$\int y'\, dx = \int \cos x + e^x\, dx = \sin x + e^x + C = y(x)$

Applying the initial condition, $5 = y(0) = \sin 0 + e^0 + C = 1 + C$.

Thus,

$$y(x) = \sin x + e^x + 4$$

Question 4.8.4

(a) For $y' + \frac{1}{2}y = 2x$, we have $F = 1/x$ and $G = 2x$. Then $\int F\, dx = \int 1/x\, dx = \ln x$ and $\int 2x \cdot e^{\ln x}\, dx = \int 2x^2\, dx = \frac{2}{3}x^3 + C$. Thus,

$$y = e^{-\ln x} \cdot \left[\frac{2}{3}x^3 + C \right] = \frac{2}{3}x^2 + C\frac{1}{x}.$$

(b) For $y' - (3/x)y = 2x^3$, we have $F = -3/x$ and $G = 2x^3$. Then $\int F\, dx = \int -3/x\, dx = -3\ln x$ and $\int 2x^3 \cdot e^{-3\ln x}\, dx = \int 2x^3/x^3\, dx = 2x + C$.

Thus,

$$y = e^{3\ln x} \cdot [2x + C] = 2x^4 + Cx^3.$$

Question 4.8.5

(a) $-\dfrac{1}{2y^2} = \dfrac{x^4}{4} + C$

(b) $\ln |y| - \arctan y = \dfrac{x^3}{3} + x + C$

Question 4.8.6

(a) $\dfrac{1}{2} \cdot L \cdot 2 \cdot \dfrac{dx}{dt} \cdot \dfrac{d^2x}{dt^2} = g \cdot \sin x \cdot \dfrac{dx}{dt}$

$$L \cdot \dfrac{d^2x}{dt^2} = g \cdot \sin x$$

(b) $L\dfrac{d^2x}{dt^2} = g\sin x \Rightarrow L\dfrac{d^2x}{dt^2} = gx$

$$\Rightarrow \dfrac{d^2x}{dt^2} = \dfrac{gx}{L}$$

$$\Rightarrow \dfrac{d^2x}{dt^2} - \dfrac{gx}{L} = 0$$

(c) $x = C\sin\left[\sqrt{\dfrac{-g}{L}}\, t\right] + D\cos\left[\sqrt{\dfrac{-g}{L}}\, t\right]$

$\dfrac{dx}{dt} = C\cos\left[\sqrt{\dfrac{-g}{L}}\, t\right] \cdot \sqrt{\dfrac{-g}{L}}$

$\qquad - D\sin\left[\sqrt{\dfrac{-g}{L}}\, t\right] \cdot \sqrt{\dfrac{-g}{L}}$

$\dfrac{d^2x}{dt^2} = -C\sin\left[\sqrt{\dfrac{-g}{L}}\, t\right] \cdot \left(\dfrac{-g}{L}\right)$

$\qquad - D\cos\left[\sqrt{\dfrac{-g}{L}}\, t\right] \cdot \left(\dfrac{-g}{L}\right)$

$$= C \sin\left[\sqrt{\frac{g}{L}}\, t\right] \cdot \left(\frac{-g}{L}\right)$$

$$+ D \cos\left[\sqrt{\frac{-g}{L}}\, t\right] \cdot \left(\frac{g}{L}\right)$$

Direct substitution shows that

$$\frac{d^2 x}{dt^2} - \frac{g}{L} x = 0.$$

(d) When $t = 0$, then $x = X$ and $x' = 0$. Substituting into the expressions from (c) provides the value of the constants as follows

$$X = C \cdot 0 + D \cdot 1 \Rightarrow D = X,$$

$$0 = C \cdot 1 \cdot \sqrt{\frac{-g}{L}} - D \cdot 0 \Rightarrow C = 0.$$

Thus, the solution is

$$x = X \cos\left[\sqrt{\frac{-g}{L}}\, t\right].$$

Question 4.8.7

$$H_2 = (-1)^2 \cdot e^{x^2} \cdot \frac{d^2}{dx^2}\left[e^{-x^2}\right]$$

$$= 1 \cdot e^{x^2} \cdot \left[(-2x)e^{-x^2}(-2x)\right.$$

$$\left. + e^{-x^2} \cdot (-2)\right]$$

$$= 4x^2 - 2$$

$$H_3 = (-1)^3 \cdot e^{x^2} \cdot \frac{d^3}{dx^3}\left[e^{-x^2}\right]$$

$$= -1 \cdot e^{x^2} \cdot \left[(4x^2)e^{-x^2}(-2x)\right.$$

$$\left. + e^{-x^2} \cdot (8x) + (-2) \cdot e^{-x^2} \cdot (-2x)\right]$$

$$= 8x^3 + 12x$$

$$H_4 = (-1)^4 \cdot e^{x^2} \cdot \frac{d^4}{dx^4}\left[e^{-x^2}\right]$$

$$= 1 \cdot e^{x^2} \cdot \left[(-8x^3)e^{-x^2}(-2x)\right.$$

$$+ e^{-x^2} \cdot (-24x^2) + (12x)$$

$$\left. \cdot e^{-x^2} \cdot (-2x) + 12 \cdot e^{-x^2}\right]$$

$$= 16x^4 - 48x^2 + 12$$

Question 4.8.8

Since $H_3 = 8x^3 - 12x$, then $H_3' = 24x^2 - 12$ and $H_3'' = 48x$. Substituting into the given differential equation results in

$$y'' - 2xy' + 6y$$

$$= H_3'' - 2xH_3' + 6H_3$$

$$= 48x - 2x \cdot [24x^2 - 12]$$

$$+ 6 \cdot [8x^3 - 12x]$$

$$= 48x - 48x^3 + 24x + 48x^3$$

$$- 72x = 0.$$

Question 4.8.9

$$\int_{-\infty}^{\infty} H_0 \cdot H_1 \cdot e^{-x^2}\, dx$$

$$= \int_{-\infty}^{\infty} 1 \cdot 2x \cdot e^{-x^2}\, dx$$

$$= \int_{-\infty}^{0} 2xe^{-x^2}\, dx + \int_{0}^{\infty} 2xe^{-x^2}\, dx$$

$$= \lim_{a \to -\infty} \int_{b}^{0} 2xe^{-x^2}\, dx$$

$$+ \lim_{b \to \infty} \int_{0}^{b} 2xe^{-x^2}\, dx$$

$$= \lim_{a \to -\infty} -e^{-x^2}\Big|_{a}^{0} + \lim_{b \to \infty} -e^{-x^2}\Big|_{0}^{b}$$

$$= \lim_{a \to -\infty} -1 + e^{-a^2} + \lim_{b \to \infty} -e^{-b^2} + 1$$

$$= -1 + 0 + 0 + 1 = 0$$

5.1 Combinatorics

Question 5.1.1

(a) order, repetition
(b) order, no repetition
(c) no order, no repetition
(d) no order, repetition

Question 5.1.2

(a) $2 \cdot 2 = 4$
(b) $6 \cdot 2 = 12$

Question 5.1.3

(a) $20 \cdot 20 \cdot 20 = 8,000$
(b) $26 \cdot 26 \cdot 26 \cdot 26 \cdot 26 = 11,881,376$

Question 5.1.4

(a) TO, OT
(b) TT, TO, OT, OO
(c) TO, TP, OT, OP, PT, PO

Question 5.1.5

(a) $P(5, 1) = \dfrac{5!}{(5 - 1)!} = \dfrac{5!}{4!} = 5$

(b) $P(5, 3) = \dfrac{5!}{(5 - 3)!} = \dfrac{5!}{2!} = 60$

(c) $P(15, 1) = \dfrac{15!}{(15 - 1)!} = \dfrac{15!}{14!} = 15$

(d) $P(15, 3) = \dfrac{15!}{(15 - 3)!} = \dfrac{15!}{12!}$
$= 2,730$

Question 5.1.6

(a) $P(20, 3) = \dfrac{20!}{(20 - 3)!} = \dfrac{20!}{17!}$
$= 6,840$

(b) $P(26, 5) = \dfrac{26!}{(26 - 5)!} = \dfrac{26!}{21!}$
$= 7,893,600$

Question 5.1.7

(a) $\{T, O\}; \{T, P\}; \{O, P\}$
(b) $\{T, O, P\}$

Question 5.1.8

(a) $C(5, 1) = \dfrac{5!}{1! \cdot (5 - 1)!} = \dfrac{5!}{1! \cdot 4!} = 5$

(b) $C(5, 3) = \dfrac{5!}{3! \cdot (5 - 3)!} = \dfrac{5!}{3! \cdot 2!} = 10$

(c) $C(15, 1) = \dfrac{15!}{1! \cdot (15 - 1)!}$
$= \dfrac{15!}{1! \cdot 14!} = 15$

(d) $C(15, 3) = \dfrac{15!}{3! \cdot (15 - 3)!}$
$= \dfrac{15!}{3! \cdot 12!} = 455$

Question 5.1.9

(a) $C(20, 3) = \dfrac{20!}{3! \cdot (20 - 3)!}$
$= \dfrac{20!}{3! \cdot 12!} = 1,140$

(b) $C(26, 5) = \dfrac{26!}{5! \cdot (26 - 5)!}$
$= \dfrac{26!}{5! \cdot 21!} = 65,780$

Question 5.1.10

(a) Using $n = 20$ and $k = 3$ in $C(n + k - 1, k)$, produces $C(20 + 3 - 1, 3) = C(22, 3) = 1,540$.
(b) Using $n = 26$ and $k = 4$ in $C(n + k - 1, k)$, produces $C(26 + 4 - 1, 4) = C(29, 4) = 23,751$.

Question 5.1.11

(a) $C(12, 2) \cdot C(3, 2) \cdot C(8, 3) = 11,088$
(b) $C(2, 1) \cdot C(4, 1) \cdot C(16, 1) = 128$
(c) $C(5, 4) + C(5, 3) \cdot C(20, 1) + C(5, 2) \cdot C(20, 2) + C(5, 1) \cdot C(20, 3)$
$= 7,805$
(d) $P(12, 4) \cdot C(4, 4) \cdot C(4, 3) \cdot C(4, 2) \cdot C(4, 1) = 1,916,006,400$

5.2   Pascal's Triangle and the Binomial Theorem

Question 5.2.1

(a) $C(4, 0) = 1$; $C(4, 1) = 4$; $C(4, 2) = 6$; $C(4, 3) = 4$; and $C(4, 4) = 1$.

(b) $C(5, 0) = 1$; $C(5, 1) = 5$; $C(5, 2) = 10$; $C(5, 3) = 10$; $C(5, 4) = 5$; and $C(5, 5) = 1$.

(c) $C(6, 0) = 1$; $C(6, 1) = 6$; $C(6, 2) = 15$; $C(6, 3) = 20$; $C(6, 4) = 15$; $C(6, 5) = 6$; and $C(6, 6) = 1$.

(d) $C(7, 0) = 1$; $C(7, 1) = 7$; $C(7, 2) = 21$; $C(7, 3) = 35$; $C(7, 4) = 35$; $C(7, 5) = 21$; $C(7, 6) = 7$; and $C(7, 7) = 1$.

Question 5.2.2

(a) $(a+b)^5 = a^5 + 5a^4b + 10a^3b^2 + 10a^2b^3 + 5ab^4 + b^5$

(b) $(2x + b)^5 = (2x)^5 + 5(2x)^4b + 10(2x)^3b^2 + 10(2x)^2b^3 + 5(2x)b^4 + b^5$
$= 32x^5 + 80x^4b + 80x^3b^2 + 40x^2b^3 + 10xb^4 + b^5$

(c) $(2x - 3y)^5 = (2x)^5 + 5(2x)^4 \cdot (-3y) + 10(2x)^3(-3y)^2 + 10(2x)^2(-3y)^3 + 5(2x)(-3y)^4 + (-3y)^5$
$= 32x^5 - 240x^4y + 720x^3y^2 - 1080x^2y^3 + 810xy^4 - 243y^5$

(d) $(a+b)^6 = a^6 + 6a^5b + 15a^4b^2 + 20a^3b^3 + 15a^2b^4 + 6ab^5 + b^6$

(e) $(3x + 2y)^6 = (3x)^6 + 6(3x)^5 \cdot (2y) + 15(3x)^4(2y)^2 + 20(3x)^3 \cdot (2y)^3 + 15(3x)^2(2y)^4 + 6(3x) \cdot (2y)^5 + (2y)^6$
$= 729x^6 + 2{,}916x^5y + 4{,}860x^4y^2 + 4{,}320x^3y^3 + 2{,}160x^2y^4 + 5{,}76xy^5 + 64y^6$

(f) $(3x - 2y)^6 = (3x)^6 + 6(3x)^5(-2y) + 15(3x)^4(-2y)^2 + 20(3x)^3(-2y)^3 + 15(3x)^2(-2y)^4 + 6(3x)(-2y)^5 + (-2y)^6$
$= 729x^6 - 2{,}916x^5y + 4{,}860x^4y^2 - 4{,}320x^3y^3 + 2{,}160x^2y^4 - 576xy^5 + 64y^6$

Question 5.2.3

$2^n = (1 + 1)^n$

$= C(n, 0) \cdot 1^n \cdot 1^0$

$\quad + C(n, 1) \cdot 1^{n-1} \cdot 1^1$

$\quad + \cdots + C(n, n) \cdot 1^0 \cdot 1^n$

$= C(n, 0) + C(n, 1)$

$\quad + \cdots + C(n, n)$

$= \sum_{k=0}^{n} C(n, k)$

Question 5.2.4

$$C_4 = \frac{8!}{5! \cdot 4!} = 14$$

Question 5.2.5

$C(2n, n) - C(2n, n+1)$

$= \frac{(2n)!}{n! \cdot (2n-n)!} - \frac{(2n)!}{(n+1)! \cdot (n-1)!}$

$= \frac{(2n)!}{n! \cdot n!} - \frac{(2n)!}{(n+1)! \cdot (n-1)!}$

$= \frac{(2n)! \cdot (n+1)! \cdot (n-1)! - (2n)! \cdot n! \cdot n!}{n! \cdot n! \cdot (n+1)! \cdot (n-1)!}$

$= \frac{(2n)! \cdot (n)! \cdot (n-1)! \cdot [(n+1)-n]}{n! \cdot n! \cdot (n+1)! \cdot (n-1)!}$

$= \frac{(2n)!}{(n+1)! \cdot n!}$

Question 5.2.6

The $n = 0$ to $n = 11$ rows are:

```
           1
          1 1
         1 2 1
        1 0 0 1
       1 1 0 1 1
      1 2 1 1 2 1
     1 0 0 2 0 0 1
    1 1 0 2 2 0 1 1
   1 2 1 2 1 2 1 2 1
  1 0 0 0 0 0 0 0 0 1
 1 1 0 0 0 0 0 0 1 1
1 2 1 0 0 0 0 0 0 1 2 1
```

Question 5.2.7

The $n = 0$ to $n = 11$ rows are:

```
           1
          1 1
         1 2 1
        1 3 3 1
       1 0 2 0 1
      1 1 2 2 1 1
     1 2 3 0 3 2 1
    1 3 1 3 3 1 3 1
   1 0 0 0 2 0 0 0 1
  1 1 0 0 2 2 0 0 1 1
 1 2 1 0 2 0 2 0 1 2 1
1 3 3 1 2 2 2 2 1 3 3 1
```

5.3 Basic Probability Theory

Question 5.3.1

(a) Flip a coin.
(b) The sum of the roll of two dice.
(c) The number of coin flips until heads is tossed.

$$P(E) = \frac{5}{36}$$

$$P(F) = \frac{6}{36} = \frac{1}{6}$$

$$P(G) = \frac{18}{36} = \frac{1}{2}$$

Question 5.3.2

(a) $D = \{(1, 2), (1, 3), (2, 2), (2, 3), (3, 2), (3, 3), (4, 2), (4, 3), (5, 2), (5, 3), (6, 2), (6, 3)\}$

(b) $E = \{(1, 5), (2, 4), (3, 3), (4, 2), (5, 1)\}$

(c) $F = \{(1, 1), (2, 2), (3, 3), (4, 4), (5, 5), (6, 6)\}$

(d) $G = \{(2, 1), (2, 2), (2, 3), (2, 4), (2, 5), (2, 6), (3, 1), (3, 2), (3, 3), (3, 4), (3, 5), (3, 6), (5, 1), (5, 2), (5, 3), (5, 4), (5, 5), (5, 6)\}$

Question 5.3.3

$$P(D) = \frac{12}{36} = \frac{1}{3}$$

Question 5.3.4

Y	1	2	3	4	5	6
$P[Y = y]$	1/6	1/6	1/6	1/6	1/6	1/6

Question 5.3.5

$P[X = n] = [1/2]^n$. Since $1/2 + 1/4 + 1/8 = 0.875$ and $1/2 + 1/4 + 1/8 + 1/16 = 0.9375$, a fair coin must be tossed four or more times until there is a better than 90% chance of tossing heads.

Question 5.3.6

$$P[X \leq 20] = \sum_{i=1}^{20} P[X = i] = 1 - (0.99)^{20} \approx 0.18209$$

Question 5.3.7

$$P[Y \geq 10] = P[Y=10] + P[Y=11]$$

$$+ P[Y=12]$$

$$= C(12,10) \cdot (0.85)^{10} \cdot (0.15)^2$$

$$+ C(12,11) \cdot (0.85)^{11} \cdot (0.15)^1$$

$$+ C(12,12) \cdot (0.85)^{12} \cdot (0.15)^0$$

$$\approx 0.735818$$

Question 5.3.8

(a) $P[S=s] = \dfrac{C(52,s) \cdot C(18, 20-s)}{C(70, 20)}$

(b) $P[S = 19] + P[S = 20]$

$$= \frac{C(52, 19) \cdot C(18, 1)}{C(70, 20)}$$

$$+ \frac{C(52, 20) \cdot C(18, 0)}{C(70, 20)}$$

$$\approx 0.00926$$

(c) $P[S=s] = \dfrac{C(49,s) \cdot C(18, 17-s)}{C(67, 17)}$

$$P[S = 16] + P[S = 17]$$

$$= \frac{C(49, 16) \cdot C(18, 1)}{C(67, 17)}$$

$$+ \frac{C(49, 17) \cdot C(18, 0)}{C(67, 17)}$$

$$\approx 0.019801$$

Question 5.3.9

$$P[-10 < X < 10] = \int_{-10}^{10} \frac{1}{60}\, dx = \frac{1}{3}$$

Question 5.3.10

$$P[1.3 < X < 1.5] \approx 0.22974$$

$$P[X \geq 1.5] = 0.5 - P[1.2 < X < 1.5] \approx$$

$$0.02275$$

Question 5.3.11

$$\mu = 0 \cdot P[X = 0] + 1 \cdot P[X = 1] =$$

$$0 + 1/2 = 1/2$$

Question 5.3.12

For the density function

$$f(x) = \begin{cases} 1/(b-a) & x \in [a, b] \\ 0 & x \notin [a, b] \end{cases}$$

$$E[X] = \int_{-\infty}^{\infty} x \cdot f(x)\, dx = \int_a^b x \cdot \frac{1}{b-a}\, dx$$

$$= \frac{x^2}{2(b-a)} \bigg]_a^b = \frac{b^2 - a^2}{2(b-a)} = \frac{a+b}{2}$$

Question 5.3.13

$$\sigma = \sqrt{\left[0 - \frac{1}{2}\right]^2 \cdot \frac{1}{2} + \left[1 - \frac{1}{2}\right]^2 \cdot \frac{1}{2}}$$

$$= \sqrt{\left[\frac{1}{2}\right]^2 \cdot \frac{1}{2} + \left[\frac{1}{2}\right]^2 \cdot \frac{1}{2}}$$

$$= \sqrt{\left[\frac{1}{2}\right]^2} = \frac{1}{2}$$

Question 5.3.14

For the density function

$$f(x) = \begin{cases} 1/(b-a) & x \in [a, b] \\ 0 & x \notin [a, b] \end{cases}$$

$$\sigma^2 = \int_{-\infty}^{\infty} (x-\mu)^2 \cdot f(x)\, dx$$

$$= \int_a^b \left(x - \frac{a+b}{2}\right)^2 \cdot \frac{1}{b-a}\, dx$$

$$= \frac{1}{b-a} \int_a^b x^2 - (a+b)x + \left[\frac{a+b}{2}\right]^2 dx$$

$$= \frac{(b-a)^2}{12}$$

Thus, $\sigma = \dfrac{b-a}{\sqrt{12}}$.

5.4 Application: Statistical Inference and Hypothesis Testing

Question 5.4.1

(a) For a good part, let $X = 1$, and for a defective part, let $X = 0$. Let p be the percent of parts assigned $X = 0$. Is $p < 0.5$ percent?

(b) Let X denote the difference in food intake for an individual fish and let $\overline{X}$ be the average difference. Is $\overline{X} > 0$?

(c) Let X denote the time to pain relief for a new medicine and let $\overline{X}$ be the average time to pain relief. Is $\overline{X} < 3.5$?

Question 5.4.2

The Chevalier wonders if a pair of dice is fair. For the experiment of rolling the dice, he defines a random variable $X = 1$ if double-sixes are rolled and $X = 0$ if not. The population is the set of all dice rolls (past, present, and future), and a sample of any size n is formed by rolling the pair of dice n times and recording the value $X = 1$ or $X = 0$ for each roll.

Question 5.4.3

Just ask voters in one party or another; just ask retirees; just ask college students; and other answers are possible.

Question 5.4.4

There are many possible such lists.

Question 5.4.5

(a) Roll a die a certain number of times and record the numbers rolled as sample elements.

(b) Randomly select a certain number of juvenile diabetes patients and record body weight loss as sample elements.

(c) Randomly select patients from a list of all juvenile diabetics that she has treated and record body weight loss as sample elements.

Question 5.4.6

(a) When $\alpha = 0.01$, the P-value of 0.0719 is greater than α and so the null hypothesis is not rejected—keeping the assumption that accounts were randomly selected.

(b) When $\alpha = 0.10$, the P-value of 0.0719 is less than α and so the null hypothesis is rejected—keeping the assumption that accounts were not randomly selected.

Question 5.4.7

Was there bias in the selection process? Did the company purposefully only select males?

1. The experiment selects one of the applicants; define a random variable $X = 1$ if male and $X = 0$ if female.
2. The population of this experiment is the set of all 35 applicants.
3. The sample of size $n = 5$ is claimed to be unbiased and consists of five males; that is, the observation about X is $\{1, 1, 1, 1, 1\}$.
4. Performing a hypothesis test at $\alpha = 0.05$.

Step 1: State the hypotheses:
H_0 : The selection is unbiased.
H_a : The selection is biased.

Step 2: $P\text{-value} = \dfrac{C(35, 5)}{C(55, 5)} \approx 0.0933$.

Step 3: The P-value of 0.0933 is greater than $\alpha = 0.05$, and so the null hypothesis is not rejected.

Question 5.4.8

(a) 2.86
(b) 2.51
(c) 3.12
(d) 3.73

Question 5.4.9

$H_a : \mu > 2$ the P-value is $P[\overline{X} \geq 3]$
$H_a : \mu \neq 2$ the P-value is $2 \cdot P[\overline{X} \geq 3]$

Question 5.4.10

Performing a hypothesis test at $\alpha = 0.05$.

Step 1: State the hypotheses:
$H_0 : \mu = 43$ and $H_a : \mu > 43$.

Step 2: Compute the P-value. Since $n = 40 > 35$, the central limit theorem applies with $n = 40$, $\mu = 43$, and $\sigma = 1.3$. P-value $= P[\overline{X} \geq 43.6] \approx 0.00175$.

Step 3: The P-value of 0.00175 is less than $\alpha = 0.05$, and so the null hypothesis is rejected. Adopting the alternate hypothesis, assume that the mean miles per gallon of one of its cars exceeds the mean EPA rating of 43 miles per gallon.

Question 5.4.11

Performing a hypothesis test at $\alpha = 0.05$.

Step 1: State the hypotheses:
$H_0 : \mu = 1$ and $H_a : \mu \neq 1$.

Step 2: Compute the P-value. Since $n = 40 > 35$, the central limit theorem applies with $n = 40$, $\mu = 1$, and $\sigma = 0.07$. P-value $= P[\overline{X} \geq 1.02] \approx 0.0707$.

Step 3: The P-value of 0.0707 is greater than $\alpha = 0.05$, and so the null hypothesis is not rejected.

5.5 Least Squares Regression

Question 5.5.1

There are many possible answers.

Question 5.5.2

$$\sum_{i=1}^{3} [\widehat{Q}_i - Q_i]^2 = [(40.627 - 12.082 \cdot 1.95) - 20]^2$$
$$+ [(40.627 - 12.082 \cdot 2.32) - 12]^2$$
$$+ [(40.627 - 12.082 \cdot 1.85) - 16]^2$$
$$\approx 13.81450475$$

$$\sum_{i=1}^{3} [Q_i^* - Q_i]^2 = [(41 - 12 \cdot 1.95) - 20]^2$$
$$+ [(41 - 12 \cdot 2.32) - 12]^2$$
$$+ [(41 - 12 \cdot 1.85) - 16]^2$$
$$= 14.9456$$

Question 5.5.3

$\widehat{\beta}_1 = 0.1266233766$ and $\widehat{\beta}_0 = 0.0876623377$

Question 5.5.4

(a) $a = n$

$$b = \sum_{i=1}^{n} 2 \cdot (\beta_1 x_i - y_i) \cdot \beta_0$$

$$c = \sum_{i=1}^{n} (\beta_1 x_i - y_i)^2$$

(b) $\beta_0 = -\dfrac{b}{2a}$

$$= -\frac{1}{2n} \sum_{i=1}^{n} 2 \cdot (\beta_1 x_i - y_i)$$

$$= -\sum_{i=1}^{n} \beta_1 \frac{x_i}{n} - \frac{y_i}{n}$$

$$= \overline{Y} - \beta_1 \overline{X}$$

(c) $\beta_1 = -\dfrac{\sum_{i=1}^{n} x_i(\widehat{\beta}_0 - y_i)}{\sum_{i=1}^{n} x_i^2}$

(d) $\widehat{\beta}_1 = \dfrac{\sum_{i=1}^{n} x_i(y_i - \overline{Y} + \widehat{\beta}_1 \overline{X})}{\sum_{i=1}^{n} x_i^2}$

$$\widehat{\beta}_1 \sum_{i=1}^{n} x_i^2 = \sum_{i=1}^{n} x_i y_i - x_i \overline{Y} + \sum_{i=1}^{n} x_i \widehat{\beta}_1 \overline{X}$$

$$\widehat{\beta}_1 \left[\sum_{i=1}^{n} x_i^2 - \sum_{i=1}^{n} x_i \overline{X} \right] = \sum_{i=1}^{n} x_i y_i - \sum_{i=1}^{n} x_i \overline{Y}$$

$$\widehat{\beta}_1 = \frac{\sum_{i=1}^{n} x_i y_i - \overline{Y} \sum_{i=1}^{n} x_i}{\sum_{i=1}^{n} x_i^2 - \overline{X} \sum_{i=1}^{n} x_i}$$

$$\widehat{\beta}_1 = \frac{\sum_{i=1}^{n} x_i y_i - (1/n) \sum_{i=1}^{n} y_i \sum_{i=1}^{n} x_i}{\sum_{i=1}^{n} x_i^2 - (1/n) \sum_{i=1}^{n} x_i \sum_{i=1}^{n} x_i}$$

$$\widehat{\beta}_1 = \frac{\sum_{i=1}^{n} (x_i - \overline{X}) \cdot (y_i - \overline{Y})}{\sum_{i=1}^{n} (x_i - \overline{X}) \cdot (x_i - \overline{X})}$$

$$\widehat{\beta}_1 = \frac{\sum_{i=1}^{n} (x_i - \overline{X}) \cdot (y_i - \overline{Y})}{\sum_{i=1}^{n} (x_i - \overline{X})^2}$$

Question 5.5.5

$\overline{X} = 6.125$

$\overline{Y} = 49.125$

$SS_X = 112.875$

$SS_Y = 4792.875$

$\widehat{\beta}_1 = 6.430786268$

$s = 4.563040246$

P-value $= 2 \cdot P[T_6 > 159.00770541] \approx 2 \cdot 0 = 0$. Since the P-value is less than $\alpha = 0.05$, the linear model is useful.

In this case,

$\widehat{\beta}_1 = 6.430786268$ and $\widehat{\beta}_0 = 9.736434109$.

Question 5.5.6

(a) $\widehat{Y} = 3 \cdot 2.2 + 7 = 13.6$, so $\$136,000$.

(b) $\widehat{Y} = 3 \cdot 1.25 + 7 = 10.75$, so $\$107,500$.

(c) $\widehat{Y} = 3 \cdot 3.1 + 7 = 16.3$, so $\$163,000$.

(d) $\widehat{Y} = 3 \cdot 4 + 7 = 19$, so $\$190,000$.

6.1 An Introduction to Graph Theory

Question 6.1.1

(a) and (b) have three vertices, while (c) has two vertices

(a) has three edges, (b) has five edges, and (c) has two edges

(b) has a loop

(b) and (c) have vertices that are connected by more than one edge

Question 6.1.2

(a) The "graph" in (a) has no vertices, and is a graph in the sense of analytic geometry rather than a graph in the sense of graph theory.

(b) The picture in (b) has no vertices.

(c) The picture in (c) has only one vertex with the right-hand of the edge not connected to any vertices.

Question 6.1.3

(a)

(b)

(c) Such a graph does not exist—simple graphs do not have loops by definition.

Question 6.1.4

(a) $d(A) = 3$; $d(B) = 3$; $d(C) = 3$; and $d(D) = 3$. The total degree of the graph is 12.

(b) $d(A) = 3$; $d(B) = 5$; $d(C) = 3$; and $d(D) = 3$. The total degree of the graph is 14.

Question 6.1.5

The total degree of the graph is even.

Question 6.1.6

If the graph has n edges, then the total degree is twice the number of edges and so is $2n$. Since $2n$ is even, the total degree of the graph is even.

Question 6.1.7

(a) a simple path: t, x

a nonsimple path: t, s, w

(b) $t, r, u, z, x.$

$v, z, w, r, s.$

(c) $u, t, v.$

$v, x, z.$

Question 6.1.8

(a)

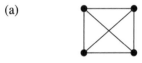

(b)

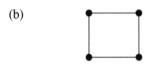

(c)

(d)

(e)

Question 6.1.9

(a) Let V and W be vertices in G. If $V \neq W$, then there exists an edge from V to W since G is connected and this is a path from V to W. If $V = W$, let U be any other vertex in the graph and consider the path consisting of traversing the edge from V to U and then the edge from U to $V = W$ (again such an edge exists since G is connected).

(b) Let $G = C_n$ be a cycle graph with vertices $X_1, \ldots, X_n$ that are adjacent based on this listing. Let V and W be vertices in G. Then for some $1 \leq i, j \leq n$, $V = X_i$ and $W = X_j$. Without loss of generality, assume $i < j$, in which case the path following the edges from X_i to X_{i+1} to $\ldots$ to X_{j-1} to X_j is the desired path.

(c) Let V and W be any two vertices in a null graph. Since the null graph does not contain any edges, there is no path from V to W in the graph.

6.2 The Explorer and the Traveling Salesman

Question 6.2.1

(a) An Eulerian circuit: s, t, x, z, w, v, y, u.

(b) There is no Eulerian circuit.
An Eulerian path: t, v, y, z, w, u, x.

Question 6.2.2

(a) There is no Eulerian circuit since not every vertex has even degree. There does exist an Eulerian path from B to D since these are the only two vertices of odd degree; one such path is: DB, BC, CA, AB.

(b) There does not exist an Eulerian circuit or path since the graph is not connected.

(c) There does not exist an Eulerian circuit or path since there are more than two (in particular, four) vertices with odd degree.

(d) An Eulerian circuit: AB, BC, CD, DB, BD, DA.

Question 6.2.3

(a) There is no Hamiltonian cycle—a cycle would need to revisit vertex C in order to move from, say, vertices A, B to vertices D, E and then back to the initial verted in vertices A, B.

A Hamiltonian path: AB, BC, CE, ED.

(b) A Hamiltonian cycle: AB, BD, DC, CA.

(c) There is no Hamiltonian cycle or path, as in example 6.2.3(c).

Question 6.2.4

(a) Since there are $n = 4 \geq 3$ vertices and for all pairs of nonadjacent vertices V and W, we have $d(V) + d(W) \geq 6$, there exists a Hamiltonian cycle.

In particular, note that:

A, D: $d(A) + d(D) = 3 + 3 = 6 \geq 4$

B, C: $d(B) + d(C) = 3 + 3 = 6 \geq 4$

A Hamiltonian cycle: AB, BD, DC, CA.

(b) Since there are $n = 6 \geq 3$ vertices and for all pairs of nonadjacent vertices V and W, we have $d(V) + d(W) \geq 6$, there exists a Hamiltonian cycle.

A Hamiltonian cycle: AB, BC, CF, FE, ED, DA.

(c) There are $n = 5 \geq 3$ vertices, but for the nonadjacent vertices D and E, $d(D) + d(E) = 1 + 1 = 2 \not\geq 3$. Therefore, theorem 6.2.2 does not apply.

6.3 Shortest Paths and Spanning Trees

Question 6.3.1

Vertices E and F are labeled $1 : A$; vertices B and C are labeled $2 : F$; and vertices D, G, and H are labeled $3 : C$. Therefore, a shortest path is that given by AF, FC, CH.

Question 6.3.2

(a) Not a tree since BC, CE, EB is a cycle.

(b) A tree. Note that this is *not* a spanning tree of the graph in (a) since edge AB is in graph (b) but not in graph (c).

(c) A tree, and a spanning tree of (a).

Question 6.3.3

(a) The graph is connected with $5 = 4 + 1$ vertices and 6 edges. Since $6 \neq 4$, this graph is not a tree by theorem 6.3.2.

(b) The graph is connected with $5 = 4 + 1$ vertices and 4 edges, and so this graph is a tree by theorem 6.3.2.

(c) The graph is connected with $5 = 4 + 1$ vertices and 4 edges, and so this graph is a tree by theorem 6.3.2.

Question 6.3.4

(a) Such a graph does not exist by theorem 6.3.2.

(b)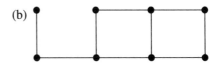

(c)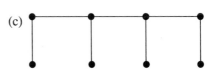

(d) Such a graph does not exist by theorem 6.3.2.

Question 6.3.5

$V_1 = A$
$V_2 = E$, so include AE
$V_3 = F$, so include AF
$V_4 = B$, so include FB
$V_5 = C$, so include FC
$V_6 = D$, so include CD
$V_7 = G$, so include CG
$V_8 = H$, so include CH

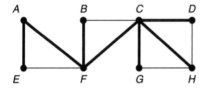

3: $E = F$, so include EF
4: $F = B$, so include FB
5: $B = C$, so include BC
6: $C = D$, so include CD
7: $D = H$, so include DH
8: $H = G$, so include HG

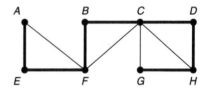

Question 6.3.6

$1 = A$
2: $A = E$, so include AE

Question 6.3.7

See the comments after question 6.3.7 in the text.

6.4   Application: Weighted Graphs

Question 6.4.1

There are many possible solutions—see the model in the text.

Question 6.4.2

Postal delivery routes; garbage and recycling pick-up routes; hiking trails; among many possible options.

Question 6.4.3

The seven possible weights are: 17, 18, 19, 19, 20, 21, and 23. The minimum weight spanning tree for the graph has weight 17 as illustrated in the following.

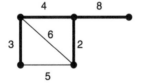

Question 6.4.4

Another possible spanning tree with minimum weight 17 is illustrated below.

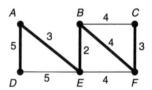

Question 6.4.5

Kruskal's algorithm identifies the following spanning tree with minimum weight 25.

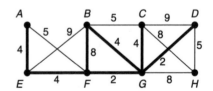

Question 6.4.6

Prim's algorithm identifies the following spanning tree with minimum weight 17.

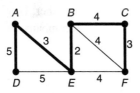

Question 6.4.7

The final Hamiltonian cycle below has the desired minimum weight of seven.

Question 6.4.8

weight : edges
1: AB, CD
2: BD
3: AE, ED, AC
4: BC
5: BE
6: CE
7: AD

The sorted edges algorithm identifies the following Hamiltonian cycle with weight 13.

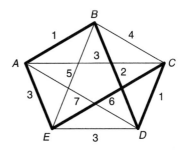

Question 6.4.9

The Hamiltonian cycle determined by the outer edges has weight 12; and so, we see that the sorted edges algorithm may not find a minimum weight Hamiltonian cycle, but just a Hamiltonian cycle with an approximately minimum weight.

Question 6.4.10

The nearest neighbor algorithm identifies the following Hamiltonian cycle with weight 13.

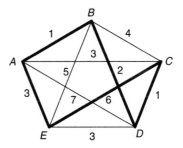

Question 6.4.11

The Hamiltonian cycle determined by the outer edges has weight 12; and so, we see that the nearest neighbor algorithm may not find a minimum weight Hamiltonian cycle, but just a Hamiltonian cycle with an approximately minimum weight.

Question 6.4.12

(a) Using vertex B as the base vertex, the nearest neighbor algorithm identifies the following two Hamiltonian cycles with weight 12.

(b) Using different bases the nearest neighbor algorithm identifies a Hamiltonian cycle with a weight of either 12 or 13, as indicated in accompanying graphs/figures.

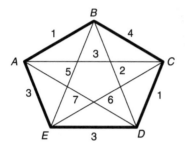

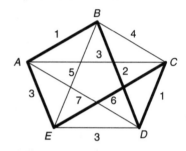

Complex Numbers and Complex Functions

Question 7.1.1

(a) $x = 1$ and $x = -1$

(b) $x = i$ and $x = -i$

(c) $x = -2$ with multiplicity 2

(d) $x = 2 + 2i$ and $x = 2 - 2i$

Question 7.1.2

(a) $Re(z) = 2$ and $Im(z) = 1$

(b) $Re(z) = -5$ and $Im(z) = 3$

(c) $Re(z) = 0$ and $Im(z) = 2$

(d) $Re(z) = 4$ and $Im(z) = -7$

Question 7.1.3

The labeled points are: $-4 - i$, $-3 + 0i$, $-3 + 6i$, $-2 - 6i$, $-1 + 2i$, $0 + 5i$, $1 - 4i$, $2 + 4i$, $4 - 4i$.

Question 7.1.4

(a) $|z| = 2\sqrt{2}$ and $\theta = \pi/4$

(b) $|z| = 4$ and $\theta = \pi/6$

(c) $|z| = 4$ and $\theta = \pi/2$

(d) $|z| = 4$ and $\theta = \pi$

(e) $|z| = 3$ and $\theta = \pi/6$

(f) $|z| = 2$ and $\theta = -\pi/6$

Question 7.1.5

(a) $t + v = 14 + 2i$

(b) $v - t = 6 - 12i$

(c) $u + w = (\sqrt{2} - \pi) + (\ln 2 - 0.5)i$

(d) $t - u = (4 - \sqrt{2}) + 6\frac{1}{2}i$

(e) $\bar{v} = 10 + 5i$

(f) $\bar{v} + v = 20$

(g) $t \cdot v = 75 + 50i$

(h) $t^2 + u \cdot v = (-\frac{71}{2} + 10\sqrt{2}) + (51 - 5\sqrt{2})i$

(i) $\dfrac{t}{v} = \dfrac{1}{15} + \dfrac{6}{5}i$

(j) $\dfrac{u}{w} = \dfrac{[-\pi\sqrt{2} - \frac{1}{2}\ln 2] + [\frac{\pi}{2} - \sqrt{2}\ln 2]i}{\pi^2 + (\ln 2)^2}$

(k) $|t| = \sqrt{65}$

(l) $|v| = \sqrt{2 + \frac{1}{4}}$

Question 7.1.6

(a) $(3\sqrt{2} + 3\sqrt{2}i)(2\sqrt{2} - 2\sqrt{2}i) = 3\sqrt{2}e^{i\pi/4} \cdot 2\sqrt{2}e^{-i\pi/4} = 12e^0 = 12$

(b) $(\sqrt{2} + \sqrt{2}i)(-3i) = \sqrt{2}e^{i\pi/4} \cdot 3e^{-i\pi/2} = 3\sqrt{2}e^{-i\pi/4} = 3 - 3i$

(c) $\dfrac{3\sqrt{2} + 3\sqrt{2}i}{2\sqrt{2} - 2\sqrt{2}i} = \dfrac{3\sqrt{2}e^{i\pi/4}}{2\sqrt{2}e^{-i\pi/4}} = \dfrac{3}{2}e^{i\pi/2} = 0 + \dfrac{3}{2}i$

(d) $\dfrac{\sqrt{2} + \sqrt{2}i}{-3i} = \dfrac{1e^{i\pi/4}}{3e^{-i\pi/2}} = \dfrac{1}{3}e^{i3\pi/4} = -\dfrac{1}{3} + \dfrac{1}{3}i$

(e) $(3\sqrt{2} + 3\sqrt{2}i)^8 = (3\sqrt{2}e^{i\pi/4})^8 = (3\sqrt{2})^8 e^{i2\pi} = (3\sqrt{2})^8 + 0i$

(f) $(\sqrt{2} + \sqrt{2}i)^{-3} = (\sqrt{2}e^{i\pi/4})^{-3} = (\sqrt{2})^{-3}e^{-i3\pi/4} = -\frac{1}{4} - \frac{1}{4}i$

Question 7.1.7

(a) $\sqrt{5}e^{i(-13)\pi/14}$ and $\sqrt{5}e^{i\pi/14}$

(b) $\sqrt{10} \cdot \sqrt[4]{2}e^{i(-7)\pi/8}$ and $\sqrt{10} \cdot \sqrt[4]{2}e^{i\pi/8}$

(c) $\sqrt[3]{10}e^{i(-7)\pi/18}$, $\sqrt[3]{10}e^{i5\pi/18}$, and $\sqrt[3]{10}e^{i17\pi/18}$

(d) $\sqrt[4]{10}e^{i(-19)\pi/24}$, $\sqrt[4]{10}e^{i(-7)\pi/24}$, $\sqrt[4]{10}e^{i5\pi/24}$, and $\sqrt[4]{10}e^{i17\pi/24}$

(e) $\sqrt[12]{8}e^{i(-17)\pi/24}$, $\sqrt[12]{8}e^{i(-9)\pi/24}$, $\sqrt[12]{8}e^{i(-1)\pi/24}$, $\sqrt[12]{8}e^{i7\pi/24}$ $\sqrt[12]{8}e^{i15\pi/24}$, and $\sqrt[12]{8}e^{i23\pi/24}$

(f) $\sqrt[5]{8}e^{i(-14)\pi/15}$, $\sqrt[5]{8}e^{i(-8)\pi/15}$, $\sqrt[5]{8}e^{i(-2)\pi/15}$, $\sqrt[5]{8}e^{i4\pi/15}$, and $\sqrt[5]{8}e^{i10\pi/15}$

Question 7.1.8

(a) Single-valued since a complex polynomial.
(b) Three-valued since a cube root.
(c) Single-valued since a complex polynomial.
(d) Four-valued since a fourth root.

Question 7.1.9

(a) $\arg(z) = \dfrac{\pi}{4} + 2k\pi$, $\operatorname{Arg}(z) = \dfrac{\pi}{4}$

(b) $\arg(z) = \dfrac{\pi}{6} + 2k\pi$, $\operatorname{Arg}(z) = \dfrac{\pi}{6}$

(c) $\arg(z) = \dfrac{\pi}{2} + 2k\pi$, $\operatorname{Arg}(z) = \dfrac{\pi}{2}$

(d) $\arg(z) = \pi + 2k\pi$, $\operatorname{Arg}(z) = \pi$

(e) $\arg(z) = \dfrac{\pi}{6} + 2k\pi$, $\operatorname{Arg}(z) = \dfrac{\pi}{6}$

(f) $\arg(z) = -\dfrac{\pi}{6} + 2k\pi$, $\operatorname{Arg}(z) = -\dfrac{\pi}{6}$

Question 7.1.10

(a) $u_x = \lim\limits_{\Delta x \to 0} \dfrac{2(x + \Delta x) - y - (2x - y)}{\Delta x}$

$= \lim\limits_{\Delta x \to 0} 2 = 2$

$u_y = \lim\limits_{\Delta y \to 0} \dfrac{2x - (y + \Delta y) - (2x - y)}{\Delta y}$

$= \lim\limits_{\Delta y \to 0} -1 = -1$

(b) $v_x = \lim\limits_{\Delta x \to 0} \dfrac{(x + \Delta x) + 2y - (x + 2y)}{\Delta x}$

$= \lim\limits_{\Delta x \to 0} 1 = 1$

$v_y = \lim\limits_{\Delta y \to 0} \dfrac{x + 2(y + \Delta y) - (x + 2y)}{\Delta y}$

$= \lim\limits_{\Delta y \to 0} 2 = 2$

(c) $f_x = \lim\limits_{\Delta x \to 0} \dfrac{(x + \Delta x)^2 y - x^2 y}{\Delta x}$

$= \lim\limits_{\Delta x \to 0} \dfrac{x^2 y + 2x\Delta x y + (\Delta x)^2 y - x^2 y}{\Delta x}$

$= 2xy$

$f_y = \lim\limits_{\Delta y \to 0} \dfrac{x^2(y + \Delta y) - x^2 y}{\Delta y}$

$= \lim\limits_{\Delta y \to 0} x^2 = x^2$

(d) $g_x = \lim\limits_{\Delta x \to 0} \dfrac{\dfrac{x + \Delta x}{y} - \dfrac{x}{y}}{\Delta x}$

$= \lim\limits_{\Delta x \to 0} \dfrac{\Delta x}{y\Delta x} = \dfrac{1}{y}$

$g_y = \lim\limits_{\Delta y \to 0} \dfrac{\dfrac{x}{y + \Delta y} - \dfrac{x}{y}}{\Delta y}$

$= \lim\limits_{\Delta y \to 0} \dfrac{-x\Delta y}{y(y + \Delta y)} = \dfrac{-x}{y^2}$

Question 7.1.11

(a) $f_x = \sin(x^2 + y^2) \cdot 2x, \ f_y = \sin(x^2 + y^2) \cdot 2y$

(b) $g_x = 20xe^{2y}, \ g_y = 20x^2 e^{2y}$

(c) $h_x = (x+y) \cdot \sin(x^2+y^2) \cdot 2x + 1 \cdot \sin(x^2+y^2)$
$h_y = (x+y) \cdot \sin(x^2+y^2) \cdot 2y + 1 \cdot \sin(x^2+y^2)$
(d) $j_x = 2xy^{-3}$, $j_y = x^2(-3)y^{-4}$
(e) $u_x = 4x^3 - 12xy^2$, $u_y = -12x^2y + 4y^3$
(f) $v_x = 12xy^2 - 4y^3$, $u_y = 4x^3 - 12xy^2$

Question 7.1.12

(a) $u_x = 2x$, $u_y = -2y$
(b) $v_x = 2y$, $v_y = 2x$
(c) It appears that $u_x = v_y$.
(d) It appears that $v_x = -u_y$.

Question 7.1.13

To prove that $\lim_{(x,y)\to(0,0)} x^2 - y^2 = 0$,
let $\varepsilon > 0$, choose $\delta = \sqrt{\varepsilon/2}$, and assume
that $\sqrt{(x-0)^2 + (y-0)^2} < \delta$. Note that
this implies $|x|, |y| < \delta$. Under this
assumption,

$$|x^2 - y^2 - 0| = |x^2 - y^2| \le |x|^2 + |y|^2$$

$$< \delta^2 + \delta^2 = 2\delta^2 = \varepsilon.$$

To prove that $\lim_{(x,y)\to(0,0)} 2xy = 0$, let
$\varepsilon > 0$, choose $\delta = \sqrt{\varepsilon/2}$, and assume
that $\sqrt{(x-0)^2 + (y-0)^2} < \delta$. Note that
this implies $|x|, |y| < \delta$. Under this

assumption,

$$|2xy - 0| = |2xy| = 2 \cdot |x|^2 \cdot |y|$$

$$< 2\delta^2 = \varepsilon.$$

To prove that $u_x = 2y$ is continuous, let
(x_0, y_0) be an arbitrary point on the complex plane and show that $\lim_{(x,y)\to(x_0,y_0)} 2y = 2y_0$. Let $\varepsilon > 0$, choose $\delta = \varepsilon/2$, and assume that $\sqrt{(x-x_0)^2 + (y-y_0)^2} < \delta$. Note that this implies $|y - y_0| < \delta$. Under this assumption,

$$|2y - 2y_0| = 2|y - y_0| < 2\delta = \varepsilon.$$

To prove that $u_y = 2x$ is continuous, let
(x_0, y_0) be an arbitrary point on the complex plane and show that $\lim_{(x,y)\to(x_0,y_0)} 2x = 2x_0$. Let $\varepsilon > 0$, choose $\delta = \varepsilon/2$, and assume that $\sqrt{(x-x_0)^2 + (y-y_0)^2} < \delta$. Note that this implies $|x - x_0| < \delta$. Under this assumption,

$$|2x - 2x_0| = 2|x - x_0| < 2\delta = \varepsilon.$$

7.2 Analytic Functions and the Cauchy–Riemann Equations

Question 7.2.1

(a) $\delta = \varepsilon/12$
(b) $\delta = \varepsilon/\sqrt{13}$

Question 7.2.2

(a) $f'(x) = (3 + 3i)z^2$
(b) $g'(z) = 4z + i$

Question 7.2.3

(a) $f'(x) = 30z^2 - 14iz$
(b) $g'(z) = 3(iz + 2)^2 i$

(c) $h'(z) = \dfrac{(5z^2 + 3i)i - (iz + 2)10z}{(5z^2 + 3i)^2}$

(d) $j'(z) = (2z + 9i)^5 \cdot 48iz^2(4iz^3 + 7)^3 + 10(2z + 9i)^4 \cdot (4iz^3 + 7)^4$

Question 7.2.4

$$\frac{d}{dz}[cf(z)] = \lim_{\Delta z \to 0} \frac{cf(z + \Delta z) - cf(z)}{\Delta z}$$

$$= cf'(z)$$

Question 7.2.5

(a) $u_x = 5x^4 - 30x^2y^2 + 5y^4 =$
v_y, $u_y = -20x^3y + 20xy^3 = -v_x$

(b) $u_x = 10x^4 - 60x^2y^2 + 10y^4 +$
$6xy = v_y$, $u_y = -40x^3y +$
$40xy^3 + 3x^2 - 3y^2 = -v_x$

Question 7.2.6

$$f'(z) = \lim_{\Delta z \to 0} \frac{e^{z+\Delta z} - e^z}{\Delta z}$$

$$= e^z \cdot \lim_{\Delta z \to 0} \frac{e^{\Delta z} - 1}{\Delta z} = e^z$$

Question 7.2.7

(a) $\text{Log}(-4) = \ln(4) + i\pi$, $\log(-4)$
$= \ln(4) + i(\pi + 2n\pi)$, for $n \in \mathbb{Z}$

(b) $\text{Log}(2\sqrt{3} + 2i) = \ln(4) + i(\pi/6)$,
$\log(2\sqrt{3} + 2i) = \ln(4) + i(\pi/6 + 2n\pi)$, for $n \in \mathbb{Z}$

(c) $\text{Log}(3e^{i\pi/6}) = \ln(3) + i(\pi/6)$,
$\log(3e^{i\pi/6}) = \ln(3) + i(\pi/6 + 2n\pi)$,
for $n \in \mathbb{Z}$

(d) $\text{Log}(2e^{i(11\pi/6)}) = \ln(2) + i(11\pi/6)$,
$\log(2e^{i(11\pi/6)}) = \ln(2) + i(11\pi/6 + 2n\pi)$, for $n \in \mathbb{Z}$

Question 7.2.8

(a) $-e^3$

(b) $\log(1 + i\sqrt{3}) = \ln(2) + i(\pi/3 + 2n\pi)$, for $n \in \mathbb{Z}$

(c) $\text{Log}(1 + i\sqrt{3}) = \ln(2) + i(\pi/3)$

(d) $e^{i\log(-2)} = e^{-(\pi+2n\pi)} \cdot [\cos(\ln 2) + i\sin(\ln 2)]$, for $n \in \mathbb{Z}$

7.3 **Power Series Representations of Analytic Functions**

Question 7.3.1

(a) Set $N = \ln(1/\varepsilon)/\ln(\sqrt{85})$. Then whenever $n > N$, $|9 - 2i|^{-n} < \varepsilon$.

(b) Set $N = \ln(\varepsilon)/\ln(.75)$, and use the fact that $n < 1.5^n$ for any $n \in \mathbb{N}$. Then whenever $n > N$, $|n/(2i)^n| < .75^n < \varepsilon$.

Question 7.3.2

(a) $(7 - 3i)(1 + i)$

(b) $4/5$

(c) $i/(2 - 3i)$

(d) $1 - i$

Question 7.3.3

(a) 0

(b) 0

(c) 0

(d) 1

Question 7.3.4

i, $2i$, $3i$, ... $8i$. $\sum_{n=0}^{\infty} i = \lim_{n \to \infty} ni$, which diverges.

Question 7.3.5

(a) $a_n = 1/(n + 3i)$, $R = 1$.

(b) $a_n = \dfrac{n!}{(3n + i)(n + 1)!}$, $R = 1$.

(c) $a_n = \dfrac{in}{(n + 2)!}$, $R = \infty$.

(d) $a_n = 1$, $R = 1$.

Question 7.3.6

(a) $\lim_{n \to \infty} |nK^n|$
$= \lim_{n \to \infty} n|K|^n = 0$, since $|K| < 1$. By theorem 7.3.2, $\lim_{n \to \infty} nK^n = 0$.

(b) For any fixed z with $|z| < r < R$, $|z/r| < 1$, and so by part (a), $\lim_{n\to\infty} |n|z/r|^n = 0$. By the definition of limit applied to $\varepsilon = |z|$, there exists $N > 0$ such that $n|z/r|^n < |z|$ whenever $n > N$; in other words, $n|z|^{n-1} < r^n$. The rest of the answer follows from multiplication on both sides by $|a_n|$ and applying the ratio test.

For $\sin z$, $\lim_{n\to\infty} |a_{n+1}/a_n|$

$$= \lim_{n\to\infty} \frac{1}{(2n+3)(2n+2)} = 0.$$

Question 7.3.9

Splitting the sum into even and odd terms, $e^{iz} = \sum_{n=0}^{\infty} (-1)^n \frac{z^{2n}}{(2n)!} + i \sum_{n=0}^{\infty} (-1)^n \cdot$

$\frac{z^{2n+1}}{(2n+1)!} = \cos z + i \sin z.$

Question 7.3.7

$f^{(n)}(z) = n!(1-z)^{-(n+1)}$, and so $f^{(n)}(0) = n!$. Therefore $f(z) = \sum_{n=0}^{\infty} z^n$. By the ratio test, $R = 1$.

Question 7.3.10

$e^{iz} + e^{-iz} = \sum_{n=0}^{\infty} \left(i^n \frac{z^n}{n!} + (-i)^n \frac{z^n}{n!} \right)$

$= \sum_{n=0}^{\infty} \left((-1)^n \cdot \frac{2z^{2n}}{(2n)!} \right) = 2 \cos z.$

Question 7.3.8

For $\cos z$, $\lim_{n\to\infty} |a_{n+1}/a_n|$

$$= \lim_{n\to\infty} \frac{1}{(2n+2)(2n+1)} = 0.$$

Similarly,

$e^{iz} - e^{-iz} = \sum_{n=0}^{\infty} \left(i^n \frac{z^n}{n!} - (-i)^n \frac{z^n}{n!} \right)$

$= \sum_{n=0}^{\infty} \left((-1)^n \cdot \frac{2iz^{2n+1}}{(2n+1)!} \right) = 2i \sin z.$

7.4 Harmonic Functions

Question 7.4.1

(a) $u_{xx} = 12x^2 - 12y^2 = -u_{yy}$. The second partial derivatives are continuous because they are polynomials.

(b) $u_{xx} = \frac{2xy}{(x^2+y^2)^2} = -u_{yy}$; these rational functions are continuous wherever the denominators are nonzero.

(c) $f(x+iy) = x^4 - 6x^2y^2 + y^4 = i(3x^3y - 3xy^3)$. Then $f_{xx} = 12x^2 - 12y^2 + i(18xy) = -f_{yy}$. The second partial derivatives are continuous because they are polynomials.

(d) $f(x+iy) = 0.5 \ln(x^2+y^2) + i \tan^{-1}(y/x)$. The first term in the sum is a harmonic function; if $u(x,y) = \ln(x^2+y^2)$, then

$$u_{xx} = \frac{2(y^2-x^2)}{(x^2+y^2)^2} = -u_{yy}$$

(and rational functions are continuous wherever the denominator is nonzero). The second term is harmonic from part (b).

Question 7.4.2

(a) $\int x^2 - y^2 \, dx = x^3/3 - xy^2 + C(y)$, and $\int x^2 - y^2 \, dy = x^2y - y^3/3 + C(x)$.

(b) $\int x^2 e^y \, dx = x^3 e^y/3 + C(y)$, and $\int x^2 e^y \, dy = x^2 e^y + C(x)$.

Question 7.4.3

(a) $\int_0^3 x^2 - y^2 \, dx = 9 - 3y^2$

(b) $\int_2^4 x^2 e^y \, dy = 2y - 56/3$

(c) $x^3/3 - xy^2$

(d) $x^2/y - y^3/3$

Question 7.4.4

(a) $u_{xx} = e^x \cos y = -u_{yy}$. Since both terms in the product are separately continuous in x and y, the product is continuous.

(b) $v(x, y) = e^x \sin y$

(c) $f(z) = e^x \cos y + ie^x \sin y = e^x e^{iy} = e^{x+iy} = e^z. f'(z) = e^z.$

Question 7.4.5

(a) u_x is computed in example 7.2.11 of section 7.2. Then

$$u_{xx} = \sum_{k=0}^{\lfloor (n+1)/2 \rfloor - 2} \frac{(-1)^k n! x^{n-2k-2} y^{2k}}{(2k)!(n-2k-2)!}$$

and

$$u_{yy} = \sum_{k=1}^{\lfloor (n+1)/2 \rfloor - 1} \frac{(-1)^k n! x^{n-2k} y^{2k-2}}{(2k-2)!(n-2k)!}.$$

Reindexing by setting $j = k - 1$,

$$u_{yy} = -\sum_{j=0}^{\lfloor (n+1)/2 \rfloor - 2} \frac{(-1)^j n! x^{n-2j-2} y^{2j}}{(2j)!(n-2j-2)!},$$

and so $u_{xx} = -u_{yy}$.

(b) Example 7.2.11 of section 7.2 gives $v(x, y)$ and shows $f(z) = z^n$.

7.5   Application: Streamlines and Equipotentials

Question 7.5.1

(a) $\langle 0, 2 \rangle$ is a vertical vector of length 2 directed upwards.

(b) In standard position, $\langle -1, -2 \rangle$ emanates from the origin and extends to the point $(-1, -2)$.

(c) $\langle 0, 0 \rangle$ is the zero vector, which has no length.

(d) $\langle -3, 0 \rangle$ is a horizontal vector of length 3 directed leftward.

Question 7.5.2

(a) 2

(b) $\sqrt{5}$

(c) 0

(d) 3

Question 7.5.3

(a) $\| \langle 0, y \rangle \| = |y|$

(b) $\| \langle \vec{F}(x, y) \rangle \| = 1$

Question 7.5.4

(a) $u(x, y) = x^2 - y^2 - x$, $\vec{\nabla} u = \langle 2x - 1, -2y \rangle$.

(b) $u(x, y) = x^3 - 3xy^2$, $\vec{\nabla} u = \langle 3x^2 - 3y^2, -6xy \rangle$.

(c) $e^{2x+i2y} = e^{2x} \cos(2y) + ie^{2x} \sin(2y)$. Hence $\vec{\nabla} u = \langle 2e^{2x} \cos(2y), -2e^{2x} \cdot \sin(2y) \rangle$.

(d) $u(x, y) = \cos x \cosh y$, $\vec{\nabla} u = \langle -\sin x \cosh y, \cos x \sinh y \rangle$.

Question 7.5.5

(a) $u(x, y) = x$, $v(x, y) = y$,

(b) $\vec{\nabla} u = \langle 1, 0 \rangle$. All vectors are directed to the right one unit.

(c) The streamlines $v(x, y) = C$ are horizontal lines through C, and the equipotentials $u(x, y) = C$ are vertical lines through C.

(d) The path extends horizontally to the right of $(3, 5)$.

Question 7.5.6

(a) Since $f(z) = y + i(-x)$, $u(x, y) = y$ and $v(x, y) = -x$.

(b) $\vec{\nabla} u = \langle 0, 1 \rangle$. All vectors are directed upward with unit length.

(c) The flux lines $v(x, y) = C$ are vertical lines through $-C$, and the isothermals $u(x, y) = C$ are horizontal lines through C.

Question 7.5.7

Setting $z = re^{i\theta}$, the imaginary part of f is $v(r, \theta) = (r - r^{-1}) \sin \theta + (\ln r)/2\pi$, and so the streamlines are of the form $(r - r^{-1}) \sin \theta + (\ln r)/2\pi = C$. The choice of $r = 1$ (which is the unit circle) satisfies $C = 0$, and so the unit circle emerges as a streamline for this fluid flow.

Answers to Odd-Numbered Exercises

1.1 **The Formal Language of Sentential Logic**

1. C

3. $M \rightarrow Q$

5. $(\sim C) \rightarrow (\sim Q)$

7. If Taylor is either a natural leader or math major, then she will be qualified for a high-paying job.

9. If Taylor is not a college student, then she will not be qualified for a high-paying job.

11. Taylor is a college student if and only if she is a natural leader.

13. $G \wedge (\sim C)$

15. $F \rightarrow (A \wedge C)$

17. X is either a field or a group.

19. X is associative does not imply that X is a group.

21. If X is a field, then X is commutative, associative, and a group.

23. $C \vee D$

25. $(B \wedge M) \rightarrow (\sim D)$

27. $\sim[D \rightarrow (\sim B)]$

29. A sequence does not diverge iff the sequence converges.

31. If a sequence is not bounded but monotonic, then it diverges.

33. If a sequence diverges, then the sequence is either not bounded or not monotonic.

35. $(p \wedge Q) \rightarrow R$

37. $[(\sim L) \wedge (\sim R)] \wedge (\sim Z)$ or $[\sim(L \vee R)] \wedge (\sim Z)$

39. $A \vee (\sim B)$

41. $[(\sim E) \wedge (\sim F)] \vee G$ or $[\sim(E \vee F)] \vee G$

43. $Y \leftrightarrow [Z \wedge (W \rightarrow X)]$

45. $(H \rightarrow J) \vee (K \wedge L)$

47. $(H \vee J) \rightarrow (K \wedge L)$

49. $(q \wedge r) \rightarrow p$

51. $(p \vee q) \rightarrow (r \leftrightarrow s)$

53. $[(\sim p) \wedge (\sim q)] \wedge (\sim r)$ or $[\sim(p \vee q)] \wedge (\sim r)$

55. Sentence.

57. Nonsentence, two adjacent connectives.

59. Nonsentence, missing parentheses.

61. Nonsentence, missing parentheses.

63. Nonsentence, adjacent sentence variables.

65. Four alternating connectives.

67. Parentheses do not actually appear in a variable or a symbol.

69. $(1 + m + n)$ left parentheses and $(m + n + 1)$ right parentheses appear in $(\mathbb{B} \wedge \mathbb{C})$.

1. Contradiction

p	$(\sim p)$	$p \leftrightarrow (\sim p)$
T	F	F
F	T	F

3. Contingency

p	$(\sim p)$	$(\sim p) \to p$	$\sim[(\sim p) \to p]$
T	F	T	F
F	T	F	T

5. Contingency

p	q	$(\sim p)$	$(\sim p) \to q$
T	T	F	T
T	F	F	T
F	T	T	T
F	F	T	F

7. Tautology

p	q	$(q \to p)$	$p \to (q \to p)$
T	T	T	T
T	F	T	T
F	T	F	T
F	F	T	T

9. Contingency

p	q	$(p \leftrightarrow q) \leftrightarrow (\sim p)$
T	T	F
T	F	T
F	T	F
F	F	T

11. Contingency

p	q	$[(p \to q) \wedge (\sim q)] \to p$
T	T	T
T	F	T
F	T	T
F	F	F

13. Tautology

p	r	$(p \vee r) \leftrightarrow \{\sim[(\sim p) \wedge (\sim r)]\}$
T	T	T
T	F	T
F	T	T
F	F	T

Note: This follows from DeMorgan's laws.

15. Contingency

p	q	r	$(p \wedge q)$	$(p \wedge q) \vee r$
T	T	T	T	T
T	T	F	T	T
T	F	T	F	T
T	F	F	F	F
F	T	T	F	T
F	T	F	F	F
F	F	T	F	T
F	F	F	F	F

17. Contingency

p	q	r	$(p \leftrightarrow q) \leftrightarrow (\sim r)$
T	T	T	F
T	T	F	T
T	F	T	T
T	F	F	F
F	T	T	T
F	T	F	F
F	F	T	F
F	F	F	T

19. Contingency

p	q	r	$\{p \to [\sim(q \wedge r)]\} \to (r \to p)$
T	T	T	T
T	T	F	T
T	F	T	T
T	F	F	T
F	T	T	F
F	T	F	T
F	F	T	F
F	F	F	T

21. Yes

p	$(\sim p)$	$\sim(\sim p)$
T	F	T
F	T	F

23. Yes

p	q	$\sim(p \vee q)$	$(\sim p) \wedge (\sim q)$
T	T	F	F
T	F	F	F
F	T	F	F
F	F	T	T

25. No, row 3

p	q	$p \vee q$
T	T	T
T	F	T
F	T	T
F	F	F

27. Yes

p	q	$p \vee q$	$q \vee p$
T	T	T	T
T	F	T	T
F	T	T	T
F	F	F	F

29. Yes

p	q	r	$(p \vee q) \vee r$	$p \vee (q \vee r)$
T	T	T	T	T
T	T	F	T	T
T	F	T	T	T
T	F	F	T	T
F	T	T	T	T
F	T	F	T	T
F	F	T	T	T
F	F	F	F	F

31. Yes

p	q	r	$p \wedge (q \vee r)$	$(p \wedge q) \vee (p \wedge r)$
T	T	T	T	T
T	T	F	T	T
T	F	T	T	T
T	F	F	F	F
F	T	T	F	F
F	T	F	F	F
F	F	T	F	F
F	F	F	F	F

33. No, rows 2 and 4

p	q	r	$p \vee (q \wedge r)$	$(p \vee q) \wedge r$
T	T	T	T	T
T	T	F	T	F
T	F	T	T	T
T	F	F	T	F
F	T	T	T	T
F	T	F	F	F
F	F	T	F	F
F	F	F	F	F

35. Yes

p	q	$(p \rightarrow q)$	$(\sim q) \rightarrow (\sim p)$
T	T	T	T
T	F	F	F
F	T	T	T
F	F	T	T

37. No, rows 2 and 3

p	q	$(p \rightarrow q)$	$(q \rightarrow p)$
T	T	T	T
T	F	F	T
F	T	T	F
F	F	T	T

39. Yes

p	q	$(p \rightarrow q)$	$\sim[p \wedge (\sim q)]$
T	T	T	T
T	F	F	F
F	T	T	T
F	F	T	T

41. Yes

p	q	$p \leftrightarrow q$	$(p \rightarrow q) \wedge (q \rightarrow p)$
T	T	T	T
T	F	F	F
F	T	F	F
F	F	T	T

43.

A	B	$\sim B$	$A \rightarrow (\sim B)$
T	F	T	T

45.

p	A	$A \rightarrow p$
T	T	T
F	T	F

47.

p	A	B	$p \rightarrow (A \vee B)$
T	T	F	T
F	T	F	T

49.

p	A	B	$A \leftrightarrow [p \vee (\sim B)]$
T	T	F	T
F	T	F	T

51.

p	q	A	B	$[\sim(B \wedge q)] \rightarrow (A \leftrightarrow p)$
T	T	T	F	T
T	F	T	F	T
F	T	T	F	F
F	F	T	F	F

53. The standard truth table for $\mathbb{B}$ has the same final column as itself. Therefore, $\mathbb{B} \equiv \mathbb{B}$.

55. If $\mathbb{B} \equiv \mathbb{C}$, the final column of the truth table for $\mathbb{B}$ is the same as the final column of the truth table for $\mathbb{C}$. Similarly, if $\mathbb{C} \equiv \mathbb{D}$, the final column of the truth table for $\mathbb{C}$ is the same as the final column of the truth table for $\mathbb{D}$. Since the final column of the truth table for $\mathbb{B}$ is the same as the final column of the truth table for $\mathbb{C}$ which is the same as the final column of the truth table for $\mathbb{D}$, the final column of the truth table for $\mathbb{B}$ must be the same as the final column of the truth table for $\mathbb{D}$. Therefore, $\mathbb{B} \equiv \mathbb{D}$.

57. If $\mathbb{B} \leftrightarrow \mathbb{C}$ is a tautology, then from the basic truth table for the connective $\leftrightarrow$, we conclude that $\mathbb{B}$ and $\mathbb{C}$ must have precisely the same final truth table column. Therefore, $\mathbb{B} \equiv \mathbb{C}$.

59. $f_\rightarrow(T, T) = T$
$f_\rightarrow(T, F) = F$
$f_\rightarrow(F, T) = T$
$f_\rightarrow(F, F) = T$

61. $f_\wedge(f_\sim(T), F) = f_\wedge(F, F) = F$

63. $f_\rightarrow(f_\vee(T, F), f_\wedge(F, T))$
$= f_\rightarrow(T, F) = F$

65. $f_\sim(f_\leftrightarrow(T, F)) = f_\sim(F) = T$

67. $f_\sim \circ f_\wedge(T, T) = F$
$f_\sim \circ f_\wedge(T, F) = T$
$f_\sim \circ f_\wedge(F, T) = T$
$f_\sim \circ f_\wedge(F, F) = T$

69. The composition is undefined because the function $f_\sim$ outputs a single value and the function $f_\vee$ is not defined for single-value inputs.

1.3 An Algebra for Sentential Logic

1. $p \rightarrow q$

3. $(\sim p) \vee q$

5. $(\sim p) \vee q$

7. $p \leftrightarrow q$

9. $[(\sim p) \vee q] \wedge [(\sim q) \vee p]$

11. $\sim\{\sim[(\sim p) \vee q] \vee [\sim[(\sim q) \vee p]]\}$

13. $\sim[(\sim p) \wedge (\sim q)]$

15. $\sim[p \wedge (\sim q)] \wedge \{\sim[q \wedge (\sim p)]\}$

17. $\sim\{p \wedge [q \wedge (\sim p)]\}$

19. $p \wedge [(\sim p) \wedge (\sim q)]$

21. $\{\sim[(p \wedge q) \wedge (\sim r)]\}\wedge$
$\{\sim[r \wedge [\sim(p \wedge q)]]\}$

23. $\sim\{\sim[(\sim p) \wedge (\sim r)] \wedge [\sim(q \wedge r)]\}$

25. $\sim[(\sim p) \vee (\sim q)]$

27. $\sim\{[\sim[(\sim p) \vee q]]\vee[\sim[(\sim q) \vee p]]\}$

29. $(\sim p) \vee [(\sim q) \vee p]$

31. $[p \wedge (\sim q)] \vee [(\sim p) \wedge q] \vee$
$[(\sim p) \wedge (\sim q)]$

33. $[p \wedge q] \vee [(\sim p) \wedge q] \vee$
$[(\sim p) \wedge (\sim q)]$

35. $p \wedge (\sim q)$

37. $[p \wedge (\sim q) \wedge r] \vee [p \wedge (\sim q)\wedge$
$(\sim r)] \vee [(\sim p) \wedge (\sim q) \wedge r]$

39. $[p \wedge q \wedge (\sim r)] \vee [(\sim p)\wedge$
$q \wedge r] \vee [(\sim p) \wedge q \wedge (\sim r)]$
$\vee [(\sim p) \wedge (\sim q) \wedge r] \vee$
$[(\sim p) \wedge (\sim q) \wedge (\sim r)]$

41. $[p \wedge q \wedge r] \vee [p \wedge (\sim q) \wedge r]\vee$
$[(\sim p) \wedge (\sim q) \wedge r] \vee [(\sim p)\wedge$
$(\sim q) \wedge (\sim r)]$

43.

$\{\sim, \vee, \wedge\}$	$\{\sim, \vee\}$
$\sim p$	$\sim p$
$p \vee q$	$p \vee q$
$p \wedge q$	$\sim[(\sim p) \vee (\sim q)]$

45.

$\{\sim, \vee, \wedge\}$	$\{\sim, \rightarrow\}$
$\sim p$	$\sim p$
$p \vee q$	$(\sim p) \rightarrow q$
$p \wedge q$	$\sim[p \rightarrow (\sim q)]$

47. $\{\sim, \wedge, \rightarrow\}$ is adequate since $\{\sim, \wedge\}$ is adequate.

49. $\{\sim, \wedge, \leftrightarrow\}$ is adequate since $\{\sim, \wedge\}$ is adequate.

51. $\{\sim, \wedge, \vee, \leftrightarrow\}$ is adequate since $\{\sim, \vee\}$ is adequate (alternatively, since $\{\sim, \wedge\}$ is adequate).

53. The following truth tables for sentences using only $\sim$ and $\leftrightarrow$ indicate that the final column of such truth tables must contain an even number of T's and F's. However, half of all two-variable truth tables contain an odd number of T's and F's in their final columns and, since $\{\sim, \leftrightarrow\}$ cannot express every truth table, this set is not adequate.

p	q	$(p \leftrightarrow p) \leftrightarrow (q \leftrightarrow q)$	$p \leftrightarrow q$
T	T	T	T
T	F	T	F
F	T	T	F
F	F	T	T

p	q	$\sim p$	$[p \leftrightarrow (\sim p)] \leftrightarrow (q \leftrightarrow q)$
T	T	F	F
T	F	F	F
F	T	T	F
F	F	T	F

p	q	$\sim q$	$p \leftrightarrow (\sim q)$
T	T	F	F
T	F	T	T
F	T	F	T
F	F	T	F

55. The connectives $\rightarrow$ and $\leftrightarrow$ cannot produce the truth value F in the first row of a truth table. Thus, the truth table for $(\sim p)$ cannot be expressed by a sentence using only the connectives $\rightarrow$ and $\leftrightarrow$, and so this set is not adequate.

57.
p	$p \mid p$
T	F
F	T

59.
p	$p \downarrow p$
T	F
F	T

61. Antecedent: p
    Consequent: q
    Contrapositive: If $\sim q$, then $\sim p$.

63. Antecedent: $p \vee q$
    Consequent: $q \vee p$
    Contrapositive: If $\sim(q \vee p)$, then $\sim(p \vee q)$.

65. Antecedent: $q = F$
    Consequent: $(p \vee q) \equiv p$
    Contrapositive: $q \not\equiv F$ when $(p \vee q) \not\equiv p$.

67. Antecedent: $n > 2$
    Consequent: $n^2 > 4$
    Contrapositive: If $n^2 \not> 4$, then $n \not> 2$.

69. Antecedent: $n > 2$
    Consequent: $n^2 > 4$
    Contrapositive: $n \not> 2$ when $n^2 \not> 4$.

1.4   Application: Designing Computer Circuits

1. $(1 \vee 1) \wedge (\sim 1) = (T \vee T) \wedge (F) = T \wedge F = F = 0$

3. $(0 \vee 1) \wedge (\sim 1) = (F \vee T) \wedge (F) = T \wedge F = F = 0$

5. $[(\sim 1) \vee 1] \wedge [1 \vee (\sim 1)] = [(\sim T) \vee T] \wedge [T \vee (\sim T)] = T \wedge T = T = 1$

7. $[(\sim 0) \vee 1] \wedge [0 \vee (\sim 1)] = [(\sim F) \vee T] \wedge [F \vee (\sim T)] = T \wedge F = F = 0$

9. $(\sim 1) \wedge (1 \vee 1) = (\sim T) \wedge (T \vee T) = F \wedge T = F = 0$

11. $(\sim 1) \wedge (0 \vee 1) = (\sim T) \wedge (F \vee T) = F \wedge T = F = 0$

13. $(\sim 0) \wedge (0 \vee 1) = (\sim F) \wedge$
    $(F \vee T) = T \wedge T = T = 1$

15. $(1 \wedge 1) \vee [(\sim 1) \wedge 1] = (T \wedge T) \vee [(\sim T) \wedge T] = T \vee F = T = 1$

17. $(1 \wedge 0) \vee [(\sim 1) \wedge 1] = (T \wedge F) \vee [(\sim T) \wedge T] = F \vee F = F = 0$

19. $(0 \wedge 0) \vee [(\sim 0) \wedge 1] = (F \wedge F) \vee [(\sim F) \wedge T] = F \vee T = T = 1$

21. $[p \wedge (\sim q)] \vee [(\sim p) \wedge q] \vee [(\sim p) \wedge (\sim q)]$

23. $[p \wedge (\sim q)] \vee [(\sim p) \wedge q]$

25. $p \wedge (\sim q)$

27. $[p \wedge (\sim q) \wedge r] \vee [p \wedge (\sim q) \wedge (\sim r)] \vee [(\sim p) \wedge (\sim q) \wedge r]$

29. $[p \wedge q \wedge (\sim r)] \vee [(\sim p) \wedge q \wedge r] \vee [(\sim p) \wedge q \wedge (\sim r)] \vee [(\sim p) \wedge (\sim q) \wedge r] \vee [(\sim p) \wedge (\sim q) \wedge (\sim r)]$

31. $[p \wedge q \wedge r] \vee [p \wedge (\sim q) \wedge r] \vee [(\sim p) \wedge (\sim q) \wedge r] \vee [(\sim p) \wedge (\sim q) \wedge (\sim r)]$

33. $(\sim p) \vee (\sim q)$

35. $\sim p$

37. $p \vee (\sim p)$

39. $q \vee [(\sim p) \wedge r]$

41. $(p \wedge q) \vee [(\sim p) \wedge (\sim q)]$

43. $q \vee r$

45. $(\sim p) \vee (\sim q)$

47. $[p \wedge (\sim q)] \vee [(\sim p) \wedge q]$

49. $p \wedge (\sim q)$

51. $[p \wedge (\sim q)] \vee [(\sim q) \wedge r]$

53. $(\sim p) \vee [q \wedge (\sim r)]$

55. $(p \wedge r) \vee [(\sim p) \wedge (\sim q)]$

57. q

59. $p \vee (\sim q)$

61. $[p \wedge r] \vee [p \wedge (\sim q)]$

63. $[(\sim p) \wedge (\sim q)] \vee (\sim r)$

65. $[(\sim p) \vee (\sim q)]$

67. $[(\sim p) \vee (\sim q)] \wedge [p \vee q]$

69. $[(\sim p) \vee (\sim q) \vee (\sim r)] \wedge [(\sim p) \vee (\sim q) \vee r] \wedge [p \vee (\sim q) \vee (\sim r)] \wedge [p \vee (\sim q) \vee r]$

1.5 Natural Deductive Reasoning

1. Modus tollens

p	q	$(p \rightarrow q) \wedge (\sim q)$	$\sim p$
T	T	F	F
T	F	F	F
F	T	F	T
F	F	T	T

3. Conjunctive simplification

p	q	$p \wedge q$	q
T	T	T	T
T	F	F	F
F	T	F	T
F	F	F	F

5. Disjunctive syllogism

p	q	$(p \vee q) \wedge (\sim p)$	q
T	T	F	T
T	F	F	F
F	T	T	T
F	F	F	F

7. Disjunctive addition

p	q	$p \vee q$
T	T	T
T	F	T
F	T	T
F	F	F

9. Hypothetical syllogism

p	q	r	$(p \rightarrow q) \wedge (q \rightarrow r)$	$(p \rightarrow r)$
T	T	T	T	T
T	T	F	F	F
T	F	T	F	T
T	F	F	F	F
F	T	T	T	T
F	T	F	F	T
F	F	T	T	T
F	F	F	T	T

11. Contradiction

p	q	r	$[\, p \wedge (\sim q)] \rightarrow [r \wedge (\sim r)]$	$p \rightarrow q$
T	T	T	T	T
T	T	F	T	T
T	F	T	F	F
T	F	F	F	F
F	T	T	T	T
F	T	F	T	T
F	F	T	T	T
F	F	F	T	T

13.

p	q	$(p \leftrightarrow q) \wedge p$	q
T	T	T	T
T	F	F	F
F	T	F	T
F	F	F	F

15.

p	q	$(p \leftrightarrow q) \wedge (\sim p)$	$\sim q$
T	T	F	F
T	F	F	T
F	T	F	F
F	F	T	T

17.

p	q	r	$(p \leftrightarrow q) \wedge (p \leftrightarrow r)$	$(\sim q) \vee r$
T	T	T	T	T
T	T	F	F	F
T	F	T	F	T
T	F	F	F	T
F	T	T	F	T
F	T	F	F	F
F	F	T	F	T
F	F	F	T	T

19. row 3

p	q	$(p \rightarrow q) \wedge (\sim p)$	$\sim q$
T	T	F	F
T	F	F	T
F	T	T	F
F	F	T	T

21. row 1

p	q	$(p \vee q) \wedge p$	$\sim q$
T	T	T	F
T	F	T	T
F	T	F	T
F	F	F	T

23. row 3

p	q	$p \vee q$	p
T	T	T	T
T	F	T	T
F	T	T	F
F	F	F	T

25. row 1

p	q	$p \wedge q$	$\sim q$
T	T	T	F
T	F	F	T
F	T	F	F
F	F	F	T

27. row 4

p	q	r	$[(p \wedge q) \rightarrow r] \wedge p$	r
T	T	T	T	T
T	T	F	F	F
T	F	T	T	T
T	F	F	T	F
F	T	T	F	T
F	T	F	F	F
F	F	T	F	T
F	F	F	F	F

29. row 6

p	q	r	$[(p \lor q) \lor r] \land (\sim p)$	r
T	T	T	F	T
T	T	F	F	F
T	F	T	F	T
T	F	F	F	F
F	T	T	T	T
F	T	F	T	F
F	F	T	T	T
F	F	F	F	F

31. row 1 or row 8

p	q	r	$(p \leftrightarrow q) \land (q \leftrightarrow r)$	$(\sim p) \land r$
T	T	T	T	F
T	T	F	F	F
T	F	T	F	F
T	F	F	F	F
F	T	T	F	T
F	T	F	F	F
F	F	T	F	T
F	F	F	T	F

33. 1. premise
    2. 1—double negation
    3. premise
    4. 2,3—modus ponens
35. 1. premise
    2. 1—conjunctive simplification
    3. premise
    4. 2,3—modus tollens
37. 1. premise
    2. premise
    3. 1,2—hypothetical syllogism
    4. premise
    5. premise
    6. 4,5—disjunctive syllogism
    7. 3,6—modus ponens
39. 1. premise
    2. 1—conjunctive simplification
    3. premise
    4. 2,3—modus ponens
    5. 4—conjunctive simplification
    6. 5—double negation
    7. premise
    8. 6,7—disjunctive syllogism
41. 1. premise
    2. premise

3. 1,2—modus tollens
4. 3—De Morgan's laws
5. 4—conjunctive simplification
6. premise
7. 5,6—modus tollens
8. 4—conjunctive simplification
9. premise
10. 8,9—disjunctive syllogism
11. 7,10—conjunctive addition

43. When $p = F$ and $q = T$, the premises $(p \rightarrow q)$ and $(\sim p)$ are true while the conclusion $(\sim q)$ is false.

45. When $p = T$ and $q = T$, the premises $(p \lor q)$ and p are true while the conclusion $(\sim q)$ is false.

47. When $p = F$ and $q = T$, the premise $(p \lor q)$ is true while the conclusion p is false.

49. When $p = T$ and $q = T$, the premise $(p \land q)$ is true while the conclusion $(\sim q)$ is false.

51. When $p = T$, $r = F$, and either of $q = T$ or $q = F$, the premises $[(p \land q) \rightarrow r]$ and p are true while the conclusion r is false.

53. When $p = F$, $q = T$, and $r = F$, the premises $[(p \vee q) \vee r]$ and $(\sim p)$ are true while the conclusion r is false.

55. When $p = q = r$ are either of T or F, the premises $(p \leftrightarrow q)$ and $(q \leftrightarrow r)$ are true while the conclusion $[(\sim p) \wedge r]$ is false.

57. modus ponens

59. inverse error

61. modus tollens

63. inverse error

65. If $\mathbb{B} \equiv \mathbb{C}$, then $\mathbb{B}$ and $\mathbb{C}$ have the same final column in their truth tables. It follows from the basic truth table for the connective $\leftrightarrow$, that $\mathbb{B} \leftrightarrow \mathbb{C}$ is true whenever corresponding rows in the final truth table columns for $\mathbb{B}$ and $\mathbb{C}$ have the same truth value. Since this happens in every row if $\mathbb{B} \equiv \mathbb{C}$, then $\mathbb{B} \leftrightarrow \mathbb{C}$ is true in every row and is therefore a tautology.

67. If $\mathbb{B} \leftrightarrow \mathbb{C}$ is a tautology, then we know that $(\mathbb{B} \to \mathbb{C}) \wedge (\mathbb{C} \to \mathbb{B})$ is a tautology because $(\mathbb{B} \to \mathbb{C}) \wedge (\mathbb{C} \to \mathbb{B})$ is logically equivalent to $\mathbb{B} \leftrightarrow \mathbb{C}$. By conjunctive simplification, we know that $\mathbb{B} \to \mathbb{C}$ is also a tautology. Then by the definition of a rule of deduction of the form $\mathbb{B} \therefore \mathbb{C}$, we know that $\mathbb{B} \therefore \mathbb{C}$ is a valid argument.

69. If $\mathbb{B} \equiv \mathbb{C}$, then from exercise 65 $\mathbb{B} \leftrightarrow \mathbb{C}$ is a tautology. Now, applying exercise 67, $\mathbb{B} \therefore \mathbb{C}$ is a valid argument.

1.6 The Formal Language of Predicate Logic

1. $L(c, p) \vee L(p, c)$

3. $L(c, p) \wedge L(p, c)$

5. $L(p, c) \wedge L(p, p)$

7. $\forall y[\sim(x = c) \to L(p, y)] \wedge [\sim L(p, c)]$

9. $\exists x[L(x, c) \to L(p, c)]$

11. $\forall x L(x, x)$

13. $\forall x \forall y\{[\sim L(x, y)] \to [\sim L(c, x)]\}$

15. $\forall x L(x, x) \to \forall x \exists y L(x, y)$

17. $P(2) \wedge E(2)$

19. $[\sim E(5)] \wedge E(2)$

21. $[(2 + 5) > 2] \wedge [(2 + 5) > 5]$

23. $P(2 + 5) \to [\sim E(2 + 5)]$

25. $\forall n\{(n > 0) \to [\sim(n = 0)]\}$

27. $\forall n \forall m\{[E(n) \wedge E(m)] \to E(n + m)\}$

29. $\forall k \forall n \forall m\{[E(k) \wedge E(n) \wedge E(m)] \to [\sim P(k + n + m)]\}$

31. $\forall n \exists m[E(n) \wedge E(m) \wedge (m > n)]$

33. A: $E(0)$
    B: $\exists x[Z(x) \wedge E(x)]$

35. A: $\exists n(n > 0)$
    B: $\exists n \forall x[Z(x) \to (n > x)]$

37. A: $\forall n(n > 0)$
    B: $\forall n \forall x[Z(x) \to (n > x)]$

39. A: $\exists n(0 > n)$
    B: $\exists n \forall x[Z(x) \to (x > n)]$

41. $\forall x \forall y[(x + y) = (y + x)]$

43. $\exists e \forall x\{[(x + e) = x] \wedge [(e + x) = x]\}$

45. $\forall x[\sim(x < x)]$

47. $\forall x \forall y\{(x < y) \to [\sim(y < x)]\}$

49. $\forall x \forall y\{(x < y) \to \exists z[(x < z) \wedge (z < y)]\}$

51. $\forall \varepsilon \exists \delta \forall x\{[[(\varepsilon > 0) \wedge (\delta > 0) \wedge [d(x, c) < \delta] \wedge [\sim(x = c)]] \to [d(f(x), L) < \varepsilon]\}$

53. $\forall M \exists \delta \forall x\{[(M > 0) \wedge (\delta > 0) \wedge [d(x, c) < \delta] \wedge [\sim(x = c)]] \to [f(x) > M)\}$

55. $\forall \varepsilon \exists N \forall x\{[(\varepsilon > 0) \wedge (N > 0) \wedge (x > N)] \to [d(f(x), L) < \varepsilon]\}$

57. $\forall y[L(c, y) \to (y = p)]$

59. $\forall x \exists y L(x, y)$

61. $\sim \exists x \forall y L(x, y)$

63. $\exists x\{E(x) \wedge P(x) \wedge \forall y[(E(y) \wedge P(y)) \to (x = y)]\}$

65. $\exists x \exists y\{(x > 0) \wedge (y > 0) \wedge$
$[\sim(x = y)]\}$

67. $\exists x \exists y \exists z\{[(x > 0) \wedge (y > 0) \wedge$
$(z > 0)] \wedge [\sim[(x = y) \vee$
$(x = z) \vee (y = z)]]\}$

69. $\exists x \exists y \exists z\{[(x > 0) \wedge (y > 0) \wedge$
$(z > 0)] \wedge [(4 > x) \wedge (4 > y) \wedge$
$(4 > z)] \wedge [\sim[(x = y) \vee$
$(x = z) \vee (y = z)]]\}$

Fundamentals of Mathematical Proofs

1. Let $n = 2i$, $m = 2j + 1$ where $i, j \in \mathbb{Z}$. Then $n + m = 2i + (2j + 1) = 2(i + j) + 1$. Since $i + j$ is an integer, $2(i + j) + 1$ is an odd integer. Therefore $n + m$ is an even integer.

3. Let $n = 2i$, $m = 2j + 1$ where $i, j \in \mathbb{Z}$. Then $n \cdot m = (2i)(2j + 1) = 2(2ij + i)$. Since $2ij + i$ is an integer, $2(2ij + i)$ is an even integer. Therefore $n \cdot m$ is an even integer.

5. Let $n = 2i + 1$, $m = 2j + 1$ where $i, j \in \mathbb{Z}$. Then $n + m = (2i + 1) + (2j + 1) = 2i + 2j + 2 = 2(i + j + 1)$. Since $i + j + 1$ is an integer, $2(i + j + 1)$ is an even integer. Therefore $n + m$ is an even integer.

7. Let $n = 2i + 1$, $m = 2j + 1$ where $i, j \in \mathbb{Z}$. Then $n \cdot m = (2i + 1)(2j + 1) = 4ij + 2i + 2j + 1 = 2(2ij + i + j) + 1$. Since $2ij + i + j$ is an integer, $2(2ij + i + j)$ is an even integer. Therefore $2(2ij + i + j) + 1$ is an odd integer. Therefore $n \cdot m$ is an odd integer.

9. Let $n = 2k + 1$ where $k \in \mathbb{Z}$. Then $n^2 = n \cdot n = (2k + 1)(2k + 1) = 4k^2 + 4k + 1 = 4k(k + 1) + 1$. Examine the case where k is an even integer: $k = 2r$ for some $r \in \mathbb{Z}$. Then $4k = 4(2r) = 8r$ and so $n^2 = 8r(k + 1) + 1$. Letting $r(k + 1) = i$, we have $n^2 = 8i + 1$. Examine the case where k is an odd integer: $k = 2r + 1$ for some $r \in \mathbb{Z}$. Then $(k + 1) = (2r + 1 + 1) = 2(r + 1)$ and so $n^2 = (4k)(2)(r +$

$1) + 1 = 8k(r + 1)$. Letting $k(r + 1) = i$, we have $n^2 = 8i + 1$. Therefore, when n is odd, $n^2 = 8i + 1$ for some integer $i \in \mathbb{Z}$.

11. Let $n + m = 2i + 1$ for some $i \in \mathbb{Z}$ where $n, m \in \mathbb{Z}$. Without loss of generality, let $n > m$ (we know $n \neq m$). Then $n - m = (n + m) - 2m = (2i + 1) - 2m = 2(i - m) + 1$. Since $i - m$ is an integer, $2(i - m) + 1$ is an odd integer. Therefore $n - m$ is an odd integer.

13. Let $x = p/q$ and $y = r/s$ for some $p, q, r, s \in \mathbb{Z}$ where $q, s \neq 0$. Then

$$x - y = \frac{p}{q} - \frac{r}{s} = \frac{ps - qr}{qs}.$$

Since $ps - qr$ is an integer and qs is a nonzero integer by the zero product property, $(x - y) \in \mathbb{Q}$.

15. Let $x = p/q$ and $y = r/s$ for some $p, q, r, s \in \mathbb{Z}$ where $p, q, r, s \neq 0$. Then

$$x \div y = \frac{p}{q} \div \frac{r}{s} = \frac{ps}{qr}.$$

Since ps is a nonzero integer and qr is a nonzero integer by the zero product property, $x \div y$ is a nonzero rational number.

17. Let $x = p/q$ for some $p, q \in \mathbb{Z}$ where $q \neq 0$. Then

$$2x = 2 \cdot \frac{p}{q} = \frac{2p}{q}.$$

Since $2p$ is an integer and q is a nonzero integer by the zero product property, $2x \in \mathbb{Q}$.

19. Let n be a nonzero even integer (if $n = 0$, then $n^2 = 0$, which is an even integer). Assume that n^2 is an odd integer. Then $n^2 = 2i + 1$ for some $i \in \mathbb{Z}$. Then

$$n = \frac{n^2}{n} = \frac{2i + 1}{n}.$$

If n is even, then

$$\frac{2i + 1}{n} = 2k$$

for some $k \in \mathbb{Z}$. Then $2i + 1 = 2kn$. Since kn is an integer, $2kn$ is an even integer. Thus the odd integer $2i + 1$ is an even integer, and we have a contradiction of the parity property of the integers. Therefore, if n is an even integer, then n^2 is also an even integer.

21. Let n be a nonzero even integer (if $n = 0$, then $n^3 = 0$, which is an even integer). Assume that n^3 is an odd integer. Then $n^3 = 2i + 1$ for some $i \in \mathbb{Z}$. Then

$$n = \frac{n^3}{n^2} = \frac{2i + 1}{n^2}.$$

If n is even, then

$$\frac{2i + 1}{n^2} = 2k$$

for some $k \in \mathbb{Z}$. Then $2i + 1 = 2kn^2$. Since kn^2 is an integer, $2kn^2$ is an even integer. Thus the odd integer $2i + 1$ is an even integer, and we have a contradiction of the Parity Property of the Integers. Therefore, if n is an even integer, then n^3 is also an even integer.

23. Choose r where $r \notin \mathbb{Q}$. Assume that $\sqrt{r}$ is rational. Then $\sqrt{r} = p/q$ for some $p, q \in \mathbb{Z}$ where $q \neq 0$. Then $(\sqrt{r})^2 = p^2/q^2$ is a rational number. Thus, since $r = (\sqrt{r})^2$, the irrational number r is rational, and we have a contradiction. Therefore,

if r is irrational, then $\sqrt{r}$ is also irrational.

25. Choose x where $x \in \mathbb{Q}$. Assume that $x^2 \notin \mathbb{Q}$ (i.e. x^2 is irrational). Then $x = \sqrt{x^2}$. From exercise 22 we know that if x^2 is irrational, that $\sqrt{x^2}$ is also irrational. Thus the rational number x is irrational, and we have a contradiction. Therefore, if x is rational, then x^2 is also rational.

27. Assume that n is the greatest integer. If n is an integer, then $n + 1$ is also an integer. We know that $n + 1 > n$. Thus $n + 1$ is greater than the greatest integer n, and we have a contradiction. Therefore there does not exist a greatest integer.

29. Assume that x is the least positive rational number. If x is rational, $\frac{x}{2}$ is also rational since $x/2 = p/(2q)$ for some $p, q \in \mathbb{Z}$ where $q \neq 0$ and thus $2q \neq 0$. Since x is positive, we know that $0 < x/2 < x$. Thus $x/2$ is a positive rational number less than the least positive rational number, and we have a contradiction. Therefore there does not exist a least positive rational number.

31. Assume that n is an odd integer. Then $n = 2k + 1$ for some $k \in \mathbb{Z}$. Then $n^3 = (2k + 1)^3 = 8k^3 + 12k^2 + 6k + 1 = 2(4k^3 + 6k^2 + 3k) + 1$. Since $4k^3 + 6k^2 + 3k$ is an integer, $2(4k^3 + 6k^2 + 3k) + 1$ is an odd integer. Thus n being odd implies that n^3 is also odd. Therefore, by contrapositive, we know that if n^3 is even, then n is also even.

33. Assume that m or n is an even integer; with loss of generality, we assume m is even, so $m = 2k$ for some $k \in \mathbb{Z}$. Then $m \cdot n = (2k)n = 2(kn)$. Since kn is an integer, $2(kn)$ is an even integer. Thus m or n being even implies that $m \cdot n$ is also even. Therefore, by

contrapositive, we know that if $m \cdot n$ is odd, then both m and n are odd.

35. Assume that $\sqrt{r}$ is a rational number. Then $\sqrt{r} = p/q$ for some $p, q \in \mathbb{Z}$ where $q \neq 0$. Then $r = (\sqrt{r})^2 = p^2/q^2$. Since p^2 and q^2 are both integers and $q^2 \neq 0$ by the zero product property, p^2/q^2 is a rational number. Thus $\sqrt{r}$ being a rational number implies that r is also a rational number. Therefore, by contrapositive, we know that if r is irrational, then $\sqrt{r}$ is also irrational.

37. $\Rightarrow$ Assume that n is an even integer. Then $n = 2k$ for some $k \in \mathbb{Z}$. Then $n^2 = n \cdot n = (2k)(2k) = 4k^2 = 2(2k^2)$. Since $2k^2$ is an integer, $2(2k^2)$ is an even integer. Thus n being even implies that n^2 is also even. Therefore, by contrapositive, we know that if n^2 is odd, then n is also odd.

$\Leftarrow$ Let $n = 2k + 1$ where $k \in \mathbb{Z}$. Then $n^2 = n \cdot n = (2k+1)(2k+1) = 4k^2 + 4k + 1 = 2(2k^2 + 2k) + 1$. Since $2k^2 + 2k$ is an integer, $2(2k^2 + 2k) + 1$ is an odd integer. Therefore n^2 is an odd integer.

39. $\Rightarrow$ Let n be an odd integer. Then $n = 2k + 1$ for some $k \in \mathbb{Z}$. Then $n + 1 = (2k + 1) + 1 = 2k + 2 = 2(k + 1)$, which is an even integer. Therefore, if n is odd, then $n + 1$ is even.

$\Leftarrow$ Let $n+1$ be an even integer. Then $n + 1 = 2k$ for some $k \in \mathbb{Z}$. Then

$n = (n + 1) - 1 = 2k - 1$. Let $k = i + 1$ for some $i \in \mathbb{Z}$. Notice that all we have done is rewritten the integer k in terms of the integer directly preceding k (e.g., $13 = 12 + 1$). Then $n = 2k - 1 = 2(i + 1) - 1 = 2i + 1$, which is an odd integer. Therefore, if $n + 1$ is even, then n is odd.

41. $3 = 2 \cdot 1 + 1$ is an odd integer.

43. 8 is even, since $8 = 2 \cdot 4$, and can be written as the sum of two distinct primes, since $8 = 3 + 5$.

45. π is an irrational number.

47. $2 = 2/1$ is a rational integer.

49. 2 is prime, but $2 = 2 \cdot 1$ is not odd.

51. The ratio $C : r = 2\pi$, but 2π is not rational.

53. The sum of the odd integers 3 and 5 is 8, and $8 = 2 \cdot 4$ is not odd.

55. $3 = 2 \cdot 1 + 1 = 3/1$ is an odd integer that is not irrational.

57. The sum of the two irrational numbers π and $-\pi$ is 0, which is not irrational.

59. For the pair of reals 2 and -2 satisfies the condition $2^2 = (-2)^2$, but $2 \neq -2$.

61. $\sqrt{4} = 2$ is a rational number.

63. 2 is a rational number that is not odd.

65. 2 is an even integer.

67. $0^2 = 0$ is not greater than zero.

69. This proof incorrectly *assumes* that the conclusion is true, then proceeds to deduce that if the conclusion is true, it must be true.

2.1 The Algebra of Sets

1. $\{w, z\}$

3. $\{x, y\}$

5. $\{x, y, z\}$

7. $\{x, y\}$

9. $\{y\}$

11. $\{(x, x), (x, y), (x, z), (y, x), (y, y), (y, z)\}$

13. $\{\emptyset, \{x\}, \{y\}, \{x, y\}\}$

15. $A^C = (-\infty, 0] \cup (2, \infty)$, or
$A^C = \{x : -\infty < x \le 0$ or
$2 < x < \infty\}$

17. $A \cap B = [1, 2] = \{x : 1 \le x \le 2\}$

19. $A \setminus B = (0, 1) = \{x : 0 < x < 1\}$

21. $A^C \cap B^C = (A \cup B)^C = (-\infty, 0]$
$\cup [3, \infty)$, or
$A^C \cap B^C = \{x : -\infty < x \le 0$ or
$3 \le x < \infty\}$

23. $1 \in \mathbb{N}$, but $1 \notin \emptyset$.

25. $\frac{1}{2} \in \mathbb{Q}$, but $\frac{1}{2} \notin \mathbb{Z}$.

27. $i \in \mathbb{C}$, but $i = \sqrt{-1} \notin \mathbb{R}$.

29. The sets $\{1, 2\}$ and $\{2, 1\}$ have precisely the the same elements and the *order* in which the elements of a set are listed is not important. Therefore $\{1, 2\} = \{2, 1\}$.

31. Let $x \in A$, then $x \in A$ and we have $A \subseteq A$. Notice that $A \not\subset A$ since $A = A$.

33. Since the empty set $\emptyset$ contains no elements, $\emptyset^C$ contains every element in the universe. Since $A \setminus \emptyset = A \cap \emptyset^C$, we have $A \cap \emptyset^C = A$ since the intersection between any set and the entire universe is the set itself. Therefore $A \setminus \emptyset = A$.

35. By definition, if some element $a \in A \cap B$, then $a \in A$ and $a \in B$. Thus, for every element $a \in A \cap B$, we have $a \in A$. Therefore $A \cap B \subseteq A$.

37. Let $a \in A$. By definition of the union, $A \cup B$ is the set of elements either in A, in B, or in both A and B. Thus, if $a \in A$, we have $a \in A \cup B$. Therefore $A \subseteq A \cup B$.

39. Let $a \in A$. Since $A \subseteq B$, we have $a \in B$. Since $A \subseteq C$, we have $a \in C$. If $a \in B$ and $a \in C$, then $a \in B \cap C$. Since for any element $a \in A$ we have $a \in B \cap C$, we then have $A \subseteq B \cap C$.

41. Let $X \in \mathbb{P}(A)$. Then $X \subseteq A$. Since $x \in X$ and $X \subseteq A$, we have $x \in A$. In addition, $A \subseteq B$, so $x \in B$. Since $x \in X$ implies $x \in B$, we have $X \subseteq B$. Then by the definition of a power set, we have $X \in \mathbb{P}(B)$. Since for every $X \in \mathbb{P}(A)$ we have $X \in \mathbb{P}(B)$, $\mathbb{P}(A) \subseteq \mathbb{P}(B)$.

43. An element is not a set and thus cannot be a subset.

45. A power set is a set of sets. So a subset of a power set must be another set of sets. Since the set $\{1\}$ is a set of numbers, not sets, $\{1\}$ cannot be a subset of a power set.

47. Let $A = \emptyset$ and B be any finite set.

49. Let $A = \{1, 2\}$ and $B = \{1, 2\}$.

51. Let $A = \{1, 2, 3\}$, $B = \{2, 3, 4\}$, and $C = \{2, 3\}$.

53. Let $A = \{1, 2, 3\}$, $B = \{1, 2\}$, and $C = \{3, 4\}$.

55. C is only disjoint from F.

57. E is disjoint from D and F.

59. The two elements are $\emptyset$ and $\{1\}$.

61. The eight elements are $\emptyset, \{1\}, \{2\}, \{3\}$, $\{1, 2\}, \{1, 3\}, \{2, 3\}, \{1, 2, 3\}$.

63. $5 = \{1, 2, 3, 4\}$

65. Each set contains a number of elements that is equal to its corresponding natural number. The set corresponding to the natural number 50 contains 50 elements.

67. The barber is only allowed to shave people who do not shave themselves. Thus, if the barber shaves himself, he must not shave himself.

69. If N is a normal set, then N cannot be in N (that is, $N \notin N$). But N contains *all* normal sets and since N is normal, N is in N (that is, $N \in N$). Contradiction!

2.2   The Division Algorithm and Modular Addition

1. $q = 5$ and $r = 4$

3. $q = -2$ and $r = 5$

5. Integers from $\{\ldots, -13, -6, 1, 8, 15, \ldots\}$

7. Integers from $\{\ldots, -10, -3, 4, 11, 18, \ldots\}$

9. $q = 4$ and $r = 1$

11. $q = -2$ and $r = 5$

13. Integers from $\{\ldots, -15, -7, 1, 9, 17, \ldots\}$

15. Integers from $\{\ldots, -11, -3, 5, 13, 21, \ldots\}$

17. $1, 4, 7$

19. $3, 10, 17$

21. $3, 13, 23$

23. $4, 6, 5, 6$

25. $7, 1, 5, 6$

27. $2, 9, 10, 9$

29. $0, 1$

31. $0, 1, 2, 3, 4$

33. $a^p \bmod p = a \bmod p$ where $a \in \mathbb{Z}^+$ and p is any prime number.

35. $\{0, 1, 2, 3, 4, 5, 6, 7\}$

37. $\{0, 1, 2, 3, 4, 5, 6, 7, 8, 9, 10, 11, 12, 13, 14\}$

39. $n = 5k + 4$ is the set $\{\ldots, -11, -6, -1, 4, 9, 14, \ldots\}$.

41. $n = 8k + 4$ is the set $\{\ldots, -20, -12, -4, 4, 12, 20, \ldots\}$.

43. $0, 1$

45. $1, 2$

47. $2, 4$

49. $4, 2$

51. $0, 7$

53. $6, 0$

55. 0 and 2 are their own inverses; 1 and 3 are inverses.

57. 0 and 4 are their own inverses; 1 and 7 are inverses; 2 and 6 are inverses; and 3 and 5 are inverses.

59. 0 is its own inverse; 1 and 14 are inverses; 2 and 13 are inverses; 3 and 12 are inverses; 4 and 11 are inverses;

5 and 10 are inverses; 6 and 9 are inverses; and 7 and 8 are inverses.

61. Since $a \equiv b \bmod n$, we have $a = n \cdot e + r$ and $b = n \cdot f + r$, for some $e, f, r \in \mathbb{Z}$. Similarly, since $c \equiv d \bmod n$, we have $c = n \cdot g + s$ and $b = n \cdot h + s$, for some $g, f, s \in \mathbb{Z}$. Taking differences, we have $a - c = n \cdot (e - g) + (r - s)$ and $b - d = n \cdot (f - h) + (r - s)$. By the division algorithm, $r - s = n \cdot p + t$ for some $p \in \mathbb{Z}$ and $t \in \mathbb{Z}_n$. Thus, $a - c$ and $b - d$ both have a remainder of t under division by n, and we have $(a - c) \equiv (b - d) \bmod n$.

63. If a is odd, then $a = 2s + 1$ for some $s \in \mathbb{Z}$. Then $a^2 = 4s^2 + 4s + 1 = 4(s^2 + s) + 1$, which is one greater than a multiple of 4. Since every number that is one greater than a multiple of 4 is equivalent to 1 mod 4, $a^2 \equiv 1 \bmod 4$.

65. Each element $a \in A \cap B$ is in both A and B. Since $B \cap A$ contains all elements common to both B and A, $a \in B \cap A$. Thus $A \cap B \subseteq B \cap A$. Each element $b \in B \cap A$ is in both B and A. Since $A \cap B$ contains all elements common to both A and B, $b \in A \cap B$. Thus $B \cap A \subseteq A \cap B$. Therefore $A \cap B = B \cap A$.

67. Counterexample: Let $A = \{1, 2\}$ and $B = \{2, 3\}$.

69. Let $a \in A \cup \emptyset$. Then, by the definition of union of sets, $a \in A$ or $a \in \emptyset$ or $a \in A \cap \emptyset$. By definition, the empty set $\emptyset$ contains no elements. Therefore, for all $a \in A \cup \emptyset$, we have $a \in A$. Thus $A \cup \emptyset = A$. By identical argument $\emptyset \cup A = A$. Therefore, since $A \cup \emptyset = A$ and $\emptyset \cup A = A$, the empty set $\emptyset$ is the identity for union of sets.

1. 0, 1

3. 2, 1

5. 6, 6

7. 8, 0

9. 9, 2

11. {0, 1}

13. {0, 1, 2, 3, 4, 5, 6, 7, 8, 9, 10}

15. {1, 3, 5, 7}

17.
$\oplus$	0	1
0	0	1
1	1	0

19.
$\oplus$	0	1	2	3	4	5	6	7
0	0	1	2	3	4	5	6	7
1	1	2	3	4	5	6	7	0
2	2	3	4	5	6	7	0	1
3	3	4	5	6	7	0	1	2
4	4	5	6	7	0	1	2	3
5	5	6	7	0	1	2	3	4
6	6	7	0	1	2	3	4	5
7	7	0	1	2	3	4	5	6

21.
$\odot$	1	2	3	4
1	1	2	3	4
2	2	4	1	3
3	3	1	4	2
4	4	3	2	1

23. 1 and 4 are their own inverses; 2 and 3 are inverses.

25. $(2 \odot 3) \odot 4 = \{[(2 \cdot 3) \bmod 11] \cdot 4\}$ $\bmod 11 = [(6 \bmod 11) \cdot 4] \bmod 11 = (6 \cdot 4) \bmod 11 = 24 \bmod 11 = 2$
$2 \odot (3 \odot 4) = \{2 \cdot [(3 \cdot 4) \bmod 11]\} \bmod 11 = [2 \cdot (12 \bmod 11)] \bmod 11 = (2 \cdot 1) \bmod 11 = 2 \bmod 11 = 2$

27. $(4 \odot 8) \odot 10 = \{[(4 \cdot 8) \bmod 11] \cdot 10\} \bmod 11 = [(32 \bmod 11) \cdot 10] \bmod 11 = (10 \cdot 10) \bmod 11 = 100 \bmod 11 = 1$
$4 \odot (8 \odot 10) = \{4 \cdot [(8 \cdot 10) \bmod 11]\} \bmod 11 = [4 \cdot (80 \bmod 11)] \bmod 11 = (4 \cdot 3) \bmod 11 = 12 \bmod 11 = 1$

29.
$\odot$	1	2	3	4	5	6	7
1	1	2	3	4	5	6	7
2	2	4	6	0	2	4	6
3	3	6	1	4	7	2	5
4	4	0	4	0	4	0	4
5	5	2	7	4	1	6	3
6	6	4	2	0	6	4	2
7	7	6	5	4	3	2	1

31. 1, 3, 5, and 7 are their own inverses. The other elements do not have inverses. The elements with multiplicative inverses are precisely those elements that are relatively prime to 8.

33. $101 \odot 48 = (101 \cdot 48) \bmod 11 = 4{,}848 \bmod 11 = 8(101 \bmod 11) \cdot (48 \bmod 11) = 2 \cdot 4 = 8$

35. $14 \odot 410 = (14 \cdot 10) \bmod 11 = 140 \bmod 11 = 8(14 \bmod 11) \cdot (10 \bmod 11) = 3 \cdot 10 = 30$
The conjecture $a \odot b = (a \bmod 11) \cdot (b \bmod 11)$ fails here.

37. $(2!) \bmod 3 = 2$; $(4!) \bmod 5 = 4$; $(6!) \bmod 7 = 6$. For the cases we have seen, $(n-1)! \bmod n = (n-1)$.

39. Since $a \equiv b \bmod n$, we have $a = n \cdot e + r$ and $b = n \cdot f + r$, for some $e, f, r \in \mathbb{Z}$. Similarly, since $c \equiv d \bmod n$, we have $c = n \cdot g + s$ and $b = n \cdot h + s$, for some $g, f, s \in \mathbb{Z}$. Taking products we have $a \cdot c = n \cdot (neg + rg + es) + rs$ and $b \cdot d = n \cdot (nfh + rh + sf) + rs$. By the division algorithm, $rs = n \cdot p + t$ for some $p \in \mathbb{Z}$ and $t \in \mathbb{Z}_n$. Thus, $a \cdot c$ and $b \cdot d$ both have a remainder of t under division by n, and we have $(a \cdot c) \equiv (b \cdot d) \bmod n$.

41. ($\Rightarrow$) If $a \equiv b \bmod n$, then $a = n \cdot q + b$ for some $q \in \mathbb{Z}$. Subtracting b, we have $a - b = n \cdot q$. Thus n divides $(a - b)$.

($\Leftarrow$) If n divides $(a - b)$, then $a - b = n \cdot q$ for some $q \in \mathbb{Z}$. Adding b, we have $a = n \cdot q + b$. Thus $a \equiv b \bmod n$.

43. If a is even, then $a = 2s$ for some $s \in \mathbb{Z}$ and we have $a^2 = 4s^2$. Since $a^2 - 0 = a^2$ is divisible 4, we apply exercise 41 to obtain $a^2 \equiv 0 \bmod 4$.

45. If a is odd, then $a = 2s + 1$ for some $s \in \mathbb{Z}$. Then $a^2 = 4s^2 + 4s + 1 = 4s(s + 1) + 1$. The parity property of integers tells us that we only have two cases to consider: when s is even and when s is odd. Examine the case when s is even. Then $s = 2k$ for some $k \in \mathbb{Z}$. Then $4s(s + 1) + 1 = 4(2k)(2k + 1) = 8k(2k + 1) + 1$, which is one greater than a multiple of 8. Examine the case when s is odd. Then $s = 2k + 1$ for some $k \in \mathbb{Z}$. Then $4s(s + 1) + 1 = 4(2k + 1)(2k + 2) + 1 = 8(2k + 1)(k + 1) + 1$, which is one greater than a multiple of 8. Since every number that is one greater than a multiple of 8 is equivalent to 1 mod 8, $a^2 \equiv 1 \bmod 8$.

47. Reflexivity: $\sim P(a, a)$.
It is not always true that a is related to itself.
Symmetry: $\sim [\, p(a, b) \to P(b, a)]$.
If a is related to b, then it is not necessarily true that b is related to a.
Transitivity: $\sim \{[\, p(a, b) \land P(b, c)] \to P(a, c)]\}$.
If a is related to b and c is related to d, then it is not necessarily true that a is related to c.

49. Yes, this is an equivalence relation.

51. No—fails all three properties.

53. No—only symmetric.

55. (Alex, Andy), (Andy, Alex), (Alex, Alex), (Andy, Andy), (Bailey, Bailey), (Chris,Chris), (Dakota, Dakota), (Morgan,Morgan).
Equivalence classes:
[Alex] = {Alex, Andy } = [Andy];

[Bailey] = {Bailey}; [Chris] = {Chris};
[Dakota] = {Dakota}; [Morgan] = {Morgan}.

57. Reflexivity: $a - a = 0 \in \mathbb{Z}$.
Symmetry: If $a - b = c \in \mathbb{Z}$, then $b - a = -c \in \mathbb{Z}$.
Transitivity: If $a - b = d \in \mathbb{Z}$ and $b - c = e \in \mathbb{Z}$, adding the two equations produces $a - c = d + e \in \mathbb{Z}$.
Examples of equivalence classes:
$[0] = \{x : x \in \mathbb{Z}\} = \{\ldots, -1, 0, 1, \ldots\}$ and
$[\frac{1}{2}] = \{x : (x - \frac{1}{2}) \in \mathbb{Z}\} = \{\ldots, -\frac{1}{2}, \frac{1}{2}, \frac{3}{2}, \ldots\}$.

59. Reflexivity: $a + a = 2a$ is even.
Symmetry: If $a + b = 2k$ for some $k \in \mathbb{Z}$, then $b + a = 2k$.
Transitivity: If $a + b = 2i$ and $b + c = 2j$ for some $i, j \in \mathbb{Z}$, then $a + c = (2i - b) + (2j - b) = 2i + 2j - 2b = 2(i + j - b)$ is even.
Examples of equivalence classes:
$[0] = \{n : n = 2k, k \in \mathbb{Z}\}$ and
$[1] = \{n : n = 2k + 1, k \in \mathbb{Z}\}$.

61. Reflexivity: For all $a \in \mathbb{R}$, $a = a$, so $(a, b) \sim (a, b)$.
Symmetry: $(a, b) \sim (x, y)$ implies $a = x$ implies $x = a$ implies $(x, y) \sim (a, b)$.
Transitivity: Assume $(a, b) \sim (x, y)$ and $(x, y) \sim (c, d)$. Then $a = x$ and $x = c$, so $a = c$ by transitivity of equality for reals. Thus, $(a, b) \sim (c, d)$.
Examples of equivalence classes:
$[(1, 7)] = \{(1, y), y \in \mathbb{R}\}$ and
$[(2, 17)] = \{(2, y), y \in \mathbb{R}\}$.

63. Assume f, g, h are differentiable.
Reflexivity: $f' = f'$.
Symmetry: If $f' = g'$, then $g' = f'$.
Transitivity: If $f' = g'$ and $g' = h'$, then $f' = h'$.

Examples of equivalence classes:
$[x] = \{f : f(x) = x + c, c \in \mathbb{R}\}$ and
$[x^2] = \{f : f(x) = x^2 + c, c \in \mathbb{R}\}$.

65. Note that slopes are real numbers and equality of real numbers is an equivalence relation.

Reflexivity: The slope of J is equal to the slope of J.

Symmetry: If the slope m of line J equals the slope n of the line K, then $n = m$ by symmetry of equality of reals.

Transitivity: If the slope m of line J equals the slope n of the line K and n equals the slope p of the line L, then $m = p$ by transitivity of equality for reals.

Examples of equivalence classes:
$[2] = \{f : f(x) = 2x + c, c \in \mathbb{R}\}$ and $[3] = \{f : f(x) = 3x + c, c \in \mathbb{R}\}$.

67. Property that holds: transitivity.

Properties that fail: reflexivity, symmetry.

69. Properties that hold: reflexivity, transitivity.

Property that fails: symmetry.

2.4 An Introduction to Groups

1. Let $p(a, b) = a \circ b$ denote the group operation.
Closure: $\forall a, b[a, b \in G \rightarrow p(a, b) \in G]$.
Associativity: $\forall a, b, c[\, p(a, p(b, c)) = p(p(a, b), c)]$.
Identity: $\exists e \forall a[\, p(a, e) = a \wedge p(a, e) = a]$.
Inverses: $\forall a \exists b[\, p(a, b) = e \wedge p(b, a) = e]$.

3. Let $p(a, b) = a \circ b$ denote the group operation.
(not) Closure: $\exists a, b[a, b \in G \wedge p(a, b) \notin G]$.
(not) Associativity: $\exists a, b, c [\, p(a, p(b, c)) \neq p(p(a, b), c)]$.
(not) Identity: $\forall e \exists a[\, p(a, e) \neq a \vee p(e, a)a]$.
(not) Inverses: $\exists a \forall b[\, p(a, b) \neq e \vee p(b, a). \neq e]$

5. Identity: $\frac{1}{1}$.
Inverses: $a/b \cdot b/a = 1$.
$0 \in \mathbb{Q}^*$ does not have a multiplicative inverse.

7. Identity: 1.
Inverses: $r \cdot 1/r = 1$

$0 \in \mathbb{R}^*$ does not have a multiplicative inverse.

9. Identity: 1.
Inverses: $(a + bi) \cdot \dfrac{1}{a + bi} = 1$ where
$$\frac{1}{a + bi} = \frac{1}{a + bi} \cdot \frac{a - bi}{a - bi}$$
$$= \frac{a - bi}{a^2 + b^2} = \frac{a}{a^2 + b^2} - \frac{bi}{a^2 + b^2}.$$
$0 \in \mathbb{C}^*$ does not have a multiplicative inverse.

11. Identity: $(0, 0)$.
Inverses: $(r, s) + (-r, -s) = (0, 0)$.

13. Identity: $(1, 1)$.
Inverses: $(r, s) \cdot (1/r, 1/s) = (1, 1)$.

15. Identity: $(1, 0)$ since $(r, s) * (1, 0) = (r - 0, 0 + s) = (r, s) = (r - 0, s + 0) = (1, 0) * (r, s)$.
Inverses: $(r, s)^{-1} =$
$$\left(\frac{r}{r^2 + s^2}, \frac{-s}{r^2 + s^2} \right) \text{ since}$$
$$(r, s) * \left(\frac{r}{r^2 + s^2}, \frac{-s}{r^2 + s^2} \right)$$
$$= \left[\left(\frac{r^2}{r^2 + s^2} + \frac{s^2}{r^2 + s^2} \right), \right.$$

$$\left(\frac{-rs}{r^2+s^2}+\frac{rs}{r^2+s^2}\right)\right] = (1,0) \text{ and}$$

$$\left(\frac{r}{r^2+s^2},\frac{-s}{r^2+s^2}\right)*(r,s)$$

$$=\left[\left(\frac{r^2}{r^2+s^2}+\frac{s^2}{r^2+s^2}\right),\right.$$

$$\left.\left(\frac{rs}{r^2+s^2}-\frac{rs}{r^2+s^2}\right)\right]=(1,0).$$

17. Axiom that fails: Inverses.
Example: 0 has no multiplicative inverse in $\mathbb{Z}$; in fact, only 1 and -1 have inverses in $\mathbb{Z}$ under standard multiplication.

19. Axiom that fails: Inverses.
Example: 0 has no multiplicative inverse in $\mathbb{R}$.

21. Axioms that fail: Closure, Inverses.
Examples: $2 + 3 = 5 \notin \{0, 1, 2, 3\}$, and 1 has no additive inverse since $-1 \notin \{0, 1, 2, 3\}$.

23. Axiom that fails: Closure.
Example: $1 + 1 = 2 \notin \{-1, 0, 1\}$.

25. Axiom that fails: Closure.
Example: $n + n = 2n \notin \{-n, \ldots, -2, -1, 0, 1, 2, \ldots, n\}$.

27. Axiom that fails: Inverses.
Example: $(0, 0)$ has no component-wise multiplicative inverse in $\{(r, s) : r, s \in \mathbb{R}\}$.

29. $(3 - 2) - 1 \neq 3 - (2 - 1)$

31. Following the text's proof of the left cancellation theorem:
If $a \circ b = c \circ b$, then we have $(a \circ b) \circ b^{-1} = (c \circ b) \circ b^{-1}$, which, using associativity, yields $a \circ (b \circ b^{-1}) = c \circ (b \circ b^{-1})$, which, using the inverse axiom, yields $a \circ e = c \circ e$, which, using the identity axiom, yields $a = c$. Therefore, if $a \circ b = c \circ b$, then $a = c$.

33. Since $1 \odot 1 = 1; 1 \odot 2 = 2; 1 \odot 3 = 3; 1 \odot 4 = 4; 1 \odot 5 = 5$, 1 is not a zero divisor.

35. For $b = 2$, we have $3 \odot 2 = 0$

37. Since $5 \odot 1 = 5; 5 \odot 2 = 4; 5 \odot 3 = 3; 5 \odot 4 = 2; 5 \odot 5 = 1$, 5 is not a zero divisor.

39. 2, 4, 6

41. In $\mathbb{Z}_9$, 3 is a zero divisor since $3 \odot 3 = 0$.

43. The number 5 is prime.

$\odot$	0	1	2	3	4
0	0	0	0	0	0
1	0	1	2	3	4
2	0	2	4	1	3
3	0	3	1	4	2
4	0	4	3	2	1

45. In $\mathbb{Z}_n$, consider the elements 0, p, and n/p, where p is a prime factor of n. We know that n is not prime because $\mathbb{Z}_n$ has zero divisors (see exercises 37–43). Then $0 \odot p = 0$ and $n/p \odot p = 0$, and $0 \neq \frac{n}{p}$. For example, consider the elements 3 and 4 in $\mathbb{Z}_{12}$. $0 \odot 3 = 0$ and $4 \odot 3 = 0$, and $0 \neq 4$.

47. 0, 1

49. 0, 1, 3, 4

51. 0, 1, 7, 8

53. If $a, b \in \mathbb{Z}_n$ are both idempotents, then, since modular multiplication is both associative and commutative, $(a \circ b)^2 = (a \circ b) \circ (a \circ b) = a \circ (b \circ a) \circ b = a \circ (a \circ b) \circ b = (a \circ a) \circ (b \circ b) = a \circ b$. We also have $(a^2 \circ b^2) = (a \circ b)$ and so $(a \circ b)^2 = a \circ b$.

55. $\alpha = \begin{bmatrix} 1 & 2 & 3 & 4 \\ 1 & 3 & 4 & 2 \end{bmatrix}$

57. $\alpha = \begin{bmatrix} 1 & 2 & 3 & 4 \\ 2 & 3 & 4 & 1 \end{bmatrix}$

59. $\alpha = \begin{bmatrix} 1 & 2 & 3 & 4 \\ 4 & 1 & 2 & 3 \end{bmatrix}$

61. $\alpha \circ \beta = \begin{bmatrix} 1 & 2 & 3 & 4 & 5 \\ 4 & 1 & 2 & 5 & 3 \end{bmatrix}$

63. $\beta \circ \alpha = \begin{bmatrix} 1 & 2 & 3 & 4 & 5 \\ 2 & 3 & 5 & 1 & 4 \end{bmatrix}$

67. $\epsilon \circ \epsilon = \begin{bmatrix} 1 & 2 \\ 1 & 2 \end{bmatrix}$

65. $\gamma \circ \alpha = \begin{bmatrix} 1 & 2 & 3 & 4 & 5 \\ 4 & 5 & 2 & 3 & 1 \end{bmatrix}$

69. $\alpha \circ \alpha = \begin{bmatrix} 1 & 2 \\ 1 & 2 \end{bmatrix}$

2.5 Dihedral Groups

1. Impossible to sketch. By definition, a polygon must have at least three sides.
3. A regular pentagon.
5. Move one vertex of exercise 3—but not too much!
7. A misshapen pentagon.
9. Start with a regular hexagon, then stretch one pair of opposite sides.
11. Number the vertices of a square as follows: upper left = 1; upper right = 2; lower right = 3; lower left = 4. We then have:

    R_0 is a counterclockwise rotation of 0 degrees.

    R_0 fixes all vertices.

    R_{90} is a counterclockwise rotation of 90 degrees.

    R_{90} moves 1 to 4, 2 to 1, 3 to 2, and 4 to 3.

    R_{180} is a counterclockwise rotation of 180 degrees.

    R_{180} moves 1 to 3, 2 to 4, 3 to 1, and 4 to 2.

    R_{270} is a counterclockwise rotation of 270 degrees.

    R_{270} moves 1 to 2, 2 to 3, 3 to 4, and 4 to 1.

    F_V is a flip about the vertical axis through the centers of the top and bottom sides.

    F_V moves 1 to 2, 2 to 1, 3 to 4, and 4 to 3.

    F_H is a flip about the horizontal axis through the centers of the left and right sides.

    F_H moves 1 to 4, 4 to 1, 2 to 3, and 3 to 2.

    F_R is a flip about the diagonal axis through the upper left and lower right vertices.

    F_R moves 2 to 4 and 4 to 2; 1 and 3 are fixed.

    F_L is a flip about the diagonal axis through the lower left and upper right vertices.

    F_L moves 1 to 3 and 3 to 1; 2 and 4 are fixed.

13. R_0 is the identity.
15. The identity R_0 is its own inverse. Every flip is its own inverse. Since there are 360 degrees in one complete rotation of the plane, the inverse of a rotation must be another rotation such that the sum of their degrees is 360; that is, the inverse of R_n is R_{360-n}.
17. $C(R_0) = D_4 = C(R_{180})$;
    $C(R_{90}) = \{R_0, R_{90}, R_{180}, R_{270}\} = C(R_{270})$;
    $C(F_V) = \{R_0, R_{180}, F_V, F_H\} = C(F_H)$;
    $C(F_R) = \{R_0, R_{180}, F_R, F_L\} = C(F_L)$.

19.

$\circ$	R_0	R_{72}	R_{144}	R_{216}	R_{288}
R_0	R_0	R_{72}	R_{144}	R_{216}	R_{288}
R_{72}	R_{72}	R_{144}	R_{216}	R_{288}	R_0
R_{144}	R_{144}	R_{216}	R_{288}	R_0	R_{72}
R_{216}	R_{216}	R_{288}	R_0	R_{72}	R_{144}
R_{288}	R_{288}	R_0	R_{72}	R_{144}	R_{216}
F_1	F_1	F_5	F_4	F_3	F_2
F_2	F_2	F_1	F_5	F_4	F_3
F_3	F_3	F_2	F_1	F_5	F_4
F_4	F_4	F_3	F_2	F_1	F_5
F_5	F_5	F_4	F_3	F_2	F_1

$\circ$	F_1	F_2	F_3	F_4	F_5
R_0	F_1	F_2	F_3	F_4	F_5
R_{72}	F_2	F_3	F_4	F_5	F_1
R_{144}	F_3	F_4	F_5	F_1	F_2
R_{216}	F_4	F_5	F_1	F_2	F_3
R_{288}	F_5	F_1	F_2	F_3	F_4
F_1	R_0	R_{288}	R_{216}	R_{144}	R_{72}
F_2	R_{72}	R_0	R_{288}	R_{216}	R_{144}
F_3	R_{144}	R_{72}	R_0	R_{288}	R_{216}
F_4	R_{216}	R_{144}	R_{72}	R_0	R_{288}
F_5	R_{288}	R_{216}	R_{144}	R_{72}	R_0

21. R_0 is its own inverse; Each flip is its own inverse; R_{72} and R_{288} are inverses; R_{144} and R_{216} are inverses.

23. There are many pairs $a, b \in D_5$ such that $a \circ b \neq b \circ a$. For example, $F_1 \circ R_{72} = F_5$, but $R_{72} \circ F_1 = F_2$.

25. The centralizer of any flip in D_5 is the set containing the identity R_0 and the flip itself. For example, $C(F_1) = \{R_0, F_1\}$.

27. For all $n \neq 0$, we have both $R_n \circ R_{360-n} = R_0$ and $R_{360-n} \circ R_n = R_0$. We know that the identity R_0 is its own inverse.

29. There are n rotations in D_n. Since all interior angles of a regular polygon are identically $360/n$ degrees, each rotation must be a multiple of $360/n$ degrees in order to move each vertex to a position previously occupied by a vertex. We need n rotations of $360/n$ degrees to complete a full 360 degree rotation of the plane

31. If n is odd, then there are n flips in D_n. Each flip is about an axis passing through a vertex and the center of the opposite side. Since no vertex of a polygon with an odd number of sides is located symmetrically opposite another vertex, each vertex corresponds to a distinct flip—so n vertices give us n flips.

33. If n is even, then there are n flips in D_n, one flip across each axis

through the center of each pair of opposite parallel sides and one flip across each axis through a pair of opposite vertices. Every polygon with an even number of sides has each of its sides parallel to exactly one other side and has each vertex symmetrically opposite one other vertex across the center point. The number of paired parallel sides is $n/2$, and the number of pairs of these symmetrically opposite vertices is also $n/2$. Therefore the total number of flips is $n/2 + n/2 = n$.

35. The order of $\mathbb{Z}_5$ is 5.

37. The order of D_5 is 10.

39. The order of $U(8)$ is 4.

41. The order of $\mathbb{Z}_{11}$ is 11.

43. The order of $U(14)$ is 6.

45. $\{R_0, R_{180}, F_V, F_H\}$

47. $\{R_0, R_{180}, F_V, F_H\}$, where F_V and F_H denote flips across the diagonals connecting opposite vertices

49. $\{R_0, R_{180}, F_V, F_H\}$, where F_V and F_H denote flips across the major and minor axes of the ellipse

51. $\{R_0, R_{90}, R_{180}, R_{270}, F_V, F_H, F_R, F_L\} = D_4$

53. $\{R_0, R_{180}, F_V, F_H\}$

55. $A + B = \begin{bmatrix} 0 & 0 \\ 1 & 0 \end{bmatrix}$

57. $A + C = \begin{bmatrix} 2 & 3 \\ 2 & 5 \end{bmatrix}$

59. $A \cdot B = \begin{bmatrix} -1 & 0 \\ 1 & -1 \end{bmatrix}$

61. $B \cdot C = \begin{bmatrix} -1 & -3 \\ -1 & -1 \end{bmatrix}$

63. Matrix addition is commutative.

65. Matrix multiplication is *not* commutative. From exercises 61 and 62, we have $B \cdot C \neq C \cdot B$.

67. $\alpha \circ \alpha = \begin{bmatrix} 1 & 2 & 3 \\ 3 & 1 & 2 \end{bmatrix}$

69. $\beta \circ \alpha^2 = \begin{bmatrix} 1 & 2 & 3 \\ 2 & 1 & 3 \end{bmatrix}$

Application: Check Digit Schemes

1. 12344
3. 12844
5. no—the check digit should be 0
7. yes
9. 12341
11. 12846
13. 21352
15. yes
17. no—the check digit should be 8
19. yes
21. 2181-2389-8824-3989
23. 1234-7898-3243-3116
25. 7678-1443-3425-7682
27. no—the check digit should be 7
29. yes
31. no—the check digit should be 9
33. 1231
35. 83549
37. 53458
39. yes
41. yes

43. yes
45.

n	0	1	2	3	4	5	6	7	8	9
$f_5(n)$	4	2	8	6	5	7	3	9	0	1

47. 847,658
49. 0,123,450
51. 2-3474-9129-6
53. 0-7167-3818-X
55. 3-3458-2134-6
57. yes
59. no—the check digit should be 3
61. no—the check digit should be 4
63. yes—the unique isomorphism from $\mathbb{Z}_2$ to $U(4)$ is given by mapping 0 to 1 and 1 to 3
65. no—D_3 is nonAbelian, while $U(7)$ is Abelian
67. no—$\mathbb{Z}_3$ has order 3, while $U(8)$ has order 4
69. no—$\mathbb{Z}_6$ is Abelian, while S_3 is nonAbelian

Prime Numbers

1. 3
3. 71
5. If m divides n with quotient q, then $-m$ divides n with quotient $-q$.
7. If $mq = n$ and $nr = k$, where $q, r \in \mathbb{Z}$, then $m(qr) = k$. Since $qr \in \mathbb{Z}$, m divides k.
9. If $mq = a$ and $nr = b$, where $q, r \in \mathbb{Z}$, then $mn(qr) = ab$. Since $qr \in \mathbb{Z}$, mn divides ab.
11. Assume m divides n. By definition, $n = mq + 0$ for some $q \in \mathbb{Z}$, and so $n \bmod m = 0$ since the remainder is 0. Similarly, if $n \bmod m = 0$, then $n = mq + 0 = mq$, and so m divides n.
13. If $2q = n$ where $q \in \mathbb{Z}$, then $n^2 = (2q)^2 = 4q^2$. Since $q^2 \in \mathbb{Z}$, 4 divides n^2.
15. If p divides n^2, then p is one of the primes in the prime power

factorization of n^2. Since the primes appearing in the prime power factorization of n^2 are exactly those that appear in the prime power factorization of n, p must appear in the prime power factorization of n, and so must divide n.
17. If $pq = m$ and $pr = n$, where $q, r \in \mathbb{Z}$, then $p^4(q^4 - r^4) = m^4 - n^4$. Since $q^4 - r^4 \in \mathbb{Z}$, p^4 divides $m^4 - n^4$.
19. For any integer n, 3 divides one of $n-1, n$, or $n+1$. Hence 3 divides the product $(n-1)n(n+1) = n^3 - n$.
21. 1 divides 3 and 1 divides 2, but $2 = 1 + 1$ does not divide $5 = 3 + 2$.
23. 1 divides 2 but 2 does not divide 1.
25. For example, let $m = 2$ and $n = 3$.
27. $123 = 3 \cdot 41$
29. $1{,}225 = 5^2 \cdot 7^2$
31. $2{,}301 = 3 \cdot 13 \cdot 59$

33. 11 is prime, $12 = 2^2 \cdot 3$, 13 is prime, $14 = 2 \cdot 7$, $15 = 3 \cdot 5$, $16 = 2^4$, 17 is prime, $18 = 2 \cdot 3^2$, 19 is prime, and $20 = 2^2 \cdot 5$.

35. $18 = 2 \cdot 3^2$ and $60 = 2^2 \cdot 3 \cdot 5$.

37. $\gcd(12, 50) = 2$

39. $\gcd(31, 32) = 1$

41. p and 1

43. $\gcd(3, 8) = 1 = 3 \cdot 3 + (-1) \cdot 8$

45. $\gcd(12, 16) = 4 = (-1) \cdot 12 + 1 \cdot 16$

47. $\gcd(12, 175) = 1$

49. $\gcd(637, 26400) = 1$

51. $\gcd(517, 31891) = 1$

53. By way of contradiction: if not, then there would be a factor q of some integer n that would also factor p, contradicting the fact that p is prime.

55. If q factored n and $n + 1$, then it would also factor $(n + 1) - n = 1$.

57. If m and n are relatively prime, then they share no common primes in their prime power factorizations. Since m^2 and n^2 have the same primes (but with doubled powers) in their prime power factorizations as n and m, respectively, m^2 and n^2 can have no common primes in their prime power factorizations either.

59. $\pi(2) = 1$, $\pi(3) = 2$, $\pi(4) = 2$, $\pi(5) = 3$, $\pi(6) = 3$, $\pi(7) = 4$, $\pi(8) = 4$, $\pi(9) = 4$, $\pi(10) = 4$

61. $\pi(100) = 25$, $\pi(200) = 46$

63. $\pi(10)/10 = .4$, $1/\ln(10) \approx .434$. $\pi(1,000)/1,000 = 0.168$, $1/\ln(10) \approx .145$. $\pi(100,000)/100,000 = 0.09592$, $1/\ln(100,000) \approx 0.087$. $\pi(10,000,000)/10,000,000 \approx 0.0665$, $1/\ln(10,000,000) \approx 0.062$. $\pi(1 \text{ billion})/(1 \text{ billion}) \approx 0.0508$, $1/\ln(1 \text{ billion}) \approx 0.0483$. About $1/\ln(n)$ of the integers less than or equal to n are prime.

65. $4 = 2 + 2$, $6 = 3 + 3$, $8 = 3 + 5$, $10 = 3 + 7$, $12 = 5 + 7$, $14 = 3 + 11$, $16 = 5 + 11$, $18 = 7 + 11$, $20 = 7 + 13$, $22 = 11 + 11$, $24 = 11 + 13$, $26 = 13 + 13$, $28 = 11 + 17$, $30 = 13 + 17$, $32 = 13 + 19$.

67. $(3, 5)$, $(5, 7)$, $(11, 13)$, $(17, 19)$, $(29, 31)$, $(41, 43)$, $(59, 61)$, $(71, 73)$.

69. $2^2 < 5 < 3^2 < 11 < 4^2 < 17 < 5^2 < 29 < 6^2 < 37 < 7^2 < 59 < 8^2 < 71 < 9^2 < 83 < 10^2 < 101 < 11^2 < 127 < 12^2 < 149 < 13^2 < 173 < 14^2 < 197 < 15^2 < 227 < 16^2 < 257 < 17^2 < 293 < 18^2 < 331 < 19^2 < 367 < 20^2 < 401 < 21^2$.

3.2 Application: Introduction to Coding Theory and Cryptography

1. 01|12|07|05|02|18|01

3. 16|05|01|03|05

5. EULER

7. NEWTON

9. SPEAK TRUTH

11. WALK WITHOUT BLAME

13. 3 digits.

15. 5 digits.

17. 01|20|01|12|31|13|15|13

19. 26|06|21|14|12|27|22|14

21. HOPE

23. CHARITY

25. 0001|5646|6299|4221|1784|0036|0001

27. 1318|4221|0001|4502|4221

29. HOPE

31. JUSTICE

33. $143 = 11 \cdot 13$

35. $12,533 = 83 \cdot 151$

37. For example, let $n = 383,993$ and choose $e = 13$.

39. $2^{11} - 1 = 2,047 = 23 \cdot 89$

41. Not defined, since $2 \neq 3$.

43. [7]

45. $[110 \quad -8]$

47. $\begin{bmatrix} a & b & c \\ d & e & f \\ g & h & i \end{bmatrix}$

49. [1 1 1 0 1 0 0] [1 1 0 1 0 1 0]

51. [0 0 0 1 0 1 1] [1 0 0 1 1 0 0]
    [1 0 1 1 0 0 1]

53. Correct.

55. Incorrect. The seventh digit in the second vector should be 0.

57. The product is [1 0 0] and it recommends changing the vector to [1 0 1 1 0 0 1].

59. The product is [1 0 1] and it recommends changing the vector to [1 1 1 1 1 1 1].

61. 11

63.
$$P = \begin{bmatrix} 1 & 1 & 1 & 1 \\ 1 & 1 & 1 & 0 \\ 1 & 1 & 0 & 1 \\ 1 & 0 & 1 & 1 \\ 0 & 1 & 1 & 1 \\ 1 & 1 & 0 & 0 \\ 1 & 0 & 1 & 0 \\ 0 & 1 & 1 & 0 \\ 1 & 0 & 0 & 1 \\ 0 & 1 & 0 & 1 \\ 0 & 0 & 1 & 1 \\ 1 & 0 & 0 & 0 \\ 0 & 1 & 0 & 0 \\ 0 & 0 & 1 & 0 \\ 0 & 0 & 0 & 1 \end{bmatrix}$$

65. 57

67. [1 0 1 0 1 0 1 0 1 0 1 0 1 0 0]

69. Correct.

3.3   From the Pythagorean Theorem to Fermat's Last Theorem

1. Yes.

3. No.

5. Yes.

7. 65

9. 9

11. $a = 3, b = 4$

13. $a = 20, c = 29$

15. Assuming $0 < a < b < c, c = 5$.

17. $n = 4$: $(3, 4, 5)$. $n = 6$: $(6, 8, 10)$. $n = 8$: $(8, 15, 17)$.

19. $n = 3$: $(3, 4, 5)$. $n = 5$: $(5, 12, 13)$. $n = 7$: $(7, 24, 25)$.

21. $m = 1, n = 2$: $(3, 4, 5)$. $m = 2, n = 3$: $(5, 12, 13)$. $m = 2, n = 4$: $(12, 16, 20)$.

23. A pythagorean triple is (ruw, stw, suv).

25. Since $p^2 + q^2 = 2^2$, p and q are at most 1, and so cannot be prime.

27. Exercises 24–26 rule out the possible cases: that p, q, and r are odd primes, or that any one of them is 2 and the others are odd primes.

29. $(AC)^2 = AH \cdot AB$

31. $(BC)^2 = AB \cdot BH$

33. Since triangle ABC is similar to ACH, the ratios of corresponding sides implies $AC/AB = AH/AC$, and so $(AC)^2 = AH \cdot AB$. Similarly, triangle ABC is similar to BCH, and so $BC/AB = BH/BC$, which means $(BC)^2 = AB \cdot BH$. Adding, $(AC)^2 + (BC)^2 = AH \cdot AB + AB \cdot BH = AB(AH + BH) = (AB)^2$.

35. Every odd number appears in the right column.

37. $(3, 4, 5)$ and $(5, 12, 13)$

39. Odd integers have odd squares.

41. There are no positive integers x and y that solve $3x + 5y = 12$.

43. There are no positive integers x and y that solve $8x + 5y = 1$.

45. $x = 21, y = 10$ and $x = 10, y = 20$ are the only two solutions.

47. $x = 1, y = 3$ is the only solution.

49. $3^2 - 2 \cdot 2^2 = 9 - 8 = 1$

51. Suppose x is even. Then $x^2 - 8y^2$ has a factor of 4, and so cannot equal 1.

53. $2^3 - 7 \cdot 1^3 = 8 - 7 = 1$

55. If x and y are both even, then $x^3 - 7y^3$ has a factor of 8, and so cannot equal 1. If $x = 2k - 1$ and $y = 2j - 1$, then $x^3 - 7y^3 = (2k - 1)^3 - 7(2j - 1)^3 = 8(k^3 - 7j^3) - 4(k^2 - 7j^2) + 2(k - 7j) + 6$, which is even and cannot equal 1.

57. For example, $a = 2, b = \sqrt[3]{56}, c = 4$.

59. If $1,701 = r^3/s^3$, then $r^3 + (bs)^3 = (cs)^3$.

61. $x = ruw$, $y = stw$, and $z = suv$. If $a^n + b^n = c^n$, where $n > 2$, then at least one of a, b, or c is irrational.

63. The prime power factorization of $(c^2 + b^2)(c^2 - b^2)$ is the same as the prime power factorization of a^2, and so all its exponential powers must be even. Since $\gcd(c^2 + b^2, c^2 - b^2) = 1$, any term of the form p^{2n} in this prime power factorization must appear in the prime power factorization of one of $c^2 + b^2$ or $c^2 - b^2$. Hence the exponential powers of the prime power factorization of each of these factors must be even, and so the factors are each expressible as squares.

65. Since $2uv = b^2$, the prime power factorization of b^2 contains a term of the form 2^{2n}, where $n \in \mathbb{N}$, and so either u or v has a factor of 2. Without loss of generality, $u = 2k$, where $k \in \mathbb{N}$. If any prime p divides both u and v, then p divides $2u + 2v = (s + t) + (s - t) = 2s$ and $2u - 2v = 2t$, and so p would divide s and t, which contradicts $\gcd(s, t) = 1$.

67. $u^2 + v^2 = c^2$ implies $v^2 = c^2 - u^2 = (c - u)(c + u) = e^2f^2$, hence $v = ef$. But $x^2 - y^2 = ((e + f)/2)^2 - ((e - f)/2)^2 = ef$, and so $v = ef = x^2 - y^2$.

69. Assume there exist $k, j \in \mathbb{N}$ such that $x = pk$ and $y = pj$. Then $v = x^2 - y^2 = p^2(k^2 - j^2)$, and so p divides v. Similarly, if p divides both e and f, then p^2 divides $2c = e^2 + f^2$, and so p divides c. Since p divides v, it divides $2uv = b^2$, and so p divides b. But then $\gcd(b, c) \neq 1$, a contradiction.

| 3.4 | **Irrational Numbers and Fields** |

1. $173/500$

3. $953/1,000$

5. $7/9$

7. $25/99$

9. If the digits 1828 repeated indefinitely without interruption, then e would be rational.

11. For example, $1, 2, 3, 4, \cdots$.

13. If $n = 2 \cdot q$ where $q \in \mathbb{Z}$, then $n^2 = 4 \cdot q^2$. Since $q^2 \in \mathbb{Z}$, 4 divides n^2.

15. The prime power factorizations of n and n^2 contain exactly the same primes.

17. The prime power factorizations of n and n^k contain exactly the same primes.

19. Proof by contradiction. Assume there exist integers m and n such that $\sqrt{7} = m/n$, where m and n have no common factors. Then $m^2 = 7n^2$, and 7 divides m^2. Hence 7 divides m by the uniqueness of prime power factorizations from the fundamental theorem of arithmetic, and $m = 7k$ for $k \in \mathbb{Z}$. Then $49k^2 = 7n^2$, and 7 divides n^2. As above, 7 divides n,

and so 7 is a factor of both m and n, a contradiction.

21. Proof by contradiction. Assume there exist integers m and n such that $\sqrt{6} = m/n$, where m and n have no common factors. Then $m^2 = 6n^2$, and 6 divides m^2. By the uniqueness property from the fundamental theorem of arithmetic, the prime power factorization of m contains the primes 2 and 3, which must both divide m; that is, we have $m = 2 \cdot 3 \cdot k$ for some $k \in \mathbb{Z}$. Then $36k^2 = 6n^2$, and 6 divides n^2. As above, both 2 and 3 divide n, and so 6 is a common factor of both m and n, a contradiction.

23. There exist integers m and n such that $\sqrt{pq} = m/n$, where m and n have no common factors. Then $m^2 = pqn^2$, and pq divides m^2. Hence the prime power factorization of m contains the primes p and q, which must both divide m; we have $m = p \cdot q \cdot k$ for $k \in \mathbb{Z}$. Then $p^2q^2k^2 = pqn^2$, and pq divides n^2. Then both p and q divide n, and so m/n is not in lowest terms, a contradiction.

25. There exist integers m and n such that $\sqrt[k]{2} = m/n$, where m and n have no common factors. Then $m^k = 2n^k$, and 2 divides m^k. Hence 2 divides m (by exercise 17) and $m = kq$ for $q \in \mathbb{Z}$. Then $2^kq^k = 2n^k$, and 2 divides n^k. Hence 2 divides n, and so 2 is a factor of both m and n, a contradiction.

27. For example, $\sqrt{p+q}$ is rational for $p = 2$ and $q = 7$, while $\sqrt{p+q}$ is irrational for $p = 2$ and $q = 3$.

29. $\sqrt{2}$

31. Does not exist—every irrational number is complex.

33. Many answers are possible. The quaternions extend the complex numbers. Alternatively, consider the modular numbers in such sets as $\mathbb{Z}_4$ or $U(6)$.

35. sum $= 7 - 2i$; product $= 13 - i$

37. sum $= 12 + 45i$; product $= 540i$

39. sum $= 2a$; product $= a^2 + b^2$

41. The additive inverse is $-7 + 2i$; the multiplicative inverse is $(7/53) + (2/53)i$.

43. The additive inverse is $-i$; the multiplicative inverse is $-i$.

45. The additive inverse is 2; the multiplicative inverse is $-\frac{1}{2}$.

47. $\left(\dfrac{1}{\sqrt{2}} + \dfrac{1}{\sqrt{2}}i\right)^2 = i$ by direct multiplication.

49. The additive inverse is $(-a) + (-b)\sqrt{3}$, where $-a$ and $-b$ are the additive inverses of a and b, respectively, in $\mathbb{Z}_5$. The multiplicative inverse is $a(a^2 - 3b^2)^{-1} - b(a^2 - 3b^2)^{-1}\sqrt{3}$, where c^{-1} stands for the multiplicative inverse of c in $U(5)$.

51. The additive inverse is $(-a) + (-b)\sqrt{3}$, where $-a$ and $-b$ are the additive inverses of a and b, respectively, in $\mathbb{Q}$. The multiplicative inverse is $\dfrac{\sqrt{3} \cdot (a - b)}{a^2 - 3b^2}$.

53.

+	0	1	2	3	4
0	0	1	2	3	4
1	1	2	3	4	0
2	2	3	4	0	1
3	3	4	0	1	2
4	4	0	1	2	3

×	0	1	2	3	4
0	0	0	0	0	0
1	0	1	2	3	4
2	0	2	4	1	3
3	0	3	1	4	2
4	0	4	3	2	1

55.

+	0	2	4	6	8
0	0	2	4	6	8
2	2	4	6	8	0
4	4	6	8	0	2
6	6	8	0	2	4
8	8	0	2	4	6

×	0	2	4	6	8
0	0	0	0	0	0
2	0	4	8	2	6
4	0	8	6	4	2
6	0	2	4	6	8
8	0	6	2	8	4

57. Since $\mathbb{N}$ does not contain 0, the additive identity property fails, which also causes the additive inverse property to fail.

59. For $k \neq 1$, there is no multiplicative identity; this also causes the multiplicative inverse axiom to fail.

61. The additive closure axiom fails.

63. Elements 2, 3, and 4 have no multiplicative inverse.

65. Elements generally do not have a multiplicative inverse.

67. If $r, s \in \mathbb{R}$, then $(-r) \cdot s = -(r \cdot s)$ and $r \cdot (-s) = -(r \cdot s)$, the unique additive inverse of $r \cdot s$. To prove $(-r) \cdot s$ is the additive inverse of $r \cdot s$, we compute $r \cdot s + (-r) \cdot s = (r + (-r)) \cdot s = 0 \cdot s = 0$. Similarly, $r \cdot (-s)$ is the additive inverse of $r \cdot s$.

69. Proof that $r < 0$ implies $-r > 0$: if $r < 0$, then the fact that addition preserves order implies $r + (-r) < 0 + (-r)$, and so $0 < -r$. Proof that $r > 0$ implies $-r < 0$: similarly, using the fact that addition preserves order implies $r + (-r) > 0 + (-r)$, and so $0 > -r$ (that is, $-r < 0$).

3.5 | Polynomials and Transcendental Numbers

1. 2

3. Not a polynomial due to the -1 exponent.

5. Not a polynomial due to the tangent term.

7. Yes.

9. No.

11. Yes.

13. 1, 2, 4, and 5 are zeros over $\mathbb{Z}_6$.

15. Since $x > 0$, $x + 3 > 0$ for any $x \in \mathbb{N}$.

17. It is a zero of $5x - 4$.

19. It is a zero of $x^3 - 5$.

21. It is a zero of $5x^4 - 14$.

23. It is a zero of $x - a$.

25. By way of contradiction: If it were not, then it would be algebraic, and so there would be a polynomial $p(x)$ with coefficients in $\mathbb{Q}$ such that $p(\pi/2) = 0$. But then the polynomial $q(x) = p(x/2)$, which has coefficients in $\mathbb{Q}$, would satisfy $q(\pi) = 0$, which would mean that π is algebraic, a contradiction.

27. If it were, then π would be a zero of $x^2 - ax - b$, which would imply that π is algebraic.

29. Let $a = e$ and $b = c = 0$.

31. 2

33. 2

35. $\dfrac{-6}{5} + \dfrac{7}{5}i$

37. $\dfrac{-5}{12} - \dfrac{3}{4}i$

39. -5 and 3

41. $\dfrac{-1 \pm \sqrt{5}i}{3}$

43. Use $x^3 - a^3 = (x - a)(x^2 + ax + a^2)$. The zeros are 2 and $-1 \pm i\sqrt{3}$.

45. 4 and $-2 \pm i\sqrt{3}$

47. -4 and 2 (which has multiplicity 2).

49. 1 (which has multiplicity 3).

51. $x^3 - 7x^2 + x - 7 = (x - 7)(x^2 + 1)$ has roots $7, i, -i$.

53. 7 and $4 \pm 3i$.

55. $x_1 = -\sqrt[3]{1/2} \cdot [\sqrt[3]{7 - \sqrt{549}} + \sqrt[3]{7 + \sqrt{549}}]$

57. $x_1 = \sqrt[3]{1 + \sqrt{2}} - \sqrt[3]{-1 + \sqrt{2}}$

59. $y^3 - (103/27)y + (848/729)$, where $y = x + 7/9$.

61. $y^3 + (167/27)y - (934/729)$, where $y = x - 2/9$.

63. $3 + 2i + 3j + 8k$

65. $-1 + 2i - 7j - 6k$

67. $3 + 12i + 40j + 10k$

69. $91 + 29i - 12j + 3k$

3.6	**Mathematical Induction**

1. $4 = 2 + 2$, $6 = 3 + 3$, $8 = 3 + 5$, $10 = 3 + 7$, $12 = 5 + 7$, $14 = 3 + 11$, $16 = 5 + 11$, $18 = 7 + 11$, $20 = 7 + 13$.

3. $4 = 2 \cdot 2$, $6 = 2 \cdot 3$, $8 = 2 \cdot 4$, $10 = 2 \cdot 5$, $12 = 2 \cdot 6$, $14 = 2 \cdot 7$, $16 = 2 \cdot 8$, $18 = 2 \cdot 9$, $20 = 2 \cdot 10$.

5. $3, 5, 7, 11, 13, 17$, and 19 are prime; 9 and 15 are composite.

7. $2^2 = 4 \le 2^2$, $2^3 = 8 \le 9 = 3^2$, and $2^4 = 16 \le 4^2$.

9. $-0 = 0$, $-1 = 6$, $-2 = 5$, $-3 = 4$, $-4 = 3$, $-5 = 2$, and $-6 = 1$.

11. Base case: $1 = 1$. If $\sum_{i=1}^{n} 1 = n$, then $\sum_{i=1}^{n+1} 1 = \sum_{i=1}^{n} 1 + (1) = n + 1$.

13. Base case: $2^1 = 2 = 4 - 2 = 2^2 - 2$. If $\sum_{i=1}^{n} 2^i = 2^{n+1} - 2$, then $\sum_{i=1}^{n+1} 2^i = \sum_{i=1}^{n} 2^i + (2^{n+1}) = 2^{n+1} - 2 + 2^{n+1} = 2^{(n+1)+1} - 2$.

15. Base case: $1 \cdot (1!) = 1 = 2! - 1$. If $\sum_{i=1}^{n} i \cdot (i!) = (n + 1)! - 1$, then $\sum_{i=1}^{n+1} i \cdot (i!) = \sum_{i=1}^{n} i + (n + 1)(n + 1)! = (n + 1)! - 1 + (n + 1)(n + 1)! = (n+1)!(1 + n + 1) - 1 = (n+2)! - 1$.

17. Base case: $1^3 = 1 = 1^2 \cdot 2^2/4$. If $\sum_{i=1}^{n} i^3 = n^2(n + 1)^2/4$, then $\sum_{i=1}^{n+1} i^3 = \sum_{i=1}^{n} i^3 + (n + 1)^3 = n^2(n + 1)^2/4 + (n + 1)^3 = (n + 1)^2((n + 1) + 1)^2/4$.

19. Base case: $1^2 = 1 = 2 \cdot 1 - 1$. If $\sum_{i=1}^{n}(2i - 1) = n^2$, then $\sum_{i=1}^{n+1}(2i - 1) = \sum_{i=1}^{n}(2i - 1) + 2(n + 1) - 1 = n^2 + 2n + 1 = (n + 1)^2$.

21. Base case: $4(1) - 3 = 1 = 1(2(1) - 1)$. If $\sum_{i=1}^{n}(4i - 3) = n(2n - 1)$, then $\sum_{i=1}^{n+1}(4i - 3) = \sum_{i=1}^{n}(4i - 3) + 4(n + 1) - 3 = n(2n - 1) + 4n + 1 = (n + 1)(2(n + 1) - 1)$.

23. Base case: $1(1 + 1) = 2 = 1(1 + 1)(1 + 2)/3$. If $\sum_{i=1}^{n} i(i + 1) = n(n + 1)(n + 2)/3$, then $\sum_{i=1}^{n+1} i(i + 1) = \sum_{i=1}^{n} i(i + 1) + (n + 1)(n + 2) = n(n + 1)(n + 2)/3 + (n + 1)(n + 2) = (n + 1)(n + 2)(n + 3)/3$.

25. Base case: $\dfrac{1}{1 \cdot 3} = 1/3$. If $\sum_{i=1}^{n} \dfrac{1}{(2i - 1)(2i + 1)} = n/(2n + 1)$,

then $\displaystyle\sum_{i=1}^{n+1} \frac{1}{(2i-1)(2i+1)} =$

$\displaystyle\sum_{i=1}^{n} \frac{1}{(2i-1)(2i+1)} + \frac{1}{(2n+1)(2n+3)}$

$= n/(2n+1) + \dfrac{1}{(2n+1)(2n+3)}$

$= \dfrac{n(2n+3)+1}{(2n+1)(2n+3)} = (n+1)/$

$(2(n+1)+1)$.

27. Base case: $5^2 = 25 < 32 = 2^5$. If $n^2 < 2^n$, then $(n+1)^2 = n^2 + 2n + 1 < 2^n + 2n + 1 < 2^n + 2^n$ by example 3. Hence $(n+1)^2 < 2 \cdot 2^n = 2^{n+1}$.

29. Base case: $3(2)+1 = 7 < 9 = 3^2$. If $3n + 1 < 3^n$, then $3(n+1) + 1 = 3n + 1 + 3 < 3^n + 3 < 3^n + 3^n$, since $3 < 3^n$ for any $n \geq 1$. Hence $3(n+1) + 1 < 2 \cdot 3^n < 3 \cdot 3^n = 3^{n+1}$.

31. Base case: $3^7 = 2187 < 5040 = 7!$. If $3^n < n!$, then $3^{n+1} = 3 \cdot 3^n < 3 \cdot n! < (n+1) \cdot n!$ for $n \geq 7$. Hence $3^{n+1} < (n+1)!$.

33. Base case: $3^{2 \cdot 0} - 1 = 1 - 1 = 0$ is divisible by 8, since 0 is divisible by any nonzero integer. If $3^{2n} - 1$ is divisible by 8, then $3^{2(n+1)} - 1 = 9 \cdot 3^{2n} - 1 = 9 \cdot (3^{2n} - 1) + 8$ is divisible by 8.

35. Base case: $7^0 - 2^0 = 1 - 1 = 0$ is divisible by 5, since 0 is divisible by any nonzero integer. If $7^n - 2^n$ is divisible by 5, then $7^{n+1} - 2^{n+1} = 7 \cdot 7^n - 2 \cdot 2^n = 7(7^n - 2^n) + 5 \cdot 2^n$ is divisible by 5.

37. Base case: $1^3 - 1 = 0$, which is divisible by 3. If $n^3 - n$ is divisible by 3, then $(n+1)^3 - (n+1) = n^3 + 3n^2 + 3n + 1 - n - 1 = (n^3 - n) + 3n^2 + 3n$ is divisible by 3.

39. Strong induction. Base case: $x^2 - y^2 = (x+y)(x-y)$ and $x^4 - y^4 = (x+y)(x-y)(x^2+y^2)$ are divisible by $x + y$. If $x^{2k} - y^{2k}$ is divisible by $x + y$ for $k = 1, 2, \ldots, n$, then

$x^{2(n+1)} - y^{2(n+1)} = (x^2 + y^2)(x^{2n} - y^{2n}) - x^2 y^2 (x^{2(n-1)} - y^{2(n-1)})$, where each term on the right is divisible by $x + y$.

41. Strong induction. Base case: $a_2 = 2$ and $a_3 = 2(1) + 2$ are even. If a_k is even for $k = 2, 3, \ldots, n+1$, then $a_{n+2} = 2a_n + a_{n+1}$ is even, since each term on the right is even.

43. Strong induction. Base case: $b_1 = 4$ and $b_2 = 8$ are even. If b_k is even for $k = 2, 3, \ldots, n+1$, then $b_{n+2} = b_n + b_{n+1}$ is even, since each term on the right is even.

45. Strong induction. Base case: $c_1 = 1 < 3^1$, $c_2 = 1 < 3^2$, and $c_3 = 3 < 3^3$. If $c_k < 3^k$ for $k = 1, 2, \ldots, n+2$, then $c_{n+3} < 3^n + 3^{n+1} + 3^{n+2} = 3^n(1 + 3 + 9) < 3^n \cdot 3^3 = 3^{n+3}$.

47. Base case: $d_1 = 2 \leq 3^1$. If $d_n \leq 3^n$, then $d_{n+1} = 3d_n \leq 3 \cdot 3^n = 3^{n+1}$.

49. Base case: $d_1 = 2$ is even. If d_n is even, then $d_{n+1} = 3d_n$ is even.

51. Base case: $e_1 = 3 = 3 + 2(1 - 1)$. If $e_n = 3 + 2(n - 1)$, then $e_{n+1} = 2 + e_n = 2 + 3 + 2(n - 1) = 3 + 2n$.

53. Strong induction. Base case: $f_1 = 1 \leq 2^1$ and $f_2 = 1 \leq 2^2$. If $f_k \leq 2^k$ for $k = 1, 2, \ldots, n+1$, then $f_{n+2} \leq 2^n + 2^{n+1} = 2^n(1 + 2) < 2^n 2^2 = 2^{n+2}$.

55. Base case: $f_1 + f_3 = 1 + 2 = 3 = f_4$. If $f_1 + f_3 + \ldots + f_{2n-1} = f_{2n}$, then $f_1 + f_3 + \ldots + f_{2n+1} = f_{2n} + f_{2n+1} = f_{2n+2}$.

57. Base case: $L_1 = 2 \leq 2^1$. If $L_n \leq 2^n$, then $L_{n+1} = L_n + L_{n-1} \leq 2^n + 2^{n-1} = 2^{n-1}(2 + 1) < 2^{n-1} \cdot 2^2 = 2^{n+1}$.

59. Strong induction. Base case: $L_3 = 3 = 2(1) + 1 = 2f_1 + f_2$ and $L_4 = 4 = 2(1) + 2 = 2f_2 + f_3$. If $L_{k+2} = 2f_k + f_{k+1}$, then $L_{n+3} = L_{n+1} + L_{n+2} = 2f_{n-1} + f_n + 2f_n + f_{n+1} = 2f_{n+1} + f_{n+2}$.

61. Base case: $6 = 2(-2) + 5(2)$. If $n = 2s + 5t$, then $n + 1 = 2s + 5t + 2(-2)2 + (1)5 = 2(s - 2) + 5(t + 1)$.

63. Base case: $1 - 1/2^2 = 3/4 = (2+1)/(2 \cdot 2)$. If $\prod_{i=2}^{n}(1 - 1/i^2) = (n+1)/(2n)$, then $\prod_{i=2}^{n+1}(1 - 1/i^2) = [1 - 1/(n+1)^2] \cdot \prod_{i=2}^{n}(1 - 1/i^2) = \dfrac{((n+1)^2 - 1)(n+1)}{(n+1)^2(2n)} = \dfrac{n+2}{2(n+1)}$.

65. Base case: If A is a set with one element, then $\mathbb{P}(A) = \{\emptyset, A\}$ has $2 = 2^1$ elements. Suppose any set A containing n elements has $\mathbb{P}(A)$ containing 2^n elements. A set B containing $n + 1$ elements may be written as $B = C \cup \{b\}$, where $b \in B$ and C has n elements. Then $\mathbb{P}(B)$ consists of the elements in $\mathbb{P}(C)$ along with sets of the form $D \cup \{b\}$, where $D \in \mathbb{P}(C)$. Hence $\mathbb{P}(B)$ has $2^n + 2^n = 2^{n+1}$ elements.

67. Base case: $(d/dx)(x) = 1 = 1x^{1-1}$. If $(d/dx)(x^n) = nx^{n-1}$, then by the product rule $(d/dx)(x^{n+1}) = x \cdot nx^{n-1} + 1 \cdot x^n = (n+1)x^n$.

69. Let n be the number of connectives in a sentence. Base case: If a sentence has $n = 0$ connectives, then it has no right parentheses. Assume that a sentence with $n = k$ connectives has k right parentheses. Then a sentence with $n = k + 1$ connectives has one of the forms $(\sim \mathbb{B})$, $(\mathbb{B} \wedge \mathbb{C})$, $(\mathbb{B} \vee \mathbb{C})$, $(\mathbb{B} \to \mathbb{C})$, or $(\mathbb{B} \leftrightarrow \mathbb{C})$, where $\mathbb{B}$ and $\mathbb{C}$ together have k connectives. Hence each of these forms adds one right parenthesis; the total number of right parentheses must be $k + 1$.

4.1 Analytic Geometry

1. The point (π, e), for example, is found by going π units (approximately 3.14) along the x axis and e units (approximately 2.718) along the y axis.

3. All but one of these points all lie collinearly along the line $y = x$; the point $(-2, -1)$ is found by going leftward 2 units along the x axis and downward 1 unit along the y axis.

5. The equation is $y = 2$.

7. The equation is $y = -x + 1$.

9. $(x - 1)^2 + (y - 2)^2 = 36$

11. $x^2 + (y - 2)^2 = 4$

13. $y = x^2/4$

15. $x^2/100 + y^2/75 = 1$

17. $x^2/4 - y^2/21 = 1$

19. The plane cuts diagonally across one of the cones at an angle steeper than the side of the cone and, say, above the common vertex.

21. The plane cuts diagonally across one of the cones at an angle less steep than the side of the cone and, say, above the common vertex.

23. The plane cuts horizontally across the double-napped cone and intersects it at the common vertex.

25. The plane cuts vertically straight down through the double-napped cone.

27. The focus is $(0, 5/4)$ and the directrix is $y = 3/4$.

29. The point $(0, 1)$ is closest.

31. Changes in e stretch the ellipse outward from a circular form either along the x axis or along the y axis. If e is close to 0, then the ellipse is close to circular. If e is close to 1, then the ellipse is stretched broadly in the y axis direction.

33. $y = 5x - 4$.

35. $y = -(5/4)x + 0.5$

37. $y = (1/4)x + 17/4$

39. $(0, 1)$ and $(1, 2)$

41. There are no intersection points.

43. $(\sqrt{(2/3)}, 1)$ and $(-\sqrt{(2/3)}, 1)$

45. $\{(-4, 4), (-3, 2), (-1.5, 1), (0, .5),$
    $(1.5, 1), (3, 2), (4, 4)\}$

47. The eight points include, for example, the point found by moving leftward three units along the x-axis and downward 54 units along the y-axis, as well as the origin.

49. $y = 2x^3$

51. They fit approximately into a circular (or elliptical) pattern.

53. $2x + 4y = 5$

55. $2x + 2y + 2z = 9$

57. $3x + 2y + 6z = 6$

59. The resulting equation is $((a^2 + c^2)z^2 + 2(by - c)z) + ((a^2 + b^2)y^2 - 2by) = a^2 - 1$. Completing the square both in z and in y will result in

either an ellipse, a circle, or a single point.

61. $a = b = 0$; the plane is then level and intersects the unit sphere at its "north pole."

63. The point $(\hat{x}, \hat{y})$ is on the line $y = mx + b$, and so $\hat{x} = 1/m(\hat{y} - b)$. The perpendicular line segment has slope $-1/m$, and so $-1/m(\hat{x} - x_0) = \hat{y} - y_0$. Substituting the expression for $\hat{x}$ and algebraically rearranging the terms produces the result. When solved for $\hat{y}$, the expression is
$$\hat{y} = \frac{m^2 y_0 + mx_0 + b}{m^2 + 1}.$$

65. $\cos\theta = $ adj/hyp $= AB/AC = AD/AE = AD/1 = AD$

67. $\tan\theta = $ opp/adj $= BC/AB = DE/AD$

69. $\csc\theta = $ hyp/opp $= AC/BC = AE/DE = 1/DE$

4.2   Functions and Inverse Functions

1. Not a function since the domain value 2 appears in two ordered pairs.

3. Yes, a function.

5. Yes, a function.

7. Not a function. For example, $x = 4$ has two output values $y = 2$ and $y = -2$.

9. Yes, a function; each x value in the domain has only one corresponding y range value given by $y = \sqrt{1 - x^2}$.

11. Not a function. For example $x = 0.5$ has two output values $y = \pm\sqrt{3}/2$.

13. For example, $f : A \rightarrow B$, where $f(1) = 4, f(2) = 5,$ and $f(3) = 6$.

15. For example, $f : A \rightarrow C$, where $f(1) = 7, f(2) = 7,$ and $f(3) = 8$.

17. The function is onto. For any $y \in \mathbb{R}$, $x = (1/2)(y - 7)$ has $f(x) = y$.

19. The function is onto. For any $y \in \mathbb{R}$, $x = \sqrt[3]{y + 1}$ has $h(x) = y$.

21. The function is not onto; for example, $y = 2$ is not in the range.

23. The inverse is $f^{-1}(x) = (1/5)(x + 2)$ with domain $= \mathbb{R}$ and range $= \mathbb{R}$.

25. The inverse is $h^{-1}(x) = \sqrt{x - 12} - 3$ with domain $= \{x \in \mathbb{R} : x \geq 21\}$ and range $= \{x \in \mathbb{R} : x > 0\}$.

27. The inverse is $k^{-1}(x) = \sqrt{\ln x}$ with domain $= \{x \in \mathbb{R} : x \geq 1\}$ and range $= \{x \in \mathbb{R} : x > 0\}$.

29. The inverse is $q^{-1}(x) = 1/(2x) - 3/2$ with domain $= \{x \in \mathbb{R} : x < 0\}$ and range $= \{x \in \mathbb{R} : x < -3/2\}$.

31. For example, restrict the domain to $D = \{x : -\pi/2 \leq x \leq \pi/2\}$.

33. For example, restrict the domain to $D = \{x : -\pi/2 \leq x \leq \pi/2\}$.

35. The graph of f is a line with slope 1 and y intercept 5, while f^{-1} is a line with slope 1 and y intercept -5.

37. With domain restricted to $[0, \infty)$, the graph of h is the right half of a parabola extending upward along the y axis, while h^{-1} is the top half of a parabola extending rightward.

39. The graph of r is a standard exponential graph; r^{-1} is its reflection across the line $y = x$.

41. $f(g(x)) = \sqrt{5/(2x+1) - 1}$, whose domain is the interval $D = (-1/2, 2]$. Similarly, $g(f(x)) = 1/(2\sqrt{5x-1} + 1)$, whose domain is $D = \{x : x \geq 1/5\}$.

43. $f(g(x)) = \ln(5/(x+2) - 1)$, whose domain is $D = (-\infty, -2) \cup (-2, 3)$. Similarly, $g(f(x)) = 1/(\ln(5x-1) + 2)$, whose domain is $D = \{x : x > 1/5 \text{ and } x \neq (e^{-2}+1)/5\}$.

45. True.

47. False; for example, $f(x) = x$ and $g(x) = -x$ are both onto $\mathbb{R}$, but $(f+g)(x) = 0$ is not.

49. False; for example, $f(x) = x$ is onto $\mathbb{R}$, but $f(x) \cdot f(x) = x^2$ is not.

51. False; for example, $f(x) = x^2$ does not satisfy this property.

53. False; the inverse of the polynomial function $p(x) = x$ is itself.

55. False; it equals x.

57. Proceed by way of contradiction; assume g is not onto. Then there exists $y \in C$ where $y \neq g(t)$ for any $t \in B$. Since $f(x) \in B$ for every $x \in A$, there cannot exist $x \in A$ with

$g(f(x)) = y$, contradicting the fact that $g \circ f$ is onto.

59. Assume $a, b \in \mathbb{R}$ with $f(a) = f(b)$. Then $g(f(a)) = g(f(b))$. Since $g \circ f$ is one-to-one, this equality implies $a = b$. Hence f is one-to-one.

61. Assume $a, b \in \mathbb{R}$ with $g(f(a)) = g(f(b))$. Since g is one-to-one, this equality implies $f(a) = f(b)$. Since f is one-to-one, this equality implies $a = b$. Hence $g \circ f$ is one-to-one.

63. For any element x in the range R of f, define a function as $g(x) = y$ when $f(y) = x$. Because f is one-to-one, any element $x \in R$ appears exactly once in the set of ordered pairs $(x, g(x))$ (which make up the function g). Hence g is defined properly as a function. For any y in the domain of f, $g(f(y)) = g(x) = y$, and for any $x \in R$, $f(g(x)) = f(y) = x$. Hence g is the inverse function f^{-1}.

65. Suppose g and f are inverse functions for f. For every y in the range R of f, there exists an x in the domain of f where $y = f(x)$. Hence $g(y) = g(f(x)) = x$ and $h(y) = h(f(x)) = x$. Thus $g(y) = h(y)$ for every $y \in R$, and so $g = h$.

67. $f^{-1}(g^{-1}(g(f(x)))) = f^{-1}(f(x)) = x$ and $g(f(f^{-1}(g^{-1}(x)))) = g(g^{-1}(x)) = x$.

69. Since the function is nonconstant, $m \neq 0$, and so $f^{-1}(x) = (x-b)/m$.

4.3 Limits and Continuity

1. 20

3. ∞

5. The limit does not exist, since the left and right limits are not equal.

7. Given any $\varepsilon > 0, 0 < |x-2| < \varepsilon/7$ implies $|(7x-8) - 6| = 7|x-2| < 7 \cdot (\varepsilon/7) = \varepsilon$.

9. Given any $\varepsilon > 0, 0 < |x-a| < \varepsilon/|m|$ implies $|(mx+b) - (ma+b)| = |m| \cdot |x-a| < |m| \cdot (\varepsilon/|m|) = \varepsilon$.

11. Given any $\varepsilon > 0$, $0 < |x| < b/a$ and $0 < |x| < \varepsilon/(2b)$ together imply $|(ax^2+bx+c) - c| = |x| \cdot |ax+b| < \varepsilon/(2b) \cdot (a(b/a) + b) = \varepsilon$.

13. Given any $\varepsilon > 0$, the definition is satisfied by choosing $\delta = \varepsilon/2$.

15. Given any $\varepsilon > 0$, the definition is satisfied by choosing $\delta = \min\{1, \varepsilon/5\}$.

17. Given any $\varepsilon > 0$, the definition is satisfied by choosing $\delta = \min\{1, \varepsilon(\sqrt{8} + 3)/3\}$.

19. 9

21. 3

23. 1/2

25. 1

27. 8

29. -5

31. $\sqrt{14}$

33. Given any $\varepsilon > 0$, the definition is satisfied by choosing $\delta = \varepsilon$.

35. Given any $\varepsilon > 0$, the definition is satisfied by choosing $\delta = \sqrt{\varepsilon}$.

37. We show f is continuous at any $a \in \mathbb{R}$. Case 1: Suppose $a > 0$. Then the definition is satisfied by choosing $\delta = \min\{a/2, \varepsilon\}$. Case 2: Suppose $a < 0$. Then the definition is satisfied by choosing $\delta = \min\{-a/2, \varepsilon\}$. See exercise 34 for the case of $a = 0$.

39. The function is continuous at all real x except $x = -1$.

41. The function is continuous at all real x except $x = 2$.

43. Given any $M > 0$, the definition is satisfied by choosing $\delta = \min\{1, \sqrt{4/M}\}$.

45. Given any $\varepsilon > 0$, the definition is satisfied by any choice of $\delta > 0$.

47. Given any $\varepsilon/2 > 0$, there exist $\delta_f > 0$ and $\delta_g > 0$ that satisfy the definition for f and g, respectively. Then $\delta = \min\{\delta_f, \delta_g\}$ satisfies the definition for $f - g$.

49. A polynomial $p(x)$ is the sum of terms that are scalar multiples of variable powers. Applying the limit of a scalar multiple rule, the limit of a sum rule, and the limit of a product rule, along with the fact that $\lim_{x \to c} x = c$, we see that $\lim_{x \to c} p(x) = p(c)$.

51. Let $a = 0$ and
$$f(x) = \begin{cases} 1 & \text{if } x \geq 0 \\ -1 & \text{if } x < 0 \end{cases}.$$

53. Assume $L > 0$ (the case $L < 0$ is similar). Given $M > 0$, the value $M(L^2 + 1)/L > 0$, and so there exists δ_f such that $0 < |x - a| < \delta_f$ implies $f > M(L^2 + 1)/L$. Setting $\varepsilon = 1/L$, there exists δ_g such that $0 < |x - a| < \delta_g$ implies $|g(x) - L| < 1/L$. This last inequality implies $1/g(x) > L/(L^2 + 1)$. Choosing $\delta = \min\{\delta_f, \delta_g\}$, we have
$$f(x)/g(x) > \frac{M(L^2 + 1)}{L} \cdot \frac{L}{L^2 + 1} = M$$
whenever $0 < |x - a| < \delta$.

55. For both parts, let $a = 0$ and $f(x) = \begin{cases} x & \text{if } x \neq 0 \\ 1 & \text{if } x = 0 \end{cases}.$

57. This fact follows from the limit of a constant rule in theorem 4.3.2.

59. Let f and g be continuous at a. By the limit of a difference rule in theorem 4.3.2, $\lim_{x \to a} f(x) - g(x) = \lim_{x \to a} f(x) - \lim_{x \to a} g(x) = f(a) - g(a)$.

61. Applying the definition of the limit and the identity $|-f(x) - (-f(a))| = |f(x) - f(a)|$, you can prove $\lim_{x \to a} -f(x) = -f(a)$.

63. The base case is that $f(x) = x$ is continuous, which follows from the definition by choosing $\delta = \varepsilon$ for any given $\varepsilon > 0$. The induction hypothesis is that $f(x) = x^k$ is continuous for $k \in \mathbb{N}$. Then $g(x) = x^{k+1} = x^k \cdot x$ is the product of two continuous functions, and so is continuous by theorem 4.3.5.

65. The functions
$$f(x) = \begin{cases} -1 & \text{if } x \geq 0 \\ 1 & \text{if } x < 0 \end{cases}$$

and

$$g(x) = \begin{cases} 1 & \text{if } x \geq 0 \\ -1 & \text{if } x < 0 \end{cases}$$

provide a counterexample.

67. The functions $f(x) = 0$ and

$$g(x) = \begin{cases} 1 & \text{if } x \geq 0 \\ -1 & \text{if } x < 0 \end{cases}$$

provide a counterexample.

69. Intuitively, when x is close to a, $f(x)$ is a very large negative number. Precisely, let $f : D \to Y$ be a function whose domain D contains all points of an open interval around $a \in \mathbb{R}$, except for a itself. Then the expression $\lim_{x \to a} f(x) = -\infty$ means: for every real $M > 0$, there exists $\delta > 0$ such that $0 < |x - a| < \delta$ implies $f(x) < -M$.

4.4 The Derivative

1. The slope is $[(4^2 + 2) - (3^2 + 2)]/(4 - 3) = 7$. The graph is a parabola with vertex at $(0, 2)$ having secant line passing through the points $(3, 11)$ and $(4, 18)$.

3. The slope is $[(3.0001^2 + 2) - (3^2 + 2)]/(3.0001 - 3) = 6.0001$. The graph is a parabola with vertex at $(0, 2)$ having secant line passing through the points $(3, 11)$ and $(3.0001, 11.00060001)$.

5. The slope is $[(0.01^3) - (0^3)]/(0.01 - 0) = 0.0001$. The graph is the standard cubic having secant line passing through the points $(0, 0)$ and $(0.01, 0.000001)$.

7. $f'(x) = \lim_{h \to 0} [(2(x + h) + 3) - (2x + 3)]/h = 2$.

9. $h'(x) = \lim_{h \to 0} [((x + h)^2 + 1) - (x^2 + 1)]/h = 2x$.

11. $p'(x) = \lim_{h \to 0} [1/(x + h) - 1/x]/h = -1/x^2$

13. $r'(t) = \lim_{h \to 0} [1/(x + h - 3) - 1/(x - 3)]/h = -1/(x - 3)^2$

15. $t'(x) = \lim_{h \to 0} (\sqrt{2(x + h) + 2} - \sqrt{2x + 2})/h = 1/\sqrt{2x + 2}$

17. For $x < 2$, $v'(x) = \lim_{h \to 0} (4(x + h) - 4x)/h = 4$. For $x > 2$,

$v'(x) = \lim_{h \to 0} (2(x + h)^2 - 2x^2)/h = 4x$. The derivative does not exist at $x = 2$.

19. $f'(x) = 37(x^9 + x^6)^{36}(9x^8 + 6x^5)$

21. $f'(x) = (3x^2 + \sqrt{6x + 5} - 4) \cdot (2 - x^{-2}) + (6x + 0.5(6x + 5)^{-1/2}) \cdot (2x + 1/x)$

23. $f'(x) = 5\sin^4(x^3 + 2x) \cdot \cos(x^3 + 2x) \cdot (3x^2 + 2)$

25. $f'(x) = \dfrac{-2\csc^2(2x)}{\ln(3) \cdot \cot(2x)}$

27. $f'(x) = (kx^5 + 2x) \cdot (1/3)x^{-2/3} + (5kx^4 + 2)\sqrt[3]{x}$.

29. $h'(3\pi/4) = 14 + \sqrt{2}$. The tangent line is $y = (14 + \sqrt{2})x + 28 + \sqrt{2} + \pi^3 - (14 + \sqrt{2})3\pi/4$.

31. $h'(\dfrac{3\pi}{4}) = \dfrac{28 - 3\pi}{72}$.

The tangent line is $y - \dfrac{20 + 3\pi}{24} = \dfrac{28 - 3\pi}{72}\left(x - \dfrac{3\pi}{4}\right)$.

33. $h'(3\pi/4) = 2\pi$. The tangent line is $y = 2\pi x - 1 - 3\pi^2/2$.

35. $f'(x) = \lim_{h \to 0} (\sqrt{x + h} - \sqrt{x})/h = \dfrac{1}{2\sqrt{x}}$.

37. $y = (1/6)x + (3/2)$

39. The area of the small triangle is $(1/2)\cos\theta \cdot \sin\theta$, the area of the pie-shaped region is $\theta/2$, and the area

of the large triangle is $(1/2)\tan\theta$. Therefore $\cos\theta \cdot \sin\theta < \theta < \tan\theta$.

41. $\cos\theta < \sin\theta/\theta < \sec\theta$. The limit as θ approaches 0 of the two outside terms is 1.

43. $\lim_{h\to 0}[(\sin x\cos h + \sin h\cos x) - \sin x]/h = \cos x \lim_{h\to 0}(\sin h)/h + \sin x \lim_{h\to 0}[\cos h - 1]/h = \cos x.$

45. $\dfrac{\cos\theta \cdot \cos\theta - \sin\theta(-\sin\theta)}{\cos^2\theta}$
$= \sec^2\theta.$

47. $\dfrac{\cos\theta \cdot 0 - 1 \cdot (-\sin\theta)}{\cos^2\theta}$
$= \dfrac{1}{\cos\theta}\dfrac{\sin\theta}{\cos\theta} = \sec\theta \cdot \tan\theta.$

49. Since the derivative is defined in terms of a limit, and the limit is unique, the derivative is unique.

51. $(d/dx)[c] = \lim_{h\to 0}(c - c)/h = 0.$

53. $\lim_{h\to 0}[f(x + h)g(x + h) - f(x) \cdot g(x)]/h = \lim_{h\to 0}[f(x + h)g(x + h) - f(x + h)g(x) + f(x + h)g(x) - f(x)g(x)]/h = \lim_{h\to 0}f(x + h) \cdot [g(x + h) - g(x)]/h + \lim_{h\to 0}g(x) \cdot [f(x+h) - f(x)]/h = f(x)g'(x) + g(x)f'(x).$

55. Assuming there are no values x for which $g(x) = g(t)$, the difference quotient $[f(g(t)) - f(g(x))]/[t - x]$ is
$$\dfrac{f(g(t)) - f(g(x))}{g(t) - g(x)} \cdot \dfrac{g(t) - g(x)}{t - x}.$$

Taking the limit as t approaches x produces the chain rule formula.

57. Apply the power rule, the scalar multiple rule, and the sum rule to a general polynomial $p(x) = \sum_{k=0}^{n} a_k x^k$, where $n \in \mathbb{N} \cup \{0\}$.

59. By L'Hopital's rule applied to h, $f'(x) = \lim_{h\to 0}[f(x + h) - f(x)]/h = \lim_{h\to 0}[f'(x+h) - 0]/1 = \lim_{h\to 0}f'(x+h).$

61. $f'(0) = \lim_{h\to 0}[(0+h)^2 \sin(1/(0+h)) - 0]/h = \lim_{h\to 0}[h \cdot \sin(1/h)] = 0$, since $|\sin(1/h)| \leq 1.$

63. A counterexample is $f'(x) = x^3$ with $a = 0$ on the interval $(-1, 1).$

65. Differentiating both sides of $f(x) - g(x) = C$ for any $x \in (a, b)$, $f'(x) - g'(x) = 0.$

67. By the definition of the derivative, $\lim_{h\to 0}[f(g(a)+h) - f(g(a))]/h = f'(g(a)).$

69. $f(g(a) + h) = f(g(a) + g(a + k) - g(a)) = f(g(a + k))$. From exercise 68, $h \cdot F(h) = f(g(a) + h) - f(g(a)) = f(g(a + k)) - f(g(a)).$ Now $k \cdot G(k) \cdot F(k \cdot G(k)) = k \cdot G(k) \cdot \dfrac{f(g(a) + k \cdot G(k)) - f(g(a))}{k \cdot G(k)} = f(g(a) + k \cdot G(k)) - f(g(a)) = f(g(a) + g(a+k) - g(a)) - f(g(a)) = f(g(a + k)) - f(g(a)).$

4.5 Understanding Infinity

1. For example, define f so that $f(a) = 1$ and $f(b) = 2$.

3. No such function exists because $|B| > |A|$.

5. No such function exists because $|B| > |A|$.

7. For example, define g so that $g(x) = a$.

9. No such function exists because $|A| > |C|$.

11. For example, define f so that $f(a) = y$ and $f(b) = z$.

13. No such function exists because $|D| > |C|$.

15. The function is not onto; for example, $2 \in \mathbb{N}$, but there exists no value $x \in \mathbb{N}$

with $2x + 1 = 2$. The function is one-to-one; if $2a + 1 = 2b + 1$, then $a = b$.

17. f is one-to-one: If $a^3 + 1 = b^3 + 1$, then $a = b$. f is onto: If $r \in \mathbb{R}$, then $x = \sqrt[3]{r - 1}$ satisfies $f(x) = r$.

19. f is one-to-one: If $a^n + 1 = b^n + 1$, then $a = b$. f is onto: If $r \in \mathbb{R}$, then $x = \sqrt[n]{r - 1}$ satisfies $f(x) = r$.

21. f is not onto $\mathbb{R}$; for example, $-1 \in \mathbb{R}$, but there exists no domain value x with $\sqrt{x} = -1$. The function is one-to-one; if $\sqrt{a} = \sqrt{b}$, then $a = b$ since $\mathbb{R}^+$ is the codomain.

23. There are two, defined by $f(a) = 1$ and $g(a) = 2$. Both are one-to-one; neither is onto.

25. The eight maps are determined by mapping each of the elements $1, 2$, and 3 to one of u and v. None are one-to-one, and all but $f(1) = f(2) = f(3) = u$ and $g(1) = g(2) = g(3) = v$ are onto.

27. Note that the cardinality of the second set is smaller than the cardinality of the first.

29. For example, define $f(1) = a, f(2) = b$, and $f(n) = n - 2$ for $n \geq 3$.

31. Note that because $\mathbb{Q}$ is countable, but $\mathbb{P}(\mathbb{N})$ is uncountable.

33. Define a one-to-one correspondence $f : (0, 2) \rightarrow (0, 4)$ by $f(x) = 2x$. This is a one-to-one, onto function on these intervals of reals.

35. $0.101010\ldots$

37. $0.01010001\ldots$

39. $D = \{2, 3, 4, \ldots\}$

41. $D = \{2, 3, 5, 6, \ldots\}$

43. Reflexivity: The identity map is a one-to-one correspondence, so $|A| = |A|$.
Symmetry: If $|A| = |B|$, there is a one-to-one correspondence $f : A \rightarrow B$. The inverse $f^{-1} : B \rightarrow A$ is also a one-to-one correspondence, so $|B| = |A|$.

Transitivity: If $|A| = |B|$ and $|B| = |C|$, there are one-to-one correspondences $f : A \rightarrow B$ and $g : B \rightarrow C$. The composite function $g \circ f : A \rightarrow C$ is also a one-to-one correspondence (see section 4.2), and so $|A| = |C|$.

45. Proceed by way of contradiction: suppose both A and B are countable. Then $A \cup B$ would be countable by theorem 4.5.1.

47. Suppose $|A| \geq 2$ and label a and b two elements of A. Define a one-to-one correspondence f so that $f(a) = b$, $f(b) = a$, and $f(x) = x$ if $x \neq a, b$. Now suppose f is a one-to-one correspondence that is not the identity function; there must exist an element $a \in A$ such that $f(a) = b$, where $b \neq a$. Thus A contains at least the elements a and b.

49. f is onto: Given any $k \in \mathbb{N}$, set m so that 2^{m-1} is the power of 2 in the prime power factorization of k. Then $k/2^{m-1}$ is odd, and so there is a value $n \in \mathbb{N}$ so that $k/2^{m-1} = 2n - 1$. f is one to one: Assume $2^{m_1-1}(n_1 - 1) = 2^{m_2-1}(n_2 - 1)$. By the uniqueness of the prime power factorization, $m_1 - 1 = m_2 - 1$, and so $m_1 = m_2$. Dividing by the equal power of 2, we then have $n_1 - 1 = n_2 - 1$, and so $n_1 = n_2$. Hence $(m_1, n_1) = (m_2, n_2)$.

51. f is onto: For $r \in (0, \frac{1}{2}]$, set $x = 1/(2(r - 1))$ and for $r \in (\frac{1}{2}, 1)$, set $x = (r + \frac{1}{2})/(1 + r)$; in both cases, $f(x) = r$. Since $f'(x) > 0$, f is increasing and so one-to-one.

53. $\mathbb{P}(A) = \{\emptyset, \{\emptyset\}\}$. $|A| = 1$ and $|\mathbb{P}(A)| = 2$.

55. $\mathbb{P}(A) = \{\emptyset, \{w\}, \{x\}, \{y\}, \{z\}, \{w, x\}, \{w, y\}, \{w, z\}, \{x, y\}, \{x, z\}, \{y, z\}, \{w, x, y\}, \{w, x, z\}, \{w, y, z\} \{x, y, z\}, \{w, x, y, z\}\}$. $|A| = 4$ and $|\mathbb{P}(A)| = 16$.

57. $|A| = 6$ and $|\mathbb{P}(A)| = 64$.

59. $|\mathbb{Q}| = \omega$ and $|\mathbb{P}(\mathbb{Q})| = 2^{\omega}$.

61. $|\mathbb{P}(\mathbb{R})| = 2^{2^{\omega}}$ and $|\mathbb{P}[\mathbb{P}(\mathbb{R})]| = 2^{2^{2^{\omega}}}$.

63. Any such polynomial $y = ax + b$ is determined by its rational coefficients a and b. Therefore the cardinality of this set is the same as $|\mathbb{Q} \times \mathbb{Q}|$, which is ω (see exercise 44). Also, $|\mathbb{P}(\mathbb{Q} \times \mathbb{Q})| = 2^\omega$.

65. The cardinality of the set of all polynomials over $\mathbb{Q}$ is ω; since each polynomial has a finite number of roots (equal to its degree), the set of algebraic numbers is of cardinality ω. Also, $|\mathbb{P}(\text{the algebraic numbers over } \mathbb{Q})| = 2^\omega$.

67. Let $A = \{a, b, c, d, e, f, g\} \cup \mathbb{N}$ represent $7 + \omega$ and set $f(a) = 1, f(b) = 2, \cdots f(g) = 7$, and $f(n) = n + 7$ for $n \in \mathbb{N}$. The Hilbert Hotel drawing is similar to figure 4.17, except that $a, b, \cdots, g$ fill the first seven rooms, and $1, 2, \cdots$ line up in the remaining rooms.

69. Let $A = \{a_n : n \in \mathbb{N}\}$, $B = \{b_n : n \in \mathbb{N}\}$, and so on until $E = \{e_n : n \in \mathbb{N}\}$. Define $f(a_n) = 5n - 4$, $f(b_n) = 5n - 3$, and so on until $f(e_n) = 5n - 0$. The Hilbert Hotel proof would begin with five copies of ω, with elements from each denoted by a_n, b_n, c_n, d_n, e_n. These copies are then folded together to obtain a final list that begins $a_1, b_1, c_1, d_1, e_1, a_2, b_2, c_2, d_2, e_2, \ldots$.

4.6 The Riemann Integral

1. 5.625

3. 11.8125

5. $\sup S = 3$, $\inf S = 1$, $s = 3$, and $t = 1.0005$.

7. $\sup S = 15$, $\inf S = 0$, $s = 14.9995$, and $t = 0.0005$.

9. $\sup S = 1/2$, $\inf S = 0$, $s = 1/2$, and $t = 1/2^{10}$.

11. $\sup S = 2$, $\inf S = 1$, $s = 2$, and $t = 1 + 1/1{,}001$.

13. $(-1, 4)$

15. No such set exists by the axiom of completeness.

17. The set of reals in the interval $[0, 1)$

19. $\{2 + 1/n : n \in \mathbb{N}\}$

21. $P = \{3, 4, 5, 7\}$, $Q = \{3, 4, 5, 5.5, 6, 7\}$.

23. $P = \{0, 2, 6, 8\}$, $Q = \{0, 2, 4, 6, 7, 8\}$.

25. $U(f, P) = 73$, $L(f, P) = 37$.

27. $U(f, P) = -46$, $L(f, P) = -181$.

29. $U(f, P) = 2\sqrt{8} + \sqrt{11} + 2\sqrt{19} \approx 17.69$ and $L(f, P) = 2\sqrt{2} + \sqrt{8} + 2\sqrt{11} \approx 12.29$

31. $M_i(f) = 4 = m_i(f)$ for $i = 1, 2, 3$, and $\int_1^6 4dx = 20$.

33. $M_1(f) = 5$, $M_2(f) = 3$, $M_3(f) = 2$. $m_1(f) = 3$, $m_2(f) = 2$, $m_3(f) = 0$. $\int_1^6 -x + 6dx = 12.5$.

35. $M_1(f) = -7, M_2(f) = -13, M_3(f) = -16$. $m_1(f) = -13, m_2(f) = -16, m_3(f) = -22$. $\int_1^6 -3x - 4\, dx = -66.5$.

37. $M_1(f) = 9$, $M_2(f) = 16$, $M_3(f) = 36$. $m_1(f) = 1$, $m_2(f) = 9$, $m_3(f) = 16$. $\int_1^6 x^2\, dx = 215/3$.

39. $M_1(f) = 16$, $M_2(f) = 26$, $M_3(f) = 52$. $m_1(f) = 2$, $m_2(f) = 16$, $m_3(f) = 26$. $\int_1^6 x^2 + 3x - 2\, dx = 685/6$.

41. $M_1(f) = 27$, $M_2(f) = 64$, $M_3(f) = 216$. $m_1(f) = 1$, $m_2(f) = 27$, $m_3(f) = 64$. $\int_1^6 x^3\, dx = 1295/4$.

43. $\lim_{n \to \infty} \sum_{i=1}^{n} 4 \cdot \dfrac{3}{n} = \lim_{n \to \infty} \dfrac{12n}{n} = 12$

45. $\lim_{n \to \infty} \sum_{i=1}^{n} \dfrac{3i}{n} \cdot \dfrac{3}{n} = \dfrac{9}{2}$

47. $\lim_{n \to \infty} \sum_{i=1}^{n} -\left(\dfrac{2i}{n}\right)^2 \cdot \dfrac{2}{n} = -\dfrac{8}{3}$

49. $\lim_{n\to\infty} \sum_{i=1}^{n} \left(\frac{2i}{n}\right)^3 \cdot \frac{2}{n} = 4$

51. The function is integrable since it is bounded on $[1, 4]$ and continuous except at $x = 2$.

53. The function is integrable, since it is bounded and discontinuous only on the measure zero countable set $\{1/2^n : n \in \mathbb{N}\}$.

55. Suppose $\sup S = s$ and $\sup S = t$. Since t is an upper bound for S, $s \geq t$. Since s is an upper bound for S, $t \geq s$. Hence $s = t$.

57. Suppose $M = \sup S$. Then M is an upper bound for S by definition. Given any $\varepsilon > 0$, $M - \varepsilon < M$, and so $M - \varepsilon$ cannot be an upper bound for S. Thus there is an element $s \in S$ such that $S > M - \varepsilon$.
Conversely, if the two conditions in lemma 4.6.1 hold, then M is an upper bound by the first condition. Suppose A is any other upper bound. If $A < M$, then we may set $\varepsilon = M - A > 0$; the second condition implies there exists a value $s \in S$ such that $s > M - \varepsilon = A$, contradicting the fact that A is an upper bound.

59. Proceed by induction as was done in the proof of lemma 4.6.2(b). Suppose Q adds one additional point x^* to $P = \{x_0, \cdots x_n\}$, where $x_{k-1} < x^* < x_k$ and set $m_1^*(f) = \inf\{f(x) : x \in [x_{k-1}, x^*]\}$ and $m_2^*(f) = \inf\{f(x) : x \in [x^*, x_k]\}$. Then $m_k(f) \leq m_1^*(f)$ and $m_k(f) \leq m_2^*(f)$. A calculation similar to that in the proof of lemma 4.6.2(b) now shows $L(f, P) \leq L(f, Q)$.

If Q is an arbitrary refinement of P, consider a sequence of partitions refining P one point at a time until obtaining Q, applying the fact just proved to see that $L(f, P) \leq L(f, Q)$.

61. This set identity follows from exercise 60 and the fact that $\{x : x > 0\} = \bigcup_{i=1}^{\infty} \{x : x \geq 1/n\}$.

63. The first term in the given sequence of inequalities is $osc(f, a)$ and the last term is $M_i(f) - m_i(f)$. Since $a \in S_n$, we have $1/n \leq osc(f, a) \leq M_i(f) - m_i(f)$.

65. Exercise 64 says that $\varepsilon/2 > \sum(x_i - x_{i-1})$, where the sum is taken over only those subintervals containing points from S_n. Since all points of S_n must be either one of the points $x_0, x_1, \cdots, x_{n-1}$ or in one of these subintervals, and since the measure of $\bigcup_{i=0}^{n-1}(x_i - \frac{\varepsilon}{4n}, x_i + \frac{\varepsilon}{4n})$ is $\varepsilon/2$, all points in S_n are contained in a collection of open intervals having total measure less than ε.

67. After three steps, the measure of the sets removed is $19/27$. The nth step removes 2^{n-1} "middle" intervals, each of width $1/3^n$. Let n go to infinity in the sum to get the total measure of the sets removed.

69. Assume all elements of the Cantor set may be listed, where the nth number in the list is $0.b_{1n}b_{2n}b_{3n} \cdots [3]$. Obtain an element $0.a_1a_2a_3 \cdots [3]$ of the Cantor set not in this list by setting $a_k = 2$ if $b_{kk} = 0$ and $a_k = 0$ if $b_{kk} = 2$, where $k \in \mathbb{N}$.

4.7 The Fundamental Theorem of Calculus

1. $f(x) = 2x - 1$

3. $c = 1.5$

5. $c = \sqrt{8}$

7. $c = \pm\frac{2}{\sqrt{3}}$

9. $c = \sqrt{5 - \dfrac{\sqrt{3345}}{15}}$

11. By the mean value theorem, the speed at some point must have equaled the average speed, which is 120 miles per hour.

13. $f(x) = x^3 + e^x + C$

15. $y = (x - 1)e^x + C$

17. $f(x) = \sin x + e^x + 5$

19. $f(x) = \sin x + e^x + 1 - e^{\pi/2}$

21. $f(x) = x^4 + 2e^x$

23. $y = (x + 1)e^x - 1 - 2\ln 2$

25. $x^4 + 3x^{5/3} + C$

27. $\dfrac{-1}{16x^2} + \dfrac{\ln |x|}{2} + C$

29. $-e^{-2x}/2 + \sin x + C$

31. $\sin(e^x) - \dfrac{\cos(2x)}{2}$

33. $x + 1/x$

35. $-x^2 - \cos x$

37. $-(\sqrt{x^5} + e^{(x^5)})(5x^4)$

39. $[xe^{x^2} + (x^2 + \ln x)^8](2x + 1/x)$

41. $3 + e$

43. $\dfrac{64\sqrt{2}}{15}$

45. $4(e^4 - e) + 31/80$

47. 0

49. $\lim\limits_{x \to 0+} \dfrac{1}{x} = \infty$

51. $\sqrt{-1} \notin \mathbb{R}$

53. Proceed similarly to the proof given for theorem 4.7.3. Given $\varepsilon = -f'(c) > 0$, there exists $\delta > 0$ such that $f(x) < f(c)$ when $x \in (c, c + \delta)$. But $f(c)$ is a relative minimum, and so $f(c) < f(x)$ for all x in some open interval about c, which is a contradiction.

55. The function $f(x) = 1/x$ has no maximum or minimum on $[-1, 1]$.

57. Set
$$2Ac + B = \dfrac{Ab^2 + Bb + C - Aa^2 - Ba - C}{b - a}$$
to get $c = (a + b)/2$.

59. Without loss of generality, assume $c \in (a, b)$. The Riemann–Lebesgue theorem implies $\int_c^b f(x)dx$ and $\int_a^c f(x)\,dx$ exist. For any partitions P_c of $[a, c]$ and P^c of $[c, b]$, $P_c \cup P^c$ is a partition of $[a, b]$ with $U(f, P_c \cup P^c) = U(f, P_c) + U(f, P^c)$ and $L(f, P_c \cup P^c) = L(f, P_c) + L(f, P^c)$. The first sum implies
$\inf\{U(f, P): P$ is a partition of $[a, b]\} \le \inf\{U(f, R): R$ is a partition of $[a, c]\} + \inf\{U(f, Q): Q$ is a partition of $[c, b]\}$, which means
$$\int_a^b f(x)\,dx \le \int_a^c f(x)\,dx + \int_c^b f(x)\,dx.$$
The second sum similarly shows
$$\int_a^b f(x)\,dx \ge \int_a^c f(x)\,dx + \int_c^b f(x)\,dx.$$

61. For any partition P of $[a, b]$, $M_i(f) = m_i(f) = r$ for $i = 1, 2, \cdots, n$. Hence $U(f, P) = L(f, P) = r(b - a)$.

63. Let $F(t) = \int_{g(a)}^{t} f(x)\,dx$ so that $F'(t) = f(t)$ and $(F \circ g)'(t) = F'(g(t)) \cdot f'(t) = f(g(t))g'(t)$. By the fundamental theorem, $\int_a^b f(g(t))g'(t)\,dt = (F \circ g)(b) - (F \circ g)(a) = F(g(b)) - F(g(a)) = \int_{g(a)}^{g(b)} f(x)\,dx.$

65. Use, for example,
$$f(x) = \begin{cases} 1 & \text{if } x \in (0, 1) \\ 0 & \text{if } x \in (1, 2) \end{cases}.$$

67. Use, for example, $f(x) = x^2$ and $g(x) = x$.

69. Use, for example,
$$f(x) = \begin{cases} 1 & \text{if } x \in \mathbb{Q} \\ 0 & \text{otherwise} \end{cases}$$
and
$$g(x) = \begin{cases} -1 & \text{if } x \in \mathbb{Q} \\ 0 & \text{otherwise} \end{cases}.$$

| 4.8 | **Application: Differential Equations** |

1. The equation is first order, linear, and separable.

3. The equation is second order, nonlinear, and not separable.

5. The equation is first order, nonlinear, and separable.

7. The equation is of the twenty-third order, nonlinear, and not separable.

9. $y = C$, where $C \in \mathbb{R}$.

11. $y = x^2 + \cos x + C_1 x + C_2$, where $C_1, C_2 \in \mathbb{R}$.

13. $y = (1/3)xe^{3x} - (1/9)e^{3x} + (5/3)x^3 - 2x + C$, where $C \in \mathbb{R}$.

15. $y = \cos x + x \sin x + (1/2)e^x(\sin x - \cos x) + C$, where $C \in \mathbb{R}$.

17. $y = C$, where $C \in \mathbb{R}$.

19. $y = Ce^{4e^x}$, where $C \in \mathbb{R}$.

21. $y = -2x^2 + Cx^3$, where $C \in \mathbb{R}$.

23. $y^2 = X/(1 - X)$, where $X = Ce^{(2/3)x^3 - x^2}$ for $C \in \mathbb{R}$.

25. $y = Ce^{(2/3)x^{3/2}}$, where $C \in \mathbb{R}$.

27. $y = -1/2 + Ce^{x^2}$, where $C \in \mathbb{R}$.

29. $y = -1/2 + (5/2)e^{x^2 - 1}$

31. $y = 2e^{-x^2/2}$

33. $y = -x^4/2 + 3x^2$

35. Assuming $y > 0$, $y = \sqrt{5x^2 - 19}$.

37. $y = (-1/2)(x\sqrt{1 + x^2} + \ln|x + \sqrt{1 + x^2}|) + 5$

39. $y = e^{(1/3)(x \sin x + \cos x - 1)}$

41. $y = Ce^{-3x} + De^x$, where $C, D \in \mathbb{R}$.

43. $y = Ce^{-x/2}\cos(x) + De^{-x/2}\sin(x)$, where $C, D \in \mathbb{R}$.

45. $y = Ce^x + Dxe^x$, where $C, D \in \mathbb{R}$.

47. $y = 9e^{-2x} - 7e^{-3x}$

49. $y = e^{-x}\cos(\sqrt{3}x) + (2/\sqrt{3})e^{-x}\sin(\sqrt{3}x)$

51. $H_0(x) = 1$ satisfies the equation because $H_0''(x) = 0$ and $H_0'(x) = 0$.

53. $48x - 2x(24x^2 - 12) + 6(8x^3 - 12x) = 0$

55. $\lim_{R \to \infty} -2Re^{-R^2} - \lim_{T \to -\infty} -2Te^{-T^2} = 0 - 0 = 0$

57. Converting to polar coordinates,

$$\left[\int_{-\infty}^{\infty} e^{-x^2}\, dx\right]^2 =$$

$$\int_{-\infty}^{\infty} e^{-x^2}\, dx \int_{-\infty}^{\infty} e^{-y^2}\, dy =$$

$$\int_{-\infty}^{\infty} \int_{-\infty}^{\infty} e^{-(x^2 + y^2)}\, dx dy =$$

$$\left(\int_{-\pi}^{\pi} d\theta\right)\left(\int_0^{\infty} e^{-r^2} r\, dr\right) =$$

$$2\pi \lim_{R \to \infty} (1 - e^{-R^2})/2 = \pi.$$

Therefore $\int_{-\infty}^{\infty} e^{-x^2}\, dx = \sqrt{\pi}$.

59. $L_0(x) = 1, L_1(x) = 1 - x$

61. Use integration by parts. $\int_0^{\infty}(1 - x)e^{-x}\, dx = \lim_{R \to \infty} -e^{-R} + Re^{-R} + e^{-R} + 1 - 0 - 1 = 0$.

63. $\int dy/y = \int k\, dt$ implies $y = e^{kt + C} = y_0 e^{kt}$.

65. Use theorem 4.8.1 with $F(t) = R/L$ and $G(t) = E/L$. $I = E/R + Ce^{-Rt/L}$, where $C \in \mathbb{R}$.

67. $b \approx 0.80537$.

69. The population is approximately 7.854 billion in 2020 in this model.

| 5.1 | **Combinatorics** |

1. 30

3. 220

5. 10

7. 6,720

9. 165

11. n

13. 64; order is important and repetition is allowed.

15. 4; order is not important and repetition is not allowed.

17. 4,096; order is important and repetition is allowed.

19. None possible; order is not important and repetition is not allowed.

21. No repetition, order: 1,816,214,400 lists.

23. No repetition, no order: 210 subsets.

25. No repetition, order: 479,001,600 ways.

27. No repetition, no order: 125,970 outcomes.

29. No repetition, no order: $462 \cdot 3,628,800 \cdot 259,459,200$ ways.

31. No repetition, no order: 25,920 options.

33. No repetition, no order: 56 ways.

35. Repetition, order: 2^{63} ways.

37. No repetition, no order: 10 ways.

39. Four of a kind occurs more often— there are 624 ways it can occur, compared with 40 ways.

41. $60 \times 60 \times 60 = 216,000$

43. $40 \times 40 \times 40 = 64,000$

45. $40 \times 40 \times 20 = 32,000$

47. $40 \times 40 \times 40 + 40 \times 40 \times 20 = 96,000$

49. $40 \times 60 \times 60 = 144,000$

51. Once one person sits down (which orients the circular arrangement by determining where the arrangement "starts"), there are $P(3, 3) = 6$ ways to order the remaining three people.

53. The five socks represent $n = 5$ "pigeons" residing in $n - 1 = 4$ "pigeonholes," which are the four different colors of socks representing each matching pair. The pigeonhole principle says that at least two of the five socks are guaranteed to be matching. In contrast, if the owner removes only four socks, all four can be mismatched.

55. The eight women separate the 18 chairs into spaces (the pigeonholes), where any group of adjacent chairs not separated by a woman sitting in between is one pigeonhole. There are at most nine such spaces (since there is a maximum of eight divisions), and so the pigeonhole principle says that at least one space holds more than one male (the 10 males are the "pigeons").

57. Think of the subsets $\{1, n\}$, $\{2, n-1\}$, $\cdots \{(n + 1)/2 - 1, (n + 1)/2 + 1\}$, $\{(n+1)/2\}$ as the $(n + 1)/2$ different pigeonholes. These sets represent the different ways two numbers can combine to sum to $n + 1$. Interpreting the numbers as the pigeons, the pigeonhole principle proves the result.

59. The five disciplines are the pigeonholes and the courses are the pigeons. Since there are more pigeons than pigeonholes, the pigeonhole principle proves the result.

61. $C(n, 0) = \dfrac{n!}{0!(n - 0)!} = 1$, since $0! = 1$.

63. $C(n, k) + C(n, k - 1)$

$$= \frac{n!}{k!(n - k)!} + \frac{n!}{(k - 1)!(n - k + 1)!}$$

$$= \frac{(n - k + 1) \cdot n!}{k!(n - k + 1)!} + \frac{k \cdot n!}{(k)!(n - k + 1)!}$$

$$= \frac{(n - k + 1 + k) \cdot n!}{k!(n - k + 1)!}$$

$$= \frac{(n + 1)!}{k!(n + 1 - k)!} = C(n + 1, k).$$

65. Follow the hint.

67. 1,487,285,800

69. 226,800

5.2 Pascal's Triangle and the Binomial Theorem

1. The $n = 8$ row is 1 8 28 56 70 56 28 8 1.
3. The $n = 16$ row is 1 16 120 560 1820 4368 8008 11440 12870 11440 8008 4368 1820 560 120 16 1.
5. 1,365
7. 924
9. 11,6280
11. They are both 1.
13. They are both 1.
15. They are both 8.
17. They are both 12.
19. All but the first and last.
21. All but the first and last.
23. The only non-one elements are 5 and 10, which are both divisible by 5.
25. The only non-one elements are 11, 55, 165, 330 and 462, which are all divisible by 11.
27. Let $n = 9$ and note that 9 does not divide $C(9, 3)$.
29. $1 + 3 = 4 = 2^2$, $3 + 6 = 9 = 3^2$, $6 + 10 = 16 = 4^2$, $10 + 15 = 25 = 5^2$, $15 + 21 = 36 = 6^2$, $21 + 28 = 49 = 7^2$, $28 + 36 = 64 = 8^2$, $36 + 45 = 81 = 9^2$, $45 + 55 = 100 = 10^2$, $55 + 66 = 121 = 11^2$.
31. The left–right symmetry of the triangle corresponds to this property.
33. $C(n - 1, n/2 - 1) + C(n - 1, n/2)$

$$= \frac{(n - 1)!}{(n/2 - 1)!(n - n/2)!}$$

$$+ \frac{(n - 1)!}{(n/2)!(n - n/2 - 1)!}$$

$$= \frac{(n/2 + n/2) \cdot (n - 1)!}{(n/2)!(n/2)!}$$

$$= \frac{n!}{(n/2)!(n - n/2)!} = C(n, n/2)$$

35. The base case is that $C(0, 0) = 1$. By reindexing and using the fact that

$C(n + 1, 0) = C(n, 0) = C(n, n) =$
$C(n + 1, n + 1) = 1$, $\sum_{k=0}^{n+1} C(n + 1, k) = C(n + 1, 0) + \sum_{k=1}^{n} \{C(n, k - 1) + C(n, k)\} + C(n + 1, n + 1) = C(n + 1, 0) + \sum_{k=0}^{n-1} C(n, k) + \sum_{k=1}^{n} C(n, k) + C(n + 1, n + 1) = 2 \sum_{k=0}^{n} C(n, k) = 2 \cdot 2^n.$

37. Use the fact that $2^0 + 2^1 + \cdots 2^{n-1} = 2^n - 1$.
39. For example, the $n = 5$ row is 1 5 20 60 120 120.
41. $P(5, 2) = 20$ and $P(5, 3) = 60$.
43. $P(9, 3) = 504$ and $P(9, 5) = 15,120$.
45. $x^4 + 2x^2y^2 + y^4 + x^4 - 4x^3y + 6x^2y^2 - 4xy^3 + y^4$.
47. $x^3 + 3x^2y^2 + 13xy^4 + y^6 + 32x^5 - 80x^4y + 80x^3y^2 - 40x^2y^3 - y^5$.
49. $78,125x^7 + 218,750x^6y + 262,500x^5y^2 + 175,000x^4y^3 + 70,000x^3y^4 + 16,800x^2y^5 + 2,240xy^6 + 128y^7$
51. $x^7y^7 + 7x^6y^6z + 21x^5y^5z^2 + 35x^4y^4z^3 + 35x^3y^3z^4 + 21x^2y^2z^5 + 7xyz^6 + z^7$.
53. $262,144t^9 + 294,9120t^8s + 1,4745,600t^7s^2 + 43,008,000t^6s^3 + 80,640,000t^5s^4 + 100,800,000t^4s^5 + 84,000,000t^3s^6 + 45,000,000t^2s^7 + 14,062,500ts^8 + 1,953,125s^9$
55. $x^{18} - 9x^{16}y^2 + 36x^{14}y^4 - 84x^{12}y^6 + 126x^{10}y^8 - 126x^8y^{10} + 84x^6y^{12} - 36x^4y^{14} + 9x^2y^{16} - y^{18}$
57. $x^{11} + 11yx^{10} + 55y^2x^9 + 165y^3x^8 + 330y^4x^7 + 462y^5x^6 + 462y^6x^5 + 330y^7x^4 + 165y^8x^3 + 55y^9x^2 + 11y^{10}x + y^{11}$

59. If $n \in \mathbb{N}$, then the numerator terms of the form $n(n-1)(n-2)\cdots$ will equal zero after a finite number of terms, since one of the factors will eventually equal $(n-n)$.

61. $C_1 = 1$, $C_2 = 2$, $C_3 = 5$, $C_4 = 14$, $C_5 = 42$, $C_6 = 132$, and $C_7 = 429$.

63. $C(2n, n) - C(2n, n+1)$

$$= \frac{(2n)!}{n!n!} - \frac{(2n)!}{(n+1)!(n-1)!}$$

$$= \frac{(n+1-n)\cdot(2n)!}{n!(n+1)!}$$

$$= \frac{(2n)!}{n!(n+1)!} = C_n.$$

65. For example, $f_{12} = 144 = 1 + 10 + 36 + 56 + 35 + 6$.

67. The $n = 0$ to $n = 6$ rows modulo four are:

1
1 1
1 2 1
1 3 3 1
1 0 2 0 1
1 1 2 2 1 1
1 2 3 0 3 2 1.

A richly colored fractal pattern version of Sierpinski's triangle appears in the colored mod 4 version of Pascal's triangle.

5.3 Basic Probability Theory

1. 8
3. $A = \{HTT, HTH, HHT, HHH\}$
5. $A' = \{TTT, THT, TTH, THH\}$
7. $B = \{TTT\}$
9. $P[B] = 1/8$ and $P[C] = 3/8$
11. $C(1000, 15)$

13. $$\frac{C(980, 15) + 20 \cdot C(980, 14) + C(20, 2) \cdot C(980, 13)}{C(1000, 15)}$$
$\approx 0.9973.$

15. $C(1000, 30)$

17. $$\frac{C(980, 30) + 20 \cdot C(980, 29) + C(20, 2) \cdot C(980, 28)}{C(1000, 30)}$$
$\approx 0.9803.$

19. $1/210$
21. $1/35$
23. $2/7$
25. $1/10$
27. $3/10$
29. For example, let $X =$ "twice the total number of dots appearing on the two dice rolled." $X = 10$ in the case given.

31. For example, let $Y =$ "the time it takes for the person to run the race."

33. For example, let $N =$ "the numerical rank of the card," where an Ace has rank 1, a King has rank 13, a Queen has rank 12, a Jack has rank 11, and all other cards have rank equal to their number. $N = 1$ in the case given.

35. $P[X = 10] = 0.0000512$ and $P[X \neq 10] = 0.9999487$

37. Both probabilities are 0.25.

39. $P[X = 1] = 0.2684$ and $P[X \leq 1] = 0.3758$

41. $P[X = 10] = 0.0961$

43. Both probabilities are 0.5.

45. $P[-1 \leq X < 5] = 0.75$ and $P[X = 3] = 0.$

47. Both probabilities are 0.1359. The graph is symmetric across 0.

49. $\int_{-\infty}^{\infty} f dx = \int_0^{\infty} e^{-x} dx = \lim_{a \to \infty} \int_0^a e^{-x} dx = \lim_{a \to \infty} -e^{-a} + e^0 = 1$, $P[X > 1] = e^{-1}$, and $P[-1 < X \leq 2] = e - e^{-2}$.

51. $P[X = 10,000] = C(4, 4)/C(20, 4) \approx 0.000206$. $P[X = 2,000] = C(4, 3)/C(20, 4) \approx 0.000824$. $E[X] = \$3.71$.

Since $E[X] = \$3.71$, for a charge of $5 we don't play the game and for a charge of $2 we do play the game.

53. The probability is about 0.6687.

55. $E[X] = 1.7$. $\sigma \approx 1.187$.

57. $E[X] = 1.8$. $\sigma \approx 1.327$.

59. $E[X] = 1$. $\sigma = 1/\sqrt{6}$.

61. $E[X] = 73/84$. $\sigma \approx 0.5939$.

63. For discrete X: $E[cX] = \sum cx \cdot P[cX = cx] = c \sum x \cdot P[X = x] = cE[X]$. For continuous X: $E[cX] = \int_{-\infty}^{\infty} cx \cdot f(x) \, dx = c \int_{-\infty}^{\infty} x \cdot f(x) \, dx = cE[X]$.

65. For discrete X:

$$\sigma_{cX} = \sqrt{\sum (cx - cE[X])^2 \cdot P[cX = cx]}$$

$$= |c|\sigma_X.$$

For continuous X:

$$\sigma_{cX} = \left[\int_{-\infty}^{\infty} (cx - c\mu)^2 \cdot f(x) \, dx \right]^{1/2}$$

$$= |c|\sigma_X.$$

67. 1

69. Approximately 3.3234.

71. Approximately 0.7096.

73. Approximately 0.5802.

5.4   Application: Statistical Inference and Hypothesis Testing

1. Let $X = $ "the number of calories consumed by a student for lunch in the school cafeteria." Is the corresponding population average $\mu > 1,000$?

3. Let $X = $ "the moisture level of a randomly selected patch of soil from the region." Is the corresponding population average $\mu < \mu_0$, where μ_0 is a selected level of moisture that would indicate severe drought?

5. Let $X = $ "the amount of time a student spends working on college-owned computers in campus labs." Is the corresponding population average $\mu = \mu_0$, where μ_0 is a selected level that constitutes Information Technology's best estimate of this time?

7. Let $X = $ "the number of times a mu-proton passes through the particle chamber in a 10 day period." Is the corresponding population average $\mu = 1$?

9. All flights on States Airline from New York to Chicago.

11. All Horse and Rider cigarettes.

13. All high school seniors.

15. The answer depends upon the random numbers generated by the computing device. If the numbers generated were 3, 8, 11, 12, 18, and 21, then the average would be $\overline{X} = [15 + 19 + 32 + 32 + 52 + 80]/6 = 38.\overline{3}$.

17. The answer depends upon the random numbers generated by the computing device. If the numbers generated were 3, 8, 11, 12, 13, 15, 16, 18, 21, and 23 then the average would be $\overline{X} = [15 + 19 + 32 + 32 + 32 + 35 + 38 + 52 + 80 + 162]/10 = 49.7$.

19. The answer depends upon the random numbers generated by the computing device. If the numbers generated were 3, 4, 5, 6, 7, 8, 11, 12, 18, and 21, then the average would be $\overline{X} = [15 + 15 + 16 + 17 + 19 + 19 + 32 + 32 + 32 + 35 + 38 + 52 + 80 + 162]/14 \approx 40.29$.

21. The P-value is 0.6470; do not reject H_0. The test concludes that the selection process is random (without bias).

23. The P-value is 0.0549; do not reject H_0. The test concludes that the selection process is random.

25. The P-value is 0.00002; reject H_0. The test concludes that the selection process was not random (it was not without bias).

27. The P-value is $P[5 \text{ or } 6 \text{ reds}] = 0.2167$. Do not reject H_0; conclude the that the selection process was random (it was without bias).

29. The P-value is $P[\text{at least 2 are red}] = 0.9640$. Do not reject H_0; conclude the that the selection process was random (it was without bias).

31. The P-value is $P[\text{at least 5 are red}] = 0.4103$. Do not reject H_0; conclude the that the selection process was random (it was without bias).

33. The P-value is $P[\text{at least 13 are red}] = 0.0115$; reject H_0 and conclude the that the selection process was not random (it was biased toward selecting the red balls).

35. The P-value is 1; do not reject H_0; conclude the that the selection process was random (it was without bias toward selecting the red balls).

37. Step 1. Will fewer than 2/3 of the customers purchase agency trips? Step 2. All agency customers. Step 3. 0 of 4 purchased an agency trip. Step 4. Test H_0: $p = 2/3$ vs. H_a: $p < 2/3$, where p is the probability that a randomly chosen customer will purchase an agency trip. If $X =$ "the number of customers purchasing agency trips," then X is binomial; $n = 4$ on the sample, and H_0 assumes $p = 2/3$. The P-value is $P[X = 0] = 0.0123$. Reject H_0 and conclude that the estimate should be lowered.

39. Step 1. Was the original set of bolts randomly selected? Step 2. All 10,000 bolts. Step 3. 1 of 200 were defective. Step 4. Test H_0: the selection was random vs. H_a: the selection was not random. If $X =$ "the number defective bolts in a sample of 200," then X is binomial; $n = 200$ and $p = 0.053$. The P-value is $P[X \leq 1] = 0.00023$. Reject H_0 and conclude that the bolts were not randomly supplied.

41. The two-tailed P-value is $2 \cdot P[\overline{X} \leq 20.5] \approx 0$ and the one-tailed P-value is also approximately 0. Reject H_0 in both cases.

43. The two-tailed P-value is $2 \cdot P[\overline{X} \geq 22.1] = 0.6733$ and the one-tailed P-value is 0.3366. Do not reject H_0 in either case.

45. The two-tailed P-value is $2 \cdot P[\overline{X} \geq 25] \approx 0$ and the one-tailed P-value is also 0. Reject H_0 in both cases.

47. The two-tailed P-value is $2 \cdot P[\overline{X} \geq 20.5] \approx 0.035$ and the one-tailed P-value is approximately 0.018. Do not reject H_0 in either case.

49. The P-value is 0.0260. Reject H_0 at $\alpha = 0.05$ but not at $\alpha = 0.01$.

51. The P-value is approximately 0. Reject H_0 at both $\alpha = 0.05$ and $\alpha = 0.01$.

53. The P-value is 0.0852. Do not reject H_0 at either $\alpha = 0.05$ or $\alpha = 0.01$.

55. The P-value is $P[\overline{X} \leq 3.75] \approx 0$. Reject H_0 and conclude that the average is less than advertised.

57. The P-value is $P[\overline{X} \geq 1,053] \approx 0$. Reject H_0 and conclude that the local high-income student average is higher than the national norm.

59. The P-value is $2 \cdot P[\overline{X} \geq 300,000] \approx 0$. Reject H_0 and conclude that the average is different than 275,000; it appears to be higher.

| 5.5 | **Least Squares Regression** |

1. The P-value is $2 \cdot P[T_2 > 5.185] = 0.035$. Reject H_0; the test confirms the usefulness of a linear model.

3. The regression line is $\widehat{Y} = 2.2X + .4$. The scatterplot consists of the four points plotted in the X–Y plane.

5. The sum of squares of vertical distances for the regression line is 3.6; it is 20 for the line $y = 2x + 3$.

7. Since the regression line based on only two points automatically runs through both points, there is no need for a test on linearity. The data set needs additional values.

9. The P-value is $2 \cdot P[T_2 > 0.192] = 0.8652$. Do not reject H_0; conclude the data do not indicate the usefulness of a linear model.

11. The P-value is $2 \cdot P[T_4 > 1.920] = 0.1273$. Do not reject H_0; conclude the data do not indicate the usefulness of a linear model.

13. The response is $4 \cdot 1.5 - 2 = 4$.

15. The response is $4 \cdot 2 - 2 = 6$. It unexpectedly matches the scatterplot point's Y value of 6 when $X = 2$. The regression line minimizes the sum of squares, providing the best fit to the data set as a whole, and so may not intersect all of the data points.

17. The response is 14, which differs from the scatterplot point's Y value of 8 when $X = 4$. The regression line minimizes the sum of squares, providing the best fit to the data set as a whole, and so may not intersect all of the data points.

19. The test for linearity's P-value is $2 \cdot P[T_3 > 4.0858] = 0.0264$; the regression line is $\widehat{Y} = 0.9692X - 1.099$.

21. The test for linearity's P-value is $2 \cdot P[T_5 > 1.210] = 0.2802$.

23. The test for linearity's P-value is $2 \cdot P[T_5 > 2.368] = 0.0640$.

25. The test for linearity's P-value is $2 \cdot P[T_3 > 2.154] = 0.1203$.

27. The test for linearity's P-value is $2 \cdot P[T_5 > 8.757] \approx 0.0003$; the regression line is $\widehat{Y} = 1.674X - 53.41$.

29. $\widehat{Y} = 0.8056X - 0.0278$.

31. $\widehat{Y} = 2.407X - 2.986$.

33. The test for linearity's P-value is $2 \cdot P[T_{18} > 0.9227] = 0.3683$; the data do not indicate a usefulness of a linear model. The regression line is $\widehat{Y} = 0.2396X + 1.34$.

35. One such scatterplot would consist of points fairly collinear with a very steep upward slope.

37. One such scatterplot would consist of points fairly collinear with a very shallow downward slope.

39. Any scatterplot that consists of collinear points along a line with a negative slope.

41. One such scatterplot would consist of points that lie fairly close along the curve $Y = X^2$.

43. One such scatterplot would consist of points that lie fairly close along the curve $Y = e^X$.

45. One such scatterplot would consist of points that lie fairly close along the curve $Y = \sin X$.

47. The test for linearity's P-value is $2 \cdot P[T_5 > 1.792] = 0.1331$; do not conclude the usefulness of the linear model.

49. The test for linearity's P-value is $2 \cdot P[T_4 > 3.828] = 0.0186$; conclude the usefulness of the linear model.

51. The test for linearity's P-value is $2 \cdot P[T_3 > 2.655] = 0.0765$; do not conclude the usefulness of the linear model.

53. The test for linearity's P-value is $2 \cdot P[T_5 > 3.89027] \approx 0.01145$; conclude the usefulness of the linear model.

55. The test for linearity's P-value is $2 \cdot P[T_5 > 6.371] = 0.0007$;

conclude the usefulness of the linear model.

57. $\sum_{i=1}^{n} x_i y_i - \widehat{\beta}_0 \sum_{i=1}^{n} x_i - \widehat{\beta}_1 \sum_{i=1}^{n} x_i^2 = 0$

59. $\widehat{\beta}_1 = \dfrac{\sum\limits_{i=1}^{n} x_i y_i}{\sum\limits_{i=1}^{n} x_i^2}$

6.1   An Introduction to Graph Theory

1. Not a graph—the lines (edges) do not join vertices.

3. A graph with one edge joining two vertices.

5. A graph with one vertex and no edges.

7. It is simple (it has no loops or parallel edges), connected (a path exists from any one vertex to another), not complete (both the top left and bottom right vertices, and the top right and bottom left vertices are not joined by an edge) and a cycle graph (it consists of a single cycle).

9. It is simple (it has no loops or parallel edges), not connected (a path does not exists from every vertex to any other), not complete (each vertex is not joined to every other vertex by exactly one edge) and is not a cycle graph.

11. It is simple (it has no loops or parallel edges), connected (a path exists from any one vertex to another), complete (each vertex is joined to every other vertex by exactly one edge) and is not a cycle graph (it contains numerous cycles).

13. For example, arrange the four vertices in a square and join them with four edges around the perimeter of the square, but do not include the

square's "diagonals" as edges in the graph.

15. For example, arrange the four vertices in a square and join them with four edges around the perimeter of the square.

17. For example, arrange the four vertices in a square and join them with four edges around the perimeter of the square.

19. No such graph exists, since a null graph has no edges, and so the vertices would be isolated.

21. (a) Labeling edges by their endpoints, a walk is AB, BE. (b) AB, BE. (c) BE, EA, AD, DE, EC, CB. (d) EA, AD, DE. (e) $d(A) = d(B) = 3$, $d(C) = d(D) = 2$, $d(E) = 4$. (f) 14.

23. (a) AE (b) AE (c) No circuit based at vertex B exists. (d) No cycle based at E exists. (e) $d(A) = d(B) = d(D) = 1$, $d(C) = 2$, $d(E) = 3$ (f) 8.

25. (a) AB, BE. (b) AB, BE. (c) BE, EA, AD, DE, EC, CB. (d) EA, AD, DE. (e) $d(A) = d(B) = d(C) = d(D) = d(E) = 4$. (f) 20.

27. Since the total degree of a graph is even, when the sum of the degrees of vertices with even degrees is subtracted, the sum of the degrees of vertices with odd degrees must also

be even. Hence there are an even number of such vertices of an odd degree.

29. If the walk has no repeated edge, then it is a path. Otherwise, for each repeated edge (say joining vertex A to vertex B) remove the portion of the walk that starts at vertex A and returns to vertex A. There will be no repeated edges after these removals, and so the resulting walk is a path.

31. Examine the portion of the graph consisting of the vertices T for which there is a path from V to T, along with any edge incident to these vertices. Since this portion of the graph is itself a graph (called a subgraph), its total degree is even. Since $d(V)$ is odd, there must be another vertex in the subgraph with odd degree. But W is the only other vertex with odd degree, and so W must be in the subgraph. By construction of the subgraph, there must then be a path from V to W.

33. K_1 is a single vertex (with no edges); K_2 consists of two vertices joined by an edge; and K_3 consists of three vertices joined by three edges that form a triangular cycle.

35. The base case states that K_1 has no edges, which is true. Assuming K_n has $n(n-1)/2$ edges, add one additional vertex V and form K_{n+1} by joining V to each of the vertices of K_n, a process that requires the addition of n edges. Hence K_{n+1} has $n(n-1)/2 + n = (n+1)n/2$ edges.

37. C_4 can be represented as four vertices joined by four edges that form a square. C_5 can be represented as five vertices joined by five edges that form a pentagon.

39. The complement consists of the four vertices with two edges: one that is incident to A and C, and one that is incident to B and C.

41. The complement consists of the six vertices with nine edges, listed according to the two vertices incident to a given edge: $AB, AC, AF, BD, BF,$ $CD, CE, DE,$ and DF.

43. The graph has six vertices and eight edges.

45. The graph is K_5.

47. $A : C, D; B : C, D; C : A, B;$ $D : A, B.$

49. $A : B; B : A, D, E, F; C : F;$ $D : B, F; E : B, F; F : B, C, D, E.$

51. $A \backsim_G C, A \backsim_G D, B \backsim_G C, B \backsim_G D$

53. $A \backsim_G B, B \backsim_G A, B \backsim_G D,$ $B \backsim_G E, B \backsim_G F$

55. The property does not hold in general; a graph consisting of a single vertex V with a loop joining V to itself is a counterexample. If the graph has no loops, then the property holds.

57. The property does not hold in general; a counterexample is a graph consisting of three vertices $A, B,$ and C, and two edges, one joining A to B and the other joining B to C. For the property to hold, all sets of three vertices $A, B,$ and C connected by a two-edged path (AB and BC) must contain a cycle.

59. The graph has three vertices that may be arranged in a triangle, where three edges form the perimeter of the triangle and one vertex has a loop joining it with itself.

61. The graph has four vertices that may be arranged in a square. Each vertex has two loops joining it with itself. Four edges form the perimeter of the square. There are two parallel edges that cut across one diagonal, and there is one edge that runs along the other diagonal.

63. $\begin{bmatrix} 1 & 1 & 1 \\ 1 & 0 & 0 \\ 1 & 0 & 0 \end{bmatrix}$

65. $\begin{bmatrix} 0 & 0 & 1 & 0 \\ 0 & 1 & 1 & 1 \\ 1 & 1 & 0 & 1 \\ 0 & 1 & 1 & 1 \end{bmatrix}$

67. The graph has n vertices but no edges.

69. In this case, the matrix element a_{ij} would be 0 when $i \neq j$.

<table><tr><td>**6.2**</td><td>**The Explorer and the Traveling Salesman**</td></tr></table>

1. By theorem 6.2.1, there exists only an Eulerian path. Labeling each edge by the vertices that are incident to it, one such path is: CA, AB, BD, DC, CB.

3. By theorem 6.2.1, there exists only an Eulerian path. Labeling each edge by the vertices that are incident to it, one such path is: $AB, BC, CE, EB, BD, DE, EA, AD$.

5. Since the graph is not connected, there is no Eulerian path or circuit.

7. By theorem 6.2.1, there exists only an Eulerian path. Labeling each edge by the vertices that are incident to it, one such path is: $CB, BA, AD, DB, BE, EC, CF, FE$.

9. Labeling each edge by the vertices that are incident to it, a Hamiltonian path is: AB, BC, CD, DE. Trial and error shows that there is not a Hamiltonian circuit.

11. Labeling each edge by the vertices that are incident to it, a Hamiltonian cycle is: AB, BD, DC, CA.

13. Labeling each edge by the vertices that are incident to it, a Hamiltonian cycle is: AB, BC, CF, FE, ED, DA.

15. Labeling each edge by the vertices that are incident to it, a Hamiltonian cycle is: AB, BD, DC, CA.

17. An example is a graph containing just one vertex having one loop joining it to itself.

19. An example is a graph with four vertices arranged in a square, with four edges forming the perimeter of the square, and one edge cutting across one diagonal.

21. An example is a graph containing just one vertex having one loop joining it to itself.

23. The graph K_4 is an example.

25. If there exists an Eulerian path from V to W in such a graph G, then any vertex A besides V and W must have even degree: the path having edge incident to and entering A must have an additional edge incident to and leaving A, since A is not a terminal vertex of the path.

27. We construct an algorithm to partition the Eulerian circuit into distinct cycles. If the Eulerian circuit does not repeat any vertex besides the base vertex, then it is a cycle. If it does repeat a vertex, then partition the path into two pieces: the portion of the path from the repeated vertex back to itself, and the rest of the path. Repeat this process for each of the two pieces, continuing to repeat the process until there are no repeated vertices besides the base vertex in any portion.

29. Prove the result by way of contradiction, supposing that G is a graph with no isolated vertices that is not connected. Then G has at least two subgraphs each containing one or more edges, where the subgraphs share no edges. Beginning at an arbitrary vertex V of any subgraph, it is impossible to construct a path from V that contains any edge in the

other subgraph(s). Hence G has no Eulerian circuit. A graph consisting of four vertices A, B, C, and D and two edges AB and CD is an example of a graph that has no isolated vertices but does not have an Eulerian circuit.

31. A cycle in G that is both Eulerian and Hamiltonian must include all the edges and all the vertices of G exactly once. Hence G is a cycle graph C_n, which has the same number of edges as vertices (see exercise 38 of section 6.1). A graph of four vertices arranged in a square with edges around the perimeter of the square and one diagonal is an example of a graph that does not have a cycle that is both Eulerian and Hamiltonian.

33. Label the graph's vertices $V_1, V_2, \cdots, V_n$. Since the graph is complete, there is an edge incident to V_k and V_{k+1} for any $k = 1, 2, \cdots, n - 1$, and there is an edge incident to V_n and V_1. This collection of edges forms a Hamiltonian cycle.

35. Labeling the vertices as in figure 6.8(b), the additional bridge is built from vertex A to D, providing another edge. Then, labeling any edge by the vertices incident to it, an Eulerian path is BC, CB, BA, AB, BD, DA, AD, DC.

37. Labeling the vertices as in figure 6.8(b), the additional bridge could be built from vertex A to C. Then, labeling any edge by the vertices incident to it, an Eulerian path is BC, CB, BA, AB, BD, DA, AC, CD.

39. Labeling the vertices as in figure 6.8(b), the additional bridges could be built from vertex A to D and from vertex B to C. Then every vertex has even degree, and so the resulting graph has an Eulerian circuit.

41. Labeling the vertices as in figure 6.8(b), the removal of the bridge corresponds to removing the edge from vertex B to D. Then, labeling any edge by the vertices incident to it, an Eulerian path is AB, BA, AD, DC, CB, BC.

43. It does not have an Eulerian circuit because $d(C) = d(F) = 1$. Add an edge incident to C and F.

45. It does not have an Eulerian circuit because $d(A) = d(D) = 1$. Add an edge incident to A and D.

47. Degrees $d(B) = d(D) = d(F) = d(H) = 3$ are all odd. Add an edge incident to B and D and one incident to F and H.

49. Graph (b) is a subgraph of (a).

51. Graph (b) is a subgraph of (a).

53. Graph (b) is a subgraph of (a).

55. The graph has three vertices A, B, and C, with a loop incident to each of A, B, and C, and two other edges joining A and B, and A and C.

57. The graph has four vertices A, B, C, and D, with a loop incident to each of A and D, and three other edges joining A and C, B and D, and C and D.

59. $\{\{A, B\}, \{A, D\}, \{A, D\}, \{A, E\}, \{B, C\},$ $\{B, D\}, \{B, E\}, \{D, E\}, \{D, E\}\}$

61. $\{\{A, B\}, \{A, D\}, \{B, C\}, \{B, E\}, \{B, F\},$ $\{C, F\}, \{D, E\}\}$

63. $A \sim_G A$, $A \sim_G B$, $A \sim_G C$, $A \sim_G C$, $B \sim_G A$, $B \sim_G B$, $B \sim_G C$, $B \sim_G D$

65. $A \smallfrown_G A$, $A \smallfrown_G B$, $A \smallfrown_G D$, $B \smallfrown_G A$, $B \smallfrown_G E$

67. False, unless the graph has no circuit (including no loops).

69. True; if there exists a path from V to W and a path from W to X, then there exists a path from V to X (construct this path from the first two, removing portions caused by repeated edges if necessary).

1. No such tree exists; it must have one more vertex than edge.

3. The tree may be represented as seven collinear vertices, where any physically adjacent two are joined by an edge.

5. Such a connected graph with no cycles would be a tree, which cannot exist because there are eight edges but only seven vertices.

7. The tree may be represented as eight collinear vertices, where any physically adjacent two are joined by an edge.

9. No such graph exists; since there are no cycles, the graph must have at least one more vertex than edge.

11. Label the four vertices A, B, C, and D. Labeling each edge in terms of the vertices incident to it, include in the graph the edges AB, BC, and AC.

13. Labeling each edge in terms of the vertices incident to it, a shortest path is AB, BC, CH.

15. A shortest path is AB, BC, CH.

17. A shortest path is AB, BF, FH.

19. A shortest path is AB, BD, DH.

21. The spanning tree consists of edges AB, BE, BF, BG, GC, GD, and GH. The algorithm labels the additional vertices in the following way: $E = V_3$, $F = V_4$, $G = V_5$, $C = V_6$, $D = V_7$, and $H = V_8$.

23. The spanning tree consists of edges AB, AE, EF, FC, FG, CD, and CH. The algorithm labels the additional vertices in the following way: $E = V_3$, $F = V_4$, $C = V_5$, $G = V_6$, $D = V_7$, and $H = V_8$.

25. The spanning tree is the graph itself. The algorithm labels the additional vertices in the following way: $G = V_3$, $H = V_4$, $D = V_5$, $C = V_6$, $F = V_7$, and $E = V_8$.

27. The spanning tree consists of edges AB, AC, AE, BD, BF, CH, and DG. The algorithm labels the additional vertices in the following way: $C = V_3$, $E = V_4$, $D = V_5$, $F = V_6$, $H = V_7$, and $G = V_8$.

29. The spanning tree consists of edges AB, BE, EF, FG, GC, CD, and GH. The algorithm labels the vertices in the following way: $B = 1 : A, E = 2 : B, F = 3 : E, G = 4 : F, C = 5 : G, D = 6 : C$, and $H = 7 : G$.

31. The spanning tree consists of edges AB, BE, EF, FC, CD, DH, and HG. The algorithm labels the vertices in the following way: $B = 1 : A, E = 2 : B, F = 3 : E, C = 4 : F, D = 5 : C, H = 6 : D$, and $G = 7 : H$.

33. The spanning tree is the graph itself. The algorithm labels the vertices in the following way: $B = 1 : A, G = 2 : B, H = 3 : G, D = 4 : H, C = 5 : D, F = 6 : C$, and $E = 7 : H$.

35. The spanning tree consists of edges AB, BD, DC, CH, HE, EF, and FG. The algorithm labels the vertices in the following way: $B = 1 : A, D = 2 : B, C = 3 : D, H = 4 : C, E = 5 : H, F = 6 : E$, and $G = 7 : F$.

37. The leaves are A, B, D, and E. Add edges AB and DE to obtain an extension with no leaves.

39. The leaves are E, F, G, H, I, and J. Add edges EF, GH, and IJ to obtain an extension with no leaves.

41. Based on the hint give a contradiction argument: since a tree is connected with finitely many vertices, such a path exists from any vertex to a vertex with degree one (otherwise there is a cycle or there are infinitely many vertices, neither of which is possible for a tree).

43. The empty graph and the graph with one vertex are the only examples of graphs with fewer than two vertices.

45. Base case: A tree with two vertices A and B has one edge AB. Now follow the hint; assume a tree with n vertices has $n - 1$ edges. A tree T with $n + 1$ vertices has at least one leaf V. Then $T \setminus \{V\}$ is formed by removing V and the one edge incident to it, and so $T \setminus \{V\}$ has n vertices; by the induction hypothesis, it has $n - 1$ edges. But then T has n edges, since only one edge was removed.

47. Proceed by way of contradiction; assume a tree with two or more vertices had no leaves. Then the degree of every vertex would be two, which contradicts the result of exercise 46.

49. No such binary tree exists; the fewest number of vertices in any binary tree with five leaves is eight.

51. From the root extend one edge to a second vertex, and from this second vertex split into two more vertices.

53. A full, binary tree has an even number of leaves.

55. There is no upper bound, since one path from root to leaf could continue indefinitely but produce only one leaf. The lower bound is 3.

57. The lower bound is 3; the upper bound is 8.

59. The tree is binary since each vertex is adjacent to at most two vertices at the next level. It is full since every vertex except leaves has two children. It is not complete since, for example, D is a leaf at level two.

61. The tree is binary since each vertex is adjacent to at most two vertices at the next level. It is full since every vertex except leaves has two children. It is not complete since, for example, C is a leaf at level one.

63. The tree is binary, full, and complete.

65. $r = 9$, $e = 15$, and $v = 8$, so $r - e + v = 2$.

67. $r = 3$, $e = 7$, and $v = 6$, so $r - e + v = 2$.

69. One such graph consists of six vertices arranged in two triangular shapes, with six edges forming the two perimeters of the disconnected triangles; this graph has $r = 3$, and $r - e + v = 3$.

6.4 Application: Weighted Graphs

1. The tree's edges, as labeled by endpoints and in the order selected by the algorithm, are Dublin-Chicago, Chicago-Rome, and Dublin-Melbourne.

3. The tree's edges, as labeled by endpoints and in the order selected by the algorithm, are BC, DE, AB, AD, and EF.

5. The tree's edges, as labeled by endpoints and in the order selected by the algorithm, are Chicago-LA, LA-Charlotte, LA-DC, New Orleans-DC.

7. The tree's edges, as labeled by endpoints and in the order selected by the algorithm, are AB, AE, GH, BF, CD, CG, and CF.

9. Starting at vertex Dublin, the tree's edges, as labeled by endpoints and in the order selected by the algorithm, are Dublin-Chicago, Chicago-Rome, and Dublin-Melbourne.

11. Starting at vertex A, the tree's edges, as labeled by endpoints and in the order selected by the algorithm, are AB, BC, AD, DE, and EF.

13. Starting at vertex Chicago, the tree's edges, as labeled by endpoints and in the order selected by the algorithm, are Chicago-LA, LA-DC, DC-New Orleans, and New Orleans-Charlotte.

15. Starting at vertex A, the tree's edges, as labeled by endpoints and in the order selected by the algorithm, are AB, AE, BF, FC, CD, CG, and GH.

17. In the order selected by the algorithm, the edges of the cycle are AB, AC, BD, and CD.

19. In the order selected by the algorithm, the edges of the cycle are AC, BD, AD, and BC.

21. In the order selected by the algorithm, the edges of the cycle are AC, DE, AB, CE, and BD.

23. In the order selected by the algorithm, the edges of the cycle are BC, BD, AF, AC, EF, and DE.

25. Using A as the base vertex, the algorithm adds edges in the following order: AB, BC, CD, and AD.

27. Using A as the base vertex, the algorithm adds edges in the following order: AC, CB, BD, and DA.

29. Using A as the base vertex, the algorithm adds edges in the following order: AC, CB, BD, DE, and EA.

31. Using A as the base vertex, the algorithm adds edges in the following order: AF, FE, EB, BC, CD, and DA.

33. Using A or D as the base vertex produces cycle weight 13; using B or C as the base vertex produces cycle weight 12.

35. Any base vertex produces cycle weight 14.

37. Using A as the base vertex produces cycle weight 11; using B, C, D, or E as the base vertex produces cycle weight 9.

39. Using B or F as base vertex produces cycle weight 54, using A or E

produces cycle weight 53, and using C or D produces cycle weight 52.

41. It is 3-regular.

43. It is 0-regular.

45. It is not regular; in particular, $d(A) = d(B) = d(D) = d(E) = 2$, while $d(C) = d(F) = 3$.

47. The graph K_4.

49. For example, a graph with two vertices and four parallel edges incident to both.

51. Each vertex of K_n is joined to each of the other $n - 1$ vertices through exactly one edge, implying that the degree of any vertex is $n - 1$.

53. By theorem 6.2.1 of section 6.2, two vertices of G must be of odd degree, while the other vertices have even degree.

55. $f : A \to V, B \to X, C \to Z,$
$D \to W, E \to Y$
$g : b \to v, c \to x, d \to z,$
$e \to w, a \to y$

57. No isomorphism exists, since G has three edges, but G^* has only two.

59. No isomorphism exists. Because isomorphisms preserve edge-endpoint relationships, two vertices mapped to each other would have to have the same degree. But $d(Z) = 1$ in G^*, while G has no vertex of degree one.

61. Follow the hint. Define f mapping a vertex of $2^{<\omega}$ to a natural number in the following way: the jth vertex (counting left to right) at level k is mapped to $n = 2^k + j - 2$. Since any path of finite length can be identified with the vertex at which the path terminates, this mapping serves as a one-to-one correspondence between the set of paths of finite length and $\mathbb{N}$.

63. $S \to x$. The parse tree has root S and one vertical edge that terminates at x.

65. $S \to -S \to -(S) \to -(S + S) \to -(x + S) \to -(x + x)$. The parse

tree is identical to the one given in the example, except the symbol $*$ is replaced with $+$.

67. $S \to S - S \to S - (S) \to S - (-S) \to x - (-S) \to x - (-x)$. The parse tree has root S and branches to three vertices S, $-$, and S at level one. Then the left S branches to x, while the right branches to vertices $(, S, \text{ and })$ at level two. Then S branches to vertices $-$ and S at level

three, and S branches to x at level four.

69. $S \to S * S \to (S) * S \to (S) * x \to (S + S) * x \to (x + S) * x \to (x + x) * x$. The parse tree has root S and branches to three vertices S, $*$, and S at level one. Then the left S branches to $(, S, \text{ and })$ while the right branches to x at level two. Then S branches to vertices S, $+$ and S at level three, and each S branches to x at level four.

7.1 Complex Numbers and Complex Functions

1. $6e^{(-5\pi/6 + 2k\pi)i}$, where $k \in \mathbb{Z}$

3. $8\sqrt{2}e^{(\pi/4 + 2k\pi)i}$, where $k \in \mathbb{Z}$

5. $2e^{2k\pi i}$, where $k \in \mathbb{Z}$

7. $-i$

9. $\dfrac{\sqrt{2}}{2} + \dfrac{\sqrt{6}}{2}i$

11. $-\dfrac{\sqrt{2}}{2} + \dfrac{\sqrt{2}}{2}i$

13. $5 + 7i$

15. $8 + 2i$

17. $26 + 28i$

19. $8 + 3i$

21. $\dfrac{26}{73} + \dfrac{28}{73}i$

23. $\dfrac{91}{73} + \dfrac{194}{73}i$

25. $4^5 e^{i\pi}$

27. $18e^{i\pi/2}$

29. $8^7 e^{-i\pi/2}$

31. $\sqrt[4]{18}e^{-i5\pi/8}$

33. $\sqrt[3]{4}e^{-i\pi/9}$, $\sqrt[3]{4}e^{i5\pi/9}$, and $\sqrt[3]{4}e^{-i7\pi/9}$.

35. $\sqrt[8]{128}e^{-i11\pi/16}$, $\sqrt[8]{128}e^{-i3\pi/16}$, $\sqrt[8]{128}e^{i5\pi/16}$, and $\sqrt[8]{128}e^{i13\pi/16}$.

37. $\bar{r} = \overline{r + i0} = r - i0 = r$

39. $\overline{z \cdot w} = \overline{(a + ib) \cdot (c + id)}$

$= \overline{(ac - bd) + i(ad + bc)}$

$= (ac - bd) - i(ad + bc)$.

Also, $\bar{z} \cdot \bar{w} = (a - ib) \cdot (c - id) = (ac - bd) - i(ad + bc)$.

41. $(z + \bar{z})/2 = (a + ib + (a - ib))/2 = a = Re(z)$.

43. Use $\cos(-\theta) + i\sin(-\theta) = \cos(\theta) - i\sin(\theta)$.

45. Use $(a + ib) \cdot (a - ib) = a^2 + b^2$. Alternatively, $z \cdot \bar{z} = |z|e^{i\theta} \cdot |z|e^{-i\theta}$.

47. $\sqrt{a^2 + b^2} \le \sqrt{a^2} + \sqrt{b^2}$, since $a^2 + b^2 \le a^2 + 2\sqrt{a^2}\sqrt{b^2} + b^2 = (|a| + |b|)^2$

49. let $z = a + ib$ and $w = c + id$. $(|z| + |w|)^2 = (a + c)^2 + (b + d)^2 = |z|^2 + |w|^2 + 2Re(z \cdot \overline{w}) = |z|^2 + |w|^2 + 2|z \cdot \overline{w}| = |z|^2 + |w|^2 + 2|z||w| = (|z| + |w|)^2$

51. The f_x difference quotient simplifies to 2. The f_y difference quotient simplifies to -4.

53. The u_x difference quotient simplifies to y^2. The u_y difference quotient simplifies to $(x+1)(2y+\Delta y)$.

55. $u_x = y^2 \; u_y = 2(x+1)y$

57. $u_x = 2(x+y) + 2 \; u_y = 2(x+y) - 1$

59. $u_x = \sqrt{1+x^2} \; u_y = x + y^3$

61. Given $\varepsilon > 0$, choose $\delta = (-1 + \sqrt{1+12\varepsilon})/6$. Then $\sqrt{x^2+y^2} < \delta$ implies $|3x^2 + y| \le |3x^2| + |y| < 3\delta^2 + \delta = \varepsilon$.

63. Given $\varepsilon > 0$, choose $\delta = (-1 + \sqrt{1+4\varepsilon})/4$. Then $\sqrt{x^2+y^2} < \delta$ implies $|x^2 - 2x + 3y^2| \le |x^2| + |2x| + |3y^2| < \delta^2 + 2\delta + 3\delta^2 = 4\delta^2 + 2\delta = \varepsilon$.

65. Given $\varepsilon > 0$, choose $\delta = \varepsilon/3$. Then $\sqrt{(x-1)^2 + (y-2)^2} < \delta$ implies $|x - 2y - (-3)| = |(x-1) - 2(y-2)| \le |x-1| + 2|y-2| < \delta + 2\delta = 3\delta = \varepsilon$.

67. Given $\varepsilon > 0$, choose $\delta = \min\{0.5, \varepsilon/2\}$. Then $\sqrt{x^2 + (y-1)^2} < \delta$ implies $|y - 1| < \delta$, which means $.5 < y$. It also means $|x/y| < \delta/|y| < \delta/0.5 \le \varepsilon$.

69. Any choice of δ satisfies the definition of continuity at any point (a, b).

71. Given $\varepsilon/2 > 0$, choose $\delta_u > 0$ and $\delta_v > 0$ so that u and v, respectively, satisfy the definition of continuity at (a, b). Then

$|u(x, y) + v(x, y) - (u(a, b) + v(a, b))| \le |u(x, y) - u(a, b)| + |v(x, y) - v(a, b)| < \varepsilon/2 + \varepsilon/2$ whenever $\sqrt{(x-a)^2 + (y-b)^2} < \delta = \min\{\delta_u + \delta_v\}$.

73. The proof is similar to the proof of example 4.3.6 in section 4.3.

75. Call $u(a, b) = L$ and assume $L > 0$; a similar proof works if $L < 0$. Given $L^2\varepsilon/2$, there exists δ_1 such that $|u(x, y) - L| < L^2\varepsilon/2$ whenever $\sqrt{(x-a)^2 + (y-b)^2} < \delta_1$. Given $L/2 > 0$, there exists δ_2 such that $\sqrt{(x-a)^2 + (y-b)^2} < \delta_2$ implies $|u(x, y) - L| < L/2$, which implies $L/2 < u(x, y)$. Hence $\sqrt{(x-a)^2 + (y-b)^2} < \delta = \min\{\delta_1, \delta_2\}$ implies

$$\left| \frac{1}{u(x, y)} - \frac{1}{L} \right| = \frac{|u(x, y) - L|}{|u(x, y)| \cdot L}$$

$$< \frac{L^2\varepsilon}{2|u(x, y)| \cdot L} < \varepsilon.$$

77. Use the fact that $|e^x \sin y| = |e^x||\sin y| \le |e^x| \cdot 1 = e^x \le e^{|x|} < e^\delta$ whenever $|x| < \delta$.

79. Choose $\delta = \varepsilon - 1$. Then $\sqrt{x^2 + y^2} < \delta$ implies $|(x+1)\sin y| \le |x+1| \le |x| + 1 < \delta + 1 = \varepsilon$.

81. $u(x, y) = \text{Re}(g) = x^3 - 3xy^2$; $v(x, y) = \text{Im}(g) = 3x^2 - y^3$. Then $u_{xx} = 6x$ and $u_{yy} = -6x$. Also $v_{xx} = 6y = -v_{yy}$.

7.2   Analytic Functions and the Cauchy–Riemann Equations

1. $3 - i$

3. $4 - i$

5. $i(\pi/2)$

7. Choose $\delta = \varepsilon$. Then $\sqrt{x^2 + y^2} < \delta$ implies $|iz + i - i| = |z| < \delta = \varepsilon$.

9. Choose $\delta = \varepsilon/|m|$. Then $\sqrt{x^2 + y^2} < \delta$ implies

$$|mz + b - b| = |m||z| < \varepsilon.$$

11. Using the fact that if $|z| < 1/2$, then $|z + 1| > 1/2$, choose

$\delta = \min\{1/2, \varepsilon/2\}$. Then $\sqrt{x^2 + y^2} < \delta$ implies

$$\left| \frac{1}{z+1} - 1 \right| = \left| \frac{z}{z+1} \right|$$

$$< \frac{\delta}{|z+1|} < 2\delta \le \varepsilon.$$

13. Assume $\lim\limits_{z \to 0} f(z) = L$ and $\lim\limits_{z \to 0} f(z) = M$, where $L > M$. Define $\varepsilon = (L - M)/2$. Examine the definition of limit in terms of L and M to find a contradiction (as there cannot exist $\delta > 0$ such that $f(z)$ is within ε of both L and M whenever $\sqrt{x^2 + y^2} < \delta$).

15. Given $\varepsilon/|c| > 0$, there exists $\delta > 0$ such that $\sqrt{x^2 + y^2} < \delta$ implies $|cf(z) - cL| = |c||f(z) - L| < |c|(\varepsilon/|c|) = \varepsilon$.

17. Given $\varepsilon/2 > 0$, there exists $\delta_f > 0$ and $\delta_g > 0$ that satisfy the definition of limits for f and g, respectively. Then $\sqrt{x^2 + y^2} < \delta = \min\{\delta_f, \delta_g\}$ implies $|f(z) - g(z) - (L - M)| \le |f(z) - L| + |g(z) - M| < \varepsilon/2 + \varepsilon/2$.

19. Write $f(z)g(z) - L \cdot M$ as $(f(z) - L)M + (g(z) - M)f(z)$. Choose δ_f so that $|f(z) - L| < \varepsilon^*$ whenever $\sqrt{x^2 + y^2} < \delta_f$ and choose δ_g so that $|g(z) - M| < \varepsilon^*$ if $\sqrt{x^2 + y^2} < \delta_g$, where $\varepsilon^*(1 + |L| + |M|) = \varepsilon$. Then prove the definition holds for for a given $\varepsilon > 0$ with the choice of $\delta = \min\{\delta_f, \delta_g\}$.

21. The difference quotient simplifies to 3.

23. The difference quotient simplifies to $18z + 9\Delta z + 6i$.

25. The difference quotient simplifies to $\dfrac{-i}{(z + \Delta z + 2i)(z + 2i)}$.

27. $10z$

29. $6(3z + i)$

31. $\dfrac{2iz(z^3 + 2i) - 3z^2(iz^2 + 2)}{(z^3 + 2i)^2}$

33. $2ze^{z^2}$

35. $2(\text{Log } i)e^{z \text{ Log } i} = i\pi i^z$

37. The uniqueness of the derivative follows immediately from the property that limits are unique; see theorem 7.2.1.

39. The proof is similar to that given in example 4.4.3 of section 4.4.

41. The proof is similar to that given in example 4.4.4 of section 4.4, but with subtraction replacing the sum.

43. Apply the chain rule to $h(g(z))$, where $h(z) = e^z$.

45. $u_x = 1 = v_y$ and $u_y = 0 = -v_x$.

47. $u_x = 18x = v_y$ and $u_y = -18y - 6 = -v_x$.

49. $u_x = 1 + e^x \cos y = v_y$ and $u_y = -e^x \sin y = -v_x$.

51. $e^4(\cos 6 - i \sin 6)$

53. $\ln 4 - i(\pi/6)$

55. $e^2(\cos 1 \cdot \ln \sqrt{32} - \sin 1 \cdot (3\pi/4 + 2k\pi)) + i[e^2(\sin 1 \cdot \ln \sqrt{32} + \cos 1 \cdot (3\pi/4 + 2k\pi))]$, where $k \in \mathbb{Z}$.

57. $-4 + 4i$

59. $5e^{-\pi/2} - 2e^{-\pi/2}i$

61. $e^{-2k\pi}(\cos(\ln 2) + i \sin(\ln 2))$, where $k \in \mathbb{Z}$.

63. $e^{\ln \sqrt{2} - \pi/4 - 2k\pi}(\cos(\ln \sqrt{2} + \pi/4 + 2k\pi) + i \sin(\ln \sqrt{2} + \pi/4 + 2k\pi))$, where $k \in \mathbb{Z}$

65. $|e^z|^2 = e^z \overline{e^z} = e^{z + \bar{z}} = e^{2\text{Re}(z)} = (e^{\text{Re}(z)})^2$.

67. $\overline{e^z} = \overline{e^x e^{iy}} = e^x(\cos y - i \sin y) = e^x(\cos(-y) + i \sin(-y)) = e^{x - iy} = e^{\bar{z}}$.

69. If $x > 0$, then $v_y = (1/x)/[1 + (y/x)^2] = x/(x^2 + y^2)$. The cases for $x < 0$ and the calculations of v_x are similar.

1. Choose $N = \ln(1/\varepsilon)/\ln(5)$.

3. Choose $N = 1/\varepsilon$.

5. For any real value n, $|e^{in}| = 1$. The sequence is constant; any choice of N works.

7. The sequence is constant since $|n + i/n - i| = 1$; any choice of N works.

9. 2

11. $8 - 2i$

13. 0

15. 0

17. $1/(1 - z)$

19. Assume $\lim_{n\to\infty} z_n = L$ and $\lim_{n\to\infty} z_n = M$, where $L > M$. Define $\varepsilon = (L - M)/2$. Examine the definition of limit in terms of L and M to find a contradiction (as there cannot exist a value N such that z_n is within ε of both L and M whenever $n > N$).

21. Given $\varepsilon > 0$, choose N so that it satisfies the definition of limit for the given value $\varepsilon/|c|$ for the original sequence.

23. Write $z_n w_n - L \cdot M$ as $(z_n - L)M + (w_n - M)z_n$. Choose N_z so that $|z_n - L| < \varepsilon^*$ if $n > N_z$ and choose N_w so that $|w_n - M| < \varepsilon^*$ if $n > N_w$, where $\varepsilon^*(1 + |L| + |M|) = \varepsilon$. Then prove the definition holds for a given $\varepsilon > 0$ with the choice of $N = N_z + N_w$.

25. The sequence $z_n = (-1)^n i$ is a counterexample.

27. Use the fact that $|z_n - 0| = |z_n| = ||z_n| - 0|$.

29. $a_n = 1/(2n - i)$; $R = 1$

31. $a_n = (2i)^n/n!$; $R = \infty$

33. $a_n = 1/(6i)^n$; $R = 6$

35. $a_n = 1/[n!(5 + i)^n]$; $R = \infty$

37. $\text{Log}(z) = \sum_{n=1}^{\infty} \frac{(-1)^{n-1}(z-1)^n}{n}$; $R = 1$.

39. $\frac{1}{z} = \sum_{n=0}^{\infty} (-1)^n(z - 1)^n$; $R = 1$.

41. $\arctan(z) = \sum_{n=0}^{\infty} \frac{(-1)^n z^{2n+1}}{2n + 1}$; $R = 1$.

43. $ie^{-z} = \sum_{n=0}^{\infty} \frac{i(-1)^n z^n}{n!}$; $R = \infty$.

45. $e^z + e^{-z} = \sum_{n=0}^{\infty} \frac{(1 + (-1)^n)z^n}{n!}$; $R = \infty$.

47. $5i\cos(iz) = \sum_{n=0}^{\infty} \frac{5i(1 + (-1)^n)z^n}{2 \cdot n!}$; $R = \infty$.

49. $\sin(iz/5) = \sum_{n=0}^{\infty} \frac{((-1)^n - 1)z^n}{2i \cdot 5^n n!}$; $R = \infty$.

51. $\frac{(e^{\pi} - e^{-\pi})}{2}i$

53. $\frac{e^{-1} + e}{e^{-1} - e}i$

55. Applying theorem 7.3.6,
$$g'(z) = \sum_{n=1}^{\infty}(-1)^n \frac{(2n)z^{2n-1}}{(2n)!}.$$
Now cancel the $(2n)$'s and reindex.

57. For $f(z) = \cos z$,
$$f'(z) = \frac{ie^{iz} - ie^{-iz}}{2} = -\sin z.$$

59. $\cos(z^2 + 1) \cdot 2\sin(5z^3 + iz) \cdot \cos(5z^3 + iz)(15z^2 + i) - \sin(z^2 + 1)(2z)\sin^2(5z^3 + iz)$

61. $-2\sec(z^3 + i)\csc^2(2iz^2 + iz) \cdot \cot(2iz^2 + iz)(4iz + i) + \sec(z^3 + i)\tan(z^3 + i)(3z^2)\csc^2(2iz^2 + iz)$

63. $\cos x(e^y + e^{-y})/2 - i\sin x(e^y - e^{-y})/2 = [(\cos x + i\sin x)e^{-y} + (\cos x - i\sin x)e^y]/2 = [e^{ix-y} + e^{-ix+y}]/2 = \cos(x + iy)$

65. Apply the triangle inequality at the induction step to obtain $|z_{n+1} + \sum_{k=0}^{n} z_k| \le |z_{n+1}| + |\sum_{k=0}^{n} z_k|$.

67. Prove $z_n = z_0^{2n}$, and study its convergence properties in each case. For (c), write $z_0 = e^{it}$ for some $t \in \mathbb{R}$.

69. Examine $|z_n|$ for $|z_0| \le 2$.

7.4 Harmonic Functions

1. $u = 4x^3 + 12x^2 - 12y^2 - 12xy^2$ implies $u_{xx} = 24x + 24 = -u_{yy}$.

3. $u_{xx} = -6xy = -u_{yy}$

5. $u_{xx} = -4(e^{-2y+i2x}) = -u_{yy}$

7. $u_{xx} = -\sin x \cos(iy) = -u_{yy}$

9. $u_{xx} = -e^{ix}\cos(iy) = -u_{yy}$

11. $u_{xx} = \dfrac{2x^3 - 6xy^2}{(x^2 + y^2)^3} = -u_{yy}$;

    these second partial derivatives are continuous in S because they are rational functions.

13. $u_{xx} = \dfrac{2(y^2 - x^2)}{(x^2 + y^2)^3} = -u_{yy}$

15. The function is the imaginary part of $f(z) = \text{Log}(z + 1)$, which is analytic in S.

17. Not harmonic.

19. Not harmonic.

21. Harmonic; $u_{xx} = -6xy = -u_{yy}$ (these polynomials are continuous).

23. Not harmonic.

25. Harmonic; $u_{xx} = 25e^{5x+3}\cos(5y + 3) = -u_{yy}$.

27. Harmonic; $u_{xx} = -\cos x \cosh y = -u_{yy}$.

29. Continuity of the second partial derivatives follows automatically from the fact that the functions are real and imaginary parts of an analytic function. $u(x, y) = 2x + 4$, so $u_{xx} = 0 = u_{yy}$. Also $v(x, y) = 2y$, so $v_{xx} = 0 = v_{yy}$.

31. $u(x, y) = x^2 - y^2 + x$, so $u_{xx} = 2 = -u_{yy}$. Also $v(x, y) = 2xy+y$, so $v_{xx} = 0 = v_{yy}$.

33. $u(x, y) = e^x \cos y$, so $u_{xx} = e^x \cos y = -u_{yy}$. Also $v(x, y) = e^x \sin y$, so $v_{xx} = e^x \sin y = -v_{yy}$.

35. $\cos z = (1/2)(e^{ix}e^y + e^{i(-x)}e^{-y})$, and so $u(x, y) = Re(\cos z) = (1/2)(e^y \cos x + e^{-y}\cos x) = (1/2)(e^y + e^{-y})\cos x = \cos x \cosh y$. Now see exercise 27. Also, $v(x, y) = Im(\cos z) = \sin x \sinh y = (1/2)(e^y \sin x - e^{-y}\sin x)$, and so $v_{xx} = -\sin x \sinh y = -v_{yy}$.

37. Proceed as in example 11 in section 7.2; since $u_x = v_y$, then $u_{xx} = v_{yx}$. Also, since $u_y = -v_x$, then $u_{yy} = -v_{xy} = -v_{yx}$ due to continuity. The result follows.

39. $u_{xx} = 0 = u_{yy}$; these functions are continuous.

41. If u and v are harmonic, then $(u + v)_{xx} = u_{xx} + v_{xx} = -u_{yy} - v_{yy} = -(u + v)_{yy}$. The second partial derivatives are continuous, since the sum of two continuous functions is continuous.

43. If u and v are harmonic, then $(au + bv)_{xx} = au_{xx} + bv_{xx} = -au_{yy} - bv_{yy} = -(au + bv)_{yy}$. The algebraic operations preserve continuity of the second partial derivatives.

45. $u(x, y) = x$ is harmonic, but $u(x, y) \cdot u(x, y) = x^2$ is not.

47. By the chain rule, $[u(ax + b, ay + c)]_{xx} = a^2 u_{xx}(ax + b, ay + c) = -a^2 u_{yy}(ax + b, ay + c) = -[u(ax + b, ay + c)]_{yy}$. The functional operations preserve continuity of the second partial derivatives.

49. $u_{xx} = 2c = -u_{yy}$.

51. $u^*(x, y) = 4xy.\ f(z) = 2z^2 + 5 =$ $u + iu^*$ is analytic.

53. $u^*(x, y) = 6xy - y.\ f(z) = 3z^2 - z =$ $u + iu^*$ is analytic.

55. $u^*(x, y) = e^y \cos x.\ f(z) = ie^{-iz} =$ $u + iu^*$ is analytic.

57. $u^*(x, y) = 2\ \text{Arg}(x + iy)$, where $-\pi < \text{Arg}(z) < \pi.\ f(z) = 2\ \text{Log}\ z =$ $u + iu^*$ is analytic.

59. $u^*(x, y) = \cos x \cosh y.\ f(z) = i \cos z$ $= u + iu^*$ is analytic.

61. $h_x = 2uu_x - 2vv_x$, so $h_{xx} = 2(u_x u_x + uu_{xx} - v_x v_x - vv_{xx})$. $h_y = 2uu_y - 2vv_y$, so $h_{yy} = 2(u_y u_y + uu_{yy} - v_y v_y - vv_{yy})$. Now apply the Cauchy–Riemann equations $u_x = v_y$, $u_y = -v_x$, and the Laplacian identities $u_{xx} = -u_{yy}$ and $v_{xx} = -v_{yy}$.

7.5 | Application: Streamlines and Equipotentials

1. $\sqrt{2}$

3. $\sqrt{2}$

5. $\sqrt{5}$

7. 4

9. 3

11. $\|\vec{F}\| = \sqrt{1^2 + 1^2} = \sqrt{2}$

13. $\|\vec{F}\| = |x|$

15. $\|\vec{F}\| = \sqrt{x^2 + y^2}$

17. $\|\vec{F}\| = |x + y|$

19. $\|\vec{F}\| = \sqrt{1 + e^{2x}}$

21. $\|\vec{F}\| = \sqrt{1 + \cos^2 x}$

23. $\|\vec{F}\| = e^y$

25. $\vec{F} = \langle 0, 0 \rangle$. Set $u^*(x, y) = 0$, then $f(z) = u + iu^* = 2$ is a constant analytic function.

27. $\vec{F} = \langle y - b, x - a \rangle$. Set $u^*(x, y) = (y^2 - x^2)/2 + ax - by$, then $f(z) = u + iu^* = -iz^2/2 - bz + iaz + ab$ is analytic.

29. $\vec{F} = \langle 6x - 1, -6y + 2 \rangle$. Set $u^*(x, y) = 6xy - y - 2x$, then $f(z) = u + iu^* = 3z^2 - z - 2iz$ is analytic.

31. $\vec{F} = \langle 3x^2 - 3y^2, -6xy \rangle$. Set $u^*(x, y) = 3x^2 y - y^3$, then $f(z) = u + iu^* = 3z^3$ is analytic.

33. $\vec{F} = \langle e^y \cos x, e^y \sin x \rangle$. Set $u^*(x, y) = e^y \cos x$, then $f(z) = ie^{-iz} =$ $u + iu^*$ is analytic.

35. $\vec{F} = \langle 2e^{x^2 - y^2}(x \sin(2xy) + y \cos(2xy)),$ $2e^{x^2 - y^2}(x \cos(2xy) - y \sin(2xy)) \rangle$. Set $u^*(x, y) = e^{x^2 - y^2}(\tan^2(xy) - 1)/$ $\sec^2(xy)$, then $f(z) = u + iu^*$ is analytic.

37. $\vec{F} = \langle \cos x \cosh y, \sin x \sinh y \rangle$. Set $u^*(x, y) = -\cosh x \cos y$, then $f(z) = -i \cos(-iz) = u + iu^*$ is analytic.

39. $\vec{F} = \langle \cos x \cosh y, \sin y \sinh y \rangle$. Set $u^*(x, y) = \cos x \sinh y$, then $f(z) = \sin(z) = u + iu^*$ is analytic.

41. $u(x, y) = 2xy$, $v(x, y) = y^2 - x^2$.

43. The streamlines are hyperbolas of the form $y^2 - x^2 = C$, where $C \in \mathbb{R}$. The equipotentials are reciprocal functions of the form $y = K/x$, where $K \in \mathbb{R}$.

45. $u(x, y) = e^x \cos y$, $v(x, y) = e^x \sin y$.

47. The streamlines are of the form $e^x \sin y = C$, where $C \in \mathbb{R}$. The equipotentials are of the form $e^x \cos y = K$, where $K \in \mathbb{R}$.

49. The fluid flow is $\vec{F} = \langle 0, 0 \rangle$, which has no velocity. The corresponding

stream function is $v(x, y) = 0$. There is no movement in this stream; a pebble dropped into it would not move.

51. The fluid flow is $\vec{F} = \langle 4, 5 \rangle$, which describes a constant velocity and direction. The corresponding stream function is $v(x, y) = 4y - 5x$ (then $f(z) = 4z - 5iz$). The streamlines are lines of the form $y = (5/4)x + C$, where $C \in \mathbb{R}$. The equipotentials are lines of the form $y = (-4/5)x + K$, where $K \in \mathbb{R}$.

53. The fluid flow is $\vec{F} = \langle e^x \cos y, -e^x \sin y \rangle$. The corresponding stream function is $v(x, y) = e^x \sin y$ (then $f(z) = e^z$). The streamlines are of the form $e^x \sin y = C$, where $C \in \mathbb{R}$. The equipotentials are of the form $e^x \cos y = K$, where $K \in \mathbb{R}$.

55. The fluid flow is $\vec{F} = \langle 0, 0 \rangle$, which has no velocity. The corresponding harmonic potential is $u(x, y) = 0$. There is no movement in this stream; a pebble dropped into it would not move.

57. The harmonic potential is $u(x, y) = x^2 - y^2 + x$; the fluid flow is $\vec{F} = \langle 2x + 1, -2y \rangle$. The corresponding stream function is $v(x, y) = 2xy + y$. The streamlines are of the form $y = C/(2x + 1)$, where $C \in \mathbb{R}$. Their

graphs have a vertical asymptote at $x = -1/2$. The equipotentials are of the form $[x - (1/2)]^2 - y^2 = K$, where $K \in \mathbb{R}$.

59. The isothermals are of the form $x = C$, where $C \in \mathbb{R}$, which are vertical lines on the $x - y$ plane. The flux lines are of the form $y = K$, where $K \in \mathbb{R}$, which are horizontal lines. The heat flux is $\vec{F} = \langle 2, 0 \rangle$, which consists of vectors directed horizontally toward the right of length 2.

61. The isothermals are circles centered at the origin of the form $x^2 + y^2 = C$, where $C \in \mathbb{R}$. The flux lines are of the form $y = Kx$, where $K \in \mathbb{R}$, which are lines through the origin. The heat flux is

$$\vec{F} = \left\langle \frac{2x}{x^2 + y^2}, \frac{2y}{x^2 + y^2} \right\rangle.$$

63. They are identical vector fields.

65. $(r + s)\langle a, b \rangle = \langle (r + s)a, (r + s)b \rangle = \langle ra + sa, rb + sb \rangle = \langle ra, rb \rangle + \langle sa, sb \rangle = r\langle a, b \rangle + s\langle a, b \rangle.$

67. Let $a = b = c = d = 1$ and $r = 2$ and note that $2 \cdot [\langle 1, 1 \rangle \cdot \langle 1, 1 \rangle] = \langle 2, 2 \rangle$, but $[2 \cdot \langle 1, 1 \rangle] \cdot [2 \cdot \langle 1, 1 \rangle] = \langle 4, 4 \rangle$.

69. The identity vector is $\langle 0, 0 \rangle$. Commutativity follows from the additive commutativity of the real numbers.

Online Resources

In the past 30 years the Internet has become a powerful and important resource for sharing and gathering information. Mathematicians were among the first people involved in the development of computing systems and a great deal of mathematical information can be found on the World Wide Web. To search for mathematical ideas on the web, choose your favorite search engine, type in a mathematical word, and see what pops up. Two well-regarded Internet search engines are *Google* at http://www.google.com and *Yahoo* at http://www.yahoo.com.

Many excellent mathematics encyclopedic references have been developed and provide quick access to diverse mathematical ideas. The following free online resources generally rely on the contributions of readers for their content, correction, and comment. The most extensive and well-regarded of the free online encyclopedias include:

- *MathWorld* at http://mathworld.wolfram.com/
  Among the first of the free online mathematics encyclopedias, this website was first created by physics and astronomy student Eric W. Wiesstein, who posted electronic notes based on various mathematical books and classes.
- *PlanetMath* at http://planetmath.org/
  A temporary shutdown of MathWorld in 2000 prompted the creation of this second well-respected online mathematics encyclopedia. This website is user-generated with individual authors "owning" their article and providing peer review and editorial rights to others in the mathematical community.
- *Wikipedia* at http://en.wikipedia.org/wiki/Mathematics
  Wikipedia is a general encyclopedia providing information on a wide range of topics, but has an extensive and well-developed presentation of mathematical ideas.

Many other focused websites provide more detailed information about particular mathematical topics. A few that may be of interest are:

- *Earliest Uses of Various Mathematical Symbols* at http://members.aol.com/Jeff570/mathsym.html
  This site presents a history of mathematical notation and other symbols compiled by Gulf High School mathematics teacher Jeff Miller—a useful and fun website for learning more about the history of mathematics.

- *Earliest Known Uses of Some of the Words of Mathematics* at http://hometown. aol.com/jeff570/mathword.html
  Also developed by Jeff Miller, this sister website surveys various mathematical terms and their history.
- *The Great Internet Mersenne Prime Search (or GIMPS)* at http://www. mersenne.org/
  Section 3.1 introduced Mersenne primes—primes of the form $2^p - 1$. GIMPS coordinates a networking of personal computers to investigate the divisibility properties of such large numbers. As of November 2008 there are 46 known Mersenne primes with the 46th equal to $2^{43,112,609} - 1$, which has 12,978,189 digits.
- *The MacTutor History of Mathematics Archive* at http://www-groups.dcs. st-and.ac.uk/ history/
  This website is an excellent resource for learning more about individual mathematicians throughout history. In addition to the many biographical sketches (accompanied by pictures and quotes), the site includes surveys about various areas and ideas of mathematics.
- *The Math Forum at Drexel University* at http://mathforum.org/
  Drexel set up this interactive website to promote the learning, teaching, and communication of mathematics through weekly problems, an "Ask Dr. Math" forum, and links to mathematical tools on the Internet.
- *The Mathematics Subject Classification* at http://www.ams.org/msc/
  This classification of the many areas in mathematics is used by the leading research journals and databases to categorize publications and presentations by mathematicians.
- *Mathwords* at http://www.mathwords.com/
  This online dictionary explains various terms and formulas used by mathematicians in courses from beginning algebra through calculus.
- *The Primes Page* at http://primes.utm.edu/
  Another fun resource for learning more about prime numbers, this website includes lists of primes, historical surveys, and links to other sites with interesting facts and applications of primes.

At various points, this text has provided helpful commands for computing devices. The following websites provide further information about these computer algebra systems and calculators.

- *Maple* at http://www.maplesoft.com/
- *Mathematica* at http://www.wolfram.com/
- *Matlab* at http://www.mathworks.com/
- *Texas Instruments Calculators* at http://education.ti.com/educationportal/sites/ US/homePage/index.html

A number of different professional societies have been established for promoting mathematics. The homepages of these organizations have links to information that is accessible to undergraduates (some publish monthly columns by well-respected authors) as well as announcements about mathematical meetings and events. Among these organizations are:

- *American Mathematical Association of Two-year Colleges* at http://www.amatyc.org/
- *American Mathematical Society* at http://www.ams.org/
- *American Statistical Association* at http://www.amstat.org/
- *Association for Computing Machinery* at http://www.acm.org/
- *Association of Symbolic Logic* at http://www.aslonline.org/index.htm
- *Mathematics Association of America* at http://www.maa.org/
- *Society of Actuaries* at http://www.soa.org/ccm/content/
- *Society of Industrial and Applied Mathematics* at http://www.siam.org/

National mathematics honor societies recognize excellence in mathematical studies by undergraduate and high school students, and promote scholarly interest and activity in mathematics. These honor societies include:

- *Kappa Mu Epsilon* at http://kappamuepsilon.org/, an honor society for undergraduates
- *Mu Alpha Theta* at http://www.mualphatheta.org/, an honor society for high school and two-year college students
- *Pi Mu Epsilon* at http://www.pme-math.org/, an honor society for undergraduates

References

1. Abbott, Stephen, *Understanding Analysis*, Springer-Verlag, Berlin–New York, 2001. ISBN: 0-387-95060-5
2. Aczel, Amir D., *Descartes's Secret Notebook: A True Tale of Mathematics, Mysticism, and the Quest to Understand the Universe*, Broadway, New York, 2005. ISBN: 0-767-92033-3
3. Aigner, Martin, and Günter Ziegler, *Proofs from THE BOOK*, 3rd ed., Springer-Verlag, Berlin–New York, 2003. ISBN: 3-540-40460-0
4. Aldous, Joan M., and Robin J. Wilson, *Graphs and Applications: An Introductory Approach*, Springer-Verlag, Berlin–New York, 2003. ISBN: 1-852-33259-X
5. Andreescu, Titu, and Zuming Feng, *A Path to Combinatorics for Undergraduates: Counting Strategies*, Birkhäuser, Basel-Boston, 2003. ISBN: 0-817-64288-9
6. Andrews, George E., *Number Theory*, Dover, New York, 1994. ISBN: 0-486-68252-8
7. Antognazza, Maria Rosa, *Leibniz: A Biography*, Cambridge University Press, Cambridge, 2004. ISBN: 0-521-80619-4
8. Archimedes and Sir Thomas Heath (translator), *The Works of Archimedes*, Dover, New York, 2002. ISBN: 0-486-42084-1
9. Axler, Sheldon, Paul Bourdon, and Wade Ramey, *Harmonic Function Theory*, 2nd ed., Springer-Verlag, Berlin–New York, 2001. ISBN: 0-387-95218-7
10. Barnier, William J., and Norman Feldman, *Introduction to Advanced Mathematics*, 2nd ed., Prentice-Hall, Englewood Cliffs, 1999. ISBN: 0-130-16750-9
11. Bashmakova, Isabella G., *Diophantus and Diophantine Equations*, The Mathematical Association of America, Washington DC, 1998. ISBN: 0-883-85526-7
12. Beckmann, Petr, *A History of Pi*, St. Martin's Griffin, New York, 1976. ISBN: 0-312-38185-9
13. Beerends, R. J., H. G. ter Morsche, J. C. van den Berg, and E. M. van de Vrie, *Fourier and Laplace Transforms*, Cambridge University Press, Cambridge, 2003. ISBN: 0-521-53441-0
14. Belhoste, Bruno, *Augustin-Louis Cauchy: A Biography*, Springer-Verlag, Berlin–New York, 1991. ISBN: 0-387-97220-X
15. Bell, Eric T., *Men of Mathematics*, Simon and Schuster, New York, 1986. ISBN: 0-671-62818-6
16. Benjamin, Arthur T., and Jennifer Quinn, *Proofs that Really Count: The Art of Combinatorial Proof*, The Mathematical Association of America, Washington DC, 2003. ISBN: 0-883-85333-7
17. Bertsekas, Dimitri P., and John N. Tsitsiklis, *Introduction to Probability*, Athena Scientific, Nashua, 2002. ISBN: 1-886-52940-X
18. Bierbrauer, Jurgen, *Introduction to Coding Theory*, Chapman & Hall/CRC, Los Angeles, 2004. ISBN: 1-584-88421-5

19. Blanchard, Paul, Robert L. Devaney, and Glen R. Hall, *Differential Equations*, 2nd ed., Brooks Cole, Pacific Grove, 2002. ISBN: 0-534-38514-1

20. Blatner, David, *The Joy of Pi*, Walker, New York, 1997. ISBN: 0-802-71332-7

21. Bollobas, Bela, *Modern Graph Theory*, Springer-Verlag, Berlin–New York, 2002. ISBN: 0-387-98488-7

22. Boole, George, *An Investigation of the Laws of Thought*, Reprint ed., Dover, New York, 1958. ISBN: 0-486-60028-9

23. Bottazini, Umberto, and Warren Van Egmond (translator), *The Higher Calculus: A History of Real and Complex Analysis from Euler to Weierstrass*, Springer-Verlag, Berlin–New York, 1986. ISBN: 0-387-96302-2

24. Box, Joan Fisher, *R. A. Fisher: the Life of a Scientist*, Wiley, New York, 1978. ISBN: 0-471-09300-9

25. Boyce, William, and Richard C. DiPrima, *Elementary Differential Equations and Boundary Value Problems*, 8th ed., Wiley, New York, 2004. ISBN: 0-471-43338-1

26. Boyer, Carl B., *History of Analytic Geometry*, Dover, New York, 2004. ISBN: 0-486-43832-5

27. Boyer, Carl B., *The History of the Calculus and Its Conceptual Development*, Dover, New York, 1959. ISBN: 0-486-60509-4

28. Boyer, Carl B., and Uta C. Merzbach, *A History of Mathematics*, 2nd ed., Wiley, New York, 1991. ISBN: 0-471-54397-7

29. Braun, Martin, *Differential Equations and Their Applications: An Introduction to Applied Mathematics*, 4th ed., Springer-Verlag, Berlin–New York, 1992. ISBN: 0-387-97894-1

30. Briggs, John, *Fractals: The Patterns of Chaos*, Touchstone, New York, 1992. ISBN: 0-671-74217-5

31. Brown, James Ward, and Ruel Vance Churchill, *Complex Variables and Applications*, 7th ed., McGraw-Hill, New York, 2003. ISBN: 0-072-87252-7

32. Browne, M. Neil, and Stuart M. Keeley, *Asking the Right Questions: A Guide to Critical Thinking*, Prentice-Hall, Englewood Cliffs, 2003. ISBN: 0-131-82993-9

33. Brualdi, Richard A., *Introductory Combinatorics*, 4th ed., Prentice-Hall, Englewood Cliffs, 2004. ISBN: 0-131-00119-1

34. Burger, Edward B., and Michael Starbird, *The Heart of Mathematics: An Invitation to Effective Thinking*, Key College, Emeryville, 2000. ISBN: 1-559-53407-9

35. Cahill, Thomas, *Sailing the Wine-Dark Sea: Why the Greeks Matter*, Nan A. Talese, New York, 2003. ISBN: 0-385-49553-6

36. Cameron, Peter J., *Combinatorics: Topics, Techniques, Algorithms*, Cambridge University Press, Cambridge, 1994. ISBN: 0-521-45761-0

37. Cantor, Georg, *Contributions to the Founding of the Theory of Transfinite Numbers*, Dover, New York, 1955. ISBN: 0-486-60045-9

38. Casella, George, and Roger L. Berger, *Statistical Inference*, 2nd ed., Duxbury Press, Pacific Grove, 2001. ISBN: 0-534-24312-6

39. Chartrand, Gary, *Introductory Graph Theory*, Dover, New York, 1984. ISBN: 0-486-24775-9

40. Christianson, Gale E., *Isaac Newton*, Oxford University Press, Oxford, 2005. ISBN: 0-195-30070-X

41. Clarke, Desmond, *Descartes: A Biography*, Cambridge University Press, Cambridge, 2006. ISBN: 0-521-82301-3

42. Colledge, Tony, *Pascal's Triangle: A Teacher's Guide with Blackline Masters*, Tarquin, Norfolk, 1992. ISBN: 0-906-21284-7

43. Consortium for Mathematics and Its Applications (COMAP), *For All Practical Purposes: Mathematical Literacy in Today's World*, 6th ed., W. H. Freeman, New York, 2002. ISBN: 0-716-74783-9

44. Comer, David J., *Digital Logic and State Machine Design, (Series in Electrical Engineering)*, 3rd ed., Saunders College Publishing, Orlando, 1997. ISBN: 0-030-94904-1

45. Conway, John H., and Richard K. Guy, *The Book of Numbers*, Springer-Verlag, Berlin–New York, 1998. ISBN: 0-387-97993-X

46. Copi, Irving M., and Carl Cohen, *Introduction to Logic*, Prentice-Hall, Englewood Cliffs, 2004. ISBN: 0-131-89834-5

47. Coxeter, Harold M. S., *Mauritus C. Escher: Art and Science* (Proceedings of the International Congress on M. C. Escher Rome, Italy, 26–28 March, 1985), North-Holland, Amsterdam, 1987. ISBN: 0-444-70011-0

48. Crilly, Tony, *Arthur Cayley: Mathematician Laureate of the Victorian Age*, The Johns Hopkins University Press, Baltimore, 2005. ISBN: 0-801-88011-4

49. Crossley, John N., C. J. Ash, C. J. Brickhill, J. C. Stillwell, and N. H. Williams, *What is Mathematical Logic?*, Dover, New York, 1972. ISBN: 0-486-26404-1

50. Cutland, Nigel J., *Computability: An Introduction to Recursive Function Theory*, Cambridge University Press, Cambridge, 1980. ISBN: 0-521-29465-7

51. D'Angelo, John P., and Douglas B. West, *Mathematical Thinking: Problem-Solving and Proofs*, 2nd ed., Prentice-Hall, Englewood Cliffs, 1999. ISBN: 0-130-14412-6

52. Dawson, John W. Jr., *Logical Dilemmas: The Life and Work of Kurt Godel*, A. K. Peters, Wellesley, 1996. ISBN: 1-568-81025-3

53. Dauben, Joseph Warren, *Georg Cantor: His Mathematics and Philosophy of the Infinite*, Princeton University Press, Princeton, 1990. ISBN: 0-691-08583-8

54. Davis, Martin, *The Universal Computer: The Road from Leibniz to Turing*, W. W. Norton, New York–London, 2000. ISBN: 0-393-04785-7

55. Davis, Philip J., Reuben Hersh, Elena A. Marchisotto, and Gian-Carlo Rota (Introduction), *The Mathematical Experience*, Study ed., Birkhäuser, Basel-Boston, 1995. ISBN: 0-817-63739-7

56. Dedekind, Richard, *Essays on the Theory of Numbers*, Dover, New York, 1963. ISBN: 0-486-21010-3

57. Derbyshire, John, *Prime Obsession: Bernhard Riemann and the Greatest Unsolved Problem in Mathematics*, Joseph Henry Press, Washington DC, 2003. ISBN: 0-309-08549-7

58. Descartes, René, and David E. Smith and Marcia Latham (translators), *The Geometry of René Descartes*, Dover, New York, 1954. ISBN: 0-486-60068-8

59. Diestel, Reinhard, *Graph Theory*, 2nd ed., Springer-Verlag, Berlin–New York, 2000. ISBN: 0-387-98976-5

60. Du Sautoy, Marcus, *The Music of the Primes: Searching to Solve the Greatest Mystery in Mathematics*, Harper Collins, New York, 2003. ISBN: 0-066-21070-4

61. DuChateau, Paul, and David Zachmann, *Applied Partial Differential Equations*, Dover, New York, 2002. ISBN: 0-486-41976-2

62. Dunham, William, *The Calculus Gallery: Masterpieces from Newton to Lebesgue*, Princeton University Press, Princeton, 2004. ISBN: 0-691-09565-5

63. Dunham, William, *Euler: The Master of Us All*, The Mathematical Association of America, Washington DC, 1999. ISBN: 0-883-85328-0

64. Dunham, William, *Journey through Genius: Great Theorems of Mathematics*, 1st ed., Wiley, New York, 1990. ISBN: 0-471-50030-5

65. Dunnington, G. Waldo, *Carl Friedrich Gauss: Titan of Science*, The Mathematical Association of America, Washington DC, 2004. ISBN: 0-883-85547-X

66. Durant, Will, and Ariel Durant, *The Story of Civilization, Volume VII: The Age of Reason Begins*, Simon and Schuster, New York, 1963. ISBN: 0-671-01320-3

67. Durant, Will, and Ariel Durant, *The Story of Civilization, Volume VIII: The Age of Louis XIV*, Simon and Schuster, New York, 1963. ISBN: 0-671-01215-0

68. Dyke, Philip P. G., *An Introduction to Laplace Transforms and Fourier Series*, Springer-Verlag, Berlin–New York, 2002. ISBN: 1-852-33015-5

69. Edwards, Anthony W. F., *Pascal's Arithmetical Triangle: The Story of a Mathematical Idea*, The Johns Hopkins University Press, Baltimore, 2002. ISBN: 0-801-86946-3

70. Edwards, Harold M., *Galois Theory*, Springer-Verlag, Berlin–New York, 1997. ISBN: 0-387-90980-X

71. Enderton, Herbert B., *A Mathematical Introduction to Logic*, 2nd ed., Academic Press, New York, 2000. ISBN: 0-122-38452-0

72. Epp, Susanna S., *Discrete Mathematics with Applications*, 3rd ed., Brooks Cole, Pacific Grove, 2004. ISBN: 0-534-35945-0

73. Euclid and Thomas L. Heath (translator), *Euclid's Elements*, Green Lion Press, Santa Fe, 2002. ISBN: 1-888-00918-7

74. Euler, Leonhard, and J. D. Blanton (translator), *Introduction to Analysis of the Infinite: Book I*, Springer-Verlag, Berlin–New York, 1988. ISBN: 0-387-96824-5

75. Euler, Leonhard, and J. D. Blanton (translator), *Foundations of Differential Calculus*, Springer-Verlag, Berlin–New York, 2000. ISBN: 0-387-98534-4

76. Eymard, Pierre, Jean-Pierre Lafon, and Stephen S. Wilson (translator), *The Number Pi*, The American Mathematical Society, Providence, 2004. ISBN: 0-821-83246-8

77. Fadiman, Clifton, *The Mathematical Magpie*, 2nd ed., Springer-Verlag, Berlin–New York, 1997. ISBN: 0-387-94950-X

78. Fadiman, Clifton (editor), *Fantasia Mathematica*, Springer-Verlag, Berlin–New York, 1997. ISBN: 0-387-94931-3

79. Farlow, Stanley J., *Partial Differential Equations for Scientists and Engineers*, Reprint ed., Dover, New York, 1993. ISBN: 0-486-67620-X

80. Feingold, Mordechai, *Before Newton: The Life and Times of Isaac Barrow*, Cambridge University Press, Cambridge, 1990. ISBN: 0-521-30694-9

81. Feller, William, *An Introduction to Probability Theory and Its Applications, Volume 1*, 3rd ed., Wiley, New York, 1968. ISBN: 0-471-25708-7

82. Fibnoacci, Leonardo, and Laurence Sigler (translator), *Fibonacci's Liber Abaci*, Springer-Verlag, Berlin–New York, 2002. ISBN: 0-387-95419-8

83. Fisher, Sir Ronald Aylmer, *The Genetical Theory of Natural Selection*, Dover, New York, 1958. ISBN: 0-486-60466-7

84. Fisher, Stephen D., *Complex Variables*, 2nd ed., Dover, New York, 1999. ISBN: 0-486-40679-2

85. Flannery, Sarah, and David Flannery, In *Code: A Mathematical Journey*, Workman Publishing, New York, 2001. ISBN: 0-761-12384-9

86. Fleiss, Joseph L., Bruce Levin, and Myunghee Cho Paik, *Statistical Methods for Rates and Proportions*, 3rd ed., Wiley, New York, 2003. ISBN: 0-471-52629-0

87. Foulds, Les R., *Graph Theory Applications*, Springer-Verlag, Berlin–New York, 1995. ISBN: 0-387-97599-3

88. Fraleigh, John B., *A First Course in Abstract Algebra*, 7th ed., Addison-Wesley, Boston, 2002. ISBN: 0-201-76390-7

89. Franklin, Benjamin, *The Autobiography of Benjamin Franklin*, Dover, New York, 1996. ISBN: 0-486-29073-5

90. Franzen, Torkel, *Gödel's Theorem: An Incomplete Guide to Its Use and Abuse*, A. K. Peters, Wellesley, 2005. ISBN: 1-568-81238-8

91. Frege, Gottlob, and John L. Austin (Translator), *The Foundations of Arithmetic: A Logico-Mathematical Enquiry into the Concept of Number*, Northwestern University Press, Evanston, 1980. ISBN: 0-810-10605-1

92. Fritsch, Rudolf, Gerda Fritsch, and Julie Peschke, *The Four-Color Theorem: History, Topological Foundations, and Idea of Proof*, Springer-Verlag, Berlin–New York, 1998. ISBN: 0-387-98497-6

93. Gallian, Joseph A., *Contemporary Abstract Algebra*, 6th ed., Houghton-Mifflin, Boston-New York, 2005. ISBN: 0-618-51471-6

94. Gamelin, Theodore W., *Complex Analysis*, Springer-Verlag, Berlin–New York, 2003. ISBN: 0-387-95069-9

95. Garland, Trudi Hammel, *Fascinating Fibonaccis: Mystery and Magic in Numbers*, Dale Seymour, Lebanon, 1987. ISBN: 0-866-51343-4

96. Garling, D. J. H., *A Course in Galois Theory*, Cambridge University Press, Cambridge, 1987. ISBN: 1-584-88393-6

97. Gauss, Carl Friedrich, and Arthur A. Clarke (translator), *Disquisitiones Arithmeticae*, Yale University Press, New Haven, 1965. ISBN: 0-300-09473-6

98. Gillispie, Charles Coulston, Robert Fox, and Ivor Grattan-Guinness, *Pierre-Simon Laplace, 1749–1827*, Princeton University Press, Princeton, 1997. ISBN: 0-691-01185-0

99. Gödel, Kurt, *On Formally Undecidable Propositions of Principia Mathematica and Related Systems*, Reprint ed., Dover, New York, 1992. ISBN: 0-486-66980-7

100. Goldberg, Richard R., *Methods of Real Analysis*, 2nd ed., Wiley, New York, 1976. ISBN: 0-471-31065-4

101. Goffman, Casper, and George Pedrick, *First Course in Functional Analysis*, Chelsea, New York, 1983. ISBN: 0-828-40319-8

102. Grabiner, Judith, *The Origin's of Cauchy's Rigorous Calculus*, Dover, New York, 2005. ISBN: 0-486-43815-5

103. Grimaldi, Ralph, *Discrete and Combinatorial Mathematics, An Applied Introduction*, 5th ed., Addison-Wesley, Boston, 2004. ISBN: 0-201-72634-3

104. Gullberg, Jan, and Peter Hilton, *Mathematics from the Birth of Numbers*, W. W. Norton, New York–London, 1997. ISBN: 0-393-04002-X

105. Gustason, William, and Dolph E. Ulrich, *Elementary Symbolic Logic*, Waveland Press, Prospect Heights, 1989. ISBN: 0-881-33412-X

106. Gutin, Giregory, and Abraham P. Punnen (editors), *The Traveling Salesman Problem and Its Variations*, Springer-Verlag, Berlin–New York, 2002. ISBN: 1-402-00664-0

107. Hahn, Alexander J., *Basic Calculus: From Archimedes to Newton to Its Role in Science*, Springer-Verlag, Berlin–New York, 1998. ISBN: 0-387-94606-3

108. Halmos, Paul R., *Naive Set Theory (Undergraduate Texts in Mathematics)*, 1st ed., Springer-Verlag, Berlin–New York, 1998. ISBN: 0-387-90092-6

109. Hamilton, Alan G., *Logic for Mathematicians*, Cambridge University Press, Cambridge, 1988. ISBN: 0-521-36865-0

110. Hamming, Richard W., Error Detecting and Error Correcting Codes, *The Bell System Technical Journal*, Volume 16, Number 2, pp. 147–160, 1950.

111. Hankins, Thomas L., *Sir William Rowan Hamilton*, Reprint ed., The Johns Hopkins University Press, Baltimore, 2004. ISBN: 0-801-86973-0

112. Hardy, Godfrey Harold, *A Mathematician's Apology*, Reprint ed., Cambridge University Press, Cambridge, 1992. ISBN: 0-521-42706-1

113. Heath, Thomas L., *Diophantus of Alexandria: A Study in the History of Greek Algebra*, Martino, Mansfield Centre, 2003. ISBN: 1-578-98403-3

114. Hellegouarch, Yves, *Invitation to the Mathematics of Fermat–Wiles*, Academic Press, Burlington, 2001. ISBN: 0-123-39251-9

115. Hill, Raymond, *A First Course in Coding Theory*, Oxford University Press, Oxford, 1990. ISBN: 0-198-53803-0

116. Hillman, Abraham P., and Gerald L. Alexanderson, *A First Undergraduate Course in Abstract Algebra*, 2nd ed., Wadsworth, Belmont, 1978. ISBN: 0-534-00525-X

117. Hodges, Andrew, and Douglas R. Hofstadter (foreword), *Alan Turing: The Enigma*, Reprint ed., Walker, New York, 2000. ISBN: 0-802-77580-2

118. Hofstadter, Douglas R., *Gödel, Escher, Bach: An Eternal Golden Braid*, Basic Books, New York, 1999. ISBN: 0-465-02656-7

119. Hogg, Robert V., and Elliot A. Tanis, *Probability and Statistical Inference*, 7th ed., Prentice-Hall, Englewood Cliffs, 2005. ISBN: 0-131-46413-2

120. Huffman, W. Cary, and Vera Pless, *Fundamentals of Error-Correcting Codes*, Cambridge University Press, Cambridge, 2003. ISBN: 0-521-78280-5

121. Hughes-Hallet, Deborah, William G. McCallum, Andrew M. Gleason, Daniel E. Flath, Patti Frazer Lock, Sheldon P. Gordon, David O. Lomen, David Lovelock, Brad G. Osgood, Andrew Pasquale, Douglas Quinney, Jeff Tecosky-Feldman, Joe B. Thrash, Karen Rhea, and Thomas W. Tucker, *Calculus*, 4th ed., Wiley, New York, 2004. ISBN: 0-471-47245-X

122. Hungerford, Thomas W., *Algebra (Graduate Texts in Mathematics)*, 1st ed., Springer-Verlag, Berlin–New York, 1997. ISBN: 0-387-90518-9

123. Hunter, Geoffrey, *Metalogic: An Introduction to the Metatheory of Standard First Order Logic*, University of California Press, Berkeley, 1971. ISBN: 0-520-02356-0

124. Ireland, Kenneth, and Michael Rosen, *A Classical Introduction to Modern Number Theory*, Springer-Verlag, Berlin–New York, 1998. ISBN: 0-387-97329-X

125. Isaacson, Walter, *Benjamin Franklin: An American Life*, Simon and Schuster, New York, 2003. ISBN: 0-684-80761-0

126. Jacobs, Harold R., *Mathematics: A Human Endeavor*, 2nd ed., W. H. Freeman, New York, 1982. ISBN: 0-716-71326-8

127. Jacquette, Dale, *Symbolic Logic*, Wadsworth Publishing, Stamford, 2001. ISBN: 0-534-53730-8

128. James, Ioan, *Remarkable Mathematicians: From Euler to von Neumann*, Cambridge University Press, Cambridge, 2003. ISBN: 0-521-52094-0

129. Jaynes, Edwin T., *Probability Theory: The Logic of Science*, Cambridge University Press, Cambridge, 2003. ISBN: 0-521-59271-2

130. Jech, Thomas, *Set Theory*, 3rd ed., Springer-Verlag, Berlin–New York, 2002. ISBN: 3-540-44085-2

131. Jeffrey, Richard, *Formal Logic: Its Scope and Limits*, 3rd ed., McGraw-Hill, New York, 1990. ISBN: 0-070-32357-7

132. Johnston, William W., *An Introduction to Statistical Inference*, Mohican Press, Perrysville, 2001. ISBN: 0-923231-40-4

133. Jones, Frank, *Lebesgue Integration on Euclidean Space*, Revised ed., Jones and Bartlett, Sudbury, 2000. ISBN: 0-763-71708-8

134. Jones, Gareth A., and Josephine M. Jones, *Elementary Number Theory*, Springer-Verlag, Berlin–New York, 1998. ISBN: 3-540-76197-7

135. Kahn, Charles H., *Pythagoras and the Pythagoreans: A Brief History*, Hackett, Indianapolis, 2001. ISBN: 0-872-20575-4

136. Kaplan, Robert, and Ellen Kaplan, *The Nothing that Is: A Natural History of Zero*, Oxford University Press, Oxford, 2000. ISBN: 0-195-14237-3

137. Karnaugh, Maurice, The Map Method for Synthesis of Combinational Logic Circuits, *Transactions of the American Institute of Electrical Engineers pt. I*, volume 72 number 9, pp. 593–599, November 1953.

138. Kelley, David, *The Art of Reasoning*, W. W. Norton, New York–London, 1998. ISBN: 0-393-97213-5

139. Kerns, David V., and J. David Irwin, *Essentials of Electrical and Computer Engineering*, Prentice-Hall, Englewood Cliffs, 2004. ISBN: 0-139-23970-7

140. Kirkwood, James R., *An Introduction to Analysis*, 2nd ed., Waveland Press, Long Grove, 2002. ISBN: 1-577-66232-6

141. Kirtland, Joseph, *Identification Numbers and Check Digit Schemes*, 1st ed., The Mathematical Association of America, Washington DC, 2001. ISBN: 0-883-85720-0

142. Kline, Morris, *Mathematics in Western Culture*, Oxford University Press, Oxford, 1971. ISBN: 0-19-500714-X

143. Krantz, Steven G., *Geometric Function Theory: Explorations in Complex Analysis*, Birkhäuser, Basel-Boston, 2005. ISBN: 0-817-64339-7

144. Kruskal, Joseph B., On the shortest spanning subtree and the traveling salesman problem, *Proceedings of the American Mathematical Society*, volume 7, 1956, pp. 48–50.

145. Kuipers, Jack B., *Quaternions and Rotation Sequences*, Princeton University Press, Princeton, 1998. ISBN: 0-691-05872-5

146. Kunen, Kenneth, *Set Theory*, Reprint ed., Springer-Verlag, Berlin–New York, 1983. ISBN: 0-444-86839-9

147. Lang, Serge, *Complex Analysis*, 4th ed., Springer-Verlag, Berlin–New York, 2003. ISBN: 0-387-98592-1

148. Lang, Serge, *Real and Functional Analysis*, Springer-Verlag, Berlin–New York, 1993. ISBN: 0-387-94001-4

149. Larson, Ron, Robert P. Hostetler, and Bruce H. Edwards, *Calculus*, 8th ed., Houghton-Mifflin, Boston, 2006. ISBN: 0-618-50298-X

150. Laugwitz, Detleff, and Abe Shenitzer, *Bernhard Riemann 1826–1866: Turning Points in the Conception of Mathematics*, Birkhäuser, Basel-Boston, 1999. ISBN: 0-817-64040-1

151. Lawler, E. L., Jan Karel Lenstra, A. H. G. Rinnooy Kan, and D. B. Shmoys (editors), *The Traveling Salesman Problem: A Guided Tour of Combinatorial Optimization*, Wiley, New York, 1985. ISBN: 0-471-90413-9

152. Leary, Christopher, *Friendly Introduction to Mathematical Logic*, Prentice-Hall, Englewood Cliffs, 1999. ISBN: 0-130-10705-0

153. Lebesgue, Henri, Sur une généralisation de l'intégrale définie, *Comptes Rendus*, April 29, 1901.

154. Lehmann, Erich L., and Joseph P. Romano, *Testing Statistical Hypotheses*, 3rd ed., Springer-Verlag, Berlin–New York, 2006. ISBN: 0-387-98864-5

155. LeVeque, William J., *Fundamentals of Number Theory*, Dover, New York, 1996. ISBN: 0-486-68906-9

156. Ling, San, and Chaoping Xing, *Coding Theory: A First Course* Cambridge University Press, Cambridge, 2004. ISBN: 0-521-52923-9

157. van Lint, Jacobus H., and R. M. Wilson, *A Course in Combinatorics*, 2nd ed., Cambridge University Press, Cambridge, 2001. ISBN: 0-521-00601-5

158. Livio, Mario, *The Equation that Couldn't Be Solved: How Mathematical Genius Discovered the Language of Symmetry*, Simon and Schuster, New York, 2005. ISBN: 0-743-25820-7

159. Locher, J. L. *M. C. Escher: Life and Work*, Harry N. Abrams, New York, 1992. ISBN: 0-810-98113-0

160. Logan, J. David, *Applied Partial Differential Equations*, 2nd ed., Springer-Verlag, Berlin–New York, 2004. ISBN: 0-387-20953-0

161. MacHale, Desmond, *George Boole: His Life and Work*, Boole Press, Dublin, 1985. ISBN: 0-906-78305-4

162. MacWilliams, Florences J., and Neil J. A. Sloane, *The Theory of Error-Correcting Codes*, North Holland, Amsterdam, 1998. ISBN: 0-444-85193-3

163. Mahoney, Michael Sean, *The Mathematical Career of Pierre de Fermat, 1601–1665*, Princeton University Press, Princeton, 1994. ISBN: 0-691-03666-7

164. Mandelbrot, Benoit B., *Fractals and Chaos*, Springer-Verlag, Berlin–New York, 2004. ISBN: 0-387-20158-0

165. Mandelbrot, Benoit B., *The Fractal Geometry of Nature*, W. H. Freeman, New York, 1982. ISBN: 0-716-71186-9

166. Mandelbrot, Benoit B., and Richard L. Hudson, *The (Mis)Behavior of Markets: A Fractal View of Risk, Ruin And Reward*, Perseus Press, New York, 2006. ISBN: 0-465-04357-7

167. Mano, M. Morris, *Computer System Architecture*, 3rd ed., Prentice-Hall, Englewood Cliffs, 1992. ISBN: 0-131-75563-3

168. Maor, Eli, *e: The Story of a Number*, Princeton University Press, Princeton, 1998. ISBN: 0-691-05854-7

169. Maor, Eli, *To Infinity and Beyond*, Princeton University Press, Princeton, 1991. ISBN: 0-691-02511-8

170. Marsden, Jerrold E., and Michael J. Hoffman, *Basic Complex Analysis*, 3rd ed., W. H. Freeman, New York, 1998. ISBN: 0-716-72877-X

171. Mathews, John H., and Russell W. Howell, *Complex Analysis for Mathematics and Engineering*, 5th ed., Jones and Bartlett, Sudbury, 2006. ISBN: 0-763-73748-8

172. Mazur, Barry, *Imagining Numbers*, Farrar, Straus, and Giroux Press, New York, 2003. ISBN: 0-374-17469-5

173. McInerny, Dennis Q., *Being Logical: A Guide to Good Thinking*, Random House, New York, 2004. ISBN: 1-400-06171-7

174. Meli, Domenico Bertoloni, *Equivalence and Priority: Newton versus Leibniz*, Oxford University Press, Oxford, 1993. ISBN: 0-198-53945-2

175. Mendelson, Elliot, *Introduction to Mathematical Logic*, Chapman & Hall/CRC, Los Angeles, 1997. ISBN: 0-412-80830-7

176. Montgomery, Douglas C., Elizabeth A. Peck, and G. Geoffrey Vining, *Introduction to Linear Regression Analysis*, 3rd ed., Wiley, New York, 2001. ISBN: 0-471-31565-6

177. Mosteller, Frederick, *Fifty Challenging Problems in Probability with Solutions*, Dover, New York, 1987. ISBN: 0-486-65355-2

178. Nagel, Ernest, James R. Newman, and Douglas Hofstadter (editor), *Gödel's Proof*, Revised ed., New York University Press, New York, 2002. ISBN: 0-814-75816-9

179. Nahin, Paul J., *An Imaginary Tale*, Princeton University Press, Princeton, 1998. ISBN: 0-691-02795-1

180. Newton, Isaac, I., Bernard Cohen, and Anne Whitman (translators), *The Principia: Mathematical Principles of Natural Philosophy*, University of California Press, Berkeley, 1999. ISBN: 0-520-08817-4

181. Nolt, John, Dennis Rohatyn, and Achille Varzi, *Schaum's Outline of Logic*, McGraw-Hill, New York, 1998. ISBN: 0-070-46649-1

182. Nolt, John, Dennis Rohatyn, Achille Varzi, and Alex M. McAllister (abridgement editor), *Schaum's Easy Outline of Logic*, McGraw-Hill, New York, 2005. ISBN: 0-071-45535-3

183. O'Connell, Marvin R., *Blaise Pascal: Reasons of the Heart*, Eerdmans, Grand Rapids, 1997. ISBN: 0-802-80158-7

184. Ostebee, Arnold, and Paul Zorn, *Calculus from Graphical, Numerical, and Symbolic Points of View*, Houghton Mifflin, Boston, 2002. ISBN: 0-616-24787-4

185. Pascal, Blaise, and Alban J. Krailsheimer (translator), *Pensées*, Reissue ed., Penguin, New York, 1995. ISBN: 0-140-44645-1

186. Peck, Roxy, Chris Olsen, and Jay L. Devore, *Introduction to Statistics and Data Analysis*, 2nd ed., Duxbury Press, Pacific Grove, 2004. ISBN: 0-534-46710-5

187. Pesic, Peter, *Abel's Proof: An Essay on the Sources and Meaning of Mathematical Unsolvability*, The MIT Press, Boston, 2003. ISBN: 0-262-16216-4

188. Posamentier, Alfred S., and Ingmar Lehmann, *Pi: A Biography of the World's Most Mysterious Number*, Prometheus, Amherst, 2004. ISBN: 1-591-02200-2

189. Powers, David L., *Boundary Value Problems*, Academic Press, New York, 1979. ISBN: 0-125-63760-8

190. Prim, Robert C., Shortest Connection Networks and Some Generalisations, *Bell System Technical Journal*, Volume 36, 1957, pp. 1389–1401.

191. Remmert, Reinhold, and R. B. Burckel (translator), *Theory of Complex Functions*, Springer-Verlag, Berlin–New York, 1998. ISBN: 0-387-97195-5

192. Ribenboim, Paulo, *Fermat's Last Theorem for Amateurs*, Springer-Verlag, Berlin–New York, 2000. ISBN: 0-387-98508-5

193. Richmond, Bettina, and Thomas Richmond, *A Discrete Transition to Advanced Mathematics*, Brooks Cole, Pacific Grove, 2003. ISBN: 0-534-40518-5

194. Riedwig, Christoph, and Steven Rendall, *Pythagoras: His Life, Teaching, and Influence*, Cornell University Press, Ithaca, 2005. ISBN: 0-801-44240-0

195. Rivest, Ronald, Adi Shamir, and Leonard Adleman, A Method for Obtaining Digital Signatures and Public-Key Cryptosystems, *Communications of the ACM*, Volume 21, Number 2, 1978, pp. 120–126.

196. Rockmore, Dan, *Stalking the Riemann Hypothesis: The Quest to Find the Hidden Law of Prime Numbers*, Pantheon, New York, 2005. ISBN: 0-375-42136-X

197. Rosen, Kenneth H., *Elementary Number Theory and Its Applications*, 5th ed., Addison Wesley, Boston, 2004. ISBN: 0-321-23707-2

198. Ross, Sheldon, *A First Course in Probability*, 6th ed., Prentice-Hall, Englewood Cliffs, 2001. ISBN: 0-130-33851-6

199. Royden, Halsey L., *Real Analysis*, 3rd ed., Prentice-Hall, Englewood Cliffs, 1988. ISBN: 0-024-04151-3

200. Rozanov, Y. A., *Probability Theory: A Concise Course*, Revised ed., Dover, New York, 1977. ISBN: 0-486-63544-9

201. Rucker, Rudy, *Infinity and the Mind: The Science and Philosophy of the Infinite*, Princeton University Press, Princeton, 2004. ISBN: 0-691-12127-3

202. Rudin, Walter, *Principles of Functional Analysis*, 3rd ed., McGraw-Hill, New York, 1976. ISBN: 0-070-54235-X

203. Rudin, Walter, *Functional Analysis*, 2nd ed., McGraw-Hill, New York, 1991. ISBN: 0-070-54236-8

204. Russell, Bertrand, *The Autobiography of Bertrand Russell*, Reissue ed., Routledge Press, Oxford, 1998. ISBN: 0-415-18985-3

205. Sainsbury, R. Mark, *Paradoxes*, Cambridge University Press, Cambridge, 1995. ISBN: 0-521-48347-6

206. Salmon, Wesley C., *Zeno's Paradoxes*, Reprint ed., Hackett, Indianapolis, 2001. ISBN: 0-872-20560-6

207. Salsburg, David, *The Lady Tasting Tea: How Statistics Revolutionized Science in the Twentieth Century*, W. H. Freeman, New York, 2001. ISBN: 0-716-74106-7

208. Scheaffer, Richard L., *Introduction to Probability and Its Applications*, 2nd ed., Duxbury Press, Pacific Grove, 1994. ISBN: 0-534-23790-8

209. Scheinerman, Edward A., *Mathematics: A Discrete Introduction*, 2nd ed., Brooks Cole, Pacific Grove, 2005. ISBN: 0-534-39898-7

210. Seber, George A. F., and Alan J. Lee, *Linear Regression Analysis*, 2nd ed., Wiley, New York, 2003. ISBN: 0-471-41540-5

211. Seidenfeld, Teddy, *Philosophical Problems of Statistical Inference: Learning from R. A. Fisher*, Springer-Verlag, Berlin–New York, 1979. ISBN: 9-027-70965-3

212. Seife, Charles, *Zero: The Biography of a Dangerous Idea*, Penguin, New York, 2000. ISBN: 0-140-29647-6

213. Simmons, George F., *Differential Equations with Applications and Historical Notes*, 2nd ed., McGraw-Hill, New York, 1991. ISBN: 0-07-057540-1

214. Simmons, George F., and Steven G. Krantz, *Differential Equations: Theory, Technique, and Practice*, McGraw-Hill, New York, 2006. ISBN: 0-072-86315-3

215. Singh, Simon, *The Code Book: The Science of Secrecy from Ancient Egypt to Quantum Cryptography*, Anchor, New York, 2000. ISBN: 0-385-49532-3

216. Singh, Simon, and John Lynch, *Fermat's Enigma: The Epic Quest to Solve the World's Greatest Mathematical Problem*, Anchor, New York, 1998. ISBN: 0-385-49362-2

217. Smith, David Eugene, *A Source Book in Mathematics, Volume Two*, Dover, New York, 1959. ISBN: 0-486-60553-1

218. Smith, Derek, and John Horton Conway, *On Quaternions and Octonions*, A. K. Peters, Wellesley, 2003. ISBN: 1-568-81134-9

219. Smith, Douglas, Maurice Eggen, and Richard St. Andre, *A Transition to Advanced Mathematics*, 6th ed., Brooks Cole, Pacific Grove, 2005. ISBN: 0-534-39900-2

220. Smith, Robin (editor), *Aristotle: Prior Analytics*, Hackett, Indianapolis, 1989. ISBN: 0-872-20064-7

221. Smithies, Frank, *Cauchy and the Creation of Complex Function Theory*, Cambridge University Press, Cambridge, 1997. ISBN: 0-521-59278-X

222. Smullyan, Raymond M., *Forever Undecided: A Puzzle Guide to Gödel*, Knopf, New York, 1987. ISBN: 0-394-54943-0

223. Smullyan, Raymond M., *Gödel's Incompleteness Theorems (Oxford Logic Guides, No. 19)*, Oxford University Press, Oxford, 1992. ISBN: 0-195-04672-2

224. Smullyan, Raymond M., *Satan, Cantor, and Infinity*, Knopf, New York, 1992. ISBN: 0-679-40688-3

225. Smullyan, Raymond M., *What Is the Name of This Book?: The Riddle of Dracula and Other Logical Puzzles*, Prentice-Hall, Englewood Cliffs, 1978. ISBN: 0-139-55088-7

226. Spiegel, Murray R., *Schaum's Outline of Complex Variables*, McGraw-Hill, New York, 1968. ISBN: 0-070-60230-1

227. Stevens, Roger, *Creating Fractals*, Charles River Media, Revere, 2005. ISBN: 1-584-50423-4

228. Stewart, Ian, *Does God Play Dice?: The Mathematics of Chaos*, Reprint ed., Blackwell Publishers, Oxford, 1990. ISBN: 1-557-86106-4

229. Stewart, Ian, *Galois Theory*, 3rd ed., Chapman & Hall/CRC, Los Angeles, 2003. ISBN: 1-584-88393-6

230. Stewart, Ian, *Letters to a Young Mathematician*, Basic Books, New York, 2006. ISBN: 0-465-08231-9

231. Stewart, Ian, *Nature's Numbers: The Unreal Reality of Mathematics*, Basic Books, New York, 1997. ISBN: 0-465-07274-7

232. Stewart, Ian, Arthur C. Clarke, Benoit Mandelbrot, Michael Barnsley, Louisa Barnsley, Will Rood, Gary Flake, David Pennock, Robert R. Prechter, Jr., and Nigel Lesmoir-Gordon, *The Colours of Infinity: The Beauty, and Power of Fractals*, Clear Books, Somerset, 2004. ISBN: 1-904-55505-5

233. Stewart, Ian, and David Tall, *Algebraic Number Theory and Fermat's Last Theorem*, A. K. Peters, Wellesley, 2001. ISBN: 1-568-81119-5

234. Stewart, James, *Calculus*, 5th ed., Brooks Cole, Pacific Grove, 2002. ISBN: 0-534-39339-X

235. Struik, Dirk J., *A Concise History of Mathematics*, Dover, New York, 1987. ISBN: 0-486-60255-9

236. Stubhaug, Arild, and Richard R. Daly (translator), *Niels Henrik Abel and His Times*, Springer-Verlag, Berlin–New York, 2000. ISBN: 3-540-66834-9

237. Swallow, John, *Exploratory Galois Theory*, Cambridge University Press, Cambridge, 2004. ISBN: 0-521-54499-8

238. Tent, Margaret B. W., *Prince of Mathematics: Carl Friedrich Gauss*, A. K. Peters, Wellesley, 2006. ISBN: 1-568-81261-2

239. Thomas, George B. Jr., Maurice D. Weir, Joel Haas, and Frank R. Giordano, *Thomas' Calculus*, 11th ed., Addison-Wesley, Boston, 2005. ISBN: 0-321-18558-7

240. Thomson, Garrett, *On Leibniz*, Wadsworth, Belmont, 2000. ISBN: 0-534-57634-6

241. Tiles, Mary, *The Philosophy of Set Theory: An Historical Introduction to Cantor's Paradise*, Dover, New York, 2004. ISBN: 0-486-43520-2

242. Trudeau, Richard J., *Introduction to Graph Theory*, Dover, New York, 1994. ISBN: 0-486-67870-9

243. Tucker, Alan, *Applied Combinatorics*, 4th ed., Wiley, New York, 2001. ISBN: 0-471-43809-X

244. Van Der Waerden, Bartel Leendart, *A History of Algebra: From Al-Khwarizmi to Emmy Noether*, Springer-Verlag, Berlin–New York, 1985. ISBN: 0-387-13610-X

245. Vilenkin, N. Ya., *In Search of Infinity*, Birkhäuser, Basel-Boston, 1995. ISBN: 0-817-63819-9

246. Wackerly, Dennis, William Mendenhall, and Richard L. Scheaffer, *Mathematical Statistics with Applications*, 6th ed., Duxbury Press, Pacific Grove, 2001. ISBN: 0-534-37741-6

247. Wahl, Mark, *Mathematical Mystery Tour: Higher Thinking Math Tasks*, Zephyr Press, Chicago, 1989. ISBN: 0-913-70526-8

248. Wallace, David Foster, *Everything and More: A Compact History of Infinity*, W. W. Norton, New York–London, 2003. ISBN: 0-393-00338-8

249. Walter, Wolfgang, and R. Thompson (translator), *Ordinary Differential Equations*, Springer-Verlag, Berlin–New York, 1998. ISBN: 0-387-98459-3

250. Watkins, Ann E., Richard L. Scheaffer, and George W. Cobb, *Statistics in Action: Understanding a World of Data*, Key College, Emeryville, 2004. ISBN: 1-931-91427-3

251. Weir, Alan J., *Lebesgue Integration and Measure*, Cambridge University Press, Cambridge, 1996. ISBN: 0-521-09751-7

252. Weisberg, Sanford, *Applied Linear Regression*, 3rd ed., Wiley, New York, 2005. ISBN: 0-471-66379-4

253. Wells, David, *Prime Numbers: The Most Mysterious Figures in Math*, Wiley, New York, 2005. ISBN: 0-471-46234-9

254. West, Douglas B., *Introduction to Graph Theory*, 2nd ed., Prentice-Hall, Englewood Cliffs, 2000. ISBN: 0-130-14400-2

255. Westfall, Richard S., *The Life of Isaac Newton*, Reprint ed., Cambridge University Press, Cambridge, 1994. ISBN: 0-521-47737-9

256. White, Michael, *Acid Tongues and Tranquil Dreamers: Eight Scientific Rivalries that Changed the World*, Harper, New York, 2002. ISBN: 0-380-80613-4

257. Wiles, Andrew, Modular elliptic curves and Fermat's Last Theorem, *Annals of Mathematics (2)*, Volume 141, Number 3, 1995, pp. 443–551.

258. Wilson, Robin, *Four Colors Suffice: How the Map Problem Was Solved*, Princeton University Press, Princeton, 2003. ISBN: 0-691-11533-8

259. Zygmund, A., *Trigonometric Series*, 3rd ed., Cambridge University Press, Cambridge, 2003. ISBN: 0-521-89053-5

Index